SOLIDS PROCESS DESIGN AND MANAGEMENT

Prepared under a Cooperative Research and Development Agreement between

U.S. Environmental Protection Agency, Office of Research and Development, National Risk Management Research Laboratory and the Office of Water, Office of Wastewater Management

and

Water Environment Federation and Water Environment Research Foundation

Under the auspices of the **Technical Practice Committee** of the **Water Environment Federation**

This manuscript was developed with the intent of revising "Technology Transfer Process Design Manual for Sewage Sludge," EPA 625/1-79-011, which was published by the U.S. Environmental Protection Agency in 1979.

U.S. Environmental Protection Agency
National Risk Management Research Laboratory
26 West Martin Luther King Drive
Cincinnati, OH 45268

U.S. Environmental Protection Agency
Office of Wastewater Management (4204M)
EPA EAST
1200 Pennsylvania Avenue, N.W.
Washington, DC 20460

Water Environment Federation
601 Wythe Street
Alexandria, VA 22314-1994
http://www.wef.org

Water Environment Research Foundation
635 Slaters Lane, Suite G-110
Alexandria, VA 22314
http://www.werf.org

About WEF

Formed in 1928, the Water Environment Federation® (WEF®) is a not-for-profit technical and educational organization with members from varied disciplines who work toward WEF's vision to preserve and enhance the global water environment.

For information on membership, publications, and conferences:

Water Environment Federation®
601 Wythe Street
Alexandria, VA 22314-1994
(703) 684-2400
http://www.wef.org

SOLIDS PROCESS DESIGN AND MANAGEMENT

Water Environment Federation
the water quality people™

WERF
Water Environment Research Foundation
Collaboration. Innovation. Results.

EPA

Water Environment Federation Alexandria, Virginia

McGraw Hill

New York Chicago San Francisco Lisbon London Madrid
Mexico City Milan New Delhi San Juan Seoul
Singapore Sydney Toronto

The McGraw-Hill Companies

Cataloging-in-Publication Data is on file with the Library of Congress.

Solids Process Design and Management

1 2 3 4 5 6 7 8 9 0 DOC/DOC 1 9 8 7 6 5 4 3 2

ISBN 978-0-07-178095-7
MHID 0-07-178095-5

Printed and bound by RR Donnelley

This book is printed on acid-free paper.

IMPORTANT NOTICE

The material presented in this publication has been prepared in accordance with generally recognized engineering principles and practices and is for general information only. This information should not be used without first securing competent advice with respect to its suitability for any general or specific application.

The contents of this publication are not intended to be a standard of the Water Environment Federation (WEF), Water Environment Research Foundation (WERF), or the U.S. Environmental Protection Agency (U.S. EPA) and are not intended for use as a reference in purchase specifications, contracts, regulations, statutes, or any other legal document.

No reference made in this publication to any specific method, product, process, or service constitutes or implies an endorsement, recommendation, or warranty thereof by WEF, WERF, or U.S. EPA.

WEF, WERF, and U.S. EPA makes no representation or warranty of any kind, whether expressed or implied, concerning the accuracy, product, or process discussed in this publication and assumes no liability.

Anyone using this information assumes all liability arising from such use, including but not limited to infringement of any patent or patents.

NOTICE

The U.S. Environmental Protection Agency, through its Office of Research and Development, funded and managed, or partially funded and collaborated in, the research described herein. It has been subjected to the Agency's peer and administrative review and has been approved for external publication. Any opinions expressed in this publication are those of the author (s) and do not necessarily reflect the views of the Agency, therefore, no official endorsement should be inferred. Any mention of trade names or commercial products does not constitute endorsement or recommendation for use.

Special Publications of the Water Environment Federation

The WEF Technical Practice Committee (formerly the Committee on Sewage and Industrial Wastes Practice of the Federation of Sewage and Industrial Wastes Associations) was created by the Federation Board of Control on October 11, 1941. The primary function of the Committee is to originate and produce, through appropriate subcommittees, special publications dealing with technical aspects of the broad interests of the Federation. These publications are intended to provide background information through a review of technical practices and detailed procedures that research and experience have shown to be functional and practical.

Contents

Chapter 1 Introduction

Chapter 2 Considerations for Planning of Biosolids Management Projects

Chapter 3 Public Outreach and Involvement

Chapter 4 Solids Production and Characterization

Chapter 5 Design Approach

Chapter 6 Conveyance of Wastewater Residuals

Chapter 7 Chemical Conditioning

Chapter 8 Thickening

Chapter 9 Sludge Minimization Technologies

Chapter 10 Anaerobic Digestion

Chapter 11 Aerobic Digestion

Chapter 12 Dewatering

Chapter 13 Composting

Chapter 14 Alkaline Treatment

Chapter 15 Disinfection and Stabilization Processes

Chapter 16 Thermal Drying of Biosolids

Chapter 17 Thermal Oxidation

Chapter 18 Pyrolysis and Gasification and Other Thermal Processes

Chapter 19 Transport and Storage

Chapter 20 Biosolids Odor Management

Chapter 21 Sidestreams from Solids Treatment Processes

Chapter 22 Instrumentation and Monitoring

Chapter 23 Land Application and Product Distribution

Chapter 24 Landfill Management Systems

Chapter 25 Emerging Technologies

Chapter 26 Treatment and Use of Green Gases

List of Figures

List of Tables

Preface

This publication represents a comprehensive and significant update of *Process Design Manual for Sludge Treatment and Disposal*, originally published by the U.S. Environmental Protection Agency in 1979. It was prepared and intended for use by professionals engaged in the design, approval, and operation of municipal solids treatment and disposal systems. This new manual expands the previous manual from 19 to 26 chapters, with contemporary new material on public outreach and involvement, waste minimization, anaerobic and aerobic digestion, odors, and treatment and utilization of green gases. While comprehensive in scope and depth in its own right, the revised manual includes many references for related subject material presented and readily available elsewhere.

The project was overseen by the following members of the Project Advisory Board:

U.S. EPA
Robert Bastian
Mark C. Meckes
James E. Smith, Jr., D.Sc., BCEEM (Retired)

WEF Technical Practice Committee
Albert B. Pincince
Todd Williams

WEF
Lorna Ernst
Melissa Jackson

WERF
Alan Hais, P.E., BCEE (Retired)
Amit Pramanik, Ph.D., BCEEM

This publication was developed under the direction of Timothy G. Shea, Ph.D., P.E., BCEE, *Task Force Co-Chair*, and Michael D. Moore, *Task Force Co-Chair*. Project management was provided by Jon Dyer and WEF as well as the following Topic Leaders:

Peter Brady (Chapters 16, 17, 18, and 22)
John F. Donovan, P.E., BCEE (Chapters 10, 11, 13, 14, and 15)
Jon Dyer and WEF (Chapters 4, 7, 8, 9, and 12)
Mark E. Lang, P.E., BCEE (Chapters 19, 20, 21, 23, 24, and 25)
Warren Uhte (Chapters 1, 2, 3, 5, 6, and 26)

Special thanks go to James E. Smith, Jr., D.Sc., BCEEM, for initiating this project and his leadership throughout the development of this publication.

The principal authors of this publication are as follows:

Chapter 1	Amit Pramanik, Ph.D., BCEEM
Chapter 2	Timothy G. Shea, Ph.D., P.E., BCEE Rick Stevens Michael D. Moore
Chapter 3	Ned Beecher Gordon Butte
Chapter 4	W. James Gellner, P.E. Peter Dold Richard Jones Wayne J. Parker, Ph.D., P.Eng. Ana Pena-Tijerina Oliver Schraa
Chapter 5	Bob Forbes, P.E. Timothy G. Shea, Ph.D., P.E., BCEE
Chapter 6	George Anipsitakis, Ph.D., P.E. Derya Dursun, Ph.D., P.E. Kenneth D. Fonda, P.E.
Chapter 7	Excerpted from *Design of Municipal Wastewater Treatment Plants* (WEF and ASCE, 2009) and *Operation of Municipal Wastewater Treatment Plants* (WEF, 2007) with support from Al Cassel
Chapter 8	Robert A. Gillette, P.E., BCEE John Bratby, Ph.D., P.E. Jeff Berk, P.E. D. Scott Joslyn, P.E. Susanna Leung, P.E. Rebecca Overacre, P.E. Tracy Stigers, P.E. Alan Straub, P.E. Steve Swanback, P.E.

Chapter 9	K. Richard Tsang, Ph.D., P.E., BCEE
	Engin Guven, Ph.D.
	Christopher Muller, Ph.D., P.E.
	Marc Roehl, P.E.
	Michael Salerno, Ph.D.
Chapter 10	David L. Parry, Ph.D., P.E., BCEE
	Cameron Clark
	Dustin Craig, P.E.
	Kudo Kilian
	Cale A. McPherson
	Perry Schafer, P.E., BCEE
	Toshio Shimado, Ph.D., P.E.
	Alex Travaglia, P.E.
	Jonathan Treadway, P.E., BCEE
	Scott Vandenburgh
	John Willis, P.E., BCEE
	Thomas E. Wilson, Ph.D., P.E., BCEE
Chapter 11	Paul Bizier, P.E., BCEE
	Harlan G. Kelly, P.E., BCEE
	Victor K. Lo
	Donald S. Mavinic, Ph.D.
Chapter 12	Angela Essner
	David C. Hagan, P.E.
	Carl Koch
	Peter LaMontagne, P.E.
	Nathan Lester
	Sarah A. Miller, BSc, MSc (Chem Eng)
	Rebecca Overacre, P.E.
	Rick Shanley, P.E.
	Julia Spicher, P.E.
	Steve Swanback, P.E.
Chapter 13	Todd O. Williams, P.E., BCEE
	Charlie Alix, P.E.
Chapter 14	James E. Smith, Jr., D.Sc., BCEEM
	Maurice Barker

	Kari Fitzmorris Brisolara, ScD
	Tim Doutt
	Harry Francis
	Greg Kester
	Terry Logan, Ph.D.
	Cheryl Minskey
	David W. Oerke, P.E.
	J. Kilren Vidrine
Chapter 15	James E. Smith, Jr., D.Sc., BCEEM
	Lewis Naylor, Ph.D.
	Bethany Acquisto, Ph.D.
	Kari Fitzmorris Brisolara, ScD
	Albert Cox, Ph.D.
	Joseph B. Farrell, Ph.D.
	Harry Francis
	Ben Gould
	Mark C. Meckes
	Fred P. Mussari
	Suresh D. Pillai, Ph.D.
	Shelly Wenger
Chapter 16	Harikishan Santha, P.E.
	Robert W. Bates
	Scott Carr, P.E., BCEE
	Edward DeLaForest, P.E.
	Eric A. Evans, Ph.D., P.E.
	Webster F. Hoener
	John Igoe
	Ajay Kasarabada, P.E.
	Neil S. Massart, P.E.
	Tazio Qubeck, P.E.
	James M. Rowan
Chapter 17	Robert P. Dominak, P.E., BCEE
Chapter 18	Stanley Chilson, P.E.
Chapter 19	Lisa Williams, CPSSc
	James L. Anderson, Ph.D., CPSS

	Amanda L. Bauner, P.E.
	Tim Chronister
	Sherrie Elliot
	A. Thomas Ferrero, Jr.
	Brooke Henderson
	Jean Malafronte, P.E.
	Magdalena Pietrzak, P.E.
	Pete Sarin
	Nicole Spieles, P.E.
	Andrew L. Thompson, P.E.
	Bill Toffey
Chapter 20	Chris Easter, P.E.
	Robert P. G. Bowker, P.E.
Chapter 21	Mario Benisch
	Michael Falk, Ph.D., P.E.
	Samuel Jeyanayagam, Ph.D., P.E., BCEE
Chapter 22	David W. Roberts
	Alex Ekster, Ph.D., P.E.
	Jeffrey S. Miller
	Patrick Moorman
Chapter 23	Lynne H. Moss, P.E., BCEE
	Jonathan Hill
	Terry Logan, Ph.D.
	Robert Pepperman
Chapter 24	Mike Sullivan, P.E.
	Robert Conway
	Michael R. Murray, Ph.D.
	James J. Walsh, P.E., BCEE
Chapter 25	K. Richard Tsang, Ph.D., P.E., BCEE
	Scott Carr, P.E., BCEE
	Alan B. Cooper, P.E., BCEE
	Engin Guven, Ph.D.
Chapter 26	Jim Schettler

In addition to the Authors, Topic Leaders, Task Force Co-Chairs, and members of the Project Advisory Board, technical material or review was provided by the following individuals:

Mohammad Abu-Orf, Ph.D.
Robert Adamski, P.E., BCEE
Layne Baroldi
Allen Baturay, P.E.
Jeanette Brown, P.E., BCEE
Peter Burrowes, P.Eng.
Stephanie Cheng, P.E.
Dick Christy
Tim Constantine, M.Eng., P.Eng.
David Cooley, P.E.
James Cox
Ron Crites
Steve Dentel, Ph.D.
Carla D. Dillon, D.P.A., P.E.
Jorja DuFresne
Lowell Eisaugle
Richard Finger
Stephan D. Frank
Kim Fries
Andrew Gall
Diane Garvey
Robert Gaudes, P.E., BCEE
Diane Gilbert
Alan Godfree, Ph.D.
Tom Granato
John Gundlach
Cecil Lue-Hing, Ph.D., P.E., BCEE
Raymond Kearney, P.E.
Michael Keleman
Robert Kelly
Michael R. Leffler, P.E., BCEE
Stephen G. Lippy, P.E., BCEE, SC
Lorrie Loder
Jim Marx
Bob Ooten
Chris Peot, P.E.
Michael Puccio
Mark Ramirez
Robert S. Reimers, Ph.D., Q.E.P.
Diego Rosso, Ph.D., D.Ch.Eng.
Thomas E. Sadick, P.E.
Kevin A. Samson
Julian Sandino, Ph.D., P.E., BCEE
Patricia Scanlan
Mike Scharp
Harold E. Schmidt, Jr., P.E., BCEE
Ralph B. "Rusty" Schroedel, Jr., P.E., BCEE
James Scisson, P.E.
James B. Slaughter, J.D.
Stephen Smith
Dave Stensel, Ph.D.
Donald C. Trueblood
Aarne Vesilind, Ph.D.
James Welp
Drury Whitlock, P.E.
Christopher Wilson, Ph.D.
James C. Young
Paula Zeller
Jianpeng Zhou
Jeff Ziegenbein

SOLIDS PROCESS DESIGN AND MANAGEMENT

Chapter 1

Introduction

1.0 BACKGROUND

This publication, *Solids Process Design and Management*, a collaborative effort between the U.S. Environmental Protection Agency (U.S. EPA), the Water Environment Federation (WEF), and the Water Environment Research Foundation (WERF), represents a comprehensive and significant update to *Process Design Manual for Sludge Treatment and Disposal* (U.S. Environmental Protection Agency, 1979). This update comes at a time of rapid change in solids management that is motivated by the need for greater cost-efficiency and improved technology performance.

More than 30 years after publication of the previous manual, wastewater requires increasing levels of treatment, resulting in even greater volumes of solids that need to be managed in an efficient and effective manner to protect human health and the environment. Additional changes include a greater understanding of the characteristics of wastewater and solids and recognition that solids resulting from wastewater treatment can be used as renewable resources if treatment methods are properly designed and managed. Finally, growing concerns have emerged about sustainability and the effects of wastewater and solids on air, water, and land. Global transfer of technologies and knowledge have led to significant advancements in equipment and process and resulting changes in planning, design, and management practices.

Table 1.1 Comparison of content of current and previous solids publications.

Solids Process Design and Management (U.S. EPA/WEF/WERF, 2012)		*Process Design Manual for Sludge Treatment and Disposal* (U.S. EPA, 1979)	
Chapter	**Topic**	**Chapter**	**Topic**
1	Introduction	1	Purpose and Scope
2	General Considerations for Planning of Biosolids Management Projects	2	General Considerations
3	Public Outreach and Involvement		
4	Solids Production and Characterization	4	Wastewater Solids Production and Characterization
5	Design Approach	3	Design Approach
6	Conveyance of Wastewater Residuals		
7	Chemical Conditioning	8	Conditioning
8	Thickening	5	Thickening
9	Sludge Minimization Technologies		
10	Anaerobic Digestion		
11	Aerobic Digestion		
12	Dewatering	9	Dewatering
13	Composting	12	Composting
14	Alkaline Treatment		
15	Disinfection and Stabilization Processes	6	Stabilization
		7	Disinfection
16	Thermal Drying of Biosolids	10	Heat Drying
17	Thermal Oxidation	11	High Temperature Processes
18	Pyrolysis and Gasification and Other Thermal Processes	13	Miscellaneous Processes
19	Transport and Storage	14	Transportation
		15	Storage
20	Biosolids Odor Management		
21	Sidestreams from Solids Treatment Processes	16	Sidestreams from Solids Treatment Processes

(*continued*)

TABLE 1.1 Comparison of content of current and previous solids publications (*Continued*).

Solids Process Design and Management (U.S. EPA/WEF/WERF, 2012)		*Process Design Manual for Sludge Treatment and Disposal* (U.S. EPA, 1979)	
Chapter	**Topic**	**Chapter**	**Topic**
22	Instrumentation and Monitoring	17	Instrumentation
23	Land Application and Product Distribution	18	Utilization
24	Landfill Management Systems	19	Disposal to Land
25	Emerging Technologies		
26	Treatment and Use of Green Gases		

In 2007, U.S. EPA began a collaboration with the WEF and WERF under a Cooperative Research and Development Agreement (CRADA) among the three organizations to significantly revise and update the 1979 manual and make it available on the Internet.

2.0 SCOPE AND PROCESS

This updated publication was prepared and intended for use by professionals engaged in the design, approval, and operation of solids treatment and disposal systems. This revised publication includes references to related material that is readily available elsewhere rather than duplicating it here.

This new publication comprises 26 chapters, compared to 19 in the 1979 edition, and presents new material on public outreach and involvement, waste minimization, anaerobic and aerobic digestion, odors, and treatment and use of green gases. Every effort has been made to include as many of the original chapters as possible; references to other sources of information are included throughout the chapters.

Solids Process Design and Management was developed using consensus-review procedures established by WEF's Technical Practice Committee (TPC). Review of WEF's peer-reviewed publications is open to anyone in the industry. Review comments are submitted in writing by established deadlines and are blinded to maintain the anonymity of reviewers, collated, and distributed to authors. Authors then revise the text accordingly, incorporating all review comments or providing written reason for not doing so. The revised text and responses to review comments are distributed to the original pool of reviewers so that they may see how their review comments have been addressed and whether their comments have been addressed to their satisfaction. These consensus-review procedures are followed until resolution of all review comments is achieved, resulting in a consensus of the industry.

It is highly recommended that design standard setting organizations embrace the changes incorporated in this design publication. Recommendations in this publication represent best practices that go well beyond the minimum design requirements necessary to comply with requirements of Standards for the Use or Disposal of Sewage Sludge (40 CFR Part 503) (U.S. Environmental Protection Agency, 1993). This is increasingly important for biosolids management to remain viable and environmentally sustainable in the future.

Finally, *Solids Process Design and Management* represents the knowledge, experience, and hard work of approximately 200 subject matter experts, staff, and members from each of the three partner organizations. The endeavor relied heavily on volunteers who served as members of the management team, topic leaders who coordinated groups of chapters, chapter lead authors, coauthors, and peer reviewers. These participants are listed in the front matter of this publication and their participation is gratefully acknowledged.

3.0 REFERENCES

U.S. Environmental Protection Agency (1979) *Process Design Manual for Sludge Treatment and Disposal*, EPA 625/1-79-011; U.S. Environmental Protection Agency, Municipal Environmental Research Laboratory and Center for Environmental Research Information and Technology Transfer: Cincinnati, Ohio.

U.S. Environmental Protection Agency (1993) Standards for the Use or Disposal of Sewage Sludge, Final Rule. *Code of Federal Regulations*, Part 503, Title 40; *Fed. Regist.*, **58**, 9248.

Chapter 2

Considerations for Planning of Biosolids Management Projects

(continued)

1.0 INTRODUCTION

Biosolids management planning is an exercise that starts with knowledge of the final use of biosolids products and then establishes what has to be done to deliver those products reliably and cost-effectively. A biosolids management plan (BMP) typically is iterative, changing as work progresses and more knowledge is gained.

This chapter is intended to preview the essential steps to create a well-planned and functioning BMP. These steps will be interpreted in multiple ways. Solids treatment and biosolids management compose approximately 50% of the annual and capital cost of wastewater treatment. This chapter is intended to identify issues of concern. It is not possible to provide all the details of a subject that has many challenges and opportunities, including significant public infrastructure, many end-users, and a significant land-use footprint.

Several subjects are addressed in the main body of the text and supported with appendices:

- Federal and state legislation and regulations,
- First considerations for planning a solids project,
- Project financing,
- Planning efforts,
- Required permits and applicable building codes, and
- Managing the completed system—an environmental management system (EMS).

2.0 FEDERAL AND STATE REGULATIONS

2.1 Background

The 1979 design manual presented an extensive review of the effect of the environmental legislation then in place for wastewater solids management (U.S. EPA, 1979). The reader is referred to this reference for the historical background, most of which is still relevant to the broader aspects of federal regulations affecting solids management. Of course, recycling biosolids to land had occurred in the United States—and remained largely unregulated—for many years before the founding of the U.S. Environmental Protection Agency (U.S. EPA) in 1970.

Of particular relevance to planning efforts is the role of U.S. EPA and its seminal Part 503 Standards for the Use or Disposal of Sewage Sludge, 40 CFR 503 (U.S. EPA, 1993). Under 40 CFR Part 503, U.S. EPA regulates the treatment, use, and disposal of biosolids (treated sludge) to protect public health and the environment from any reasonably anticipated adverse effects of pollutants that may be present in the biosolids. The regulations at 40 CFR 503 establish requirements for the final use and disposal of biosolids in three circumstances: (1) when applied to land for a beneficial purpose (e.g., to fertilize plants or to condition the soil, including products that are sold or given away for use in home gardens); (2) when disposed of by being placed on surface disposal sites (including dedicated land disposal sites and monofills); (3) when incinerated. State and local governments make the decision whether to recycle the biosolids as a fertilizer, to incinerate it, or to bury it in a landfill. Regardless of end use or disposal, however, state and local governments must comply with federal standards and requirements under the Clean Water Act (P.L. 92–500). The U.S. EPA does not promote land application over other options.

2.2 Part 503 Standards—Class A and Class B Biosolids

The standards for each end use and disposal practice consist of general requirements, numeric limits for up to ten metals concentrations, management practices, and operational standards. The final rule also includes monitoring, recordkeeping, and reporting requirements.

These requirements are designed to generate two classes of biosolids: (1) Class A, which can be used without restrictions (e.g., on farms, home lawns and gardens); and (2) Class B, which can be used with restrictions on use, including for crop harvesting, grazing animals, and public access. State biosolids regulations have played an increasingly significant role in biosolids management and use, including requiring inspection and approval of specific sites before biosolids application can occur.

Fecal matter that may contain pathogenic microorganisms enters community wastewater collection systems from sources such as hospitals, embalming facilities, animal slaughtering operations, and dwellings. Although these wastewaters are cleaned and treated in wastewater treatment plants (WWTPs), surviving pathogenic microorganisms will become concentrated in sludge. If sludge is to be beneficially used on land and possibly come in contact with humans, then it must be further treated so that levels of microorganisms and pathogens are significantly reduced and metals levels fall below certain thresholds. For Class B biosolids, harvesting, grazing, and public access restrictions also must be used to protect public health.

2.3 Pathogen Reduction

The U.S. EPA's rules put multiple lines of defense in place to reduce risk to negligible levels (U.S. EPA, 1993, 2003). The commonly used disinfection "processes to significantly reduce pathogens" (PSRP) for Class B biosolids and "processes to further reduce pathogens" (PFRP) for Class A biosolids are shown in the Table 2.1. This is only a core list of established technologies, and the later chapters of this manual contain many examples on other technologies that are categorized as established, innovative, or emerging.

TABLE 2.1 Typically used sludge disinfection processes (PSRP = process to significantly reduce pathogens; PFRP = process to further reduce pathogens).

PSRP	PFRP
Anaerobic digestion	Heat drying
Aerobic digestion	Composting
Lime stabilization	Thermophilic aerobic digestion

These processes are validated, benchmarked, and well described (U.S. EPA, 2003). The intent of PFRP processes is to reduce pathogenic organisms to below their analytical detection limits; for PSRP processes, it is to partially reduce the number of pathogens. Because pathogens can be present with the use of PSRP processes, the rules mandate site-access restrictions for Class B biosolids once land applied to undergo further pathogen reduction by natural attenuation. Thus public access, crop harvesting, and grazing restrictions are applied. These restrictions are detailed in Part 503 and the literature (U.S. EPA, 2003).

2.3.1 *Pathogen Reduction to Achieve Class A*

The requirements for a biosolids product to meet Class A for pathogens can be met with any one of six alternatives, which includes use of a PFRP process. Common to every alternative is the necessity for the density of *Salmonella* species to be reduced to less than 3 MPN per 4 g of dry sludge solids or for the fecal coliforms to be less than 1000 MPN/g of dry sludge solids. Because levels of fecal coliforms in untreated sludge often approach 10^8 MPN/g dry weight of sludge solids, it is expected that there will be at least a 5 log reduction to achieve Class A status. The four typically used alternatives are summarized below.

- Alternative 1. Time (D = days) and temperature (T = °C) are related by the equation: $D = 31\,700\,000/10^{0.14T}$ or $50\,070\,000/10^{0.14T}$. The first equation applies when total solids > 7%; T > 50 °C, and time is > 20 minutes. If the sludge particles are small and are heated by warmed gases or an immiscible liquid, then the minimum time is 15 seconds. This equation also applies when the solids < 7%, T > 50 °C, and time is > 15 s to < 30 min. The second equation applies for total solids < 7 %; T > 50 °C, and D > 30 minutes. These requirements were established from U.S. Food and Drug Administration (FDA) requirements for eggnog, data from German sources, and other data collected during composting experiments in the United States (U.S. EPA, 1992).

Many facilities approach Alternative 1 without realizing that it was derived from experience with fluids. With fluids it is not difficult to ensure that all particles meet the requirements for time and temperature. So, when a facility submits a plan to the permitting authority, the facility staff is often shocked when asked for a monitoring plan that will ensure that all parts of the sludge will meet the appropriate time and temperature requirements. In general these processes are limited to batch or plug-flow reactors rather than continuous-flow reactors. It is critical that these sludge treatment processes are operated without the occurrence of short circuiting.

- Alternative 2. This alternative is based on disinfection research done by the N-Viro Energy Systems, Inc. in the late 1980s (U.S. EPA, 2003). The pH is raised to greater than 12 for more than 72 hours, the temperature is greater than 52 °C, and, after the 72 hours, the treated sludge is air-dried to 50% solids or greater.
- Alternatives 3 and 4 are not discussed here. They depend on monitoring for the presence of enteric viruses and viable helminth ova. This is only useful when significant numbers of enteric viruses and helminth ova are present in the raw sludge and monitoring is done solely to measure the effectiveness of the treatment process. Otherwise, the monitoring is meaningless because the presence or absence of enteric viruses and helminth ova says nothing about the presence or absence of other pathogens.
- Alternative 5. The sludge is treated by a PFRP process. The three processes most frequently used are composting, pasteurization, and heat drying. With composting it is critical that all or most of the sludge particles pass through a zone where they can be held for no less than three days at 55 °C and then be removed without contamination. Pasteurization is holding the temperature of a fluid at 70 °C or above for at least 30 minutes. A good example is milk pasteurization. Achieving the same results by mixing a powder, such as quicklime or similar reagent, with semisolid material, sludge, is not easy. The mixing has to be thorough to ensure that all or most of the sludge particles come into contact with the alkaline material; the temperature gets elevated to 70 °C or above and is held there for at least 30 minutes. Some minimal level of moisture is necessary

and critical to ensure that thorough mixing can occur. These systems need to be engineered and carefully monitored to ensure that they are properly operating.

- Alternative 6. The sludge is treated by a process equivalent to a PFRP. Some technologies found to be equivalent include two-stage sludge stabilization (a thermophilic aerobic digester followed by a mesophilic anaerobic digester), autothermal thermophilic aerobic digestion, two-phase thermo-meso anaerobic digestion, and OxyOzonation (U.S. EPA, 2003). Readers are referred to U.S. EPA's Pathogen Equivalency Committee Website at http://www.epa.gov/nrmrl/pec/basic.html for further information about processes equivalent to PFRP and how to apply for equivalency.

2.3.2 *Pathogen Reduction to Achieve Class B*

Three alternatives can be used to meet the requirements for biosolids to be Class B for pathogens:

- Alternative 1. The geometric mean fecal coliform density [either most probably number (MPN) or colony-forming units (CFU) per gram of dry sludge solids] of seven samples shall be less than 2 000 000. Here, because levels of fecal coliforms in the untreated sludge can approach 10^8 MPN/g dry weight sludge solids, at least a 2 log reduction is expected.
- Alternative 2. The sludge is treated by a PSRP process. As previously noted, the typically used processes are aerobic digestion, air drying, anaerobic digestion, composting (less stringent thermal requirements than PFRP), and lime stabilization (U.S. EPA, 2003).
- Alternative 3. The sludge is treated by a process equivalent to a PSRP. Refer to http://www.epa.gov/nrmrl/pec/basic.html for further information about processes equivalent to PSRP and how to apply for equivalency.

2.4 Vector Attraction Reduction

The U.S. EPA regulates vector attraction reduction with processing or physical barrier alternatives (see Appendix A). Both Class A and Class B processes are required to minimize the attractiveness of biosolids to vectors, such as flies and birds. Within the United States, concerns beyond vector attraction reduction to achieve public acceptance are the responsibility of state and local authorities. It is recognized, though, that if a material is no longer putrescible (i.e., a minimal amount of biodegradable organic material is present), then it is not likely to be odorous.

There are specific requirements to be met: 38% volatile solids reduction number; demonstrating a specific oxygen uptake rate (SOUR) at 20 °C of 1.5 mg oxygen/hr/g total sludge solids; and aerobically treating sludge for at least 14 d at > 40 °C with an average temperature of > 45 °C. These requirements indicate removal of most putrescible material from sludge so that it is not attractive to vectors. Not surprisingly, these test values are not optimal. Adjustments are needed. For example, we know that biological digestion processes can achieve volatile solids reductions of 20 to 70% as a function of the specific sludge being digested. Thus what is needed is research that may address and find alternatives for replacing the 38% value with a formula into which the parameters for a specific wastewater sludge can be inserted. The SOUR number can only be used within a narrow temperature range and with a relatively low solids concentration. These conditions need to be expanded and include the situation of thermophilically digested wastewater sludge. Better tests are already available for composted materials, such as measuring the evolution of CO_2, which should be considered by regulatory authorities.

2.5 Numeric Limits for Heavy Metals

The heavy-metal concentration of biosolids to be land applied is also regulated. Two parameters are used to characterize the ten heavy metal concentrations (Table 2.2). These include pollutant "ceiling concentrations" and pollutant concentrations (the latter is labeled "high quality" pollutant concentrations in Table 2.2). To qualify for land application, biosolids or material derived from biosolids must meet or fall below the pollutant ceiling concentration limits and meet Class B requirements for pathogens and vector attraction reduction requirements.

Cumulative pollutant loading rates are imposed on biosolids that meet pollutant ceiling concentrations, with the exception of biosolids that meet the criteria for "exceptional quality." Exceptional quality biosolids must meet three criteria: pollutant concentration limits, Class A pathogen requirements, and vector attraction reduction requirements.

The minimum and maximum observed values for the ten heavy metals from the Targeted National Sewage Sludge Survey (TNSSS) also are presented in Table 2.2. The maximum values for several of the metals exceed the Part 503, Table 3 criteria (Section 503.13) for high-quality pollutant concentration limits. Only three metals

Table 2.2 Numeric criteria and Targeted National Sewage Sludge Survey (TNSSS) results.

	Part 503 Numeric Criteria[a] Land Application Tables in Section 503.13				TNSSS Results[b] Observed Dry-Weight Concentration	
Pollutant	**Table 1[c] mg/kg[g]**	**Table 2[d] kg/ha**	**Table 3[e] mg/kg**	**Table 4[f] kg/ha/yr**	**Minimum mg/kg**	**Maximum mg/kg**
Arsenic	75	41	41	2	1.18	49.2
Cadmium	85	39	39	1.9	0.21	11.8
Chromium	–	–	–	–	6.74	1160
Copper	4300	1500	1500	75	115	2580
Lead	840	300	300	15	5081	450
Mercury	57	17	17	0.85	0.17	8.3
Molybdenum	75	–	–	–	2.5	132
Nickel	420	420	420	21	704	526
Selenium	100	100	100	5	1.1	24.7
Zinc	7500	2800	2800	140	216	8550

[a] Includes updates to Part 503: 1) A February 25, FR Notice deleting the values for molybdenum in Tables 2–4 (59 FR 9095), and 2) An October 25, 1995 FR Notice dropping Chromium from coverage and raising the Selenium value for Table 3 (60 FR 54764).

[b] Targeted National Sewage Sludge Survey (http://www.epa.gov/waterscience/biosolids/tnsss-overview.html)

[c] Ceiling Concentrations

[d] Cumulative Pollutant Loading Rates

[e] "High Quality" Pollutant Concentrations

[f] Annual Pollutant Loading Rates

[g] Dry weight basis

in the TNSSS exceeded the land application ceiling concentrations: molybdenum, nickel, and zinc (Table 1, Section 503.13). The maximum observed concentrations for all other regulated metals were less than the land application ceiling concentrations. None of the facilities that actually dispose of sludge by land application exceeded the limits.

2.6 State Regulations

The U.S. EPA's regulations at 40 CFR Part 503 establish requirements for final disposal or use of biosolids that are land applied to condition the soil for vegetation, placed in a surface disposal site, or burned in a sludge incinerator. States may establish more stringent standards than the Federal requirements. For example, states may prohibit application to sensitive areas, such as areas with shallow groundwater. Most importantly, states typically inspect and test application sites before they are approved and issue a permit for land application, thereby allowing evaluation of soil quality, topography, and population density. State and local governments make the decision whether to recycle the biosolids as a fertilizer, to incinerate it, or to bury it in a landfill. The U.S. EPA allows the choice to be a local one, as long as the local decision complies with standards and requirements under the Clean Water Act (33 U.S.C. § 1251). The U.S. EPA does not promote land application over other options.

Most states have regulations that impose additional restrictions and require additional management practices, treatment processes, and testing requirements (North East Biosolids and Residuals Association, 2007). States typically assign biosolids regulatory oversight to the water/wastewater program, solid waste program, or the combined water/wastewater and solid waste program of state environmental agencies. In a few states, responsible agencies also include the public health department (e.g., Arkansas); the air quality division (e.g., Connecticut and other states with sludge incinerators); or the environmental agency (e.g., Virginia) (North East Biosolids and Residuals Association, 2007). Because the assignments and priorities for biosolids management shift from time to time, maintaining a current database of state regulations is a continuing process. The Residuals and Biosolids Committee of WEF has agreed to undertake the periodic update of this manual to include new developments, including state regulations. A similar process will be used to add important regulatory developments at the local level that may ultimately influence state and national levels.

Since 1993, when the Part 503 rule was adopted, the U.S. EPA has offered states the option of becoming delegated to administer the rule. Delegation requires that states have regulatory and enforcement programs that ensure compliance with Part 503. Delegation allows states some autonomy, which can streamline the regulatory process by avoiding involvement of, and reporting to, both state and federal authorities. As of 2010, relatively few states have pursued delegation: Arizona, Michigan, Ohio, Oklahoma, South Dakota, Texas, Utah, and Wisconsin.

3.0 BASIS FOR A BIOSOLIDS MANAGEMENT PLAN

This section reviews background factors to consider in the preparation of a BMP. Ownership of a WWTP can be vested in a city, a county, an authority, a sanitary district, or other such entities. The owner of the WWTP may have agreements in place with other user jurisdictions to provide wastewater treatment and other services. In such cases, the user jurisdictions will be part of the rate base of the owning agency and, as a result, may expect a voice in the selection and implementation of the solids processing systems at the treatment plant. The result is that the ownership and user jurisdictions for each treatment plant will result in a group or groups of stakeholders with specific interests in the solids processing and biosolids management programs.

A second factor in planning of solids processing facilities is the treatment plant itself including the processes, operations, and practices and end uses or disposal outlets for biosolids that are in use. The existing program will influence the next solids project in multiple ways, from the historical views of agency management and the personal experiences of the operating and maintenance personnel, to the availability of tankage that can be retrofitted for other purposes in newer processing schemes.

Several drivers for a project may exist:

- Legal and regulatory actions requiring new, expanded, or upgraded treatment and end-use management.
- Public or political support for recycling or, conversely, public misunderstanding of biosolids and opposition to a project.
- Changing end uses or outlets for biosolids because of urban sprawl or as exemplified by developments such as, for example, a reduction in agronomic loading rates when phosphorus rather than nitrogen is the limiting nutrient; the acceptance of pelletized biosolids as a home garden fertilizer; and the acceptance of pellets into cement manufacturing facilities as a substitute fuel.
- Community growth that will overload existing facilities.
- Wastewater master plans that add new sources of solids such as tunnel pumpouts from treatment of combined sewer overflows and the direct acceptance of fat-oil-grease (FOG) and food wastes at the solids processing site of the treatment plant; or sidestream treatment systems to alleviate nutrient loadings to enhanced nutrient reduction (ENR) projects.

- Global climate change regulations and inventories that provide incentives or mandates that the carbon footprint or greenhouse gas emissions from a solids processing scheme be minimized.
- Availability of new technology that has been demonstrated to the satisfaction of a client and that offers significant technical, economic, or environmental advantages.

As the process unfolds, some institutional difficulties may be encountered if a completely new approach is advocated. It can be difficult financially for a WWTP owner to walk away from an operating solids processing facility, especially if fully amortized. There can be a "continental divide" that exists to the present day that was created decades ago by an agency's decision to favor anaerobic digestion or incineration. This divide means that an agency wishing to change from incineration to anaerobic digestion will have to pay full price for new facilities. This economic burden can be insurmountable, especially because many agencies are also paying for court-mandated programs to end sanitary and combined-sewer overflows. The incentive to make such changes has to be either a strong economic advantage or a legal ruling against continuation of the present program. As a result, a solids project will often become discretionary (subject to deferred funding).

4.0 PROJECT FINANCING

Solids processing and biosolids management projects are uniquely amenable to a range of public or private financing alternatives. A role for the private sector is practical because certain components of a solids processing train are at the end of the line, allowing clear assignment of responsibilities for the quality of the material transferred from upstream (the owner) and from the privatized sector downstream (the contractor). It should be noted that a WWTP generating the solids should never transfer full liability to a third party and must realize that they have cradle-to-grave responsibility. Many states do not allow the legal transfer of liability to a third party and would require the contractor to act merely as an agent to the generator. Moreover, experience has shown, with land application in particular, that it is prudent business for the generator to closely monitor the operations of a contractor to ensure long-term viability of the program. Lack of oversight of third-party management has been the source of many areas of controversy that periodically surface. Because Part 503 mandates that agencies are responsible for the treatment of biosolids and for certifying to a contractor that the biosolids are fit for land application, agencies and contractors should have a clear understanding of who will defend allegations if the biosolids have caused injury or nuisance.

The financing of solids projects can come from conventional sources, such as bond issues and State Revolving Fund monies, from operating and maintenance budgets, or from private sources. In some cases, the funding can come from a program that creates an income stream for repayment of the private source. The latter approach has been used to finance the construction of biogas treatment and combined heat-and-power facilities, in return for the income stream associated with the revenue from the electricity generated with the biogas. The incentive to the private sector is the opportunity to receive an acceptable return on its capital investment in a high-technology system.

The delivery of a solids project has typically been done by conventional design-bid-build procurement. As solids processing has become more complex in nature, specialty contractors have risen to the need. Private sector involvement has been at the end of the processing line and involved terminal operations such as thermal drying with removal and marketing of the heat dried biosolids product or receiving, dewatering, transport, storage, and land application of stabilized or Part 503 compliant biosolids. The amount of private sector involvement in a solids processing program will depend upon the availability of funds within an agency and the view of management toward contract operations.

There are now many treatment plants where the entire solids processing train is operated by contractors experienced in this type of work. In some cases, the private contractor will have financed the upgrade or expansion of the facility to meet current standards in return for a payment stream that incentivizes the contractor to greater efficiency.

5.0 THE PLANNING EFFORT

The planning effort is the essential starting point for any solids processing and biosolids management program (BMP). The planning process begins with the end uses of the biosolids, leading to development of the supporting process technologies, risk management procedures, project phasing, and other considerations. The authors of the BMP have to create a road map for the utility manager that is based on clear, defensible decisions, provides diversification and redundancy, addresses risks and liabilities, is sustainable, incorporates continual improvement, and has stakeholder and public support.

In starting with the end use or uses for biosolids, a BMP should establish the preferred alternative, or the highest ranked alternatives, and the issues to be resolved for selection. The BMP should be a road map for the development of the program. It should provide the core information for the capital improvement program (CIP), including a prioritized project list, implementation schedule, capital cost estimates, financial plan, and documentation of the role played by the governing board in reaching consensus of the final plan of the governing body of the owner. The effort must be based upon a well-considered and systematic decision and implementation process that is responsive to the legal and financial responsibilities of the owner. Roadblocks may surface because of organizational issues such as insufficient direction from the owner, unclear responsibilities, differing perspectives, conflicting objectives, limited information, and inadequate alternatives. These are all issues to be resolved at the onset, not after a facilities design program has been started.

There are multiple approaches to development of a BMP, one of which is the "six-step process." The six-step process is used as an example because it provides important organizational benefits to the process. These include the establishment of leadership and commitment, identification of conflicting objectives, provision of tools for conflict resolution, creation of a clear track record of decision-making, process auditing and improvements, and implementation guidance. The six-step process also focuses on identification of the correct problems, identifies the important alternatives, addresses risk (uncertainty), and provides a template for collecting essential information. Meaningful stakeholder involvement must occur at each step and typically includes a workshop that allows for significant input from stakeholders.

The six steps and questions to be addressed in each are presented below.

1. Develop leadership and commitment. Why is the project to be done? Who are the decision-makers? Who are the key stakeholders? What roles will individuals play? Answers to these questions are the foundation of a successful decision process, providing the vision and boundaries for the project, ensuring the credibility of resources with milestones, and involving the right people in effective team partnerships.
2. Frame the problem. What are the core issues to be addressed? What decisions, uncertainties, and values must be clarified? How do these relate to

each other? What policies, strategies, and tactics are known? What general types of information are needed to evaluate the alternatives?

3. Build the evaluation model. How will it be clear when the stated objectives have been achieved? What is the hierarchy of objectives to be met by the successful alternative? How can the outcomes associated with each alternative be measured? What are the primary drivers for the outcomes?
4. Collect meaningful, reliable data. What data is needed? In what format? How will expert opinions be incorporated? Is the data acceptable to a technical peer review? Is the data credible to the decision-makers and stakeholders?
5. Evaluate alternatives and make decisions. What is the optimal strategy? The best alternative strategies? What risks exist? What factors would have to change to select an alternative strategy? Is there enough data to select a strategy and what additional data are needed? Was the correct problem solved?
6. Develop the implementation plan. How is the decision process communicated? What tasks are required and what barriers to success exist? How to the barriers to be overcome? Who is responsible for each task? What contingencies are required? What "meaningful stakeholder involvement" should be included in the decision-making process?

The process requires transparency and involvement of stakeholders throughout the effort. There is considerable latitude required from one planning effort to the next (i.e., there is no standard template). The successful BMP will include a plan for the short-term, interim, and long-term projects and operations. The BMP will include prioritization of projects and the basis for the CIP and preparation of the financing and facility planning.

6.0 REQUIRED PERMITS AND APPLICABLE BUILDING CODES

It is prudent at the planning stage to identify all the required permits and applicable building codes for a project, especially those with long lead times and extensive information preparation requirements. The permits of interest will include the National Pollutant Discharge Elimination System (NPDES) and applicable state permits with sections pertaining to solids processing and biosolids management, permits to construct and to operate, and permits for construction in waterways and wetlands.

If an air permit is required, then the application process may require air quality modeling, consideration of the most advanced technology in use for a given application, the new source review procedure, and other demonstrations of the adequacy of a proposed air pollution control strategy. The initial step will be development by the applicant of an air permit acquisition strategy, to be submitted for review by the permitting agency as a component of the work program. The application also will require any odor mitigation that the project will employ. Upon approval of the air application strategy, the project can begin as defined in the application.

The requirements for a building permit can vary, with the local fire inspector likely to be the ultimate judge of acceptability. These requirements can get complex for solids processing systems such as thermal dryers. For example, the building housing a thermal dryer system must be classified according to the local building code, and this step requires the selection of the correct electrical classification. The building code review should be done by an architect familiar with dryer facility design, and the electrical code classification review should be carried out by an electrical engineer experienced in dryer facility design. These requirements are explained further in Chapter 17, Thermal Oxidation.

7.0 EMERGING ISSUES

An emerging issue is defined as a factor that may influence the viability of a specific project, even if not fully understood. Several examples of emerging issues that should be considered from the start of a project are:

- The selection of a solids processing technology that yields a byproduct that is difficult to treat with an existing liquid treatment train.
- Identification of potentially new pathogens of wastewater origin that can be transmitted from human to human or from animal to human in public drinking water supplies, treated effluents, and biosolids.
- Identification of "microconstituents" (i.e., anthropogenic elements and compounds that are being detected in water, wastewater, and biosolids including pharmaceuticals, personal care products, household cleaning products, flame retardants, and pesticides as limits of detection decrease) and their sources. The WERF report, *Technical Brief: Trace Organic Compounds and Implications for Wastewater Treatment*, is an excellent summary of available scientific data on

trace organic compounds in wastewater and the receiving environment that puts these concerns into perspective (Anderson, 2008).

- Nuisance factors associated with biosolids management practices, especially as related to the land application of biosolids.
- Management systems for excessive nutrient loadings to the land, such as the "phosphorus index," which is a measure of the relative mobility of phosphorus from soil to water column. The phosphorus index is now being incorporated at the state level into farm-by-farm nutrient management plans (NMPs). Nutrient management plans with phosphorus-index control measures should include water extractable phosphorus (WEP) in their calculations. This is a measure of the soluble portion of phosphorus, which is of prime environmental concern. Research has shown that the WEP in most biosolids is a fraction of that in manures and commercial fertilizer (Brandt et al., 2002). If the WEP is not considered for biosolids, then phosphorus-index based control measures can result in a reduction of the allowable loadings of biosolids on phosphorus-sensitive lands by 80% or more—greatly altering the cost-effectiveness and perhaps feasibility of land application.
- Global climate change is being recognized for what it is—a long-term threat to the planet. The public reaction to global climate change is happening without regulatory support from the federal level, but with significant support from the states of California, Oregon, Pennsylvania, and others, and with firm support from organizations like the U.S. Conference of Mayors. This subject is addressed in more detail in Chapter 3.

8.0 ENVIRONMENTAL MANAGEMENT SYSTEMS

8.1 The Challenges

The challenges—such as assuring management support, documenting the standard operating procedures, and ensuring compliance with regulatory requirements—of creating a viable BMP are just the beginning of a successful BMP. Implementation is often more challenging. The National Biosolids Partnership (NBP) has responded to this challenge by developing elements of an environmental management system (EMS) to build the management infrastructure for a successful, sustainable biosolids program. The elements of an EMS are set forth in the references cited at the end of this chapter, from which this section has been developed.

8.2 Management Overview (or Ensuring Management Support)

If management does not support or understand development and implementation of a management system, then there will never be staff-level support. Management needs to review and agree to the process required to develop the EMS, including staff time and meaningful stakeholder involvement and commit to the continuous improvement philosophy that is central to an effective EMS. Management also needs an understanding of the cost-benefit ratio of an EMS.

8.2.1 Management Policy and Support

Experience has shown that the top management of an organization must first determine, commit to, and embrace an EMS as a way to organize and implement a continuous improvement program for biosolids operations. Understanding, buy-in, and ownership are critical for those who will develop and implement the EMS, not only for managers but all the way through the organization. It is essential to reach out to key staff and educate them on the benefits of an EMS. This will avoid springing it on them or handing down a directive, because this has caused problems to occur time and again. The EMS needs full initial support to serve as a foundation on which to build success. Staff can build support among senior management and help them champion the effort by explaining how an EMS can help link up and strengthen existing practices and ensure compliance, improve performance, engender greater public acceptance, and develop greater efficiency.

An environmental policy is top management's declaration of its commitment to the environment. This policy should serve as the foundation for an EMS and provide a unifying vision of environmental concern by the entire organization. Given its importance, the policy should be more than just flowery prose and should serve as the basis for the management system. Because it serves as the framework for setting environmental objectives and targets, the policy should be brought to life in plans and deeds. Everyone in the organization should understand the policy and what is expected of them to achieve the organization's objectives and targets.

The policy should contain three key commitments: to continual improvement, to pollution prevention, and to compliance with relevant laws and regulations. This does not mean that improvements need to be made at once in all areas. Rather it means that the policy should drive the organization's efforts to continually improve environmental management (and the improved performance that results from these efforts).

8.2.2 *Goals and Objectives for Continual Improvement*

Goals and measurable objectives and targets help an organization translate purpose into action. These environmental goals and objectives should be factored into strategic plans. This can facilitate the integration of environmental management with the organization's other management processes.

Goals can be applied organization-wide or to individual units, departments, or functions, depending on where implementing actions will be needed. In setting objectives for an environmental policy, it is important to keep in mind three pillars of environmental, social, and financial considerations. Consideration should be given to significant environmental aspects, applicable legal and other requirements, the views of interested parties, technological options, and financial, operational, and other organizational considerations.

There are no "standard" environmental objectives that make sense for all organizations. Objectives and targets should reflect what the organization does, how well it is performing, and what it wants to achieve.

8.2.3 *Roles and Responsibilities*

For an EMS to be effective, roles and responsibilities must be clearly defined and communicated. The commitments of all employees are needed for an EMS to live up to its full potential. Top management plays a key role by providing resources needed to implement the EMS, one of its most important responsibilities. In some organizations, top management might be a single individual; in other organizations, it might be a group of people (such as a board of directors).

An effective management system needs an advocate, and to this end top management should appoint a management representative. It should be the duty of this representative to ensure that (1) the EMS is established and implemented; (2) reports are prepared on its performance over time; and (3) management is involved with others to modify the EMS as needed. A business owner, plant or shop manager, or any number of other people might serve as an effective EMS management representative.

8.2.4 *Management Review*

Just as a person should have periodic physical exams, an EMS must be reviewed periodically by top management to stay "healthy." Management reviews are one key to continual improvement and for ensuring that the EMS will continue to meet the organization's needs over time. These reviews are an opportunity for management to gauge the overall effectiveness of the system and make adjustments to improve the system and respond to emerging challenges. Management reviews also offer a great opportunity to keep the EMS efficient and cost-effective. For example, some organizations have found that certain procedures and processes initially put in place were not needed to achieve their environmental objectives or to control key processes. If EMS procedures or other activities did not add value, then they can be eliminated.

The management review seeks to answer: Is the system working? Is it suitable, adequate, and effective given our needs?

8.3 Quality Management Practices

Many types of management practices exist: some good but not sustainable, some are poor but persist for the time being, and some are of the highest quality and are sustainable. This section will discuss the types of management practices that can be acceptable, both from a public and regulatory perspective, if everything is done right.

8.3.1 *Best Management Practice Guidelines*

A best management practice can be defined as a practice, or combination of practices, that is the most effective and feasible means of controlling a process or source of pollutants for the attainment of air, land, or water quality objectives. The guidelines for determining best management practices are dependent on what needs to be managed.

8.3.2 *Safe Practices Manual Guidelines*

Guidelines for safe practices are the safety procedures for each of the best management practices and standard operating procedures to ensure that employees remain safe and that public health and the environment are protected.

8.3.3 *Environmental Aspects and Critical Control Points*

Identification of critical control points along an organization's biosolids value chain is fundamental to effectively operating an EMS. By identifying critical control points and associated environmental effects, an organization can plan and implement proactive steps and operational controls. These efforts will ensure that the environmental products/discharges material characteristics are consistent with intended or actual final use or disposal and will manage or change environmental effects.

8.3.4 *Operational Controls*

Operational controls are needed to ensure that the commitments in the environmental policy are satisfied and that certain operations and activities be controlled. Where operations or activities are complex and/or the potential environmental impacts are significant, controls should include documented procedures. Procedures can help the organization to manage its significant environmental aspects, ensure regulatory compliance and achieve environmental objectives. Procedures can also play a prominent role in employee training.

Procedures should be documented where needed to prevent deviations from the environmental policy (including the commitments to compliance and pollution prevention) or from the objectives and targets. Determining which operations should be documented and how those operations should be controlled is a critical step in designing an effective EMS. The reader should keep in mind that operational controls might be needed to manage significant aspects of the regulatory or legal requirements, regardless of whether there are established objectives and targets for each of them.

In determining which operations and activities need to be controlled, one must look beyond routine production or service. Activities such as equipment maintenance, management of onsite contractors, and services provided by suppliers or vendors could affect the organization's environmental performance significantly.

8.3.5 *Monitoring and Measurement*

An EMS must have effective monitoring and measurement processes. Monitoring and measurement enables an organization to

- Evaluate environmental performance;
- Analyze root causes of problems;
- Assess compliance with legal requirements;
- Identify areas requiring corrective action; and,
- Improve performance and increase efficiency.

Monitoring helps manage an organization better. Pollution prevention and other strategic opportunities are identified more readily when current and reliable data are available. Historic data typically are helpful in performing forensic evaluations and trouble-shooting tasks.

The organization should develop procedures to

- Monitor key characteristics of operations and activities that can have significant environmental effects or compliance consequences;
- Track performance (including progress in achieving objectives and targets);

- Calibrate and maintain monitoring equipment; and,
- Evaluate compliance with applicable laws and regulations through periodic internal audits.

8.3.6 Documentation and Record Keeping

People in the organization probably use various documents (procedures, work instructions, forms, drawings) as they perform their duties. An organization must provide personnel with the proper tools to ensure that they are consistently performing their jobs correctly. In the context of an EMS, the "tools" needed are correct and up-to-date procedures, instructions, and other documents. Without a mechanism to manage these EMS documents, the organization cannot be sure that people are working with the right tools.

To ensure that everyone is working with the proper EMS documents, the organization should have a procedure that describes how such documents are controlled. Implementation of this procedure should ensure that

- EMS documents can be located (staff knows where to find them);
- They are periodically reviewed (staff checks to make sure they are still valid);
- Current versions are available where needed (staff ensures the right people have access to them); and,
- Obsolete documents are removed (staff does not use the wrong documents by mistake).

Procedures should designate responsibility and authority for preparing documents, making changes to them, and keeping them up-to-date. In other words, make it clear who can actually generate and change documents and the process for doing so.

The value of records management is fairly simple—to demonstrate that the organization is actually implementing the EMS as designed. Although records have internal value, over time they may be needed to provide evidence of EMS implementation to external parties (such as customers, a registrar, or the public). Records management is sometimes seen as bureaucratic, but it is difficult to imagine a system operating effectively and consistently without accurate records.

The basics of records management are straightforward: decide what records will be kept, how they will be kept, and for how long. The organization also should consider how records will be disposed of when they are no longer needed. If the organization has an ISO 9001 (or other) management system, then there may be a process in place for managing records. This process could be adapted for EMS purposes.

8.3.7 *Auditing, Nonconformance, Preventive and Corrective Action*

Once an organization has established its EMS, verifying implementation of the system through an internal audit will be critical. To identify and resolve EMS deficiencies, an organization must actively seek them out.

In a smaller organization, periodic internal audits can be particularly valuable. Managers are often so close to the work performed that they may not see problems or bad habits that have developed. Periodic internal EMS audits will help determine whether all of the requirements of the EMS are being carried out in the specified manner.

For the EMS internal audit program to be effective, an organization should

- Develop audit procedures and protocols;
- Determine an appropriate audit frequency;
- Select and train auditors; and,
- Maintain audit records.

Results of the EMS audits should be linked to the corrective and preventive action process described earlier. Internal audits can be performed by existing staff or by a qualified outside organization. Although they can be time-consuming, EMS audits are critical to effectiveness. Systematic identification and reporting of EMS deficiencies to management provides a great opportunity to maintain focus on the environment, to improve the EMS and its performance, and to ensure the system's cost-effectiveness.

8.3.8 *Emergency Preparedness and Response*

Despite an organization's best efforts, the possibility of accidents and other emergency situations still exists. Effective preparation and response can reduce injuries, prevent or minimize environmental effects, protect employees and neighbors, reduce asset losses, and minimize downtime.

An effective emergency preparedness and response program should include provisions for

- Assessing the potential for accidents and emergencies;
- Preventing incidents and their associated environmental effects;
- Plans/procedures for responding to incidents;

- Periodic testing of emergency plans/procedures; and,
- Mitigating the effects of these incidents.

Consistent with the focus on continual improvement, it is important to review the emergency response performance after an incident has occurred. This review can be used to determine if more training is needed or if emergency plans/procedures should be revised.

8.4 Relationships with Interested Parties

8.4.1 Public Participation

Planning through final use or disposition of product is a requirement of the National Biosolids Partnership's EMS Program that has been invaluable in gaining the public's acceptance.

Securing and maintaining public acceptance of the biosolids management program, particularly with respect to beneficial use, should be part of the policy. Information from public participation also should be considered in setting EMS goals and objectives because public acceptance (or lack thereof) likely will have an effect on overall success. (See Goals and Objectives/Targets in the NBP Manual of Good Practice, at http://www.wef.org/biosolids/) Although the requirements are intended to be flexible, there are specific requirements for implementing a proactive approach and should be consistent with the degree of current public interest, historic public involvement, the method of biosolids management, and other local circumstances. A "proactive public participation approach" is one that creates avenues for interested parties, as identified by the organization, to communicate their perceptions and concerns regarding biosolids management activities near the beginning of the EMS planning process and during the periodic review of EMS program goals and objectives.

Specifically, this process applies to identifying environmental effects and setting performance goals and objectives for the biosolids program. A "proactive" participation approach also enables the organization to understand the needs, concerns, and perspectives of interested parties before problems arise or incidents occur. Public participation requires that the input from interested parties be considered but clearly leaves decision-making up to the operating agency.

8.4.2 Communication and Outreach (Internal, External, Regulatory)

The importance of employee involvement in developing and implementing the EMS has been discussed earlier. In addition, there may be parties with an interest in the organization's environmental performance and management efforts outside the organization. Effective environmental management requires effective communications, both internally and externally.

Effective communications will help to

- Motivate the workforce;
- Gain acceptance for the plans and efforts;
- Explain the environmental policy and EMS and how they relate to the overall organizational vision;
- Ensure understanding of roles and expectations;
- Demonstrate management commitment;
- Monitor and evaluate performance; and,
- Identify potential system improvements.

Effective internal communication requires mechanisms for information to flow top-down, bottom-up, and across functional lines. Because employees are on the "front lines," they can be an excellent source of information on issues, concerns, and ideas.

Proactive, two-way communication with external parties is also important for an effective EMS. Taking steps to obtain the views of stakeholders, which can include neighbors, customers, community groups, and regulators, will help the organization better understand how it is perceived by others.

These stakeholders also can bring important environmental issues to the organization's attention that should be addressed by the EMS. The organization also should consider ways to get specific advice from these stakeholders when developing critical elements of the EMS such as setting objectives and targets. Involving these parties, however, does not mean the organization should cede control of the EMS or decision-making to them, but rather use their input to make the EMS stronger and more responsive to community concerns. Doing so typically will provide long-term benefits to the organization.

Thus, an effective EMS should include procedures for communicating internally (between levels and functions within the organization) and soliciting, receiving, documenting, and responding to external communications.

8.4.3 Training

There are two excellent reasons for training employees about environmental management and the EMS: (1) every employee can have potential effects on the environment and (2) any employee can have good ideas about how to improve environmental management efforts.

Each person and function within the organization can play a role in environmental management. For this reason, the training program should be broad. Every employee and manager should be aware of the environmental policy, the significant environmental effects of their work activities, key EMS roles and responsibilities, procedures that apply to their work, and the importance of conformance with EMS requirements. Employees also should understand the potential consequences of not following EMS requirements (such as spill, releases, and fines or other penalties).

All personnel should receive appropriate training. Such training should be tailored to the different needs of various levels or functions in the organization. Training is just one element of establishing competence, however, which typically is based on a combination of education, training, and experience. For certain jobs, particularly tasks that can cause significant environmental effects, the organization should establish criteria to measure the competence of individuals performing those tasks.

8.4.4 Regulatory Compliance/Legal Requirements

To comply with laws and regulations that apply to the organization, employees must first know what the rules are and how they affect what they do. Compliance with legal requirements is a critical part of the basis of an environmental policy. The potential costs of non-compliance (possible damage to the environment, revenue loss, and damage to public image, for example) can be high. Thus, an effective EMS should include processes to

- Identify and communicate applicable legal and other requirements;
- Ensure that these requirements are factored into the organization's management efforts; and,
- Respond promptly to instances of non-compliance, investigate the root causes of the non-compliance, and ensure that appropriate practices and procedures are in place to prevent future non-compliance.

New or revised legal requirements might require modification of the environmental objectives or other EMS elements. Anticipating new requirements and making changes to operations may avoid some future compliance obligations and their costs.

Organizations should consider the following factors in setting measurable objectives and targets:

- Prevention of noncompliance;
- Prevention of pollution at the source;
- Minimization of cross-media pollutant transfers; and,
- Environmental performance improvement.

9.0 REFERENCES

Anderson, P. D. (2008) *Trace Organic Compounds and Implications for Wastewater Treatment: Technical Brief*. Water Environment Research Foundation: Alexandria, Virginia.

Brandt, R. C.; Elliott, H. A.; O'Connor, G. A. (2002) Comparative Evaluation of Water Extractible P in Biosolids and Livestock Manures. *Proceedings of the Water Environment Federation Specialty Conference—16th Annual Residuals and Biosolids Conference*, Austin, Texas, March 3–6; Water Environment Federation: Alexandria, Virginia.

North East Biosolids and Residuals Association (2007) *A National Biosolids Regulation, Quality, End Use and Disposal Survey*, Final Report; North East Biosolids and Residuals Association: Tamworth, New Hampshire (http://www.nebiosolids.org).

U.S. Environmental Protection Agency (1979) *Process Design Manual for Sludge Treatment and Disposal*, EPA 625/1–79-011; U.S. Environmental Protection Agency, Municipal Environmental Research Laboratory and Center for Environmental Research Information and Technology Transfer: Cincinnati, Ohio.

U.S. Environmental Protection Agency (1992) *Technical Support Document for Reduction of Pathogens and Vector Attraction in Sewage Sludge*, EPA/822/R-93–004; U.S. Office of Water, Environmental Protection Agency: Washington, D.C.

U.S. Environmental Protection Agency (1993) Standards for the Use or Disposal of Sewage Sludge, Final Rule. *Code of Federal Regulations*, Part 503, Title 40; *Fed. Regist.*, **58**, 9248.

U.S. Environmental Protection Agency (2003) *Environmental Regulations and Technology: Control of Pathogens and Vector Attraction in Sewage Sludge*, EPA/625/R-92/013; U.S. Environmental Protection Agency: Cincinnati, Ohio. http://water.epa.gov/scitech/wastetech/biosolids/upload/2007_05_31_625r92013_625R92013.pdf (accessed October 2011).

10.0 SUGGESTED READINGS

Commission for Environmental Cooperation Guidance Document: Improving Environmental Performance and Compliance: 10 Elements of Effective Environmental Management Systems; Enforcement Cooperation Program of the Commission for Environmental Cooperation, Commission for Environmental Cooperation, June 2000, http://www.epa.gov/compliance/resources/policies/incentives/ems/cecguidedoc.pdf (accessed June 2011).

Public Entity EMS Resource Center homepage http://www.peercenter.net/ (accessed June 2011).

U.S. Environmental Protection Agency, EMS Key References http://www.epa.gov/osw/inforesources/ems/ems-101/otherresources.htm (accessed June 2011).

U.S. Environmental Protection Agency (2009) *Targeted National Sewage Sludge Survey*. U.S. Environmental Protection Agency, Office of Water, Office of Science and Technology: Washington, D.C. http://water.epa.gov/scitech/wastetech/biosolids/biosolids_index.cfm (accessed October 2011).

U.S. Environmental Protection Agency Voluntary Environmental Management Systems/ISO 14001 http://www.epa.gov/owm/iso14001/index.htm (accessed June 2011).

Chapter 3

Public Outreach and Involvement

(*continued*)

1.0 IMPORTANCE OF THIS CHAPTER

Other chapters of this manual are about excellence in the design, construction, and operation of wastewater solids treatment and management facilities. Those chapters are written by leading technical experts with extensive experience. Information found there, when applied to a project, typically leads to predictable outcomes in terms of engineering and operations.

In contrast, this chapter addresses factors that confound such predictable outcomes: people's understanding and judgments regarding the risks and benefits associated with biosolids production and use. The topics in this chapter should become familiar to all of those developing wastewater solids management projects: engineers, solids program leadership and staff, operators, wastewater treatment plant managers and governing boards, and the staff of regulatory agencies.

2.0 PUBLIC OUTREACH AND INVOLVEMENT BASICS

Those involved in wastewater solids management are providing a vital public service that protects the environment and public health. But people can easily ignore this fact when faced with a new proposal regarding wastewater and sludge, about which they may know little.

Managing wastewater solids often involves bringing people's attention to a waste they find distasteful. It may involve transport or use of biosolids near their homes or workplaces. It involves real and perceived risks. Done poorly, the communications around biosolids management programs can lead to uncertainty, unease, negative judgments, and conflict. And such outcomes cannot be entirely overcome by improved engineering design or technical planning.

In addition to excellence in design and engineering, the most effective way to increase the likelihood of a biosolids management program's success is through proactive public outreach and involvement, at the core of which is effective communication about risks and benefits.

Alternatively, if there is no proactive public outreach and involvement about a biosolids management program, or if outreach fails, then the common alternative method of dealing with public opposition is legal action. Going to court may eventually yield a desired outcome, but it is unlikely to improve relationships with communities. Such relationships are vital to long-term success. And legal action may cost much more in dollars, time, and frustration, and its outcome is by no means certain.

2.1 When Are Public Outreach and Involvement Needed?

Public outreach and involvement are needed whenever a significant, noticeable change in solids management is planned. It is especially important if any of the following are involved:

- Construction;
- Change of land use;
- New and visible activities;
- Public funding; or
- Wastewater solids management outside of the wastewater treatment facility.

The right time to start public outreach and involvement is as the project *concept* is being developed and during initial *project planning*, even before formal design begins. By beginning early in the development of a project, public outreach and involvement help to minimize the risk of misunderstandings that can lead to concerted opposition. Technical professionals tend to delay public outreach until their design is "perfected"; at that point, it may be too late. Beginning outreach early and continuing it throughout the project process, right through start-up and operations, is critical.

Public outreach and involvement is especially critical—and inevitable—before and during the permitting process. In fact, permitting may be the only time public outreach and involvement is required. Many jurisdictions require formal public notices, comment periods, or public hearings for permitting of biosolids facilities, products, or land application sites. The federal National Environmental Policy Act (NEPA) requires environmental impact statements for some projects that may have federal effects or funding.

The minimum required public outreach and involvement should not be considered adequate. Communicating and consulting with people for the first time at a formal public hearing can result in serious miscommunications, strong emotions, and long-lasting negative communications, which can bring a project to a halt.

Starting public outreach and involvement early and continuing it through the process will help to ensure widespread public awareness of the program, ownership and support by diverse stakeholders, and approvals of permits and funding.

2.2 Siting a Biosolids Program and Community Acceptance

A biosolids management program is built on an unavoidable premise: wastewater solids are produced as a necessary part of wastewater treatment. They exist and must be responsibly managed. This can involve introducing change to one or more communities.

As a new or changed biosolids management operation is introduced, it introduces new perceived risks and benefits faced to a community. Each person and group, and the community as a whole, must absorb this change, just as it must absorb any change. Different people will adapt in different ways and with different levels of ease or consternation. This is to be expected. Whether or not the biosolids program succeeds depends on whether the public outreach and involvement process enables diverse community members to integrate the change into a new status quo for the community.

2.3 The Dangers of Poor Outreach

The history of wastewater solids management since the 1970s is littered with abandoned projects stopped by neighbors, community groups, politicians, and skeptical public officials. Here is an example provided by Dan Thompson, Ph.D., Division Manager at the City of Tacoma, Washington (adapted from Beecher et al., 2004):

Case Study: Everett and Seattle, Washington's Land Application at Port Susan

The City of Everett entered into a partnership with Seattle Metro (now King County Department of Natural Resources and Parks, Seattle, Washington) to site a regional biosolids recycling facility on a large tract of forested land (925 acres) on the Tulalip Indian Reservation near Port Susan, Washington. The city negotiated a complex "win-win" agreement with the Tulalip tribal government, which agreed to support the city in its efforts to obtain permits to recycle biosolids on a property donated to the tribes.

Public outreach and education were recognized as crucial elements in successfully establishing this biosolids recycling facility. The city and the tribes collaborated in an outreach program, contacting neighbors and inviting interested citizens to informational meetings and to participate in selecting a consultant to do the project design. Unfortunately, a very short period of time was allotted for this important outreach, and several neighborhoods were not contacted. In addition, input—public outreach and involvement in decision-making—was not a priority.

When news of the proposed project spread, opposition quickly organized. The city began implementing the project in May 1990, and early that month the first letters to the editor objecting to the project were published in the local paper. On May 23, the city council meeting was dominated by a large contingent of concerned neighbors. The council was regaled with fears that water quality would be jeopardized by heavy metals and pathogens. Citizens requested information on several aspects of the project, including the effects of sludge on humans and wildlife. The city had no answers, and they did not have any consultants or technical experts on board at the time. The council was left with the impression that city staff did not know what they were doing. Neighbors of the proposed project formed an activist group that organized a letter-writing campaign directed at the city and elected officials.

The controversy intensified through mid-1991, as permitting and environmental evaluations were conducted. The city toiled for some time after that, but never released a final scoping document. Early in 1994, a new city administration took office. The new mayor saw the property as far more valuable for other uses, and the wastewater solids project ended.

Technically, the Tulalip property was an excellent choice for a biosolids recycling facility. Still, the project failed despite recognition of the need to engage stakeholders. Outreach efforts came too late and were not integrated into the processes of decision-making and project design. Proponents were unprepared, were forced into a defensive position, and ended up playing catch-up. The result was a painful and expensive negative experience.

2.4 Managing Public Involvement

Since the 1980s, the public has developed a strong appetite for involvement in projects affecting their communities—especially those that are publicly funded. Today, no significant wastewater solids management project proposal can escape public scrutiny. Project proponents—the design engineers and utility or company managers—cannot avoid public outreach and involvement. Project proponents can help direct the timing and nature of that public outreach and involvement and the effectiveness of the communications. As in the example above, involving the public late in the project after millions of dollars have been spent could jeopardize the entire project and result in the loss of millions of dollars. It is far better to integrate public involvement at each step of the planning, design, construction, and operations, which will make it less likely that the public's inevitable involvement will cause an expensive redirection of the project.

2.5 Responsibility to Protect the Public's Investment

Given that most wastewater solids management projects are supported by public funds, it is irresponsible for designers and proponents to not provide ample opportunities for public outreach and involvement throughout the process. If the public was not involved, and the project is stopped by public outrage, then the designers and proponents of the project who did not reach out to the public are responsible for wasting public funds.

For these reasons, the U.S. Environmental Protection Agency and other agencies and companies recognize the need for public outreach and involvement and require specific formal steps be taken to ensure it happens. Managers should pay attention to these requirements for public hearings, reviews of drafts, and other steps and integrate them into their project planning and schedules.

The process cannot stop there, however. These required actions will not ensure public support. Voluntary steps indicated by an assessment of the needs of the community might also be required. Questions to ask include "What steps will the community need to go through to get from their current situation to the new situation that is envisioned?" and "What will help them absorb the change?" Several concepts and actions discussed below (Figure 3.1). The system of public outreach and involvement must be designed to meet the specific needs of the program. The plan should also include any steps required by regulations and regulatory agencies.

2.6 Public Outreach and Involvement Requires Time and Money

Public outreach and involvement requires planning and plenty of time for implementation. This means it costs staff time and money. To ensure this staff time and money is allocated, it needs to be written into contracts and to be part of someone's job description. Money needs to be budgeted to support meetings, create project literature, and pay for travel and other expenses. These expenditures are money well spent on insurance against the potential for significant project disruption.

It is common for project proponents to underestimate the time and cost of effective public outreach and involvement. Conducting the formal and informal research needed to find, contact, and understand stakeholders and their thinking, interests, and priorities can take months. It cannot be skipped without serious risk of failure of the outreach and involvement effort.

2.7 How Much Public Outreach and Involvement Is Needed?

All key stakeholders should be involved in public outreach and involvement efforts. Stakeholders may be neighbors, community members, local leaders, public officials, community groups, and environmental advocates. Stakeholders can be individuals, groups, and, to a lesser degree, the public around the proposed project. Deciding who is a stakeholder can be difficult. Typically, a stakeholder is anyone who thinks he or she is one. It is better to err on the side of considering more people to be stakeholders and offering them the opportunity to be involved. Trouble comes from leaving people out.

How each person should be engaged is difficult to decide. Most of the public will not pay much attention to the project and will be satisfied if they feel that the development process was open and fair and that *they could have been more involved if they had wanted to be.* Other stakeholders, such as direct neighbors, public officials, and

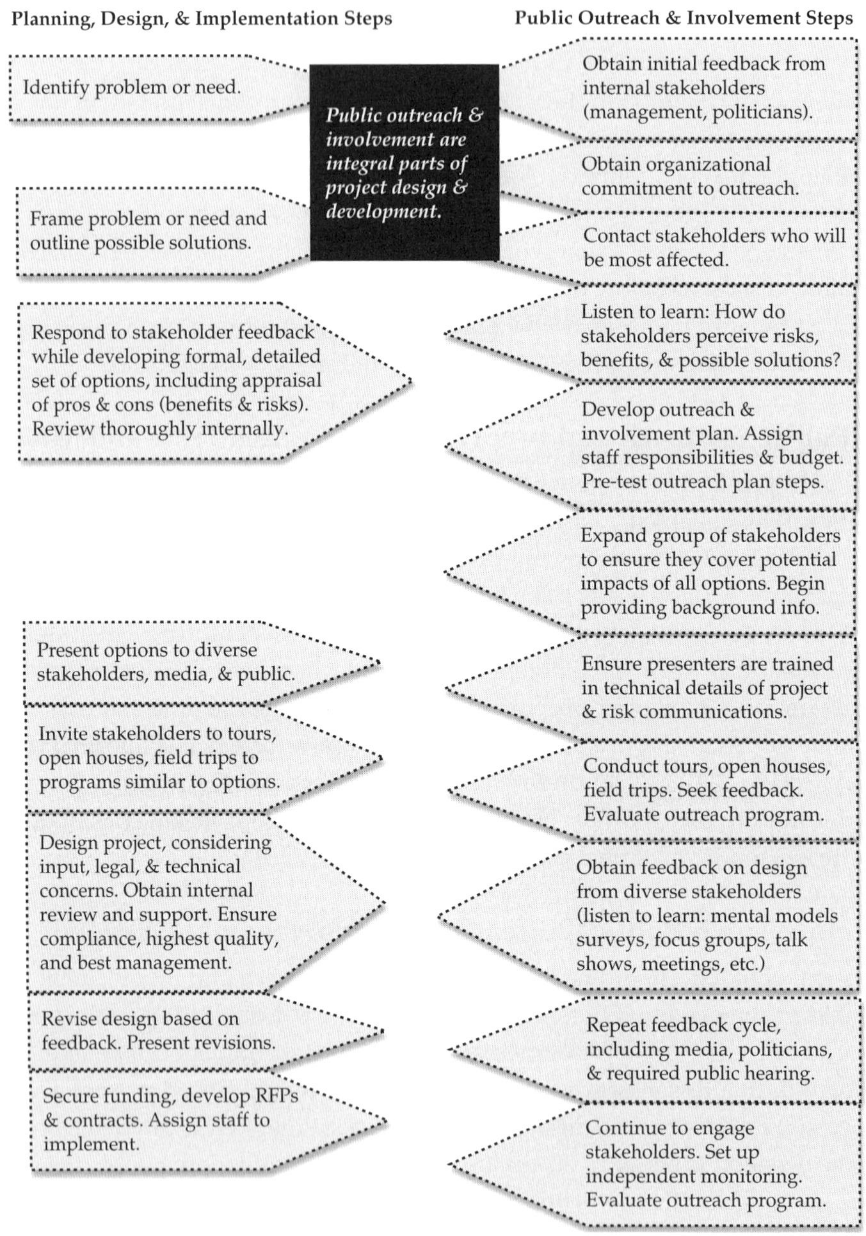

FIGURE 3.1 Project Public Outreach and Involvement Flow Chart.
Integrate public outreach and involvement steps (right) into the timeline for project planning, design, and implementation (left). Note that time and money for public outreach and involvement are spent at different times than the time and money spent on engineering and construction.

community group leaders, will need to be more involved. Ideally, the public outreach and involvement program should go far enough so that the project team knows what all key stakeholders are thinking and feeling about the project. If a project team is surprised by a citizen's sudden involvement or attack on the project, the public outreach and involvement program was not comprehensive enough. Various levels of public outreach and involvement are discussed in Table 3.1.

Fischhoff (1995) notes that "deciding how much effort to make requires a comparison of the costs and benefits of perfunctory communications. In some ways, communication is like an insurance policy. It is a fixed cost that can prevent larger damage. In evaluating a particular policy, one needs to decide how complete the coverage is, how much protection one can afford, and how much the attendant peace of mind is worth."

He notes that understanding of risk perception and communication theory helps guide the decision regarding how much public outreach and involvement is needed. Risk communication literature shows that biosolids recycling to land triggers a large increase in perceived risk, which is discussed further below. The experience of biosolids professionals dealing with public skepticism corroborates this fact (Beecher

Table 3.1 Levels of public outreach and involvement. The level of communication and involvement will vary for different stakeholders. For a particular stakeholder, it will depend on that stakeholder's level of interest and the degree to which he or she can influence decision-making. With limited resources, it makes sense to provide opportunities for more extensive two-way communications and involvement to some groups of stakeholders. All stakeholders should be provided some level of outreach and the option of some involvement.

Less involved	One-way communication	Fact sheets, news releases, white papers, letters to editors
↑	Some two-way communication	Informational meetings, public hearings, surveys
	Active two-way communication	Facility and site tours, workshops, door-to-door visits, focus groups
↓	Two-way communication and involvement	Citizen advisory groups
More involved	Extensive two-way communication and involvement	Joint fact-finding, consensus building, collaboration in decision-making, shared implementation

et al., 2004). Therefore, the level of public outreach and involvement around a biosolids management project is likely going to have to be greater than for most other projects.

By starting public outreach and involvement early, project teams can learn from stakeholders and from the process itself to decide how extensive the involvement will need to be.

2.8 Benefits of Public Outreach and Involvement

Public outreach and involvement can result in more stakeholders providing support for the project team's decisions and actions. It means the project team—the project design and management team—is not making decisions alone, in a vacuum. Even more importantly, public outreach and involvement can lead to better decisions and designs and improved environmental performance and efficiency. This is because public outreach and involvement engages more minds and brings more diverse experience (including local knowledge) to an issue.

There are several additional benefits to public outreach and involvement:

- Increases the likelihood of avoiding costly mistakes that will be judged, in retrospect, to have been stupid;
- Helps to ensure the biosolids management technology or program will work well within the particular community;
- Reduces the vulnerability of the program to criticism;
- Helps expand the group of knowledgeable allies who will voice their support to peers, elected officials, and others;
- Provides some confidence that the assumptions, design, and decision-making are appropriate and workable.

It is quite possible that a robust public outreach and involvement process will result in a decision that is not the initial first choice of the biosolids engineers and program managers. If the final choice is a good choice, however, then that is okay. Public outreach and involvement requires flexibility and an open mind.

At the same time, public outreach and involvement does not take away the responsibility of the wastewater solids manager and the design engineers for meeting contractual obligations and legal requirements.

Note also that there is no guarantee that public outreach and involvement will result in selection of the best possible option—that depends on how careful, fair, and equitable the public input process is. There are examples in the literature regarding public participation in environmental decision-making that resulted in poor and wasteful choices. To avoid bad decisions, it is critical that the biosolids engineers and managers—along with other participants in the decision-making process—strongly state their interests and knowledge. Experts need to clearly state their best professional judgment and be willing and able to deny any decision that they know is not technically viable, environmentally sound, or economically possible. This may require that the biosolids engineers and managers learn more about decision-making in a contentious atmosphere. Necessary public infrastructure should not be derailed by third parties who are not interested in good-faith negotiations or who are pursuing agendas completely at odds with the project's public health and environmental goals. At some point, political and legal decisions may have to be made to proceed with a project despite the vocal opposition of a few people.

2.9 Risk Assessment and Risk Perception

"Public resistance to new technology remains more a response to loss of control, respect, justice, or courtesy than to the technology itself" (McAvoy, 1999).

Decades of research into how humans perceive risks, what influences their perceptions, and how they make decisions support McAvoy's statement (Butte and Thorne, 2007). Gaining public support for a new program is about helping a community absorb change. When citizens react negatively to a proposed project, it is likely because of their perceptions of risks. Their perceptions can be dramatically influenced by project's development process, distrust of the proponents of the project, and the levels of respect and courtesy the proponents display.

A flawless technical design will not, by itself, win public support. Good design is, of course, important; but equally important is the process by which those potentially affected by the project—the community—are informed and involved.

Scientists, engineers, and other experts assess risk through mathematical models that evaluate hazards and the likelihood of something going wrong. In contrast, those not expert in a particular field tend to assess risk through social and political factors. Social scientists have developed lists of factors that influence this risk *perception*. A person who experiences a proposed change in any of the following ways will perceive greater risk if the project is

- Run by people who are not trustworthy (trust is often considered the most significant factor);

- Involuntary (i.e., it is foisted on them);
- Artificial and industrial (e.g., a biosolids incinerator, or a compost facility, or land-application machinery);
- Exotic or unfamiliar (most people do not have experience with wastewater treatment plants or biosolids management systems);
- Hard to understand (it is not self-explanatory);
- Memorable (e.g., odors or other nuisances);
- Dreaded (the "yuck" factor of the origins of biosolids creates dread);
- Unknowable (there is a greater level of uncertainty regarding biosolids land application than, for example, for animal manures; biosolids have more diverse inputs from municipal sewers, so its precise constituents are more unknowable);
- May affect them, creating a personal stake (e.g., neighbors to biosolids incinerators may believe they will be affected by emissions);
- Controlled by others or "the system" (wastewater treatment and associated systems can appear to be large, inevitable, uncontrollable systems);
- Involves a closed process (i.e., lack of transparency);
- Gets a lot of media attention (media stories about a biosolids projects heighten local interest and, if they report opposition, public concern tends to increase); and,
- Provides limited or no visible benefits (communities receiving biosolids from somewhere else often perceive little benefit to anyone but, perhaps, the farmer).

Social science research has demonstrated that these risk perception factors apply in diverse situations and disciplines, from health care to hazardous waste management to nuclear energy.

A careful review of the list shows that most of the factors apply to wastewater solids programs. Some biosolids programs involve all of these factors. In addition, wastewater and solids management is an uncomfortable topic of discussion for most people. Humans do not like to dwell on human waste. On top of all that, wastewater solids can create malodors, and malodors naturally trigger unease or fear.

Add together the incremental increases in perception of risk caused by each of these factors and the total perceived risk reaches a high level; fear results and people react when they feel fear.

2.10 Risk Communication

Risk communication developed in response to understanding of risk perception. The practice of strategic risk communication aims to ensure that diverse people share a common understanding of the level of risk to ensure "policy decisions and public discussion based on the best information available" (Powell, 1996). Risk communication is "an interactive process of exchange of information and opinion...". (U.S. Department of Health and Human Services, 2002). It is *not* a substitute for risk management. And it is *not* a way of hiding something or manipulating opinions.

Today, risk communication is widely applied in many fields and increasingly is being applied in wastewater and biosolids management programs.

Risk communication aims to directly address the factors in the list above that increase the perception of risk. Public outreach and involvement using risk communication makes a biosolids management project less involuntary, less exotic and unknowable, and less controlled by others. It provides opportunities for trust to develop. The goal of a risk-communication-driven public outreach and involvement program is to remove—or diminish the effects of—as many risk perception factors as possible.

2.11 Building Trust

Because trust is critically important to developing public support for a program or project, and because trust is easily lost and difficult and time-consuming to regain, any public outreach and involvement program should focus on building trust with key stakeholders and the public (Table 3.2).

Experience with risk communications suggests that trust can be earned by being open and direct with people, showing respect, sharing information, and addressing concerns in a caring manner. It is also gained by being knowledgeable and competent. Quality projects, people, products, and operations generate trust.

Trust does not come quickly or easily. Project teams should not assume that people will develop trust, although many will give the benefit of the doubt at first. Then, trust builds slowly and steadily. It can take less than a minute for trust to be lost, however. For example, some inappropriate remarks at a public meeting can begin the slide into distrust that can take years to reverse.

Table 3.2 U.S. Environmental Protection Agency's seven rules of risk communication (U.S. Environmental Protection Agency, 2007).

1. Accept and involve the public as a legitimate partner.
2. Listen to the audience.
3. Be truthful, honest, frank, and open.
4. Coordinate, collaborate, and partner with other credible sources.
5. Meet the needs of the media.
6. Speak clearly and with compassion.
7. Prepare, plan carefully, and evaluate communication performance.

Well-designed public outreach and involvement initiatives create a favorable environment in which trust is built. For example, by beginning the process of developing a plan for wastewater solids management by hosting a tour of a wastewater treatment facility, project teams can create an informal atmosphere where diverse stakeholders can learn together, begin to know and respect each other, and find some common ground.

3.0 ACTION ITEMS FOR EFFECTIVE PUBLIC OUTREACH AND INVOLVEMENT

3.1 Consulting Available Resources

Numerous resources exist on public outreach, involvement, and risk communication. These should be used. It is not worth repeating here the detailed instructions available (see 5.0 References and Suggested Readings).

What follows here are the most essential concepts and action steps for developing a public outreach and involvement program tailored for biosolids management (Figure 3.1).

3.2 Starting Public Outreach and Involvement Right Away

As discussed above, public outreach and involvement should be integrated into the planning and design process from the beginning. Public outreach and involvement is challenging and takes time, so it is best to get started on it right away. It will then be in place and working when it is really needed (for example, during permitting).

Case Study: Siting a Biosolids Composting Facility

In some parts of southern California, biosolids management was a contentious topic in the media and in politics throughout the late 1990s and early 2000s. Nonetheless, a large biosolids composting facility that manages wastewater solids from southern California cities was successfully sited in a rural community. Developers of the project not only designed a facility with the most advanced odor control and other technical systems, they also spent a lot of time with local citizens, talked with the local newspaper, and looked at what they could do for the town (e.g., processing local biosolids, helping with a local beautification program, and supporting town birthday celebrations). They met with local citizen groups at those groups' convenience. They ensured that significant oversight and enforcement—five layers in all—were visible to everyone. They talked about the necessity of recycling biosolids for sustainability. They focused on the end product (quality compost) rather than on the inputs (biosolids), thus reducing reactions of disgust. The facility's outreach person said, "We need to listen to communities; when you hear outrage, listen! We need to integrate into communities, find symbiotic relationships, offer something to communities. Look at technologies that will work in concert with communities. We need to be willing to compromise, work through disagreements."

3.3 Ensuring That the Project Team Is Committed to the Process

Worse than not having a public outreach and involvement program at all is to have one with no follow-through. People quickly lose trust if their input is requested and then ignored, or if they expect public outreach and involvement and it does not occur.

Planning a public outreach and involvement program begins with ensuring that the project team and their facility and any other organizations involved in the biosolids program planning have committed the resources, including staff time and money, to fulfill public outreach and involvement promises. The project team and the organizations involved need to ensure the necessary level of commitment. If resources for the public outreach and involvement program are limited, then the program and expectations should be scaled down.

For projects where one group of proponents is involved in all three stages (design, build, operate), having a consistent outreach and involvement component is easier. Special attention is required, however, if more than one team is involved such as when there is one team doing project design and another conducting operations. The public outreach and involvement plan should address all stages of the project, including how to deal with changing project staff.

3.4 Identifying Diverse Stakeholders

The next step is to identify stakeholders. These will certainly include the organizations involved in the project team, such as design engineers, wastewater treatment agency management, municipal administrators, and facility operators. Figure 3.2 provides suggestions of other groups and individuals to include.

The creation of this list of potential stakeholders should be thorough because it could be detrimental to leave out anyone by mistake. It is better to err on inviting more people

City of Tulsa: Stakeholder Map of Sustainable Biosolids Business

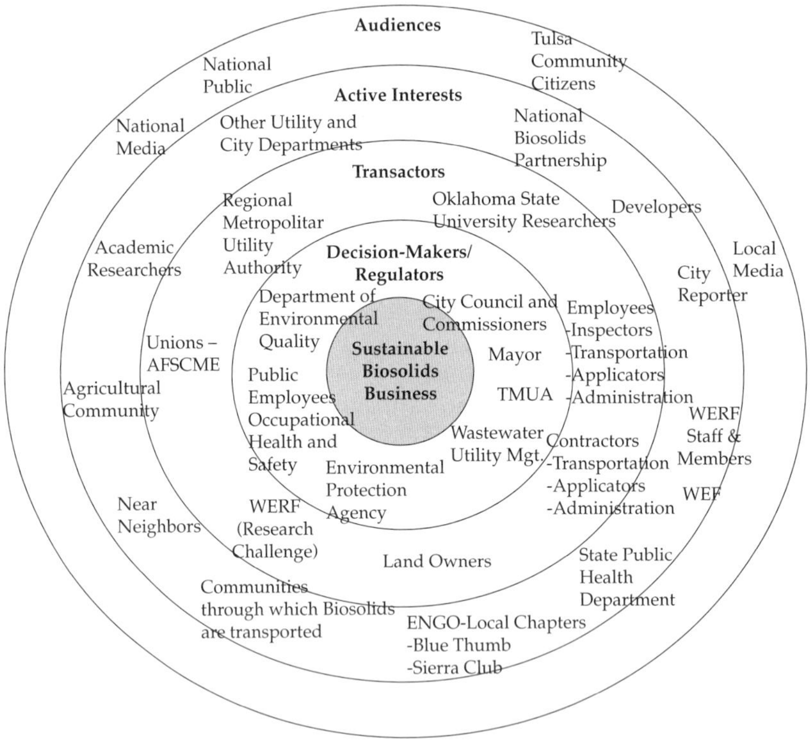

FIGURE 3.2 Mapping stakeholders. This map of stakeholders was created as part of a public outreach effort by the City of Tulsa, Oklahoma. Biosolids managers worked with a risk communications consultant to identify all stakeholders (interested parties) and place them in the appropriate circle on the diagram. Stakeholders closest to the center tend to have the most direct influence on decisions affecting the biosolids program. This does not mean that more distant stakeholders should be ignored by the outreach and involvement program; rather, each should be given opportunities to be involved to the extent he or she chooses.

than fewer. Potential stakeholders can then make the decision whether to be involved. This makes it their choice, which builds a sense of openness, credibility, and trust.

Stakeholders can be arranged on a diagram like the one shown in Figure 3.2. When making assumptions about how involved each stakeholder should be, one question to ask is, "Whose decision-making and behavior are most important to moving the biosolids project forward?" These stakeholders should be closer to the center of the diagram and should receive more attention and a greater allocation of outreach resources. This is a working diagram, and stakeholders will be moved around as more is learned about the level of involvement each stakeholder chooses.

3.5 Listening to Learn from Stakeholders and the Community

Managers and communicators involved with design and operations often ask, "Why take the time and trouble to listen-to-learn from stakeholders? Why not just start with messages, or go right into "messaging" mode?" The simple answer is listening is important to prevent potentially negative outcomes of communication.

People are complicated and their thinking can be unpredictable. In the absence of empirical evidence, no one can confidently predict people's information needs and how they will interpret messages they receive.

Listening to learn from stakeholders is the primary means for generating empirical evidence about people's thinking and communications needs. If communications are to have their intended effect on people's decision-making and behavior, then they must reflect: (1) an understanding of people's information needs for decision-making and (2) an understanding of the decision-relevant topics, such as the risks and benefits associated with biosolids management options. Messages prepared without using a framework that addresses both of these may be counterproductive.

This section summarizes some ways to gain insight into people's thinking. Specifically, it covers opinion polling, focus groups, and focused mental models interviews, which allow effective design of public outreach and involvement programs and risk communications. There are resources available to assist with these and other methods (Eggers et al., 2011; Deeb and Means, 2009; Frank and Sigler, 2004; International Association for Public Participation, 2000; Morgan et al., 2001). In addition to formal methods of research, there are informal methods such as semi-structured conversations with individuals or a small number of stakeholders in a group.

The first step in any research into people's perceptions, thinking, and communication needs is to recognize that assumptions about them are just that: assumptions. These assumptions can serve as hypotheses that will be tested in the formal and informal research.

3.5.1 *Mental Models (Focused Interviews or Dialogue Method)*

Decades of research have shown that tacit webs of beliefs—beliefs that are sometimes below the surface of consciousness—guide people's decision-making. These are called mental models (Morgan et al., 2001). People draw on their mental models to make inferences about issues or opportunities that come to their attention through various forms of communication. To address people's attitudes, beliefs, and behaviors, one must understand their mental models and address them through communication.

Analysis typically seeks better understanding:

- What people know that is correct;
- What people do not know that is consequential;
- What people might misunderstand (about the issue or topic);
- What they want to know;
- Whom they trust (as an information source); and
- What communication methods they prefer.

The mental models research method typically involves individual, one-on-one, confidential interviews done in person or by telephone. During the interview, the participant is led through an agenda of topics. He or she is encouraged to elaborate on issues or topics that come to mind in the course of the conversation. The method allows for free expression and encourages elaboration on topics to reveal individual perspectives at considerable depth. From the interviews, analysts can identify what people believe and, more significantly, *why* they believe it. They are also able to compare analyses over time and provide insights into why beliefs may have changed.

Properly done, the mental models method can produce rich results more efficiently than can the equivalent time and effort invested in opinion polls or focus groups. This method can also be used effectively in combination with opinion polls and focus groups.

3.5.2 *Group Discussion/Focus Groups*

A focus group moderator leads a small group of individuals through an agenda of topics, formulated to allow free expression within the constraints of group dynamics. Opinions are often interpreted and summarized impressionistically. The focus group method is most useful when studying beliefs that are shaped and expressed in groups that have the same composition. That is, the members of the focus group should be people who work or otherwise interact together outside of the research grouping.

By allowing free expression of beliefs, focus groups can get closer than opinion polling to understanding what a group believes and why it holds those beliefs. The approach is inherently limited by the effects that specific group interactions have on what participants do and do not say and on which lines of thought can and cannot be fully pursued by individual members. In focus groups, the opinions that participants express depend on the skills and the agenda of the moderator. The moderator typically will interpret and summarize these exchanges impressionistically.

3.5.3 *Opinion Polling*

Polling methods are well-suited to identifying opinions in a large sample of people and measuring their distribution across the sample. Polling typically offers a fixed set of questions to a selected group. Answers to these questions are subjected to statistical analyses, which must then be interpreted in terms of how respondents are *believed* to have interpreted the questions.

Opinion polling typically tries to capture *what* respondents believe, not *why* they believe it. Such an approach is useful when only discrete answers to particular questions are needed. It is also useful when precise estimates of the frequency of specific beliefs in a population are required.

3.5.4 *Informal Methods for Listening and Learning*

Informal research is research that can easily be conducted by members of a project team and typically focuses on semi-structured conversations with a few key stakeholders. It is typically limited to issues of program management options and their benefits and risks. Informal research might also be done when time is short or if sufficient resources are not available to conduct formal research. Examples of informal research techniques are outlined below:

- **One-on-one dialogue**. Test a hypothesis with one or more members of a target stakeholder group through one-on-one dialogue, following a dialogue script and documenting the conversation. For example, this could be done as door-to-door visits around a biosolids management site. Members of a project team can then share the notes from their dialogue sessions and impressionistically analyze the results. This is a simple and effective way to test a hypothesis and build or reinforce relationships with important individuals.
- **Breakfast roundtables, advisory task groups, advisory panels**. A breakfast roundtable, advisory task group, or an existing stakeholder advisory panel can provide an excellent way to get input from members of target stakeholder

groups fairly quickly. Comments can be solicited in the context of a meeting, giving the biosolids project team immediate insight or data to work with.

- **"5 × 5" dialogues**. The "5 × 5" dialogue technique is an ideal method for the team to use to gather insight from a relatively large number of people in a short period of time. As a research technique, each team member must have a script and commit to talk to five (or whatever number is appropriate) members of the priority stakeholders by a certain date. It is critical that each of these conversations is documented at the time by the team member. The team then gets back together and shares its data.

Any of the informal methods above can be quickly implemented when time is tight. A simple protocol of 6 to 10 questions with prompts to enable the team to understand not just what people are thinking but, more importantly, why they are thinking it, can be designed and implemented in a matter of hours. If the topic is confidential, the interviews can be conducted with 12 nontechnical people within the organization (people not involved in the biosolids program).

The key for listening to learn from people is to gain insight about stakeholders and what they are thinking and why in the time that is available, rather than relying on guesswork about their perceptions and opinions. Taking this step will ensure that public outreach and involvement efforts—and dollars—are not wasted or counterproductive.

3.6 Continue Public Outreach and Involvement

Don't stop listening to learn as the project moves ahead. Outreach and involvement efforts should continue using some of the tools discussed in Table 3.3.

TABLE 3.3 Outreach and involvement options.

Tool	Benefits	Limitations
One-way (passive) communications tools		
Brochures Bill stuffers Fact sheets Newsletters Advertising	Printed material is needed to convey information; can reach large audience; should be visually pleasing.	Limited amount of information; may not be read; depends on quality of mailing list or target audience list.
Technical reports	Best if written by independent authority; conveys detailed information; should be written clearly.	If too detailed, then may not be read by many.

(continued)

Table 3.3 Outreach and involvement options (*Continued*).

Tool	Benefits	Limitations
News releases	Keeps editors/media outlets aware of project developments; reaches wide audience; controls message to some degree.	Requires follow-up calls to ensure it is printed; often poorly placed in papers; may not be read.
News articles	Shows greater interest on part of media and public; gives independent perspective.	Requires working with reporter and editor to help ensure accuracy; requires an "angle."
Op-eds/letters to the editor	Allows control of the message; makes editors and readers aware that expertise is available.	May not be widely read; space limitations make it hard to explain complexities; can lead to winless back-and-forth debate.
Radio and television stories or programs	Cable and internet options make it more accessible and cost-effective; is more accessible to most people than written media.	Can be expensive; is more difficult to get air-time than to get into print; important to have experienced speakers.
Public service announcements	Can reach a large audience effectively; broadcasters may be motivated to air or print these.	Must be concise; no room for much detail; can be costly; little control about when aired.
Websites	Moderately priced; can provide lots of information, both summary and detailed, including visuals.	Must be kept updated and organized.
Blogs	Opportunity to express commentary quickly; often includes room for comments in response.	Part of the information overload of the Internet; may not be widely read.
Podcasts	Can be an accessible way for people to get lots of information; not expensive.	Requires technical savvy and planning; has to be relevant to audience to get any attention; part of the information overload; unlikely to be used by most people.
Handheld device applications	Most convenient systems for busy people to obtain up-to-date information; holds loyal audience.	Requires programming; can be expensive; must be updated regularly; unlikely to be used by most people.

(*continued*)

TABLE 3.3 Outreach and involvement options (*Continued*).

Tool	Benefits	Limitations
Two-way (interactive) communications and involvement		
Informational meetings/expert panels	In-person interactions are a benefit; useful in reaching coherent communities (e.g., neighborhoods).	Venue and presenters must be open and welcoming; setup should encourage dialogue, not lecturing; requires thorough preparation.
Surveys	Simple, cost-effective way to get input from diverse stakeholders; Internet surveys are especially efficient.	Challenging to write questions properly to receive meaningful responses; typically a low rate of response.
Complaint/feedback phone line	Gives people another simple option for providing input and expressing themselves.	Has some costs; requires someone to respond to input in a timely manner.
Open houses, tours, and field days	Best simple, cost-efficient option for conveying lots of information about a program; provides chances for questions and answers and dialogue; builds credibility and trust.	Arranging a time that works for intended audience can be a challenge; need to ensure everything is safe; may need to offer food and transportation.
Public hearings	Provide formal process for public input in accordance with regulations.	Typically required by regulations; should not be the only outreach and involvement effort; if it is formal, can be contentious; requires thorough preparation.
Participating in community events (e.g., fairs)	Builds relationships with community, media, and groups; provides formal and informal opportunities to share information.	May not be able to control the message; may be seen as a publicity stunt; no control of who attends.
Coffee klatches and other community meetings	Excellent way to bring the information-sharing session to the audience in a setting familiar to them; builds dialogue and relationships.	Takes time and money and may not reach critical stakeholders (depends on who shows up); presenters must be experienced, amicable.

(*continued*)

TABLE 3.3 Outreach and involvement options (*Continued*)

Tool	Benefits	Limitations
Door-to-door visits	Conveys information and builds relationships.	Time-consuming; presenters must be experienced, amicable.
Call-in radio and TV talk shows	Can create excellent dialogue and convey plenty of information.	Interviewee must be experienced and well prepared.
Email or listservs	Simple, inexpensive way of regularly providing timely information to an interested audience.	Often not considered high-quality information; information overload.
Social media (e.g., Twitter), Internet comment and chat areas	Opportunity for information exchange and dialogue; quick, timely, and inexpensive.	Quality of information and dialogue widely varies; may not be used by many people.
Mediation/ facilitation	Addresses specific issues with specific audience; typically a last resort to address conflict.	Costly, takes a lot of time; outcome is not guaranteed.
Consensus building process	Useful when the level of conflict and distrust is high; can overcome impasses, avoiding litigation.	Costly; requires experienced facilitators; takes lots of time; outcome is not guaranteed.
Focus groups	Good way to obtain detailed, nuanced feedback (or test messages) from a selected group of people in a short amount of time.	Requires careful planning and group selection; experience required, which costs money.
Advisory committees	Excellent for maintaining relationships and knowledgeable stakeholders over long periods of time; a good way to obtain diverse input and advice and reach effective compromises	Requires facilitation and clearly defined purpose and process; requires give-and-take; time and labor intensive.
Independent oversight/quality assurance	Improves credibility and trust, especially when developed with diverse stakeholder input.	Time and labor intensive, with ongoing significant costs.
Environmental management system	Improves credibility and trust, especially when developed with diverse stakeholder input; leads to more effective, efficient operations.	Time and labor intensive, with ongoing significant costs; requires training and change in organizational culture.

Approaches should be selected that will help to build a relationship with each stakeholder. The project team also should consider how the communications will be received and use approaches that will make stakeholders feel welcome and at ease, for example:

- Bringing the meeting to them, in their community, at a time that works for them;
- Offering food to create a comfortable, informal time for sharing;
- Going door to door can make people feel personally involved, if it is done in a respectful way and at an appropriate time;
- Offering people several opportunities and ways to be involved gives them a choice and a sense of control.

As project team members begins to know people and see that some people are growing more interested and involved, they will be able to ask them what the best way is to obtain their ongoing involvement. For example, would they like monthly meetings in person or by conference calls? Where would they like to meet? Do they prefer to communicate in person, by phone, or by email? Team members may have to make some of these decisions, but they will appreciate being asked for their input.

As the stakeholder involvement grows, it is a good idea to ensure that representation is included from all of the most important stakeholder groups. It is likely that some group is not represented or is underrepresented. In which case, it might be necessary to contact people who declined involvement and find out why. For example, was it the time involved, or the distance to meetings, or the timing of meetings? Some individual will not be willing or able to be involved; others will need to be accommodated to make their involvement possible. In some cases, to gain diverse representation from various stakeholder groups, a stipend or mileage may be necessary (small citizen or environmental groups often cannot afford to donate staff time). Such arrangements, however, must be fair and equitable among all stakeholders.

The complexity of the public outreach and involvement program depends on the scope of the project and the community. For smaller projects, it may not be possible to hold many or any meetings, and these may rely on newsletters, other mailings, newspaper articles, telephone hotlines, interactive websites, and a few tours. For larger projects, more interactions and time are necessary.

Finally, in the field of wastewater and biosolids management, it is often difficult to get people interested in the topic. As noted above, this is not something most people like to think about. It will be necessary to use innovative public outreach and involvement tools and be determined and persistent to attain involvement (see Table 3.3). Not having this involvement is not an option.

3.7 Communicating

From listening to learn from stakeholders, it will become clear what information needs to be communicated and the best format. For example, it is likely that for biosolids land application, people will need information on how biosolids are made, what it contains, and its risks and benefits, and on relevant regulations. There is extensive information on communicating effectively regarding proposed or operating programs (Beecher et al., 2004; Eggers et al., 2011; Haworth et al., 2005; Water Environment Federation, 2002). The program should make use of these resources and training in risk communications, crisis communications, and working with the media.

Throughout the process, communication should

- Be open and honest (deception is easily spotted and kills trust);
- Offer two-way communications and create dialogue;
- Tailor information to the intended audience (i.e., avoid technical jargon); and,
- Be accurate.

It is also important to always be aware of nonverbal cues during public interactions. For example, standing distantly up on a stage and preaching is not recommended. Good communication includes a lot of listening and interaction.

As communications are developed, all project team members and their management, employees, and organizations need to be kept informed too. These internal communications should not be forgotten. If management and employees remain informed, some of them will become excellent public spokespersons.

The effects of communications technologies and practices should be considered before their adoption. Haworth et al. (2005) recommend, for example, that wastewater treatment facilities should abandon automated answering machines and have people answer instead. Who greets visitors and how also should be considered. Finally, the facility's buildings and grounds should be inviting and pleasing (New Zealand Ministry for the Environment Urban Design Toolkit, 2009).

There are myriad ways to communicate (Table 3.3). The final selection should be based on the targeted audiences, the cost of each option, and how each option helps to increase public outreach and involvement and relationships with key stakeholders.

Perhaps the most important statement about good communications is, "Say what you mean and mean what you say." All statements should be crafted so that the intended audience will readily understand. It is good practice to collect and structure thoughts first, followed by an effort to "open the door to a productive exchange" (Haworth et al., 2005).

Another important statement is, "Say what you do and do what you say." This is a hallmark of the National Biosolids Partnership's EMS program: "A well-run program provides the foundation upon which public support can be built. It gives you the confidence to go out to your community and emphasize the benefits of your program to win public support" (National Biosolids Partnership, 2005).

3.8 Quality

The best public outreach and involvement efforts will have little effect in helping a project gain public support if there is one thing missing: quality. Sometimes, public outrage and the "public veto" of a project are deserved because the project was poorly conceived, designed, or operated. If there had been good public outreach and involvement from the beginning of the project, then it is unlikely that the project would get far without being redirected toward achieving a higher level of quality. Stakeholders will force it to happen or will force the project to be abandoned before much time and money are wasted. This is a good thing.

As Draman (1995) urged, wastewater solids programs must develop strong communications, but "equally important is the performance of everyone involved with beneficial use projects—it must be done right from top to bottom."

Case Study: New York City Biosolids in Colorado

A portion of New York biosolids has been land applied in Colorado. The company managing the land application of New York biosolids created a unique way of doing business. It incorporates numerous special management measures developed with local public input to address local farmer and public concerns. For example, each rail car carrying the pelletized biosolids to Colorado has eight security seals that ensure the contents are free from tampering during transit. The company pays $1 per ton to county government and more than $2 per ton to the state (as required for all Colorado biosolids land appliers) to fund independent monitoring. Inspectors visit land application sites regularly, often unannounced. Extensive testing data and other information are provided regularly to farmers and area citizens, helping ensure their comfort with the program. And the Colorado farm communities are experiencing demonstrated benefits. As the program manager wrote: "Ultimately, the farmers are our best communicators of the value of biosolids utilization. Our role is to deliver not just a product, but a service that meets the farmers' needs by becoming a critical component of their fertilization program" (Scharp, 2000). All of these extra measures cost considerable money but, because they were developed with local input, the money is well spent ameliorating potential issues. The result has been years of success.

3.9 Independent Evaluation and Monitoring

Showing the public that things are being done right can be difficult, unless there is some form of independent monitoring or oversight. Independent monitoring helps to provide essential confidence to local communities. Often, governmental regulatory agencies are under-funded and do not have the staff available to conduct comprehensive monitoring and enforcement. In such cases, biosolids program designers and managers can take control of their program's destiny by working with local communities to establish and help fund a monitoring program. Similarly, when design engineers put forth a project plan that may be challenged, having an independent evaluation, a second opinion, by credible experts chosen by a consensus of stakeholders will generate trust and support. A biosolids EMS can involve independent, third-party auditing.

3.10 Evaluating a Public Outreach and Involvement Program

Like any program, it is essential to evaluate the effectiveness of public outreach and involvement efforts, both as they are being implemented and after the process is complete. This provides correction and improvement, both as the process advances and in future efforts.

Case Study: Denver's Biosolids Land Application Program

Frank and Sigler (2004) wrote an excellent overview of a proactive public outreach and involvement program. The Wastewater Reclamation District in Denver, Colorado, produces Class B biosolids that are hauled to the east of Denver to a 21 000-ha (52 000-ac) farm in Deer Trail, Colorado, that is owned by Metro. In 1995, after startup of land application at the farm, some neighbors became upset. Tensions and conflict brewed for two years, and in mid-1997, the county issued a cease-and-desist order, forcing the land application program to stop.

Frank and Sigler noted that, "Metro was then faced with a decision: Fight the order in court or find a way to work things out peacefully with the neighbors, other concerned citizens, and the two counties in which it owns land. Metro chose the latter option and began an effort to repair its relationships with its neighbors, concerned citizens and the counties."

Over the next several years, Metro focused on addressing the concerns the Deer Trail community had raised. For example, the Metro farm changed its tilling practices to reduce soil erosion by wind, something recommended by local experts. Metro opened up dialogue with neighbors and public officials and, together, they conceived an independent monitoring program. The U.S. Geological Survey conducted studies beginning in 1999. In 2000, a survey of neighbors and other key stakeholders gave Metro's program "about a D," noted a project leader. Metro had improved somewhat, but there was plenty of skepticism as to its sincerity and willingness to follow through on perceived promises. This survey was used as a benchmark; in 2003, it was repeated, and the grade given Metro was "about a B."

In the time between the surveys, Metro had taken the results of the survey seriously and implemented many improvements, such as hosting an annual picnic and information session at its farm, providing tours of the wastewater treatment facility, sharing road maintenance costs with the county, and working with the local soil conservationists to improve practices. "Key Metro staff members have worked hard to make sure community members know they are accessible and the community's questions will be answered," Frank and Sigler noted.

4.0 SPECIAL CIRCUMSTANCES

4.1 Public Outreach and Involvement in Ongoing Operations

Many biosolids management programs have operated with the assumption that it is best to avoid being noticed; they have been "under the radar" for many years, and they like it that way. This practice is dangerous. Not communicating is communicating. Practicing avoidance is a decision. People are always watching and will interpret—often negatively—what they do not see happening and what they are not hearing. A program that is "under the radar" may be a public upset waiting to happen. Although it can be scary, coming out from "under the radar" is good.

When working to increase public outreach and involvement, the best way to reduce the potential for creating public upset is to ensure the program is in compliance and uses the latest best management practices to reduce real and perceived risks, including odors. Many wastewater solids management programs focus only on compliance with regulations, giving less attention to odor because it is not directly measured or regulated. But, in practice, odor is often the most significant trigger for public outrage.

As a biosolids management program increases its visibility, it may trigger initial negative reactions and, possibly, conflict. Any conflict must be handled in a patient and professional manner. People need time to learn, and the community must absorb what it perceives as change, even though the change might simply be that it now knows what has been going on for some time. Some of the feedback received will improve the biosolids management program.

The single most effective way to involve the public in ongoing operations is by offering open houses and tours and by bringing people into the facility or to the land application site and showing them what is done. Tours provide opportunities for dialogue, questions and answers, and shared understanding.

Ongoing wastewater solids management operations should also focus attention on how complaints and concerns are addressed, especially because odor complaints can play a tremendous part in creating public outcry against a program. Responding in a timely manner with clear and honest evaluations of the issues is critical to building trust. Concerns should be discussed with the complainant and then he or she can be involved in evaluating and monitoring the problem. This can be followed by implementation of corrective actions and then asking for more input. Regular performance reports and updates on construction activities will let people know what to expect next.

Case Study: Philadelphia Land Application

Philadelphia, Pennsylvania, had Class B land application and Class A composting programs for many years. Both programs have run into public outrage, however, often triggered by malodors at biosolids land application or processing sites—as is typical of many programs. To address the decline in public acceptance, a consultant for Philadelphia took one of the application programs—the revegetation of some gamelands—in a new direction by engaging stakeholders and developing local relationships. What worked, she noted, was "being on a first-name basis, taking it to the one-on-one relationship level. We are now invited to participate in community events, conferences and seminars, environmental improvement projects, and planning sessions. We have an e-mail network to disseminate information quickly. The newspaper reporters are knowledgeable." She admitted that the start was rough, with some difficult public meetings. "We had to learn that we are not just there to 'educate the public.' What worked was apologizing for not providing information sooner, honesty—even if it meant saying 'I don't know,' explaining research that is underway, testimonials from respected stakeholders, showing the benefits of reclaiming the gamelands, presenting biosolids as part of the big picture, and working with a watershed specialist and providing grant support to the watershed group. We learned we must develop relationships within the community and look for opportunities for shared experiences. Information will be absorbed only when trust is achieved."

4.2 Take Seriously Public Reactions to Odors

Wastewater solids management operations tend to create malodors.

Some involved in the biosolids management profession are able to ignore odors, or they have come to perceive a particular odor as minimal, perhaps because of their constant exposure and adjustment. And it is likely that those working in the field are members of a self-selected group that are less sensitive to wastewater and solids odors than some people.

A significant portion of the population is highly sensitive to various sensory inputs, including odor. Biosolids management odor control systems must meet the needs of these people to minimize the chance of odors causing public upset. Fortunately, objective odor measurement systems are available that can help pinpoint the level of odor control that is needed for a particular project (e.g., nasal rangers, odor panels). And many technologies can help to avoid, contain, mitigate, or treat malodors.

Toffey (2007) noted, "The biosolids manager [can influence whether] his biosolids communicate with neighbors in a way that builds support for the recycling program." Managers should pay attention to odors and other nuisance factors throughout the planning, design, construction, and operations of any biosolids management project. Malodors plus lack of response can equal eroded trust.

4.3 Working with the Media

Part of involving the public requires working with the media. Ample information exists on working constructively with the media, and water quality professional associations and other organizations regularly offer workshops on the topic (Goodman and Goodman, 2005; U.S. Environmental Protection Agency, 2007; Water Environment Federation, 2002). From the media, the public learn about the world around them; as a result, the media also can help the public understand a biosolids management program. The media, on the other hand, needs the expertise and accurate information provided by technical experts, such as biosolids managers, that gives their work meaning and credibility.

By being open, honest, and helpful, biosolids managers can expect, in return, journalists' adherence to their professional standards of truthfulness, accuracy, objectivity, impartiality, fairness, and public accountability.

The media's understanding of biosolids may not advance much over time because biosolids are a topic that is mostly hidden and is covered only sporadically (Goodman and Goodman, 2005). But with patience, more occasions can be created to discuss the topic. With social media, there are now many more ways available to a biosolids management program to share information that should be used.

4.4 Crisis Communications

A subdiscipline of risk communications focuses on communications during emergencies. Most often, crisis communications are aimed at calming fears, but there are instances when its goal is to heighten fear to instigate protective actions. In both cases, the communications also must provide accurate information regarding what happened, the potential risks involved, and, in many cases, what actions people can take to reduce their exposure to risk.

In biosolids management, crisis communications are most likely to be needed in the event of an accident, fire, or spill. In such situations, communications should "first, do no harm" (U.S. Department of Health and Human Services, 2002). Before any type of communication, it is important to know the facts of what has happened, what people know (whether accurate information or rumor), and how the potential risks are being perceived (are people worried?). The choice of wording will affect the perception of risk and the level of concern. If enough information does not yet exist, then it is acceptable to say that more information will be provided at some specified future time.

Crisis communications should be an integral part of an organization's emergency response plan. It requires forethought and planning before a crisis. A competent, trained individual should be available throughout a crisis to conduct communications and work with the media. There is excellent guidance on crisis communications available (U.S. Department of Health and Human Services, 2002).

4.5 Working with Elected Officials

Working with elected officials requires knowing what they need and how these busy people and their worlds function. Haworth et al. (2005) summarized "basic rules that are essential for working effectively with elected officials." These include, "[B] e informed, tell the truth, make connections, don't burn bridges, speak from experience, stay on message, keep in contact, and continue to build the relationship."

It is important to remember that elected officials are swayed mostly by their constituents. If biosolids management team members are not constituents, then they likely will not be on the same side. In which case. allies should be found that are also their constituents; those allies can then carry the team's messages.

4.6 Dealing with Dedicated Opponents and Other Agendas

Individuals and groups who oppose biosolids land application or biosolids incineration or even biosolids generation (some oppose centralized wastewater treatment) will always exist. Some may use opposition to biosolids as a way to generate publicity or support for their organization. But whatever their motivations, they feel their concerns are justified, and they want to be taken seriously. Being dismissed as extreme only makes them angrier and more motivated.

The debates that dedicated opponents help generate are healthy, ensuring that wastewater and biosolids management professionals do not become complacent. Pressure from voiced public concerns has led to many improvements. Nonetheless, the statements and tactics of dedicated opponents can be frustrating. Sometimes, even a biosolids management program that has conducted an excellent, open public outreach and involvement process will have its program halted by a last-minute intervention by a few people.

Although it is not possible to control the actions of others, the project team and organizations involved can follow the recommendations of this chapter. By identifying and engaging stakeholders, and by having robust public outreach and involvement throughout the project, the program will be more immune to anything anyone throws at it. Dedicated opponents harvest public outrage; they do not create it. Their effectiveness in generating opposition depends on their ability to instigate uncertainty and fear. They can only do that when stakeholders and the public around the biosolids program have little information or when the program staff are not perceived as credible or trustworthy. If program staff members have developed honest relationships with diverse stakeholders who have become informed and involved, then they will have their support when unexpected challenges arise.

4.7 Public Outreach and Involvement Are Part of a Strong Legal Defense

Sometimes, when intractable conflicts develop, project proponents may need to look to the courts to defend their rights and investments. During the 1990s and 2000s, biosolids management programs were stopped or challenged by local ordinances, regulatory changes, and direct lawsuits based on toxic tort, personal injury, or nuisance claims. Ultimately, if legal action does occur, good project design, strong public outreach and involvement, and reliance on sound science will lay the foundation for a successful legal defense of a solids management program. Courts look favorably on public projects that involve professionalism, transparency, and public outreach and involvement. A summary of the current state of legal battles and precedents around the land application of biosolids appears in *Biosolids Management: Options, Opportunities, and Challenges* (National Association of Clean Water Agencies, 2006).

4.8 Conclusion: Building Relationships for the Long Term

As a solids management program is developing, it is vital to integrate public outreach and involvement into the entire progression of design, construction, and ongoing operations and to build lasting relationships with key stakeholders.

5.0 REFERENCES AND SUGGESTED READINGS

Beecher, N.; Connell, B.; Epstein, E.; Filtz, J.; Goldstein, N.; Lono, M. (2004) *Public Perception of Biosolids Recycling: Developing Public Participation and Earning Trust*. Water Environment Research Foundation: Alexandria, Virginia.

Butte, G.; Thorne, S. (2007) Personal communication.

Decision Partners Home Page http://www.decisionpartners.com (accessed June 2011).

Decision Partners (2011) *Conducting Effective Community Outreach and Dialogue on Biosolids Land Application: Primer for Biosolids Professionals*. Water Environment Research Foundation: Alexandria, Virginia.

Deeb, R.; Means, E. (2009) *Communication Principles and Practices, Public Perception and Message Effectiveness*. Water Environment Research Foundation: Alexandria, Virginia.

Draman, G. A. (1995) Public Perception is Key to Biosolids Acceptance. *BioCycle*, **36** (9), 82–83.

Eggers, S.; Thorne, S.; Butte, G.; Sousa, K. (2011) *A Strategic Risk Communications Process for Outreach and Dialogue on Biosolids Land Application*; Water Environment Research Foundation: Alexandria, Virginia.

Federation of Canadian Municipalities InfraGuide Best Practices Reports (2005) *Communication and Public Consultation for Biosolids Management*. Green Municipal Fund, Federation of Canadian Municipalities: Ottawa, Ontario. http://www.sustainablecommunities.fcm.ca/Infraguide/Storm_and_Water.asp (accessed June 2011).

Fischhoff, B. (1995) Risk Perception and Communication Unplugged: Twenty Years of Process. *Risk Anal.*, **15** (2), 137–145.

Frank, S. D.; Sigler, L. A. (2004) It's What You Do, Not What You Say That Counts: Denver Metro Wastewater Reclamation District Learns Valuable Stakeholder Lessons. *The Environmental Horizon*, **12** (1), 1–9.

Goodman, J. R.; Goodman, B. P. (2005) Beneficial or Biohazard: How the Media Frame Biosolids. Unpublished work.

Haworth, J.; Beekman, C.; Brewer, L.; Sandul, D. (2005) Great Communication Ideas for Utilities. *Water Environ. Technol.*, September, 2005.

Health Canada (2000) *Health Canada Policy Toolkit for Public Involvement in Decision Making;* Cat. H39–549/2000E. Minister of Public Works and Government Services Canada: Gatineau, Quebec. http://www.hc-sc.gc.ca/ahc-asc/pubs/_public-consult/2000decision/intro-purpose-but-eng.php (accessed June 2011).

International Association for Public Participation (2000) *IAP2 Public Participation Toolbox*. http://www.iap2.org (accessed June 2011).

McAvoy, G. E. (1999) *Controlling Technocracy: Citizen Rationality and the NIMBY Syndrome*. Georgetown Univ. Press: Washington, D.C.

Morgan, M. G.; Fischhoff, B.; Bostrom, A.; Altman, C. (2001) *Risk Communication: A Mental Models Approach*. Cambridge University Press: New York.

National Association of Clean Water Agencies (2006) *Biosolids Management: Options, Opportunities and Challenges*. National Association of Clean Water Agencies: Washington, D.C.

National Biosolids Partnership (2005) *National Manual of Good Practice for Biosolids*. National Biosolids Partnership: Alexandria, Virginia. http://www.wef.org/Biosolids/page.aspx?id=7767 (accessed June 2011).

New Zealand Ministry for the Environment Urban Design Toolkit (2009) http://www.mfe.govt.nz/publications/urban/urban-toolkit-2009/ (accessed June 2011).

Powell, D. (1996) *An Introduction to Risk Communication and the Perception of Risk.* University of Guelph: Guelph, Ontario.

Sandman, P. (2000) Dealing with Outrage: A Key Communication Tool for Biosolids Professionals. *Proceedings of the Water Environment Federation Specialty Conference—14th Annual Residuals and Biosolids Conference,* Boston, Massachusetts, March; Water Environment Federation: Alexandria, Virginia.

Scharp, M. (2000) Why Colorado Farmers Want New York City Biosolids. *BioCycle,* **41** (7), 71.

Susskind, L.; Field, P. (1996) *Dealing with an Angry Public: A Mutual Gains Approach to Rebuilding Trust and Improving Long Term Relationships.* The Free Press: New York.

Toffey, W. E. (2007) Biosolids Odorants Emissions as a Cause of Somatic Disease: What Ought to be Our Profession's Response? *Proceedings of the Water Environment Federation Specialty Conference—21st Annual Residuals and Biosolids Conference*; Water Environment Federation: Alexandria, Virginia.

U. S. Department of Health and Human Services (2002) *Communicating in a Crisis: Risk Communication Guidelines for Public Officials.* U.S. Department of Health and Human Services: Rockville, Maryland.

U. S. Environmental Protection Agency (2007) *Risk Communication in Action: The Risk Communication Workbook.* EPA/625/R-05/003; U.S. Environmental Protection Agency: Washington, D.C. http://www.epa.gov/nrmrl/pubs/625r05003/625r05003.pdf (accessed June 2011).

U. S. Environmental Protection Agency (2008) Public Involvement. http://www.epa.gov/publicinvolvement (accessed June 2011).

Water Environment Federation (2002) *Survival Guide: Public Communications for Water Professionals.* Alexandria, Virginia: Water Environment Federation.

Water Environment Research Federation (2011) *A Strategic Risk Communications Process for Outreach and Dialogue on Biosolids Land Application*; WERF Research Challenge SRSK2R08; © Decision Partners, LLC.

Chapter 4

Solids Production and Characterization

(continued)

4.3 Secondary Treatment

Biological treatment processes generate secondary sludges. The characteristics of these sludges vary with type and with the specific operating conditions of the treatment processes.

4.3.1 *Conventional Activated Sludge*

Conventional activated sludge processes are typically preceded by primary sedimentation; hence, the sludges are primarily a product of biological growth. Table 4.7 summarizes typical characteristics of waste activated sludges from conventional processes. The concentration of solids in waste activated sludge depends on the settling characteristics of the solids and the loading rate to the secondary settler process. Sludge settleability is affected by

- **Biological solids characteristics**. Characteristics are highly dependent on the SRT and operating conditions. Shorter retention times and low dissolved oxygen in aerobic environments can encourage the growth of filamentous organisms. These types of organisms adversely affect settling properties. Sludge bulking caused by inadequate nutrients or other conditions can also adversely affect settling.
- **Temperature**. Increased water density at low temperatures that reduces the rate of sludge settling can result in reduced solids concentrations. Low temperatures can also affect the physical characteristics of the sludge.
- **Mechanical limits of the sludge collection system**. High solids concentrations cannot be conveyed by some solids handling equipment because of the pseudo-plastic and viscous nature of these streams.
- **Presence of dense suspended solids**. The addition of chemicals for phosphorus removal can result in the formation of dense precipitates that have increased settling properties. In facilities that do not use primary sedimentation, the presence of dense fixed solids in the sludge will increase solids concentrations.
- **Type of biological process**. The settleability of solids in the final clarifiers is strongly influenced by the biological processes upstream. For instance, plants with selectors at the head of the activated sludge system often produce better settling and much more consistent solids. As another example, biological phosphorus removal processes also yield solids that settle more quickly than other processes.

Table 4.7 Typical data for the physical characteristics and quantities of sludge produced from various wastewater treatment operations and processes (adapted from Metcalf and Eddy, Inc.,2003).

Treatment operation or process	Specific Gravity of Solids	Specific Gravity of sludge	Dry solids, lb/10^3 gal		Dry solids kg/10^3 m^3	
			Range	Typical	Range	Typical
Primary sedimentation	1.4	1.02	0.9–1.4	1.25	110–170	150
Activated sludge (waste biosolids)	1.25	1.005	0.6–0.8	0.7	70–100	80
Trickling filter (waste biosolids)	1.45	1.025	0.5–0.8	0.6	60–10	70
Extended aeration (waste biosolids)	1.30	1.015	0.7–1.0	0.8[a]	80–120	100[a]
Aerated lagoon (waste biosolids)	1.30	1.01	0.7–1.0	0.8[a]	80–120	100[a]
Filtration	1.20	1.005	0.1–0.2	0.15	12–24	20
Algae removal	1.20	1.005	0.1–0.2	0.15	12–24	20
Chemical addition to Primary tanks for Phosphorus removal						
Low lime (350–500 mg/L)	1.9	1.04	2.0–3.3	2.5[b]	240–400	300[b]
High Lime (800–1600 mg/L)	2.2	1.05	5.0–11.0	6.6[b]	600–1300	800[b]
Suspended growth Nitrification	–	–	–	–	–	–[c]
Suspended growth Denitrification	1.20	1.005	0.1–0.25	0.15	12–30	18
Roughing filters	1.28	1.02	–	–[d]	–	–[d]

[a] Assuming no primary treatment.

[b] Solids in addition to that normally removed by primary sedimentation.

[c] Negligible.

[d] Included in biosolids production from secondary treatment processes.

4.3.2 *Extended Air/Oxidation Ditch*

Extended aeration processes typically are not preceded by primary sedimentation and are operated at higher SRTs. Therefore, the solids generated in these processes typically have higher levels concentrations of inert solids than those generated in conventional activated sludge processes.

4.3.3 *Biological Nutrient Removal*

Biological nutrient removal (BNR) processes typically use alternating environments (aerobic, anoxic, anaerobic) to facilitate enhanced removal of nitrogen and phosphorus by biological means. Biological processes can also be combined with chemical addition when low final effluent phosphorus concentrations are required. The operation of BNR processes can affect the properties of the generated waste sludge streams. Nitrogen present in waste activated sludges from BNR processes is associated with the cellular matter and typically constitutes 8 to 12% of the dry weight. Phosphorus is either taken up by the biomass during cell synthesis (1.5 to 2% dry weight) or is stored as long chained polyphosphates in enhanced biological phosphorus removal processes (EBPR). The latter storage mechanism can result in a total P content of 3.5 to 6%. The polyphosphates are subject to ready release to a soluble form if the sludges are exposed to anaerobic conditions (i.e., secondary settlers with deep sludge blankets, gravity thickeners, anaerobic digestion). Release of this soluble phosphorus can result in significant recycling of phosphorus back to the wastewater treatment process in thickening and dewatering return streams. In some facilities, biological sludges are wasted from the aeration basin rather than the secondary clarifier underflows to eliminate phosphorus solublization associated with anaerobic underflow conditions. This practice will result in the production of high volumes of dilute secondary sludges, which will likely require thickening during sludge handling.

4.3.4 *Fixed-Film/Hybrid Processes*

Typically, fixed-film processes have included trickling filters and rotating biological contactors (RBCs). More recent innovations have included moving bed biofilm reactors (MBBRs), integrated fixed-film activated sludge (IFAS), and biological aerated filters (BAFs). Sludges from fixed-film processes tend to contain increased concentrations of higher organisms including algae, fungi, protozoa, and worms. Sludges produced by biofilm systems typically have slightly improved settling characteristics compared to suspended growth sludges. When removed from secondary clarifiers, therefore, the solids content of these may be slightly higher compared with other biological processes. In addition, observed yields from fixed systems are slightly lower than conventional biological systems because of higher solids retention time in the biofilms.

4.3.5 *Membrane Bioreactors*

Membrane bioreactors (MBRs) are increasingly being used in wastewater treatment. Although the principals of biological treatment are similar for MBRs as for other types of biological treatment, they have several unique features that affect sludge properties:

- Membrane bioreactors require the use of fine screens (i.e., 1- to 2-mm openings) before the secondary process to minimize membrane fouling. Therefore, more fixed solids are removed upstream and solids generated will have a higher volatile fraction.
- Membrane bioreactors are often operated at extended SRTs or low F/M values. The higher volatile fraction resulting from finer screening upstream can be offset by more stabilized solids at higher SRTs. In addition, sludge yield is also likely lower than in other biological systems.

4.3.6 *Tertiary Treatment*

As effluent requirements become more stringent, tertiary processes are being used more often to further enhance wastewater quality. Typical tertiary processes include filtration, chemical treatment for additional phosphorus removal, membrane processes for wastewater reuse, and post-nitrification and denitrification facilities. The characteristics of the solids are highly dependent on the processes used and the upstream processes. For example, solids removed by filtration will mimic the upstream biological process but will typically be more dilute before thickening. Chemical solids generated from phosphorus precipitation will have a higher inert solids concentration and can be thickened in the reactor to higher total solids concentrations.

5.0 ESTIMATION OF WASTEWATER SOLIDS QUANTITIES

5.1 Criteria Used for Solids Processing Design

A designer must choose the most appropriate design conditions for the solids treatment process. Unlike with the liquid stream, solids treatment systems are not covered by daily hydraulic maximums. Often, solids streams are designed for a combination of annual-average and maximum-month solids production conditions estimated for a future design year, typically 10 to 20 years depending on client needs. Typically, different design conditions are also applied to different steps in the treatment process.

5.1.1 *Annual Average*

Annual average solids production is typically described as the daily average solids production for a treatment process or system when the data is averaged over a calendar year period. In most cases, annual average solids production is used for two purposes:

1. Determination of normal operating conditions of a treatment process. When considered during startup conditions, annual average conditions are used to determine typical operating duration of a solids treatment system. Annual average solids production can also be used as a measure when phasing of additional construction of solids treatment systems may be needed in future years.
2. Determination of costs of operation. Annual average quantities are used to estimate present and future operating costs associated with solids treatment systems.

5.1.2 *Maximum Month*

In most cases, solids treatment systems are designed to accommodate maximum-month or peak 30-day solids production conditions. Unit process sizing is typically tied to an expected maximum month sludge production condition in a future design year.

Maximum month solids production is typically defined as the highest average daily solids production rate determined for a calendar month for a given calendar-year period. Peak 30-day solids design criteria, which corresponds to the highest rolling 30-day average of daily solids production, can also be used. The use of either maximum-month or peak 30-days should be consistent with the criteria chosen for the liquid stream (i.e., if peak 30-day criteria are established for the liquid stream, then it recommended that a peak 30-day criteria also be used for design of the solids treatment system).

5.1.3 *Peak Week*

Peak-week solids production is typically used as a check of the solids treatment system for adequate performance under extreme solids production conditions. Peak-week solids production is defined as the maximum daily solids production when averaged over a weekly period and when a calendar-year period is considered. A peak week can either be a calendar week or a peak seven-day period. Under a peak-week solids production condition, the solids treatment process should be able to accommodate expected production, either by storage or by increased hours of operation for thickening and dewatering systems.

5.1.4 *Peak Day*

The peak-day criterion is typically not used for design purposes for a solids treatment system. Peak-day solids production represents an extremely high solids production condition that is typically not sustained for a long period. As such, this criterion does have applicability in the solids removal portion of the liquid stream (i.e., solids pumping). Typically, this criterion is used to check a pumping system to make sure that all the solids can be removed under this peak production condition. In most cases, pumps are designed for intermittent operation and can accommodate higher solids production through longer operation. Nevertheless, it is recommended in design that solids pumping systems and conveyance systems used for solids removal from the liquid stream be checked for capacity to accommodate these potential spikes in solids production. Similar to peak week, storage or attenuation in the liquid stream are also integral to management of solids production.

5.1.5 *Hours of Operation*

Hours of operation are design criteria established for thickening, dewatering, or other solids treatment process where the presence of an operator is required for startup/shutdown and for more hands-on operation. The design criteria chosen are highly dependent on the specific process and often correspond to specific shift hours used by a WWTP. The hours of operation should be considered carefully in final sizing of unit processes and may offer flexibility to accommodate greater-than-average solids production. For instance, a belt filter press dewatering operation may be designed for maximum-month solids production and a solids concentration range of 3 to 4%. If design of the system was based around a period of operation shorter than 24 hours, 7 days per week, then there is additional capacity for treatment/handling of larger unexpected flows. A facility designed to be operated for 5 to 16 hours per day could be operated for a longer period of time each day or for more days to accommodate spikes in production. Additionally, if a solids treatment process is down for unexpected repairs, additional hours of operation can be added to catch up and to reestablish steady-state operating conditions.

5.2 Typical Quantities of Solids

Many factors influence the amount of solids produced in a WWTP including influent characteristics, size of primary and secondary treatment systems, and operating conditions. Therefore, it is recommended that plant-specific data be considered wherever possible and that a mass balance be established for the plant.

If a conceptual level review/design of a system is desired, then typical rules of thumb can be applied. A conservative rule of thumb for approximation of solids production is 1 Mg dry solids per 4 ML treated. As with any rule of thumb, application of this criterion should only be used for cursory review and for conceptual level consideration. The total amount can vary widely as can the distribution between primary and waste activated solids.

Typical quantities of solids for different unit processes are shown in Table 4.7. These characteristics should also be used with caution. Wherever possible, actual plant data should be used to establish specific design criteria for the unit processes.

5.3 Methods for Estimating Solids Quantities

5.3.1 Overview

The quantity of sludge to be handled must be estimated when designing and retrofitting solids processing, treatment, and disposal facilities. Estimating solids production is also important when evaluating the performance of WWTPs and determining their sludge handling capability. Solids quantities can be estimated using historical data (where available), empirical methods, and model-based methods. The method used will depend on the context (e.g., design versus performance evaluation), the level of detail required, and the available information.

Historical data, where available, are a valuable source of information on solids production. Analysis of actual solids production data can be used to project solids production for similar plants for the purposes of design, or to analyze the performance of an existing plant for the purposes of retrofitting or optimization. Historical data should be checked for reliability, where possible, using solids production estimates developed using an empirical or model-based approach.

Empirical estimation of solids production involves using the actual or projected plant loading and performance data, typical chemical addition dosages, and typical sludge yields for the processes found within the plant. For example, using the mass of five-day biochemical oxygen demand (BOD_5) removed across an activated sludge process and a published yield or sludge production ratio, it is possible to estimate the expected mass of secondary sludge produced.

The model-based approach involves using fundamental mass-balance-based models to predict sludge production. These models may be steady-state design models such as those published by the Water Research Commission (1984), Wentzel et al. (1990), and Tchobanoglous et al. (2003), or dynamic simulation models that include more detailed models of the activated sludge process such as ASM1, ASM2d, or ASM3 (Henze et al., 2000). Both model-based approaches are based on chemical oxygen demand (COD) mass balances, incorporate biological growth kinetics, and give comparable estimates of sludge production when used consistently. Model-based methods are the most comprehensive and reliable methods of estimating solids production when used appropriately but require more input information than empirical methods.

5.3.2 *Analysis of Historical Data*

Sources of solids production data include plant historical records, sampling program data, and literature data. Most plants typically measure the flow rates of sludge streams or the volumes of sludge wasted by measuring volume changes in holding tanks or use pump curves (e.g., pump strokes for a positive displacement pump are often used). They will also typically measure the solids concentrations in the sludge streams.

Measurements may be based on grab or composite samples. Composite samples are preferred with flow-weighted composite samples being the most representative. Municipal permits typically require composite samples for influent and effluent streams but sampling of sludge streams often is based on grab samples. These samples are also taken less frequently. Grab samples represent the state of the wastewater stream sampled at an instant in time. Grab sampling data should be used carefully because they are dependent on when the samples are collected during the sludge removal cycle. When all the required data are not available, a sampling study can be initiated to augment the historical data.

Before using historical data, it should be analyzed using statistical techniques to detect outliers, to identify trends, and to select representative data for calculating average sludge production rates. Typical analyses conducted would include calculating the average, median, minimum, maximum, standard deviation, and variance of the data and plotting time-series data. Calculation of the 95th percentile of the dataset can be helpful in determining a representative maximum value. Plots are useful for visually identifying outliers, checking for trends and seasonal effects, and selecting representative periods for the calculation of average values used for further analysis.

Sludge data may be unreliable because of poor sampling techniques, inaccurate flow measurements, and unmeasured sludge flow streams (Ministry of the Environment, 1984; U.S. Environmental Protection Agency, 1989). Flow measurements of sludge streams are a common source of error.

Solids mass balances should be performed, where possible, to check the consistency and reliability of historical data. For example, the mass of solids entering a clarifier in the influent should equal the mass of solids exiting the clarifier in the overflow and underflow streams for stable periods of plant operation (i.e., no significant change in clarifier storage or blanket level). If the difference is more than 10 to 15%, then further investigation of the plant data, sampling procedures, and operational procedures is warranted.

The historical sludge production data should also be compared against projected sludge production estimates calculated using empirical or model-based techniques. Discrepancies greater than 10 to 15% suggest the need for further analysis and investigation of the data. Plant influent loading data may also contain errors and should be verified by comparison to typical per capita wastewater flow rate and constituent data.

5.3.3 *Empirical Sludge Estimation Techniques*

Empirical estimation techniques are based on using typical process performance characteristics, sludge yields, and chemical addition stoichiometry to calculate sludge production. For primary clarifiers, the sludge production from suspended solids removal can be calculated as follows:

$$PS_M = \frac{Q \times TSS_{\text{inf}} \times E}{1000} \qquad 4.1$$

where

PS_M = mass of sludge, kg/d,

Q = primary influent flow, m^3/d,

TSS = primary influent total suspended solids, mg/L, and

E = removal efficiency, fraction.

Typical primary clarifier removal efficiencies for BOD_5, total suspended solids (TSS), and total phosphorus are given in Table 4.8 with and without chemical coagulation (Water Environment Federation, 2009). If the actual TSS removal is known, then it can be used to calculate the primary sludge production instead of a typical efficiency factor. Once the mass of solids is known, historical average concentrations can be used to calculate volumes of material sent to solids treatment. If data on solids concentrations is not known, then typical values such as those discussed in previous sections can be used.

Typical sludge yields for secondary treatment systems are presented in Table 4.9 (U.S. Environmental Protection Agency, 1989). These yields can be used along with plant performance data to estimate sludge production. Additional information on

Table 4.8 Typical primary clarifier removal efficiencies (BOD_5 = five-day biochemical oxygen demand; TSS = total suspended solids; TP = total phosphorus) (adapted from Water Environment Federation et al., 2009).

Parameter	Without chemical coagulation	With chemical coagulation
Removal efficiency (%)		
BOD_5	25–40	40–70
TSS	40–70	60–90
TP	5–10	70–90

Table 4.9 Typical sludge yields for suspended growth processes (BOD_5 = five-day biochemical oxygen demand) (adapted from U.S. Environmental Protection Agency, 1989).

Process type	Sludge production ratio (kg TSS (sludge)/ kg BOD_5 removed)
Activated sludge after primary clarification	0.7
Activated sludge without primary clarification	
Conventional (tapered aeration, complete-mix, plug-flow, step feed)*	0.85
Extended aeration and oxidation ditch	0.65
Contact stabilization	1.0

*For systems with hydraulic retention times less than 10 hours.

observed yields as a function of influent BOD removed is provided in recent texts (Water Environment Federation et al., 2009). Typical yields are only approximate because secondary sludge production depends on many factors including the following:

- Mass of organic matter applied to the biological system, expressed as the mass of five-day BOD or COD in kg/day;
- Mass of TSS applied to the biological system, expressed in kg/day;
- Effective SRT of the biological process, expressed in days;
- Wastewater temperature; and
- Characteristics of the influent organic matter (BOD_5 or COD), TSS, and the biological wastewater treatment system.

The mass of organic matter affects biological solids production because biological solids are generated from the applied organic matter. The mass of TSS affects biological solids production because a portion of the applied TSS is nonbiodegradable. Included in this are not only the influent fixed (or nonvolatile) suspended solids (FSS), but also a portion of the volatile suspended solids (VSS) that are not biodegraded in the biological treatment system. Typically, 30 to 40% of the VSS in municipal wastewaters are not biodegraded in biological wastewater treatment systems.

The SRT affects biological solids production through its effect on the overall degradation of influent organic matter. Increased SRT allows increased endogenous respiration, which leads to greater degradation of synthesized biological solids. Endogenous respiration results in the conversion of synthesized biological solids to carbon dioxide (CO_2) and water (H_2O). Synthesized biological solids cannot be fully oxidized, however, because a nonbiodegradable residue remains. Thus, biological system solids production cannot be reduced to zero through extended SRT. Solids retention time also affects the degradation of influent particulate matter. At short SRT (less than approximately three days), little degradation of influent particulate organic matter occurs. At SRTs in excess of about 3 days, biodegradable particulate matter in the influent wastewater is degraded, leaving only the nonbiodegradable residue. Wastewater and biological system characteristics affect biological solids production directly. The degradability of the influent organic matter affects the conversion of organic matter to biological solids and its subsequent oxidation as a result of increased SRT. Likewise, biological system characteristics affect biological solids production and degradation rates.

Membrane bioreactor and fixed-film systems will have solids yields that are lower than the values shown for conventional activated sludge. Therefore, when applying the values shown to those systems, the solids estimates should be viewed as slightly conservative.

Chemical solids are typically produced by the addition of chemicals to remove phosphorus. The quantity of chemical sludge produced can be estimated using the metal salt dosage, the phosphorus removed, and the chemical reaction stoichiometry. The traditional procedure for estimating the chemical sludge formed involves first calculating the quantity of metal phosphates formed and then calculating the metal hydroxides formed with the excess metal ions. The mass of metal phosphates formed is calculated using the stoichiometry of the chemical reaction and the mass of phosphorus removed. The excess metal ion is then calculated based on the metal dosage. Using reaction stoichiometry, the mass of metal hydroxide formation is then estimated. The mass of chemical sludge determined using this method is often increased by 35% because the stoichiometry is only an approximation of the chemical reactions that occur, and larger quantities are often required in practice (U.S. Environmental Protection Agency, 1987). More recent developments in chemical phosphorus removal theory have been developed relating to different types of solids.

If there is a significant amount of soluble organics in the wastewater when chemicals are added, precipitation of some dissolved solids can also occur. The quantity of sludge formed because of the removal of dissolved solids (i.e., soluble organics) can be estimated using the following formula and an assumed soluble organics removal efficiency of 30% (U.S. Environmental Protection Agency, 1987):

$$DS_M = STOC_{inf} \times E_{sol} \times 1.6 \times 1.18 \tag{4.2}$$

$$DS_M = SCOD_{inf} \times E_{sol} \times 1.6 \times 1.18 \tag{4.3}$$

$$DS_M = SBOD_{inf} \times E_{sol} \times 1.6 \times 1.18 \tag{4.4}$$

where

DS_M = mass of sludge resulting from removal of dissolved solids, kg/d;
$STOC_{inf}$ = soluble TOC (total organic carbon) in influent, mg/L;
$SCOD_{inf}$ = soluble COD (chemical oxygen demand) in influent, mg/L;
$SBOD_{inf}$ = soluble BOD (biochemical oxygen demand) in influent, mg/L; and
E_{sol} = soluble organic removal efficiency, %.

Biological excess phosphorus removal (BEPR) processes would be expected to have an increased secondary sludge yield compared to a conventional activated sludge process. This occurs because of the storage of large quantities of inorganic polyphosphate and its associated counter-ions and the lower decay rate of phosphorus accumulating organisms (Wentzel and Ekama, 1997). The secondary sludge mass in a BEPR system can increase by as much as 20 to 25% for raw wastewater and by 45 to 55% for settled wastewater, compared to a conventional activated sludge process according to the steady-state model of Wentzel et al. (1990).

5.3.4 *Theoretical Model-Based Estimation*

Wastewater treatment in activated sludge systems results in biosolids from waste activated sludge (WAS) and primary sludge if the plant incorporates primary clarifiers. Quantifying sludge production is important for sizing the activated sludge system and secondary clarifiers and for evaluating solids treatment and disposal options and costs. Also, the composition of the excess sludge is important because it affects the selection and design of solids treatment options.

Many factors affect sludge production and composition: (1) the influent wastewater composition; (2) primary clarifier performance; (3) activated sludge system SRT; and (4) temperature.

5.3.4.1 *Influent Wastewater Characteristics*

Estimating sludge production for both primary sludge and WAS requires detailed information on influent wastewater characteristics to identify the different components making up the total organic load. The influent wastewater characterization protocol developed by the Water Environment Research Foundation (2003) identifies the division of the total COD into soluble, colloidal (nonsettleable), and particulate components and the biodegradable and inert soluble and particulate components (see Figure 4.1). [Methods for measuring the different components are outlined in WERF (2003). This characterization is critical to estimating sludge production. An additional important factor is the inorganic suspended solids (ISS) content (ISS = TSS – VSS). Also, the COD/VSS ratio of influent organic solids has an effect on sludge production estimates and the composition of WAS and primary sludge.

Division of the influent wastewater COD into the four primary subcomponents can be defined by three fractional parameters (f_{US}, f_{UP}, and f_{BS}), typically expressed as a decimal value (i.e., unitless):

f_{US} = fraction of the total influent COD, which is nonbiodegradable soluble
f_{UP} = fraction of the total influent COD, which is nonbiodegradable particulate
f_{BS} = fraction of the total influent COD, which is readily biodegradable

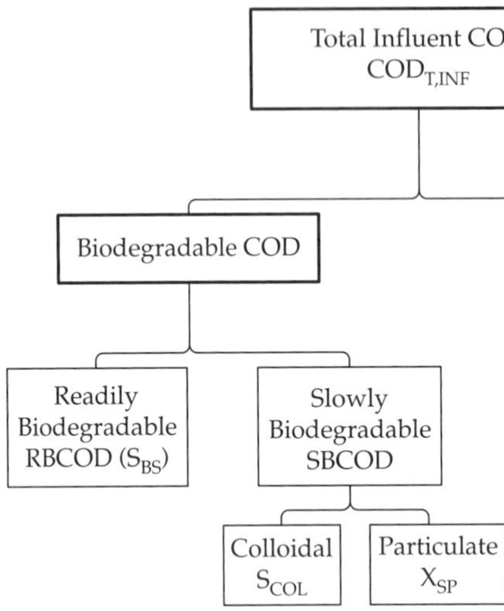

FIGURE 4.1 Division of influent chemical oxygen demand (COD) into characteristic fractions (SBCOD = slowly biodegradable COD; RBCOD = readily biodegradable COD) (WERF, 2003).

Using these fractions, the division of the total influent COD into concentrations of the four primary sub-components may be expressed as follows:

$$S_{us} = f_{us} \times S_{T,INF} \quad 4.5$$

$$X_I = f_{up} \times S_{T,INF} \quad 4.6$$

$$S_{BS} = f_{bs} \times S_{T,INF} \quad 4.7$$

$$SBCOD = (1 - f_{BS} - f_{US} - f_{UP}) \times S_{T,INF} \quad 4.8$$

where

S_{US} = soluble nonbiodegradable COD concentration (mg/L);
X_I = particulate nonbiodegradable COD concentration (mg/L);
S_{BS} = soluble readily biodegradable COD concentration (mg/L);
$SBCOD$ = slowly biodegradable COD concentration (mg/L); and
$S_{T,INF}$ = total influent COD concentration (mg/L).

For modeling the organics removal in a primary settling tank and the COD composition and load of the stream passing to the activated sludge system, it is important to distinguish between the portions of the slowly biodegradable COD (SBCOD) that are colloidal and particulate because only a portion of the slowly biodegradable material is settleable. This division of SBCOD into colloidal and particulate portions is defined by another fraction (unitless):

f_{XSP} = fraction of the SBCOD that is particulate (not colloidal)

$$X_s = f_{XSP} \times \text{SBCOD} \quad 4.9$$

$$S_{COL} = (1 - f_{XSP}) \times \text{SBCOD} \quad 4.10$$

where

S_{COL} = slowly biodegradable colloidal COD concentration (mg/L) and
X_S = slowly biodegradable particulate COD concentration (mg/L).

In summary

$$S_{T,INF} = S_{BS} + S_{US} + X_I + X_{SP} + S_{COL} \quad 4.11$$

$$X_{ISS} = TSS - VSS \quad 4.12$$

Typical organic components for a high-strength municipal wastewater are shown in Table 4.10, based on a total influent COD of 600 mg/L and using the methodology outlined above for derivation of components. The table includes quantification of the different components as fractions of total COD and the corresponding concentration of each component. The approach of defining total COD and fractions of the subcomponents is useful because although total COD may vary from plant to plant, the values of the fractions will be more consistent. The table includes typical BOD_5, VSS, and TSS concentrations for the wastewater, and the COD/VSS ratio for influent solids.

5.3.4.2 Primary

A simple mass balance approach can be applied using a spreadsheet to quantify and characterize primary sludge production and the composition of the primary effluent, based on the soluble/particulate COD division and an assumed percent solids removal. Typical primary effluent concentrations for high-strength wastewater characteristics are also shown in Table 4.10, assuming 60% solids removal efficiency in primary clarifiers. In the primary clarifiers, it is assumed that no soluble components are removed. Therefore, the fractions describing COD also change as a result of primary treatment. The primary effluent fractions that result are also shown in the table.

TABLE 4.10 Typical raw influent and primary effluent composition for a high-strength wastewater (primary effluent values developed based on 60% solids removal efficiency and assuming a 2% primary solids concentration) (COD = chemical oxygen demand; BOD_5 = five-day biochemical oxygen demand; VSS = volatile suspended solids; TSS = total suspended solids; ISS = inert suspended solids).

Component	Raw influent	Primary effluent	Units
Total COD ($S_{T,INF}$)	600	378	mg /L
Fractions			
f_{US}	0.05	0.079	
f_{UP}	0.13	0.084	
f_{BS}	0.16	0.254	
f_{XSP}	0.75	0.550	
Concentrations			
S_{BS}	96	96	mg/L
S_{US}	30	30	mg/L
X_I	78	32	mg/L
S_{COL}	99	99	mg/L
X_{SP}	297	121	mg/L
Other typical characteristics			
BOD_5	295	202	mg/L
COD / BOD_5	2.04	1.87	mg/mg
VSS	235	96	mg/L
TSS	290	118	mg/L
X_{ISS} = ISS = TSS-VSS	55	22	mg/L
$f_{CV,P}$	1.6	1.6	mg/mg

Primary volatile sludge production per unit influent COD load can be estimated based on a fractional primary clarifier mass removal and the definition of influent COD fractions as follows:

$$\frac{PS_{VSS}}{MS_T} = \frac{f_{REM} \times Q_{INF} \times (X_S + X_I) / f_{CV,P}}{Q_{INF} \times S_{T,INF}}$$
$$= f_{REM} \times \left[f_{XSP} \times (1 - f_{US} - f_{UP} - f_{BS}) + f_{UP} \right] / f_{CV,P} \qquad 4.13$$

where

PS_{VSS} = primary sludge VSS production rate (g/day);
MS_T = raw influent total COD mass rate (g/day);
Q_{INF} = raw influent flow rate (m^3/day);
f_{REM} = PST fractional mass removal (unitless); and
$f_{CV,p}$ = COD/VSS.

Primary total sludge production per unit influent COD load can be estimated by adding the ISS component as follows:

$$\frac{PS_{TSS}}{MS_T} = \frac{PS_{VSS}}{MS_T} + \frac{f_{REM} \times Q_{INF} \times ISS}{Q_{INF} \times S_{T,INF}}$$
$$= \frac{PS_{VSS}}{MS_T} + f_{REM} \times \frac{X_{ISS}}{S_{T,INF}} \qquad 4.14$$

where

PS_{TSS} = primary sludge TSS production rate (g/day) and
X_{ISS} = raw influent ISS concentration (mg/L).

5.3.5 *Biological*

The development of equations for quantifying the behavior and production of solids in a biological system dates back to the early 1950s. A detailed history of development is included in Dold (2007). Quantifying sludge production in an activated sludge system treating municipal wastewater, and operated at a given solids retention time, requires the following information:

- Mass of substrate utilized for biomass growth;
- Yield coefficient (*Y*);
- Endogenous respiration rate (*b*);
- Amount of endogenous residue generated (quantified by the factor, *f*, defined below);
- Inert (nonbiodegradable) particulate organics from the influent; and
- Inorganic suspended solids from the influent (ISS = TSS – VSS).

This section summarizes steady-state equations for quantifying waste activated sludge quantity and composition as a function of SRT (and temperature). It should be noted that the definition of wastewater COD fractions is *with respect to the activated sludge influent stream.* In cases with primary settling these fractions differ from those for the raw influent stream.

The mass of COD used for biomass growth per day can be calculated using the following:

$$\begin{aligned} M\Delta S &= Q_{INF} \times S_{T,INF} \times (1 - f_{US} - f_{UP}) \\ &= MS_{T,INF} \times (1 - f_{US} - f_{UP}) \end{aligned} \qquad 4.15$$

where

$M\Delta S$ = mass of substrate utilized per day (g COD/d);
Q_{INF} = activated sludge influent flow rate (m^3/d);
$S_{T,INF}$ = Total COD of activated sludge influent (mg COD/L);
$MS_{T,INF}$ = mass of substrate in influent (g COD/d);
f_{US} = fraction of $S_{T,INF}$, which is nonbiodegradable soluble (unitless); and
f_{UP} = fraction of $S_{T,INF}$, which is nonbiodegradable particulate (unitless).

The total amount of solids produced in a biological system can be estimated using the following equations.

$$MX_T = MX_V + MX_{ISS} \qquad 4.16$$

$$\begin{aligned} MX_V &= MX_H + MX_E + MX_I \\ &= MS_{T,INF} \times \theta_X \times \left\{ \frac{Y \times (1 - f_{US} - f_{UP})}{1 + b \times \theta_X} \times (1 + f \times b \times \theta_X) + \frac{f_{UP}}{f_{CV,P}} \right\} \end{aligned} \qquad 4.17$$

$$MX_{ISS} = Q_{INF} \times X_{ISS} \times \theta_X \qquad 4.18$$

where

MX_T = mass of total solids (g/d);
MX_H = mass of active biomass (g/d);
MX_E = mass of endogenous residue (g/d);
MX_I = mass of inert solids (g/d);
$S_{T,INF}$ = total COD of activated sludge influent (mg COD/L);

f_{US} = fraction of $S_{T,INF}$, which is nonbiodegradable soluble (unitless);
f_{UP} = fraction of $S_{T,INF}$, which is nonbiodegradable particulate (unitless);
Y = yield (mg VSS/mg COD) – 0.47 mg VSS/mg COD;
B = decay coefficient (d^{-1}) – 0.24 $(1.029)^{T-20}d^{-1}$
F = fraction of cells remaining as inerts after endogenous respiration (unitless) –0.2 (unitless).

Other useful relationships are the active fraction of the VSS and the sludge production per unit influent COD removed.

Biomass (active) fraction of volatile solids:

$$f_{AV} = \frac{MX_H}{MX_V} = \frac{1}{1+f\times b\times\theta_X+(f_{UP}/f_{CV,P})\times((1+b\times\theta_X)/[Y\times(1-f_{US}-f_{UP})])} \quad 4.19$$

The VSS production per unit influent COD removed (influent COD minus filtered effluent COD):

$$\frac{M\Delta X_V}{M\Delta S} = \frac{MX_V/\theta_X}{M\Delta S} = \frac{\left(\{[Y\times(1-f_{US}-f_{UP})]/(1+b\times\theta_X)\}\times(1+f\times b\times\theta_X)+(f_{UP}/f_{CV,P})\right)}{(1-f_{US})} \quad 4.20$$

The TSS production per unit influent COD removed (influent COD minus filtered effluent COD):

$$\frac{M\Delta X_T}{M\Delta S} = \frac{MX_T/\theta_X}{M\Delta S} = \frac{\left(\{[Y\times(1-f_{US}-f_{UP})]/(1+b\cdot\theta_X)\}\times(1+f\times b\times\theta_X)+(f_{UP}/f_{CV,P})\right)+(X_{ISS}/S_{T,INF})}{(1-f_{US})}$$

4.21

6.0 COMPREHENSIVE SIMULATION MODELS

The majority of biological models applied in widely used simulators are based on ASM-type models (Henze et al., 2000). In these models, organism decay is based on the death-regeneration approach as an alternative to the endogenous respiration approach used in the equations above. The endogenous respiration method has been used here because it enables development of simple steady-state equations. Both approaches use two parameters to quantify organism decay: (1) a decay rate, b; and (2) an endogenous residue fraction, f. Values of the parameters differ significantly for the two approaches, however, which has caused confusion. The two approaches should give identical predictions if corresponding b and f values are applied.

6.1 Additional Factors Affecting Sludge Production

Many factors may cause the observed sludge production at a particular plant to differ somewhat from those predicted using the simplified equations. For example,

- Yields for anoxic growth are lower than for aerobic growth; therefore, sludge production in systems with extensive denitrification should be lower. (Anoxic decay rates are also assumed to be lower, however, and this would have a compensating effect).
- In systems with carbon addition for gentrification, the substrates typically are simple organic compounds (e.g., methanol). Yield coefficients for these may be lower than for the mix of more complex organics in municipal wastewater.
- In systems with chemical addition for phosphorus precipitation, the generation of chemical sludge must be considered.
- In biological phosphorus removal, an apparent degree of stabilization occurs in anaerobic zones, reducing VSS production. Stored P, however, increases the apparent ISS.
- The calculations do not account for nitrifier sludge growth; however, this is a small component in systems with typical influent TKN/COD.

The designer should be aware of the above factors and take these influences into account in the consideration of solids design criteria.

7.0 REFERENCES AND SUGGESTED READINGS

American Public Health Association; American Water Works Association; Water Environment Federation (2005) *Standard Methods for the Examination of Water and Wastewater*, 21st ed.; American Public Health Association: Washington, D.C.; American Water Works Association: Denver, Colorado; Water Environment Federation: Alexandria, Virginia.

Dold, P. L. (2007) Quantifying sludge production in municipal treatment plants. *Proceedings of the 80th Annual Water Environment Federation Technical Exposition and Conference* [CD-ROM]; San Diego, California, Oct 15–17; Water Environment Federation: Alexandria, Virginia.

Henze, M.; Gujer, W.; Mino, T.; van Loosdrecht, M. C. M. (2000) *Activated Sludge Models ASM1, ASM2, ASM2d, and ASM3*, IWA Scientific and Technical Report No. 9; IWA Publishing: London, United Kingdom.

Metcalf and Eddy, Inc. (2003) *Wastewater Engineering: Treatment and Reuse*, 4th ed.; Tchobanoglous, G.; Burton, F. L.; Stensel, H. D., Eds.; McGraw-Hill; New York.

Ministry of the Environment (1984) *Guidelines for the Design of Sewage Treatment Works*. Ontario Ministry of the Environment: Toronto, Ontario, Canada.

Pepper, I.; Gerban, C.; Brusseau, M. (1996) *Pollution Science*. Academic Press: New York.

Tchobanoglous, G.; Burton, F.; Stensel, D. (2003) *Wastewater Engineering: Treatment and Reuse*. McGraw-Hill: New York.

U.S. Environmental Protection Agency (1987) *Design Manual: Phosphorus Removal*, Report No. EPA/625/1-87/001; U.S. Environmental Protection Agency, Office of Research and Development, Center for Environmental Research Information and Water Engineering Research Laboratory: Cincinnati, Ohio.

U.S. Environmental Protection Agency (1989) *Handbook: Retrofitting POTWs*, Report No. EPA/625/6-89/020; U.S. Environmental Protection Agency, Office of Research and Development, Center for Environmental Research Information and Water Engineering Research Laboratory: Cincinnati, Ohio.

Water Environment Federation (2005) *Wastewater Sampling for Process and Quality Control*; Water Environment Federation: Alexandria, Virginia.

Water Environment Federation; American Society of Civil Engineers; Environmental and Water Resources Institute (2009) *Design of Municipal Wastewater Treatment Plants*, 5th ed., WEF Manual of Practice No. 8; ASCE Manual and Report on Engineering Practice No. 76; McGraw-Hill: New York.

Water Environment Research Foundation (2003) *Methods for Wastewater Characterization in Activated Sludge Modelling*, Project 99-WWF-3; Water Environment Research Foundation: Alexandria, Virginia.

Water Research Commission (1984) *Theory, Design and Operation of Nutrient Removal Activated Sludge Processes*, Wiechers, H. N. S., Ed.; Water Research Commission: Gezina, South Africa.

Wentzel, M. C.; Ekama, G. A. (1997) Principles in the Design of Single-Sludge Activated-Sludge Systems for Biological Removal of Carbon, Nitrogen, and Phosphorus. *Water Environ. Res.*, **69** (7), 1222–1231.

Wentzel, M. C.; Ekama, G. A.; Dold, P. L.; Marais, G.v.R. (1990) Biological Excess Phosphorus Removal—Steady State Process Design. *Water SA*, **16** (1), 29–48.

Chapter 5

Design Approach

(continued)

1.0 INTRODUCTION AND SCOPE

This chapter presents a methodology for the design of wastewater solids management systems. Topics discussed include systems approach, process selection logic, mass balance calculations, concept of sizing equipment, contingency planning, and other design considerations such as energy conservation and cost-effective analysis.

Many factors and considerations beyond the process selection, design criteria, and their application influence the financial, environmental, and social effects of a solids processing project. Each of these factors needs to be adequately defined and considered. These additional project-defining considerations include, but are not limited to, incorporation of key project stakeholders in the planning process, adoption of comprehensive life-cycle cost methodologies, and definition of project implementation sequences and procurement mechanisms relevant to the specific technical, financial, and institution capabilities of the agency. This chapter discusses the overall design considerations to be taken into account in the design of a successful solids process design project.

Overall, wastewater treatment plant (WWTP) performance is the sum of the combined performances of each linked component. The actions of one component affect the performance of all the others. For example

1. Materials not captured in solids treatment processes will be returned in the sidestreams to the wastewater treatment system as a recirculating load. This load may cause degradation in effluent quality, an increase in wastewater treatment costs, and process upsets.
2. Failure to remove and treat solids at the same rate as they are produced within the wastewater treatment system will eventually cause effluent degradation and may increase operating costs.
3. Hydraulic overloads resulting from inadequate solids thickening can cause downstream solids treatment processes (such as anaerobic digestion) to operate less effectively.
4. The addition of chemicals to the wastewater treatment process for purposes of nutrient and suspended solids removal will increase the quantity and alter the characteristics of solids to be disposed of or treated.

It is important to understand the relationship between process parameters and the performance of processes such as, for example, how thickener feed rates affect thickener performance. It is equally important to understand how individual processes affect one another when combined into a system such as, for instance, how the performance of the thickener affects digestion and dewatering. Interactions between the processes in a system are described in this chapter.

2.0 DESIGN PROJECT PHASES

The size and complexity of the project will make a difference to the design planning, design processes and the design documenting requirements. There is no "one size fits all" approach to the design process. The design approach is customized on a project-by-project basis. It all begins with a design plan where the project is defined and the design activities are carefully mapped out.

Design process flowcharts are shown in this chapter as illustrative examples. They are example flowcharts that provide a guide for most projects. They can be modified to suit other, specific project requirements. Further details on the design activities can be found below.

2.1 Small Design Projects

If there is a design component within the project, then some form of design plan is required. Several questions should be considered in development of the design plan:

- What are the inputs to the design (client, performance, regulatory, standards, and specifications, outside services)?
- What is the process for incorporating safety in design activities?
- What design tools will be used (software, spreadsheets, computer-aided design and drafting, etc.)?
- What processes will be followed for design chartering, optioneering, multicriteria analysis, innovation, value engineering, constructability, operability, and risk management?
- What review processes are required (checking of calculations, discipline reviews, inter-discipline reviews, reports, drawings, documents) and what level of approval or authority is required?
- What verification activities are required (qualification of verifier, external, internal, etc.)?
- What form will be the design outputs take (drawings, specifications, procurement documents, etc.)?
- How will design records be handled (records of review, verification, correspondence, request for information, engineering change request, etc.)?

2.2 Large Design Projects

For larger projects with a design component, the design plan will be more detailed than for a smaller project, and it may be necessary to develop a project-specific design management plan. All of the considerations above (see small design project) need to be considered but they will be more detailed. With larger projects, work may need to be done by others. If so, then a subconsultant agreement will need to be drafted.

Activities such as chartering, optioneering, safety in design, constructability, operability, value engineering, and risk management may involve workshops with multiple parties involved.

2.3 Typical Phases of a Design Project

Most solids-processing design projects can be partitioned into the following phases, with percentages referring to the percent of project completion:

- **Project definition phase (0–5%).** The design is planned and documented. The aim is for the design to meet the functional needs of the project cost effectively within the budget assigned. The scope of the work is determined. A design and engineering strategy for delivery of that work is established (within the overall project management framework). The design management plan is drafted. The design criteria, such as solids characterization and quantity estimation, are determined.
- **Concept or schematic design phase (5–30%).** Design is taken to the full-concept stage. All options and alternatives have been reviewed and analyzed and the final outcomes have been determined. This stage includes development of preliminary arrangement drawings, flow sheets, sketch plans, design criteria, general arrangements, piping and instrumentation diagrams, equipment lists, discipline criteria, notes, descriptive schedules of materials, single line diagrams, and preliminary functional specifications. The design is frozen at the completion of this stage and any changes to the documented design require a formal design change process.
- **Design development or preliminary design phase (30–60%).** The selected concept (or schematic) design is further developed, costed, and reviewed. It includes preparation of detail drawings, specifications, schedules, calculations, and suppliers' information. At the end of this phase, the design is again frozen.
- **Construction documents or final design phase (60–100%).** All of the drawings, specification, data sheets, and details are ready to be issued for construction stage. During this phase, each of the disciplines (civil, electrical, mechanical, etc.) should be able to work relatively independent of each other and freely without further client direction as they complete their detail drawings and related documentation.

- **Design Support or Construction Services Phase.** This phase is where support is provided during the construction and commissioning phases of the work to deal with unforeseen issues.

2.4 Design Processes and Activities

Additional design processes and activities are used on many projects. A selection of these design activities is outlined below.

- **Safety and environment in design.** During the design process, consideration needs to be given to ways of eliminating hazards and reducing risks associated with the construction, operation, and maintenance of the item being designed. On most projects, this is a mandatory requirement.
- **Design innovation.** Are there new and better ways to achieve the performance and expected outcomes through an innovative process? This is applicable to all types and sizes of projects. It should be a part of the early chartering and optioneering sessions.
- **Design chartering.** At the commencement of design, the project team (which may also include construction personnel, operations personnel, and client representatives), reviews the design processes and all aspects of the way the project design will be carried out.
- **Optioneering.** A formal process at the concept stage of design during which all of the feasible options and alternatives are considered and evaluated. Typically occurs on larger projects; however, the principles can be applied to any size projects. It relies on a team approach to look for the different configurations and their benefits, often using multicriteria analysis (MCA).
- **Constructability reviews.** These reviews are used especially on design/build or engineer-procure-construct (EPC) projects/programs. The construction manager and other construction personnel provide this review early in project design to ensure constructability and clarity of the design documents and cost optimization. Constructability reviews performed in the early stages of a project have proven extremely effective. Typically, constructability reviews provide further assurance that technical requirements are met while minimizing redesign, reducing construction costs, expediting project completion, and reducing environmental effects and long-term maintenance costs.

- **Operability reviews.** Operations staff typically undertake operability reviews during design of a project. Operability reviews address factors such as equipment selection, facility layout, maintenance access, safety, and ease of operation. The focus is on assuring that the constructed facility will satisfy its intended function and purpose at the best life-cycle combination of benefits and costs.
- **Value engineering.** A formal process undertaken to optimize the value of projects. It is typically performed by a panel of experts not associated with the design team. The panel suggests design changes that are typically to reduce construction cost but which also may be focused on reliability or external factors. This process occurs in a formal, documented manner on larger projects and typically takes place at either the 30 or 60% completion stage of the design process.
- **Design risk management.** A supplementary activity similar to the safety and environment in design risk assessment, during which other design risks and uncertainties are reviewed and mitigation measures put in place. It can also include Monte Carlo type simulations to determine the degrees of uncertainty in design capacities and redundancies of unit processes. Monte Carlo simulations involve a mathematical analysis of all foreseeable operating scenarios with projections for the probability of each scenario to occur. For estimates, a cost/risk approach using Monte Carlo simulations can provide confidence levels.

2.5 Required Design Processes

In summary, there are a consistent set of design requirements regardless of the size of work:

- Design planning, which ranges from flowchart to full design management manual depending on project size.
- Design inputs that need to be defined and agreed to by those involved. Typically uses some form of design input sheet and include client, code, and special requirements.
- Design outputs that meet the input requirements and are in a form that is appropriate for purchasing, construction, and other phases.
- Design review, including a hold point for review and checking by an independent part. This is often done at a key stage of the design such as concept

completion, 30% completion, 60% completion (and which may be part of the value engineering process discussed above). It can also apply to individual design outputs such as drawings, calculations, and design documents.

- Design verification, which is an extensive overview of the design process. It will involve independent, often external, checks of the design process. For example: Are the outputs a function of the inputs? Has the proper process been used? Design verification must be performed by individuals who are not directly involved in the design of the project.
- Design validation is related to design providing and commissioning and is a consequence of effective design processes being in place.
- Design changes that include a progressive freezing of design throughout the design stages. This applies to the concept design, 60% design, 90% design etc. If anything needs to be changed to a frozen concept after approval of a prior phase, then a formal change process is required. The purpose of the design freeze, and the design change process, is to ensure that the full implications of any change are understood, properly costed, and signed off by the design coordinator and project manager.
- Document control for all documents. The primary design documents are drawings, but there are also many other documents that will need to be controlled with regard to their issue level, authorization, and approval (e.g., specifications, operating procedures, data sheets, unit process guidelines, vendor data, design plans, procedures, forms, etc.). For large projects with large number of documents and multiple review and approval workflows, it is often necessary to provide a structured document management system.

3.0 PROCESS SELECTION

Wastewater treatment and wastewater solids and disposal systems must be put together to ensure the most efficient use of resources, such as money, materials, energy, and work force, in meeting treatment requirements. Logic dictates what the process elements must be and the order in which they go together.

A methodical process of selection must be followed in choosing a resource-efficient and environmentally sound system from the myriad of treatment and disposal options available. The most elementary selection mechanism is the "principle of successive elimination," an iterative procedure in which less effective options are progressively culled from the list of candidate systems until only the most suitable system or systems for the particular site remain. The process is described in this chapter, and other, more sophisticated processes are introduced.

The concept of the "treatment train" has been developed as a result of a systems approach to problem solving. This concept is useful in the design process, however, only if all components of the train are considered. This includes not only solids treatment, reuse, and disposal components, but wastewater treatment options and other critical linkages such as solids transportation, storage, and sidestream treatment. The successful development of a treatment train from a collection of individual components depends on a rigorous system selection procedure, or logic. For large plants, system selection is complex, and a methodical approach is required. Progressive and concurrent documentation of the procedure is mandatory because it prevents a cursory dismissal of options. For smaller plants (< 4 ML/d or 1 mgd) the system choices are necessarily more obvious and the selection procedure is typically shorter and less complex.

3.1 Relevant Criteria for Solids Process Selection

Criteria for system selection must be pinpointed before system synthesis. A listing of potential criteria for consideration is shown in Figure 5.1. The list is not necessarily complete, and planners may find other criteria they wish to include. The relative importance of each criterion will vary from site to site. For example, reliability may be most important at one site, whereas minimizing costs may be paramount at another. Criteria deemed relevant for each site in question are subsequently used in the system selection procedure.

3.2 Identification of System Options

Candidate systems are synthesized from an array of components, such as these shown on Figure 5.2. Wastewater and solids management components are listed as a reminder that all components of the train must be considered. Figure 5.3 illustrates how Figure 5.2 can be used to develop a specific flow sheet. Process streams can be drawn on copies of the master drawing. Relevant information such as solids concentrations and mass flow rates can be entered directly on the flow sheet, if desired. The advantages of using arrays such as Figure 5.2 to develop specific project flow sheets are that nearly all potential options are identified and process streams are clearly displayed.

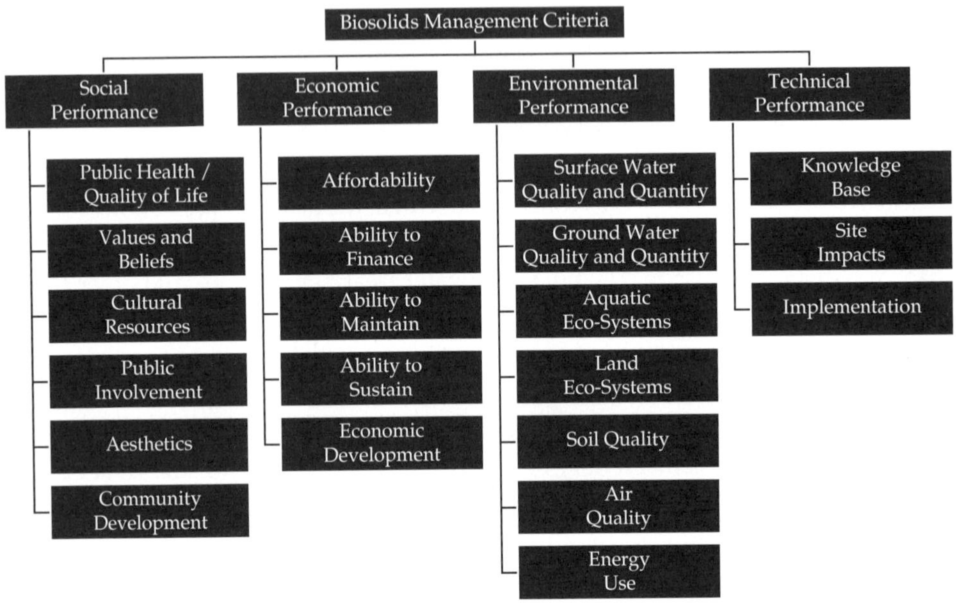

FIGURE 5.1 Potential criteria for solids process selection.

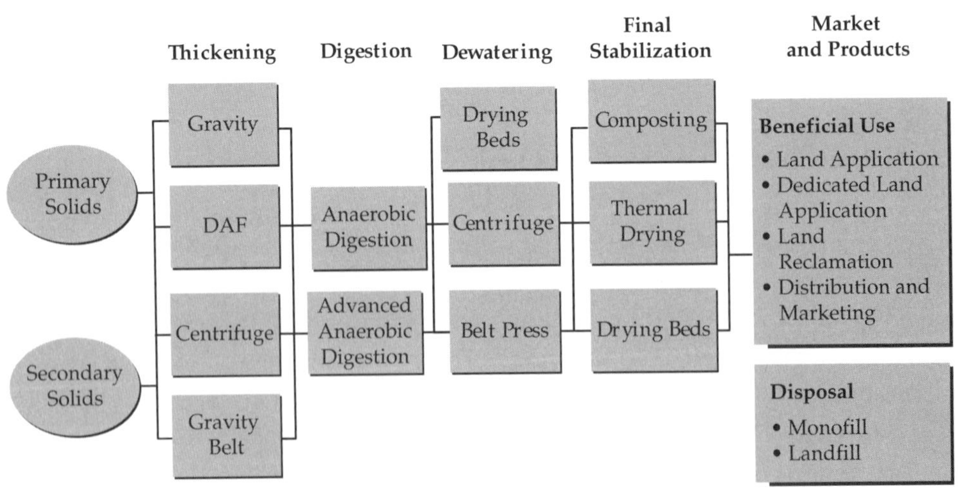

FIGURE 5.2 Potential array of components for biosolids processing.

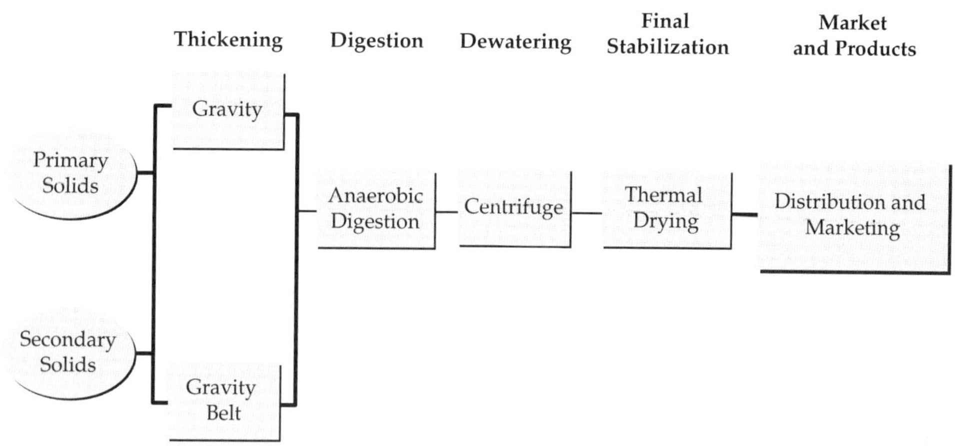

FIGURE 5.3 Example flow sheet for a biosolids processing train.

3.3 Biosolids Process Selection—Evaluation

Biosolids process evaluations involve the integration of three essential components:

1. Incorporating stakeholder views on the successful process for decision making and outcome of the design. This is typically accomplished by facilitating one or more meetings with project stakeholders during which consensus is reached on the project vision and the process for achieving it.
2. Using a decision process for narrowing the choices to a solution. This process is developed and endorsed by the project stakeholders as described above.
3. Setting the stage for the design process to follow. The stage is set for the design process when the stakeholders create a project vision and endorse the decision-making process.
4. Ensuring that the activities involved in process selection are based upon six steps for ensuring decision quality and implementation of the resulting program that were covered in Chapter 2.
5. Achieving several additional goals and objectives when the planning and process selection phases have been completed successfully:

   - Identification of the pertinent issues;
   - Selection of a preferred alternative;

- Enhancement of the credibility of the alternative;
- An auditable process with deliverables that have been quality controlled and "frozen" for purposes of proceeding with design;
- Best use of resources; and
- Sustainable program for maintaining the selected process.

Figure 5.4 provides a workflow diagram for the alternatives screening and selection process. The work is initiated with an identification and evaluation of the project delivery approaches and construction phasing. A preferred project delivery approach is selected from this effort. The project phasing is used to establish intermediate milestones and a project schedule for all the projects in the selected program.

3.3.1 *Developing Alternatives*

Biosolids alternative selection seeks to develop consensus on a "best" alternative, where the best alternative is not necessarily the best technical alternative but rather one that can be implemented and is technically rational. Each alternative can consist of one or more unit processes, some of which are common to most or all of the alternatives because of immediate needs, and others that may be phased into operation on a "just-in-time" basis. The first building blocks for the decision tool are the descriptions of each alternative and of its elements. Each element must be described at an appropriate and consistent level of detail that is sufficient to define the process involved and differentiate among the alternatives.

With this guidance, a typical alternative will describe a conceptual process train that produces single or multiple products or disposal routes for biosolids. The

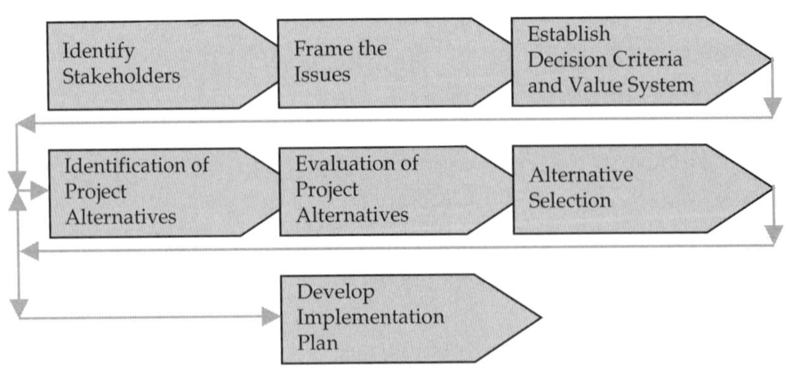

FIGURE 5.4 Workflow diagram for alternatives selection and screening process.

conceptual process train will have a defined process origin and endpoint. Each train will consist of a sequence of unit processes, and each unit process may have several functionally equivalent options (technologies and/or vendor proprietary processes). Those unit process options should then be evaluated for the preferred alternative.

Biosolids process selection typically involves alternatives with a common starting point and different endpoints. Typical endpoints might include, for example, incineration (process train component) with ash (product) disposal to landfill (disposal route), land application (use) of Class A or B biosolids (product), distribution and marketing (use) of pellets (product) from thermal drying (process train component).

It is important when identifying or developing candidate alternatives that the use disposal routes and the products be identified and developed as part of the alternative. There is no value to an agency in having a great process train, but no sustainable use or disposal route for the biosolids product produced. It is also important to note that final strategies could include combinations of alternatives to provide diversity, reliability, and sustainability.

3.3.2 *Screening Alternatives*

Each alternative is characterized in terms of its performance using screening criteria that are selected by project stakeholders. A "value hierarchy" is established, by which a set of scalar values and weights are placed on the screening criteria and used to assess the advantages and disadvantages of each alternative. A value hierarchy allows the stakeholders to agree that more weight should be placed on one measure over another. The hierarchy is structured by polling the stakeholders to obtain their weighting preferences. The key rule of the exercise is that there should be stakeholder commitment from all to the weightings that are used.

At least one dozen in-plant residuals and biosolids treatment processes are in use, with well more than 40 different associated technologies. More than one dozen use and or disposal routes are used, and several more are emerging as agencies look for new and innovative ways of managing residuals. Before selecting candidate alternatives, it is necessary to reduce these alternatives and their various combinations into a manageable quantity so that the candidate alternatives selected are meaningful and relevant to the agency. In most projects, there should be between 5 and 10 candidate alternatives screened for more detailed evaluation.

The key to effective screening is to have a set of simple and clear screening criteria that can be applied to uses, products, and processes/technologies. Thus, screening criteria for uses and products may be different for processes and technologies. For example, existence of markets is important to the viability of a product such as pellets, and reliability and demonstrated experience at a similar scale are important to the selection of a dewatering process and a particular technology. These criteria are usually "pass/fail" or "must have". Following are examples of each and related questions that should be answered.

1. Use/disposal
   - Sustainability—would the use endure over time in an environmentally safe manner while meeting all regulations and would it have the capacity to handle all biosolids expected to be produced at the WWTP?
   - Reliability—would the use be reliable and enforceable within the agency's current framework?
   - Flexibility—would the use be adaptable to a variety of processes or treatment technologies under changing circumstances?
2. Processes/technologies
   - Operational compatibility—can the process or technology accommodate the operating schedule for the facility where it will be located?
   - Residuals removal—how effective is the process or technology at separating the residuals from the liquid or recycle stream?
   - Reliability—would the process/technology meet regulations and has it been successfully implemented at this scale?

Once the screening criteria have been established, they are applied to each process or technology, product, and use. The remaining items are then used to develop the candidate alternatives. In selecting the candidate alternatives, use alternatives are selected, followed by the desirable products, followed by the process units that would produce these products.

The second set of building blocks for the decision tool consists of the screening criteria and the value hierarchy used to assess the advantages and disadvantages of each alternative. The overarching goal of the program should be sustainability, and this goal is reflected in all the criteria discussed in this section. For example, a situation in which stakeholders reach consensus on a goal to maximize ratepayer benefits would result in identification of several criteria to attain the goal.

- Product marketability, to be maximized by providing a consistent product and by using proven technology.
- Health and safety risk, to be minimized through regulatory compliance and worker protection.
- Flexibility, to ensure the ease of phasing and to be maximized by selecting options and products that provide flexibility.
- Credibility, to be maximized by such measures as delivery on time and on budget and by developing stakeholder support.

The value hierarchy allows the stakeholders to agree that more weight should be placed on one measure over another. The hierarchy is structured by polling the stakeholders to obtain their weighting preferences. The key rule of the exercise is that there be commitment from all stakeholders to the weightings that are used.

Table 5.1 contains example criteria for biosolids master planning, and Table 5.2 contains a generic listing of criteria that are applicable to resource master planning based upon sustainability. The criteria in each table are presented in relation to an objective category. The criteria of Table 5.1 are specific to biosolids processing, and are measured or scaled in terms of the extent to which each criterion is maximized or minimized.

The selection of criteria should be tailored to the specific project, and alternative selection must be specific enough to allow differentiation among the alternatives under evaluation. In this regard, the criteria presented in Table 5.2 are a starting point for any presentation, but participation of stakeholders must be taken to a more detailed level resulting in scalar measures of attainment as illustrated in Table 5.1.

3.3.3 Collecting Meaningful, Reliable Data

The data acquisition program will have to be tailored to the needs of the project. With this qualification, the following types of data are identified as being potentially meaningful to a biosolids process selection. The data and information-gathering steps described below are discussed in Chapter 4 or earlier in this chapter:

- Solids production by source, in wet and dry tons/day and in terms of volatile solids destruction (VSD) content (minimum of two years of daily observations).
- Solids composition by source, including volatile solids content (minimum of two years of daily observations).

TABLE 5.1 Example criteria for biosolids master planning.

Objective	Example criterion	Relative score (0–10) versus measure
Environmental protection	Regulatory compliance (i.e., are standards being met as set at the federal, state and local levels?)	Meets federal and state requirements = 1 Meets federal, state, and local requirements = 5 Exceeds federal, state, and local requirements = 7 Further exceeds federal, state and local requirements (as Class A biosolids with minimal odor) = 10
Environmental protection	Conservation of materials (i.e., to what degree are materials being conserved?)	No reduction achieved = 0 0 to 25% reduction achieved = 2 26 to 50% reduction achieved = 5 > 50% reduction achieved = 10
Environmental protection	Conservation of nonrenewable energy sources (i.e., to what degree is the measure effective?)	No reduction achieved = 0 0 to 25% reduction achieved = 2 26 to 50% reduction achieved = 5 > 50% reduction achieved = 10
Environmental protection	Preservation of water quality (WQS) (i.e., how frequently are water quality standards exceeded?)	> 20 days/year WQS exceedances = 0 5 to 20 days/year WQS exceedances = 3 < 5 days WQS exceedances = 5 No WQS exceedances = 10
Environmental protection	Preservation of air quality (AQS) (i.e., is air quality of site improved or impaired?)	> 20 days/year AQS exceedances = 0 < 5 to 20 days/year AQS exceedances = 3 < 5 days AQS exceedances = 5 No WQS exceedances = 10
Environmental protection	Preservation of ecology of site (i.e., is biodiversity improved or hindered?)	> 50% reduction in biodiversity = 0 25–49% reduction in biodiversity = 2 5–24% reduction in biodiversity = 5 <5% reduction in biodiversity = 10
Environmental protection	Difficulty obtaining land application permits (P) and administrative complexity (AC)	High P/high AC = 1 Medium P/high AC = 2 High P/medium AC = 3 Medium P/medium AC = 6 Low P/high AC = 7 High P/low AC = 8 Medium P/low AC = 9 Low P/medium or low AC = 10

Table 5.1 (*Continued*)

Objective	Example criterion	Relative score (0–10) versus measure
Technical	Worker protection (i.e., number of people exposed and product quality)	Client staff, contractors, and public exposed to Class B or raw sludge = 1 Client staff, contractor, and public exposed to Class A = 3 Client staff and contractors only exposed = 7 Client staff only exposed = 10
Technical	Ease of phasing of project elements to minimize plant effects	One construction phase = 1 Two construction phases = 4 Three construction phases = 6 No effect to onsite operations = 10
Technical	Diversity of processes and use options	One process; one product = 1 Multiple processes; one product = 4 One process; multiple products = 7 Multiple processes; multiple products = 10
Technical	Operations (i.e., what is the ease of operation of the alternative?	Multiple processes; multiple products = 1 One process; multiple products = 4 Multiple processes; one product = 7 One process; one product = 10
Technical	Operations (i.e., how well can technology fit within facility operations?)	Not compatible with existing facility operations = 1 Compatible except for minor exceptions = 4 Compatible with existing facilities = 7 Will match facility operations perfectly = 10
Technical	Technical basis (i.e., what types of issues about the alternative merit phased or deferred decision making?)	Issues are ill-defined and require basic research = 1 Issues are defined and resolvable by pilot investigation = 4 Issues are understood but facilities do not exist at same scale as intended application = 7 No technical uncertainties exist = 10
Technical	Quality of biosolids product	Must be landfilled = 1 Can be land applied as Class B = 4 Can be distributed and marketed at Class A = 10
Economics	Sales demand (i.e., what is the anticipated market demand for the product?)	Pay to land apply the product = 1 Product given away free = 4 50% of product sold = 6 100% of product sold = 10

(*continued*)

Table 5.1 Example criteria for biosolids master planning (*continued*).

Objective	Example criterion	Relative score (0–10) versus measure
Economics	Contracts (i.e., what is the number and complexity of the service contracts required?)	Numerous complex contracts = 1 Single complex contract = 3 Multiple simple contracts = 6 No contracts = 10
Community	Inclusiveness (i.e., to what degree have stakeholder interests been considered?)	No stakeholders consulted = 0 Local stakeholders only considered = 3 Agricultural communities also considered = 7 All communities considered = 10
Community	Public health and safety (i.e., are the health and safety of stakeholders protected?)	Water supplies are protected = 4 Water supplies and air quality are protected = 7 Water supplies and air quality are protected and agreed upon course of action if health concerns arise after implementation = 10

Table 5.2 Sustainability-based criteria and related questions.

Objective	Criteria	Evaluation considerations and performance measures
Environmental protection	Water: To what degree is the water quality and quantity being preserved or improved for current and future users and uses (i.e., recreational, irrigation, drinking)?	Will groundwater quality be affected by any land application or landfill program? Will the quality of any surface water be affected by any land application or landfill or ash disposal program? Can program be modified in future if justified by currently unforeseen research or knowledge?
	Land: To what degree is the land use being preserved for current and future uses and users?	Is site area minimized for plant structures and equipment? Is the beneficial reuse of biosolids nutrients desired by the community? Will agricultural or recreational land be improved or degraded by the land application, landfilling, or ash disposal operations?

Table 5.2 Sustainability-based criteria and related questions (*Continued*).

Objective	Criteria	Evaluation considerations and performance measures
	Energy: To what degree are nonrenewable energy sources being conserved for future users and uses?	Quantify system energy requirements in terms of net electrical power consumption per unit flow, net natural and digester gas consumption, and heat energy recovery.
	Materials: To what degree are materials being preserved for current and future users and uses?	Quantify nonrecoverable use of chemicals and supplements.
	Air: To what degree is the air quality being preserved for current and future uses and users?	What is the current state of the air quality, including odor, products of combustion, and toxic emissions, and what is the projected state?
	Watershed management: To what degree is the ecology of the project site and the watershed being preserved for current and future users and uses? Project sites include biosolids management, storage, land application, and landfill locations.	Are constituents being transferred between soil, air and water, or beyond the watershed? With positive or negative effects? Is biodiversity in the watershed improved or hindered? Are there measurable effects on global resources (agricultural nutrients, greenhouse gases)?
Technical	Sound science: To what degree does the alternative reflect current scientific understandings and research?	Are contaminants of concern identified? Has information regarding their potential effects been made available? Are the potential or claimed effects of the alternative supported by accepted studies, tests, and research? Are there uncertainties or research needs that could be addressed to improve the option, and are those uncertainties being addressed?
	Engineering: To what degree has sound engineering practice been included?	Have engineering standards been followed? Have contaminants been identified and quantified such as by creation of a mass balance?

(*continued*)

TABLE 5.2 Sustainability-based criteria and related questions (*Continued*).

Objective	Criteria	Evaluation considerations and performance measures
		Has redundancy and contingency planning been incorporated? Has the maintenance and control of the operation been considered? Can an option be readily modified in future if justified by currently unforeseen research or knowledge? Has a responsible biosolids management program been developed, including public involvement, auditing, and adherence to a biosolids master plan? Has metal content and energy dependency of commercial fertilizers been considered for landfill and incineration options?
	Operations: What is the relative ease of operation of the alternative?	Will the technology or process match the hours of operaton for the facility where it will be located?
	Regulatory: To what degree does the alternative comply with, or exceed, regulatory requirements?	Does the option comply with current and anticipated regulations? Will the permit ensure or require that the option will be modified as future regulatory changes are implemented?
Economic	Cost: What is the relative life cycle cost?	Quantify full lifecycle costs of system, including operation, replacement, maintenance, and materials consumption. Consider associated costs to society for use of nutrients, greenhouse gas emission impacts, and landfill siting requirements.
	Economics: Does the alternative improve or degrade the local economy and the prospects for economic development?	Are resources and infrastructure used effectively to promote current and future local prosperity? To what extent will the use or release of products and resources (including air emissions, energy consumption, and nutrient management) affect the economies of other areas?

Table 5.2 Sustainability-based criteria and related questions (*Continued*).

Objective	Criteria	Evaluation considerations and performance measures
Community	Quality of life: Is the quality of life preserved and/or enhanced for current and future residents and stakeholders?	Does the option define a vision of the future, or is the option an incremental advancement of the current condition? Are these incremental changes that restrict future options, or is this a master plan for the future? Are potential sources of agricultural nutrients being protected or created? Are the natural habitats for outdoor recreation, and the conditions for agricultural production, being protected? Would additional communications and involvement of communities improve the perception of the quality of life associated with an option? Can that be implemented? Will the community feel a sense of achievement of environmental responsibility? Can the quality of life be measured using the general reporting framework? Is the affected watershed and farming community being considered as a whole?
Community	Sharing: Are best practices determined from other similar communities and will they be shared with future communities?	Will the technologies or outcomes have benefit to, and be shared with, other communities? Do the affected urban and farming communities understand the science and benefits related to each option? Is research needed that will also benefit other communities, such as health effects studies? Can that research be implemented? What additional information is needed to improve a solution, and can it be obtained? If other communities have successfully implemented an option, have best efforts been made to incorporate their experience and information?

(*continued*)

Table 5.2 Sustainability-based criteria and related questions (*Continued*).

Objective	Criteria	Evaluation considerations and performance measures
Community	Inclusiveness: To what degree are the interests of all stakeholders, including future residents, considered?	Has the public been consulted and involved in the decision-making process? Has collaboration and consensus been achieved with all contributing municipal areas and any affected agricultural communities? Have industrial, commercial, and agricultural interests been included in the decision-making process? Have some participants claimed to represent classes of stakeholders? If so, has feedback been obtained to verify their representation? Have nonresident stakeholders been considered? Have the interests of future generations been considered, with respect to affordability, land use, and quality of life? Will continuing stakeholder communications and involvement occur after the project is complete? Are any or all barriers and incentives visible to the participants?
Community	Measurement: To what degree will the beneficial results be measurable and evident?	Have measurable air, soil, and water quality performance criteria been defined? Is there an agreed-upon course of action if the measurable performance criteria are not satisfied?
Community	Public health and safety: Has the health and safety of the current and future stakeholders been protected?	Will recreational areas, agricultural production, and water (groundwater and surface) supplies be protected? Is air quality protected? Will there be measurable effects on living and working conditions? Is research needed that will reduce any health effects, such as by improving a technology? Can that research be implemented? Is there an agreed-upon course of action if unforeseen public health concerns arise after implementation?

- Process performance data, showing the liquid and solids flow rates through the process train and the composition in terms of volatile solids content (minimum of two years of daily observations).
- Final production of processed biosolids, in terms of both quantity and quality parameters (minimum of two years of daily observations).
- Operating procedures that cause or induce diurnal, weekly, monthly, and seasonal variability of solids production rates and variability in sludge composition by source.
- Maintenance procedures that cause or induce diurnal, weekly, monthly, and seasonal variability of solids production rates and variability in sludge composition by source.
- Operating cost experience.
- Maintenance cost experience.
- Pertinent operating contracts (e.g., with a land application contractor).
- User requirements for biosolids products.
- Odor sources in the existing process train.
- Characterizations of odor generation and release by odor source.
- Technical reports on any aspect of the facility that can shed insight on technical problems in the plant, odor sources and characteristics, geotechnical problems, and outstanding recommendations for upgrade or improvements.

This information can be used for purposes of

- Collation and tabulation of data bases.
- Statistical analysis and interpretation of trends and investigation of causes of trends and variability in solids production rates and process-by-process lading rates and variability.
- Development of recommendations for database upgrades.

The final data bases can be used for purposes of

- Evaluating the existing facilities and conditions.
- Identifying future conditions.
- Identifying pertinent criteria, scalar values, and uncertainties.
- Detailed development of the alternatives for further evaluation.

3.3.4 *Evaluating Screened Alternatives*

Enough information will have been gathered at this juncture to permit comparative analysis of the alternatives, leading to the selection of a preferred alternative and the preparation of a risk profile on the recommendation. The risk profile is based on an assessment of the effect of the uncertainties on the outcomes, and is usually covered in qualitative terms as part of the description of each alternative. There are some statistical tools such as Monte Carlo analysis that can be use to help quantify and compare risk levels among alternatives. After a quality control review, the deliverable can be advanced for approval by the governing body or management authorities.

After quality control and stakeholder reviews, it is anticipated that a consensus will have been reached on the biosolids process selection and alternative rankings. If not, then another iteration should be undertaken to verify the alternatives analysis and related information. If consensus is reached, then the preferred alternative will be selected and a risk profile developed. A quality control review will then be conducted, and the deliverables documented. A final check on consensus should be made, and, if successful, the program can move on to process selection and design; otherwise, additional iterations will need to be made as necessary to attain consensus. Alternative evaluation proceeds along the following steps:

- Presenting updated results of the economic analyses of the alternatives, taking into account the new information from the stakeholders and other sources;
- Presenting results of the alternatives screening step based on the relative benefits of each alternative (Tables 5.1 and 5.2 provide suggested criteria by which to quantify benefits);
- Ranking alternatives in accordance with the ratio of benefit to expected net present worth for each; and
- Developing alternative implementation scenarios, presenting how the selected process might be phased to maintain operation while upgrading the facilities needed.

Documentation of a biosolids selection process typically consists of the following components

- Ranking of the alternatives and
- Selecting and presenting the preferred alternative and its risk profile.

3.3.5 *Developing Plan to Proceed with Selected Alternative*

The preferred alternative is further developed and refined in this step. The refinements are those needed to outline the biosolids management program associated with the selected alternative, to the extent that the project can be taken to the governing board or management body for approval. What is sent to the governing body for approval is a product that can be financed and implemented within the capabilities of the agency in the time frame allotted.

The financial feasibility is tested by assessing the effect of the project on the utility's rate structure. The implementation schedule is tested against milestones in the client's permits to operate to ensure that the selected process alternative can be executed on a timely basis. In conjunction with the project delivery approach, permit requirements are identified for the projects. Also, a contingency plan and a monitoring plan are developed to provide reliability and accountability for the program. These could be part of the agency's biosolids environmental management system (EMS). The financing of the project is analyzed, taking into account the funding sources available and the effect of the financing on the user rate structure. In the event that the agency is unable to finance the project according to schedule, an alternate project delivery approach and schedule may have to be developed. The end product is a program containing a series of projects that the governing or management body will approve.

The key deliverables from the selection and implementation process are as follows:

1. Recommended project delivery approach and construction phasing;
2. Recommended contingency plan;
3. Recommended monitoring plan;
4. Project financing plan;
5. Project schedule;
6. Implementation plan combining the above deliverables; and
7. Documentation of the selected process for the approval of the governing body.

4.0 MASS BALANCES AND PROCESS SCHEMATICS

This section describes mass balance concepts that are integral to solids processing at all WWTPs. Although most mass balance calculations today are performed by computer-based process modeling, a basic understanding of mass balance fundamentals is necessary to review and apply "reality checks" to process modeling results.

4.1 Mass Balance Approach to Solids Processing

The quantities of solids generated during wastewater treatment are important parameters in wastewater treatment because they determine the sizing of solids treatment processes and all solids-processing equipment used. Solids generation also affects the sizing of many liquids treatment processes. For example, the sizing of a secondary treatment process to maintain desired solids retention times (SRT) is governed by the quantity of waste activated solids (WAS) produced by the treatment process in addition to the assumptions for influent and effluent loadings of biochemical oxygen demand (BOD) and total suspended solids (TSS).

Although the importance of solids production in wastewater treatment is recognized, a great deal of plant-to-plant variability exists that can cause difficulty in accurately estimating solids quantities. The best source of information for estimating solids production is plant-specific data that reflect the nature of wastewater being treated and the types of wastewater treatment processes being used. Whenever typical value models are used to estimate solids quantities, safety factors should be applied that reflect the degree of uncertainty to which the wastewater characteristics and treatment processes are known.

An initial approximation of solids generated by a typical municipal WWTPs is a range of 0.20 to 0.25 dry kg of solids per m^3 (0.8 to 1.2 dry tons per million gal) of wastewater treated. Solids production will likely be on the low end or lower than this range for WWTPs with solids minimization or reduction processes. Solids production will be on the high end of this range or higher for WWTPs using significant amounts of chemical addition or treating high-strength wastes.

4.1.1 Generation of Wastewater Treatment Residuals

The first step in evaluating the solids processing system at an existing WWTP is to estimate the quantity of solids generated. In some instances, it will be possible to directly measure the quantity generated by measuring solids flow rates and concentrations. Experience indicates, however, that this will not always be possible, particularly in a plant with an inadequate solids processing system. In such a facility, measured solids production rates do not reflect actual solids production rates because the quantity of solids that can be processed is limited by the capability of the solids processing system.

Solids produced in excess of the capacity of the solids handling system often are inventoried in the liquid processing train until their presence begins to affect effluent quality. Alternatively, the solids processing system is operated in an overloaded condition, resulting in the recycle of excessive quantities of pollutants to the liquid processing train. In situations such as these, theoretical calculations must be relied upon to estimate solids generation rates. Chapter 4 provides a discussion covering the characterization and estimation of wastewater treatment residuals.

4.1.2 *Interfaces between Biosolids Handling and Treatment*

The recommended approach for estimating solids production is to provide a mass balance for the entire WWTP that relates it to the design parameters for the wastewater treatment processes. The mass balance should show the key constituents such as process flows, TSS, and biochemical oxygen demand (BOD). Sometimes chemical oxygen demand (COD) and solids production from nitrogen and phosphorus removal processes should also be considered in the mass balance.

Recycle streams within the WWTP must be included in the mass balance. These recycle streams are estimated based on the treatment rates and capture efficiencies of the individual solids handling processes, which are found in other chapters of this manual. For example, a centrifuge with 90% solids capture efficiency will recycle 10% of its feed solids back to the discharge point of the recycle stream, which is usually the head of the WWTP or the aeration basins. Hours of operation for solids-separation equipment are also important to consider. For example, the recycle stream from a centrifuge operating eight hours per day five days a week will have intermittent effects on upstream processes; a centrifuge operating 24 hrs per day, 7 days a week will have more consistent effects on upstream processes.

It is necessary to separately estimate the quantities of primary, secondary, and chemical solids. It is also important to take into account expected fluctuations in wastewater characteristics that occur because of changes in industrial contribution, stormwater flows, seasonal variations, and changes in the wastewater collection and conveyance systems. Chapter 4 provides guidance on how to estimate residuals resulting from typical wastewater treatment processes that can be customized for each specific case.

4.2 Process Schematic Approach

The first step in developing an understanding of the solids handling processes is to develop a process schematic diagram, also known as a process flow diagram. This can be as simple as a box and arrow diagram where boxes represent individual processes and arrows represent the piping or channel flow lines between the individual processes. Each process flow should be represented by only one flow line. Many unit processes will have more than one flow line entering and leaving the process, such as the overflow and underflow lines leaving clarifiers or thickeners, or the dewatered cake and centrate recycle lines leaving centrifuges.

The best way to illustrate the typical conventions for process flow diagrams is by way of specific examples, such as the ones that follow.

4.2.1 *Example 1—Secondary Treatment with Activated Sludge, Dewatering of WAS, and Landfilling of Raw Solids*

This process schematic is shown in Figure 5.5, and is the simplest way of handling solids from wastewater treatment. Because of the lack of solids stabilization in this type of process, however, options for disposal of waste solids are limited to landfilling or incineration. In this example, WAS from the activated sludge process is wasted from the secondary clarifiers to a dewatering centrifuge, and after dewatering, the unstabilized solids are transported to a municipal solid-waste landfill or monofill for ultimate disposal. Stabilization of biosolids is not typically required for landfilling disposal. Most landfills require that the solids not have free-draining

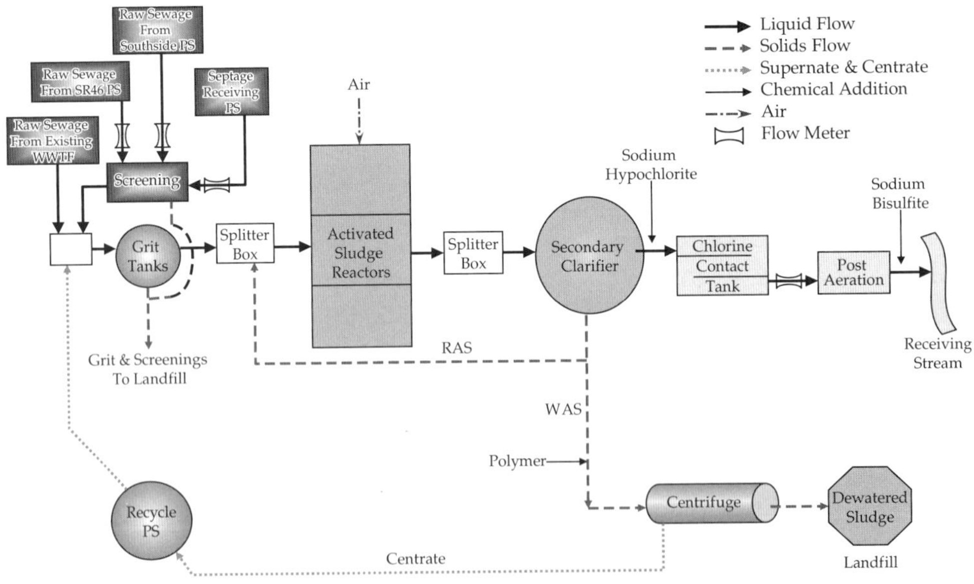

FIGURE 5.5 Example 1: Activated sludge, dewatering of WAS, and landfilling of raw solids.

liquid (accomplished by most dewatering processes) and are not hazardous or toxic. Landfilling of unstabilized solids is not traditionally regarded as a beneficial use, but biological solids from wastewater treatment can help augment the rate of decomposition of other solid waste and increase the quality and quantity of landfill gas, which can be used to generate power. This type of process is also frequently used in WWTPs that incinerate unstabilized, dewatered biosolids, which can produce energy and reduce the volume of solids requiring disposal (i.e., incinerator ash). Usually the ash is landfilled, but in some locations, options for beneficial use of incinerator ash are available, such as in brick manufacturing or concrete production.

4.2.2 *Example 2—Primary Clarification with Activated Sludge, Thickening of WAS, Anaerobic Digestion, and Land Application of Digested Biosolids*

This process schematic is shown in Figure 5.6. It incorporates primary sedimentation (producing primary sludge) and waste activated (secondary) solids thickening, and then combines those two solids streams into the feedstock for anaerobic digestion.

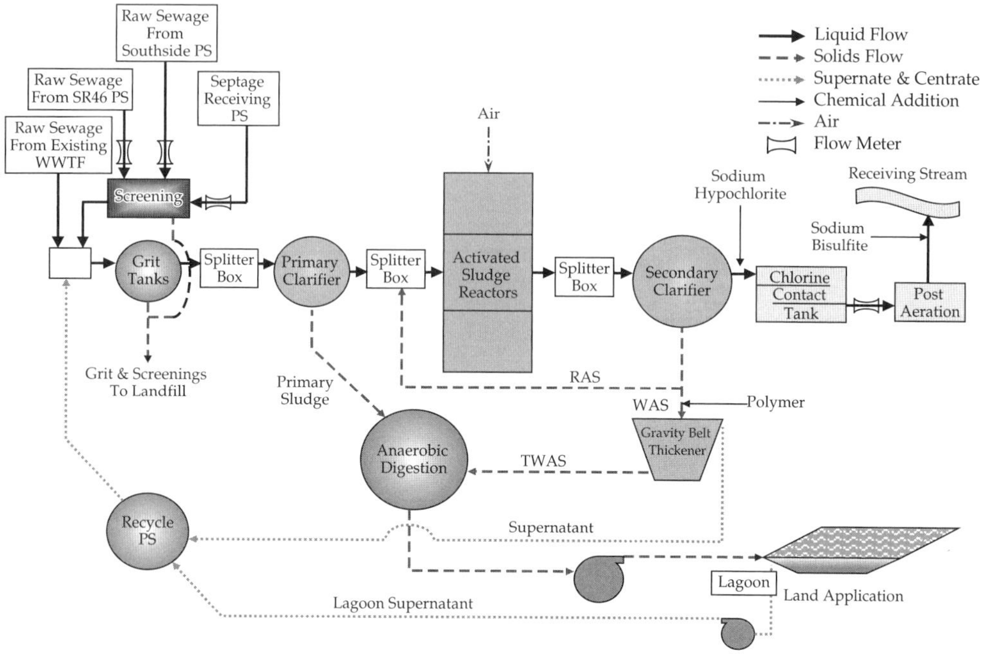

FIGURE 5.6 Example 2: Primary clarification, anaerobic digestion, and land application of thickened, Class B biosolids.

Anaerobic digestion is a proven stabilization process that will reduce pathogens to Class B levels and will produce substantial quantities of methane gas that can be used for heating and energy production. After anaerobic digestion, the biosolids is post-thickened in lagoons, where it is stored for seasonal land application on agricultural sites. Several WWTPs in the Midwest, such as Chicago, Illinois, where land application of biosolids in liquid form is often preferred, use this type of process.

4.2.3 *Example 3—Oxidation Ditch, Aerobic Digestion, Dewatering, and Land Application of Class B, Cake Biosolids*

This process schematic is shown in Figure 5.7. It represents another type of "Class B" level of biosolids stabilization and pathogen reduction achieved by aerobically digesting the waste-activated solids produced from an activated sludge process such as an oxidation ditch. Aerobic digestion is also a proven stabilization process that will reduce pathogens to Class B levels; however, unlike anaerobic digestion, aerobic digestion requires oxygen and is an energy consumer rather than an energy producer.

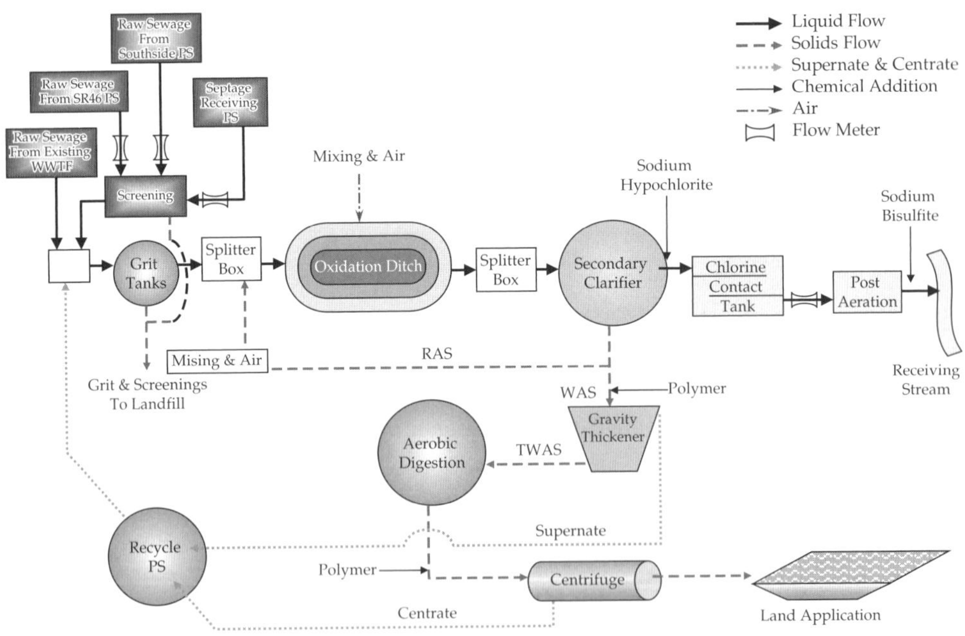

FIGURE 5.7 Example 3: Oxidation ditch, aerobic digestion, dewatering and land application of Class B, cake biosolids.

The capital investment for aerobic digestion is much less than for anaerobic digestion; therefore, this type of process schematic is often practiced at smaller WWTPs in warmer climates such as the U.S. Southeast or Southwest. After aerobic digestion, the digested biosolids are post-thickened in lagoons, where they can be stored for seasonal land application on agricultural sites.

In larger WWTPs, a mechanical dewatering process is used to reduce the volume of biosolids that have to be transported to land application sites. When lagoons are not used, biosolids storage facilities must be provided that are adequate to last through seasonal conditions that are unsuitable for land application. When land application of liquid, digested solids is practiced, additional storage needs to be provided in the form of post-digestion storage tanks. When mechanical dewatering is used, cake storage bunkers are typically provided. Caution must be exercised in the location and design of storage facilities, because dewatered cake solids are usually quite odorous. Centrifuge dewatering tends to produce more odors in the cake than other dewatering equipment that exerts less shear, such as belt-filter presses and rotary presses.

These three examples briefly describe representative process schematics for three types of solids processing and handling systems. An infinite number of combinations and permutations for other types of solids process schematics are possible.

5.0 EQUIPMENT SELECTION AND SIZING

5.1 Evaluating Equipment

The process schematics described in the previous section are typically used to develop mathematical models of the wastewater treatment process. The models are should be run under several future wastewater-flow scenarios to project the quantities and characteristics of biosolids produced during the expected life of the water reclamation facility. This allows the designer to establish such factors as operating costs and quantities of solids for ultimate disposal or beneficial reuse. The process models must be further calibrated to establish solids loadings that each equipment item must be capable of processing. To properly size solids-handling equipment, it is necessary to know the projected ranges of solids production at minimum and maximum loads. It is important to have a good idea of the operating conditions for which the equipment is being designed such as, for example, 8 hours a day, 5 days a week; 16 hours a day, 7 days a week; or 24 hours a day, 7 days a week. Given these three examples, it is evident that operating conditions and hours for solids-processing equipment are at least as important as the ranges of flows or loadings the equipment will encounter.

5.2 Integration of Equipment into Selected Process

An example will help to understand how solids handling equipment such as described above are integrated into the selected solids processing scheme. In this case, a centrifuge needs to be designed to dewater anaerobically digested primary and secondary solids at a medium-sized, 38 to 76 ML/d (10 to 20 mgd capacity), WWTP. The flow scheme is similar to that shown on Figure 5.6 (2nd process schematic). The WWTP is fully staffed only one shift per day, seven days a week. The digesters are complete-mix units equipped with floating covers. Because of the floating covers, digester volume varies. During the day shift, when operators are available to staff solids-handling processes, secondary sludge is wasted from the clarifiers to a gravity-belt thickening process before digestion. Primary solids are wasted continuously from the primary clarifiers.

The average loadings to the centrifuge are defined by the overall WWTP process model, but the necessary centrifuge capacity is calculated by considering the load dampening effect of digestion in conjunction with the WWTP operating schedule. During periods of higher-than-average solids production, loads to the dewatering units may be attenuated by storing portions of the peak loadings within the digester. Thus, peak solids loadings after digestion will be lower than peak solids production. The degree of load dampening is a direct function of the size and operating configuration of the digester.

Because the dewatering centrifuges will only operate during the day shift when solids-handling staff is available, the design capacity for dewatering must account for the handling of a full day's solids production in an eight-hour shift. Therefore, the design capacity of dewatering centrifuges would need to be calculated: $24 \div 8 = 3 \times$ average daily rate of digested biosolids production.

Further, the centrifuges require time for startup and for a cleaning cycle at shutdown. The startup and shutdown activities typically take an hour each, which needs to be considered in the sizing of the centrifuge. In addition, the effect of centrate recycle on upstream processes needs to be considered, especially when it is intermittent such as in this example.

5.3 Specifying Equipment Features and Options

The previous example shows the influence of solids peaking, storage volume, and operating strategy on the selection of design loadings for a particular sludge handling process. Several other factors are important in selecting the capacity a unit must have:

- Uncertainties. When systems are designed without the benefits of pilot- or full-scale testing, actual solids quantities and characteristics and efficiencies of solids-handling system components may not be known with certainty. The degree and potential significance of the uncertainties must be considered when developing design criteria. It is often necessary to introduce a safety factor into the design so that reliable performance can be obtained no matter what conditions are encountered in the full-scale application. The magnitude of the safety factor must be determined by the designer, based on judgment and experience.
- Equipment reliability. Greater capacity or a redundant unit must be specified if there is reason to believe that downtime for any essential equipment item will be more than a day or two. Redundant units are typically preferred because they provide backup capacity when one equipment unit is down. Also, multiple units of equipment allow for wider operating ranges to be accommodated under normal and peak conditions.
- Sensitivity of downstream components. If losses in the efficiency of a solids-processing component at peak loading conditions would cause an overload or other problem for downstream processes, then upstream process must be designed more conservatively. Conversely, if reduced efficiency of an upstream equipment unit can be tolerated by downstream equipment, then its design does not need to be as conservative.

5.4 Established, Innovative, and Embryonic Equipment

Established equipment has been proven by years of experience in full-scale WWTP applications in the United States. Examples of established equipment are as follows:

- For solids thickening processes—gravity thickeners, gravity belt thickeners, thickening centrifuges, and dissolved-air flotation thickeners;
- For digestion processes—conventional aerobic digesters or conventional high-rate, mesophilic anaerobic digesters;
- For dewatering—belt-filter presses or dewatering centrifuges.

Innovative equipment or processes are defined as technologies that have been tested at the demonstration scale of have seen some degree of initial use in full-scale demonstrations. New and emerging technologies with only one or a few full-scale applications with limited track records also fall into the innovative category. Because of their limited application, some of the scale-up issues may remain unresolved. Examples of innovative equipment include recuperative thickening and dewatering using rotary presses or screw presses.

Embryonic equipment or processes are defined as technologies that have been tested at laboratory- or bench-scale levels with limited pilot testing. Typically this category includes technologies that are still at the conceptual level. Examples of embryonic technology include high-pressure anaerobic digestion, vermicomposting, electro-acoustic dewatering, or microwave drying.

Most WWTP designs will include or consider only established equipment. There are situations where innovative equipment may be considered and specified, if the innovative equipment has been successfully demonstrated or pilot-tested onsite. Rarely is embryonic equipment considered for full-scale design applications. Such equipment needs to undergo several successful pilot tests, after which it may be upgraded to the innovative category, before being considered for full-scale applications.

6.0 CONTINGENCY PLANNING

6.1 Designing for Emergency Conditions

As indicated previously, flexibility to cope with unforeseen problems and emergency conditions is necessary in any wastewater solids management system. Examples of such situations and contingencies for coping are listed below.

- Equipment breakdowns. Downtime may be minimized by having maintenance staff on call, by advance purchase of key spare parts, by providing redundant equipment units, and by making use of excess storage capacity in upstream processes.
- Solids disposal problems. Such problems may include temporary closures of landfills, unwillingness of current users to accept biosolids, failure of a process to adequately stabilize or treat biosolids, labor strikes, or extended inclement weather. Disposal problems such as these are addressed by providing

long-term storage capacity, or more often, being capable of using more than one disposal or beneficial reuse alternative.

- Sudden increases in solids production. The most common way of addressing this problem is to operate for more hours per day or per week than normal, or by using redundant equipment units temporarily to increase solids handling capacity. In the latter case, redundancy must be available at all points in the solids processing train being used.

6.2 Process and Equipment Redundancy

Because of the emergency conditions described above, it is desirable to have more than one disposal or beneficial reuse alternative available. Backup plans typically involve higher costs, such as higher operating costs for alternative solids handling and disposal or the cost of using specialty contractors to address the emergency conditions.

6.3 Developing a Contingency Plan

Contingency plans for both design and operation of biosolids management processes are recommended for all WWTPs. Even the best facilities will encounter unexpected conditions that may radically change modes of operation and significantly affect the quality of biosolids. In such cases, the normal methods of biosolids and options for biosolids disposal or reuse may not be allowed because changes in conditions or quality of biosolids.

Contingency plans for biosolids management are strongly recommended to address unforeseen conditions and circumstances. Examples of unforeseen conditions might include extreme weather events such as hurricanes, power outages or unit process failures, or the presence of contaminants in biosolids that might cause them not to qualify for the usual and traditional biosolids management options, such as agricultural land application.

Contingency plans are best when they are developed in a workshop setting involving management, operations, and engineering staff, who brainstorm to come up with emergency conditions that require contingency planning, then work together to develop approaches for handling them. Figure 5.8 follows as an example flow chart for a contingency plan that was developed for a WWTP that normally produces Class A, thermophilically digested biosolids.

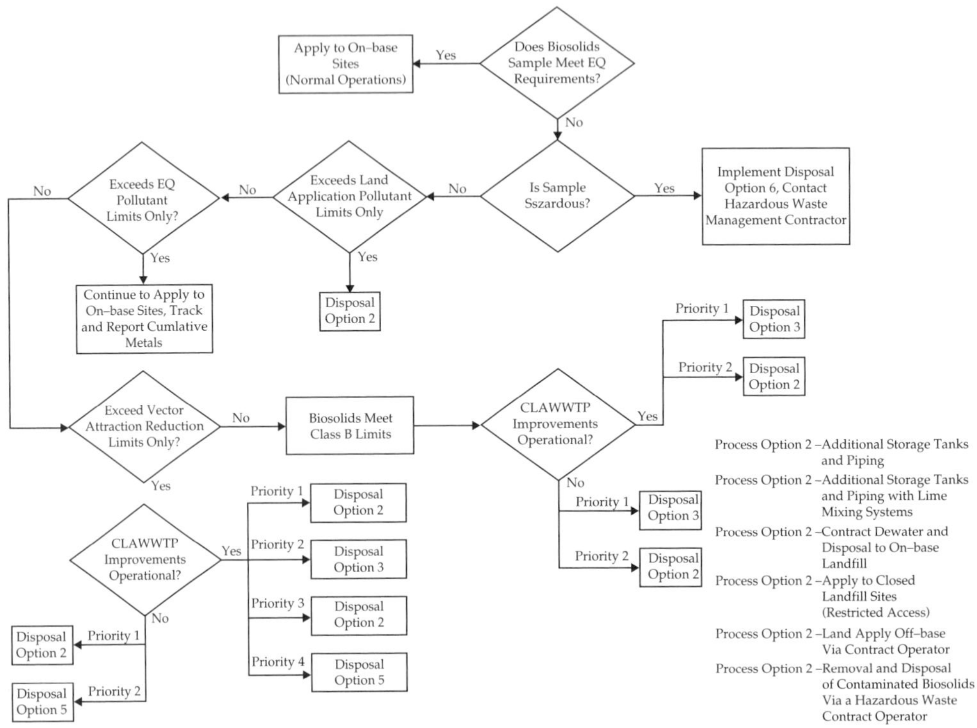

FIGURE 5.8 Example flow chart for biosolids contingency plan.

7.0 OTHER DESIGN CONSIDERATIONS

7.1 Site Variations

Characteristics such as size and location of the WWTP and solids disposal or beneficial use sites strongly influence the nature and cost of solids treatment and handling systems as outlined below.

- If sufficient land application sites are available within close proximity of the WWTP, then land application of thickened and digested biosolids may be practiced, thereby negating the need for solids dewatering.
- If the WWTPs is located near railroads or waterways then opportunities for rail or barge transportation of solids may be available at much lower costs than trucking.

- Moratorium on connections. An effective tool to catalyze stakeholders to support an improvement program in an area.
- Avoiding a repetition of history. The lessons learned from history are the legacy upon which a management approach will be developed in an agency. Often, the legacy can impede positive, progressive thinking
- Incineration as an option. Incineration is often dismissed even though it offers advantages and sustainability in a growing number of locations.
- Public relations and outcry. Maintaining continuous contact with stakeholders is vital for a successful program in a changing environment, especially to maintain a sense of urgency and risk
- Influential neighbors who can impose inordinate opposition to a proposal, especially related to land application programs in rural areas that are gentrifying.
- Structuring reliable and competitive contracts. The strategy for contracting has evolved to the point where public agencies recognize that the lowest cost is not always the most advantageous contract, especially when it results in the concentration of all land application sites in a single or a few jurisdictions.
- Overlapping capacity is desirable to ensure 100% availability, and diversity of outlets is vital for reliability.
- Querulous and unpredictable attitudes of local and state jurisdictions. Political winds can change rapidly and unpredictably, mandating the need and meriting the expense of multiple outlets for biosolids.
- Credibility concerns. This is the price of ignoring the public in the past, takes years to regain, and has often been the reason for a management change.

Today's designer of solids processing and management systems need to maintain a broad perspective and awareness of all relevant issues affecting a project, including economic, environmental, and social issues, if the design approach and alternatives analyses to follow are going to have a reasonable chance of a successful outcome.

8.0 REFERENCES AND SUGGESTED READINGS

CH2M HILL (2005) *Biosolids Master Planning Toolkit Manual*; CH2M HILL: Englewood, Colorado.

CH2M HILL (2007) *Biosolids Contingency Plan for Camp Lejeune Marine Corps Base Advanced WWTP*; CH2M HILL: Englewood, Colorado.

U.S. Environmental Protection Agency (1979) *Sludge Manual*, EPA/625/1–74-006; U.S. Environmental Protection Agency: Washington, D.C.

U.S. Environmental Protection Agency (1984) *Report on Use and Disposal of Sludge*, EPA/625/10–84-003; U.S. Environmental Protection Agency: Washington, D.C.

U.S. Environmental Protection Agency (1987) *Dewatering Municipal Wastewater Sludges*, EPA/625/1–87/014; U.S. Environmental Protection Agency: Washington, D.C.

U.S. Environmental Protection Agency (1994) A Plain English Guide to the U.S. Environmental Protection Agency Part 503 Biosolids Rule EPA/832/R-99/003; http://www.epa.gov/owm/mtb/biosolids/503pe/index.htm (accessed July 2011).

U.S. Environmental Protection Agency (1995) *Process Design Manual for Land Application of Sewage Sludge and Domestic Septage*, EPA/625/R-95/001; U.S. Environmental Protection Agency: Washington, D.C.

U.S. Environmental Protection Agency (1999) *Biosolids Generation, Use, and Disposal in the United States*, EPA/530-R-99–009; U.S. Environmental Protection Agency: Washington, D.C.

U.S. Environmental Protection Agency (2003) *Environmental Regulations and Technology Publication: Control of Pathogens and Vector Attraction in Sewage Sludge*, EPA/625/R-92/013–2003 edition; U.S. Environmental Protection Agency: Washington, D.C.

U.S. Environmental Protection Agency (2006) *Emerging Technologies for Wastewater Biosolids Management*. http://www.epa.gov/owm/mtb/index.htm (accessed July 2011).

Water Environment Research Foundation (1998) *Biosolids Management: Assessment of Innovative Processes*; Project No. 96-REM-1; Water Environment Research Foundation: Alexandria, Virginia.

Water Environment Research Foundation (2003) *Full-Cost Accounting Protocol for Biosolids Management*, Project No. 00-PUM-7; Water Environment Research Foundation: Alexandria, Virginia.

Chapter 6

Conveyance of Wastewater Residuals

1.0 INTRODUCTION

The fundamental objective of all wastewater treatment operations is to remove undesirable constituents present in wastewater and consolidate these materials for further processing and disposal. Solids removed by wastewater treatment processes include screenings, grit, scum, and residuals (sludge). This chapter discusses the conveyance of solids from point to point for treatment, storage, or disposal.

In addition to this chapter, the reader may also refer to the following references on residuals conveyance:

- Conveyance of Residuals from Water and Wastewater Treatment (American Society of Civil Engineers, 2000);
- System Design for Sludge Pumping in *Pumping Station Design* (Anderson and Hanna, 1998); and
- Solids Storage and Transport in *Design of Municipal Wastewater Treatment Plants* (Water Environment Federation et al., 2009).

2.0 PUMPING AND PIPELINES

Unless sludge has been dewatered or dried, it can be transported most efficiently and economically by pumping through pipelines. Thin sludge, typically defined as less than 5% solids in primary sludge and less than 2% solids in waste activated sludge, is subject to the same physical laws as other fluids. Work put into a fluid by a pump alters velocity, hydraulic grade line, and pressure and overcomes friction loss. Typically, the thicker the sludge (i.e., the higher its solids content), the more difficult it is to predict its flow behavior in pipelines. Uncertainties arising from deviations from the typical Newtonian flow behavior of water-like fluids complicate design and specification of the appropriate pumping equipment.

Nevertheless, sludge has been successfully pumped through short pipelines with more than 30% solids by weight and in pipelines of more than 16 km (10 miles) long at up to 8% solids concentration.

Certain principles and enough operational experience exist to allow a designer to successfully select and specify the appropriate pumping system. Those principles and experiences are presented below.

2.1 Sources and Properties of Residuals

Knowledge of the sources, composition, and characteristics of sludge is extremely important in understanding its behavior. Primary sludge behaves differently than a biological or chemical sludge, for example, and anaerobically digested sludge might behave differently from aerobically digested sludge. To predict how sludge behaves during conveyance, its flow properties, including viscosity and shear stress/shear rate relationships, have to be fully understood.

Secondary sludge is the product of the aerobic, biological conversion of biochemical oxygen demand (BOD). It also contains nonbiodegradable particulate matter not captured in the primary settling process. Secondary sludge is produced in activated sludge processes and in fixed-film processes such as trickling filters, rotating biological contactors (RBC), membrane bioreactors (MBR), or integrated fixed-film activated sludge systems (IFAS).

Mixed sludges are typically a combination of primary and secondary sludges. Mixture proportions vary from one plant to another and from one time to the next. A typical activated sludge treatment plant with primary settling will produce a mixture of approximately 50% primary and 50% activated sludge on a dry weight basis. The characteristics of this mixed sludge are not proportional to the composition of the individual type of sludge. For instance, mixtures of primary and waste activated sludge behave more like activated sludge in thickening and dewatering processes.

When sludge undergoes treatment, its properties and, therefore, its behavior may change significantly. Changes in behavior primarily will affect downstream processing, but also will affect sludge flow characteristics. Chemical properties that are relevant to residuals include pH and particle surface charge. Physical properties such as water distribution, specific gravity, particle size, particle hydration, compressibility, and septicity also significantly affect the behavior of sludge. Other physical properties that influence the behavior of sludge include the degree of flocculation, types of microorganisms, and age.

Chemical conditioning alters the characteristics of the sludge in terms of specific gravity, pH, particle size and distribution, and particle surface and surface charge. These changes affect the behavior of sludge and may affect the choice of transport devices. Anaerobic and aerobic digestion also alter sludge characteristics. In fact, digested solids typically are thinner than the raw solids fed to the digestion process and can be pumped more easily than raw or undigested solids with the same moisture content.

2.2 Application of Rheology to Sludge Pumping Problems

Water, oil, and most other common fluids are "Newtonian". This means that the pressure drop is directly proportional to the velocity and viscosity under laminar flow conditions. As velocity increases past a critical value, the flow becomes turbulent. The transition from laminar to turbulent flow depends on the Reynolds number, which is inversely proportional to the fluid's viscosity. Viscosity is a constant for the fluid at any given temperature. Formulas to solve for flow and pressure loss of Newtonian fluids are available in fluids mechanics textbooks.

Wastewater sludge, however, is a non-Newtonian fluid. The pressure drop under laminar conditions is not simply proportional to flow, so the viscosity is not a constant. Special procedures may be used, however, to determine headloss under laminar flow conditions and the velocity at which turbulent flow begins. These procedures use at least two constants to describe the fluid instead of a single constant (the viscosity), which is used for Newtonian fluids.

The behavior of wastewater sludge is compared with the behavior of water on Figure 6.1. This figure is based on steady-state behavior, after thixotropic breakdown. Thixotropic breakdown is discussed below. Several features are notable concerning the behavior of wastewater sludge:

- Essentially no flow occurs unless the pressure is high enough to exceed a yield stress, τ_y.
- Turbulent flow may occur, but a much higher velocity is needed for sludge than for water.
- In fully developed turbulent flow, the pressure drop is roughly that of water.
- For a laminar plastic flow region, sludge follows approximately the flow behavior of a Bingham plastic. A Bingham plastic is described by two constants, which are the yield stress, τ_y, and the coefficient of rigidity, η.
- It is also possible to consider sludge to be a "pseudoplastic" material. In that case, two other constants are used, and the formulas are different.

Besides possibly being dependent on the shearing rate, the flow resistance of liquids can depend on the length of time of shearing or on some function of both the time and intensity of shearing. The most commonly encountered time-dependent change in viscosity is a drop that occurs with time of shearing, followed by a gradual recovery when shearing is stopped. This behavior is called thixotropy. A familiar example is an

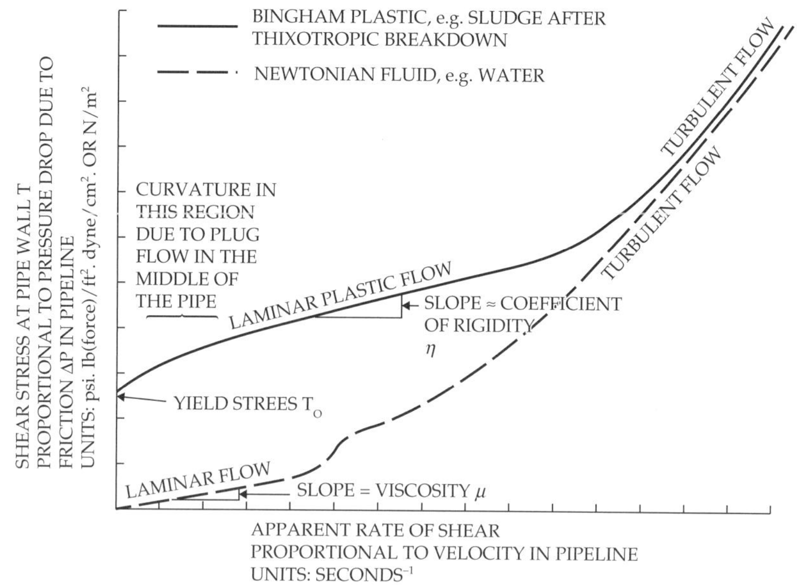

FIGURE 6.1 Comparison of flows behaviors between wastewater sludge and water in circular pipes (U.S. EPA, 1979).

ice cream milkshake, which "sets up" in its container and will only flow out when the container is rapped or jarred several times. The structure rebuilds when the rapping is stopped. Paints typically are not only Bingham plastics, but could also behave as thixotropic fluids. They will flow for a short time after being "worked" by the paint brush so brush lines tend to disappear. Their "plastic" characteristics rebuild quickly after shearing stops so the paint does not flow downward on vertical surfaces.

Wastewater sludge is thixotropic. The effect is increasingly evident as the percent solids and percent volatile solids increase. Thixotrophy has three significant effects:

1. It complicates the measurement of constants such as the yield stress, τ_y.
2. It makes pump suction conditions important. In one case, a centrifugal pump produced ample pressure to move the sludge through a hose. The pump was suspended in a lagoon, but the sludge would not flow into the pump suction. It was found that mixers next to the pump caused thixotropic breakdown sufficient for satisfactory pumping (Rimkus and Heil, 1975a, 1975b).
3. It raises the pressure needed to start flow in a pipeline that has been shut down. In one installation, this effect was found to be significant when

> shutdowns exceeded one day. These effects can be mitigated through proper operating procedures such as purging the line of sludge before extended (i.e., more than 8 hours) shutdowns (Rimkus and Heil, 1975a, 1975b).

Permanent degradation of yield stress can occur with time of shearing. Intense shearing produces this result in high-molecular-weight polymers. This phenomenon can be expected in wastewater sludges, when shear levels are sufficiently high to physically disrupt a portion of the particles making up the sludge. If this occurs, then it may be difficult to thicken or dewater the sludge later.

Sometimes, the reduction in viscosity that occurs with time of shearing is actually the effect of temperature increase produced by the energy delivered to the liquid. The effect of an increase in temperature with both Newtonian and non-Newtonian liquids typically is a reduction in viscosity. For sludge, however, the main effect of temperature is that low temperatures may cause the grease fraction of the sludge to harden. Other temperature effects appear to be unimportant, at least up to 70 °C (160 °F) (Kenny, 1969; Rimkus and Heil, 1975a).

An additional unusual effect occur in wastewater sludge pipelines: slippage and seepage (Rimkus and Heil, 1975a). Essentially, the sludge is riding on a thin film of water next to the wall of the pipe. This effect is noticeable at low velocities when starting a sludge pipeline and partially offsets the thixotropic effect. Seepage and slippage are hard to calculate, but are useful in reducing the initial pressures when restarting the flow of sludges in pipelines (Rimkus and Heil, 1975b).

The reader is referred to *Conveyance of Residuals from Water and Wastewater Treatment* (American Society of Civil Engineers, 2000) for a discussion on sludge rheology, a listing of the basic equations that govern the different fluid flow models, and an explanation on how the theory of viscosity applies to sludge behavior.

2.3 Headloss Calculations for Pumping of Non-Newtonian Sludges

2.3.1 *Introduction*

This section presents both a theoretical approach and field data for designing systems to pump concentrated residuals.

The need to pump residuals long distances has increased in the last 20 years. As a result, researchers have been developing methods to predict site-specific friction losses in pumping systems more accurately (Carthew et al., 1983; Honey and Pretorius 2000; Mulbarger et al., 1981; Murakami et al., 2001; Wagner, 1990). Study results have shown that, once rheological properties have been determined, the Bingham plastic model (Carthew ***et al.***, 1983; Mulbarger et al., 1981) or the pseudoplastic model (Honey and Pretorius, 2000; Murakami et al., 2001) for non-Newtonian fluids may describe how wastewater residuals flow.

2.3.2 *Headlosses from Pumping Dilute Residuals*

Primary and secondary clarifiers often produce a relatively dilute settled sludge (maximum concentrations of 5% are typical for primary sludge, 2% for activated sludge). At velocities greater than 0.3 to 0.6 m/s (1 to 2 ft/sec), such solids are in the turbulent flow regime and have a headloss essentially equal to that of water (Mulbarger, 1997). At lower velocities, the flow becomes laminar, and headlosses increase sharply. Engineers should design dilute-residual pumping systems to maintain a minimum velocity of 0.6 to 0.76 m/s (2 to 2.5 ft/sec) whenever possible to ensure turbulent flow.

2.3.3 *Headlosses from Pumping Thickened Residuals*

This subsection is devoted to methods for estimating the headloss in a pipe conveying homogenous slurries. The solids concentration will vary from a low of 3% to a maximum of 12% for wastewater sludges, but will depend on the composition of the specific sludge or slurry. Difficulties arise, however, when using a classical formula to defining headlosses in ranges greater than 12% total solids for wastewater sludges if the apparent viscosity is high. Several different methodologies are available when designing pumping systems for thickened solids.

2.3.3.1 *Simplified Headloss Calculations*

In water piping, flow is almost always turbulent. Formulas for calculating the friction loss of the flow of water, such as the Hazen Williams and Darcy-Weisbach equations, are based on turbulent flow. Dilute sludge also may flow turbulently, in which case, the friction loss may be roughly that of water. Sludge, however, is unlike water in that laminar flow also is typical. When laminar flow occurs, the friction loss may be much greater than for water. Furthermore, laminar flow laws for ordinary Newtonian fluids, such as water, cannot be used for laminar flow for sludge because sludge follows different flow laws.

Grit slurries are typically dilute; also, grit particles do not stick to each other. Therefore, ordinary friction formulas for water are usually adequate. Typically, a velocity of approximately 1.5 m/s (5 ft/sec) is used. Low velocities may cause deposition of grit within the pipe; high velocities may cause erosion.

2.3.3.2 *Using Known Equations for Water*

First, design engineers can use the Darcy-Weisbach equation (eq 6.1) for water to determine headloss, regardless of the solids' flow regime (laminar, transition, or turbulent), and apply a solids correction factor to the final calculation. The correction factor for residuals up to 12% in solids may be found in the figures of pages 577 and 578 in Anderson and Hanna (1998).

$$h_f = f \frac{L}{D} \frac{V^2}{2g} \qquad 6.1$$

where

h_f = dynamic headloss, m,
f = Darcy-Weisbach friction factor determined from a Moody Diagram,
L = pipeline length, m
D = pipe diameter, m
V = velocity, m/s, and
g = gravitational acceleration, m/s^2.

Under turbulent flow conditions, which occur at velocities greater than the upper critical velocity, engineers can use the Hazen Williams equation for water and multiply the solids correction factor with the final result (Anderson and Hanna, 1998; Metcalf and Eddy, 2003). When using the Hazen-Williams equation with the multiplier for solids concentration, design engineers should assume that *C* equals 140, regardless of whether a different coefficient is applicable for water.

2.3.3.3 *Using Yield Stress and Coefficient of Rigidity*

The flow characteristics of wastewater sludge at concentrations greater than 2 to 3% total solids are different from that of water. As result, design approaches used for pumping water and wastewater typically are not applicable for design of sludge piping systems. Sludge viscosity is significantly higher than that of water because of the presence of suspended solids. The solids not only increase the fluid viscosity, but are also responsible for the fact that sludge exhibits non-Newtonian behavior. This characteristic adds complexity to the design of sludge pumping systems. Despite the fact that sludge flow is complex, most sludge flows can be mathematically expressed as either Bingham plastic with a yield stress or a pseudoplastic type of flow. To use such models, design engineers need to know two special system properties: the yield stress and the coefficient of rigidity.

2.3.3.4 *Obtaining the System Properties*

Mulbarger et al. (1981) reported a relationship between yield stress, coefficient of rigidity, and sludge solids concentration, which was reproduced on pages 98 and 99 of *Conveyance of Residuals from Water and Wastewater Treatment* (American Society of Civil Engineers, 2000). Battistoni (1997), Guibaud et al. (2004), Laera et al. (2007), and Mori et al. (2007) have created similar graphs.

These parameters can also be calculated experimentally for a given system. Several types of instruments are available for viscosity measurements. Only two of these types are suitable for sludge: test pipes and rotational viscometers. Some instruments, such as capillary viscometers, are unable to handle the relatively large particles in sludge; other instruments, such as ball-drop viscometers, are not suited to strongly non-Newtonian fluids such as sludge.

Flow curves from test pipes are directly scalable to full-scale pipes provided flow is laminar. The onset of turbulence in a large pipe cannot be predicted directly from small-pipe tests. It is necessary to use the yield stress and coefficient of rigidity, compute Reynolds and Hedstrom numbers, and use Figure 6.2 to predict the onset of turbulence. The flow curves obtained with test pipes do not provide fundamental rheological data because at a given flow rate, shear stress and rate of shear vary across the radius of the pipe. By using the Rabinowitsch equation, the flow curve can be transformed into a rheologically correct shear stress versus rate of shear curve (Rabinowitsch, 1929). An offsetting disadvantage of test pipes is that a high degree of experimental skill is required to get reliable data. Also, these installations are relatively expensive and cumbersome and require large sample volumes.

For sludge, the best instrument to determine rheological properties appears to be a rotational viscometer. In this type of machine, the test liquid is placed between two concentric cylinders, one of which rotates. The torque on a cylinder is measured as a function of rotational speed. Such machines can produce approximately uniform shear rates at given shear stresses, provided the space between the bob (inner cylinder) and cup (outer cylinder) is small compared to the bob radius. Viscometers in which the bob rotates and the twisting force on the cup is measured are relatively easy to design mechanically, but turbulence occurs at low shear rates for low viscosity materials. Turbulence onset does not occur until much higher shear rates for viscometers in which the cup rotates and the twisting force on the bob is measured. In both types of viscometers, end effects become substantial if the bob and cup are not long relative to the clearance.

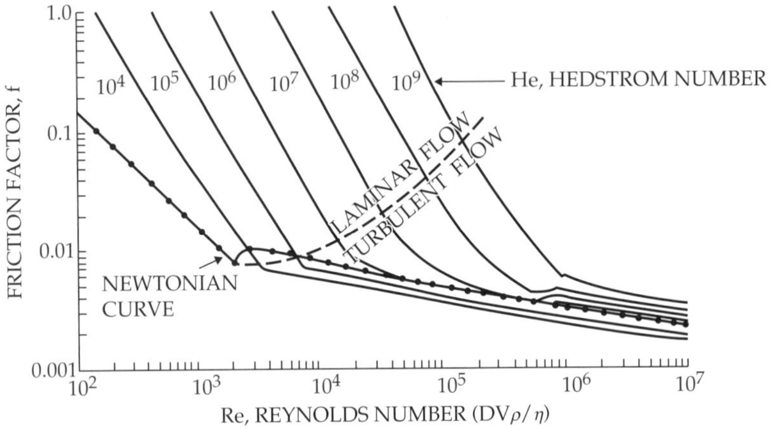

FIGURE 6.2 Friction factor for sludge analyzed as a Bingham plastic (U.S. EPA, 1979).

Several viscometers feature rotational movement, but either do not have constant clearances between an inner and an outer cylinder or do not control or measure shearing rate or shear stress. These devices are of little value for obtaining consistency curves for non-Newtonian liquids.

The nearly uniform shear rate achievable in rotational viscometers allows direct measurement of the fundamental shear stress-rate of shear curve, which is a significant advantage for complex flow relationships. Rotational viscometers are simple to operate. Their primary disadvantage is that close clearances between outer and inner cylinders are needed to give uniform shear rates across the gap between cylinders. A too-small clearance will give erroneous results for sludges. Gap size should not be reduced to less than 1.0 mm (0.04 in.). Sludge must be screened to remove large particles. This creates no substantial error because a few large particles do not strongly affect the coefficients.

A representative test curve adapted from Rimkus and Heil (1975a) is shown in Figure 6.3. In this test, the viscometer speed was gradually increased from 0 to 100 rpm and then decreased. Torque was measured and converted to shear stress, providing "consistency curves". The upper curve (increasing speed) shows thixotropy; the lower curve (decreasing speed) shows behavior of the fluidized sample. The lower curve is appropriate for pipeline design because the sludge is fluidized

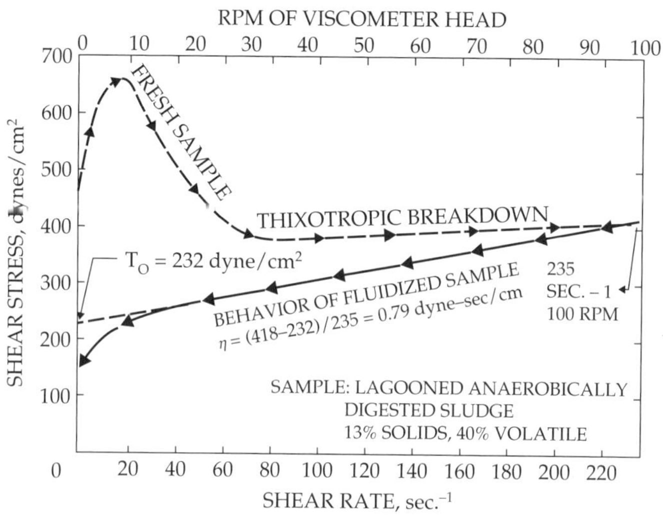

FIGURE 6.3 Viscometer test of wastewater sludge (U.S. EPA, 1979).

by passing through a pump. In this case, the shear stress projected to zero rpm (232 dynes/cm^2) is the yield stress, τ_0; the coefficient of rigidity, η, is the slope of the straight part of the lower curve. Even when fluidized, sludge is not exactly a Bingham plastic, as shown by curvature in the lower curve at low rpm. This departure from Bingham plastic conditions can be used to refine the pressure drop calculations. The viscometer for this test was a Haake Model RV-3 Rotoviso with sensor head MV-I.

Once the rigidity coefficient and yield stress are known, designers can use the following two equations to calculate the upper and lower critical fluid velocities:

$$V_{uc} = 1500\ Rc/\rho D + 1500/\rho D\ (Rc^2 + S_y\ \rho D^2/4500)^{1/2} \tag{6.2}$$

$$V_{lc} = 1000\ Rc/\rho D + 1000/\rho D\ (Rc^2 + S_y\ \rho D^2/3000)^{1/2} \tag{6.3}$$

where

V_{uc} = upper critical velocity (m/s),
V_{lc} = lower critical velocity (m/s),
ρ = fluid density (kg/m^3),
D = pipe diameter (m), and
Rc = rigidity coefficient (N-s/m^2).

2.3.3.5 *Laminar Flow*

At velocities less than the lower critical velocity, the residuals flow will be in the laminar range, and designers can follow the Carthew et al. (1983) methodology (Buckingham equation) described in both *Conveyance of Residuals from Water and Wastewater Treatment* (American Society of Civil Engineers, 2000) and *Design of Municipal Wastewater Treatment Plants* (Water Environment Federation et al., 2009).

2.3.3.6 *Laminar, Transition, and Turbulent Flow*

Regardless of flow regime, designers also can combine the Reynolds and Hedstrom numbers to determine the Fanning friction factor, which can be used in the following modified Darcy-Weisbach equation:

$$H_f = 2\,f\,L\,V^2/g\,D \tag{6.4}$$

The Reynolds number (dimensionless) is determined using Eq 6.5:

$$Re = \rho VD/Rc \tag{6.5}$$

The Hedstrom number is calculated as follows:

$$He = D^2\,\rho\,Sy/Rc^2 \tag{6.6}$$

The friction factor, f, is determined using Figure 6.2.

A few adjustments must be made to correctly use these equations. First, the Reynolds number in eq 6.5 is not the same as a Reynolds number based on viscosity. In plastic flow, an effective viscosity may be defined but is variable and can be much greater than the coefficient of rigidity. Consequently, the two Reynolds numbers can differ by factors of more than 10 under some conditions. Second, design engineers should make sure they use the correct diagram to determine the appropriate friction factor. The Darcy-Weisbach friction factor is four times larger than the Fanning friction factor. Most charts or tables indicate the type of friction factor or at least provide the formula for the friction factor with laminar flow. If the formula for laminar flow is $f = 16/Re$, then the chart provides the Fanning factor. If the formula for laminar flow is $f = 64/Re$, then the chart provides the Darcy-Weisbach factor. Third, care is required with units. For U.S. customary units, it is not possible to use pounds (mass) in density at the same time as pounds (force) in stress without introducing the conversion factor (g_C) into eqs 6.4 and 6.6. Alternatively, the "slug" U.S. customary mass unit could be used.

These equations apply to the entire range from virtually zero velocity to the fully turbulent range, except that Figure 6.2 does not allow pipe roughness. To allow for pipe roughness, ordinary water formulas may be used. If, for example, the Hazen-Williams formula gives a higher pressure drop than eq 6.4, then pipe roughness is dominant, the flow is fully turbulent, and the pressure drop will be given by the ordinary water formula to a sufficiently good approximation for engineering purposes.

Figure 6.2 also shows whether flow is laminar or turbulent. The friction factor f is located by the intersection of the Reynolds and Hedstrom numbers (*Re* and *He*). If this point is above the dashed line on Figure 6.2, or if the Reynolds number is less than 2000, then flow is laminar; otherwise it is turbulent. For example, at $Re = 10^4$, a Hedstrom number of 10^4 gives turbulent flow, and a Hedstrom number of 10^6 gives laminar flow.

Interpolation on logarithmic graphs such as Figure 6.2 is somewhat difficult. This is particularly true for the Hedstrom number curves on Figure 6.2. If the logarithmic (base 10) of *He* is calculated, then interpolation between lines will be linear. Alternatively, if flow is laminar, then the Buckingham equation may be used (Buckingham, 1921; Caldwell and Babbitt, 1941). Figure 6.2 incorporates the Buckingham equation in the laminar region.

2.3.3.7 *Field Pumping Studies and Long Distance Pumping*

Murakami et al. (2001) developed practical equations for long-distance pumping of waste activated sludge (WAS), thickened residuals, and digested biosolids in the 2 to 5% solids range. Their experimental setup and the equations can be found in the original work or in *Design of Municipal Wastewater Treatment Plants* (Water Environment Federation et al., 2009). Other studies that deal with the subject of long-distance pumping include Setterwall (1972), Spaar (1972), Carthew et al. (1983), and Honey and Pretorius (2000).

2.3.3.8 *Solved Example 1: Laminar Flow*

A comparison of headloss values using two different methods is developed below: (1) the Darcy-Weisbach equation for water with a multiplication factor and (2) the Buckingham equation with headlosses shown on Page 100 of *Conveyance of Residuals from Water and Wastewater Treatment* (American Society of Civil Engineers, 2000).

Known parameters include 10% sludge, $D = 0.30$ m, $L = 25$ m, $\rho = 1040$ kg/m^3, $V = 0.7$ m/s (less than lower critical velocity of 2.5 m/s), $Rc = 0.1$ N s/m^2, $Sy = 15$ N/m^2.

Referring to pages 577 and 578 in Anderson and Hanna (1998) for a 10% sludge with a 0.7 m/s velocity, the multiplication factor for routine operation is 10 and for worst-case is 21.

The *Re* for water at 15 °C is:

$$\text{Re} = \rho\, V\, D/\mu = 999.1 \times 0.7 \times 0.3/(1.14 \times 10^{-3}) = 1.83 \times 10^{6}$$

The "ε" coefficient for ductile iron pipe ranges from 0.13 to 0.33 mm. Taking the average value, $\varepsilon = 0.23$ mm and then $\varepsilon/D = 0.23/300 = 0.0008$.

From a Moody diagram, the Darcy-Weisbach friction factor is then determined as follows:

$$f = 0.0185$$

The Darcy-Weisbach equation is then solved for water as follows:

$$H_f = f\, L\, V^2/2\, g\, D = 0.0185 \times 25 \times 0.7^2/(2 \times 9.81 \times 0.3) = 0.0385 \text{ m of water column}$$

For sludge the headloss will then be as follows:

For routine operation, $10 \times 0.0385 = 0.385$ m of water column

For a worst-case scenario, $21 \times 0.0385 = 0.809$ m of water column

Using the Buckingham equation or referring to page 100 of *Conveyance of Residuals from Water and Wastewater Treatment* (American Society of Civil Engineers, 2000), the headloss is 0.297 m/10 m, so for 25 m, the loss is 0.743 m of water column.

2.3.3.9 *Solved Example 2: Transition/Turbulent Flow*

The preceding example was in the laminar flow range. The example below is a comparison of headloss values, in the turbulent range, using two different methods: (1) the Darcy-Weisbach equation for water with a multiplication factor and (2) the modified Darcy-Weisbach equation using *Re*, *He*, and the Fanning friction factor.

Known parameters include 10% sludge, $D = 0.30$ m, $L = 25$ m, $\rho = 1040$ kg/m^3, $V = 3.0$ m/s [fluid is in the transition flow because value is between the lower critical velocity (2.5 m/s) and the upper critical velocity (3.2 m/s)], $Rc = 0.1$ N*s/m^2, $Sy = 15$ N/m^2.

1. According to Anderson and Hanna (1998), for a 10% sludge with a 3-m/s velocity, the multiplication factor for routine operation is 1 and, at worst, is 1.3.

   Re for water at 15 °C is

$$Re = \rho\, V\, D/\mu = 999.1 \times 3 \times 0.3/(1.14 \times 10^{-3}) = 7.82 \times 10^{6}$$

The "ε" coefficient for ductile iron pipe ranges from 0.13 to 0.33 mm. Taking the median value, $\varepsilon = 0.23$ mm and then $\varepsilon/D = 0.23/300 = 0.0008$

From a Moody diagram, the friction factor is then determined as follows:

$$f = 0.0185$$

The Darcy-Weisbach equation for water is then solved as follows:

$$H_f = f\,L\,V^2/2\,g\,D = 0.0185 \times 25 \times 3^2/(2 \times 9.81 \times 0.3)$$
$$= 0.707 \text{ m of water column}$$

For sludge, the headloss will then be as follows:

For routine operation, $1 \times 0.707 = 0.707$ m of water column

For a worst-case scenario, $1.3 \times 0.707 = 0.919$ m of water column

2. $Re = 0.3 \times 3 \times 1040/0.1 = 9360$

$He = 0.3^2 \times 1040 \times 15/0.1^2 = 140\,400$

From Figure 6.2, f is approximately 0.007. The flow is close to the laminar/turbulent borderline, but appears to be turbulent.

The modified Darcy-Weisbach equation gives

$$H_f = 2\,f\,L\,V^2/g\,D = 1.070 \text{ m of water column}$$

Of the two results, the modified Darcy-Weisbach equation using the Reynolds and Hedstrom numbers is considered more accurate and the result from that method should be considered for design (Anderson and Hanna, 1998).

2.3.3.10 *Solved Example 3: Laminar and Turbulent Flow*

In this example, the designer wants to transport anaerobically digested sludge 9.7 km (6 mi.) from one plant to another plant where there are dewatering facilities. If transported a 5% solids, the sludge quantity is 378 m^3/d (100 000 gpd). The sludge may be diluted or thickened, if desired, to improve economics. All of the sludge must be pumped in a four-hour period each day to accommodate dewatering schedules at the receiving plant.

It is assumed that the sludge can be considered as a Bingham plastic using the data shown in Table 6.1 from Canton, Ohio (Hanks and Dadia, 1971). For a comparison, water has a yield stress of 0 and a coefficient of rigidity of approximately 0.01 g/cm·s. The pipe is assumed to be unlined steel pipe, schedule 40: nominal pipe sizes of 100 to 250 mm (4 to 10 in.) in diameter will be considered.

TABLE 6.1 Sludge data from Canton, Ohio (Hanks and Dadia, 1971).

Case	Solids concentration,%	Yield stress τ_y, dyne/cm²	Coefficient of rigidity η, g/cm·s
1	7.12	100	0.40
2	5.34	30.5	0.24
3	3.56	5.8	0.13

The calculation is illustrated in detail for the 200-mm (8-in.) pipe and 7.12% sludge. First, the flow rate is needed. If the sludge were at 5%, 378 m³ (100 000 gal) of sludge would be transported daily. Because the sludge is at 7.12%, the volume is

$$378 \times (5/7.12) = 266 \text{ m}^3 \text{ (70 224 gal)}$$

And the flowrate is

$$(378 \text{ m}^3/\text{d})/(4 \text{ hours flow/day} \times 60 \text{ min/h}) = 18.46 \text{ L/s (18 460 cm}^3\text{/sec or 292.6 gpm)}$$

The internal diameter of a 200-mm (8-in.) Schedule 40 pipe is 20.27 cm (7.981 in.) and the cross sectional area is 322.7 cm². The velocity, V, is the flowrate divided by the area:

$$V = (18\,460 \text{ cm}^3/\text{sec})/322.7 \text{ cm}^2 = 57.2 \text{ cm/sec}$$

The Reynolds number is calculated as follows:

$$Re = \rho VD/\eta = 1.0 \times (57.2 \times 20.27)/0.40 = 2898 \text{ (dimensionless)}$$

The Hedstrom number is calculated as follows:

$$He = (D^2\, \tau_y g_C\, \rho)/\eta^2 = (20.27)^2 \times 100 \times 1.0 \times 1.0)/(0.40)^2 = 256\,800 \text{ (dimensionless)}$$

Referring to Figure 6.2, f is approximately 0.08. The flow is laminar, not turbulent.

The length, L, is needed:

$$L = 6 \text{ miles} \times 5280 \text{ ft/mi.} \times 30.48 \text{ cm/ft} = 965\,600 \text{ cm}$$

$$\text{(miles} \times 1.609 = \text{km; ft} \times 0.3048 = \text{m)}$$

Pressure drop because of friction using eq 6.7, which is a variation of eq 6.4:

$$\begin{aligned} \Delta p &= (2f\,\rho\, LV^2)/(D\, g_C) \\ &= (2 \times 0.08 \times 1.0 \times 965\,600 \times (57.2)^2\,)/(20.27 \times 1.0) \\ &= 24\,940\,000 \text{ dyne/cm}^2 \quad \text{or} \quad 2496 \text{ kPa} \end{aligned} \qquad 6.7$$

To put it into perspective, this value may be compared to the value for water for the same conditions, calculated from the Hazen-Williams equation:

$$V = 1.318\ C\ R^{0.63}\ S^{0.54} \qquad 6.8$$

where
V = average velocity, ft/sec,
C = friction coefficient,
R = hydraulic radius = ¼ of diameter, ft
S = hydraulic gradient, ft/ft.

This equation may be rearranged and solved on a calculator, or tables or nomographs may be used. In the present case, V = 57.2 cm/sec = 1.88 ft/sec and R = 0.166 ft. With a C of 100, S is 0.00310, indicating a pressure drop of 290 kPa (42 psi). The drop with this sludge is 2496 kPa (362 psi) or about 9 times higher than the drop for water.

For the various cases, calculations are summarized in Table 6.2. Friction factor plots from Figure 6.2 are shown on Figure 6.4.

A precaution that is useful for detection of computational error is to check to see whether the pressure drop across the pipe calculated by the above procedure produces a sufficient shear stress at the pipe wall to exceed the yield stress of the sludge. If the yield stress is not exceeded, then sludge will not flow. The pressure drop needed is calculated by setting the calculated shear stress at the wall equal to yield stress:

$$\Delta p_0\, g_C\, D/4L = \tau_y g_C \qquad 6.9$$

where Δp_0 = pressure drop needed to exceed yield stress.

Results of the calculation are shown for Case 1 and Case 2 in Table 6.3. Equation 7.9 is also useful as a screening test. If τ_y, D, and L are known, it is possible to quickly calculate the minimum pressure drop that could occur, regardless of velocity of flow rate. If Δp_0 is excessive, then diameter, D, should be increased. Impractical pipe sizes could be quickly eliminated as requiring too high a pressure drop for consideration.

Values from Tables 6.2 and 6.3 are plotted on Figure 6.5. Selection of the optimum pipe diameter and solids content requires an economic analysis. It is evident, however, that at the more reasonable pressure drops (below 1400 kN/m^2 or 200 psi), the 7.12% solids has a higher pressure drop at a given pipe diameter even though the volumetric flow rate is much lower than for the other two cases. At 200 mm (8 in.), the pressure drops are approximately the same for the 5.34 and the 3.56% sludges.

As noted in Table 6.2, however, the flow is not in the turbulent regime for the 5.34% sludge. This is a disadvantage because small changes in the rheological

Table 6.2 Summarized calculations for non-Newtonian flow example problem (1 psi = 6.9 kN/m^2 = 69,000 dyne/cm^2).

Case	Diameter, in.	Diameter, cm	Average velocity, cm/s	Reynolds number, *Re*	Hedstrom number, *He*	Fanning friction factor, *f*	Pressure drop, kPa: Sludge	Pressure drop, kPa: Water[a]	Pressure drop, psi: Sludge	Pressure drop, psi: Water[a]
1	4.03	10.2	225	5750	65000	0.010[b]	9515	8205	1380	1190
	5.05	12.8	143	4580	103000	0.019[b]	5343	2717	775	394
	6.06	15.4	99.1	3820	148000	0.03[b]	3682	1117	534	162
	7.98	20.3	57.2	2900	257000	0.08[b]	2496	290	362	42
	10.02	25.4	36.3	2310	405000	0.20[b]	1999	97	290	14
2	4.03	10.2	300	12780	55300	0.008 3	14052	13927	2038	2020
	5.05	12.8	190	10150	87000	0.008 5	4640	4599	673	667
	6.06	15.4	132	8480	126000	0.009 0	1965	1896	285	275
	7.98	20.3	76.3	6440	218000	0.019[b]	1048	496	152	72
	10.02	25.4	48.4	5130	343000	0.035[b]	62[c]	165	90[c]	24
3	4.03	10.2	450	35400	36000	0.006 6	25166[c]	29510	3650[c]	4280
	5.05	12.8	286	28200	56000	0.007 0	8619[c]	9811	1250[c]	1423
	6.06	15.4	198	23500	82000	0.007 2	3537[c]	4013	513[c]	582
	7.98	20.3	114	17800	141000	0.007 5	931[c]	1048	135[c]	152
	10.02	25.4	72.6	14200	222000	0.008 0	317[c]	345	46[c]	50

[a]Calculated from Hazen-Williams equation with a friction coefficient (C) of 100.

[b]Flow is not in the turbulent region.

[c]Note that pressure drop for sludge, by equation, is less than the pressure drop for water if $C = 100$. The pressure drops would be approximately the same if $C = 110$.

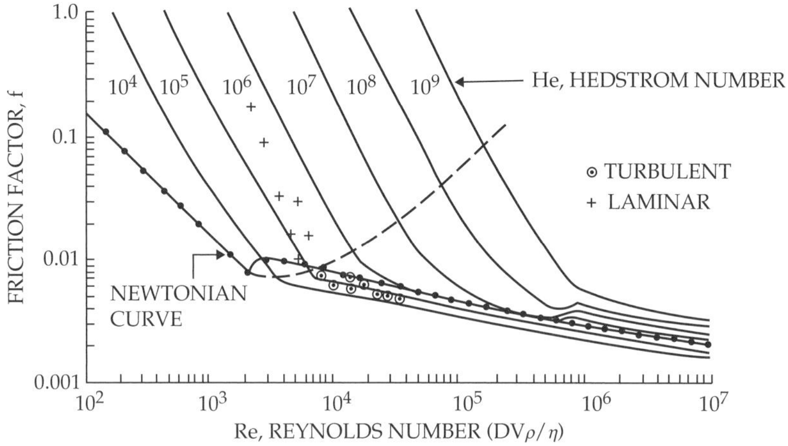

Figure 6.4 Friction factors for the example problem (U.S. EPA, 1979).

Table 6.3 Pressure required to exceed yield stress—example problem.

	Pressure drop Δp_0, kPa*		Pressure drop Δp_0, psi*	
Diameter, cm (in.)	**Case 1** τ_y **= 100 dyne/cm²**	**Case 2** τ_y **= 30.5 dyne/cm²**	**Case 1** τ_y **= 100 dyne/cm²**	**Case 2** τ_y **= 30.5 dyne/cm²**
10.2 (4.03)	3778	1151	548	167
12.8 (5.05)	3013	917	437	133
15.4 (6.06)	2503	765	363	111
20.3 (7.98)	1903	579	276	84
25.5 (10.02)	1517	462	220	67

*Pressure drop to cause the shear stress at pipe wall to exceed the yield stress τ_y. Higher pressures may be needed to start the pipeline because of thixotropic effects not considered in Figure 6.2.

constants τ_y and η could cause changes in *f*. The 3.56% solids content is probably a better selection based on the likelihood of more stable operation. At 250 mm (10 in.), the value of *f* is significantly higher for the 5.34% sludge than for the 3.56% sludge. The choices between 200- and 250-mm (8- and 10-in.) diameter and 3.56 and 5.34% sludge would have to be made on the basis of minimum overall cost. The 5.34%

sludge will be more expensive to transport, but this cost increase may be offset by more economical dewatering at the plant receiving the sludge.

Note that the pressure drop for Cases 1 and 2 is greater in all cases than the minimum drop Δp_0 (see Figure 6.5).

2.3.4 *Headlosses from Pumping Dewatered Residuals*

The hydraulic characteristics of dewatered cake (with more than 12% solids) have not been extensively studied or widely reported. At such solids concentrations, dewatered cake may exhibit both plastic and pseudoplastic (thixotropic) behavior (Barbachem and Payne, 1995; Bassett et al., 1991; List et al., 1998). Unlike liquid residuals, where pumping design equations are available for solids concentrations up to 12%, design approaches for pumping dewatered cake are more site-specific. Field testing is highly recommended, especially in relatively long-distance applications. A pipe that conveys cake containing more than 30% solids should not be more than 152 m (500 ft) long, and line lubrication is highly recommended.

In most dewatered cake applications, headlosses are high; often in the range of 141 to 704 meters of water column (200 to 1000 psi). Headloss depends on the

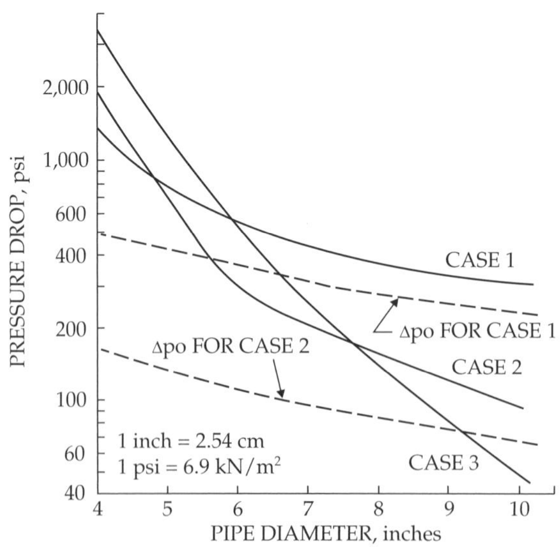

FIGURE 6.5 Pressure drops for example problem (U.S. EPA, 1979).

length, diameter, and configuration of the discharge piping. It also depends on cake type, solids concentration, and the conditioning and dewatering methods used to produce the cake. Current experience indicates that headlosses typically are sensitive to velocity and piping constrictions, particularly if the dewatered cake's solids concentration exceeds 30%. Typical piping headlosses in cake-pumping applications range from 1.15 to 8.07 m of water /m (0.5 to 3.5 psi/ft). Design engineers often use these values as a guideline during preliminary design. Ideally, the final design would keep headlosses at less than 4.61 m/m (2 psi/ft). Also, theoretical cake flow velocities should never exceed 0.15 m/s (0.5 ft/sec)—although maximum velocities of 0.08 m/s (0.25 ft/sec) are preferred, especially if dewatered cake solids concentrations exceed 30%.

A compilation of useful headloss data from pumping different dewatered solids was reported by a progressing cavity pump manufacturer (Bourke, 1997) and reproduced on pages 21–49 in *Design of Municipal Wastewater Treatment Plants* (Water Environment Federation et al., 2009). The data seem somewhat liberal, however, so designers should use more conservative values or require performance warranties from the pumping-system supplier.

Results, analysis, and comparison of four peer-reviewed case studies with actual field data were published in the 1990s are available in two of the manuals referred to in this chapter (American Society of Civil Engineers, 2000; Water Environment Federation et al., 2009). These studies explored pipeline headlosses from different types of cake ranging in thickness from 20 to 34% in dry solids. An example of calculating headlosses developed from the pumping of dewatered cake can be found in *Design of Municipal Wastewater Treatment Plants* (Water Environment Federation et al., 2009).

2.4 Overview of Residuals Pumping Equipment

Sludges can range in consistency from watery scum to thick, paste-like slurry. A different type of pump may be required for each type of sludge. Pumps used for sludge transport include centrifugal and positive displacement pumps. Centrifugal pumps are further divided into nonclog, recessed impeller, combination screw/centrifugal, and disc pumps. Positive displacement pumps include plunger, piston, piston/hydraulic diaphragm, progressing cavity, rotary, diaphragm, ejector, peristaltic, and air-lift pumps. Water eductor pumps sometimes are used to pump grit from aerated grit removal tanks. Because of space availability, the reader is referred to the three references cited in the beginning of this chapter for a description of the different pumps for residuals duty.

2.5 Application of Sludge Pumps

Suction conditions require special attention when pumping sludge. When pumping water or other Newtonian fluids, calculations of net-positive suction head (NPSH) can be used to determine permissible suction piping arrangements. Sludge, however, is a non-Newtonian fluid, especially at high solids concentrations. This behavior may drastically reduce the available NPSH. Consequently, long suction pipelines should be avoided, and the sludge pump should be several feet below the liquid level in the tank from which the sludge is to be pumped (flooded suction). If these conditions are not met, then the pump will not be able to handle sludge at high concentrations.

Special precautions typically are required to reliably pump screenings and grit. Screenings should be ground up and pumped by pumps that can pass large material. Torque flow pumps are ideal for this application. Grit pumping requires special consideration of both abrasion and clogging issues. Both screenings and grit pumps should be easy to disassemble with quick access to the volute and impeller. Table 6.4 presents an application matrix that identifies the various types of sludges or solids typically encountered in wastewater applications and provides a guide for the suitability of each type of pump in that service.

2.6 Pipe, Fittings, and Valves

Materials for wastewater solids pipelines include steel, cast and ductile iron, pretensioned concrete cylinder pipe, thermoplastic, fiberglass-reinforced plastic, and other materials. Steel and iron are most common. With steel or iron, external corrosion may occur in unprotected buried lines; corrosion may be adequately controlled under most conditions by coatings and, where needed, cathodic protection. Inside the pipe, a lining of cement, plastic, or glass may be used to protect the pipe from internal corrosion and abrasion. With raw sludges and scum, linings have an additional function: they provide a smooth surface that greatly retards accumulations of grease on the pipe wall (Sparr, 1971; Williams, 1977). With anaerobically digested sludge, linings may be useful to prevent crystals of struvite from growing on the pipe wall. Smooth linings are especially valuable in pump suction piping and in key portions

Table 6.4 Application of sludge pumps.

Pump*	Screen-ings	Grit	Septage	Scum	Primary	Return acti-vated	Waste acti-vated	Mixed	Float thickened	Gravity thickened	Digested	Dewatered	Notes
Non-clog mixed flow submersible			x		x	x	x	x	x	x	x		Low solids content
Axial flow						x							
Recessed impeller	x	x	x		x		x	x	x	x	x		< 2.5% solids for PS, < 4% for digested
Screw centrifugal					x	x	x	x	x	x	x		High efficiency, < 6% solids
Grinder				x							x		Low efficiency
Parallel disc		x					x		x	x	x		< 6% solids
Vertical turbine solids handling						x							
Plunger			x	x	x		x	x	x	x	x		High maintenance, < 15% solids
Diaphragm	x	x	x	x	x		x	x	x	x	x		< 10% solids
Peristaltic				x	x		x	x	x	x	x		High suction lift, < 12% solids
Rotary lobe				x	x	x	x	x	x	x	x		< 10% solids
Progressing cavity	x		x	x			x	x		x	x	x	< 20% solids

(*continued*)

TABLE 6.4 Application of sludge pumps (*Continued*).

Pump*	Screen-ings	Grit	Septage	Scum	Primary	Return acti-vated	Waste acti-vated	Mixed	Float thickened	Gravity thickened	Digested	Dewatered	Notes
Piston				x			x		x	x	x	x	High cost, 15-40% solids
Pneumatic ejector	x	x		x	x		x	x	x				
Air-lift						x	x						Low lift & efficiency
Water eductor		x											
Archimedes screw						x							

*This is a non-exhaustive list and includes variations of pumps that can perform the type of work indicated. The table, however, does not provide recommendations in selecting one type of pump over another. The considerations that go into selecting a pump for a particular sludge-pumping application are complex. Each application should be reviewed carefully as to the type and source of sludge, the downstream application, and the benefits one pump may provide over another before final selection is made.

of piping (header pipes and the like) where maintenance shutdowns would cause process difficulties.

Fittings and appurtenances must be compatible with the sludge and pipe. Long sweep elbows are preferred over short radius elbows. Grit piping may be provided with elbows and tees made of special erosion-resistant materials.

Valves of the nonlubricated eccentric plug type have proven reliable in sludge pipeline service. Care must be taken if a cleaning tool is to pass through the valves. Grit pipelines typically are equipped with tapered lubricated plug valves.

Wastewater solids piping should be designed for reasonable and convenient maintenance. Even under good conditions, pipe occasionally may have erosive wear, grease deposits, or other difficulties. Pipe in tunnels or galleries is more accessible than buried pipe. An adequate number of flanged joints, mechanical couplings, and takedown fittings should be provided. It is recommended that 10 to 15 cm (4 to 6 in.) be considered the minimum diameter for wastewater solids pipelines to minimize grease clogging or particle blockage and facilitate maintenance. Blind flanges and cleanouts should be provided for ease of line maintenance. Gas formation by wastewater solids left for long periods in confined pipe or equipment can create explosive pressures; therefore, provision should be made for flushing and draining all pipes, pumps, and equipment.

The pressure rating of wastewater solids pipelines and fittings should be adequate for unusual and routine operating pressures. Unusual pressures occasionally will occur because of high solids concentrations, pipe obstructions, gas formation, water hammer, and cleaning operations. For instance, all isolation valves should be able to withstand the expected pressures developed in the system or they could be the first point of failure. Standard pinch valves of 152 mm (6 in.) or less are rated the same, 1034 kPa (150 psi), as knife gate valves. At larger diameters, standard knife gate valves sustain their rating of 1034 kPa (150 psi), whereas the rating on pinch valves reduces to 826 kPa (125 psi), 203 mm (8 in.), and 690 kPa (100 psi), 254 mm (10 and 12 In.). Higher pressure pinch valves are available for installations where higher pressures are expected, but those are twice as expensive as standard pinch valves of the same diameter. Resilient seat gate valves can also withstand high pressure.

Temperature changes may cause stress in the pipe. Temperatures are changed by heated material as it enters cold pipe, flushing, and the use of hot fluids during cleaning to remove grease. Pipe should be designed to accommodate such stresses.

2.7 Inline Grinding

Inline grinders are used to reduce the size of sludge solids to prevent problems with the operation of downstream processes. Grinders require high maintenance; therefore, they should not be installed unless absolutely necessary. For locations where a grinder may be installed in the future, removable spool pieces should be inserted into the pipeline to facilitate the later installation of a grinder. Grinders may be applicable to streams carrying debris, rags, or stringy materials, but typically are not needed for streams carrying only secondary (biological) sludge. Grinders often have been installed preceding equipment with ball or flapper check valves. Using dual-check valving, proper stroke seating can be obtained and the grinders can often be eliminated. Grinders remain a necessity upstream from small diameter, high-pressure positive displacement pumps.

Sophisticated, slow-speed, hydraulic or electric grinders that can sense blockages and clear themselves by reverse operation are now available. Special combination centrifugal pump-grinders are available for use as digester circulation pumps and are effective in preventing rag balls. Experience indicates such pumps require as much maintenance as grinders.

2.8 Line Lubrication

In addition to the larger-diameter piping for cake transfer, installation of slip injection rings for boundary layer water or polymer injection is recommended in "high-solids" pumping applications. The method involves injecting a liquid lubricant into the discharge pipe via a ring, which distributes the liquid equally around the pipe's perimeter to create a "boundary layer". There are two types of boundary layer injection rings: the original configuration, which has individual injection points, and a newer design, in which lubricating media is injected through an annular groove around the pipe perimeter (Figure 6.6). Typically, the annular groove design is preferred because it requires less lubricant and has minimal effect on the cake's percent solids content. The lubricant could be water-, polymer-, or oil-based.

The quantity of water is small, typically 0.5 to 1.5% of the flow in the pipe. This has only a nominal effect on the delivered solids concentration (American Society of Civil Engineers, 2000). For example, if 0.5, 1.0, and 1.5% (v/v) of the

FIGURE 6.6 Annular ring for line lubrication.

material flow as water were added to a 35% sludge cake, the resulting concentrations would be 34.8, 34.7, and 34.5%, respectively. The effect of the lubricating water addition could be a 50 to 75% reduction in the pipeline pressure. The effect of the thin layer (0.20 to 0.25 mm) of water on the interface is to produce a hydrodynamic film, which is independent of the high-viscosity material flowing through the pipe. The film reduces the frictional forces and yield stress at the wall and, hence, the headlosses.

The lowering of the headloss without a significant reduction in the solids content will enhance reliability of the pump operation, reduce pump maintenance costs, and significantly lower power requirements.

The addition of the water may be initiated by the level of pressure in the discharge line. A process controller can control the amount of water added to limit the pressure to a maximum value. The injection system will consist of a high-pressure pump, pump controller, and injection ring. In a recent project in San Diego, California, the controls of a cake pumping operation had to be modified to include running the lubrication pump whenever the cake pump was operating (Fonda and Rupik, 2006). Therefore, contrary to the recommendation of pacing the line lubrication system based on the pressures experienced in the line, there may be cases where line lubrication should be provided continuously for as long as the cake pumps are running.

Willis et al. (2002) reported significant reduction in pipeline pressures encountered in a cake pumping installation in Canada where water was added for lubrication. In that installation, a single piston pump was operated at a 19–22 m^3/h capacity, conveying biosolids of approximately 28% dry solids. A 55 to 65% reduction in pipeline pressures was observed along the entire pipe when the lubrication rate was increased from 1350 to 5400 mL/min, representing 0.4% and 1.6% respective increases in cake volume because of the added lubrication. This increased volume corresponded to a decrease in cake concentration from 28.0% to 27.89% and 27.55%, respectively.

Additional studies have documented the even greater benefits of boundary layer injection using polymer instead of water (American Society of Civil Engineers, 2000; List e al., 1998). Incorporating slip injection water or polymer solution into the pumping system will eliminate many of the uncertainties of conveying 20 to 35% total solids cakes using high-pressure positive displacement pumps.

2.9 Experience and Design Guidelines for Long-Distance Pumping of Liquid Residuals

As wastewater treatment plants (WWTPs) are expanded to handle higher flows, limited onsite space can force organizations to relocate sludge treatment to remote facilities. Space that was once available adjacent to a WWTP may have been taken over by residential and commercial development. Because of increased urbanization surrounding treatment plants, the distance between the plant and the solids handling facilities can become significant. The method chosen to transport solids from one facility to the other is a decision that requires careful consideration. Because of greenhouse gas emissions, transporting solids by truck or rail may not be an environmentally sound option. Sludge can be pumped for miles. As discussed earlier, the pumping characteristics of sludge change as the total solids concentration increases. Pipelines frequently are the least expensive and most environmentally sound option. Although this may be the case, pumping sludge long distances can present its own challenges and considerations. Long-distance pumping typically creates high-pressure losses, so design engineers should choose pumps that can generate the high pressures needed. In the United Kingdom, for example, a vertical, positive-displacement, hydraulically driven ram pump transfers primary and activated sludge over a 2.2-km-long pipeline, working against pressures of up to 2600 kPa (377 psi) (World Pumps, 1999).

2.9.1 *Experience*

Table 6.5 presents case history data from seven long distance sludge pumping applications that may be helpful to engineers that design similar projects. A review of this table shows that

- Both centrifugal and positive displacement pumps are used;
- Velocities are typically greater than 0.6 m/s (2 ft/sec) to minimize solids accumulation;
- Pigging or flushing systems are provided on most long distance pipelines;
- Pressures up to 3930 kPa (570 psi) are encountered, but most are below 690 kPa (100 psi);
- Raw sludge becoming septic in the pipeline results in gas production in the pipeline and high odors at receiving end.

2.9.2 *Design Guidance*

Proper preplanning of a pipeline installation is critical. For example, a pump breakdown or a plugged pipeline can significantly affect plant operation. Negative effects can be minimized by good initial design and equipment selection.

If digestion will be part of the system, then digesters may be located either before or after the long sludge pipeline. Sludge, however, is much easier to pump after it has been digested. In addition, raw sludges may cause problems related to thickening, odors, and corrosion at the receiving point because septic conditions may develop in the pipeline. If raw sludge is to be pumped long distances, then the least environmental effects will result if the pipeline contents are discharged directly into anaerobic digesters.

Sludge that has been piped for a long distance may experience floc breakdown. If this occurs, then thickening and dewatering may be impaired. Chemical conditioning may require a higher chemical dose; thermal conditioning may produce sludge with poorer dewatering properties. Several special design features should be considered for long-distance pipelines:

1. Two pipes should be provided unless a single pipe can be shut down for several days without causing problems in wastewater treatment system.
2. External corrosion and pipe loads should be considered just as for any other utility pipeline, for example, water or natural gas. External corrosion has

Table 6.5 Long-distance pipeline installations conveying dilute or thickened residuals (WWTP = wastewater treatment plant; WRP = water reclamation plant; SHF = solids handling facility; BMF = biosolids management facility; WRF = water reclamation facility; WAS = waste activated sludge).

	Indianapolis, Indiana	Mesa Arizona	Metropolitan Water Reclamation District of Greater Chicago	Colorado Springs, Colorado	Malibu, California	San Diego, California	Eugene, Oregon
	Southport to Belmont Interplant Connection	Southeast WRP to Greenfield WRF	Chicago Northside to Stickney Plant Pipeline	Las Vegas WWTP to Clear Springs Ranch SHF	Rancho Los Virgenes	Point Loma to Metro Biosolids Center	Central Oregon Metropolitan Waste Management Commission
Summary description	Southport is a liquids-only facility and Belmont is a full-treatment plant (both liquids and solids processing); solids from Southport are pumped to Belmont.	The Southeast WRP is a liquids-only treatment facility; solids from Southeast WRP are sent to Greenfield WRF for processing. Greenfield WRF has liquids and solids processing.	Solids from the Northside Plant are sent to the Stickney Plant (liquids and solids processing).	Clear Spring Ranch SHF receives solids from the Las Vegas WWTP and the JD Philips Water Reclamation Facility (JDPWRF)	Tapia WWTP has a 37.9 ML/d (10 mgd) capacity but no solids processing and is "zero discharge" seven months per year, used to inject until 1970s, which built the Rancho composting facility (solids only). Solids from Tapia are sent to Rancho Los Virgenes for processing and disposal.	Point Loma Plant is space constrained to only include digesters but no post-digestion processing. Digested sludge is sent to the Metro Biosolids Center (MBC); MBC is a solids-only facility and receives raw sludge from the North City WRP.	Digested sludge is pumped from the WWTP to the BMF where it goes into four facultative lagoons; BMF also has drying beds and belt press dewatering.
Pumped fluids	Two lines: primary sludge and WAS. No scum.	Mixture of primary sludge and WAS.	Sludge from one plant to another.	Sludge, WAS, primary sludge and scum; 3.0–3.2% solids.	Combined sludge, 1° and 2° mixed to 3% solids.	Digested sludge, 1.8– 3.2% solids.	2% solids sludge from digesters.
Thickened before transport?	No	No, degritted, 1% solids	No, 1.5–2.5%	Mixed and thickened, heated, dewatered. Specific mixing protocol for primary scum.	Yes, decanted	No	No

Other systems or utilities?	There is a parallel sewer line, not part of the sludge transport system.	No	No	No	Two sludge force mains, older one is now backup; dedicated centrate line back to WWTP.	An 8-km (5-mi) transmission line runs from the North City Plant. It conveys primary scum mixed into the sludge line. A 400-mm (16 in.) line with 3785 L/min. (1000 gpm).	Two: 300-mm (12-in.) recycled water from the WWTP; supernatant line from BMF facultative sludge lagoons goes back to WWTP (uses a previously existing pipe).
Piping Age	1967	2007 but pipe was pre-existing from 1998; previously used for raw wastewater between plants.	40–50 years	2005	1 = 1970; 2 = 1980	12 years old	20 years old
Piping material	Either ductile or cast iron	Ductile iron	Ductile iron and asbestos cement	Fiber-glass-reinforced plastic (FRP)	Ductile iron, slip lined and wrapped	Ductile iron with push on restrained joints	Ductile iron. (supernatant and reclaimed are plastic)
Piping length	11 km (7 miles)	18 km (11 miles)	45 km (28 miles)	29 km (18 miles)	8 km (5 miles)	27 km (17 miles)	9.7 km (6 miles)
Piping diameter	Both 350 mm (14-in.)	250 mm (10 in.)	450–600 mm (18–24 in.)	350 mm (14 in.)	25–150 mm (1–6 in.) to 50–200 mm (2–8 in.) (newer)	300–350 mm (12–14 in.)	200 mm (8 in.)
Piping profile	Pretty flat	Drops 21 m (68 ft) to more than 18 km (11 mi)	Flat	Hilly	Hilly, goes through National Parkland and Indian Territory.	Sea level to 122 m (400 ft)	Flat
Fluid velocities	> 0.6 L/s (2 ft/sec) to minimize settlement	Calculated as < 0.3 L/s (1 ft/sec)	0.9–1.5 L/s (3–5 ft/sec)	0.3–0.6 L/s (1–2 ft/sec)	755 L/min (200 gpm); 0.4 L/s (1.2 ft/sec)	2 L/s (6 ft/sec) in 300 mm (12 in.); 1.3 L/s (4.4 ft/sec) in 350 mm (14 in.)	Unknown
Flowrates	4543 L/min (1200 gpm) each line	568 L/min (150 gpm)	Not available, they produce 35 000 dry tons per year	Approximately 871 L/min (230 gpm)	212 L/min (56 gpm)	7949 L/min (2100 gpm) peak	1703 L/min (450 gpm) is typical

(*continued*)

Table 6.5 Long-distance pipeline installations conveying dilute or thickened residuals (*Continued*).

Construction method	Direct buried, open cut through area that was undeveloped at the time. Currently some residential.	Direct buried	Direct buried	Direct buried, encased in sand and Flowfill.	Direct buried	Direct buried	Direct buried
Pressures	Approximately 414 kPa (60 psig)	< 552 kPa (80 psig)	517 kPa (75 psi) but because of leaks, 310 kPa (45 psig)	1207 kPa (175 psi) at plant, 207 kPa (30 psi) at SHF	1482 kPa (215 psi) operational limit because of failures, 2413 kPa (350 psi) pipe rating but not the fittings.	3930 kPa (570 psig); tested to 6205 kPa (900 psig)	Unknown, not high
Air vacuum valves?	Yes	Yes	Yes	Yes	Yes	Yes	Yes
Corrosion?	Some corrosion, occasional line breaks, determined to be caused by corrosive soils. rather than pumped fluids.	Unknown	Yes, and cracks from failure of bedding.	None. Damaged by unrelated contractor once but repaired	Yes, from poor soil conditions and drainage.	No, a cathodic protection system.	No
Instrumentation	No	Yes	Pressure transmitters only	Pressure and flow	Mag meter at both ends with alarm; if they do not match, then leak. System has worked well.	Yes	Indicating gauges in vaults.
Supervisory control and data acquisition	No	Operators at either end can shut off the pumps.	Pressure transmitters only.	Full package	Mag meters only	Full controls and alarms and automatic controls.	No

Communication and control	Operators at Southport control the pumps, communicate with operators at Belmont before making changes. Operators at Belmont control discharge.	Can trend flowrates.		Alarms and automatic shutdowns on high and low pressure.	Telephone communication between operators at both ends about pumping rates and solids content.	Seismic valves, etc.	Pumps controlled from main plant, distribution controlled at BMF.
Operation and maintenance issues?	Occasional line breaks from corrosive soils.	Occasionally flushed when pressure gets high.	None	Gas build-up occurs in pipe, requires burping.	Operation demands a lot of attention, mostly to keep pressures down by controlling solids content (diluting).	No	Low velocities cause settling.
Grit or debris problems?	Grit in primary sludge but does not cause a problem in transport pipeline.	No	None	None	Grit in stream does not create problems.	No	Struvite is big problem.
Pigging?	Yes, once a week.	Available but never used	No	Not available but flushing Water is available.	About twice a year	Available and tested but Never needed.	Installed, worked on Settlement for a while but then struvite problem made pigging impossible.
Additional comments	Primary sludge goes septic over the 11 km (7 miles) transport; creates high H_2S at discharge.				Isolation and check valves along route to allow for repairs when necessary.		Struvite remains a big problem, redundant pipes recommended if this might be an issue.
Type of pumps	Centrifugal Chicago pumps	Two progressing cavity pumps, 3028 L/min (800 gpm) monsters "Netzsch".	Two centrifugal pumps. No problems specific to sludge pumping.	ABEL-diaphragm pumps	Moyno	ABEL-diaphragm pumps, four in parallel.	Unknown

(continued)

Table 6.5 Long-distance pipeline installations conveying dilute or thickened residuals (*Continued*).

Scum processing	Southport: scum towers, skim, hopper, dump truck to landfill. Belmont has a scum concentrator but also goes to landfill.	Digesters, centrifuges, 91 ML/d (24 mgd) solids treatment capacity.	Centrifuge to 5–6%		1° scum screened and mixed with sludge and pumped with combined sludge to Rancho.		From clarifiers to digesters.
Filtrate and centrate disposal	Recycle stream to headworks.	To aeration basins.	Back to plant via interceptor.	No dewatering.	Dedicated centrate line back to Tapia; some SCADA on this line.	To sewer.	To lagoons, or back to sewer.
Dewatered biosolids disposal	Incineration on site.	To land applications.	Agricultural	Subsurface injection	Centrifuge, then mix solids with wood to make compost. Compost is sold to public or to landscaping companies and used all over the area.	Trucked off site, nonfood agricultural or alternate daily cover at landfills.	Compost program, grassland agricultural, or "Biocycle Farm"; 60% farms, 40% trees.
Solids handling	Gravity thickening tanks, gravity belt thickeners, filter presses	Hoppers	25% cake to farmland	Anaerobic digesters and facultative sludge.	Sludge wetwell receives stream from Tapia, to thickener, to WAS, and decant in old digester.		
Why remote? Other options?	Belmont was the original facility built in 1927 with solids processing. Southport was built in 1965 with the original design intent of sending sludge to Belmont.	The Southeast plant never had solids treatment; when Greenfield was built it was designed to handle solids from SEWRP as well.	Agreement with city not to have biosolids facility at plant site.	The current FRP pipe replaced a leaking dual 250-mm (10-in.) ductile iron pipe pair (installed 1982) that was used for the same purpose. Real estate issues at Colorado Springs; needed more space.	Real estate issues, lack of space and hilly terrain prevented adding biosolids treatment to Tapia.	WWTP is on Navy base, no additional space available. Previously used solar drying on Fiesta Island and that was no longer available because of Coastal Commission.	BMF is large facility 10 ha (25 ac); no room at main plant.

Survey conducted in 2009 for the San Francisco Public Utilities Commission by Brown and Caldwell engineers Lloyd Slezak, Pete Bellows, Kevin Kai, and Scott Parr.

been a problem on some long sludge pipelines. Electrical return currents, acid soils, saline groundwater, and other factors may cause serious difficulty unless special corrosion control measures are used. Consultation of specialists on the need for cathodic protection is advised.

3. Adding controlled amounts of water is advised to dilute the sludge or flush the line. Primary effluent may be used in raw sludge pipelines; disinfected final effluent may be preferred for digested sludge pipelines. The water connection should have a flow rate indicator. The flushing water should flow at approximately 0.9 m/s (3 ft/sec).
4. Access for inserting and removing a cleaning tool ("pig") should be provided. This can be through the line if needed. Such cleaning may be required frequently if unstabilized sludge is pumped, even if scum is handled separately. If tool cleaning is to be used, then some additional recommendations apply:

    a. Valves must provide an unobstructed waterway to pass the tool.
    b. Flushing water pressure should be sufficient to push the tool through the full length of the pipeline.
    c. Pipe bend fittings should be 45 deg or, if possible, 22.5 deg. Some cleaning tools will pass 90-deg bends, but such bends are likely to be trouble spots. Length/radius of bends should be checked with the tool supplier.
    d. A recording or totalizing flowmeter should be provided. If the tool gets stuck in the line, the flow record can be used to compute the number of liters pumped since the tool was inserted. Thus, the tool can be located and retrieved.

5. The pipeline route should be selected for ease of maintenance.
6. At high points, air or gas relief valves should be provided. With care, automatic relief valves can be made reliable on digested sludge lines; however, in unstabilized sludge lines, grease and debris cause automatic valves to be unreliable. Simple manual blowoff valves typically are better for unstabilized sludge. Air and gases from sludge pipelines may be odorous. In confined spaces, the air or gas may also be toxic, flammable, explosive, and corrosive.

7. If sludge is to be pumped at more than about 3% solids, then pumps and pipeline should be designed for high and variable friction headlosses. Sludge may flow more like a Bingham plastic than an ordinary Newtonian fluid.
8. If centrifugal pumps are used with thin sludge, then flow rates will be somewhat unpredictable because of the varying flow resistance properties of the sludge. Storage provisions should be made for these variations. Pumps should be capable of operating at shutoff head with low flow during pipeline startup. Often with thin sludge, a variable frequency drive (VFD) and a magnetic flowmeter will allow slowing the speed of the pump to the desired flowrate.
9. Positive displacement pumps may experience difficulty when starting a long sludge pipeline. The thixotropic nature of sludge may cause high resistance to flow during startup. Consequently excessive pressures may be generated by positive displacement pumps. To avoid this problem, VSDs should be provided, and the pumps should be started at low speeds. An air chamber may be installed on the discharge side of the pumps; the chamber will assist in startup and will dampen pulsations. With digested sludge, a relief valve piped back to the digesters may be used near the pumps. Also, a high-pressure switch to stop the pump from running will prevent possible pump or piping damage from excessive pressures due a partial or total plug.
10. For long lines, a booster pumping station may be required. If positive displacement booster pumps are used, then a holding tank should be provided. It is practically impossible to match booster pumping rates to the sludge flow reaching the booster station unless centrifugal pumps are used.
11. Waterhammer is best controlled by limiting velocity. Unless a special evaluation is made, velocities should not exceed approximately 0.9 m/s (3 ft/sec). Even lower velocities may be required in some cases.

2.10 Case Studies of Dewatered Cake Pumping

Table 6.6 presents two installations where dewatered cake is pumped to storage or truck loading facilities. Both facilities use slip injection with water to reduce pipeline pressures.

3.0 CONVEYOR SYSTEMS

Dewatered (containing greater than 15% solids) or dried residuals, screenings, ash, and grit can be conveyed by most forms of industrial materials handling equipment, including belt, tubular and screw conveyors; slides and inclines; elevators; and

TABLE 6.6 Pumping installations conveying dewatered biosolids residuals.

	San Diego, California Metro Biosolids Center	Toronto, Canada Ashbridges Bay Treatment Plant
Description	Dewatered cake pumped to storage and truck load out facilities	Dewatered cake pumped to storage and truck load out facilities
Pumped fluids	28% anaerobically digested and dewatered (centrifuge) cake	28% anaerobically digested and dewatered cake
Other systems or utilities	Water injection lubrication rings; one just after the pump and one approximately 61 m (200 ft) downstream of the pump.	Water injection for lubrication
Piping age	1997	2001
Piping material	Unknown	Carbon steel
Piping length	122 m (400 ft) farthest	177 (580 ft) farthest
Piping diameter	Unknown	300 mm (12 in.)
Fluid velocities	Unknown	0.01 to 0.2 m/s (0.03 to 0.6 ft/sec)
Flowrates	Unknown	5 to 30 m^3/h
Pressures	13.8 kPa/m (2 psi/ft); has gone as high as 4826 kPa (700 psi) total pipeline pressure before lubrication kicked in; lubrication now starts when the pumps start.	12 to 60 kPa/m (0.53 to 2.65 psi/ft) w/o lubrication; with water lubrication 55 to 65% reduction is achieved
Operation and maintenance issues	Original control logic of the pumps was creating excessive wear. Now pump rebuilding occurs every 1500 hours	Unknown
Additional comments	Varying pump speed to match feed-chute level rather than cyclical operation; keeping the feed screw running avoids false low level readings in the feed chute by maintaining a more even level; hydraulic bypass mode is only necessary for pump safety controls and should not be used during normal operation; running the lubrication pump all the time when the cake pump is operating also reduces line pressure.	Unknown
Type of pumps	Hydraulically driven dual piston by Schwing	Hydraulically driven piston pumps by Schwing

pneumatic systems. Each may be used to advantage in certain applications. Because the consistency of wastewater solids is highly variable and because the solids are often difficult to move and may tend to flow, the design of this equipment must consider the most severe conditions that may be expected. Discussion about the different types of conveyors may be found in *Design of Municipal Wastewater Treatment Plants* (Water Environment Federation et al., 2009) or in *Conveyance of Residuals from Water and Wastewater Treatment* (American Society of Civil Engineers, 2000).

4.0 OVERVIEW OF DESIGN AND CONTROL CONSIDERATIONS FOR RESIDUALS CONVEYANCE SYSTEMS

When designing a residuals conveyance system the engineer should consider and determine parameters and properties that pertain to (1) the material being conveyed; (2) the conveyance system (e.g., piping); and (3) the type of conveyance equipment (e.g., pump, conveyor).

Design and control of a conveyance system could be complicated because of the wide variety of pumped materials and their characteristics. The overall conveyance system selected is strongly dependent on the type of material that needs to be transferred. Some properties of the material should, therefore, be monitored regularly regardless of the process selected. These properties include viscosity, solids content, moisture content, bound water content, specific gravity, and temperature and particle size, shape, and compaction rate. In addition to the material properties, parameters such as distance, elevation, pipe diameter, pipe losses and pressures, flow rate, velocity, and flow regime (continuous versus intermittent) should also be considered early on in the design of pumping systems.

Design considerations and process control measurements would differ based on the selected pumping and conveyance equipment. For instance, design and process control of a centrifugal pump that is used to transfer dilute residuals would be different than the design and process control of a belt conveyor that is used to convey granular and compactable residuals. Table 6.7 presents design and monitoring parameters for various types of pump and conveyance equipment that should be considered early on in the selection process and later during operation of the system. Parameters highlighted are monitored throughout the life of the residuals system. Additional information on process control may also be found in the various sections describing the different residuals conveyance equipment throughout this chapter.

Table 6.7 Design considerations and process control parameters for pumping and conveying equipment (**bold** = parameters that need to be monitored continuously) (NPSH = net-positive suction head).

Category	Type	Parameters
1. Pumping equipment	• Centrifugal pumps	• Pump capacity • Pump efficiency • NPSHr • NPSHa • Total dynamic head • Suction velocities should not exceed 1.5 m/s (5 ft/sec) and should not be less than 0.6 m/s (2 ft/sec) • Flushing connections • **Wet-well levels** • **Suction and discharge pressures** • **Operating speeds** • **Flowrate** • **Vibration**
	• Positive displacement pumps	• Pump capacity • Pump efficiency • Isolation valves on both the suction and discharge sides of the pump • Pretreatment of materials (such as screening or grinding) • Clearing debris from valves • **Flowrate** • **Discharge pressure** • **Operating speeds** • **Temperature** • **Wet-well levels, hopper levels, and material weight**
2. Conveyor equipment	• Belt conveyors	• Capacity • Appropriate cleaning system • Emergency-stop switches • Speed switches • Hose station and drains • **Traveling speeds** • **Level of oil and grease of pulley and sheave axles or shafts**

(*continued*)

TABLE 6.7 Design considerations and process control parameters for pumping and conveying equipment (**bold** = parameters that need to be monitored continuously) (NPSH = net-positive suction head) (*Continued*).

Category	Type	Parameters
	• Chain or drag conveyors	• Strength of the conveyor chain • Chain tension • Inspection ports, access doors • Bolted top covers • Wear bar thickness • Bolt torques • Alignment of head and tail shafts • **Traveling speeds** • **Weight of the material being conveyed** • **Bearing and chain lubrication level**
	• Bucket elevators	• Capacity • Periodic removal of accumulated materials • Chain strength and wearing qualities • **Traveling speeds**
	• Screw conveyors	• Speed switches • Internal hangers • Torsional capacity of the center shaft and couplings • All covers and lids • **Traveling speeds**
	• Pneumatic conveyors	• Capacity • Material to air/mass ratio • Bulk density of the material • **Conveying air velocity**

5.0 ODORS

Open sludge conveyance can be a source of odors. Adding an air extraction hood to capture odors at the emission point will prevent release of odors to a building and significantly reduce the volume of air needing scrubbing. All solids transporting facilities should be well-ventilated and, if necessary, provided with odor control for the vented air. Even with stabilized sludges, if large holding or equalization tanks are required for the pumping system, then covers or special odor-control facilities for venting tank air should be provided when the detention time is greater than several hours. Chapter 19 provides information on sludge storage and Chapter 20 on information on odor management.

6.0 REFERENCES

American Society of Civil Engineers (2000) *Conveyance of Residuals from Water and Wastewater Treatment*; ASCE Manuals and Report on Engineering Practice No. 98; American Society of Civil Engineers: Reston, Virginia.

Anderson, C. N.; Hanna, D. J. (1998) *System Design for Sludge Pumping, in Pumping Station Design*, 2nd ed.; Butterworth-Heinemann: Boston, Massachusetts.

Barbachem, M. J.; Pyne, J. C. (1995) Pipeline Hydraulics of Dewatered Non-Newtonian Cakes. *Proceedings of the 68th Annual Water Environment Federation Technical Exposition and Conference*; Miami, Florida, Oct 21–25; Water Environment Federation: Alexandria, Virginia.

Bassett, D. J.; Howell, R. D.; Haug, R. T. (1991) Hydraulic Properties Evaluation for Sludge Cake Pumping. *Proceedings of the 64th Annual Water Environment Federation Technical Exposition and Conference*; Toronto, Canada, Oct 10–14; Water Pollution Control Federation: Washington, D. C.

Battistoni, P. (1997) Pretreatment, Measurement Execution Procedure and Waste Characteristics in the Rheology of Sewage Sludges and the Digested Organic Fraction of Municipal Solid Wastes. *Water Sci. Technol.*, **36** (11), 33–41.

Bourke, J. D. (1997) *Handling High Solids Content Non-Newtonian Fluids.* Moyno Industrial Products: Springfield, Ohio.

Buckingham, E. (1921) On Plastic Flow through Capillary Tubes. *Proc. Am. Soc. Test. Mater.*, **21**, 1154.

Caldwell, D. H.; Babbitt, H. E. (1941) The Flow of Muds, Sludges and Suspensions in Circular Pipe. *Trans. Am. Chem. Eng.*, **37**, 237.

Carthew, G. A.; Goehring, C. A.; van Teylingen, J. E. (1983) Development Of Dynamic Head Loss Criteria for Raw Sludge Pumping. *J. Water Pollut. Control Fed.*, **55**, 472.

Fonda K., Rupik B. (2006) Have Your Cake and Pump It Too: Lessons learned from the San Diego Metro Biosolids Center Cake Pumping Operations.

Proceedings of the Water Environment Federation Specialty Conference—22nd Annual Residuals and Biosolids Conference; Water Environment Federation, Alexandria, Virginia.

Guibaud, G.; Dollet, P.; Tixier, N.; Dagot, C.; Baudu, M. (2004) Characterisation of the Evolution of Activated Sludges Using Rheological Measurements. *Process Biochem.*, **39** (11), 1803–1810.

Hanks, R. W.; Dadia, B.H. (1971) Theoretical Analysis of the Turbulent Flow of Non-Newtonian Slurries in Pipes. *Am. Inst. Chem. Eng. J.*, **17**, 554.

Honey, H.C.; Pretorius, W.A. (2000) Laminar Flow Pipe Hydraulics of Pseudoplastic-Thixotropic Sewage Sludges. *Water SA*, **26** (1), 19–25.

Kenny, J. P. (1969) Bulk Transport of Waste Slurries to Inland and Ocean Disposal Sites, Vol. III; Bechtel Corporation; National Technical Information Service: Washington, D. C.

Laera, G.; Giordano, C.; Pollice, A.; Saturno, D.; Mininni, G. (2007) Membrane Bioreactor Sludge Rheology at Different Solid Retention Times. *Water Res.*, **41** (18), 4197–4203.

List, E. J.; Hannoun, I. A.; Chiang, W. -L. (1998) Simulation of Sludge Pumping. *Water Environ. Res.*, **70**, 197–204.

Metcalf and Eddy, Inc. (2003) *Wastewater Engineering: Treatment and Reuse*, ***4th ed.;*** Tchobanoglous, G.; Burton, F. L.; Stensel, H. D., Eds.; 4th ed.; McGraw-Hill Inc.: New York, New York.

Mori, M; Isaac, J.; Seyssiecq, I.; Roche, N. (2007) Effect of Measuring Geometries and of Exocellular Polymeric Substances on the Rheological Behaviour of Sewage Sludge. *Chem. Eng. Res. Des.*, **86** (6), 554–559.

Mulbarger, M.; Samuel, C.; Copas, R.; Kordic, J. R.; Cash, F. M. (1981) Pipeline Friction Losses for Wastewater Sludges. *J. Water Pollut. Control Fed.*, **53** (8), 1303–1313.

Mulbarger, M. C. (1997) Selected Notions about Sludges in Motion, and Movers. *Proceedings of the Central States Water Environment Association Education Seminar*, Madison, Wisconsin; Water Environment Association: Madison, Wisconsin.

Murakami, H.; Katayama, H.; Matsuura, H. (2001) Pipe Friction Head Loss in Transportation of High-Concentration Sludge for Centralized Solids Treatment. *Water Environ. Res.*, **73** (5), 558–566.

Rabinowitsch, B. (1929) Über die Viskosität und Elastizität von Solen" *Z. Phys. Chem.*, **145** (A), 1.

Rimkus, R. R. Heil, W. R. (1975a) The Rheology of Plastic Sewage Sludge. *Proceedings of the 2nd National Conference on Complete Water Reuse: Water's Interface with Energy, Air and Solids*, Chicago, Illinois, May 4–8.

Rimkus, R. R. and Heil, W. R. (1975b) Breaking the Viscosity Barrier. *Proceedings of the 2nd National Conference on Complete Water Reuse: Water's Interface with Energy, Air and Solids*, Chicago, Illinois, May 4–8.

Setterwall, F. (1972) Discussion/ Communication on Pumping Sludge Long Distances. *J. Water Pollut. Control Fed.*, **44** (1), 648.

Sparr, A. E. (1971) Pumping Sludge Long Distances. *J. Water Pol. Control Fed.*, **43** (8), 1702–1711.

U.S. Environmental Protection Agency (1979) *Process Design Manual for Sludge Treatment and Disposal*, EPA 625/1-79-011; U.S. Environmental Protection Agency, Municipal Environmental Research Laboratory and Center for Environmental Research Information and Technology Transfer: Cincinnati, Ohio.

Water Environment Federation; American Society of Civil Engineers; Environmental and Water Resources Institute (2009) *Design of Municipal Wastewater Treatment Plants*, 5th ed.; WEF Manual of Practice No. 8;ASCE Manual and Report on Engineering Practice No. 76; McGraw-Hill: New York.

Wagner, R. L. (1990) *Sludge Digester Heating*. Alfa-Laval Thermal Co.: Ventura, California.

Williams, M. L. (1977) A Guide to the Specification of Glass Lined Pipe. *Water Sewage Works*, **124** (10), 76–81.

Willis J.; Nazareth, V.; Rupke, M.; Saknenko, V.; Witzgall, R. (2002) Field Data and Experience Summary from Ashbridges Bay Treatment Plant Biosolids Loadout Facility Expansion in Toronto, Canada. *Proceedings of the Water Environment Federation Specialty Conference—16th Annual Residuals and Biosolids Conference*; Austin, Texas; Water Environment Federation, Alexandria, Virginia.

World Pumps (1999) Ram Pumps Take the Drudgery out of Sludge Transfer. February 1999, 18–19.

Chapter 7

Chemical Conditioning

(continued)

1.0 INTRODUCTION

Conditioning does not reduce the water content of solids; it alters the physical properties of solids to facilitate the release of water during thickening and dewatering. Mechanical thickening and dewatering techniques would not be economical for a utility without solids conditioning beforehand.

Conditioning is chemical or thermal treatment used to improve the efficiency of thickening or dewatering processes. Chemical conditioning processes use inorganic chemicals, organic polymers, or both to improve thickening and dewatering characteristics of solids. Thermal conditioning uses heat to condition and stabilize solids.

This chapter discusses chemical conditioning and includes design considerations. Some thickening and most dewatering of wastewater residuals—particularly those containing solids from biological treatment processes (e.g., fixed-film and suspended growth activated solids treatment systems)—typically are not practical without some type of conditioning. This conditioning step can take the form of either a chemical or physical process.

Conditioning can significantly reduce the moisture content of the solids, depending on the thickening or dewatering process used, the characteristics of the incoming solids, and the method for thickening or dewatering. Typically, chemical conditioning can increase the residuals' solids content from approximately 1% to between 15 and 30%. Conditioning not only removes water but also increases the thickening or dewatering rate significantly by adjusting the chemical and physical properties of solids.

In the United States, the most typically used inorganic chemicals for conditioning solids are iron salts (e.g., ferric chloride) and lime. In Great Britain, the typical practice is to use ferrous sulfate with lime.

The use of organic polyelectrolytes (polymers) in wastewater treatment plants (WWTPs) was introduced during the 1960s and was rapidly adopted for both thickening and dewatering processes. The primary advantage of polymers is that they do not significantly increase solids production; every kilogram of inorganic chemicals added during conditioning will produce a kilogram of extra solids (if there is no chemical reaction, such as formation of calcium carbonate from lime). Inorganic chemicals (e.g., iron salts and lime) can increase the final product by 20 to 30% (dry solids), and fly ash can increase the final product by 50 to 100% (dry solids).

The conditioning method used must be compatible with the proposed thickening or dewatering process. For example, centrifuges use pressure to compact the solids, and belt filter presses (BFPs) permit the water to pass through void spaces; therefore, a single type of conditioning agent cannot be used for all applications (Water Environment Federation, 2003).

Often, management focuses on the wet end of the plant because the purpose of the plant is to receive wastewater and produce clean water. Dewatering operations are often treated as a consequence of producing a satisfactory plant effluent. To minimize the overall cost of treatment, the wet end needs to be operated with consideration for the dry end.

2.0 DIFFICULTY IN DEWATERING BIOSOLIDS

Sludge varies greatly regarding the ease with which it can be conditioned and dewatered. Research has explored how the various ways that water is held in sludge affects its dewaterability. Vesilind (1994) understood the complexity of sludge and presented several ways that water could be held by sludge. This model can be used to understand how sludge is generated and which treatment alternatives result in more difficult-to-dewater solids. More recent research suggests that sludge is more complicated than Vesilind's model; however, for the purposes of understanding why some sludges are more difficult to dewater than others, Vesilind's model is convenient to use.

2.1 Free Water

Free water is not connected in any way with sludge. This water is the easiest of all to remove. Typically, free water is removed in clarifiers and, therefore, does not reach the dewatering process. The exception is a plant where no attention is paid to a process, such as a clarifier skimmer pumping clear water to dewatering.

2.2 Bound Water

Bound water is water held in the particle either by chemical bonds or within a membrane. An example of this might be aluminum hydroxide, which precipitates as a hexahydrate. Each molecule of aluminum has six molecules of water attached to it.

Bacteria consist of a membrane encasing mostly fluid. Mechanical dewatering cannot break the membrane to release the fluid. Primary sludge can be dewatered to 35 to 40% solids, whereas the amount of activated sludge in the same plant rarely exceeds 15 to 18% solids. From this, it can be concluded that anything that increases the percentage of bacteria in the sludge will result in wetter sludge. Similarly, although eliminating primary clarifiers saves capital costs and simplifies the flow diagram, this practice greatly increases the difficulty of dewatering the resultant sludge. In effect, easy-to-dewater primary sludge is broken down by bacteria and, thus, increases the volume of bacteria going to the dewatering process.

Minneapolis Wastewater Treatment Plant, Minnesota, has published data that showed that the polymer dose increases and cake dryness decreases as the ratio of primary to secondary sludge goes from 50:50 to 30:70 (Minneapolis Wastewater Treatment Plant, 1999).

2.3 Vicinal Water

Vesilind (1994) postulated that water in the vicinity of the surface of a solid particle is attached to the particle by electrostatic and Van der Waals forces. As a result of these forces, this water is difficult or impossible to remove by mechanical means. From this, it can be concluded that the greater the surface area per kilogram (pound) of solids, the more vicinal water will remain in the dewatered cake. Under a microscope, it can be observed that waste activated sludge (WAS) has a lot of surface area—remarkably more than primary sludge. This profusion of surface area is another reason why secondary sludge is harder to dewater than primary sludge.

If a stone is broken into pieces, then the weight of stone is unchanged, but the surface area clearly increases. Typically, larger particles have less surface area per kilogram (pound) than smaller particles and thus have less vicinal water associated with them. Therefore, it is not good practice to make little particles from big particles, which occurs when primary clarifiers are absent or not operating properly.

2.4 Interstitial Water

Interstitial water is water trapped in the spaces between sludge particles. There are several mechanisms to explain the reduction in the amount of interstitial water. Anyone filling a canister with cereal has observed that shaking the canister causes the contents to move closer together and displace some of the interstitial air from between the cereal flakes. In this example, cereal does not change shape, and the packing density increases somewhat to reach a terminal packing density. If the cereal was flexible, then applying pressure to it would press the flakes closer together, causing them to conform to one another, and press the interstitial air out. In wastewater treatment, as sludge particles form a blanket in the clarifiers, the weight of one particle upon another squeezes the sludge particles together, and squeezes water out. As the sludge blanket becomes deeper, the cumulative weight of the particles on the lowest layer becomes greater, progressively more water is removed, and the sludge blanket becomes thicker. Following sludge to the dewatering process, filters or centrifuges apply a great deal more pressure to press the solids further together, expressing even more water from the sludge. In dewatering sludge in high solids centrifuges, various techniques are used to make the cake layer in the centrifuge as deep as possible. This increases the pressure of the sludge upon itself. Polymer can affect cake dryness (to obtain dryer cakes may require an increase in polymer dosage); however, high solids centrifuge operations should not consider polymer application the primary means to affect cake dryness. Simply adding more polymer without increasing cake depth does not significantly alter cake dryness.

2.5 Capillary Water

Capillary water is water held in crevices within sludge and is contained by capillary attraction. Typically, mechanical means cannot remove this water; although perhaps if the sludge particles are not too rigid, then squeezing the particles may crush the capillaries and extrude the water, much like squeezing toothpaste from a tube. Certainly, adding sufficient pressure to nonrigid particles could express capillary water out of the sludge.

Some water is easier to remove than other water, and considering this allows one to profitably speculate on ways to make sludge easier to dewater. Unfortunately, efforts to quantify the amount of water held in these various states have not been immediately useful, although the concept itself is quite helpful. For example, in observing that dewatering sludge from 98% water to a cake containing 72% water, the dewatering process has failed to remove the great majority of the water in the sludge. By expressing cake solids as a percent, it is implicitly suggested that 100% dryness is possible. Vesilind (1994) has pointed out that it is not possible to mechanically remove 100% of the water, or 50% of the water, or perhaps even as little as 33% of the water in the sludge. As with nearly everything, as efficiency improves, further improvements are progressively harder to achieve.

3.0 FACTORS AFFECTING CONDITIONING

The type and dosage of conditioning agent needed depends on the characteristics of the residuals, solids handling and processing before and after conditioning, and the mixing process after agent addition.

3.1 Residuals Characteristics

The following solids characteristics affect conditioning requirements:

- The source of residuals;
- Solids concentration;
- Alkalinity and pH;
- Biocolloids and biopolymer production;
- Particle size and distribution;
- Degree of hydration;
- Particle surface charge; and
- Volatile suspended solids (VSS) content.

3.1.1 Source of Residuals

To some extent, the conditioning method depends on the type of solids that must be treated. Municipal primary, secondary, combined, and digested solids are a good indicator of the range of probable doses of conditioning agents that are required for subsequent solids treatment. Cationic polymers work best on primary, secondary, and digested solids, and anionic polymers may be most effective with inorganic solids.

An examination of published data for a variety of thickening and dewatering devices suggests that primary solids require lower doses than secondary solids, and that fixed-film secondary solids require lower doses than suspended-growth secondary solids. Depending on the thickening or dewatering method used, aerobically and anaerobically digested solids require conditioning doses comparable to those for secondary solids. Similarly, combined solids (primary and secondary solids) have properties that are closer to those of secondary solids, although they are affected by the composition of each type. More importantly, the characteristics of solids from the same source vary from plant to plant and can vary seasonally. The conditioner dose depends on the specific conditioning agent used and the goal (thickening or dewatering solids), and may need to be changed seasonally or with variations in characteristics.

Chemical solids are solids that have been mixed with an inorganic conditioning agent (e.g., the addition of aluminum or iron salts and lime to precipitate phosphorus or to improve suspended solids removal). These types of solids are more variable than traditional primary and secondary solids and, as a result, dosing is hard to categorize. In addition, their conditioning requirements are often qualitatively different from those for primary and secondary solids. For example, adding lime to mixed-liquor suspended solids (MLSS) before secondary clarification may improve suspended solids removal; however, the resulting solids may require an anionic polymer, not the cationic one typically used for primary or secondary solids (i.e., the positively charged calcium has neutralized some of the negative surface charge). Although charge neutralization is a fundamental part of the process, an equally significant piece is the interconnection of solids particles via the polymer chain.

3.1.2 Solids Concentration

In many applications, conditioning neutralizes the colloidal surface charge by adsorbing oppositely charged organic polymers or inorganic complexes. The residuals' solids concentration will affect dosage and dispersal of the conditioning agent. Therefore, for a given particle-size distribution, increasing the suspended solids concentration increases the required coagulant dose for effective surface coverage (on a volumetric basis).

The suspended solids concentration also affects two additional aspects of conditioning agents. First, the process is less susceptible to overdosing at higher solids

concentrations. Second, the solids and the conditioning agent are more difficult to mix at higher solids concentrations.

3.1.3 *Alkalinity and pH*

When inorganic conditioning agents are used, alkalinity and pH are the most important chemical parameters affecting conditioning. Coagulation occurs when coagulant interacts with the surface of solids' colloids, and the nature of the charged surface and coagulant charge are both pH-dependent. When inorganic conditioners are used, the solids' pH determines which chemical species are present.

Iron and aluminum salts behave like acids when added to water (i.e., these conditioners reduce pH), so the pH of the conditioning process will depend on the solids' alkalinity and the dosage of iron or aluminum salts. This pH, in turn, determines the predominant coagulant species and the nature of the charged colloidal surface. The high alkalinity typically associated with anaerobically digested solids is one reason for the higher coagulant doses required. Low-molecular-weight coagulants tend to be more effective over a broader pH range than inorganic conditioning agents.

3.1.4 *Biocolloids and Biopolymers*

Although more commonly measured by researchers than by design engineers, these fundamental parameters provide some insight to the conditioning process. Biopolymers in activated solids flocs seem to affect the physicochemical properties of flocs (e.g., floc density, floc particle size, specific surface area, charge density, bound water content, and hydrophobicity).

Other studies have shown that cations can affect bioflocculation and change the settling and dewatering properties of activated sludge flocs (Bruus et al., 1992; Eriksson and Alm, 1991; Higgins and Novak, 1997a, 1997b). Divalent cations bridge across negatively charged biopolymers to form a dense, compact floc structure. Monovalent cations tend to prevent proper flocculation by forming a much weaker structure. As a result, divalent cations promote bioflocculation and produce subsequent improvements in settling and dewatering properties. Monovalent cations tend to degrade settling and dewatering properties. It seems that settling and dewatering properties are further improved when the divalent cations are added to the feed rather than superficially added to the settling tank (Higgins and Novak, 1997b).

A series of laboratory-scale studies was conducted using WAS to gain insight to the floc-destruction mechanisms that alter solids conditioning and dewatering properties after anaerobic or aerobic digestion. The data indicated that biopolymer was released from solids under both anaerobic and aerobic conditions, but much more was released under the former. In particular, four to five times more protein was released into solution under anaerobic conditions compared with aerobic conditions (Novak et al., 2003). Both the dewatering rate (as characterized by the specific resistance to filtration) and the polymer dose depend directly on the amount of biopolymer (protein + polysaccharide) in solution.

3.1.5 *Particle Size and Distribution*

Particle-size distributions affect the total particle surface area and the porosity of cakes formed from these particles. These properties affect required coagulant doses and dewaterability. Several researchers studied the effect of particle size on dewaterability and concluded that particle size was one of the most important parameters in determining dewaterability (Karr and Keinath, 1978; Novak et al., 1988; Sorensen and Sorensen, 1997). Smaller particles (colloidal and supracolloidal) can blind filters and solids cakes and deter the release of water from the solids cake (Novak et al., 1988; Sorensen and Sorensen, 1997). Another study suggested that an increase in floc density improves dewatering properties because of a decrease in bound water associated with the flocs (Kolda, 1995). These studies concluded that dewaterability improvements often associated with other factors (i.e., pH, mixing, biological degradation, and conditioning) could be explained by the effects of these factors on particle-size distributions.

3.1.6 *Degree of Hydration*

Excessive bound water has been suggested as the cause of dewatering difficulties. The percent of bound water associated with the floc also indicates the maximum dryness that can be achieved in the solids cake by mechanical means (Robinson, 1989). Additionally, Vesilind (1979) reviewed the work of several investigators on water distribution in activated sludge. This water was described as free, floc, capillary, and bound (particle) water. These categories are defined based on the amount of centrifugal acceleration required to release a given portion of water. Vesilind suggested that water distribution in a given solids could determine the applicability of a specific thickening or dewatering operation.

3.1.7 *Particle Surface Charge*

Solids particles (e.g., subcolloidal and macromolecular constituents) typically have a negative surface charge, so they tend to repulse each other. The resulting spaces between these constituents are occupied by cations and water. If the charge can be eliminated, then thickening or dewatering will improve. This is why chemical conditioners are positively charged or become positively charged when added to water.

In most cases, polymer conditioning is optimal when the charge is neutralized. Measuring charge, such as with a streaming current detector (zeta meter), can be useful in laboratory comparisons of polymers and doses. Meters can also be used to monitor or control polymer dose real-time in dewatering processes (Dentel et al., 1995).

Shear during mixing or dewatering tends to open up new negative surfaces in the biocolloids, thereby undoing the charge effects of cationic polymers. As a result, increases in mixing shear increases the required polymer dose (Dentel, 2001).

3.1.8 Wastewater Cations

Numerous studies have suggested that cations interact with the negatively charged biopolymers in activated solids to change the structure of the floc (Bruus et al., 1992; Eriksson and Alm, 1991; Higgins, 1995; Novak and Haugan, 1979; Tezuka, 1969). One study indicated that monovalent cations tend to deteriorate settling and dewatering characteristics, and divalent cations tend to improve them (Higgins, 1995). The effects of charge density on activated sludge properties could decrease the polymer dose needed to condition secondary solids.

Researchers have studied influent concentrations of cations (e.g., aluminum, ammonium, calcium, iron, magnesium, potassium, and sodium) extensively and have postulated that they significantly affect bioflocculation. Cations have been found to influence the thickening and dewatering characteristics of biological solids. For example, high concentrations of sodium typically resulted in poor dewatering; however, if the floc contained enough aluminum and iron concentrations, it typically offset the deleterious effects of sodium (Park et al., 2006). The data on aluminum revealed that WAS with low aluminum levels contained high concentrations of soluble and colloidal biopolymer (protein + polysaccharide), resulting in high effluent chemical oxygen demand (COD) concentrations, a need for larger doses of conditioning chemical, and poor solids dewatering properties (Park et al., 2006). Studies have shown that iron may contribute to floc strength and that the reduction and solubilization of iron during anaerobic digestion may explain why digested solids dewater poorly. Also, the presence of proteins in solution contributes to poor dewatering and larger conditioning chemical doses (Novak et al., 2001).

3.1.9 Rheology

Many studies have focused on the rheological properties of wastewater solids in an attempt to correlate solids properties with chemical conditioning requirements (Ormeci et al., 2004). For example, yield strength and viscosity have been used to optimize chemical conditioning. One study demonstrated that mixing considerably affected the rheological characteristics of conditioned solids (Abu-Orf and Dentel, 1999). Another showed that solids conditioning could be improved by monitoring centrate or filtrate viscosity (Bache and Dentel, 2000). In another study, both laboratory- and full-scale testing showed that the network strength of the sludge could be used to optimize chemical conditioning and achieve drier solids (Abu-Orf and Ormeci, 2005). Typically, the type of conditioner (e.g., polymers or fly ash) and solids (e.g., chemical or WAS) involved will determine which rheological parameters should be used.

3.2 Inorganic Chemical Addition

3.2.1 *Ferric Chloride and Alum*

Ferric and aluminum salts are often added in the wet end of the plant, because both react with the phosphate in the water to form an insoluble precipitate. These inorganic salts may also be added in chemically enhanced primary treatment to improve solids capture, including some colloidal solids.

The use of these chemicals in the wet end will have a noticeable effect on dewatering; the use of ferric chloride will improve the dewatering process, whereas aluminum will have either no change or result in sludge that is more difficult to dewater. The effect on dewatering will be proportional to the amount added; therefore, the dewatering operation needs to be advised of any changes in the dosing rate. Likewise, when comparing the cost of ferric chloride and alum, the effect on dewatering also needs to be considered. The addition of ferric chloride and aluminum salts causes the suspended particles to aggregate and settle out of the water. Unlike polymers, alum and ferric chloride do not require aging time, and special mixing devices are not used. Ferric chloride also reacts with the bicarbonate alkalinity in the solids to form hydroxides that also act as flocculants.

3.2.2 *Struvite Control*

Some anaerobic digesters have a problem with struvite, which is a crystal composed of ammonium magnesium phosphate. Struvite can precipitate out in the digester anywhere carbon dioxide can flash off or where there is a lot of turbulence. The crystals sometimes are a serious problem in the centrate/filtrate lines, where they can block a pipeline in one week. In theory, struvite can be controlled by altering the amounts of component ingredients or to bind up phosphate—a key ingredient of struvite. Ferric chloride is typically added because it reacts with phosphates and lowers the pH of the sludge. Ferric chloride does improve dewatering; however, it is expensive and adds to the weight of material hauled away.

Another approach to struvite control is to acidify the feed to dewatering, lowering the pH to the point where struvite will not form. Any acid will do (sulfuric, hydrochloric, or nitric). Sometimes spent acid can be found at a good price—even free—but it is important to check it to see if it contains anything, such as heavy metals, that might be a problem to the plant. Typically, adding acid to the sludge provides little benefit to dewatering. Ferric chloride typically improves dewatering, whether it is added in the primary treatment process or just ahead of the dewatering device. It is rarely cost effective to add it solely to improve the dewaterability of the biosolids. Ferric chloride is more expensive than other, more common acids, however, so its benefits must be weighed against its higher cost.

3.3 Handling and Processing Conditions before Conditioning

The efficiency of any conditioning process depends to a large degree on the chemical and physical characteristics of the solids (e.g., origin, concentration, inorganic content, chemistry, storage time, and mixing) before conditioning. The physical characteristics are a function of the physical stresses the solids were exposed to before conditioning. For example, any process that damages the flocculent nature of solids particles typically either increases chemical conditioning demand or reduces performance in the final treatment stage. The extent of mixing and shear stress before and after conditioning can significantly affect conditioning efficiency and, ultimately, solids treatment performance.

3.3.1 Storage

There are two types of storage for liquid residuals: long- and short-term. Long-term storage may occur in stabilization processes with long detention times (e.g., aerobic and anaerobic digestion) or in specially designed tanks. Short-term storage may occur in the wastewater treatment process (e.g., increasing solids inventory) or in smaller, specially designed tanks. Storage helps smooth out fluctuations in solids production, making the solids feed rate more uniform. It also provides a place to keep solids during equipment downtime. Long-term storage, however, has been reported to negatively affect the dewaterability of solids.

Unstabilized solids that have been stored for long periods typically require more conditioning chemicals than fresh solids because the degree of hydration and percentage of fine particles increased. Storage of mixed primary and secondary sludge for any significant time results in rapid development of anaerobic conditions, which leads to higher polymer demand and increases soluble organics and odors.

Storing activated sludge increased its specific resistance to filtration and, subsequently, conditioning requirements (Karr and Keinath, 1978). Storing aerobically or anaerobically stabilized solids for extended periods typically lowers temperature significantly and can change pH and alkalinity. Temperature drops typically increase conditioning requirements. If, however, the temperature decrease is small, then the negative effect on conditioning may be more than offset by an increase in solids concentration.

3.3.2 *Pumping*

Pumping subjects solids to shear forces; the level of shear depends on the type of pump and the flowrate. Solids particles are fragile, and pumping typically causes some of them to fragment. Researchers have shown that the significant demand for chemical conditioning is associated with the fraction of particles in the colloidal and supracolloidal range (Karr and Keinath, 1978; Roberts and Olsson, 1975), so any process that reduces particle size will increase conditioning chemical requirements.

Conditioned solids should not be pumped, because pumping introduces shear forces that tend to break down flocs. If required, however, then the pump should be designed to minimize shear.

3.3.3 *Mixing*

During conditioning, the solids and added chemicals must be mixed enough to ensure that the chemical is evenly dispersed throughout the solids. The mixer must not break the floc once it has formed, however. Design engineers should optimize the mixing time with these two goals in mind. Mixing requirements depend on the thickening or dewatering method used. Inline mixers typically are used with most modern thickening and dewatering units. A separate mixing-and-flocculation tank is provided with some older thickening and dewatering devices.

For many municipal solids, intense mixing should be followed by much gentler agitation to allow fine particles to flocculate into aggregates that settle or can be readily filtered. Once the solids and conditioning chemical have been mixed thoroughly, a hydraulic retention time of 15 to 45 seconds in the pipeline or flocculation tank will complete the flocculation process before solids enter the thickening or dewatering system.

3.4 Ultimate Disposal or Use of Solids

Title 40 of the *Code of Federal Regulations* addresses the use and disposal of solids generated during the treatment of domestic wastewater (40 CFR 253) (U.S.

Environmental Protection Agency, 1990). Wastewater solids disposed in municipal landfills or used as landfill cover material must comply with federal, state, and local requirements. For example, it is becoming increasingly common for local or state authorities to require that residuals contain 35 to 40% solids before they can be codisposed with municipal solid waste. Landfilled solids may have to meet certain levels of biological stability, soil engineering properties, or both. A sufficiently high dose of lime can both stabilize and condition solids, and a dose of lime, fly ash, or other bulking materials can improve a cake's mechanical properties. The mechanical strength of solids can be measured by a slump test (similar to that used for concrete).

The need to achieve a Class A biosolids through lime addition is more common in recent years. Lime and other alkaline agents, such as fly ash and kiln dust, require proper addition. There are two ways to admix lime.

The simplest method is the wet method, in which the pulverized lime (calcium oxide) is mixed with water (though ammonia can be released using this method), and the lime slurry is pumped into the solids feed line ahead of dewatering. This is a neat, trouble-free, low-maintenance system.

Alternately, the pulverized lime is conveyed by screw conveyor or air to a paddle mixer, where it is blended directly into the dewatered solids. This releases large amounts of ammonia, dust, and sulfides. Because operators cannot work under such conditions, an odor-control system is mandated.

From both an operating and an equipment point-of-view, the wet method is a much less expensive option than mixing powdered lime directly into the dewatered cake.

On the other hand, if biosolids will be land-applied or otherwise beneficially used, then the material must meet the pollutant concentrations, pathogen-reduction requirements, and vector-attraction reduction standards outlined in 40 CFR 503. Because of continuing public concerns about biosolids use, many farmers will only accept Class A solids, which affects overall conditioning and treatment options. Also, some farmers only accept solids that were treated with specific conditioners. In addition, biosolids characteristics and site conditions (e.g., groundwater and soils) may limit the use of certain conditioners and treatment methods. For example, certain crops are better cultivated in acidic soils, and land-applying lime-treated biosolids to such fields would not help the overall agricultural operation. The effects of liming before and after dewatering are outlined below.

- Liming before dewatering

1. When the lime (calcium oxide) reacts with solids, it absorbs one molecule of water to form calcium hydroxide and calcium carbonate. This water is taken from the feed and is water that otherwise would leave the dewatering system in the filtrate, but now moves to the cake and must be hauled away as solids.
2. The lime has surface area and, therefore, will carry surface moisture with it through the dewatering process.
3. Relatively few polymers react well with high-pH solids, and, when they do, polymer dosages are higher. This alone will complicate polymer selection and increase polymer costs.
4. The addition of lime before dewatering will increase the amount of lime needed because it must also increase the pH of the filtrate/centrate that recycles back to the plant.

- Liming after dewatering

1. When the lime (calcium oxide) reacts with solids, it absorbs one molecule of water to form calcium hydroxide and calcium carbonate. This water comes from the dewatered solids and will have to be hauled away.
2. Solids to the dewatering process are unchanged, so there is no problem with polymer selection and use.
3. Only the pH of the dewatered solids is raised, no lime is wasted on the filtrate/centrate.

In filtration devices, the addition of lime reduces the specific resistance to filtration, which, in turn, improves performance. This is an additional benefit not present for centrifuges. In water and wastewater centrifuges, filtration is not a separation mechanism; therefore, there is no such mitigation. In both cases, the amount of material to be hauled is greater with preliming—in part, because the lime must be hauled away, but also because the surface area of the biosolids is covered with vicinal or surface water. Adding lime increases the amount of surface area and, therefore, adds to the surface water that moves to the solids.

Cost should be the significant factor driving this decision. It is expensive to buy special equipment to add chemicals to sludge, especially if the chemicals increase the volume and weight of the sludge, and then the sludge must be discarded. Obvious upfront costs need to be balanced by disposal costs. Regardless of how the lime is added, the product is of value for most agriculture reuse. In addition, limed sludge can be stored for a long time without odor problems. This is a great benefit when the recycle of biosolids to land is not possible during the winter or when crops are in the field.

4.0 TYPES OF CHEMICAL CONDITIONING

Solids can be conditioned via chemical, heat, and freeze–thaw methods. The most popular is chemical conditioning using polymers, inorganic chemicals, or both. Heat treatment and freeze–thaw conditioning have been used to a limited extent, but their use has declined in recent years because of problems.

In the mid-20th century, wastewater treatment professionals used various inorganic chemicals and natural organics to condition solids. The most common inorganic chemicals used were lime and iron salts.

4.1 Inorganic Chemicals

Inorganic chemical conditioning is primarily associated with recessed-plate filter presses, although it also has been used for BFPs. Ferric chloride and lime are most typically used. Compared to polymers, larger doses of inorganic chemicals are required to condition solids, and this affects the volume and mass of solids to be managed. For example, adding iron salts and lime can increase the solids mass (and volume) (Water Environment Federation, 2003).

Lime and liquid ferric chloride are the two most widely used inorganic conditioning agents for recessed-chamber filter presses. These conditioning agents are readily available and can condition a wide range of solids. In addition, the resulting biosolids are suitable for land application or composting.

Less commonly used inorganic coagulants include liquid ferrous sulfate, anhydrous ferric chloride, aluminum sulfate, and aluminum chloride. Other inorganic materials (e.g., fly ash, power plant ash, cement kiln dust, pulverized coal, diatomaceous earth, bentonite clay, and sawdust) have been used to improve dewatering; increase cake solids; and, in some cases, reduce the required dosage of other conditioning agents.

In addition to increasing the volume of solids to be managed, inorganic chemical conditioners reduce the heat value of the solids. Cake combustibility depends, however, on the ratio of water to dry volatile solids, not the level of chemical precipitates in the cake.

4.1.1 *Lime and its Characteristics*

Some plants use lime to control pH and improve settling in wastewater treatment processes and to condition and stabilize solids. Lime is commercially available in two main dry forms: pebble quicklime (CaO); and powdered hydrated lime [$Ca(OH)_2$]. In either form, lime is caustic, tends to produce dust, and tends to precipitate when slurried, forming a calcium-carbonate scale on conveyance equipment.

As a conditioner, lime typically is used to raise the pH, which was lowered by ferric-chloride addition. It also forms calcium-carbonate and calcium-hydroxide precipitates, which improve dewatering by acting like bulking agents, increasing porosity while resisting compression. Dissolved calcium hydroxide is less available at high pH levels.

Quicklime is typically 85 to 95% pure and often is called calcined lime because it is manufactured by burning crushed limestone (calcium carbonate) in high-temperature kilns to drive off carbon dioxide, leaving calcium oxide (quicklime). It typically is purchased in pebble form to minimize dust problems during handling. It rarely is applied in dry form, however, except to stabilize solids. Instead, it is mixed with water and converted to the more reactive hydrated form (calcium hydroxide) before application. This hydration reaction (called slaking) emits heat as part of the reaction:

$$CaO + H_2O \rightarrow Ca(OH)_2 + \text{heat} \tag{7.1}$$

The quicklime pebbles rupture during slaking, splitting into microparticles of hydrated lime that have a large total surface area and are highly reactive. A high-grade quicklime produces a quick-slaking, highly reactive, calcium-hydroxide slurry; a low-grade quicklime produces a slow-slaking, less-reactive slurry (see Table 7.1). Low-grade quicklime requires critical water control to maximize slaking efficiency and to minimize calcium-hydroxide particle size.

Quicklime must be stored under controlled conditions because prolonged contact with carbon dioxide in moist air causes quicklime to air-slake, cake, and become less reactive. Likewise, air or excessively hard water (alkalinity more than 180 mg/L as calcium carbonate) in a hydrate slurry encourages carbonate scaling, which can lead to plugging problems in conveyance pumps and piping.

For quality control, quicklime should be highly reactive, quick-slaking, and able to disintegrate without producing objectionable amounts of dissolved or unslaked products. Medium-slaking limes are not preferred. Low-slaking and run-of-kiln quicklimes are unacceptable.

TABLE 7.1 Relative slaking ability of quicklimes (National Lime Association, 1982).

CaO content, %	Degree of burn	Slaking ability
High	Soft	Very quick
	Normal	Medium
	Over to hard	Medium to slow
Medium	Soft	Quick to medium
	Normal	Medium
	Over to hard	Slow
Low	Soft	Quick to medium
	Normal	Medium
	Over to hard	Slow to very slow

Hydrated lime is a powdered form of calcium hydroxide; its composition and characteristics depend on the quality of its parent quicklime (see Table 7.2). Hydrated lime typically costs 30% more than quicklime with the same calcium-oxide content because of its higher production and transportation costs. At small plants where daily requirements for lime are intermittent or minimal, hydrated lime is preferred because it does not require slaking. The storage and mixing operations are relatively simple (e.g., typically a dedicated storage area and minimal manual labor). Hydrated lime is more stable than quicklime, and storage precautions are satisfied more easily. Because of its dusting characteristics, however, handling is more difficult.

4.1.2 Ferric Salts

Both ferric chloride and ferric sulfate react with the bicarbonate alkalinity in solids to form ferric hydroxide precipitates. The precipitate can lead to both charge neutralization and floc aggregation. The chemical reaction may be written as follows

$$Fe + 3H_2O \rightarrow Fe(OH)_3 + 3H \qquad 7.2$$

$$2FeCl_3 + 3Ca(HCO_3)_2 \rightarrow 2Fe(OH)_3 + 3CaCl_2 + 6CO_2 \qquad 7.3$$

The acidity in the product and chemical reactions cause the pH to drop to 6.0. Adding lime raises the pH as high as pH 8.5, thus allowing the ferric chloride reaction to be more efficient in forming hydroxides. Lime also reacts with bicarbonate to

TABLE 7.2 Characteristics of quicklime and hydrated lime (gal × 3.785 = L and lb × 0.4536 = kg) (Wang, et. al., 2007).

Material	Available forms	Containers and requirements	Appearances and properties	Weight	Commercial strength	Solubility in water
Quicklime	Pebble, 6–19 mm	80–100 lb moisture-proof bags, barrels, and container cars. Store dry, maximum 60 days in tight container and 3 months in moisture-proof bag.	White (light grey to tan). Lumps to powder. Unstable caustic irritant slakes to hydroxide slurry evolving heat. Saturated volume pH approximately 12.5.	3.4–4.7 kg/m^3 sp gr: 3.2–3.4	70–96% CaO	Reacts to form $Ca(OH)_2$; 1 lb of quicklime will form 1.16 to 1.32 lb of $Ca(OH)_2$ with 2–12% grit, depending on the purity.
Hydrated lime	Powder, <200 mesh	50-lb bags, 100-lb barrels, and container cars. Store dry, maximum 1 year.	White. Powder free of lumps. Caustic dust irritant absorbs H_2O and CO_2 to form $Ca(HCO_3)_2$. Saturated volume pH approximately 12.4.	1.6–2.5 kg/m^3 sp gr: 2.3–2.4	82–98% $Ca(OH)_2$ 62–74% CaO	10 lb/1000 gal at 70 ºF 5.6 lb/1000 gal at 175 ºF

form calcium carbonate, a granular structure that provides the porosity needed to increase the water-removal rate during pressure filtration. This chemical reaction is as follows

$$Ca(OH)_2 + Ca(HCO_3)_2 \rightarrow 2CaCO_3^- + 2H_2O \quad (7.4)$$

Depending on the type of solids involved, the ferric-chloride dosages range from 2 to 10% (dry solids basis), and lime dosages range from 5 to 40% (dry solids basis). Activated sludge requires high ferric-chloride dosages, anaerobically digested solids require midrange dosages, and fresh raw primary solids require low dosages (see Table 7.3). Typically used to flocculate solids, ferric chloride is sold as an orange-brown liquid containing between 30 and 45% (by weight) ferric chloride. At 30 °C (86 °F) and 1.39 specific gravity, a 30% ferric-chloride solution typically contains 1.46 kg (3.24 lb) of ferric chloride per 3.8 L (1 gal). Liquid ferric chloride is corrosive, so it must be handled and stored properly. In colder climates, for example, shipping strength is reduced to prevent a crystalline hydrate from forming on cold rail cars.

Ferric-chloride coagulation is pH-sensitive and works best above pH 6. Below pH 6, floc formation is weak and dewaterability can be poor. Lime is used to adjust the pH to optimize ferric-chloride use and solids dewatering. There may be situations in which ferric chloride is effective at a pH of less than 6, but these exceptions depend on solids type and cake-dryness requirements.

Liquid ferric sulfate typically is sold as a reddish-brown liquid containing 50 to 60% ferric sulfate. It is a cationic coagulant and flocculant typically used with another conditioning agent (e.g., lime or polymers). It has been reported that using ferric sulfate before solids thickening or dewatering will reduce the amount of polymer needed and improve the filtrate or centrate quality. Its use as a solids conditioner, however, is limited; it primarily is used in water and wastewater treatment to remove turbidity, color, suspended solids, and phosphorus.

Ferrous sulfate (also called *copperas*) is similar to ferric chloride in terms of handling, storage, and stoichiometry; however, its use as a solids conditioner has been limited in the United States. Ferrous sulfate ($FeSO_4 + 7H_2O$) is available in granular form in bags, barrels, and bulk. The product has a bulk density of approximately 1000 to 1100 kg/m^3 (62 to 66 lb/cu ft). Dry ferrous sulfate will begin to cake when stored at temperatures higher than 20 °C (68 °F) and will further oxidize and hydrate in moist and humid conditions. Ferrous sulfate should be stored in a dry area, and care should be taken to control dust, which can stain and irritate skin, eyes, and the respiratory tract. Ferrous sulfate forms an acidic solution, and manufacturer precautions should

TABLE 7.3 Typical dosages of ferric chloride and lime for dewatering municipal wastewater sludges (U.S. EPA, 1979) (WAS = waste activated sludge; TF = trickling filter).

Application	Sludge type	Ferric chloride, g/kg*	Lime g/kg*
Vacuum filter	Raw primary	20–40	70–90
	Raw WAS	60–90	0–140
	Raw (primary + TF)	20–40	80–110
	Raw (primary + WAS)	22–60	80–140
	Raw (Primary + WAS + septic)	25–40	110–140
	Raw (Primary + WAS + lime)	15–25	None
	Anaerobically digested primary	30–45	90–120
	Anaerobically digested (primary + TF)	40–60	110–160
	Anaerobically digested (primary + WAS)	30–60	140–190
Recessed-chamber filter press	Raw primary	40–60	100–130
	Raw WAS	60–90	180–230
	Anaerobically digested (primary + WAS)	40–90	100–270
	WAS + TF	40–60	270–360
	Anaerobically digested WAS	70	360
	Raw primary + TF + WAS	75	180

* All values shown are mass of either $FeCl_3$ or CaO per unit mass of dry solids pumped to the dewatering unit.

be followed when storing, feeding, and transporting the material. Ferrous sulfate in granular (dry) form may be fed using gravimetric or volumetric feeding equipment; it also may be fed as a solution.

The effectiveness of ferric coagulation depends on pH and alkalinity. A lower pH favors the formation of positively charged hydroxoiron (III) complexes; a higher pH favors the solid species $Fe(OH)_3(s)$ (see Figure 7.1). Because hydroxoiron (III) complexes are effective coagulants, a lower pH should produce better results (see Figure 7.2). Tenney et al. (1970) found that ferric iron was most effective between pH 5 and 8, which is near the pH of maximum precipitation (Figure 7.1).

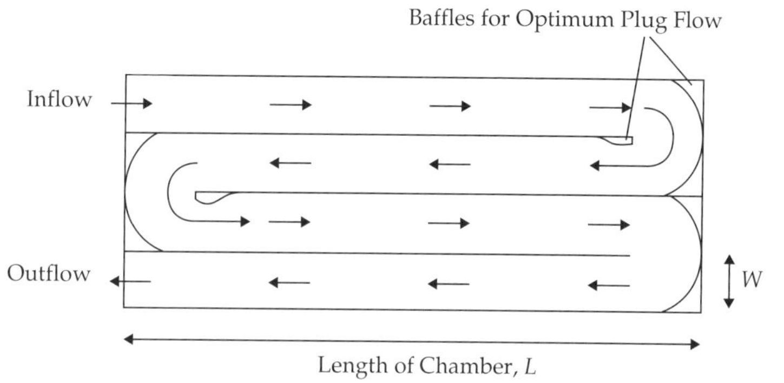

FIGURE 7.1 Equilibrium concentrations of hydroxoiron (III) complexes in a solution in contact with freshly precipitated $Fe(OH)_{3(s)}$ at 25 °C (From *Water Chemistry*, Snoeyink, V. L., and Jenkins, D., Copyright © 1980. Reprinted by permission of John Wiley & Sons, Inc.).

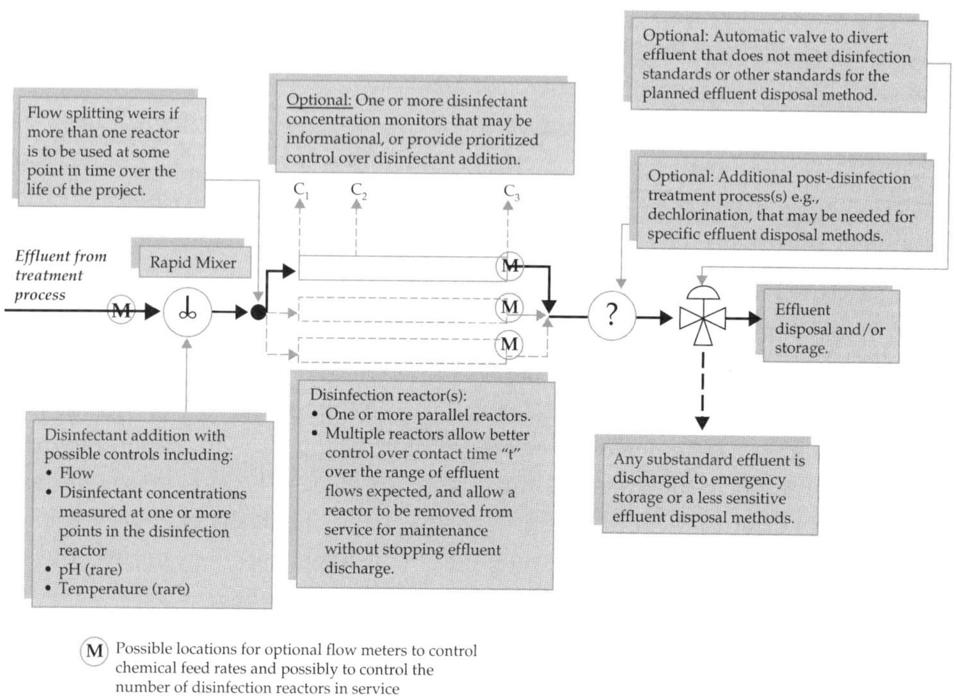

FIGURE 7.2 Effectiveness of ferric iron as a function of pH (Tenney et al., 1970).

Alkalinity is important in ferric-solids conditioning because it controls pH during conditioning. The ferric ion functions as an acid, lowering pH, while alkalinity maintains the existing pH. For a given solids, the pH decreases as ferric doses increase.

4.1.3 *Ferric Salts with Lime*

Precipitation of $Fe(OH)_3$ can neutralize charge and lead to effective aggregation and filtration in pH range 6 to 8. Practically, however, the majority of wastewater solids cannot be adequately conditioned unless lime is added after ferric salt (Christensen and Stulc, 1979). Full-scale filtration involves much more pressure than that typically applied in laboratory tests, which is why lime is required at full scale (Figure 7.2). The key ingredients are a pH of 11 to 12, a high calcium ion concentration (10^{-2} M), and the presence of solid ferric species (Christensen and Stulc, 1979). Conditioning with ferric salts and lime is not practiced in centrifugation because the solids cannot resist the imposed shear stresses but corrode and abrade metallic surfaces. Ferric should be added before lime at a separate point because adding ferric and lime to thickened solids together in the same tank (or in close proximity) adversely affects ferric conditioning (Christensen and Stulc, 1979; Webb, 1974). When conditioning with both ferric and lime, it typically takes 2 to 4 times more quicklime than ferric chloride to reach a pH between 11 and 12.

The choice of ferric salt for conditioning is more significant when the ferric salt is followed by lime (see Table 7.4). Ferric sulfate followed by lime deteriorates more rapidly and produces a poorer result than ferric chloride followed by lime. The deterioration of solids dewaterability seems to be associated with the formation of insoluble calcium sulfate.

4.1.4 *Aluminum Salts*

Aluminum salts typically are not used for solids conditioning in the United States, although they have been used at some facilities with limited degrees of success. New age coagulants (e.g., polymerized aluminum chloride and aluminum chlorohydrate) that are widely used in the water treatment industry also are being used in the wastewater treatment industry for phosphorus removal; for chemically enhanced primary treatment (CEPT) and ballasted sedimentation systems; and, to a limited degree, solids conditioning. Although aluminum salts are not widely used as a solids conditioner in the United States, aluminum chlorohydrates have been a popular conditioner in Great Britain for some time.

Table 7.4 Comparison of iron conditioners used with and without *lime* (from Lewis, C. J., and Gutschick, K. A. [1988] *Lime in Municpal Sludge Processing*. Natl. Lime Association, Washington, D.C., with permission).

Sludge[a] total solids, %	Iron conditioner	Iron dose % Fe	CST[b] after iron, seconds	Lime dose, % CaO	Specific resistance after lime, Tm/kg[c]
5.5	$FeSO_4{\cdot}7H_2O$	1.72	208	15	1.40
5.5	$FeCl_2{\cdot}4H_2O$	1.72	157	15	0.79
5.5	$Fe_2(SO_4)_3{\cdot}6H_2O$	1.72	41	15	0.50
5.5	$FeCl_3{\cdot}6H_2O$	1.72	26	15	0.26
5.5	$FeSO_4{\cdot}7H_2O$	3.44	180	30	0.60
5.5	$FeCl_2{\cdot}4H_2O$	3.44	139	30	0.29
5.5	$Fe_2Cl_4{\cdot}6H_2O$	3.44	27	30	0.23
5.5	$FeCl_3{\cdot}6H_2O$	3.44	19	30	0.12
7.0	$FeSO_4{\cdot}7H_2O$	3.44	480	20	1.1
7.0	$FeCl_2{\cdot}4H_2O$	3.44	—	20	0.56
7.0	$Fe_2(SO_4)_3{\cdot}6H_2O$	3.44	117	20	0.53
7.0	$FeCl_3{\cdot}6H_2O$	3.44	58	20	0.18

[a] Mixture of raw primary sludge and waste activated sludge.

[b] Capillary suction time.

[c] Terameters/kg (Tm/kg) = 10^{12} m/kg

The primary differences between aluminum and ferric chemistry are the relative solubility of aluminum above pH 7 and the relative insolubility of ferric above pH 7. The practical significance is that ferric hydroxide is relatively insoluble at the highest pH values used in ferric and lime conditioning (i.e., pH 12 to 12.5), whereas aluminum hydroxide is soluble above pH 10. Aluminum salts, therefore, are unlikely to be effective with the same lime doses often used with iron salts.

4.1.5 *Process Design Considerations*

The following subsections describe the use of inorganic coagulants for thickening and dewatering. Because the inorganic conditioners discussed in this chapter increase the total solids to be managed by approximately 20 to 40%, their use in thickening and dewatering applications is limited. Therefore, only one thickening and two dewatering applications that may use inorganic chemicals are discussed.

4.1.5.1 *Gravity Thickening*

Gravity thickening characteristics depend on the concentration and flocculant nature of the solids being thickened. In many cases, conditioning agents are not used; it depends on the type of solids being thickened. Although polymers are the first choice if chemical conditioning is required, alum and ferric salts—with or without lime—also could be used (see Table 7.5). Lime also could be important for odor control because it acts as a bactericide on fresh solids before land application.

The primary mechanism when using these inorganic chemicals is coagulation and flocculation. Efficient flocculation increases solids loading rates, improves solids capture, improves supernatant clarity, and may increase underflow concentrations from conventional gravity solids thickeners. Inorganic conditioning agents also increase the dry solids volume by approximately 20 to 30%. When designing any thickener, engineers should determine (whenever possible) appropriate coagulants and their dosage rates by using bench-scale tests to evaluate the effectiveness of conditioning agents during thickening operations.

4.1.5.2 *Recessed-Plate Filter Press Dewatering*

Recessed-plate filter presses are one of the oldest dewatering devices and can produce the highest cake solids concentration of any mechanical dewatering equipment (Kemp, 1997). They are used more often in industrial applications than in municipal WWTPs. Unless the inorganic content of the feed solids is high, conditioning

TABLE 7.5 Typical chemical dosages for gravity sludge thickening (WAS = waste activated sludge).

	Nature of solids/dosage of chemical			
	Raw		Anaerobically digested	
Solids	**$FeCl_3$ (mg/L)**	**CaO (mg/L)**	**$FeCl_3$ (mg/L)**	**CaO (mg/L)**
Primary	1–2	6–8	1.5–3.5	6–10
Primary + trickling filter	2–3	6–8	1.5–3.5	6–10
Primary + WAS	1.5–2.5	7–9	1.5–4	6–12
WAS	4–6	No data	No data	No data

chemicals are required for successful filter press dewatering (Kemp, 1997). Recessed-plate filter presses used to rely on lime and ferric chloride for conditioning. Although these chemicals typically produced a dewatered cake containing more than 40% solids, they increased the mass to be stored, transported, and used or disposed. Lime also is associated with ammonia releases, which must be considered in overall facility design, especially ventilation and odor-control requirements.

To produce a low-moisture cake via a recessed-plate filter press, biological solids first must be conditioned with lime and ferric chloride, polymers, or a polymer combined with either inorganic chemical. (Using only polymer tends to decrease unit performance.) Proper conditioning to a specific resistance of approximately 1×10^{12} m/kg or less typically is required for good dewatering. The specific-resistance test should be used to determine whether the solids are properly conditioned and to evaluate and improve various combinations of conditioning chemicals. This test is reliable but time-consuming. If a quick field test is needed, then the Buchner funnel test can be conducted onsite; if 200 mL can be dewatered in 100 seconds or less, then solids should be filterable.

Thin, poorly conditioned solids will approach design pressure rapidly (approximately 5 to 10 minutes); at this point, the sticky, hard-to-handle cake will not release, and the filtrate is inconsistent and of poor quality. However, thin cake continues to dewater if pressed for an extended period, yielding a good discharge from the filters. A properly conditioned sludge builds up pressure slowly, which indicates little resistance to filtration (U.S. Environmental Protection Agency, 1979). The most efficient use of chemicals in the dewatering process occurs when thickening before dewatering has been optimized.

Most researchers have found that pumping, storage, and application methods for conditioned solids significantly affect the amount of conditioning chemical needed and the performance of the dewatering process. Conditioned solids may be excessively agitated in either the conditioning tank or the surge tank. The conditioning tank should provide good mixing without shearing the floc. The surge tank should equalize conditioned solids within a maximum detention time of 30 minutes before feeding to the press. Engineers also should design the solids feed pump to minimize floc shearing.

The chemicals typically used to condition solids for a recessed-plate filter press unit are lime and ferric, either alone or with fly ash or polymers. The required dosage is approximately 3:1 [70 to 150 g/kg (140 to 300 lb/ton) of lime and 20 to 50 g/kg (40 to 100 lb/ton) of ferric chloride]. Ferric sulfate may be substituted for ferric chloride;

however, it typically requires a higher dosage. Actual dosage requirements for inorganic conditioners depend on the ratio of secondary solids to primary solids and on the percentage of solids fed to the dewatering process. Cassel and Johnson (1978) have shown that once the ratio of secondary to primary solids is more than 1:1, secondary solids become the controlling factor in dewaterability. Because the percentage of feed solids directly affects dewatering, it is important to perfect the thickening process.

Table 7.6 summarizes the performance of various types of domestic wastewater solids conditioned with 10 to 30% lime and 5 to 7.5% ferric chloride (dry weight basis). Although dewatered cake is a drier cake containing up to 45% dry solids, when lime and ferric chloride are used as conditioners, a significant portion (15 to 40%) of cake solids will offset the weight reduction of high water-removal efficiency.

4.1.5.3 Belt Filter Press Dewatering

Inorganic chemicals typically are not used to condition solids before dewatering in a BFP, and it is not recommended because of chemical deposits that can "blind" the belt and cause excessive wear on the rollers and belt, reducing the equipment's life expectancy. As a result, information is limited on using inorganic conditioners with belt filter presses. Alum is sometimes used, and other inorganic chemicals (e.g., lime) may be important for chemical stabilization before land application.

As with other dewatering processes, the optimal dose depends on feed solids concentration and type, mixing intensity, and mixing time. The limited information available indicates that the chemical dosage required varies directly with the ratio of secondary to primary solids. At a secondary-to-primary ratio of 1:1, an approximate dosage of 5% for ferric chloride and 15% for lime can be expected. Doubling the

TABLE 7.6 Recessed-plate filter press dewatering experience (WAS = waste activated sludge) (U.S. EPA, 2000).

Type of wastewater solids	Feed total solids (%)	Typical cycle time (hours)	Dewatered cake total solids (%)
Primary + WAS	3–8	2–2.5	45–50
Primary + WAS+ trickling filter	6–8	1.5–3	35–50
Primary + WAS + $FeCl_3$	5–8	3–4	40–45
Primary + WAS + $FeCl_3$ (digested)	6–8	3	40
Tertiary with lime	8	1.5	55
Tertiary with alum	4–6	6	36

secondary-to-primary ratio to 2:1 could double the required lime and ferric-chloride dosage.

Although using inorganic coagulants may be advantageous for routinely dewatering a widely varying sludge, this will cause a significant increase in solids needing disposal, leading to greater hauling, handling, and use or disposal costs. Also, design engineers should consider ventilating the belt filter press rooms because of the strong ammonia odor that may result from lime addition.

4.2 Organic Flocculants

4.2.1 Background

Organic flocculants are widely used in many industries and processes involving the separation of solids from liquids. These liquid-solid separation applications may involve processes for the recovery of finished products, clarification or purification of liquids, and volume reduction of waste materials. Several significant applications are identified in Figure 7.3, which also indicates the relative electronic charge and molecular weight of organic flocculants typically used in the respective application.

Figure 7.4 shows the various application points for organic flocculants within the typical municipal WWTP operation.

Initially, ferric chloride and lime were used to condition wastewater residuals before thickening and dewatering. In today's sophisticated municipal dewatering operations, however, these products are rarely used, with the exception of any remaining vacuum filtration and recessed-plate filtration processes.

Organic polyelectrolytes are used in liquid-solid separation. Thickening and dewatering processes, in particular, are completely dependent on their use. Several factors have contributed to the widespread use of cationic polyacrylamide (PAM) flocculants as conditioners for sludge dewatering:

- Development of more sophisticated dewatering equipment, such as BFPs and centrifuges;
- Technological advances in polymer chemistries to provide more convenient, user-friendly, and high-performance products;
- Development of convenient and effective polymer make-down units; and
- Increased emphasis on the costs of solids processing and disposal.

Because of the complex and proprietary nature of PAM flocculants and their varying levels of performance or effectiveness with different sludges, each manufacturer

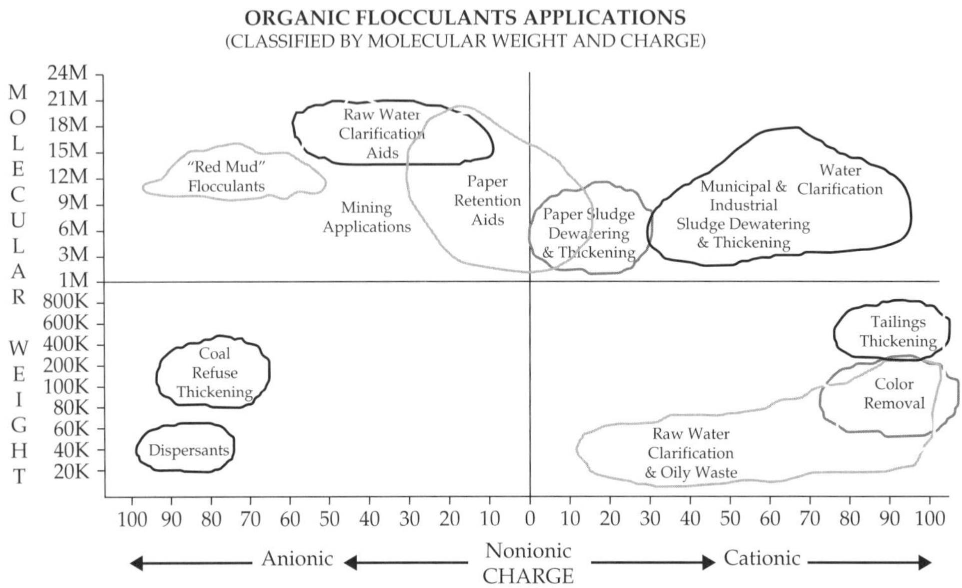

FIGURE 7.3 Typical points of addition of organic flocculants in wastewater treatment (WEF, 2009).

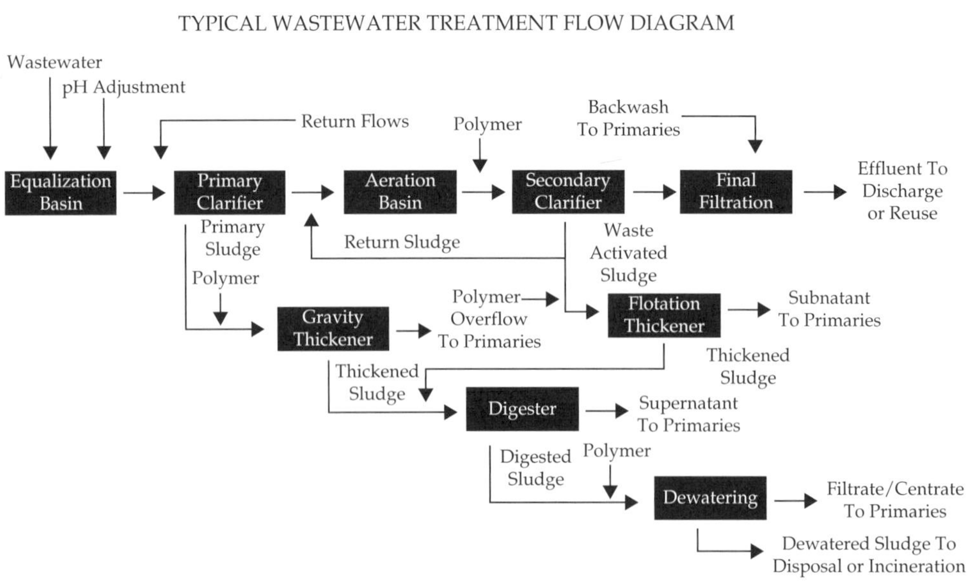

FIGURE 7.4 Typical wastewater treatment flow diagram (WEF, 2008).

offers a full range of products based on product characteristics and forms. The remainder of this discussion will focus on cationic PAMs, because they represent 95 to 100% of the organic flocculants for conditioning sludge before dewatering.

4.2.2 *Polymer Characteristics*

The product characteristics of these complex and proprietary PAM flocculants may vary according to the following:

- Electronic charge (anionic, non-ionic, or cationic);
- Charge density;
- Molecular weight (standard viscosity); and
- Molecular structure.

As shown in Figure 7.5, the "families," or types, of organic flocculants are depicted according to molecular weight and electronic charge and charge density. It is useful to review Figure 7.5 with Figure 7.3 to identify the family of organic flocculants that is frequently used in each application.

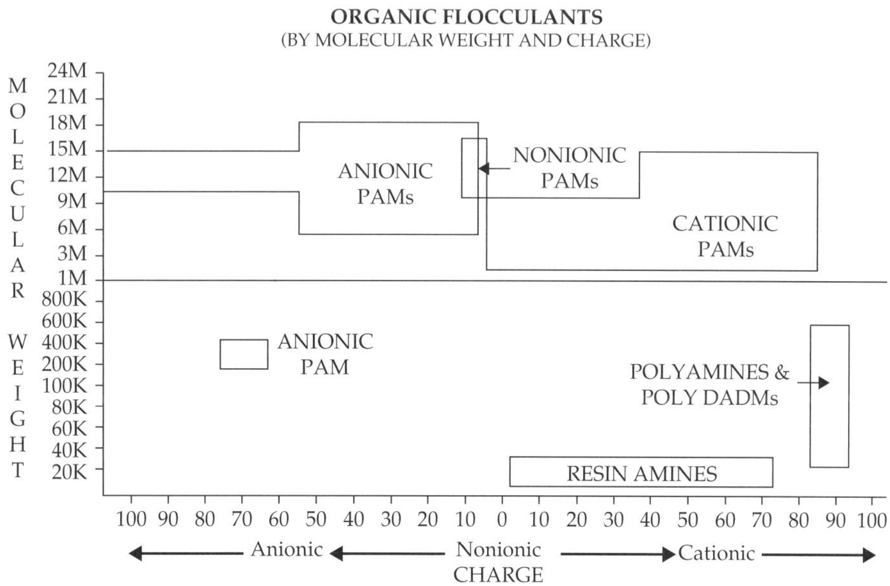

FIGURE 7.5 Types of organic flocculants according to electronic charge density and molecular weight (WEF, 2008).

4.2.2.1 *Electronic Charge (Type)*

A variety of organic flocculants are manufactured with three distinct types of available electronic charge—anionic (negative), cationic (positive), and non-ionic (no charge)—each suitable for specific liquid-solid separation applications. For conditioning wastewater residuals before dewatering, virtually all organic flocculants used are cationic PAMs. An unlimited number of cationic PAM flocculants can be produced. A variety of cationic monomers may be copolymerized with acrylamide in varied molar ratios to acrylamide to produce flocculants with a range of charge density.

Cationic PAMs are formed either by copolymerizing the acrylamide monomer with a positively charged monomer to create the cationic polymer or by modifying the acrylamide monomer itself following polymerization. The latter method is performed by reacting formaldehyde and dimethylamine with acrylamide, which yields aminomethylated PAM, the traditional Mannich solution PAM, named after the inventor of the chemical reaction.

A variety of cationic monomers may be incorporated to the PAM backbone at various molar ratios, thereby giving rise to various degrees of cationic charge or charge density. Typical molar ratios of acrylamide monomer to the cationic monomer are 80:20, 60:40, and 45:55. The weight ratios differ substantially from the molar ratios as a result of the higher molecular weights of cationic monomers to that of the uncharged acrylamide. The names and chemical structures of these cationic monomers may be found in *Guidance Manual for Polymer Selection in Wastewater Treatment Plants* (Water Environment Research Foundation, 1993).

4.2.2.2 *Charge Density*

Manufacturers of organic flocculants refer to charge density in relative terms. Although there is no standardized representation for charge density, the information in Table 7.7 may be used as a typical convention for relative charge density.

The charge density of cationic flocculants used for dewatering municipal sludges falls into the broad range of 10 to 80%. The majority of polymers, which perform effectively to condition most municipal sludges composed of blends of digested primary and WAS for dewatering in centrifuges or on BFPs, fall into the category of medium to high charge—typically 20 to 60%.

Although there is no substitute for laboratory testing and full-scale machine evaluations to determine effective products, the following general guidance is provided:

- Higher concentrations of primary sludge will require lower-charge products;

TABLE 7.7 Charge density.

Relative charge density	Charge density (mol %)
Very high	> 70 to 100
High	> 40 to 70
Medium	> 10 to 40
Low	< 10

- Higher concentrations of biological solids or WAS will require higher charge products;
- Sludge with smaller particle sizes typically will require a higher polymer dose and possibly higher charge products; and
- Older septic sludge will require higher charge products and higher polymer dose.

4.2.2.3 Molecular Weight

Similar to that for charge density, most manufacturers of organic flocculants refer to molecular weight in relative terms. Although there is no standardized representation for molecular weight, the information in Table 7.8 may be used as a typical convention for relative molecular weight.

Variations in the chain length of the polymer may result in products with an extremely wide range of molecular weight. Further, these polymers may be chemically modified for improved performance by crosslinking the polymer chain or incorporating branching to the polymer backbone. The molecular weight data can only be estimated. As an indicator of molecular weight, many manufacturers provide standard viscosity data. Unfortunately, standard viscosity measurements are not comparable across different product chemistries. As a side note, viscosity is not a measure or an indicator of polymer content.

Lower-molecular-weight products tend to be more soluble and less viscous. The majority of polymers, which perform effectively to condition most municipal sludges composed of blends of digested primary sludge and WAS for dewatering in centrifuges or on BFPs, fall into the category of medium to high molecular weight, ranging from an estimated 0.8 to 6.0 million units and represented by standard viscosities ranging from 0.0025 to 0.0035 Pa (2.5 to 3.5 cP).

TABLE 7.8 Molecular weight.

Relative molecular weight	Molecular weight
Very high	> 6 M to 18 M
High	> 1 M to 6 M
Medium	> 200 K to 1 M
Low	< 200 K

4.2.3 *Polymer Forms and Storage and Handling*

There are three physical forms of cationic PAMs that are typically used in conditioning municipal sludge for thickening and dewatering applications:

1. Dry PAMs (DPAMs);
2. Emulsion PAMs (EPAMs); and
3. Solution PAMs (SPAMs), sometimes called *liquid PAMs*.

There is a fourth form, the gel log, which is rarely used in wastewater treatment. Several factors affect the determination of which form is most suitable for an operation:

- Cost-effectiveness;
- Storage and handling of neat product;
- Polymer make-down and aging equipment and solution feeding capability; and
- Personnel safety considerations.

4.2.3.1 Dry Polyacrylamide

Dry PAM may be provided in granular (pellet), flake, or bead (pearl) forms. These may affect storage and handling considerations, as each has different moisture adsorption (caking), flowability, and wetting for polymer make-down characteristics. Dry PAMs remain effective for more than 1 year, provided that they remain free of moisture. Most manufacturers suggest a maximum shelf life of one year.

Dry PAMs are not 100% active polymer, as is frequently reported. Dry PAMs typically contain 88 to 96% active polymer, with the balance being residual moisture and inert salts. Dry PAMs, therefore, are advantageous for having the lowest shipping cost per unit weight of active ingredient. In many cases, additional salts are added to enhance product life or flowability. Accordingly, a simple solids determination probably does not accurately reflect the active polymer content.

The effective make-down of dry polymers is critical for efficient product use. Aging is another issue, especially with dry polymers. Sufficient time is required between dissolution and use to ensure that the polymer "string" has completely "unraveled" to ensure maximum effectiveness.

Each polymer particle must be wetted to ensure complete dissolution into the polymer solution and to avoid product agglomeration and gels. Any gels, or "fish-eyes," that do not properly dissolve into the polymer solution are not useable and, therefore, are wasted.

Storage and handling of DPAMs must be considered carefully. Dehumidified areas are recommended for product storage and make-down, as excessive humidity or moisture may cause caking and handling problems in the storage and conveying equipment. Conversely, dust must also be controlled, and personnel may be required to wear protective respiratory equipment. Some DPAM products have a tendency to cause more dust than others.

Spills of DPAM must be contained and promptly swept or vacuumed away for proper disposal. Moisture and water cause the polymer to become hazardous, gelatinous, and slippery, and this condition should be strictly avoided.

Polymer manufacturers should be consulted for storage and handling, humidity and moisture control, dust control, product make-down and solution concentration, and cleanup recommendations.

4.2.3.2 Emulsion Polyacrylamide

Emulsion PAMs (emulsion or dispersion) appear as a milky liquid, although some emulsions or dispersions with smaller particle sizes may appear somewhat clearer or cloudy. The emulsion product provides for a higher concentration of active polymer in a convenient, easy-to-handle, liquid form. The EPAMs typically contain 25 to 60% of active polymer.

Most emulsion products have a recommended shelf-life of six months when stored under typical conditions. Samples stored in a refrigerator have considerably longer shelf lives. Emulsions consist of two distinct phases—oil and water; there is the potential for phase separation or layering. Users should consider slow mixing or recirculation of the polymer storage tank to maintain uniformity of the neat polymer. Similarly, drum or semibulk containers of emulsion polymers should be checked for uniformity and mixed, if necessary, before use. The manufacturers have mixing recommendations for maintaining neat polymer homogeneity and can supply mixers for drums and semibulk containers.

In an emulsion, the organic polymer is actually contained in a mineral-oil phase of droplets, which are stabilized in an aqueous media by emulsifiers or surfactants. These additives also assist in the inversion of the polymer during make-down. Additional cost is associated with the oils and surfactants used to make the convenient product package.

Emulsion PAMs typically are shipped and stored in bulk tanks or semibulk, returnable containers (totebins). The EPAMs also may be available in 208-L (55-gal) steel or fiber drums. Handling takes place within pipes and hoses, which are part of a closed system. The make-down of emulsion polymers is convenient and reliable, using any of several readily available, packaged systems offered by numerous manufacturers. The bulk viscosity of EPAMs may vary considerably, thereby affecting pumping capability and calibration. This means that the equipment must be recalibrated when polymers are changed.

When lines, hoses, or other equipment containing EPAM must be disassembled, they should be flushed with lightweight machine oil, such as SAE 10 to 30, to remove residual polymer. If water is used to flush the neat emulsion lines, then polymer will gel into a sticky mass.

Similarly, spills should be quickly contained and adsorbed using clay-based adsorbents. Several commercial cleaners are available for cleaning spills and belt washing. Any polymer supplier can recommend specific types. Water use should be avoided during spill cleanup outside of a contained and properly drained area because it will cause a slippery, gelatinous mass.

The polymer manufacturer should be consulted for specific recommendations regarding product uniformity and mixing, storage conditions and product stability, neat product viscosity and pumping, product make-down and solution concentration, and cleanup.

4.2.3.3 Solution Polyacrylamide

As the name implies, SPAM is a true solution of polymer in water, appearing as a clear to cloudy, highly viscous syrup—almost gel-like. These products are also referred to as liquid PAMs or Mannichs, named after the inventor of the chemical process. Solution PAMs have a very high molecular weight and range from approximately 3 to 8% active polymer content. The neat SPAM product also may contain inert salts or diluents, so that a simple solids determination may not accurately reflect active polymer content.

These products are dilute and are typically sold in bulk tanker loads. The user will need storage tanks to hold the large-volume inventory. The storage time should be limited to one month or less, even though some manufacturers suggest a shelf life of three months or more. The SPAM products are more susceptible to biological and chemical degradation than other forms of polymers and may continue to react in storage, increasing in viscosity even to the extent of gelling. At best, this may cause pumping and product make-down difficulties; at worst, the tank has to be cleaned out and the product discarded.

Bulk storage and handling are convenient and typically are conducted in closed systems, although the pump that is used for emulsion probably will not be large enough to deliver the volume of SPAM that is needed. Because the polymer exists in an aqueous solution, traditional make-down, such as that required for dry and emulsion forms, is not necessary. Simple dilution of the neat product with water to a recommended concentration for use, followed by proper mixing, should be adequate.

Because their active polymer concentration is so low—typically 5% or less—SPAMs have high transportation costs and typically are not cost-effective beyond a 400-km (250-mi) radius of the manufacturing plant.

Spills of SPAM should be readily contained and adsorbed in appropriate media for disposal. Several commercial cleaners are available for cleaning spills and for belt washing. Water should be avoided during spill cleanup outside of a contained and properly drained area because it will cause a large slippery, gelatinous mass.

The polymer manufacturer should be consulted for specific recommendations regarding neat polymer storage conditions and product stability, neat product viscosity and pumping, product make-down and solution concentration, and cleanup.

4.2.4 General Safety

The manufacturer's product literature and material safety data sheet for each polymer should be reviewed for detailed handling and safety precautions. Although polymers used in sludge conditioning/dewatering are not particularly toxic, it is prudent to avoid contact. Safety problems are primarily related to slip hazards and dusting. Of course, personnel should take appropriate precautions to avoid getting dust into their eyes, ears, nose, and mouth. Any dust that attaches to a moist body part will become sticky and slippery and difficult to wash off. Accordingly, protective equipment and disposable clothing should be used when working in dusty polymer conditions.

Dust from dry polymers is extremely hazardous when wetted. Simply walking in a minor amount of polymer dust with wet shoes can cause a fall. Similarly, dust can be tracked or carried by drafts into an unexpected area and linger there for months. Polymer dust should be carefully contained and thoroughly swept or vacuumed and discarded before it can result in a hazardous, slippery condition.

Similar slippery conditions may result from leakage or spillage of liquid solution or emulsion polymer or from tracking liquid polymer into an unexpected area. When a liquid polymer dries, it is relatively harmless. Once rewetted, however, it quickly becomes a slippery hazard. Cleanup of liquid polymer is difficult and must be thorough. Adsorbents may be used, and steam-cleaning can be effective. The addition of rock salt or bleach (sodium hypochlorite) may help to break the polymer chain before washing with water. Several commercial cleaners are available for cleaning spills and for belt washing. Any residual polymer will result in the same hazardous, slippery condition when in contact with water.

The best precaution is to avoid polymer leaks, spills, and dusting. Prevention is much easier than containment and cleanup.

4.2.5 Polymer Specifications and Quality Control

Along with the product identification and type and form of product, the following standard product specifications should be obtained to determine storage conditions, pumping requirements, and potential hazards:

- Total solids (percent);
- Specific gravity;
- Bulk viscosity;
- Flash point;
- Freezing point; and
- Shelf-life.

The product specifications should also include related quality control limitations or variability of the manufacturer, each of which is critical to the use and performance of the product in the conditioning/dewatering process:

- Active polymer content (percent);
- Charge density (mole percent); and
- Standard viscosity.

4.2.5.1 Active Polymer Content

The user should obtain the active polymer content (percent) with total solids and should receive from the manufacturer the quality control capability, in the form of the minimum and maximum active polymer (percent), to be contained in any individual shipment.

Active polymer content is the specific measure of the real product that is responsible for dewatering performance. This measurement of active polymer eliminates inert solids, which are included in total solids, but do not contribute to conditioning/dewatering performance. The procedure for determining active polymer involves solvent extraction and is quite involved. Unfortunately, there is no industry-wide standardized procedure. It is prudent to ask for a copy of the manufacturer's procedure and consider spot-checking deliveries.

4.2.5.2 Charge Density

Similarly, cationic charge density should be obtained for each product. The charge density should be a quantitative assessment expressed as the mole percent of cationic charge, rather than a relative reference, such as low, medium, or high. Also, the end-user should receive, from the manufacturer, the quality-control capability, in the form of the minimum and maximum cationic charge density (mole percent), to be contained in any individual shipment. Charge density is an indicator of the ratio of positive charges relative to the polymer backbone. For any given application or sludge type, if the charge density falls below a certain level, then more polymer will be needed to achieve the same performance. Therefore, it is important to identify the minimum charge density that will be provided by the manufacturer for a selected optimum product. This information was identified through full-scale field evaluations, and the actual charge density of the products was tested to obtain the bid information.

4.2.5.3 Standard Viscosity

Standard viscosity is also important to the end user in conjunction with the manufacturer's relative classification of molecular weight. Standard viscosity may be used as an alternative, quantitative comparison for molecular weight.

Standard viscosity is different than bulk viscosity or the viscosity of the neat product. Bulk viscosity is determined on the neat product and is useful for determining pumping and flow and related equipment requirements. Standard viscosity is determined at various make-down concentrations of the neat polymer product and is useful as an indicator of molecular weight. For standard viscosity, there are no industry-wide standardized procedures; manufacturers, however, can provide copies of the analytical procedures they use.

4.2.6 *Product Make-Down*

A variety of effective, packaged polymer make-down systems for handling emulsion or dry polymers is available from independent equipment manufacturers and polymer producers. Unfortunately, most facilities are set up to effectively handle one form of polymer and are not readily able to convert to another product form without a lot of reengineering

Flexibility in a polymer make-down system and related polymer handling equipment is desirable. It is impossible to predict what form of polymer will be the most cost-effective in the future. Make-down systems are available that provide universal capability for dry, emulsion, and solution polymer.

To be truly effective, a universal make-down system must include the appropriate ancillary equipment, such as metering pumps and storage tanks capable of handling varied polymer concentrations and related volumes and flowrates. As such, universal product make-down systems are not always practical and are more expensive. Installation of equipment to handle dry and emulsion products is a good compromise between price and value. Sometimes the form of product being used dictates the make-down system, and built-in capabilities of ancillary equipment preclude easy conversion from one product form to another. Accordingly, several issues should be carefully considered:

- Neat product storage and handling;
- Personnel and safety;
- Make-down capability and space limitations;
- Convenience;
- Dewatering performance and economic criteria;
- Changes in the polymer market and competitive landscape;

- Potential for sludge changes, and
- Anticipation of future developments.

All of these factors should be considered before any final decisions are made relative to the form of polymer and type of make-down unit to be used. At a minimum, every system should include the make-down device and a separate agitated storage (day) tank for feeding diluted polymer solution to the conditioning/dewatering operation. The day tank should be sized to hold the equivalent volume of polymer to satisfy the maximum demand for at least one hour, thereby providing effective aging time with gentle mixing of 30 minutes minimum. Tanks should be fitted with clear and calibrated sight tubes for monitoring polymer solution draw-down.

Low-head centrifugal pumps are acceptable for one-time transfer of the polymer solution from the make-down unit to the day tank. Rotary gear pumps and progressive cavity pumps are better for metering the dilute polymer solution to the conditioning/dewatering process. Inline flow meters are important for each polymer solution feed pump, and each metering pump should be outfitted with an appropriately sized calibration cylinder for confirmation of the feed rate. Sufficient water pressure (minimum of 280 kPa [40 psi]) and flow are required to overcome system head pressure and operate an atmospheric eductor for aspirating dry polymer.

4.2.6.1 Dry Polyacrylamide

For dry polymers, care must be taken to ensure that the neat polymer remains dry and free from humidity. The polymer make-down system or dispersing unit must meter the dry polymer at a controlled rate and effectively wet individual dry polymer particles to produce a uniform dilute polymer solution without polymer agglomeration, gels, or "fish-eyes."

Typically, cationic dry polymers are made-down to a concentration ranging from 0.1 to 0.5%, with a target of 0.2%, polymer as supplied by the vendor. The dilute solution should be mixed with a low-speed mixer (400 to 500 rpm) for at least 30 minutes for proper aging and uncoiling of the polymer structure before feeding to the dewatering process. Higher concentrations of polymer and colder makeup water will probably need longer mixing time for proper aging of the polymer.

4.2.6.2 Emulsion Polyacrylamide

Emulsion polymers may separate or layer over time. Products in storage or containers should be checked for uniformity and mixed, if required, before make-down.

Emulsion polymers must be activated before they can be used. The first step is called inversion and occurs relatively quickly. During inversion, the make-down system imparts significant energy, in the form of turbulence or shear, to the emulsion and water mixture, to invert the polymer from the oil-droplet to the aqueous phase. The second step, termed aging, occurs over up to 30 minutes, during which quiescent mixing allows the polymer structure to thoroughly uncoil to become completely effective.

There are some emulsion products that may be fully effective without aging, and some equipment manufacturers claim that their make-down units are so effective that aging is not required. Nonetheless, adequate aging capability is strongly recommended in every system to provide the flexibility to use many products.

Cationic emulsion polymers are typically made-down to a concentration of 0.5 to 2.0% of polymer as supplied by the vendor, which typically amounts to approximately 0.2% on an active basis. As with dry products, the storage (day) tank should be sized for dilute polymer solution to hold the polymer solution for a minimum of one hour before refilling.

4.2.6.3 Polymer Degradation

Intense mixing, pumping, and shearing of the dilute polymer solution will reduce the molecular weight and increase the dosage. Temperatures higher than 50 °C (120 °F) should be avoided. The quality of water used for polymer make-down is important. Suspended and dissolved solids found in nonpotable water result in increased polymer demand. It is also important to check the pH of the water to ensure that it is close to 7.0.

Cationic flocculants perform effectively over the typical range of pH (6.5 to 7.5) found in municipal sludge conditioning/dewatering; however, as pH increases above this level, polymer effectiveness typically decreases. In the case of Mannich SPAM, a pH of 7.5 and higher may result in decomposition of the polymer, thereby releasing odor-causing amines. Other substances that may interfere with performance or cause a need for higher-than-normal doses of cationic PAMs include sulfides, hydrosulfides, phenolics, and chlorides. For a more detailed discussion of make-down systems, including materials of construction, schematics, and pump-and-valve specifications, the reader may consult *Design of Municipal Wastewater Treatment Plants* for polymer suppliers and manufacturers of specific make-down systems (Water Environment Federation et al., 2009).

4.2.7 *Process Design Considerations for Thickening and Dewatering*

This section includes summaries of design considerations for several types of thickening and dewatering processes. Typical polymer dosages for thickening and dewatering applications are provided in Tables 7.9 and 7.10, respectively.

4.2.7.1 *Gravity Thickening*

Conventional gravity thickening typically does not require the use of organic polymers. That said, using these chemicals increases solids and hydraulic loading rates by a factor of 2 to 4 and improves solids capture. They have minimal effect, however, on the resultant underflow solids concentration. Using polymer also increases the overall cost of gravity thickening. As a result, it only should be used to prevent operating problems caused by solids carryover.

Bench tests and other laboratory or field investigations should be performed to test the relative effectiveness of flocculating aids (either alone or in combinations). Also, care should be taken during feeding and mixing to prevent the overfeeding or poor mixing that causes "islands" to form (Water Pollution Control Federation, 1980).

Using approximately 2 to 4.5 g of active polymer/kg of dry solids (4 to 9 lb/ton) can produce a solids-loading rate of approximately 22 to 34 kg/m^2·d (4.5 to 7.0 lb/d/sq ft) when thickening primary solids. Adding approximately 4.5 to 6.0 g/kg (9.5 to 12.5 lb/ton) of polymer can increase the thickener's loading rate to 12 to 16 kg/m^2·d (2.4 to 3.2 lb/d/sq ft) when treating WAS (Ettlich et al., 1978; U.S. Environmental Protection Agency, 1978).

4.2.7.2 *Dissolved Air Flotation Thickening*

Chemical conditioning is unnecessary for dissolved air flotation (DAF) thickening if low hydraulic and solids-loading rates are used. If high loading rates are required or compaction is poor and the sludge volume index is high, then chemical conditioning will improve solids capture and can increase the float solids concentration. Although the increase in float solids is typically small (approximately 0.5%), polymers may be required for WAS if a 4% float solids concentration is to be achieved. Float solids can routinely be 6% or greater when cothickening mixtures of primary solids and WAS. Unless problems exist, solids capture without polymers is typically approximately 95%. With polymers, solids capture can increase to 97 or 98%, thereby improving subnatant quality and lessening the effect of recycle solids on plant performance. Also, with polymer addition, it is possible to as much as double the solids loading rate. [A typical rate is 10 kg/m^2·h (2 lb/sq ft/hr) with polymers.]

Table 7.9 Polymer dosages associated with various solids thickening processes (WWTF = wastewater treatment facility; WWTP = wastewater treatment plant; WRP = water reclamation plant; WPCF = water pollution control facility) (lb × 0.4536 = kg).

Facility name	Type of sludge	Influent feed solids (%)	Method of thickening	Polymer dosage (lb/ton)	Thickened solids (%)	Method of stabilization
Water Reclamation Facility. Bend, Oregon	Secondary (100%)	0.4–0.6	Gravity belt thickening	8–10	4–5	Anaerobic digestion
Irwin Creek WWTF Charlotte, North Carolina	Primary (60%); secondary (40%)	3.2 0.5–1.0	Conventional gravity thickening	0	4–5	Anaerobic digestion
City of Greeley WPCF Greeley, Colorado	Secondary (100%)	0.6–0.8	Centrifuge	1.7–2.5	4.5–6.5	Anaerobic digestion
JEA Buckman Street WWTP, Jacksonville, Florida	Primary (60%); secondary (40%)	1–3	Gravity belt thickening	9–12	3–5	Anaerobic digestion
Greenfield Road WRP Mesa, Gilbert and Queen Creek, Arizona	Primary and secondary	1–1.25	Centrifuge	0.5	5	Anaerobic digestion
Iron Bridge Water Reclamation Facility, Orlando, Florida	Secondary (100%)	0.4–0.6	Gravity belt thickening	7–9	2–3	Lime stabilization
Northwest Water Reclamation Facility, Orange County Utilities, Orlando, Florida	Secondary (100%)	0.7–0.8	Conventional gravity thickening	0	1–2	Contract lime stabilization

Water Conserv I Orlando, Florida	Secondary (100%)	0.7–1.2	Gravity belt thickening	10–12	4–5	Transported to larger facility for further stabilization
Water Conserv II Orlando, Florida	Secondary (100%)	0.4–0.8	Gravity belt thickening	10–12	2–4	Anaerobic digestion
Orange County Utilities, South WRF, Orlando, Florida	Secondary (100%)	0.4–1.25	Gravity belt thickening	2.4–6.5	3.1–6.4	Anaerobic digestion
91st Avenue Wastewater Treatment Plant Phoenix, Arizona	Secondary (100%)	1.0–1.5	Centrifuge	2.0–3.0	5.7–6.2	Anaerobic digestion
McAlpine Creek WWTF Pineville, North Carolina	Primary (50%); secondary (50%)	1.0	Conventional gravity thickening, centrifuge	0 3–5	3–5 0.9–1	Anaerobic digestion
Roger Road Wastewater Reclamation Facility	Secondary	0.2–0.4	Gravity belt thickening	9–10	5–6	Anaerobic digestion
Pima County, Tucson, Arizona	Primary	0.4–0.6	Conventional gravity thickening	0	4–5	Anaerobic digestion
Columbia Boulevard WWTP Portland, Oregon	Secondary (100%)	0.5–1.0	Gravity belt thickening	5–9	4–6	Anaerobic digestion

Table 7.10 Polymer dosages associated with various solids dewatering processes (WWTF = wastewater treatment facility; WWTP = wastewater treatment plant; WRP = water reclamation plant; WPCF = water pollution control facility; WAS = waste activated sludge).

Facility name/location	Type of sludge	Type of stabilization	Influent feed solids (%)	Method of dewatering	Polymer dosage (g/kg)	Dewatered cake solids (%)
Water Reclamation Facility Bend, Oregon	Primary (55%); secondary (45%)	Anaerobic digestion	1.8–2.1	Belt filter press	3.75–6	12–15
Irwin Creek WWTF Charlotte, North Carolina	Primary (60%); secondary (40%)	Anaerobic digestion	1.4	Belt press	3.5–4	18.75
City of Greeley WPCF Greeley, Colorado	Primary (60%); secondary (40%)	Anaerobic digestion	1.5–2.0	Centrifuge	5–8	19–22
Buckman Street WWTP, Jacksonville, Florida	Primary (60%); secondary (40%)	Anaerobic digestion	2–4	Centrifuge	3.75–6.25	19–22
Greenfield Road WRP Mesa, Gilbert and Queen Creek, Arizona	Primary and secondary	Anaerobic digestion	2.75–3.0	Centrifuge	5.75	22–23
Water Conserv II Orlando, Florida	Secondary	Anaerobic digestion	2.0–2.5	Belt filter presses	3.5–4	12
Eastern Water Reclamation Facility Orange County Utilities Orlando, Florida	Secondary (100%)	Contract lime stabilization	< 1	Belt filter presses (three belt)	3–3.5	16–17
Iron Bridge Water Reclamation Facility Orlando, Florida	Secondary	Lime stabilization	2.0–3.0	Belt filter presses	3.5–4	17

Iron Bridge Water Reclamation Facility Orlando, Florida	Secondary	Lime stabilization	0.4–0.6	Belt filter presses (three belt)	3.5–4	17
Northwest Water Reclamation Facility Orange County Utilities Orlando, Florida	Secondary (100%)	Contract lime stabilization	1–2	Belt filter presses	3.75–6.25	14–16
South Water Reclamation Facility Orange County Utilities Orlando, Florida	Secondary	Anaerobic digestion	2.3–3.7	Belt filter presses	0.5–1.1	9.3–19.7
McAlpine Creek WWTP Pineville, North Carolina	ORC reported a 1:1 ratio of primary to secondary solids	Anaerobic digestion	2.4	Centrifuge	3.75	20
Columbia Boulevard WWTP Portland, Oregon	Freshly digested (65–75%); thickened WAS (20%); primary (80%) and previously digested, lagoon stabilized thickened WAS (25–35%)	Anaerobic digestion	1.5–2.0	Belt filter presses	3.75–5	19–22
Thomas P. Smith Water Reclamation Facility Tallahassee, Florida	Secondary (100%)	Anaerobic digestion	2.81–4.39	Screw presses	3.75–6.75	13–20

Typically, a cationic polymer with a moderate charge and high molecular weight is used; however, lower-charge cationic polymers are starting to show better performance. Typical dosages are from 2 to 5 g/kg (4 to 10 lb/ton) up to 7.5 g/kg (15 lb/ton) of dry solids.

A problem typically encountered when conditioning solids is improper mixing of conditioner and solids. To mitigate this problem, a more dilute polymer solution (0.25 to 0.5%) should be used, or the flocculant should be mixed with pressurized recycle before contacting the solids (Ettlich et al., 1978; U.S. Environmental Protection Agency, 1978).

4.2.7.3 *Centrifugal Thickening*

A solid-bowl conveyor centrifuge has been used to thicken a variety of solids. Centrifugal thickening typically does not require polymer addition when treating biological and aerobically stabilized solids. Well-digested solids, however, have little natural flocculating tendency and require polymer additions to achieve acceptable solids recovery levels. Engineers should make provisions for polymer addition in the initial design, even if chemical conditioning is not planned. The design should be flexible enough to allow conditioning chemical to be added at one of several points in the influent piping.

Dry or liquid high molecular weight cationic polymers are effective thickeners. When dry polymers are used, a 0.05 to 0.1% feed solution is used; liquid polymers can range in concentration up to 0.5% on an active basis. It is important that a solids capture of at least 95% is obtained to prevent recycling filamentous bacteria and fines to the wastewater treatment process. Waste activated sludge produces a weak floc that tends to shear inside the centrifuge; a dose of up to 4 g/kg (8 lb/ton) of polymer can be used to formulate a tougher floc. Aerobically and anaerobically digested solids have little natural floc and, therefore, require approximately 4 to 8 g/kg (8 to 16 lb/ton) of polymer.

4.2.7.4 *Gravity Belt Thickening*

Gravity thickening works well with many types of solids. Difficult-to-thicken solids only require minor modifications of polymer dosages and solids loading rates to keep the effluent solids concentration and percent solids capture high. Gravity thickeners have been used to treat solids containing as little as 0.4% solids or as much as 10% solids with polymer addition. Polymer dosages ranged from between 1.5 and 3 g/kg (3 and 6 lb/ton) (dry weight basis) for raw primary solids and up to between 4 and 6 g/kg (8 and 12 lb/ton) for anaerobically stabilized solids. In all cases, solids capture remained above 95%.

4.2.7.5 Rotary Drum Thickening

Rotary drum thickeners work much like gravity belt thickeners: in both systems, a moving, porous media retains conditioned solids while allowing free water to drain through. A polymer is injected to the feed line and mixed with incoming solids before entering the flocculation tank. Once inside the tank, the mixture is exposed to a low-shear rotary mixer to ensure maximum flocculant development. Conditioned solids then flow onto the distribution tray, where they are directed onto the rotating drum. Free water passes through the openings in the drum, and captured solids remain on the drum surface for further dewatering. Radial flights inside the drum slowly transport thickening solids toward the discharge end of the drum. Thickened solids exit the unit and fall through a discharge chute into a storage tank, pump hopper, or other suitable receiving device.

Drum speed, mixer speed, and spray water cycling are adjustable to ensure maximum performance with minimal polymer and water use. Polymer requirements are approximately 10 to 20% greater than those associated with gravity belt thickeners. Rotary drum thickeners are suited for high-fiber solids and for raw and digested solids with a significant fraction of primary solids. Their success with municipal WAS is variable and depends on solids characteristics. Sludge typically can be thickened to 5 to 7% total solids (in some cases, to more than 10% total solids), with up to 99% capture of feed solids at polymer dosages ranging from 4 to 6 g/kg (8 to 12 lb/ton) (dry weight basis).

4.2.7.6 Centrifugal Dewatering

Polymers have been used with solid-bowl conveyor centrifuges to increase machine throughput without lowering cake dryness, to improve solids recovery, or both. Typically, a moderate-to-high charge, high-molecular-weight cationic polymer is used. Pilot studies are needed to determine the correct conditioning agent and dosage. Designs should include facilities for feeding both dry and liquid polymers. Polymer use typically increases solids capture; however, too much polymer can lead to a wetter cake because more fines are captured. Therefore, the relationship between recycled solids and cake dryness determines the dosage of polymer to be used.

4.2.7.7 Belt Filter Press Dewatering

The performance of BFPs depends on proper conditioning, and organic polymers traditionally are used. A properly conditioned product has a 95 to 98% solids recovery rate. The quantity of polymer required for proper conditioning varies widely; it depends on solids type, solids concentration, and ash content (typically, less polymer is needed when the ash content is high). For example, primary solids require a polymer dose of 3.5 to 5 g/kg (7 to 10 lb/ton), anaerobically digested solids require a dose of 7 g/kg (14 lb/ton on a dry weight basis), and autothermal thermophilic aerobically digested solids require a dose between 18 and 23 g/kg (37 and 47 lb/ton).

Insufficient conditioning causes inadequate dewatering in the initial sections of the press, which, in turn, can cause solids to extrude from the press section, overflow in the drainage section, or blind the belt. Over-conditioning can cause belt-blinding and over-flocculation, which can cause solids to drain too fast and mound on the belt, resulting in poor dewatering. The goal is to remove as much water as possible in the gravity section of the press. Over-flocculation may be mitigated by using deflection plates to even out the mounds before pressing begins, or by selecting a BFP with an extended gravity table.

Because of the shearing action between belts, Novak and Haugan (1980) have suggested using turbulent mixing when adding polymers for conditioning before dewatering. The best dosage and overall system performance depend on solids concentration, mixing intensity, and mixing time.

4.2.7.8 Screw Press Dewatering

The screw press is a simple, slow-moving mechanical device that gradually compresses conditioned, thickened solids that move through the unit. Dewatering is continuous. Proper screw design is critical because different solids require different polymer dosages, screw speeds, and configurations to maintain a desired dewatered cake concentration and solids capture rate.

Proper solids conditioning is essential to produce a consistent dewatered cake. Slower operations will produce a dryer cake but also will reduce solids throughput. Therefore, it is important that a relationship between polymer dosage, solids throughput, and cake dryness be established. Depending on the influent solids characteristics, the polymer dosage may range from 8 to 12 g/kg (16 to 24 lb/ton) to produce a cake containing between 12 and 25% dry solids and a solids capture rate of 90 to 95%.

4.2.7.9 Rotary Press Dewatering

The rotary press is sometimes confused with the screw press when, in fact, it operates quite differently. Another misconception is that the dewatering channels on the rotary press somehow incorporate a converging or narrowing channel. Its operating principle is relatively simple. After dosing with polymer to promote flocculation, solids are pumped into a hollow cavity between porous screens. Free water (filtrate) passes through the screens, and a cake begins to form inside the cavity. The screens constantly, slowly rotate and "grip" the dry cake (via frictional force) near the outlet, extruding it continuously through a pressure-controlled port. A septum separates the inlet side of the cavity from the outlet. The polymer dosage depends on the type of solids to be dewatered.

4.2.7.10 Drying Beds

Conditioning solids before sending them to drying beds is not widely practiced. In fact, *Recommended Standards for Wastewater Facilities* and other design guidelines do not consider using conditioning chemicals for this application, even though they may reduce drying time and, therefore, required bed area (Great Lakes Upper Mississippi River Board of State Sanitary Engineers Health Education Services, Inc., 2004).

The conditioning system must be designed to avoid rupturing the conditioned floc during transport to the drying beds. Rupturing typically occurs during pumping; it can be overcome by locating the flocculation tank close to the drying beds and allowing conditioned solids to flow by gravity from the flocculating chamber. Excessive holding increases the percentage of fines, which will impair flocculation and dewatering.

Both wedge-wire and vacuum-assisted drying beds use polymers to coagulate fines and promote rapid cake formation. The polymer is injected to the solids in the inlet line or in a flocculation tank next to the bed.

5.0 PERFORMANCE OPTIMIZATION

The economics of conditioning/dewatering performance is the bottom line. There are typically four economic factors that compose overall conditioning/dewatering performance:

1. Polymer cost (polymer dose times unit price);
2. Cost of sludge disposal or downstream processing (dependent on cake solids percent);
3. Cost of recycle (function of capture percent); and
4. Throughput (feed rate).

It is important that operating personnel have an understanding of these cost factors and can calculate these costs with real-time operating data. If operations personnel do not have real-time analytical data for feed solids (percent), cake solids (percent), capture (percent), and polymer dosage, then they cannot maintain optimum performance. Laboratory results for cake solids, sludge feed solids concentration, and suspended solids in recycle, which may be available 24 hours later, do not provide the information necessary to make immediate operational adjustments.

Similarly, it is important that sample points for collecting feed sludge, recycle (filtrate/centrate), and dewatered sludge cake are convenient for operators. It should not be necessary to walk down two flights of stairs into the basement to collect a sample of sludge feed or of dewatered cake on a different level than recycle (centrate). Ideally, the operator can see the centrate/filtrate running into a drain while walking along a well-traveled route.

5.1 Cost of Polymer

Simply stated, polymer cost is polymer dose in weight as supplied by the vendor per weight of solid processed times the unit price of the polymer (price per weight delivered). Polymer dose typically is determined from formal polymer evaluations under standardized conditions. This is discussed in further detail under Section 7.0 *Outline for Conducting Effective Polymer Evaluations and Product Selection*.

5.2 Cost of Sludge Disposal or Downstream Processing

The cost of sludge disposal or downstream processing is a function of the cake solids (percent) achieved in dewatering times the unit cost of disposal, such as incineration or transportation and landfilling. In the case of further solids processing for reuse, such as composting or lime stabilization, it would include these processing costs plus the cost for transportation and land application. More broadly, the basic disposal cost is the sum of the polymer cost plus the disposal cost. If the plant supervisory control and data acquisition (SCADA) system does not calculate and display these costs, operators have no way of determining if they are operating at the minimum cost per megagram (ton).

5.3 Cost of Recycle (Capture)

There is a quantifiable cost of solids recycled to the front end of the treatment operation. A simple gross method would be to assign to recycled solids the same cost assessed to industrial sewer customers who have excess suspended solids. Alternately, the plant budget could be divided by the amount of suspended solids and biochemical oxygen demand processed. The result is a fair representation of the cost to treat the recycled solids.

5.4 Throughput (Feed Rate)

Dewatering equipment is specified with sludge throughput parameters based on hydraulic loading (volume) or on solids loading mass of dry solids per hour or per meter per hour for BFPs. For effective process control, operators should know the throughput on a mass basis, typically in dry tons of solids per hour. Accordingly, operators must have access to accurate real-time data for hydraulic loading (flow) and solids concentration (%) in the feed sludge. Throughput varies significantly with the type of sludge being dewatered. Specific data related to throughput rates for various sludges are referenced in standard wastewater engineering manuals and Water Environment Federation publications. Primary sludges typically yield the highest throughput and digested WAS the lowest. Blends of raw or digested primary sludge with WAS or digested WAS typically yield maximum throughputs between the two extremes. Equipment capability is limited by the following:

- Intrinsic design of the equipment;
- Equipment set up and operation;
- Type and quality of sludge; and
- Strength of the reaction between polymer and sludge.

From an operational perspective, the best economics are achieved when the equipment is used to its maximum throughput, while achieving the desired performance for cake solids and solids capture. Similarly, this approach would optimize economics relative to power consumption, labor, and maintenance costs. This can only be achieved, however, with carefully obtained process data that is evaluated with good cost numbers.

Nevertheless, many plants established their dewatering throughput on the basis of the quantity of sludge required for downstream processing rates or to satisfy hauling conditions at some time in the distant past. Another factor often used to determine the throughput rate is the amount of hours necessary to fulfill the personnel schedule. On occasion, multiple units are all being operated at less than maximum throughput, even when the total volume could be readily handled by fewer machines running at a higher throughput, thereby saving power and equipment wear.

Several factors should be considered when determining the optimum throughput at a particular facility. Operations personnel should spend time and effort determining optimum performance. Once optimum performance is established, the operators need training so that they know how to keep operations at the optimum throughput rate and achieve desired performance while minimizing the costs associated with electricity, equipment wear, labor, and maintenance.

6.0 AUTOMATION

There has been much technological development on a variety of automated process sensors, controllers, and related software for managing conditioning/dewatering operations for optimum performance and polymer usage. The best success occurs when the instrument, installation parameters, and application are considered together. Several publications provide more detailed information (Gillette and Scott, 2001; Pramanik et al., 2002; Water Environment Research Foundation, 1995, 2001).

7.0 OUTLINE FOR CONDUCTING EFFECTIVE POLYMER EVALUATIONS AND PRODUCT SELECTION

Many details must be considered to conduct effective polymer evaluations for municipal sludge dewatering. A few principal suggestions can help ensure more efficient and effective polymer evaluations:

1. Identify proper performance criteria and define economic values. Considerations may include (1) evaluating the combined cost of polymer and its recycle, transport, and application or (2) using a more expensive polymer (based on dose) to reduce recycle costs and produce a drier product that reduces transport and application costs. Each facility will need to establish the relative costs for these factors.
2. Minimize the time period to complete all product evaluations. This includes allowing sufficient time for vendors to conduct informal laboratory testing and possibly informal full-scale testing to recommend their best product. The entire formal, full-scale trial period can be cut from four to six weeks to one week.

3. Standardize sludge conditions and ensure that all variations affecting sludge characteristics are thoroughly understood and controlled throughout the evaluation period.
4. Eliminate the participation of all unnecessary personnel from formal trial activity.

    a. Conduct all formal product evaluations solely with in-house staff, supplemented with independent consultants, as necessary.

    b. Minimize the number of staff personnel involved in product trials and provide thorough training and written procedures, especially for the polymer make-up calibrations to ensure consistency.

    c. Have vendor personnel onsite and providing input, but any adjustments should be made by plant personnel. When going out for bid, the vendor should be provided with the defined performance criteria and their performance data, but not that of other vendors.

5. Standardize sampling and analytical activities and generate real-time performance data. Real-time analysis and data for feed solids, cake solids, and capture is imperative to make appropriate operational adjustments to identify the product's optimum performance and operating range. If the plant does not have a microwave oven, then one should be purchased or borrowed.
6. Analyze performance data in a manner that assesses actual product performance rather than trial conditions.

    a. Determine optimum performance based on a dose–response curve, such as cake dryness versus polymer dose.

    b. Do not average individual data points to determine performance.

    c. Do not determine polymer dosage based on total polymer usage over the trial period for the respective product. Do not use totalizer readings.

By following these suggestions for more effective polymer trials, operations personnel and process engineers will be better equipped to select the most cost-effective polymer based on overall product performance and related economics, established with criteria determined by the municipality. Also, the product performance will be reproducible under day-to-day operating conditions.

8.0 REFERENCES

Abu-Orf, M. M.; Dentel, S. K. (1999) Rheology as Tool for Polymer Dose Assessment and Control. *J. Environ. Eng.*, **125** (12), 1133–1141.

Abu-Orf, M. M.; Ormeci, B. (2005) Measuring Sludge Network Strength Using Rheology and Relation to Dewaterability, Filtration, and Thickening—Laboratory and Full-Scale Experiments. *J. Environ. Eng.*, **131** (8), 1139–1146.

Bache, D. H.; Dentel, S. K. (2000) Viscous Behaviour of Sludge Centrate in Response to Chemical Conditioning. *Water Res.*, **34** (1), 354–358.

Bruus, J. H.; Nielsen, P. H.; Keiding, K. (1992) On the Stability of Activated Sludge Flocs with Implication to Dewatering. *Water Res.*, **26,** 1597–1604.

Cassel, A. F.; Johnson, B. P. (1978) Evaluation of Filter Presses to Produce High-Solids Solids Cake. *J. New Eng. Water Pollut. Control Assoc.*, **12,** 137.

Christensen, G. L.; Stulc, D. A. (1979) Chemical Reactions Affecting Filterability in Iron–Lime Sludge Conditioning. *J. Water Pollut. Control Fed.*, **51,** 2499.

Dentel, S. K. (2001) Conditioning. In *Sludge into Biosolids*; Spinosa, L., Vesilind, P. A., Eds.; IWA Publishing: London, United Kingdom.

Dentel, S. K.; Abu-Orf, M. M.; Griskowitz, N. J. (1995) *Polymer Characterization and Control in Biosolids Management*; Water Environment Research Foundation: Alexandria, Virginia.

Eriksson, L.; Alm, B. (1991) Study of Bioflocculation Mechanisms by Observing Effects of a Complexing Agent on Activated Sludge Properties. *Water Sci. Technol.*, **24,** 21–28.

Ettlich, W. F.; Hinrichs, D. J.; Lineck, T. S. (1978) *Operations Manual: Sludge Handling and Conditioning*, EPA-68/01-4424; U.S. Environmental Protection Agency: Washington, D.C.

Gillette, R. A.; Scott, J. D. (2001) Dewatering System Automation: Dream or Reality? *Water Environ. Technol.*, **13** (5), 44–50.

Great Lakes Upper Mississippi River Board of State Sanitary Engineering Health Education Services Inc. (2004) *Recommended Standards for Wastewater Facilities*; Great Lakes Upper Mississippi River Board of State Sanitary Engineering Health Education Services Inc.: Albany, New York.

Higgins, M. J. (1995) The Roles and Interactions of Metal Salts, Proteins, and Polysaccharides in the Settling and Dewatering of Activated Sludge. Ph.D.

dissertation, Virginia Polytechnic Institute and State University, Blacksburg, Virginia.

Higgins, M. J.; Novak, J. T. (1997a) Dewatering and Settling of Activated Sludges: The Case for Using Cation Analysis. *J. Water Environ. Res.*, **69,** 225–232.

Higgins, M. J.; Novak, J. T. (1997b) The Effect of Cations on the Settling and Dewatering of Activated Sludge: Laboratory Results. *J. Water Environ. Res.*, **69,** 215–224.

Karr, P. R.; Keinath, T. M. (1978) Influence of Particle Size on Sludge Dewaterability. *J. Water Pollut. Control Fed.*, **50,** 1911.

Kemp, J. S. (1997) Just the Facts on Dewatering Systems: A Review of the Features of Three Mechanical Dewatering Technologies. *Water Environ. Technol.*, **9,** (12), 47–55.

Kolda, B. C. (1995) Impact of Polymer Type, Dosage, and Mixing Regime and Sludge Type on Sludge Floc Properties. Master's Thesis, Virginia Polytechnic Institute and State University, Blacksburg, Virginia.

Lewis, C. J.; Gutschick, K. A. (1988) *Lime in Municipal Sludge Processing*; National Lime Association: Washington, D.C.

Minneapolis Wastewater Treatment Plant (1999) *Full-Scale Centrifuge Demonstration Project Centrifuge Test Program Final Report*, June 30; Minneapolis Wastewater Treatment Plant: Minneapolis, Minnesota.

National Lime Association (1982) *Lime Handling, Application, and Storage in Treatment Processes*, 4th ed.; Bulletin 213; National Lime Association: Arlington, Virginia.

Novak, J. T.; Goodman, G. L.; Pariroo, A.; Huang, J. C. (1988) The Blinding of Sludges During Filtration. *J. Water Pollut. Control Fed.*, **60** (2), 206–214.

Novak, J. T.; Haugan, B. E. (1979) Chemical Conditioning of Activated Sludge. *J. Environ. Eng.*, **105** (EE5), 993.

Novak, J. T.; Haugan, B. E. (1980) Mechanisms and Methods for Polymer Conditioning of Activated Sludge. *J. Water Pollut. Control Fed.*, **52** (10), 2571–2580.

Novak, J. T.; Miller, C. D.; Murthy, S. N. (2001) Floc Structure and the Role of Cations. *Water Sci. Technol.*, **44** (10), 209–213.

Novak, J. T.; Sadler, M. E.; Murthy, S. N. (2003) Mechanisms of Floc Destruction During Anaerobic and Aerobic Digestion and the Effect on Conditioning and Dewatering of Biosolids. *Water Res.*, **37,** 3236.

Ormeci, B.; Cho, K.; Abu-Orf, M. M. (2004) Development of a Laboratory Protocol to Measure Network Strength of Sludges Using Torque Rheometry. *J. Residuals Sci. Technol.*, **1** (1), 35–44.

Park, C.; Muller, C. D.; Abu-Orf, M. M.; Novak, J. T. (2006) The Effect of Wastewater Cations on Activated Sludge Characteristics: Effects of Aluminum and Iron in Floc. *Water Environ. Res.*, **78**, 31–40.

Pramanik, A.; LaMontagne, P.; Brady, P. (2002) Automatic Improvements, Installing an Integrated Control System Can Improve Sludge Dewatering Performance and Cut Costs. *Water Environ. Technol.*, **14** (10), 46–50.

Roberts, K.; Olsson, O. (1975) The Influence of Colloidal Particles on the Dewatering of Activated Sludge with Polyelectrolyte. *Environ. Sci. Technol.*, **9,** 945.

Robinson, J. K. (1989) The Role of Bound Water Content in Designing Sludge Dewatering Characteristics. Master's Thesis, Virginia Polytechnic Institute and State University, Blacksburg, Virginia.

Snoeyink, V. L.; Jenkins, D. (1980) *Water Chemistry*; Wiley and Sons: New York.

Sorensen, B. L.; Sorensen, P. B. (1997) Applying Cake Filtration Theory on Membrane Filtration Data. *Water Res.*, **31** (3), 665–670.

Tenney, M. W.; Echelberger, W. F., Jr.; Coffey, J. J.; McAloon, T. J. (1970) Chemical Conditioning of Biological Sludges for Vacuum Filtration. *J. Water Pollut. Control Fed.*, **42,** R1.

Tezuka, Y. (1969) Cation-Dependent Flocculation in *Flavobacterium* Species Predominant in Activated Sludge. *Appl. Microbiol.*, **17,** 222.

U.S. Environmental Protection Agency (1978) *Sludge Treatment and Disposal, Sludge Treatment, Vol. 1*; EPA-625/4-78-012; U.S. Environmental Protection Agency: Cincinnati, Ohio.

U.S. Environmental Protection Agency (1979) *Evaluation of Dewatering Devices for Producing High-Solids Sludge Cake*; EPA-600/2-79-123; U.S. Environmental Protection Agency, Water Resources Management Administration, Municipal Environmental Research Laboratory: Cincinnati, Ohio.

U.S. Environmental Protection Agency (1990) Guidelines for Federal Procurement of Products Containing Recovered Materials. *Code of Federal Regulations*, Part 553, Title 40, 1993; *Fed. Regist.*, **55** (192), 40384–40390.

U.S. Environmental Protection Agency (2000) *Biosolids Technology Fact Sheet Recessed-Plate Filter Press*; EPA-832/F-00-058; U.S. Environmental Protection Agency, Office of Water: Washington, D.C.

Vesilind, P. A. (1979) *Treatment and Disposal of Wastewater Sludges*; Ann Arbor Science Publishers: Ann Arbor, Michigan.

Vesilind, P. A. (1994) The Role of Water in Sludge Dewatering. *Water Environ. Res.*, **66,** 4–11.

Wang, L. K.; Pereira, N. C.; Hung, Y. T. (2007) *Handbook of Environmental Engineering Biosolids Treatment Processes*, 6th ed.; Humana Press: Totowa, New Jersey.

Water Environment Federation (2003) *Wastewater Treatment Plant Design*; Water Environment Federation: Alexandria, Virginia; IWA Publishing: London.

Water Environment Federation (2009) *Operation of Municipal Wastewater Treatment Plants*, 6th ed., WEF Manual of Practice No. 11; McGraw-Hill: New York.

Water Environment Federation; American Society of Civil Engineers; Environmental and Water Resources Institute (2009) *Design of Municipal Wastewater Treatment Plants*, 5th ed., WEF Manual of Practice No. 8; ASCE Manual and Report on Engineering Practice No. 76; McGraw-Hill: New York.

Water Environment Research Foundation (1993) *Guidance Manual for Polymer Selection in Wastewater Treatment Plants;* Water Environment Research Foundation: Alexandria, Virginia.

Water Environment Research Foundation (1995) *Polymer Characterization and Control in Biosolids Management*; Water Environment Research Foundation: Alexandria, Virginia.

Water Environment Research Foundation (2001) *Thickening and Dewatering Processes: How to Evaluate and Implement an Automation*; Water Environment Research Foundation: Alexandria, Virginia.

Water Pollution Control Federation (1980) *Sludge Thickening*; Manual of Practice No. FD-1; Water Pollution Control Federation: Washington, D.C.

Webb, L. J. (1974) A Study of Conditioning Sewage Sludges with Lime. *J. Water Pollut. Control Fed.*, **73,** 192.

Chapter 8

Thickening

(continued)

(*continued*)

1.0 THICKENING PROCESS SELECTION

The most common types of thickening processes are gravity thickening, DAFT, GBT, centrifuge, rotary thickening, and membrane thickening. Thickening in clarifier and storage ponds and fan press thickening are also briefly discussed. Selection of the most applicable method depends on the materials that are being thickened and other factors such as space limitations and power costs. Factors that should be evaluated during the process to select the type of thickening used include:

- Desired thickened-sludge concentration,
- Capture,
- Power use,
- Operator attention,
- Space limitations, and
- Chemical use.

1.1 Materials Thickened

1.1.1 Primary Sludge

Primary sludge has traditionally been thickened in primary clarifiers or gravity thickeners. There is some movement away from thickening in primary clarifiers because the added detention time can lead to solubilization of the organic matter, which reduces the solids removed in the primary clarifier and increases aeration requirements. If this was not a significant concern, then thickening in primary clarifiers could save the cost of a separate thickening process. Other processes that are in being used for primary sludge include rotary thickening; dissolved air flotation thickening (DAFT) and gravity belt thickening (GBT), which are typically used in combination with waste activated sludge (WAS); and centrifuges for large plants.

1.1.2 Waste Activated Sludge

Designers of WAS thickening systems have moved away from gravity thickening and toward the DAFT, GBT, and centrifuges (for some large institutions). The main disadvantage of gravity thickening of WAS is that it is typically limited to a maximum concentration of 2 to 3% solids. The use of rotary thickeners is also becoming more popular. The main disadvantage of the centrifuge for WAS thickening is the higher power use and for some sludges, difficulty in achieving the desired thickened-solids concentration, and capture without the sludge becoming so thick that it is hard to pump or mix. The higher power use is also a reason that fewer DAFTs are being designed today.

1.1.3 Fats, Oils, and Grease

Typically, FOG is best thickened using gravity and water/oil separation. Although polymer has been shown to assist in thickening FOG, DAFT and gravity thickeners are the primary methods used.

1.2 Pretreatment

Sludge conditioning before thickening is a vital part of the process, and is typically needed either to improve capture or solids concentration. Sludge conditioning is discussed in detail in Chapter 7.

Plastic materials, rags, and other particles that are collected in the scum and sludge removal processes can plug and affect the ability to thicken. In addition, these materials typically do not degrade in the digestion process and can cause problems in the dewatering processes and denigrate the quality of the final biosolids material. If these materials are not removed in the pretreatment stage of the treatment plant (i.e., bar screens), then it is desirable to remove them before thickening.

The screening devices typically consist of rotary screens with wire mesh or punch plate with small holes or step screens with small openings. Because these screens are typically used on primary sludge, they are usually covered and foul air is removed and scrubbed.

1.3 Washwater, Filtrate, Subnatant, Centrate

The recycle stream from thickening processes can significantly affect overall wastewater treatment effectiveness and efficiency.

Recycle streams from thickening processes can have high concentrations of biochemical oxygen demand (BOD), total suspended solids (TSS), ammonia, and phosphorous. This load should be taken into account when designing and operating the main wastewater treatment processes. In addition, main processes should be metered to minimize the effects of recycle stream loads or to provide treatment before it is returned to these processes.

If high-strength recycle streams will affect process performance, then storage of these streams and either metering at slow constant rates or at low loading times can significantly attenuate any effects on the secondary treatment system. Equalization can also be achieved where batch wasting is practiced at low loading so that the thickening recycle stream hits at off-peak loading times.

The treatment of thickener effluent is increasing as effluent requirements for nitrogen and phosphorous become more stringent. It can also be more cost effective to treat a concentrated recycle stream than to add more treatment to the entire process flow. Thickener recycle streams can be treated either by themselves or with dewatering recycle streams.

Processes that have been used for recycle streams should be tailored to remove the constituents that will affect the process or compliance. Systems that have been used include conventional activated sludge and several proprietary processes.

2.0 SEDIMENTATION BASINS

2.1 Primary Sedimentation

A primary clarifier can be used as a thickener under certain conditions. Primary sludge thickens well, provided the sludge is reasonably fresh, solids of biological origin (i.e., WAS) are kept to a minimum, and the wastewater is reasonably cool. If sludge of 5 to 6% solids content is to be recovered from a primary sedimentation system, then the sludge transport facilities must be able to move them. This will require short suction piping, adequate net-positive suction head on the primary sludge pump, suction sight glass inspection piping, and a positive means of ascertaining the quantity pumped and the concentration of the slurry (U.S. Environmental Protection Agency, 1974). If anaerobic decomposition begins and gas production floats the sludge, then any thickening benefit is lost and primary performance can be seriously degraded.

2.2 Secondary Sedimentation

Thickening in secondary or intermediate clarifiers has not been successful in the past because biological sludges are difficult to thicken by gravity. Thickening has been improved by using sidewater depths of 4 to 5 m (14 to 16 ft), suction sludge withdrawal mechanisms rather than plow mechanisms, and gentle floor slopes (1:12). If denitrification or anaerobic decomposition begins and gas production floats the sludge, then any thickening benefit will be lost. Thickening in the secondary clarifier can degrade secondary treatment by causing solids carry over. Although thickening within a sedimentation basin can be beneficial under certain conditions, separate thickening is typically recommended.

3.0 GRAVITY THICKENERS

3.1 Principles of Operation

Separate, continuously operating gravity thickening for municipal wastewater sludges was conceptualized in the early 1950s (Torpey, 1954). Until that time, thickening had been carried out within the primary clarifier. Operating problems such as floating sludge, odors, dilute sludge, and poor primary effluent led to the development of the separate thickening tank. Gravity thickeners became the most used sludge concentrating device. Now, however, their use is being challenged by other thickening processes.

3.1.1 *Major Elements*

The major elements of a gravity thickener include the tank, typically circular, with bottom or thickened-sludge collection and surface skimming, overflow weirs for the separated liquid, and sludge pumping. Most gravity thickeners also include the ability to add elutriation water and some form of odor control.

3.1.2 *System Design Considerations*

Circular concrete tanks are the most typical configuration for continuously operating gravity thickeners, although circular steel tanks and rectangular concrete tanks have also been used.

A typical solids concentration profile for municipal wastewater sludges within a continuously operating gravity thickener is shown in Figure 8.1. Sludge moving into the thickener partially disperses in water in the sedimentation zone and partially flows as a density current to the bottom of the sedimentation zone. The solid phase of the sludge both dispersed and, in the density current, creates flocs that settle on top of the thickening zone. Flocs in the thickening zone lose their individual character. They have mutual contacts and thus become a part of the matrix of solids compressed by the pressure of the overlying solids. The displaced water flows upward through interstitial spaces between the particles.

In decision-making about thickener size, the settling process in the sedimentation zone and the consolidation process in the thickening zone should be evaluated. The process (sedimentation or thickening) that requires greater surface area dictates the size of the thickener. For municipal wastewater sludges, the thickening zone area required is almost always greater than that for the sedimentation zone.

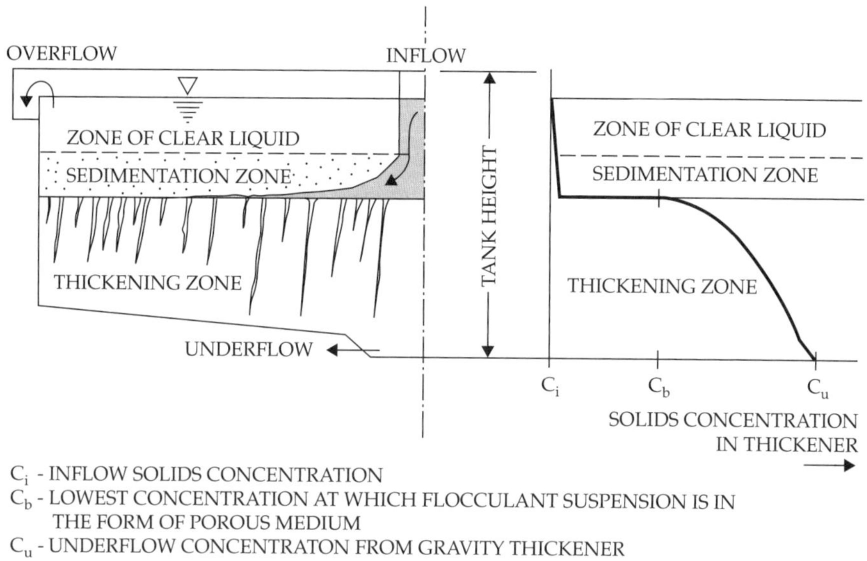

FIGURE 8.1 Typical concentration profile of municipal wastewater sludge in continuously operating gravity thickener.

3.2 Design Criteria

The two major criteria for gravity-thickener design are hydraulic loading and mass loading.

3.2.1 *Surface-Area Requirements*

Surface area of a gravity thickener is important for two reasons. First, it is related to mass loading. The quantity of solids entering the thickener is equal to the product of the flowrate and solids concentration. Upper limits that exist for mass loading will create upper limits for hydraulic loading. Secondly, high hydraulic loading causes excessive carryover of solids in the thickener effluent.

3.2.2 *Mass Loading*

If sludge from the particular facility is available for testing, then required surface area can be found by using a settling column, developing a settling flux curve, and calculating the critical flux (mass loading) for that particular sludge (Adams and Eckenfelder, 1974). In most cases, however, the sludge is not available, and the designer must resort to other methods.

The designer must specify the sludge type (for mixtures, the approximate proportions should be known), the range of solids concentrations that are expected in the thickener inflow, and the underflow concentration required for downstream processing.

3.2.3 *Hydraulic Loading*

Typical maximum hydraulic loading rates of 1200 to 1600 $L/m^2{\cdot}h$ (25 to 33 gph/sq ft) have been used in the past but mainly for primary sludges. For sludges such as waste-activated or similar types, much lower hydraulic loading rates, 200 to 400 $L/m^2{\cdot}h$ (4 to 8 gph/sq ft) are more applicable (Noland and Dickerson, 1978).

Using the typical maximum hydraulic loading rates mentioned above, maximum velocities for primary sludges are 1.0 to 1.3 m/h (3.3 to 4.4 ft/h) and for WAS are 0.2 to 0.3 m/h (0.5 to 1.1 ft/h).

Several researchers have related overflow rates to odor control, but odor is caused by excessive retention of solids and can be better controlled by removing the thickened sludge from the thickener at an increased frequency and by collecting and treating the odors.

3.2.4 *Drive-Torque Requirements*

Sludge on the floor of a circular thickener resists the movement of the solids rake and thus produces torque. Calculation of torque for a circular drive unit is based on the simple cantilevered beam equation represented by Eq. 8.1:

$$T = WR^2 \tag{8.1}$$

where

T = torque, kg/m;
W = uniform load (sludge specific; see Table 8.1); and
R = tank radius, m.

Note that there are several levels of torque that must be specified for a circular gravity thickener (Boyle, 1978).

3.2.5 *Total Tank Depth*

The total vertical depth of a gravity thickener is based on three considerations: tank free board, settling zone (zone of clear liquid and sedimentation zone), and compression and storage zone (thickening zone).

3.2.6 *Freeboard*

Tank freeboard is the vertical distance between tank liquid surface and top of vertical tank wall. It is a function of tank diameter, type of bridge structure (half or full

Table 8.1 Typical uniform load (W) values (1 lb/ft = 1.49 kg/m).

Sludge type	Truss arm W, lb/ft*
Primary sludge only (little grit)	30
Primary sludge only (with grit)	40
Primary sludge + lime	40 to 60
Waste activated sludge (WAS)	
Air	20
Oxygen	20
Trickling filter (TF)	20
Thermal conditioned	80
Primary sludge + WAS	20 to 30
Primary sludge + TF	20 to 30

*Rake arms typically have a tip speed of 3 to 6 m/min (10 to 20 ft/min).

bridge), type of influent piping arrangement, and whether skimming is provided. Freeboard will be at least 0.6 to 0.9 m (2 to 3 ft), although distances up to 2 to 3 m (7 to 10 ft) have been used by some designers.

3.2.7 *Settling Zone*

This zone encompasses the theoretical zone of clear liquid and sedimentation as shown in Figure 8.1. Typically, 1.2 to 1.8 m (4 to 6 ft) is necessary, with the greater depth being for difficult sludges, such as waste-activated or nitrified sludge.

3.2.8 *Compression and Storage Zone*

Sufficient tank volume must be provided so that the solids will be retained for the period of time required to thicken the slurry to the required concentration. In addition, sufficient storage is necessary to compensate for fluctuations in solids loading rate.

Another consideration is that gas may be produced because of anaerobic conditions or denitrification. Development of these conditions depends on the type of sludge, liquid temperature, and the length of time sludge is kept in the thickener. Plant operating experience has indicated that the total volume in this zone should not exceed 24 hours of maximum sludge wasting.

3.2.9 *Floor Slope*

The floor slopes of thickeners are typically greater than 17 cm of vertical distance per meter of tank radius (2 in. vertical distance/ft tank radius). This is steeper than the floor slopes for standard clarifiers. The steeper slope maximizes the depth of solids over the sludge hopper, allowing the thickest sludge to be removed. The steeper slope also reduces sludge raking problems by allowing gravity to do a greater part of the work in moving the settled solids to the center of the thickener.

3.2.10 *Odor Control*

Depending on the temperatures and sludge detention times, gravity thickeners can be the source of strong odors. Elutriation and continuous sludge removal can help, but often covering and odor removal and scrubbing are necessary.

3.2.11 *Pickets*

Stirring with pickets in gravity thickeners is thought by some to help consolidate sludge in the thickening zone (Dick and Ewing, 1967). Support rake mechanisms, however, typically can provide sufficient sludge mixing to make special pickets unnecessary. Others have indicated that stirring in general and pickets in particular are detrimental to thickening and unnecessarily add to power use (Water Environment Federation et al., 2009).

3.3 Performance

3.3.1 *Chemical Requirements*

Addition of polymer to gravity-thickener feed has been practiced at several plants. Results indicate that addition of polymers improves solids capture but has little or no effect on increasing solids underflow concentration (see Chapter 7 for further discussion).

3.3.2 *Power Requirements*

Annual power consumption for a continuously operating gravity thickener as a function of gravity-thickener surface area is shown in Figure 8.2. As an example, for a gravity-thickener surface area of 93 m^2 (1000 sq ft), a designer would include a yearly power usage of 4500 kW·h (16.2 GJ) in the cost analysis. Figure 8.2 does not include accessories such as pumps or polymer feed systems.

3.3.3 *Process-Control Strategy*

Process elements that can be controlled for gravity-thickener operation include feed sludge, thickened-sludge pumping, polymer feed, skimmer speed and elutriation water, sludge thickness, and thickened-sludge flow.

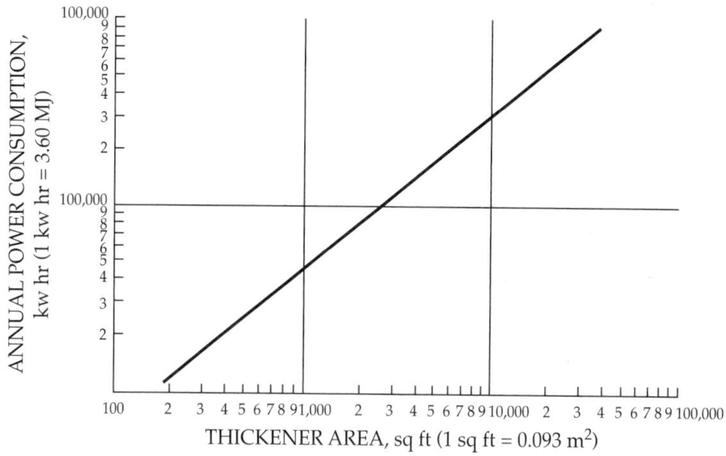

FIGURE 8.2 Annual power consumption of continuously operating gravity thickeners.

3.3.4 Gravity-Thickener Control Strategy

The goal in the gravity thickening process is to provide the digesters with primary sludge with consistent density and known sludge mass, which ensures steady-state conditions and ultimately optimizes digester performance. This strategy includes microwave density meters (MDM) on the discharge side of the thickened-sludge pumps and flow meters, one on each suction line to the thickened-sludge pumps from the gravity thickeners.

The thickening process will be controlled by adjusting the speed of the thickened-sludge pumps to draw thickened sludge with a uniform sludge density, based on MDM readings on the discharge lines of the pumps. The use of this system should be able to control the sludge density to within ±0.5% solids.

After the pumps initialize, they begin pumping thickened sludge to the digesters at the last set speed or at an operator input speed. After a preset time delay to allow the sludge from the thickener to get to the MDM, the automated speed control will begin controlling the process.

The MDM downstream of each thickened-sludge pump operates to maintain the density within a range of setpoints. When the metered thickened-sludge density approaches the upper set point, the variable frequency drive (VFD) on the pump incrementally increases the pump speed to maintain the density below the upper setpoint or until the pump has reached its upper speed limit. The pump speed remains constant until the thickened-sludge density approaches the lower setpoint, when the VFD on the pump incrementally decrease the pump speed so that the metered sludge remains above its lower density setpoint, in which case the pump stabilizes, or the pump has reached its minimum speed limit. If the thickened-sludge pump reaches its minimum speed, then it will continue to pump at that speed until either the sludge concentration increases to the upper setpoint, at which point the pump speed will increase, or the total solids concentration falls below the lower setpoint and remains there for a preset time, at which point the pump will shut off.

4.0 DISSOLVED AIR FLOTATION THICKENING

4.1 Principles of Operation

4.1.1 Major Elements

A DAFT system consists of a flotation unit and a saturator (Figure 8.3). The flotation unit separates the solid from the liquid phase and the saturator dissolves air into water under pressure. Pressure-saturated water (from the saturator) is introduced to the flotation unit through a reducing valve. Downstream from the valve, the pressure is virtually atmospheric and the saturator feed is blended with wastewater feed and becomes highly supersaturated with air. The air precipitates out of solution in the form of small bubbles that become attached to wastewater particles forming bubble-particle agglomerates with a density lower than water.

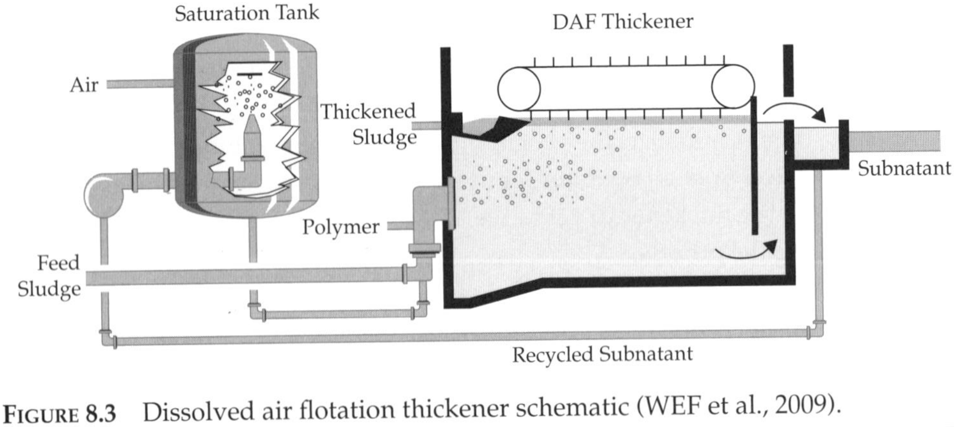

FIGURE 8.3 Dissolved air flotation thickener schematic (WEF et al., 2009).

Flotation thickening can be carried out with or without chemical aids such as polymers. If they are used, however, chemicals typically enter at the mixing point of the saturator feed and the sludge feed. Proper mixing to ensure chemical dispersion, while avoiding excessive shearing forces, produces the best aggregation of sludge particles and air bubbles. Experience has shown that, if used, introduction of polymer just downstream of the point of blending of the sludge feed and the precipitating saturator feed produces the best results. Particular advantages of DAFT systems include the following (Bratby et al., 2004)

- Thickens wastewater sludges with or without polymer;
- Co-thickens primary and secondary sludges;
- Thickens scum from both the primary and secondary scum collection systems;
- Allows scum and sludge to be transported to the thickening process with maximum water flow;
- Allows the separation and capture of grit from a continuous bottom sludge removal system when co-thickening primary and secondary sludges;
- Continuously produces a homogeneously mixed thickened-sludge product that is of ideal quality for feeding digesters;
- Allows all solids processing recycle loads to be concentrated into one stream; and
- Achieves significant soluble BOD reduction in the DAFT liquid stream.

4.1.2 Clarification and Thickening Functions

In the flotation unit, buoyancy causes the bubble-particle agglomerates to rise to the water level and accumulate as a float, achieving separation of the solid phase from the liquid phase. Because of the difference in density between float and water, a stage is reached where the top of the float, continuously rising above the water level, reaches the top of the unit where solids are continuously removed by scraping. Drainage of interstitial water from the float above the water level increases the solids concentration. This stage of the process is termed thickening.

Flotation, then, has basically two functions: (1) clarification or separation of the solid phase from the liquid phase; and (2) thickening or dewatering of the separated (accumulated) solids.

Process variables that influence the clarification aspect of DAFT are the downflow rate and the air/solids ratio. The downflow rate is analogous to the overflow rate used for sedimentation tanks. It is defined as the total flow into the unit divided by the effective plan area of the flotation unit. The air/solids ratio is defined as the mass of air precipitated per unit time divided by the mass of solids introduced per unit time. The downflow rate and the air solids ratio are related by the following empirical expression (Bratby, 1978)

$$v_L = K_1 \times (a_S)^{K2} - K_3 \quad 8.2$$

where

K_1, K_2, and K_3 = empirical constants for a particular waste or sludge;
v_L = downflow rate (m/d or gpd/sq ft); and
a_S = air/solids.

Process variables influencing thickening are the solids loading rate and the height that the float extends above the water level. Thickening of float solids occurs principally by drainage of interstitial water from the float above the water level. Both the solids loading rate and the height of float above the water level directly influence the time allowed for drainage of the float. The solids loading rate, defined as the mass of solids introduced per unit area of flotation unit per unit time, governs the rise rate of the float above the water level; the height of float above the water level governs the distance through which the float travels before being removed from the unit. The following empirical relationships describe thickening by DAFT (Bratby, 1978; Bratby and Ambrose, 1995).

$$C_F = K_4 \times (d_W)^{K5} \times (Q_S)^{-K6} \quad 8.3$$

$$d_T = (d_B + d_W) = d_W[(a_S)^{K7} + K_8] \times a_S^{-K7} \quad 8.4$$

where

$K_4 - K_8$ = constants for a particular waste or sludge;
C_F = float solids concentration (percent) at the top of the unit (i.e., where float solids are removed from the unit);
Q_S = solids loading rate (kg/m^2·d);
d_W = height of float above water level (m);
d_B = depth of float below the water level (m); and
d_T = total float depth (m).

Note that for a given set of parameters, adjusting the air/solids does not influence the float solids concentration. The air/solids does, however, affect the total depth of float solids.

The accuracy of the expression for total float depth ($d_B + d_W$) becomes uncertain at air/solids below a certain value. At lower air/solids, the float depth, in practice, tends to be greater than predicted by Eq. 8.3. The addition of polymer reduces the float depth (Bratby, 1978). With activated sludge and either with or without polymers, the limiting air/solids ratio is approximately 0.02; with raw wastewater and using metal coagulants such as alum, the limiting air/solids ratio is approximately 0.06; with stabilization pond algae separation and using metal coagulants, the limiting air/solids ratio is approximately 0.03; with flotation applied to thickening the sludge from highly colored waters using polymer, the limiting air/solids ratio is approximately 0.03.

Bratby and Ambrose (1995) expanded the equations to include the actual time spent by the float above the water level before being removed to the sludge hopper by the scraper. An operator can control the scraper speed and periodicity of operation. An effective time, t_E, spent by the float above the water level is defined as (Bratby and Ambrose 1995):

$$t_E = (t_C/t_{ON}) \times (L/v) \qquad 8.5$$

where

t_C = float scraper cycle time = $t_{ON} + t_{OFF}$ (minutes);
t_{ON} = time that the scraper is ON during a given cycle;
t_{OFF} = time that the scraper is OFF during a given cycle;
L = effective length of rectangular DAFT or periphery of circular DAFT (m or ft); and
v = velocity of travel of scrapers (peripheral speed for circular units) (m/d or ft/d).

Therefore, the following relationships incorporate the effects of t_E

$$C_F = K^* \times t_E^{[K5/(1+K5)]} \qquad 8.6$$

$$d_T = (d_B + d_W) = Q_S \times t_E^{[1/(1+K5)]} \bullet (1 + K_8/a_S^{K7})/(10 \times K^*) \qquad 8.7$$

where

$K^* = [K_4/(Q_S^{(K6-K5)} \times 10^{K5})]^{[1/(1+K5)]}$

At long effective times, t_E, more opportunity is provided for interstitial water in the float solids to drain and, consequently, the float solids concentration will be higher. A practical upper limit exists, however, for the value of t_E. For a given d_W, a greater value of d_B is required to supply the buoyancy to support the height of float above the water. Haarhoff and Bezuidenhout (1999) showed that the depth of float below the water level, d_B, is proportional to the square root of t_E.

Therefore, just as increasing t_E increases d_W, it also increases the total float depth. The limitation in t_E, however, is to avoid increasing the float depth to the extent that float solids are dragged through with the underflow, causing deterioration of effluent quality.

4.2 Equipment Description

4.2.1 *Dissolved Air Flotation Thickening Tanks*

Flotation thickeners are constructed of steel or concrete and are either rectangular or circular tanks. Typically, the smaller rectangular units up to 2- to 3-m (8- to 10-ft) wide are steel, and the larger units are concrete. Only the small, circular units in municipal plants are constructed with steel (Water Pollution Control Federation, 1980). Steel tanks come completely assembled and only require a concrete foundation pad and piping and wiring hookups.

Although the purchase price for steel tanks is much higher than concrete, they do not require the considerable field labor and expensive equipment installation. Structural and shipping problems limit steel DAF units to smaller sizes, up to approximately 40.5 m^2 (450 sq ft) or less for rectangular units, and 9 m^2 (100 sq ft) for circular units. For a large installation requiring multiple tanks or large tanks, concrete tanks are more economical.

Rectangular and circular DAFT tanks each have advantages and disadvantages. In rectangular units, space is minimized. A further advantage is that the straight-through flow regime may be better suited to higher-rate treatment. One objection to rectangular units relates to the chain-and-flight float scraper and, if used, the bottom sludge scraper mechanisms. If designed, specified, and installed correctly, however, then these tanks can operate relatively trouble-free.

Rectangular DAFTs have some advantages related to float removal. Skimmers, for example, can easily be closely spaced, and they can be designed to skim the entire surface. Because of the side-walls, float does not easily move around the end of the skimmers. Bottom sludge flights are typically driven by a separate unit and, hence, can be operated independently of the skimmer flights.

The main advantage of circular units is their lower cost in terms of both structural concrete and mechanical equipment. For example, two 18-m (60-ft) diameter circular units are equivalent to three 6-m by 27-m (20-ft by 90-ft) rectangular units. The rectangular units require approximately 11% more structural concrete and more drives and controls, which increase maintenance requirements.

The relative positions of the float removal and effluent removal points can be an issue with rectangular tanks. The float depth is typically greatest in the direction that the float is scraped, with maximum depth at the point of float removal. Therefore, some designs scrape the float from the effluent end to the feed end to avoid excessive float depths at the point where DAFT subnatant is removed from the unit. The feed end of the DAFT unit is where the likelihood of greatest float disturbance occurs. Therefore, designs that scrape to the feed end may suffer the greatest float disturbance and dilution of the final float concentration.

Scraping to the effluent end allows the thickened float solids to remain relatively undisturbed up to the point they are removed from the unit. In this case, care must be taken in design and operation to limit the float depth to avoid scouring of the underside of the float and deterioration of the DAFT subnatant. If bottom sludge scrapers are included with the float scrapers, then sludge typically is not scraped to the effluent end, which avoids increasing subnatant suspended solids. Therefore, it is possible that the thickened float pumps may be at the effluent end and bottom sludge pumps at the feed end of the DAFT tanks.

Float and bottom sludge scraper mechanisms tend to be simplest in circular units. In some successful designs, one or more float hoppers are situated at the periphery, so that float is scraped away from the influence of the DAFT inlet. In these designs, if included, bottom sludge is scraped to bottom sludge hoppers at the center, and sludge pumps are located at the outside periphery of the tank.

In rectangular designs, float must be scraped along the whole length of the unit. This could increase the possibility of rotation and dilution of the uppermost thickened float layers. The advantage of circular designs is that two or more float hoppers can be distributed around the periphery, thereby minimizing the distance over which thickened float needs to be transported before discharge.

The same degree of efficient blending of inlet flows can be achieved in both circular and rectangular designs. However, in rectangular tanks, expansion of the feed point into the tank cross-section is required, which may promote hydraulic short-circuiting unless particular care is taken in inlet baffling to mitigate this effect.

4.2.2 *Saturation System*

The heart of a DAFT system is the saturation system, where air is dissolved into water under pressure then released at near-atmospheric pressure in the form of microscopic bubbles. The saturation system typically includes a recycle pressurization pump, an air compressor, an air saturation tank, and a pressure release valve (Figure 8.3). The flow through the pressurization pump typically is recycle flow, but can be makeup water. The pressure release valve controls the pressure loss and distributes the gas saturated pressurized flow into the feed sludge as dissolved air emerges from solution. The rapid reduction in pressure causes dissolved air (under pressure) to emerge from solution or "effervesce" into minute bubbles. Several principles are important in saturation system (Bratby et al., 2004):

- Dissolution of a gas into a solute is a simple process that is dependent upon pressure, temperature, the solubility of the gas in question in the solute (in this instance, water) and the surface area of the liquid available for gas transfer.
- The solubility of nitrogen in water is roughly half that of oxygen.
- At 78% nitrogen and 21% oxygen, water will more readily absorb oxygen. The volume represented by the oxygen is left for addition of more gas in the unvented headspace typically provided in standard saturation tank designs; the volume represented by the nitrogen is absorbed by the water during the critical contact period in the headspace.

With an unvented headspace, the saturation efficiency, as measured by the volume of gas absorbed in the pressurization flow stream, rapidly deteriorates to approximately two-thirds that if the headspace gas is at or near atmospheric concentrations. This represents a 33% loss of process capacity. It follows then that it is necessary to vent excess nitrogen continuously to maintain process efficiency. Because more than 80% of the energy consumed by the DAFT process is required for pressurizing the saturation system pumps, this represents an opportunity to fine-tune the process to reduce energy consumption.

The aspirating pumped system is one alternative system that is found in several DAFT installations. The apparent advantage of such systems is that the saturation vessel is not required, and the saturation system is typically simpler, without the saturator level controls required in more conventional saturation systems.

Air is aspirated into the suction side of a specially designed centrifugal pump and saturation occurs in the pressurized side of the pumped line. Some installations include a small retention vessel in the line; whereas others use only the retention time in the pressure line for air saturation. Recent systems appear to have overcome the air binding problem common with older systems, which caused aspirating pump systems to lose favor. For example, Bratby and Marais (1975) tested an air aspirating centrifugal pump for flotation and found that only 23% of the air required for saturation could be dissolved by pump suction air aspiration and air binding of the pump occurred at 2% by volume air injection at 1000 kPa (145 psig) and at 11% by volume at 170 kPa (25 psi).

Some have claimed that recent developments in pump technology specifically for this purpose have aspirated 12% air by volume without air binding. To date, however, independently established performance data appears to be lacking. In addition, there have been reports of pump failures because of cavitation eroding the impeller.

4.2.3 *Float Collection*

All DAFT tanks are equipped with float skimming equipment. The surface skimmers remove floating material from the thickening tank to maintain a constant average float blanket depth.

Preceding discussions have pointed out that the principal mechanism affecting thickening during flotation is drainage of float layers above the water level. For this reason, float solids concentration profiles through the float depth will be at a maximum at the top of the float and will decrease markedly with depth down to the position of the water level. The concentration will then reduce through subsequent float layers until a minimum is reached at the bottom of the float.

Figure 8.4 illustrates this occurrence (Bratby, 1978). Figure 8.4 shows results obtained in a pilot flotation unit in which float solids concentrations were taken at various depths. The solids loading rate was set at 55 kg/m^2·d (11.2 lb/sq ft/d) and the air/solids at 0.036. The pilot-scale unit was fitted with a horizontal action scraper at the top of the unit. Minimal disturbance of the float occurred, and concentration profiles presented in Figure 8.4 are those obtained under ideal conditions. In contrast, Figure 8.4 presents results obtained with a full-scale proprietary flotation unit with the same sludge. In this case, the solids loading rate was approximately 60 kg/m^2·d (12.3 lb/sq ft/d) and the air/solids was 0.036 to 0.10. A marked difference is evident in the profile. The full-scale unit does not show the sharp increase in concentration above the water level achieved in the pilot-scale unit operating under similar conditions.

The float solids concentration achieved in the pilot unit was approximately 6.5%; whereas, in the full-scale unit, it was about 5%. The difference in results is because of differences in the methods by which float solids are removed from the two units. In the case of the pilot unit, the float scraper removes only the top-most slivers of float solids that reach the top. In the full-scale unit, the scraper blades dig into the float to a depth of about 178 mm (7 in.). This arrangement typifies many DAFT units.

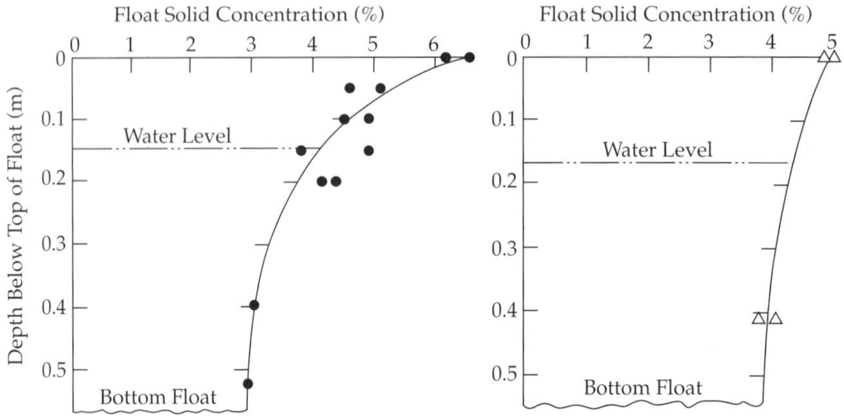

FIGURE 8.4 Float solids concentration profiles (WEF, 2008).

In the case of the full-scale unit, scraper arrangement had two effects: (1) it maintained the top of the float at the level of the float take-off weir (i.e., it fulfilled its function of removing float solids from the unit); and (2) it had the undesirable function of mixing the float. The influence of this disturbance will extend beyond the scraper depth. The effect of this mixing action is to replace the upper layers of the float, which have attained maximum drainage, with lower, wetter layers.

These results indicate that a float scraper should disturb the float as little as possible and remove only the topmost layer of the float, because solids concentration decreases markedly with depth. Minimizing disturbance requires that the bottom of the scraper be situated above the water level, as far as practicable, and that float solids do not build up excessively in front of the scraper as it travels around or along the top of the unit. A further important consideration is the rigidity of the float scraper material. Excessive flexing of the scraper material is not conducive to removing only the topmost thickened slivers of float, as described previously.

Circular tank designs provide the opportunity to minimize the distance that float is scraped by providing two or more float hoppers around the periphery.

4.2.4 Bottom Sludge Collection

Bottom sludge scrapers are often considered essential for removing nonfloatable, heavier solids that settle to the bottom of the flotation thickener. Unlike float scrapers, however, not all DAFT tanks are equipped with bottom sludge collection. For example, plants that use DAFT thickeners for WAS only could dispense with bottom sludge scrapers, unless special conditions require them.

4.2.5 *Other Equipment*

Many units are baffled and equipped with an overflow weir. In these units, clarified effluent passes under an end baffle (rectangular units) or peripheral baffle (circular units) and then flows over the weir to an effluent launder. The weir controls the liquid level within the flotation tank. A disadvantage to this, however, is that the liquid level cannot be changed and will vary depending on the total feed flow into the unit. Some successful designs have used submerged launders, which allows selection of the liquid level to maximize drainage of the float layers. A level sensor provides a signal to either an effluent pump or a modulating valve. In either case, the operator sets the operating level and, independent of feed flow, the liquid level is maintained at the setpoint by modulating the pump speed or valve position.

The float handling and pumping system is also an important part of the DAFT system. After removal of the float from the DAFT unit, float solids are deposited in a hopper and then pumped for further processing. This aspect of the operation requires special considerations because of the following characteristics of float solids:

- Floats on the surface of the liquid;
- Does not flow as easily as water because it is concentrated; and
- Does not flow easily on flat or gently sloping surfaces, especially if the surfaces lack a smooth surface that reduces skin friction.

In addition, conventional sump level controls make the situation more difficult because pumps are typically controlled on a high/low level basis, with the former used to start the pumps and the latter used to stop the pumps. The effect of this control scheme is to concentrate the floating material by decantation, rather than by pumping the solids away. If left on the surface of a sump for a protracted period and exposed to the air, solids dry out and become a dry cake, making the task of trying to move the solids toward the pump intake that much more difficult. Operators are frequently forced to spend considerable time directing high pressure hose streams to the floating mass, trying to force the material under, thereby wetting it (and decreasing its concentration) until the material can find its way into the pump.

A preferable arrangement is to position the pump intake connection in a confined sump located well below the nominal floor of the sump to concentrate the floating solids in a small surface area right over the pump inlet. The pump is then controlled to shut down not on level, but rather when the pump breaks suction, guaranteeing that the floating material has reached the pump inlet (Bratby et al., 2004).

4.3 Design Criteria

4.3.1 *Air/Solids*

The air/solids (g air/g solids) affects the sludge rise rate and, as importantly, the total float depth. The necessary air/solids, typically between 0.02 and 0.04, depends mostly on sludge characteristics (Bratby et al., 2004).

4.3.2 *Hydraulic Loading*

The hydraulic loading rate for a DAFT should be expressed as the total flow (influent plus recycle) through the unit, divided by the surface area. Units are typically $m^3/m^2{\cdot}h$ (gpm/sq ft). The maximum hydraulic rate must always be less than the minimum rise rate of the sludge/air particles to ensure that all the particles will reach the sludge float before the particle reaches the effluent end of the tank.

The DAFT process is relatively insensitive to hydraulic loading, and rates as high as 14 $m^3/m^2{\cdot}h$ (5.7 gpm/sq ft) have been applied for activated sludge thickening with no detrimental effect on effluent suspended solids. These rates were obtained using polymers. When polymers are not used, the corresponding values may be lower, although it is probable that the critical parameter for sizing the DAFT unit will still be the solids loading rate.

4.3.3 *Solids Loading*

The solids loading rate for a DAFT should be expressed as the total feed solids load to the unit divided by the surface area. Units are typically $kg/m^2{\cdot}d$ (lb/sq ft/d). Typically, increasing the solids loading rate decreases the float concentration. The addition of polymer typically will allow an increased solids loading rate for a given float solids concentration up to a given dosage (see below). The design solids loading rate depends on the characteristics of the waste and the desired float solids concentration. Design values typically are based on peak day solids loads and have ranged from less than approximately 100 to 1150 $kg/m^2{\cdot}d$ (20 to 235 lb/sq ft/d).

4.4 Performance

Through experience with specific flotation thickening units, operators may determine the optimum ranges of performance for their systems. Table 8.2 provides typical operating data from several WWTPs for comparing system performance.

TABLE 8.2 Typical operating data from dissolved air flotation thickening (DAFT) treating a variety of sludges.

Installation	Sludge type[a]	Polymer dosage, g/kg dry solids	Solids loading rate, kg/m^2·d	Float concentration, %	Subnatant suspended solids, mg/L	Ref.
Boise, Idaho	A	0	117	4.0		1
Levittown, Pennsylvania	A	0	63	6.5		1
Wayne County, Michigan	A	0	97	4.6		1
Athol, Massachusetts	A	1.0	376	4.0	50	4
Boulder, Colorado	A	2.0	89	4.0		2
Denver, Colorado	A	2.4	614	4.3	21	2
Warren, Michigan	A	7.5	247	5.3		5
Westgate, Virginia	A	1.5	822	7.3	20	5
Athol, Massachusetts	EA	1.0	376	4.0	50	5
Biddeford, Maine	CS	2.2	375	6.6		5
Philadelphia, Pennsylvania	A + RBC	0	30	4.0	200	4
Fairfax, Virginia	A[b]	1.2	820	7.3	20	4
Frankenmuth, Michigan	A[c]	0	68	3.0	750	1
Xenia, Ohio	AD	0	176	4.0		5
Xenia, Ohio	AD	3.0	705	8.0		5
Levittown, Pennsylvania	PRI + A	0	117	8.6		1
Dalton, Georgia	PRI + A	0	88	6.1		1

(continued)

Table 8.2 Typical operating data from dissolved air flotation thickening (DAFT) treating a variety of sludges (*continued*).

Installation	Sludge type[a]	Polymer dosage, g/kg dry solids	Solids loading rate, $kg/m^2 \cdot d$	Float concentration, %	Subnatant suspended solids, mg/L	Ref.
Seattle, Washington	PRI + A	1.2	111	6.2	300	6
San Jose, California	PRI + A[d]	0	223	7.1		1
San Jose, California	PRI + A[e]	0	188	5.3		1
Eugene, Oregon	PRI + TF	0	147	4.8	500	4
Springdale, Arizona	PRI + \TF	3.5	294	6.5	200	4
Eugene, Oregon	PRI + TF	0	147	5.0		5
Englewood, Colorado	PRI + TF/SC	4.0	79	6.2	210	2

[a] PRI = Primary sludge

A = Waste activated sludge (plant with primary clarifiers)

CS = Contact-stabilization activated sludge

EA = Extended aeration activated sludge

TF = Trickling filter sludge

TF/SC = Trickling filter/solids contact process

RBC = Rotating biological contactor

AD = Aerobically digested waste activated sludge

[b] Oxygen plant

[c] Considerable brewery waste

[d] Noncanning season

[e] Canning season

1 $kg/m^2 \cdot d$ = 0.20 lb/sq ft·d

1 g/kg = 2 lb/ton dry solids

1. U.S. Environmental Protection Agency, 1979 (courtesy of Envirex, Inc.).
2. Plant data, 2008.
3. Bratby and Ambrose, 1995.
4. Komline, 1976.
5. Butler et al., 1997.

4.4.1 *Thickened-Solids Concentrations*

A float total solids content of 4% represents a typical minimum for flotation thickeners handling solids without primary sludge. Depending on the solids loading rate, however, a 5 to 6% solids content can be expected.

One factor influencing the float solids concentration is sludge characteristics, often expressed in terms of the sludge volume index (SVI) of the sludge. This is probably not the principal factor influencing the thickened concentration achieved. Other factors such as design and operation of the float scraper, proper drainage of the float layers, and solids loading rate are equally, if not more, important.

Higher overall solids concentrations can be achieved by co-thickening primary and secondary sludges using DAFT, compared with separate thickening of primary sludge in a gravity thickener and secondary sludge in a DAFT (Butler et al., 1997). A float total solids content of 6% approximates a typical minimum for flotation thickeners handling a 50/50 mixture of secondary and primary sludge. Under optimum conditions, minimum solids contents ranging from 6 to 8% can be expected from thickening such mixtures (Bratby et al., 2004).

4.4.2 *Capture Rates*

Capture rates can vary widely depending on the particular installation. Even the most poorly operating DAFT systems, however, can typically produce a capture rate of at least 90%. This is calculated as the difference between feed suspended solids and DAFT effluent suspended solids. Therefore, if feed solids to the DAFT were 5000 mg/L, then a 90% capture would produce effluent suspended solids of 500 mg/L. Polymer addition tends to reduce effluent solids and improve capture rates.

Well-operated DAFTs treating WAS only have produced capture rates of approximately 99% with polymer addition, even with solids and hydraulic loadings as high as 1150 $kg/m^2{\cdot}d$ (235 lb/sq ft/d) and 13.2 $m^3/m^2{\cdot}h$ (5.4 gpm/sq ft), respectively (Bratby and Jones, 2007).

Results from DAFT systems thickening WAS show that effluent suspended solids are relatively insensitive to polymer dosage (beyond a minimum value), hydraulic loading, and air/solids ratio (beyond a minimum value).

4.4.3 *Chemical Requirements*

Dissolved air flotation thickening can be performed with or without chemical addition. The following operating advantages may occur with the use of polymers: the size of the DAFT may be reduced; solids capture may be improved, thus reducing the amount of solids recycled back to the liquid handling system; and an existing, overloaded facility in which polymers are not being used may be upgraded. Polymers also act as a surfactant, thus potentially allowing better attachment to air bubbles. The major disadvantage of polymers is cost (including polymer cost and operation and maintenance of polymer feed equipment) when calculated over the useful lifetime of the plant. Improvement in float solids concentration occurs up to a certain minimum concentration. For a DAFT facility co-thickening primary sludge and trickling filter/solids contact biological sludge, no perceptible effect of polymer dosage on float solids concentration is evident beyond approximately 2.5 g/kg dry solids (5 lb/ton dry solids).

4.4.4 *Power Requirements*

The biggest power draw with a DAFT system is the centrifugal pressurization pump. Other power requirements in the overall system are relatively minor in comparison. For this reason, Bratby et al. (2004, 2006) and Bratby and Jones (2007) pointed out that the pressurization pumps should be designed to always operate at the best efficiency point (BEP). Based on several recent installations, power requirements for DAFT systems are approximately 3 to 5 W per kg/d (0.0020 to 0.0033 hp per lb/d) design dry solids load to the DAFT system. The design dry solids load to the DAFT is typically computed as the peak-day value.

4.4.5 *Process-Control Strategy*

Strategies for DAFT operation include controlling feed sludge, skimmers, saturator pressure and feed rate, polymer feed, and thickened-sludge pumping.

Proper operation requires reducing the variations of the feed rate and concentration. A feed sludge holding and mixing tank helps with intermittent operation. Most units are operated continuously. Some are operated with a short shutdown period during weekends. Others are operated only during certain hours of the day.

Operating the skimmer so that float is only skimmed above the water level as described previously typically is achieved by manually adjusting the water level (if adjustment is included in design) to a predetermined level.

The speed and number of the float skimmer arms control the rate at which floated sludge is pushed into the sludge trough, as explained earlier. The speed and the ON-OFF times of the float skimmers should be set to maximize the float solids concentration but not too slow as to cause excessive float depth accumulation. Operation of float skimmers typically occurs through manual adjustments, based on float depth monitoring.

Pressure in the saturator should be controlled within the 350 to 590 kPa (50 to 85 psig) range, and should not be operated at less than 350 kPa (50 psig) because this can cause disproportionate reductions in efficiency and increases in total float depth (see Bratby, 1978; Bratby and Jones, 1995). The pump discharge pressure will be higher than the saturator pressure because of line and appurtenance losses. Depending on the design, pump discharge pressure should correspond with the best efficiency point (BEP) of the particular pump.

The saturator feed rate should maintain the required air/solids. This is typically set manually in DAFT systems to balance the pumped feed flow to the saturator with the feed rate to the DAFT unit from the saturator.

Polymer feed is often automatically coupled with feed-sludge rate. By monitoring the flowrate and solids content of the feed sludge, the polymer feed pumps can be automatically adjusted to maintain a predetermined polymer dosage, in terms of g polymer/kg dry sludge solids (lb/t dry solids).

Thickened-sludge pumping is often automatically controlled by the level in the float solids hopper. Initiation of the pump would result from high level in the hopper; some designers recommend stopping the pump when it breaks suction, rather than at low level, to avoid separating the sludge in the hopper (Bratby et al., 2004). During operation, DAF thickeners should be sampled and analyzed at a minimum for the following:

- Influent for TSS, total solids, and flow (typically using online instrumentation for both solids and flow).
- Effluent for TSS, total solids, and flow. Effluent solids instrumentation is a possibility, although fouling with residual polymer is an issue that has prevented their use at some installations. Continuously monitoring DAFT effluent flow and solids is an advantage in terms of conducting plant-wide mass balances and the possibility of controlling polymer feed by effluent suspended solids. This latter control strategy has been proposed as a possibility, but experience on its effectiveness needs to be demonstrated.
- Float (sludge withdrawal) for total solids and flow.
- Depth of blanket (DOB). Traditionally, float depth has been determined using a "sludge judge" type of apparatus; this has often produced unsatisfactory results, however, because of the high float concentrations and the slow passage of thickened float into the cylinder. Better results have been obtained using an optical interface detector. The sample location should be near the beach where sludge is scraped into the float trough.

Although instrumentation to automatically monitor total float depth has been attempted for both circular and rectangular DAFT units, long-term operating success has not been demonstrated yet. Availability of such instrumentation would be valuable because scraper speed could be automated based on maintaining a predetermined total float depth, thereby ensuring the maximum float concentration possible. It is not unusual to see free water at the surface of the DAFT during operation; this is not conducive to efficient operation, however, because drainage of water from the accumulated float is minimized.

If applicable, polymer quality assurance tests should be conducted. This may comprise simple polymer total solids determinations and/or jar tests or more sophisticated quality control testing such as Fourier transform infrared spectroscopy.

Occasionally, the mass of precipitated air in the saturator feed should be determined to evaluate the operating air/solids ratio. The methodology has been documented elsewhere (Bratby and Marais, 1975, 1977).

5.0 CENTRIFUGAL THICKENING

5.1 Theory of Operation

A centrifuge is often compared to a clarifier when discussing theory of operation. With a clarifier, a mixture of solids and liquids are added to a tank. With the help of gravity and time, the solids settle to the bottom of the tank while the clear liquid stays on top. This gravitational force is always occurring on earth and is equivalent to 1 G.

A centrifuge is essentially a clarifier wrapped around an axis spun at a high rate of speed. This high speed creates centripetal force that acts like gravitational force on the solids but is significantly greater. This force not only speeds up the separation process compared to a typical clarifier, but can compress the solids, which can increase the concentration.

5.2 Principles of Operation

The main components of a centrifuge are the bowl and the scroll. The bowl is mounted horizontally and turned at a high rate of speed to create centrifugal force. The scroll is mounted inside the bowl and conveys the solids from one end of the bowl to the other.

The bowl consists of a cylindrical section and a conical section. The sections are typically constructed of a stainless-steel casting but can also be constructed of rolled stainless-steel plate. The sections are bolted together in the factory, machined, and balanced at a high speed.

The scroll consists of a stainless-steel screw conveyor that is mounted on a hollow shaft. The scroll can be either an open design mounted to the shaft by a spoke arrangement or a closed design mounted directly to the shaft. The entire scroll assembly is mounted inside the bowl and can turn independently.

To thicken solids, the bowl and scroll typically operated at more than 1500 rpm and can generate more than 3000 Gs of force at the bowl wall. The scroll is rotated approximately 10 rpm to 20 rpm faster (or slower) than the bowl, creating a differential speed.

Sludge and polymer are injected into the hollow shaft of the scroll and discharged into the spinning bowl. The centripetal force created by the spinning bowl causes the solids to settle along the bowl wall. The scroll (screw conveyor) conveys the solids up the conical section of the bowl where they are discharged. The liquid is discharged at the opposite end of the bowl through openings in the end plate.

5.3 Cocurrent/Counter-Current Designs

The two basic centrifuge designs used in municipal solids separation are the countercurrent design and the cocurrent design. Dewatering centrifuges are almost exclusively countercurrent designs where thickening centrifuges can be either countercurrent or cocurrent.

5.4 Physical Features

Each centrifuge has unique physical features that can affect performance including throughput, capture efficiency, polymer dosage, cake solids concentration, and power. These features include

- Bowl geometry,
- Bowl diameter,
- Bowl length,
- Discharge diameter,
- Cone angle,
- Scroll geometry (open or closed scroll),
- Scroll configurations, and
- Materials of construction.

Design of Municipal Wastewater Treatment Plants provides additional information regarding these physical features (Water Environment Federation et al., 2009).

5.5 Mechanical Features

All manufacturers use SKF or FAG type roller bearings with labyrinth seals to support the scroll and bowl. The bearings are oil lubricated and have an L10 life of 100 000 hours.

Two types of lubrication systems are typically used: oil and grease. Oil lubrication provides a controlled flow of filtered and cooled oil to each main bearing and consists of a reservoir, pump, oil filter, water/oil cooler, and instrumentation. The oil lubrication system is interlocked with the centrifuge control panel so the oil pump has to be running before the centrifuge can be started. The grease lubrication system provides grease to the bearings.

5.6 Electrical Features

All major manufacturers provide a scroll-drive motor and VFD and a main drive motor and VFD. The motors are AC-type motors that received current directly from the VFD.

5.7 Scroll-Drive Features

The scroll- or back-drive system turns the scroll relative to the bowl creating a differential speed, which can range from 15 rpm to 1 rpm. The higher the differential speed, the faster the solids are removed from the centrifuge and the lower the discharged solids concentration. The lower the differential speed, the slower the solids are removed from the centrifuge and the higher the discharged solids concentration.

5.8 Scroll-Drive Options

The two basic scroll drives used on modern centrifuges are hydraulic and mechanical scroll drives. The hydraulic scroll drive uses a hydraulic motor to turn the scroll and the mechanical scroll drive uses an electric motor. The hydraulic scroll-drive system is comprised of three main components: motor, pump system, and control system. The mechanical scroll-drive system also is composed of three main components: gear box, scroll motor, and variable frequency drive.

5.9 Control Features

Producing a consistent predetermined, thickened-solids concentration while maintaining reasonable capture efficiency and minimizing polymer dosage is a difficult task to accomplish without instrumentation and controls. Dewatering centrifuges use an automatic torque feature to change the speed of the scroll relative to the bowl to maintain a predetermined torque setpoint, which helps maintain consistent cake solids. Thickening centrifuges operate at a much higher differential speed and do not produce a lot of torque, which makes torque-based control somewhat useless for maintaining a consistent solids concentration.

In the past, the only way to control the output of a thickening centrifuge was to adjust differential speed, sample and analyze the thickened solids, and repeat as necessary until the desired thickened solids was obtained. If the feed solids changed (which can be frequent when co-thickening primary and waste activated sludge), then the trial-and-error process of changing the differential speed and potentially changing the pond depth was onerous.

Today, all of the major centrifuge manufactures offer instrumentation and control packages that automatically change centrifuge operating parameters (such as bowl speed and pond depth) based on continuous measurement of feed solids, or discharge solids, or both.

5.10 Theoretical Performance

5.10.1 *g*-Volume Description

The *g*-volume is a widely accepted industry calculation used to determine centrifuge performance based on bowl geometry and bowl speed. Once *g*-volume is calculated, it can be used to compare centrifuges from different manufacturers. It can also be used to predict throughput of various sized centrifuges based on past performance.

The *g*-volume is calculated based on the following formula

$$g\text{-}volume = (g\text{-}force_{\text{bowl wall}}) \times (usable\ bowl\ volume_{\text{cylinder}}) \quad 8.9$$

5.10.2 *Usable Bowl Volume*

The usable bowl volume is the volume of the cylindrical section of the bowl minus the volume of air space associated with the discharge diameter. This cylindrical volume is sensitive to the cone length and discharge diameter and is shown graphically in Figure 8.5.

The cylindrical volume can be increased if the cone section is shortened by increasing the beach angle. The steeper the beach angle, however, the more torque required to scroll the solids up the conical section.

The cylindrical volume can also be affected by the pond depth. When the maximum pond depth is controlled by the discharge diameter, it is known as "neutral

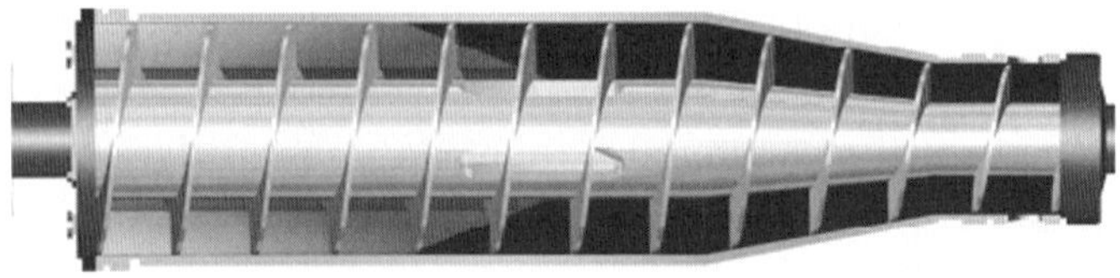

FIGURE 8.5 The scroll conveyor is mounted inside the bowl and rotates at a slightly different speed (courtesy of GEA Westfalia Separator Group, Oelde, Germany).

pond depth." Some centrifuge manufacturers use a baffle disk on the scroll to create a dam that forces the solids through a small annular space between the disk and the bowl. This allows the centrifuge to be operated with a deeper pond (negative pond), which can increase the cylindrical volume. This is shown graphically in Figure 8.5.

5.10.3 g-Force

The *g*-force is the centripetal force at the bowl wall and is associated with both the bowl speed and diameter of the bowl. For a given bowl speed, the larger the bowl diameter then the higher the *g*-force is at the bowl wall. Conversely, for a given bowl diameter, the higher the bowl speed is and the greater the *g*-force is at the bowl wall.

5.10.4 Predicted Throughput Based on g-Volume

If a particular centrifuge has a known throughput based on actual performance, throughput can be predicted for other size units based on comparing *g*-volumes.

6.0 GRAVITY BELT THICKENERS

A typical gravity belt thickener GBT is shown in Figure 8.6. It is a self-contained unit constructed of a table-like steel structure and a rotating belt. The GBT thickens sludge through sludge conditioning and subsequent gravity-forced filtration through a woven nylon belt. Advantages include low capital and operation costs, good solids capture and reliable operation (provided proper sludge conditioning is achieved), and a small footprint compared to other thickening technologies; odors and variable performance with some sludge types are the main disadvantages. With GBT, thickened sludge with concentrations between 3.5 and 13% can be achieved depending on sludge type and performance goals. Gravity belt thickeners are typically sized for hydraulic and solids loading by belt width and are available with belt widths between 0.5 and 3.0 m.

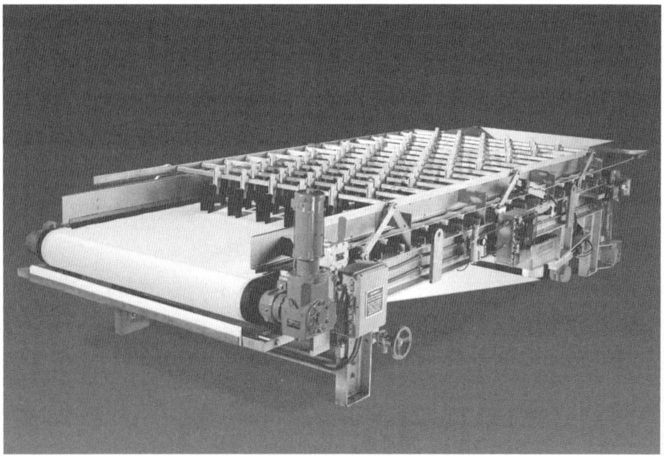

FIGURE 8.6 Gravity belt thickener (courtesy of Charter Machine Corp., Metuchen, New Jersey).

6.1 Principles of Operation

The GBT thickening process is based on liquid/solids separation through polymer addition and gravity-forced filtration as shown schematically in Figure 8.7 and 8.8.

The GBT uses polymer to achieve coagulation and flocculation of the feed sludge then distributes it on to a woven nylon belt to filter the liquid from the solids by gravity. The sludge remains on the belt and is conveyed to a hopper at the end of the GBT where the filtrate is collected in troughs and conveyed to a drain underneath the machine. Critical to the process is proper solids conditioning. Thus, GBT technology is more applicable to sludges that readily flocculate such as WAS more than those which do not, such as primary sludge. The rate at which liquid is filtered is a function of the belt area available for filtering and sludge conditioning. The larger the belt area, the more surface available for filtration and thus the higher the hydraulic and solids loading capacity of the GBT. Maintaining the filtering area and performance requires that the belt be evenly tensioned to prevent pooling of sludge and properly washed to prevent fouling. Therefore, belt wash, tensioning, and belt guide systems are included as subsystems of a GBT. The speed of the belt is also adjustable and controls, if all other operational parameters remaining constant, the time allowed for liquid/solids separation. Typically, a slower belt speed will produce thicker solids. Plows are also used to winnow the sludge and improve the filtration and thus

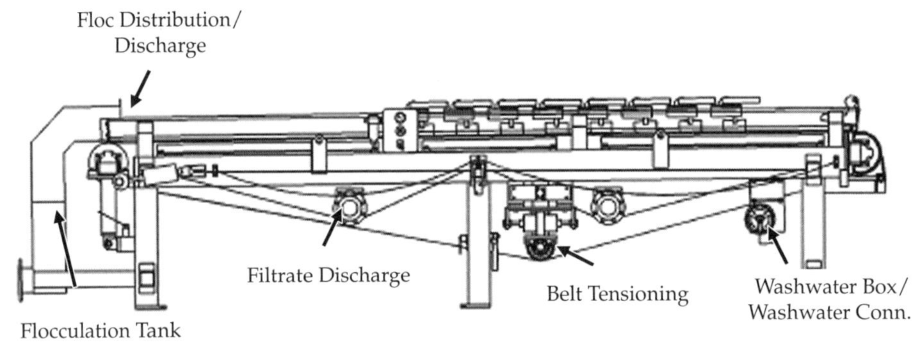

FIGURE 8.7 Elevation of a gravity belt thickener (courtesy of BDP Industries, Greenwich, Connecticut).

increase the thickened-sludge concentration. Gravity belt thickeners can experience

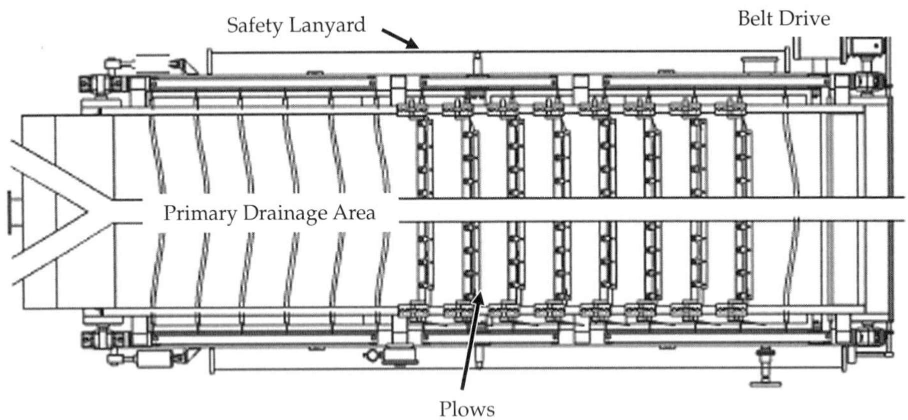

FIGURE 8.8 Plan view of a gravity belt thickener (courtesy of BDP Industries, Greenwich, Connecticut).

washout, which is where the liquid/solids separation does not effectively occur and unthickened sludge is conveyed on the belt to the thickened-sludge hopper. Improper sludge conditioning and overloading or reduced effective belt filter area can all cause washout.

6.1.1 *Major Elements of a Gravity Belt Thickener*

The GBT consists of the following elements as described below and shown in Figures 8.7 and 8.8.

- Sludge conditioning and distribution function, which uses polymer dilution, conveyance, mixing, and subsequent coagulation, flocculation tank, and floc distribution.
- Liquids/solids separation is the gravity-forced filtration through the belt and uses the belt and belt drive, sludge hopper, and filtrate collection and drainage system.
- Belt tensioning function uses the belt-tensioning mechanism and the belt-guide system.
- Belt-wash function uses belt-wash sprayers and booster pumps, as required.
- Control systems including a master control panel and various alarms.

6.1.2 *Sludge Conditioning and Distribution*

Effective flocculation of the sludge before distribution onto the belt is critical for gravity liquid/solid separation because this is a passive process. Polymer is added to the sludge stream to eliminate the negative charge on particles in the sludge and to allow flocculation by polymer bridging to occur upstream at the flocculation tank. A passive mixing device such as a Venturi/orifice plate mixer is typically used to distribute and mix the polymer into the feed sludge. This passive mixer provides good distribution of the polymer through equally spaced radial taps and can be manually adjusted to vary mixing energy. An inline mixer specifically designed for use with sludge could also be used. The time required for effective coagulation/flocculation to be achieved is not predictable and is dependent on the polymer and feed-sludge properties. Therefore, the design should incorporate up to three possible points spaced 5.6 to 9 m (15 to 30 ft) apart in the feed-sludge piping to locate the polymer mixer. Once the GBT is brought into service, the location of the polymer mixer can be finalized.

Flocculation of the sludge occurs in a small (typically 400- to 800-L or 100- to 200-gal) flocculation tank attached to the GBT. The tank slows the flowrate of the sludge to ensure flocculation and limit the shear on the floc particles. The tank is typically sized to provide a minimum of 20 seconds of hydraulic retention time based on maximum hour flows. The conditioned sludge enters the bottom of the tank and the flocculated sludge overflows onto the belt. The floc must be both evenly and gently distributed onto the belt to ensure maximum usage of belt area and prevent floc break-up.

Numerous suitable polymer types are available for thickening applications as described in Chapter 7 of this manual. Polymer selection must consider sludge thickening design requirements, costs, existing polymer systems, and jar test results. Jar testing of the feed sludge, when possible, is essential to evaluate the suitability of different polymers for the application. The amount of polymer required to thicken the feed sludge is typically described in terms of grams of dry polymer per kilogram of feed-sludge solids. Polymer requirements may vary from 1.5 to 5.5 g/kg (3 to 11 lb/ton). The majority of sludges can be thickened with less than 4.5 g/kg (9 lb/ton).

6.1.3 Liquid/Solids Separation

The GBT belt is fabricated of multiple weaves of nylon to limit affinity with fats, oils, and greases and to provide good wear and minimal stretch under tension with a consistent porosity. The belt is the filter mechanism and allows liquids but not the flocculated sludge to pass through. Belts are available in widths of 0.5 to 3 m. Belt width is a measure of the capacity of a GBT, although the actual operating area of machines vary between manufacturers. Fouling of the belt occurs when portions of the belt are coated or plugged and unable to effectively separate liquid and solids. Fouling occurs because of an improperly operating wash water system, sludges that contain high concentrations of fats, oils, and greases, or from use over time. Belt replacement typically occurs every three to five years depending on the operating conditions and can be cleaned intermittently to remove fouling.

After the filtrate drains through the belt, it gathers in drain pans and discharges below the GBT onto the drainage slab. This slab should be sized to the outline of GBT drainage area and sloped to accommodate collection of the filtrate. The maximum filtrate flow plus the washwater flow must be considered in designing the hydraulics of the drainage slab and filtrate conveyance system.

The GBT belt is rotated by an electric motor powered drive. The drive is configured like a roller and uses friction to impart torque on the belt. The motor is mounted on the GBT and ranges from 0.8 kW (1 hp) for a 0.5-m belt to 2.4 kW (3 hp) for a 2-m belt. The rotational speed of the drive is manually adjustable and controls the time allowed for liquid/solids separation. Proper gear design in conformance with American Gear Manufacturers Association (AGMA) standards is important in selecting a GBT manufacturer and drive.

The location of the GBT in the building must accommodate removal of the drive, which is typically the width of the belt plus 0.6 to 1.0 m (2 to 3 ft) and is removed perpendicular to the unit.

6.1.4 *Belt-Tensioning Function*

The belt of the GBT must be kept in tension to maintain the spacing between belt fibers, to retain an even distribution of sludge along the belt, to keep the filter surface perpendicular to gravity, and to maintain friction on the belt-drive roller. Slack in the belt will cause areas of sludge ponding, which inhibits consistent and efficient thickening. Belt slipping may occur if sufficient friction does not exist between the belt and the belt drive. Belt tracking ensures tension is applied evenly over the belt and prevents the belt from dislodging from the rollers.

Both hydraulic and pneumatic belt-tensioning systems are common throughout installations in North America. The pneumatic system uses an air compressor mounted on or near the GBT and pneumatic tubing mounted on the GBT frame to actuate the belt tensioning and tracking system. The hydraulic system is similarly configured and includes a hydraulic pump and the associated conveyance tubing.

Both manual and automatic belt tensioning and tracking systems are in-use. With the manual system a manufacturer-specified air or hydraulic pressure setpoint is maintained by the operator and adjusted as required. This adjustment typically occurs on a weekly or monthly basis or, for GBTs run less than 24 hours per day, at each startup. Manual adjustment of the belt-tracking system occurs at the same time and requires the operator to visually inspect the belt and adjust the tracking. Automated belt tensioning and tracking is available as a standard feature with most manufacturers. Both systems provide monitoring and alarming of the belt tensioning and tracking systems.

6.1.5 *Belt Wash Function*

After the thickened sludge is transferred to the sludge hopper, the belt must be cleaned before rotating back to the sludge flocculation tank. High pressure spray is applied to the belt through wash water nozzles, and the liquid stream is conveyed directly to the GBT drain. Wash water demands vary between 3.1 and 4.0 L/s (50 and 65 gpm) and 7.7 kilopascals (kPa) and psi and 9.8 kPa (55 to 70 psi) for a typical 2-m belt application. Typically, the higher the pressure, the better the cleaning, which will reduce belt plugging and ensure better sludge thickening. Too much pressure can reduce belt life. Three different sources of washwater are typically used: recycled plant water, municipal nonpotable water, and recycled filtrate.

Recycled plant water is the most common source and should have a TSS of less than 100 mg/L. Washwater nozzles should be designed to accommodate twice the maximum anticipated solids concentration (TSS) in the plant water. Plugged nozzles lessen the efficiency of the belt wash process and require excessive maintenance. A booster pump may be required to bring plant water to the necessary pressure at the nozzle.

Municipal nonpotable water does not contain solids, and nozzle plugging is not a concern; however, the cost associated with water used during the GBT operation can be prohibitive. Again, a booster pump may be required to elevate the pressure of the nonpotable water as well.

The GBT filtrate can be recycled through a booster pump and used as washwater. The TSS concentration in the filtrate must be below 100 mg/L, however, and sufficient volume of filtrate must exist during all operational scenarios. An additional source of washwater must be available during GBT startup and shut-down, when the GBT is not producing filtrate. This source of washwater is typically adopted when a satisfactory source of plant water is not available

6.2 Design Criteria

6.2.1 *Primary Sludge versus Waste Activated Sludge*

The thickening effectiveness of the GBT is closely related to the floc properties of the conditioned sludge. Gravity belt thickeners are used more often to thicken WAS, aerobically digested sludge or combined WAS and primary sludge than solely primary sludge. Manufacturers often will not guarantee performance with primary sludge because it does not reliably flocculate because of variability in the particle type. As discussed in Chapter 7, primary sludge does not consistently flocculate with polymer addition. Primary sludge can contain grit and other smaller particulates and higher concentrations of fats, oils, and greases than WAS. Grit particles, fats, oils, and grease increase the potential for belt plugging and may reduce belt life. A GBT, unlike a gravity thickener, does not have the capacity to equalize sludge properties through retention. Any momentary failure of the sludge conditioning process leads to flooding of the belt. Under this condition, there is not suitable liquid solids separation by flocculation before reaching the belt, and both solids and liquids are conveyed to the hopper or runoff the side of the belt.

6.2.2 *Hydraulic and Solids Loading Rates*

The GBT is sized to accommodate both peak hydraulic events and peak solids loadings because once belt washout occurs, the system must be shut down and restarted. Guidelines for loadings of WAS based on belt width are shown in Table 8.3. The guidelines must be adjusted for consideration of sludge type (WAS, primary sludge, etc.) and the target thickened-sludge concentration. Typically, the capacity of a GBT is limited hydraulically with WAS and by solids loading with primary sludge. Primary sludge is not thickened as effectively as WAS and commonly has a higher thickened sludge. Primary sludge and WAS combined, however, can produce thicker solids through higher solids loading on the belt. Typically, loadings are limited by the dynamics of the liquid/solids separation process, which is characterized solely by manufacturer empirical testing and operational experience.

6.2.3 *Feed-Sludge Systems*

Feed-sludge systems can be any sludge conveyance system suitable for the sludge being conveyed. The design must consider the hypothetical variation in system pressure loss because of the Venturi/orifice mixer operating under different mixing intensities, however, and the hydraulic load limitations of the GBT. A feed-sludge system that either cannot deliver the required flow to the GBT or hydraulically overloads the GBT and causes wash-out is unacceptable. Feed-sludge flow must be measured to pace the addition of polymer, especially in systems where there is more than 10% variability in flow during an operational period. Finally, considerations for feed-sludge sampling should be incorporated into the design for polymer addition calculations and controls interties to shut down the feed-sludge system upon GBT.

TABLE 8.3 Guidelines for peak hydraulic and solids loading.

Belt Width (meters)	Hydraulic Loading (L/s)	Hydraulic Loading (gpm)	Solids Loading (kg/hr)	Solids Loading (lb/hr)
0.5	7.9	125	255	561
1	15.7	250	520	1144
1.5	23.6	375	765	1683
2	31.4	500	1040	2288
3	47.2	750	1560	3432

6.2.4 *Thickened-Sludge Systems*

The thickened-sludge system typically consists of a hopper at the discharge of the GBT and pumps that can convey sludge at the solids concentrations produced by a GBT. Typically, positive displacement pumps are used because of their reliability and relatively flat flow versus discharge pressure curve. Centrifugal pumps, which are cheaper and require less maintenance, may also be used with lower concentrations depending on the application. Both the hopper and pump are typically external to the GBT, however, they are often supplied by the GBT manufacturer to ensure system compatibility. The capacity of the pump should be designed for the minimal thickened-solids concentration under the maximum hydraulic condition.. The hopper size is determined by the type of control to be used to modulate pump discharge flowrates. On/off control is based on sludge level in the hopper. This requires that the hopper be sized to limit the number of start and stops required to less than 10 per hour when the GBT is operating at any point in the flow range. More typically, pump-discharge flow is controlled by modulating the speed of the pump based on hopper sludge level through a variable frequency drive. The hopper for this mode of operation must be sized large enough to keep the rotational speed of the pump within manufacturer's recommendations during periods of minimal sludge production from the GBT. The thickened-sludge pump typically cannot be designed for flows associated with GBT belt flooding, and a high hopper level switch must be incorporated in the design to shut down the GBT upon high hopper level.

6.2.5 *Filtrate System*

The filtrate passes through the belt and is collected in the drainage slab. The slab should be sloped on all sides toward a central collection point with a minimum slop of 1:8 to ensure efficient conveyance of filtrate to the drain without puddling or substantial splashing. The drain should be sized for the peak hydraulic flow to the GBT and be between 200 and 360 mm (8 and 14 in.). The filtrate must be conveyed back upstream of the primary clarifiers to settle out solids and treat soluble nutrients before discharge. The BOD load from the filtrate must be included in any design criteria for primary and secondary processes.

6.2.6 *Chemical Feed Systems*

Polymer mixing and delivery systems and the theory of sludge conditioning are also described in Chapter 7 of this manual. The polymer system should consist of two mixing/dilution tanks each with the volumetric capacity to supply polymer to the GBT for a minimum of one operational shift at maximum-hour hydraulic and solids loading rate. Polymer must be diluted to between 0.1 and 0.5% thickened sludge to ensure good mixing and coagulation upon injection to the feed sludge. Dilution of the polymer is achieved by either manual or automated addition of water to liquid or dry polymer. Flow measurement of both water and polymer must be incorporated in the design for the process to be automated. A mixer, sized to achieve a g of between of between 50 and 80 s^{-1}, is included for each tank. Two positive displacement pumps (one active, one back-up) with automated variable speed control should convey the polymer to venture/orifice plate mixer. Pumps should be sized to convey estimated flowrate required at maximum hour loading. Polymer should not be fed at a concentration greater than 0.5% to ensure good mixing. Flow-pacing of the polymer with the feed sludge to ensure adequate but not overdosing requires either polymer flow measurement or calculated flow based on pump speed. The latter is less expensive but less reliable.

6.2.7 *Odor Control*

Gravity belt thickeners distribute conditioned sludge on the belt, which is open to the atmosphere and stores thickened sludge upon discharge in a hopper. Both processes release odor-producing compounds. Although not required by NFPA 820 (National Fire Protection Association, 2003), exhausting 12 to 20 air changes per hour is recommended to limit odors. The exhaust air intakes should be located directly above the GBT and sludge hopper to limit the dispersion of odors. Some installations enclose the GBT in smaller rooms to limit the required exhaust airflows and fan sizing. More recently, an enclosed model of a GBT has been developed to lessen the volume of exhaust air required and better contain odors.

6.3 Performance

6.3.1 *Thickened Solids*

The thickening capacity of the GBT, as mentioned earlier, ranges from 4 to 13% TSS. The final sludge concentration will be a function of the feed-sludge concentration and type and the amount of polymer added, belt speed, and whether plows were used. Belt speed and polymer dosage are the most effective ways to control the thickened-sludge concentration.

6.3.2 *Capture*

Solids capture is the percentage of solids in the feed sludge, which are thickened and discharged from the belt. Capture is a function of hydraulic loading, solids loading, and feed-sludge type; for WAS, capture should be greater than 98%.

6.3.3 *Chemical Requirements*

Chemical requirements for the GBT are dependent on sludge type, concentration and thickened-sludge design criteria, and polymer type. Typically, 4 g polymer/kg dry feed solids (8 lb/ton) is given as the maximum. As mentioned earlier, however, jar testing when possible should be undertaken to determine the required dosage of the selected polymer and WAS. Variation in polymer doses with polymer type can be as great as 40%.

6.3.4 *Power Requirements*

The GBT is energy efficient as a liquid/solids separation process. An estimate of 1.9 kW (2.5 hp) per meter of belt width is conservative. If a washwater booster pump is required, then an additional 3.8 kW (5.0 hp) load should be included in power estimates.

6.3.5 *Estimated Capital and Operation and Maintenance Costs*

Capital costs vary between manufacturers, based on specified features for the GBT. Similarly, operation and maintenance costs are dependent on polymer usage, electrical rates, and operator labor rates. For the costs parameters listed in Table 8.4, capital costs were averaged for reporting manufacturers. Solids loadings to calculate the costs per kilogram are taken from Figure 8.9.

The capital costs per kilogram of feed solids increase with decreasing belt size (see Figure 8.9). Each GBT has a base cost and master control panel associated with it that is not related to size. Operating costs increase with decreasing belt size, but only slightly, which is because of similar labor time required for both larger and smaller units that treat less feed-sludge solids.

6.3.6 *Process-Control Strategy*

Process systems for the GBTs include

- Feed-sludge system;
- Polymer system;
- Gravity belt thickener; and
- Thickened-sludge system.

The GBT control system comprises the master control panel (MCP) and the belt tensioning mechanism, belt drive, and appurtenances. The MCP provides control

TABLE 8.4 Values of parameters used for cost analysis.

Power (kW/m belt width)	Power cost ($/kW)	Polymer (g/kg feed solids)	Polymer cost ($/kg)	Labor time (h/d)	Labor rate ($/h)
2.5	0.1	3.3	6	0.75	40

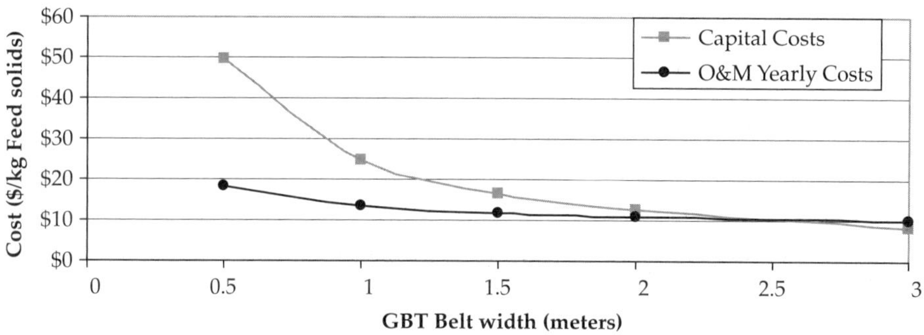

FIGURE 8.9 Capital and operation and maintenance costs per kilogram of feed solids (GBT = gravity belt thickener).

and monitoring of all the equipment that interfaces with the GBT. It typically consists of one or more programmable logic controllers with interfaces to displays and analog and digital inputs and outputs. The MCP also controls the startup and shut-down of the GBT and peripheral equipment and monitors the operation of the GBT itself. On the face of the panel, there is typically a human-machine interface (HMI) display. It also includes remote-local-off switches for the different functions of the GBT and, sometimes, for the peripheral equipment.

The startup and shut-down can be either automatic or manual. In automatic mode, the GBT MCP starts and wets the GBT and then starts and monitors peripheral equipment, engaging the control systems. In manual mode, each peripheral must be started independently. The GBT MCP monitors the following functions of the GBT and alarms when operator-determined limits are exceeded:

- Belt tension;
- Belt alignment;
- Air-compressor or hydraulic-pump pressure;

- Air compressor or pump status;
- Washwater flow; and
- GBT status.

The GBT MCP will alarm with tripping of the lanyard or emergency shutdown through the emergency shut-down button on the GBT.

The feed-sludge system consists of the following equipment and sensors:

- Feed-sludge pumps, which convey the feed sludge to the GBT and
- Feed-sludge magnetic-electric flowmeter to measure the feed-sludge flow for monitoring and flow pacing of the polymer.

The feed-sludge pump is controlled through a remote-local-off switch.

In remote on/off, control will be through the GBT MCO. Upon a signal from the GBT panel, the pump will start. The pump will be turned off upon a signal from the GBT, which could occur during shut-down or emergency shut-down of the GBT. It is critical that the feed-sludge pump is turned off while the GBT is not running to prevent washout of the GBT. In local mode, on-off pump control will be from a local field panel or from a switch at the GBT MCP.

The status of the feed-sludge pump and the flow through the feed-sludge flow-meter should be displayed at the GBT MCP and at supervisory control and data acquisition (SCADA). The following alarms should be displayed as well, at both the MCP and SCADA:

- Pump fails to start,
- Pump fails to stop, and
- Flowmeter alarm (internal to the flowmeter).

The polymer system in this example is simpler than those at larger municipal installations and may be more prevalent at smaller facilities. It consists of two polymer tanks, each sized to store enough diluted polymer to run the GBT for one shift. Polymer is added and diluted in the tanks by plant operations. The two polymer pumps are positive displacement pumps and polymer addition to the feed sludge at the venture orifice meter is accomplished by varying the speed of the polymer pump to match the measured feed-sludge flow and in accordance with a calibrated speed versus flow curve for the pump. This system precludes the need for a polymer flowmeter or an automated polymer addition and dilution system:

- Polymer mixers;
- Polymer tanks;
- Positive displacement polymer pumps; and
- Polymer tank level sensor.

The polymer mixers should be operated by a remote-local-off switch at the polymer tanks. The operator turns on the mixer after the polymer is added and diluted. The polymer pumps are operated by a remote-local-off switch.

In remote, they receive a signal from the GBT MCP to start during the GBT startup cycle. As mentioned earlier, the speed of the pumps will vary in proportion to the feed-sludge flow. The polymer pumps should be turned off whenever the feed-sludge pump is not operating to prevent excessive polymer on the belt, which can lead to fouling. In local, the pumps should be turned on/off from a switch at the GBT MCP or at a local field panel. The status and speed of the polymer pumps should be displayed at the GBT MCP and at SCADA. The following alarms should be displayed at the MCP and SCADA:

- Pump fails to start;
- Pump fails to stop; and
- Low polymer level in tank.

The thickened-sludge system consists of the following pieces of equipment:

- Hopper;
- Level sensor (either pressure or ultrasonic);
- Pump; and
- High-pressure discharge switch.

The thickened-sludge pump is controlled through a remote-local-off switch at the GBT MCP. In remote, the thickened-sludge pump is engaged during the GBT startup when the feed-sludge pump is started. The thickened-sludge pump is controlled from hopper level through speed control and a VFD. The speed will be varied to maintain a thickened-sludge level in the hopper within a certain band, typically 0.3 m (1 ft). If the thickened-sludge pump is sized for projected 20-year flows, then the operational band may need to be increased and the pump turned off when operating at less than 10% of capacity to prevent damage to the starter/motor and VFD. The thickened-sludge pump should disengage when the belt on the GBT stops.

In local mode, the pump can be turned on and off at the GBT MCP and should not be controlled by the level in hopper. Care must be taken by operators to not turn on thickened-sludge pump in local with an empty hopper.

Thickened-sludge hopper level and pump status and speed should be monitored at the GBT MCP and SCADA. The follow alarms are also monitored at the GBT MCP:

- High-high hopper level alarm; this will shut-down the GBT to prevent the hopper from overflowing.
- Thickened-sludge pump start fail.
- Thickened-sludge pump stop fail.
- Thickened discharge pump high-pressure discharge; this will be directly wired to the pump starter and prevent over-pressurization of the thickened-sludge discharge piping.

7.0 ROTARY THICKENING

Progressive equipment companies developed rotary thickeners in the 1980s but were unable to find a receptive market. They were reintroduced in the late 1990s and early 2000s with the recognition that enclosed-drum thickening provides an odor control advantage. In cases where fibrous sludge can be conditioned to form a strong floc structure, capture rates were good, and thickened-sludge concentrations were slightly higher than possible with GBTs. Rotary thickeners can accommodate stringy material with less operational challenges and should be considered if thickening primary or digested sludge is required. Because building new digester capacity is getting increasingly expensive, the concept of feeding a digester thicker solids warrants extra investigation.

Manufactures disagree on the proper naming of this thickening technology. Some manufactures refer to their equipment as rotary drum thickeners; others use the term rotary screen thickeners. There are differences in the design, and both are discussed in this section.

7.1 Principles of Operation

Rotary thickeners operate in a similar manner to GBTs. Coagulated and flocculated solids enter the process drum, and free water is released through the screen, leaving the solids to be conveyed to the discharge chute. Free water is the clear bulk liquid found in between floc chunks of the feed sludge. The rotary thickening process works as shown in Figure 8.10. The significant difference between rotary thickeners and GBT is the manipulation the sludge receives while in the thickening process.

Rotary thickeners are designed for a more aggressive manipulation of the sludge while in process so that free water is released in a smaller area. The more aggressive manipulation means that more free water has access to the drain screen. During normal operation, feed sludge can get dumped into the center of the process, and filtrate will find outlets through the screen because of the rotation of the equipment and tumbling action of the sludge. Floc shear from aggressive manipulation, however, could cause sludge particles to escape with filtrate, resulting in poor capture. Rotary-drum thickeners typically require less space than other thickening methods. Operators can control the thickening process by adjusting feed rate, polymer dose rate, and drum speed.

Rotary thickeners comprise a low-shear floc tank and the screen. The low-shear floc tank has a mixing motor so that flocculation can be controlled to develop a strong

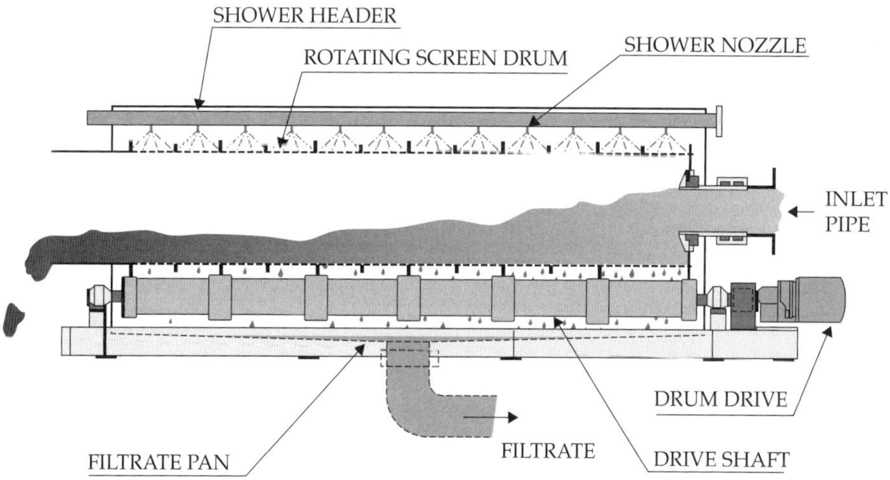

FIGURE 8.10 Rotary screen thickener process (courtesy of FKC).

floc without over-mixing or breakage. The drum screen is typically fabricated of stainless steel with a wedge wire, orifice plate, or polyester screen and VFD motor. The screen typically has spray bars to spray backwards through the screen into the thickened-sludge area to maintain porosity. This water is rereleased as more filtrate is discharged from the process. The equipment may be fully enclosed, which is recommended for odor control. Thickened sludge accumulated inside the screen is discharged through the end of the drum to the downstream pumping process.

Because of the solids concentration possible from the rotary-drum thickeners (up to 12% depending on feed sludge), sludge pumps should be specifically designed for thickened sludge or cake.

7.2 Design Criteria

Design criteria are equipment specific according to the details of each manufacturer's design. Because rotary thickening is still relatively new, there are a variety of designs. Cost-effective thickening can be accomplished with per-unit design capacities in the following range:

- Hydraulic feed rates of 95 to 1500 L/min (25 to 400 gpm), which is lower than normal GBT loading rates but ideal for smaller plants or 24-hour-per-day operation.
- Mass feed rates of 45 to 900 kg/h (100 to 2000 lb/h).
- Turndown is approximately 50% of the nominal feed rate listed to prevent excessive detention time in the drum and floc shear.
- Polymer dose rates of 2.5 to 7.5 g/kg (5 to 15 lb/dry ton) (active) are consistent with GBT operations.
- Thickened-sludge concentrations of 5 to 7% are typical, although concentrations up to 12% based on application may be possible.
- Capture rates of 98% can be specified. This is better than GBTs because spray bars clean screens back into the thickening process.
- Washwater use of 10 to 40 L/min (2.5 to 10 gpm). Some designs only require intermittent belt wash.
- Successfully used in pulp and paper, WAS thickening, and WAS plus primary sludge thickening.

7.3 Performance

Operations and process monitoring of rotary thickeners are similar to GBTs. Equipment operating variables that influence rotary thickener performance include hydraulic and mass loading feed rates, polymer dose, polymer flash mixing, sludge flocculation speed, and drum speed. Process optimization consists of testing feed sludge with different polymer doses and equipment speeds. In some cases, changes of filter media are possible, but typically, laboratory testing before fabrication identifies the best filtration media.

7.3.1 *Thickened Solids*

Thickened-sludge concentrations of 5 to 7% are typical, although concentrations of up to 12% based on application may be possible.

7.3.2 *Capture*

During equipment selection, the engineer should consider the effect the filtrate quality on the sidestream flow returning to the liquid treatment processes. Capture rates of 98% for rotary thickening can be specified, which is better than GBTs because spray bars clean screens back into the thickening process.

7.3.3 *Chemical Requirements*

As with most solids handling, process performance depends on the coagulation of the feed sludge. Chapter 7 discusses specifics of coagulation chemistry, including process operation in terms of rotary thickeners. Most manufacturers comment that the polymer dose requirements for a thickened sludge are similar to that of a GBT at an active polymer dose of 2.5 to 7.5 g/kg. (5 to 15 lb/dry ton fed). One manufacturer has suggested that an additional polymer dose of up to 30% could yield thickened-sludge concentrations of 8 to 12% depending on feed sludge. Although guaranteed thickened-sludge concentrations are much more reasonable at 5 to 7% total solids.

7.3.4 *Power Requirements*

Drive units are contained in local control panel with control PLCs. Drive motors are typically 1490 to 3728 W (2 to 5 hp), and some units have a 373-W (0.5-hp) flocculator. The drum-washer booster pump, if required, is typically 2237 to 3728 W (3 to 5 hp). Washwater flow and pressure requirements vary by manufacturer, but range from an intermittent 38 L/min (10 gpm) to a constant 150 L/min (40 gpm) at 275 to 415 kPa (40 to 60 psi.)

7.3.5 *Process-Control Strategy*

Process elements that can be controlled by process-control strategy include feed sludge, thickened-sludge pumping, polymer feed, air-to-solids ratio, and drum rotation. Automatic process-control programming would be well suited for this type of physical treatment process if an analysis determines that the functions would be cost effective for a specific project. Both hydraulic (flowrate setpoint) or mass-feed rate (flow times concentration) control algorithms could be used to maintain a consistent process feed. Automatic polymer adjustment to maintain streaming current charge setpoint or filtrate suspended solids quality could also be useful to minimize polymer dose.

8.0 MEMBRANE THICKENING

8.1 Principles of Operation

Interest in this technology has surged with the growing use of membrane bioreactors (MBR) in wastewater treatment over the last two decades. The MBR system combines an activated sludge process with solids separation by microfiltration or ultrafiltration membranes. They operate either as dedicated units or during periods of reduced flow to the WWTP. The MBR membranes are capable of integrating solids thickening into the overall plant operation. Most operating membrane-thickener systems have been designed for small-scale WWTP applications with less than 4 ML/d (1 mgd) capacity. Membrane thickeners have become a more cost-competitive option for larger facilities, however, as MBRs have become more affordable.

The membrane fibers are manufactured using a proprietary polymeric membrane bonded to a porous support fiber. Dependent on the manufacturer, these fibers are produced into one of two membrane configurations: flat-plate or hollow-fiber with an average pore size of 0.4 to 0.04 μm, respectively. Both membrane configurations maintain activated sludge between the tank and the fiber while the filtered water is pulled into the center chamber of the fiber through a slight vacuum using a permeate pump. The filtered water then manifolds to a central header and is collected. Each series of membranes are organized into cartridges or modules and then further organized together into cassettes.

The basic components of a membrane thickener system may include a dedicated membrane tank, membranes, filtrate (also known as permeate) pumps, thickened-solids pumps, and membrane aeration blowers. Membrane aeration is provided by coarse bubble diffuses to scour the membrane surface. The scouring effect minimizes the accumulation of solids on the membrane surface. The aeration also provides oxygen to ensure the thickened sludge remains aerobic. From the biological process, waste activated sludge is sent to the membrane tank where it is thickened by the action of pulling water (permeate) through the membrane fibers in a slight vacuum using a pump to the desired solids concentration. A separate pump then pulls the thickened solids for stabilization. Figure 8.11 illustrates a typical membrane thickener installation operating in batch mode.

The membrane thickener can also be configured to operate continuously by looping the thickener with a stabilization process such as aerobic digestion. This allows the thickener to operate in more of a steady-state environment. Integrating the thickener system with a sludge stabilization process also allows the thickener to produce a higher solids concentration with stabilized sludge. At this time, only a single flat-plate membrane thickener manufacturer has packaged the two processes together. Figure 8.12 illustrates a typical membrane thickener installation operating in a looped continuous mode.

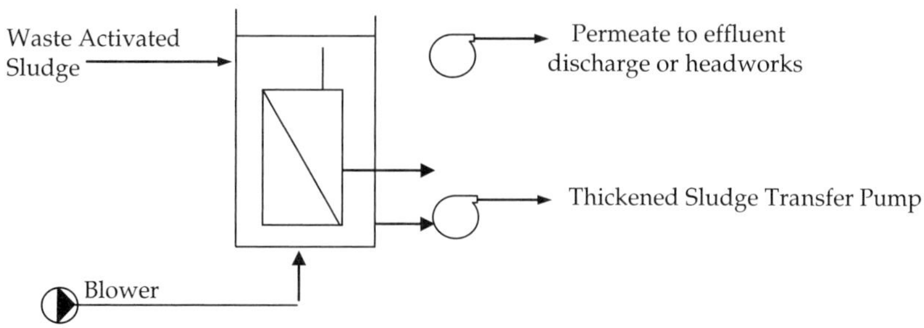

FIGURE 8.11 A simplified process flow drawing of a batch membrane thickener.

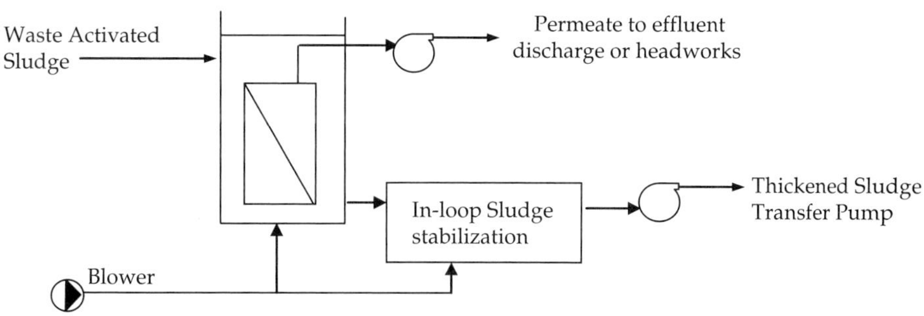

FIGURE 8.12 A simplified process flow drawing of a continuous membrane thickener.

8.2 Design Criteria

8.2.1 *Membrane Flux*

The primary controlling membrane separation design parameter is the filtration rate, commonly referred to as the membrane flux. Flux is expressed as the flowrate transferred through a specific membrane surface area such as liters per square meter per hour (gallons per day square foot = gpd/sq ft). The typical flux range for hollow-fiber sludge thickening applications is 5 to 7 $L/m^2 \cdot h$ (3 to 4 gpd/sq ft). The operating flux for flat-plate sludge thickening applications varies dependent on the feed WAS concentration. For dilute WAS, the membranes initially may operate up to 20 $L/m^2 \cdot h$ (12 gpd/sq ft) and decrease to 7 $L/m^2 \cdot h$ (4 gpd/sq ft) as the solids concentration increases.

8.2.2 *Hydraulic Loading Range*

Membranes have limited ability to sustain higher fluxes because of an increase in the rate of membrane fouling, resulting from increased transmembrane pressure (TMP). Design of a membrane thickener is dependent on the maximum design volume of WAS processed per time period being operated. Operating as a solids thickener system, the typical loading range can be designed to be narrow because of equalization upstream of the thickener. Without equalization, the thickener design focuses on peak-flow condition or sustained flux, potentially requiring more membrane area.

8.2.3 *Concentration of Waste Activated Sludge*

Membrane thickeners typically are designed in conjunction with, but are not exclusive to, MBR systems. Although MBRs typically have prethickened concentrations of 0.8 to 1.5% solids, prethickened WAS concentrations for gravity-settling plant processes range from 0.1 to 0.5% (Metcalf and Eddy Inc., 2003). Practical membrane thickening limits are up to 4.5% solids with a target post-thickened WAS concentration of 2 to 4%. Combined with sludge stabilization, membrane thickeners can reach up to 5.5% solids.

8.2.4 *Process Temperature*

The normal temperature range of the WAS will influence the thickener design through both operation and membrane longevity. Conversely, membrane permeability (permeability = flux/TMP) drops as the feed temperature decreases. This decrease is not necessarily indicative of membrane fouling, rather a function of the filtration characteristics of the mixed liquid. The membrane throughput design must be adjusted to accommodate the increase in WAS viscosity as a function of the minimum liquid temperature.

8.2.5 *Redundancy Requirements*

Redundancy requirements for membrane thickeners are similar to other thickening processes and are influenced by user philosophy. Optimum thickener operation is dependent on the plant's sludge treatment capacity for other processes, allowable membrane flux, periods of desired operation, possible peaking factors, and mode of operation selected. Additional conservatism may be added by requiring up to 25% spare membrane cartridges or modules in the provided cassette.

8.3 Performance

8.3.1 *Comparing Different Manufacturers*

In the United States, both flat-plate and hollow-fiber membrane configurations have been integrated into membrane thickening applications.

Modular in design, the systems are typically sold as core packages with optional ancillary equipment. Smaller steel-tank membrane thickeners may arrive on site as plug-and-play. With higher capacities, the systems may range from steel-tank skid-mounted (up to 15 800 L/h, or 100 000 gpd) to cassettes placed in concrete basins provided separately. The smallest package units can produce up to 300 L/h (1900 gpd) permeate with a minimum footprint of 1.3 m^2 (14 sq ft). For planning level purposes, packaged flat-plate membrane tankage footprints range from 3.5 to 4 m^2 required per 1000 L/h of permeate. For planning level purposes, packaged hollow-fiber membrane tankage footprints range from 0.7 to 1.5 m^2 required per 1000 L/h of permeate. These footprint ranges do not include the associated pumps, blowers, and other ancillary equipment. The footprint for larger custom designs may decrease with larger systems using concrete basins.

8.3.2 *Thickened Solids*

The solids thickening capability of membranes are comparable to that of other thickening processes. With the physical barrier separating solids and liquids during the thickening process, the membrane thickener is not dependent on coagulants and coagulant aids for conditioning the sludge. Similar to other mechanical processes, the energy usage and system strain increases exponentially in relationship to higher targeted post-thickened concentrations.

8.3.3 *Capture*

A significant advantage of the physical membrane barrier over conventional activated sludge is the high-quality effluent. Unlike the secondary clarifiers typically used in activated sludge treatment systems, more than 99% of suspended solids and most colloidal material are retained regardless of the settling characteristic of the mixed liquor. Therefore, permeate from a membrane thickener has low total suspended solids and turbidity. Dependent on the effluent permit water quality limitations, the permeate may potentially be blended and discharged with the effluent or sent back to the headworks for reprocessing.

8.3.4 *Chemical Requirements*

Membranes require both chemical cleaning and routine air scouring to prevent fouling from occurring. During normal operation, mineral salts, iron, biological matter, and insoluble organics such as fats, oils, and grease can foul membranes. Deposits that build up on the membrane surface during operation can cause a loss in permeability (permeability = flux/TMP). Once the TMP increases to the maximum value recommended by each manufacturer, cleaning must be performed by applying chemicals to the membranes and allowing the solution to soak through the membrane for a period of time. Chemicals such as sodium hypochlorite may be used in the cleaning process to remove organic fouling constituents. Chemicals such as citric acid or phosphoric acid may be used to remove inorganic fouling constituents. Combined with air scouring and permeate, the chemical-cleaning sequences can be categorized into two procedures: maintenance cleans and recovery cleans (only in hollow-fiber membrane).

For hollow-fiber systems, maintenance cleans are routinely scheduled to occur in conjunction with and between more intensive chemical cleanings. Hollow-fiber membrane maintenance cleanings are shorter in duration (approximately 30 minutes) and include a final 250 mg/L (0.025%) sodium hypochlorite solution that is introduced into the membrane tank. Fully automated maintenance cleans may occur between one to three times per week. Recovery cleaning for hollow-fiber membranes is an operator-initiated, intense cleaning process that involves soaking the membranes in cleaning solution for a longer period of time (several hours). Hollow-fiber membrane cleans include injection of a final 1000 mg/L (0.1%) sodium hypochlorite solution to fill the membrane tank. Fully automatic once initiated by the operator, recovery cleaning may occur one to four times per year.

Flat-plate membrane maintenance cleans occur at similar intervals as the hollow-fiber membrane recovery cleans. Flat-plate membrane cleans involve injection of 5000 mg/L (0.5%) of sodium hypochlorite solution directly into the center of each plate. This chemical solution is held within each plate for several hours.

8.3.5 *Power Requirements*

Unlike most other thickener applications, a blower mechanically aerates the membrane thickener system. This aeration energy through the coarse bubble diffusers accounts for up to 95% of the energy usage required for the system, with the remainder required for pumping. The estimated energy required to thicken a 1% WAS solids concentration to approximately 4% WAS concentration ranges from 0.05 to 0.1 kWh per kilogram of dry solids processed. Typical types of blowers and pumps used for the thickener system are as follows: aeration blower—positive displacement or turbo, permeate pump—lobe or centrifugal, and thickened WAS drain pump—progressive cavity.

8.3.6 *Process-Control Strategy*

Process elements that can be controlled for membrane thickening operation include WAS feed rate, membrane cleaning sequences and associated chemical dosages, aeration rate, permeate rate, and thickened-solids pumping. The membrane thickener incorporates the following components and features:

- Membrane tank to house the membrane cassettes;
- Dedicated membrane aeration blowers with optional online redundant blower;
- Thickener feed pump (if necessary);
- Variable speed process pump [backpulsing (hollow-fiber only) and permeation];
- Thickener drain pump (if necessary);
- Air valves providing air to the membrane tank;
- Membrane-tank-level transmitter;
- Magnetic flowmeter for WAS;
- Magnetic flowmeter for TWAS;

- Permeate turbidimeter (if necessary);
- Pressure transmitter, switches, and gauges;
- Pipe header with chemical injection locations; and
- Chemical system for cleaning (tank, metering pump, accessories).

8.3.7 *Mode of Operation*

Membrane thickeners can be designed with three modes of operation: (1) dedicated membrane thickener train operating in modified batch mode; (2) dedicated membrane thickener train operating in a continuous mode; and (3) thickening with MBR process trains during low-flow standby conditions. All three modes of operation use the same basic core components.

Dedicated membrane thickeners can be designed as a standalone modified batch operation. In this mode of operation, the membrane thickener is fed with WAS until the thickener tank is full. At that time, permeate pumps begin drawing supernatant at the same rate that WAS continues to be fed to the thickener. Continued operation in this manner will progressively concentrate solids in the membrane tank. Upon reaching the desired concentration or recovery, the WAS feed and permeate pumps stop, and the thickened sludge is drained or pumped from the tank to the next process. The process is repeated with the next WAS batch.

Dedicated membrane thickeners can be combined in with an anoxic tank and aerobic digestion to produce Class B biosolids. In this mode of operation, the membranes serve as a continuous mechanical thickener immediately upstream of the aerobic digester. Permeate pumps draw supernatant out of the membrane tank at the same rate that WAS continues to be fed to the thickener. The biological effect of inloop thickened solids on aerobic digestion is further discussed in Chapter 11.

Finally, MBR plants can incorporate the use of a standby train for membrane-thickener purposes. This mode of operation takes advantage of the membranes not required for effluent production during average or low-flow conditions. In lieu of placing the membrane train in standby mode, the train is prevented from recycling the mixed liquor. At that time, permeate pumps begin drawing supernatant at the same rate that WAS continues to be fed to the thickener. Permeation is continued until the desired solids concentration or recovery is achieved. When the target recovery is reached, the WAS feed and permeate pumps stop, and the thickened sludge is drained or pumped from the tank to the next process. When the cycle is complete, the train switches back to normal MBR production mode. If an increase to the influent plant flow requires use of the membrane train during the production of any batch, then the batch will be aborted, and the system will revert to MBR operation.

8.3.8 *Equipment Monitoring*

The degree of equipment monitoring is dependent on owner preference and level of automation desired. With the number of ancillary equipment such as process blowers, pumps, and instrumentation, equipment monitoring become extensive. Conversely, the membrane thickener can be manually operated during normal operator shift hours. Standard life and safety alarms and equipment protection monitoring, such as high temperature or overtorque, is recommended at a minimum.

8.3.9 *Process Monitoring*

Process monitoring is dependent on the mode of operation implemented. For the membrane thickener, prethickened and thickened TSS (percent solids), temperature, and pH should be recorded in a process-control log on a daily basis to assist in operations. The recommended monitoring parameters and analysis frequency should be revisited during startup and during those times when significant changes are made to operating conditions (sludge flow, sludge source, aeration rate, etc.).

If the membrane thickener is combined with aerobic digestion, then other water-quality parameters typical to aerobic digesters should be measured weekly such as dissolved oxygen, TSS, and VSS to track performance and volatile solids reduction. Other analyses, such as ammonia, nitrate, and alkalinity, are recommended for startup and troubleshooting. *Fecal coliform* and specific oxygen uptake rate (SOUR) measurements are also required periodically to satisfying Class B biosolids treatment requirements.

8.3.10 *Levels of Automation*

The recommended levels of automation and alarms for a membrane thickener can vary widely dependent on owner and manufacturer preferences. Operation can be dependent on whether the process is initiated as batch or continuous. Typically, the level of operator interaction is complemented with automated controls. For example, although an operator may manually initiate membrane cleaning, the sequence is typically automatic.

9.0 REFERENCES

Adams, A. E., Jr.; Eckenfelder, W. W., Jr. (1974) Gravity Thickening. In *Process Design Techniques for Industrial Waste Treatment*; Enviro Press: Smith field, New South Wales, Australia.

Beliew, G.; Fileccia, M. (1986) Optimization of Dissolved Air Flotation Thickeners, *Proceedings of the California Water Pollution Control Association Conference*; California Water Pollution Control Association: Oakland, California.

Boyle, W. H. (1978) Ensuring Clarity and Accuracy in Torque Determinations. *Water Sew. Works*, March.

Bratby, J. (1978) Aspects of Sludge Thickening by Dissolved Air Flotation. *Water Pollut. Control Fed.*, **77** (3), 421–432.

Bratby, J.; Ambrose, W. (1995) Design and Control of Flotation Thickeners. *Water Sci. Technol.*, **31** (3–4), 247–261.

Bratby J.; Jones G. (2007) How Far Can You Take Flotation Thickeners? *Proceedings of 5th International Conference on Flotation in Water and Wastewater Systems*, International Water Association, Sep 11–14, Seoul, Korea; IWA Publishing: London, England, United Kingdom.

Bratby, J.; Marais, G.vR. (1975) Saturator Performance in Dissolved-air Flotation. *Water Res.*, **9,** 929–936.

Bratby, J.; Marais, G.vR. (1977) Flotation. In *Solid/Liquid Separation Equipment Scale-Up*; Purchas, D. B., Ed.; Uplands Press Ltd.: London, England, United Kingdom.

Bratby, J.; Jones, G.; Uhte, W. (2004) State-of-Practice of DAFT Technology—Is There Still a Place for It? *Proceedings of the 77th Annual Water Environment Federation Technical Exposition and Conference* [CD-ROM]; New Orleans, Louisiana, Oct 2–6; Water Environment Federation: Alexandria, Virginia.

Bratby, J.; Jones, G.; Uhte, W. (2006) Maximising the Performance of Flotation Thickeners; *Water 21*, **Oct** (8.5), 29–31.

Butler, R. C.; Finger, R. E.; Pitts, J. F.; Strutynski, B. (1997) Advantages of Cothickening Primary and Secondary Sludges in Dissolved Air Flotation Thickeners. *Water Environ. Res.*, **69** (3), 311–316.

Dick, R. I.; Ewing, B. B. (1967) Evaluation of Activated Sludge Thickening Theories. *J. Environ. Eng.*, **93** (4), 9.

Haarhoff, J.; Bezuidenhout, E. (1999) Full-Scale Evaluation of Activated Sludge Thickening by Dissolved Air Flotation, *Water SA*, **25** (2), 153–166.

Jordan, V. J.; Scherer, C. H. (1970) Gravity Thickening Techniques at a Water Reclamation Plant. *J. Water Pollut. Control Fed.*, **42,** 180.

Komline, T. (1976) Flotation for Water and Wastewater Treatment. *Proceedings of the Water Research Centre Conference*, Felixstowe, United Kingdom; Water Research Center: Wiltshire, England, United Kingdom.

Metcalf and Eddy, Inc. (2003) *Wastewater Engineering: Treatment and Reuse*, 4th ed.; Tchobanoglous, G.; Burton, F. L.; Stensel, H. D., Eds.; McGraw-Hill: New York.

National Fire Protection Association (2003) *NFPA 820 Standard for Fire Protection in Wastewater Treatment and Collection Facilities*; National Fire Protection Association: Quincy, Massachusetts.

Noland, R. F.; Dickerson, R. B. (1978) *Thickening of Sludge*. EPA-MERL, Technology Transfer Seminar on Sludge Treatment and Disposal; U.S. Environmental Protection Agency: Cincinnati, Ohio.

Torpey, W. N. (1954) Concentration of Combined Primary and Activated Sludges in Separate Thickening Tanks. *J. Sanit. Eng. Div., Proc. Am. Soc. Civ. Eng.*, **80,** 443.

U.S. Code of Federal Regulations (1991) 29 CFR, Section 1910.147.

U.S. Environmental Protection Agency (1974) *Process Design Manual for Upgrading Existing Wastewater Treatment Plants*; EPA-625/1-71-004a; Technology Transfer; U.S. Environmental Protection Agency: Cincinnati, Ohio.

U.S. Environmental Protection Agency (1978) *Field Manual for Performance Evaluation and Troubleshooting at Municipal Wastewater Treatment Facilities*. U. S. Environmental Protection Agency: Washington, D.C.

U.S. Environmental Protection Agency, EPA, 1979: *Process Design Manual for Sludge Treatment and Disposal*, 625/1-79-011; U.S. Environmental Protection Agency: Washington, D.C.

Water Environment Federation (2008) *Operation of Municipal Wastewater Treatment Plants*, 6th ed.; WEF Manual of Practice No. 11; McGraw-Hill: New York.

Water Environment Federation; American Society of Civil Engineers; Environmental and Water Resources Institute (2009) *Design of Municipal Wastewater Treatment Plants*, 5th ed.; WEF Manual of Practice No. 8; ASCE Manuals and Reports on Engineering Practice No. 76; McGraw-Hill: New York.

Water Pollution Control Federation (1980) *Sludge Thickening*, Manual of Practice No. FD-1; Water Pollution Control Federation: Washington, D.C.

Water Pollution Control Federation (1980) *Sludge Thickening*; Manual of Practice No. FD-1; Water Pollution Control Federation: Washington, D.C.

Water Pollution Control Federation (1987) *Operation and Maintenance of Sludge Dewatering Systems*, Manual of Practice No. OM-8; Water Pollution Control Federation: Alexandria, Virginia.

Chapter 9

Sludge Minimization Technologies

(continued)

1.0 INTRODUCTION

Sludge minimization can be defined as the optimum reduction of sludge produced at a wastewater treatment plant (WWTP). The interest in sludge minimization technologies results in the desire of utilities to reduce capital cost and operation and maintenance costs associated with sludge management and increased federal and local restrictions on land application. Processes like Cannibal and multistage activated biological process (MSABP) reduce the waste sludge production at the WWTPs, whereas sludge disintegration improves the efficiency of downstream solids processes such as anaerobic digestion, which results in higher volatile solids removal and higher biogas production. Some of the processes presented here had few or no municipal full-scale installations at the time this section was prepared.

2.0 PROCESSES THAT REDUCE BIOMASS PRODUCTION

2.1 Cannibal Solids Reduction Process

2.1.1 Process Description

The Cannibal solids reduction process (Siemens, Warrendale, Pennsylvania) is an excess activated sludge minimization process. It reduces the yield at an activated sludge plant by (1) removing trash and inerts from the mixed liquor and (2) converting the excess biological solids back to a degradable carbon source by hydrolysing bacterial cell material through environmental conditioning. The conditioned sludge is fed back to the activated sludge process for biodegradation under aerobic conditions. Cannibal is not applicable for primary sludges or fixed-film sludges. Chemical sludges should also be kept out of the flow stream of the Cannibal process.

A process flow diagram for the Cannibal process is presented in Figure 9.1. It consists of a physical separation process called solids separation module (SSM) followed by a biological treatment process called side-stream interchange bioreactor (Roxburgh et al., 2006).

The SSM consists of a very fine drum screen (0.25 mm opening) where activated sludge from the secondary clarifier is treated on a continuous basis to remove inert organic matter. A set of hydrocyclones are also part of SSM and operated on a periodic basis to remove heavy organic material, grit, and dense inorganic particles that accumulate because of not wasting the sludge. The material removed by SSM is typically stabilized and does not require further stabilization via digestion. The trash, grit, and inert material are compressed to 20 to 40% solids (with 30% being typical) and are hauled from the plant to a landfill.

The flow that typically would be wasted from a conventional plant is wasted to the side-stream interchange bioreactor where minimal aeration is applied to maintain

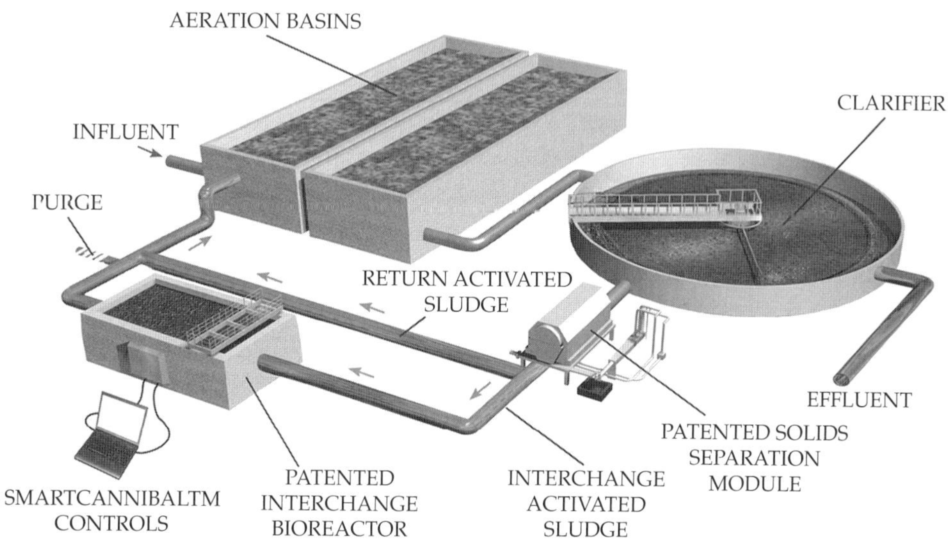

FIGURE 9.1 Flow diagram for the Cannibal process (courtesy of Siemens Water Technologies Corp.).

the tank at the border of anoxic and anaerobic conditions. Oxidation reduction potential (ORP) is monitored in the interchange bioreactor so that the air is only supplied for short periods of time when the ORP becomes too low. The cyclic conditions provide a suitable environment for conditioning of the biosolids and for the development of different microbial communities.

A portion of the interchange bioreactor is transferred back to the main treatment process, where biodegradation is completed. Because the solids are allowed to settle between aeration periods, a mixer is activated before transferring the contents from the interchange bioreactor to the activated sludge plant. As a result of this cycle between the aerobic and nonaerobic environments, solids are destroyed and the effective biological solids yield is reduced, which results in less solids production from the plant. A reduction in biological solids in the range of 40 to 60% is typical when the Cannibal process is used.

2.1.2 *State of the Technology*

In the early 1990s, several technologies were being developed that collectively now make up the Cannibal Solids Reduction Process. In the Southeast United States, Micronair, Inc. attempted to oxidize all the organic sludge at a plant. In the Midwest, sequencing facultative digestion (SFD) was developed by ATS, Inc. and now is the heart of the interchange portion of the Cannibal process. In January 2003, USFilter (now Siemens Water Technologies) acquired the patent rights to these technologies and began marketing the Cannibal Solids Reduction Process.

As of 2010, there were more than 30 WWTPs using the Cannibal process in the United States ranging in size from 1900 to 60000 m^3/d (0.5 to 16.0 mgd) in size. Several international installations were also in the works. The Cannibal process has been applied to many different types of activated sludge processes, but has proven to work best on activated sludge systems with solids retention times (SRTs) greater than 10 days.

Cannibal is applicable for activated sludge plants without primary clarifiers and with a design flow of 1900 to 95 000 m^3/d (0.5 to 25 mgd); although there are no technological reasons that it could not be applied to larger facilities.

Advantages of the Cannibal process include simplified plant operation, low biosolids yield, reduced energy consumption for aerobic digestion, and reduced plant odors. A noticeable odor likely will occur at the start of the aeration cycle in the interchange bioreactor, however (Roxburgh et al, 2006). Disadvantages of the Cannibal process include increased potential for operating at higher mixed-liquor suspended solids (MLSS) concentrations and SRTs in the activated sludge process. In addition, the inerts removal hydrocyclone operation has not been perfected.

From a microbiology perspective, the Cannibal process is compatible with activated sludge treatment facilities configured for biological nitrogen removal (BNR) and biological phosphorous removal (BPR) processes. One disadvantage occurs when the Cannibal process is applied to facilities that have phosphorus limits. In those cases, the potential for solids reduction becomes minimized as a result of needing to remove phosphorus from the treatment process in a solid form. Unlike carbon and nitrogen, phosphorus is not converted to a gas and thus stays in the biosolids. Reducing the biosolids yield ultimately reaches a limitation where no additional phosphorus can be concentrated in to the biosolids stream.

The Cannibal process typically is operated in a conventional semibatch operation with daily interchange between the main activated sludge process and the interchange bioreactor. The Cannibal process can also be setup to operate in a conventional continuous feed operation from the activated sludge system with batch interchange activated sludge (IAS) back to the activated sludge process. If phosphorus must be removed, then an additional option can be used. The process involves phosphorus scalping from the clear decant of the interchange tanks that is sent to a chemical phosphorus (P) removal system for precipitation of soluble P (typically 100 to 200 mg/L). In this case, chemical sludges are sent to disposal and kept out of the biological process flow stream.

2.1.3 *Design Considerations*

Plant type is one of the most important design criteria for achieving optimal solids reduction with the Cannibal process. University research and experience in the field has shown that the greatest amount of solids reduction is achieved when the Cannibal process is combined with activated sludge processes that have a plug-flow type of configuration. Sequential batch reactors also tend to be a good fit with the technology. Returning the interchange solids back to a nonaerobic zone along with good control over the activated sludge aeration system to prevent zones of over-aeration also are beneficial for promoting effective solids reduction.

The solids separation module of Cannibal includes rotary drum fine screens (0.25-mm openings), which perform best on lower solids concentrations (< 10 000 mg/L). Consideration for MLSS screening in lieu of return activated sludge (RAS) screening should be investigated. Ample prefiltered wash water is a must, and good drainage around the SSM is also required.

The interchange bioreactors are equipped with coarse bubble diffused aeration supplied by positive displacement blowers, independent mixing, floating decanters, and instrumentation including pH, ORP, total suspended solids (TSS), and level sensors. Interchange bioreactors can be incorporated into a wide array of aerobic digester-like tanks including both circular and rectangular tanks. Floating mixers are the preferred mixing approach, but submersible mixers can also be used if preferred. Interchange bioreactor sludge can be wasted to a separate aerobic digester and then aerated for 7 to 30 days to achieve a Class B product.

Trash (lint, hair, fibers, and paper) is screened and compacted (20 to 40% total solids). For the purposes of measuring performance, a trash yield can be calculated for given plant. Trash yield varies from community to community but ranges from 0.10 to 0.20 kg/kg five-day biochemical oxygen demand (BOD_5) (0.10 to 0.20 lb/lb BOD_5). Trash tends to be between 80 and 95% volatile solids. Fine inerts typically account for less than 0.04 kg/kg BOD_5 (0.04 lb/lb BOD_5) and have a specific gravity of approximately 1.5 and are 20 to 40% volatile solids.

Typical biological solids yield after reduction with the Cannibal process range from 0.20 kg/kg (lb/lb) BOD_5 to 0.40 kg/kg (lb/lb) BOD_5, with some facilities achieving less than 0.10 kg/kg (lb/lb) BOD_5 biological solids yield (biosolids yield values do not include trash yield stated above).

Interchange activated sludge is the product returned to the activated sludge process from the interchange tanks and contains food for the process. Concentrations of ammonia can range from 60 to 80 mg/L, total phosphorus can range from 100 to 200 mg/L, and oxygen demands can be more than 100 mg/L·h. The IAS should be returned to the anoxic portion of the activated sludge process or should be mixed with the raw wastewater before being returned.

Purged biosolids from the interchange tanks contain a high degree of biodegradable material and, therefore, a high oxygen demand. Class B biosolids typically can be easily achieved through typical aerobic digestion.

Design example:

- Average daily flow = 19 m^3/d (5.0 mgd);
- BOD_5 = 200 mg/L, 3785 kg/d (8345 lb/d);
- TSS = 240 mg/L, 4542 kg/d (10 014 lb/d);
- Total Kjeldahl nitrogen (TKN) = 40 mg/L, 757 kg/d (1669 lb/d);
- Total phosphorus = 6 mg/L, 113 kg/d (250 lb/d);
- Biological process = four-stage activated sludge system; and
- Organic loading rate (OLR) = 0.3 kg BOD/m^3 (20 lb BOD/1000 cu ft).

The process consists of two trains of treatment with four tanks in series and the ability to run anoxic conditions in tanks 1, 2, and 3 in each train. Each tank is 29 m by 9 m with a side water depth (SWD) of 5.5 m (95 ft × 30 ft × 18 ft sidewater depth). Traditional yield is 0.8 kg TSS/kg BOD removed (0.8 lb TSS/lb BOD removed).

Two 0.076 m^3/s (1200 gpm) fine screens are provided to screen the MLSS. Trash removed is approximately 0.15 kg TSS/kg BOD (0.15 lb TSS/lb BOD), leaving a biosolids yield before Cannibal of 0.60 kg TSS/kg BOD removed (0.60 lb TSS/lb BOD removed).

Two interchange tanks are provided at 1900 m^3 (0.50 mil. gal) each. Waste sludge is sent to interchange tanks on a daily basis and held there for a 10-day SRT. The interchange of sludge between the main activated sludge process and the interchange tanks results in an overall biosolids yield of 0.20 kg TSS/kg BOD removed (0.20 lb TSS/lb BOD removed).

The overall disposal consists of 0.15 kg TSS/kg BOD (0.15 lb TSS/lb BOD) of dewatered fine screenings (approximately 40% dry solids) that can be sent to a sanitary landfill or incinerated, and 0.20 kg TSS/kg BOD (0.20 lb TSS/lb BOD) of waste sludge that will need to be stabilized by aerobic digestion if the ultimate disposal method is land application.

2.2 Multistage Activated Biological Process

2.2.1 Process Identification and Description

2.2.1.1 Identification of Process

The multistage activated biological process (MSABP) is a biological municipal and industrial wastewater treatment process in which waste activated sludge (WAS) is consumed within the process, which eliminates the sludge handling. The manufacturer lists the advantages of MSABP as

- No return activated sludge or mixed liquor suspended solids control required;
- No sludge handling, inventory, or treatment;
- Compact footprint;
- Ability to handle hydraulic and organic peaks;
- Simple operation and automatic control; and
- Energy efficient.

2.2.1.2 *Scientific Basis*

The MSABP process is based on spatial microorganism successions, which is a sequence of one organism feeding the other, and trophic hydrobionts chains. In a spatially segregated trophic chain, primary microorganisms consume microbes, and then they are consumed by higher organized vulture filtrators of different trophic levels. Such spatial microorganism succession provides treatment by means of aerobic and anaerobic destructor microorganisms together with conventional copiotrophs and oligotrophs providing for high levels of organic and inorganic substrate removal. This trophic chain design leads to a total consumption of WAS in a MSABP tank.

2.2.1.3 *Process Description*

The MSABP consists of series of reactors as presented in Figure 9.2. Each reactor has removable media racks and air diffusers. The first stage in MSABP is the log-growth phase. In this stage, a high substrate-to-microorganism ratio results in microorganisms rapidly absorbing the substrate and, hence, removing the BOD. The volatile suspended solids-to-TSS ratio is 75 to 85%. The majority of the microorganisms are attached growth biomass (6000 to 10 000 mg/L), but there are also suspended growth biomass in the reactors (100 to 150 mg/L).

The second stage is the declining-growth phase. In this stage, microorganisms oxidize the absorbed organic materials and dissolved organic solids producing new microorganisms. Slow microorganism growth occurs, and the mass quantity of organic matter entering the final stage decreases.

The third and final stage is the endogenous phase. In this stage, the higher-order microorganisms consume the lower-order ones because of lack of any other food source. Total mass of new biomass and the organic products produced will be lower than the mass quantity of the raw organic material. In this way, no WAS will be generated from the MSABP. In this stage, biomass concentration does not increase because there is not enough substrate. Therefore, the attached growth biomass is less than 1000 mg/L, and the suspended growth biomass is less than 10 mg/L.

Because MSABP uses fixed microflora and zoning of organic material, the nitrification and denitrification processes occur simultaneous without alternating aerobic and anoxic zones.

2.2.1.4 *Design Data*

Depending on the level of treatment, 16 to 24 hours hydraulic retention time (HRT) is used as design criteria. For higher efficiency treatment, longer HRT is preferred. The BOD loading should be kept at less than 0.65 kg/m^3·d (40 lb/d/1000 cu ft) at 8 to 10 °C.

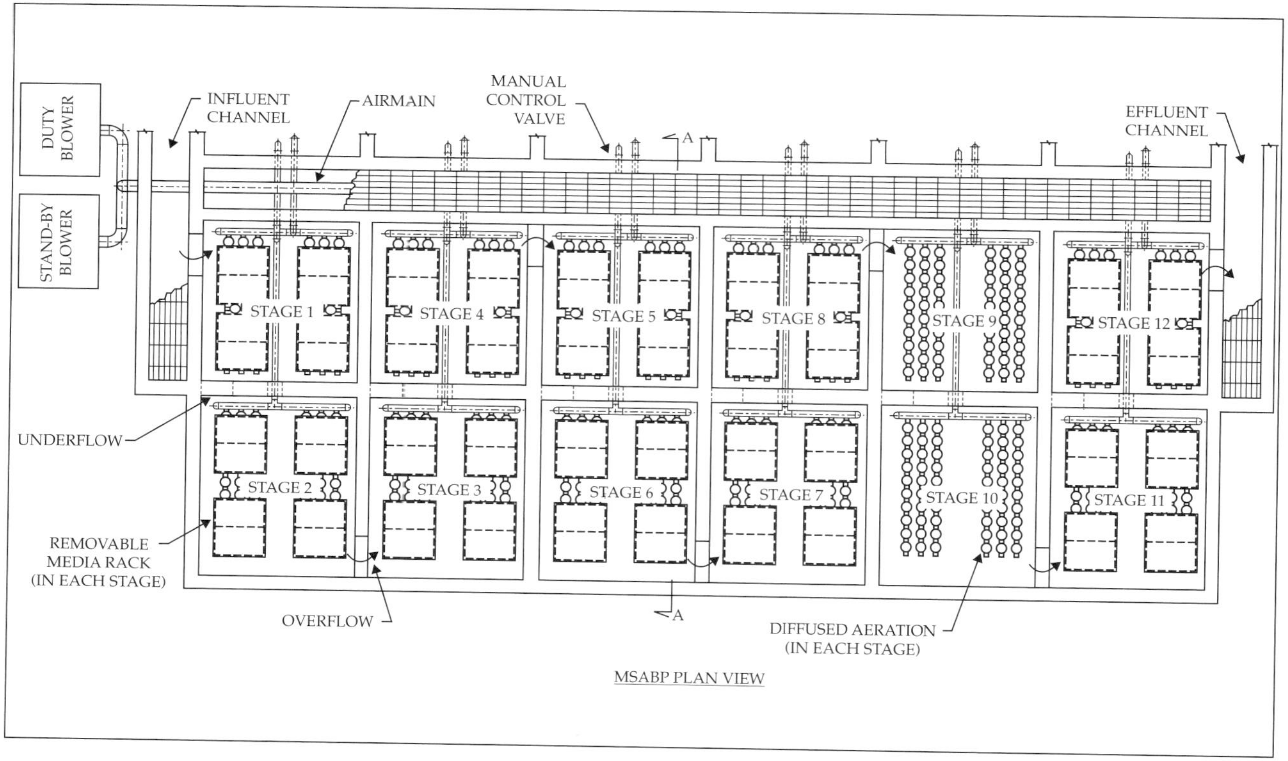

FIGURE 9.2 Arrangement of the Multi-Stage Activated Biological Process (courtesy of Aquarius Technologies, Inc.).

The bioreactor is divided into 8 to 12 stages. The number of stages is determined based on the level of treatment required. Each stage contains fine- or coarse-bubble diffused aeration and fixed-film media made from polyamide and polypropylene material. Support frames are made from 304 stainless steel. Air is provided by blowers, which are controlled by overall oxygen gradient, and blower output feedback.

2.2.1.5 *Pre- and Post-Treatment*

Removal of inert material before MSABP is important for efficiency of the system and is provided as part of the manufacturer's scope of supply. Inert removal consists of fine screens with 1-mm opening followed by standard grit removal for municipal applications.

2.2.1.6 *Environmental Effect*

Environmental effects from this process are similar to the ones from any other WWTP.

2.2.2 *Potential Applications*

The multistage activated biological process can be used for both industrial and municipal wastewater treatment.

2.2.3 *Process Evaluation*

2.2.3.1 *Levels of Removal of Contaminants*

When used for municipal wastewaters, the MSABP can produce an effluent with less than 10 mg/L BOD and less than 10 mg/L TSS without producing any WAS. If lower BOD and TSS concentrations are desired, then a tertiary filter should be used.

2.2.3.2 *Status of Development*

The MSABP has been tested at the bench-, pilot-, and full-scale levels. The pilot plants were located in Gijon, Asturias, Spain; Northern Moraine Utility Commission, Glenbeulah, Wisconsin; Devlin Wastewater Treatment Plant, Roselle, Illinois; and St. Vrain Sanitary District, Firestone, Colorado. The process has been used for both municipal and industrial wastewaters to eliminate sludge handling and disposal. It should be noted, however, that these full-scale installations have influent flows less than 800 m^3/d (0.21 mgd).

There are approximately 16 full-scale installations for industrial wastewater around the world with flows ranging from 20 to 300 m^3/d (5200 gpd to 80 000 gpd).

For municipal wastewater, there are approximately seven installations with flows ranging from 95 to 800 m^3/d (25 000 to 210 000 gpd). Several full-scale plants are listed below:

- Kvuzat Yavne, Israel: 800 m^3/d (0.210 mgd) municipal wastewater;
- JH Ranch, CA: 170 m^3/d (0.045 mgd) municipal wastewater;
- Teva Pharmaceutical Industries Ltd., Israel: 300 m^3/d (0.080 mgd) industrial wastewater; and
- Kohay Shahar, Israel: 400 m^3/d (0.108 mgd) municipal wastewater.

2.2.4 *Limitations*

2.2.4.1 *Site/Matrix Considerations*

The MSABP eliminates sludge production while producing an effluent comparable to secondary effluent from conventional activated sludge processes.

Like a conventional wastewater treatment plant, MSABP would require odor control at the headworks based on the site conditions. The process does not produce any primary or secondary sludge. Therefore, odor control would not be required for domestic wastewater applications.

2.2.4.2 *Residue Treatment*

The manufacturer states that MSABP eliminates sludge production. Therefore, there will be no sludge handling or disposal. Furthermore, the process does not produce any sidestreams except disposal of screenings and grit.

2.2.4.3 *Process Risks*

The MSABP has similar risks as operating a municipal WWTP with membrane bioreactors.

2.2.4.4 *Health and Safety Considerations*

The MSABP does not require any special health and safety considerations compared to other municipal WWTPs with membrane bioreactors.

2.2.4.5 *Reliability*

As stated above, the performance of MSABP is not known if the average daily flow is more than 800 m^3/d (0.21 mgd). Additionally, chemical phosphorous removal may decrease the performance of the MSABP because of chemical sludge accumulation in the bioreactors.

2.2.5 *Technology Prognosis*

The MSABP is an effective activated sludge technology to eliminate WAS production and, hence, sludge handling from WWTPs. Based on full-size installations, this process is shown to be effective for treatment plants that have average daily flow of 800 m^3/d (0.21 mgd) or less. A cost-benefit analysis should be performed if this process is being considered for a facility with higher influent flow rates.

3.0 SLUDGE DISINTEGRATION TECHNOLOGIES

Sludge disintegration is "the destruction of sludge by external forces" according to Müller et al. (2004). Sludge disintegration technologies, therefore, enhance the digestion process by applying external energy to increase the bioavailability of sludge being digested. The form in which the energy is applied varies based on the technology used. The technologies used to apply external forces on the sludge are quite diverse: chemical disruption (Chiu et al., 1997); pulsed electric fields (Kopplow et al., 2004); gamma-irradiation (Laffite-Trouque´ and Forster, 2002); thermal destruction (Li and Noike, 1992); mechanical shear (Muller et al. 2007; Nah et al., 2000); sonication (Neis et al., 2000; Tiehm et al., 2001); and ozonation (Weemmaes et al., 2000). The point of energy application is also variable. Primary sludge, secondary sludge, and digested solids have all been reviewed as a means of increasing sludge digestibility (Muller et al., 2006; Tiehm et al., 2001). By far, the most common point of technology application is to WAS because it is thought to be the most refractory portion of the raw sludge loaded to the digester. Although a variety of technologies are available, those reviewed in this chapter represent processes that have been developed to full-scale implementation.

Regardless of energy input location, these technologies typically provide several advantages and some disadvantages to the digestion process. Several process changes are associated with the application of disintegration technologies:

- Additional total and volatile solids destruction;
- Increased biogas production; and
- Reduced mass of material for disposal.

These changes in the digestion process and solids handling are typical; some technologies have reported additional benefits not reported by others and will be discussed with the individual technologies in later sections.

Sludge disintegration technologies improve the digestion process by increasing the bioavailability of sludge. Bioavailability is enhanced by one of two mechanisms: cellular lysis and particulate dispersion. Cellular lysis improves sludge digestion by breaking cells open releasing intracellular materials, which is often regarded as readily bioavailable. The lysis products are consumed by undamaged organisms, which generates additional biogas and solids destruction. Although cell lysis is possible, little direct evidence of a high degree of cellular lysis has been offered in the literature at this time.

Particle dispersion also improves overall sludge digestibility. A reduction in particle size increases the overall surface area for microbial colonization and hydrolytic enzyme attack. The latter is more significant because hydrolysis is considered the rate-limiting steps of anaerobic sludge stabilization (Vavlin et al., 1997). Increasing the surface area of particles has been shown to improve the observed rate of hydrolysis of a particular substrate (Sanders et al., 2000). Hydrolysis, particularly of WAS, typically is reported as the rate-limiting step in anaerobic digestion. Thus, any improvement in the hydrolysis rate will increase digester performance.

The improvement in digester performance, as noted by increased solids destruction and gas production, can result in some detriment to the treatment process. Increased volatile solids destruction will result in the release of phosphorous, ammonium, and sulfide through the biodegradation of organic matter. Additional ammonium and phosphorous release will affect the liquid and solids streams, especially if nutrient removal is implemented at the facility. These additional nutrient loads will result in an increase in oxygen demand and chemical requirements for both nitrogen and phosphorus removal, respectively.

The effect on the solids stream is in the form of chemical precipitants, vivianite, or struvite. Struvite formation is affected by flow conditions and the relative concentrations of ammonium, phosphorous, and magnesium. Increasing the concentration of ammonium and phosphorous may lead to conditions where struvite is formed and becomes an issue. Controlling solution cations concentrations is critical to the overall success of digestion enhancement. Formation of struvite because of enhanced digestion may not have a negative effect if the process is coupled with a struvite precipitation process for recovery of nutrients.

Disintegration processes also can affect the dewatering and biosolids disposal processes. Muller et al. (2007) reported increased polymer demand with sludge disintegration. Novak et al. (2007) reported that digestion enhancement using the MicroSludge process decreased volatile organic sulfur emissions from dewatered biosolids.

Based on these observations, the installation of a digestion enhancement process must consider more than the added biogas and reduced solids mass to assess the complete effect and to process cost/benefit. For example, some of the manufacturers are also claiming the use of conditioned WAS as an external carbon alternative to methanol for BNR applications.

Enhancement of anaerobic digestion can provide several added benefits depending on the degree of enhancement. Additional biogas production can be used to generate electricity or heat or can be sold as a fuel source to outside entities. The utility can realize these benefits through a reduction in electricity purchased from the grid, reduced need for natural gas, and a source of revenue if the gas is sold. When put to beneficial use, biogas can also offset carbon emission from fossil fuel sources. Enhancements can also lead to a reduction in hauling associated with improved solids destruction. Reduced hauling because of the reduction in the mass of solids can reduce the number of haul trips and reduce wear on equipment and use of fuel. Although not guaranteed, these benefits should be considered in selecting a technology.

The addition of energy to unstabilized sludge will improve process performance typically in the form of improved biogas production and solids destruction. The economics of the system, however, require other revenues to offset the costs of implementation. The following sections will discuss several representative technology types that are at different points of development.

3.1 MicroSludge: High-Pressure Homogenization Process

3.1.1 Process Identification and Description

3.1.1.1 Identification of Processes

The MicroSludge process is a high-pressure homogenization process developed by Paradigm Environmental of Vancouver, British Columbia, Canada. The process uses a caustic pretreatment combined with a high-pressure homogenizer (815 atm or 12 000 psi) to disintegrate thickened WAS (TWAS) before anaerobic digestion. Figure 9.3 shows the complete MicroSludge unit, high-pressure homogenizer, and skid. Figure 9.4 is a simplified process schematic showing the location of the MicroSludge process in a solids treatment system.

3.1.1.2 Scientific Basis

In the two-step process, caustic (NaOH, 25% wt/wt) is added to TWAS and then, following a short reaction period, the solution is processed through a high-pressure

a) MicroSludge skid—contains control elements, reaction tank, hydrocyclones, screens, and shear pump.

b) Microsludge high-pressure homogenizers.

FIGURE 9.3 Primary elements of the MicroSludge process for sludge disintegration (courtesy of Paradigm Environmental).

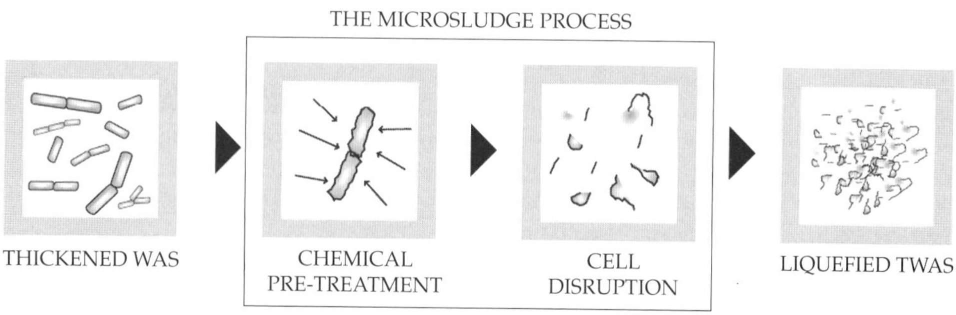

FIGURE 9.4 Simplified process flow schematic of the primary elements of the MicroSludge process for sludge disintegration (courtesy of Rabinowitz and Stephenson, 2006).

homogenizer to disintegrate the WAS. The addition of caustic to the WAS breaks down the floc structure and weakens the cell walls of the bacteria, rendering them more susceptible to lysis by the homogenization process. Figure 9.4 provides a simplified flow schematic of the disintegration process.

The homogenization process applies several different stresses on the microorganisms passing through the valve. As the sludge is pumped through the valve, the pressure increases to 815 atm (12 000 psi) and is immediately released as it passes through. The instantaneous pressure drop results in the formation of cavitation bubbles and the generation of heat when the bubbles collapse. As the pressure is relieved on the exiting sludge, the velocity of the waste stream increases to 305 m/s (682 mph) in approximately 2 microseconds (Stephenson et al., 2005). This stream affects a ring to place further mechanical stress on the organisms (Figure 9.5).

The disintegration process results in cellular lysis and particle-size reduction, which can lead to enhanced degradability in the anaerobic digester. Figure 9.6 show electron micrographs of the supernatant of TWAS before and just after processing through MicroSludge. The increase in organic matter is indicative of sludge disintegration.

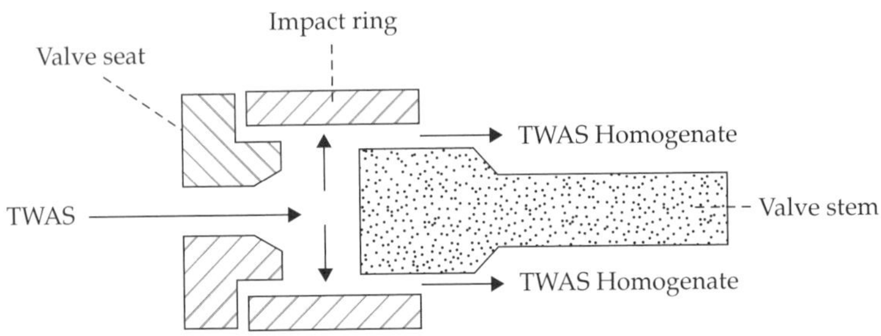

FIGURE 9.5 Schematic of the valve assembly used to generate shear forces on the caustic treated thickened waste activated sludge (TWAS) (courtesy of Rabinowitz and Stephenson, 2005).

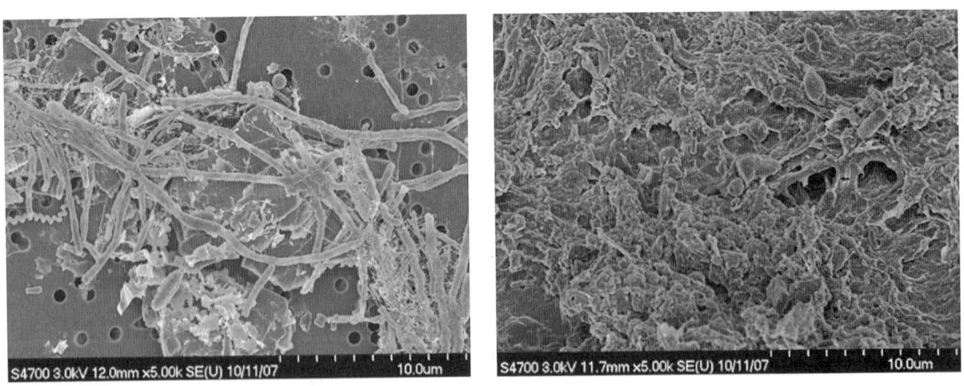

a) TWAS supernatant prior to processing

b) TWAS supernatant following processing

FIGURE 9.6 Effect of MicroSludge processing on supernatant of thickened waste activated sludge (TWAS) before digestion (courtesy of Paradigm Environmental).

3.1.1.3 Process Description

As noted in the previous section, MicroSludge couples caustic pretreatment with high-pressure homogenization to render thickened WAS more biodegradable. The process is installed on the discharge of the WAS thickening process. The thickened WAS is passed through a course screen, combined with 25% (by weight) sodium hydroxide, and homogenized in an inline shear pump. Following homogenization, the caustic treated material is passed through a second in-line screen before introduction to the age tank.

Screen size can vary to accommodate site-specific sludge conditions. The treated material exiting the caustic tank passes through two hydrocyclones to remove gas, which can damage the homogenizer valve and fines, especially grits and silts, which can result in the premature wear of the impact ring. The pretreatment steps are all contained on the primary skid of the MicroSludge unit. Figure 9.7 shows critical pretreatment elements of the unit installed as part of the demonstration project at the Des Moines Water Reclamation Facility in Des Moines, Iowa.

Prescreens remove hair and other debris from the sludge stream that would otherwise damage the homogenizer valve. The reject material from the screenings is discharged into the final product line, after the homogenizer, for introduction to the

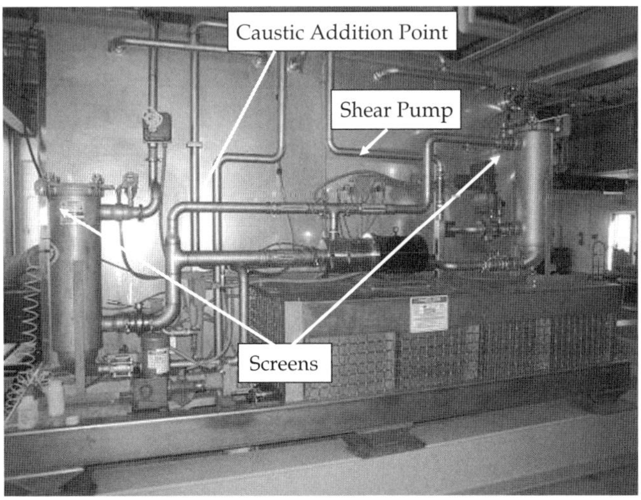

FIGURE 9.7 Preprocessing elements located on the primary skid of the MicroSludge unit (courtesy of Paradigm Environmental and Brown and Caldwell).

digester. This closes the mass balance and ensures that the maximum gas potential of the waste stream is realized.

Following the second screening, the WAS is held in the hold tank to chemically weaken the floc structure and cell wall. The hold tank is monitored to ensure a sufficient pH increase has been achieved to disrupt the floc and cell-wall structure. The pH-treated WAS is passed through a hydrocyclone before the homogenizer to ensure that any inerts such as grit or fine silts are removed because this material shortens the life span of the homogenizer.

The homogenizer is the primary disruptive force within the MicroSludge process. A 100-HP motor drives the piston pump to pump the WAS through the disrupter valve. The pressure on the inlet side of the valve increases to 815 atm (12 000 psi). As the solids pass through the valve, the pressure rapidly decreases to near ambient, which accelerates the solids to greater than the speed of sound. The high-velocity jet of WAS affects the impact ring, further disrupting the solids. The disrupted solids are then loaded to the digester for biodegradation.

The disintegration process through the homogenizer generates a high degree of cavitation as the pressure decreases from 815 atm to 3.4 atm (12 000 psi to 50 psi) almost instantaneously. The resulting collapse of the cavitation bubble imparts heat to the liquid stream. Stephenson et al. (2005) reported that the temperature of the raw sludge increased from 10 to 16 °C (50 to 60 °F) before disruption to 32 to 38 °C (90 to 100 °F) after. This could significantly reduce heating requirements, depending on the WAS fraction of the sludge stream to the digester.

3.1.2 Potential Applications

The MicroSludge process is specifically designed to liquefy or disintegrate TWAS for anaerobic digestion enhancement. Through this process, WAS bioavailability is increased, and consumption in an anaerobic digester is more complete. The process does not appear to have limitations related to digester type (thermophilic/mesophilic), though process loading, effluent quality, and energy balances need to be considered. Furthermore, it should be noted that this process may not be suitable for facilities that combine WAS and primary sludge before thickening.

3.1.3 Process Evaluation

3.1.3.1 Levels of Contaminant Removal

Varying levels of improvement in volatile solids destruction and increased gasification have been reported and are discussed in later sections. Additional benefits such as contaminant removal have not been reported. Reductions in fecal coliforms (pathogen indicator organisms), organic chemicals, or metals have not been reported to date; though with further process development and research, added benefits may be realized.

3.1.3.2 Status of Development

The MicroSludge process is an embryonic technology that has been developed to the point of full-scale application and demonstration with minor modifications to meet the specific operating conditions that vary from facility to facility. Demonstrations of the technology have been conducted at several locations including

- Chilliwack Wastewater Treatment Plant, Chilliwack, British Columbia, Canada;
- Los Angeles County Sanitation District, Joint Water Pollution Control Facility (JWPCF), Los Angeles, California; and
- Des Moines Water Reclamation Facility, Des Moines, Iowa (commenced 2008).

- At the time of publication, there were no permanent installations of this technology. The full-scale demonstrations have provided data on the effectiveness of the technology at enhancing the digestion process. Tables 9.1 and 9.2 summarize the process data for the Chilliwack 9 and Los Angeles demonstrations that were available during preparation of this manual.
- The effect of the process on overall biosolids stability has been investigated as well. Novak et al. (2007) reported that processing of WAS through the MicroSludge at the Los Angeles JWPCF resulted in a net reduction in total volatile organic sulfur compounds released from the biosolids. These findings suggest a more stable biosolids product. Adams et al. (2003) indicated that the primary odorant associated with anaerobically digested biosolids was volatile sulfur compounds.
- Additional benefits of the technology that have been reported include a reduction in sludge viscosity through disintegration, which may improve mixing, pump efficiency, and a reduction in digester foaming, which increases the potential for more stable operation.

3.1.3.3 Design Data

MicroSludge is scalable based on plant capacity and size of the community. The manufacturer markets the smallest unit at approximately 95 m^3/d (25 000 gpd);

TABLE 9.1 Summary of Chilliwack, British Columbia, Canada, demonstration (Rabinowitz and Stephenson, 2005).

Parameter	Value*	
Raw solids		
Waste activated sludge (%)	0.4	
Thickened waste activated sludge (%)	4–5	
Primary-secondary ratio (percentage basis)	65:35	
System		
Digester number	2	
Operation	Series	
Solids retention time (primary/secondary/total)	10/3/13	
Operating temperature (°C)	37	
Mixing	Draft tube/gas lances	
Process performance	**Control condition**	**MicroSludge process**
Volatile solids reduction (%)	60	78
Volatile acids (digester 1/digester 2) (mg/L as acetate)	60/5	18/1
Biogas production	n/r	n/r
Biogas methane composition (%)	n/r	58.1
Total Kjeldahl nitrogen (digester1/digester 2) (mg/L as N)	1342/802	1415/748
NH_3N (digester 1/digester 2) (mg/L as N)	420/338	706/440

*n/r = not reported.

the largest unit has a capacity of 760 m^3/d (200 000 gpd), which would service plants from 53 000 to 405 000 m^3/d (14 to 107 mgd), respectively, processing all of the WAS at the plant. For larger utilities, multiple units can be used to meet solids processing demands.

3.1.3.4 *Secondary Environmental Effects*

The MicroSludge process does not produce an additional waste product for disposal other than those commonly associated with anaerobic digestion. The process does introduce sodium cations to the digester and, eventually, to the liquid stream.

TABLE 9.2 Summary of the Los Angeles County (California) Joint Water Pollution Control Facility demonstration (SRT = solids retention time) (Rabinowitz and Stephenson, 2006).

Parameter	Value*	
Raw solids		
Waste activated sludge (%)		
Thickened waste activated sludge (%)	5–6	
Primary-secondary ratio (percentage basis)	75:25	
System		
Digester number, SRT and mixing	n/r	
Operation	parallel	
Operating temperature (°C)	mesophilic	
Process performance	**Control**	**MicroSludge process**
Volatile solids reduction (%)	51–54	54–58
Volatile acids (mg/L as acetate)	14	16
Alkalinity	3629	3912
pH	7.18	7.22
Biogas production	n/r	n/r
Biogas methane composition (%)	61.6	62.3
Nitrogen	n/r	n/r

*n/r = not reported.

Sodium, a monovalent cation, has been reported to affect activated sludge settling. Increased biological activity can result in additional ammonium and phosphorous release into the liquid stream, which can affect the overall liquid treatment processes, particularly if nutrient removal is a requirement.

3.1.4 Limitations

3.1.4.1 Site/Matrix Considerations

The process requires increasing or decreasing footprints depending on the size of the facility using the equipment. The System 25 unit requires a footprint of 6 m × 12 m (20 ft × 39 ft) to process 95 m^3/d (25 200 gpd) of TWAS as shown in Figure 9.8. The

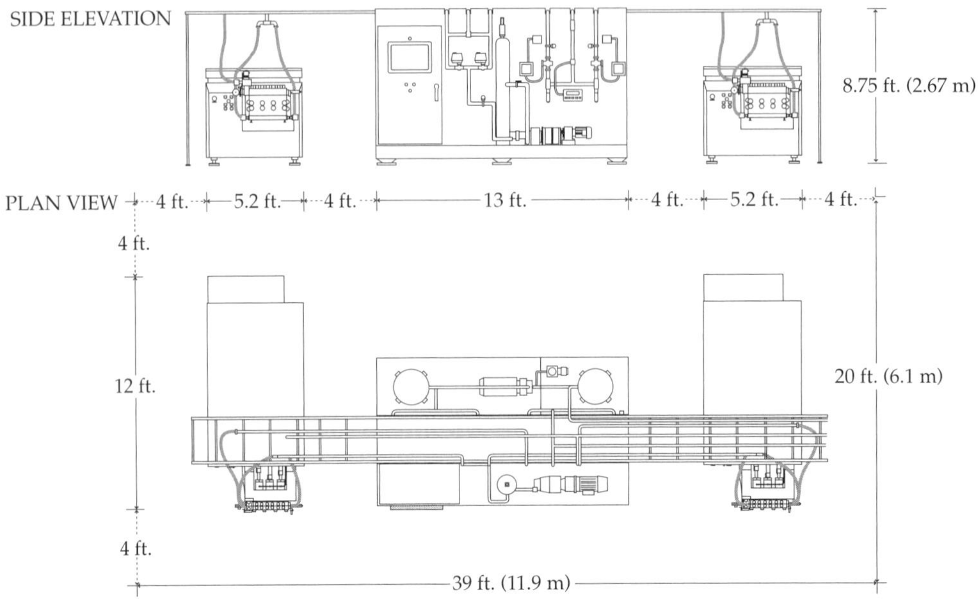

Figure 9.8 Plan and section view of the MicroSludge System 25 (courtesy of Paradigm Environmental).

system has an upper limit of 8% total solids, below which it is able to operate normally. The presence of extremely high-pressure elements and caustic should be considered when planning for installation to meet safety requirements.

Additional site consideration should include the amount of secondary sludge generated at the plant relative to primary sludge (Roxburgh et al., 2006). Because the process only affects secondary sludge, plants with high proportions of primary sludge may not realize the same benefits as a plant with a low primary sludge.

As noted previously, the economic viability of the process will be affected by beneficial use of the biogas. Co-generation or sale of biogas may be needed to offset the capital and operating cost of the unit; reduced solids production alone may not be sufficient.

3.1.4.2 Residue Treatment

The process itself does not produce a residue as long as screening and cyclone discharges are reintroduced. Handling and disposal of spent chemical containers need to be considered and appropriate safety practices adhered to. All other materials associated with the process are incorporated into the biosolids matrix and should be handled per regulatory requirements.

3.1.4.3 Process Risks

Appropriate safety measures should be practiced and incorporated into the installation to mitigate risks associated of caustic use and high pressures generated in the system must be considered.

Process effects to be considered include generation of additional ammonium-N and phosphorus through increased biological activity in the digesters. These nutrients may affect liquid-stream performance and permit levels depending on the requirements of the facility. Ammonium and phosphorous are also key elements in struvite and vivianite formation. Considerations for digester inorganic chemistry should be made to mitigate potential precipitate issues.

3.1.4.4 Reliability

At this time, there are no full-scale permanent installations for this process. As a result, long-term reliability data is unavailable.

3.1.5 Technology Prognosis

The technology is embryonic. The process does require some minor modification to operate at different facilities, as any piece of equipment does. At the time this chapter was authored, no permanent full-scale installations existed.

As process efficiency, energy recovery, and sludge minimization become more significant to utilities, the process has the potential to gain greater application.

3.2 Ultrawaves: Ultrasonic Sludge Treatment

3.2.1 Process Identification and Description

3.2.1.1 Identification of Processes

Ultrasonic pretreatment of WAS has been reported to be one of the more widely applied sludge minimization technologies that is commercially available. The process exposes TWAS to ultrasonic energy, typically 20 kHz, before digestion to breakdown or weaken the biomass cell walls. Several technologies are commercially available; the Ultrawaves technology will be reviewed as the representative technology of the group.

3.2.1.2 Scientific Basis

Ultrasonic technologies for sludge minimization can be applied to either the WAS feed of an anaerobic digester or in the RAS line of the secondary treatment system as shown in Figure 9.9. The enhancement of anaerobic digestion began to gain full-scale implementation in Europe in the mid to late 1990s. The technology developed as a means to increase biogas production while reducing the mass of biosolids for disposal—essentially increasing the gasification rate of the digester.

Enhancement of anaerobic digestion, WAS in particular, is achieved through generation of transient acoustic cavitation. Acoustic cavitation is formed in liquid medium as the ultrasonic waves place the liquid in compression and rarefaction. During rarefaction, sufficient energy may be applied so that intermolecular forces are exceeded and a void or cavitation bubble forms (Mason and Lormer , 1988). The subsequent collapse of the cavitation bubble generates significant amounts of heat (> 4000 °C or > 7200 °F) and pressure (approximately 1000 atm or 14 700 psi), forming shear forces from liquid jets (Christi, 2003). Particles located near or within a collapsing bubble can be exposed to one or more of these forces.

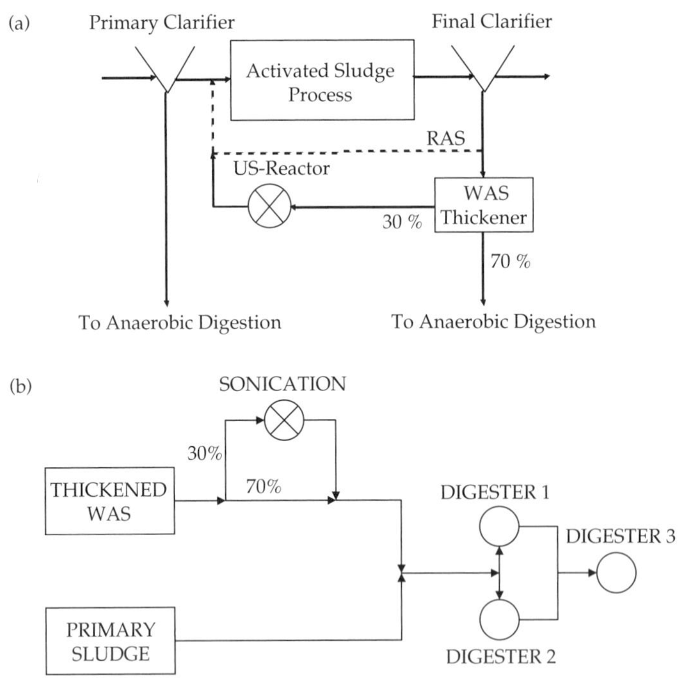

FIGURE 9.9 Process flow schematics for sludge minimization with ultrasonic technologies: (a) secondary treatment alternative and (b) anaerobic digestion alternative (WAS = waste activated sludge) (courtesy of EIMCO).

Pressure, temperature, and shear forces represent the primary forces responsible for the disintegration of WAS. Joyce et al. (2003) reported that lower frequency (20 and 38 kHz) ultrasonic energy was more effective at lysing cells than higher frequencies (512 and 820 kHz), which primarily disperse materials. Portenlänger and Heusinger (1997) reported similar results. Scherba et al. (1991) reported that 26 kHz ultrasonic treatment was capable of lysing both Gram positive and negative bacterial cell, suggesting that cell-wall composition does not provide additional protection. Furthermore, amplitude can improve the effectiveness of ultrasound for lysis (Raso et al.; 1998).

Chemical transformations are theoretically possible, as observed in the field of sonochemistry. The environment within the cavitation bubble can also lead to the formation of free radicals, which can interact with different elements in the surrounding fluid. Although sonochemical effects are possible, they are typically considered minimal relative to the mechanical effects of sonication on the WAS particles.

Other factors affect the generation of cavitation bubbles and their properties, including solids concentration (Mao et al., 2004), particle size (Tuziuti et al., 2005), liquid-gas saturation, and liquid temperature (Mason and Lormer, 1988). Optimizing conditions for all of these parameters is unlikely, however, because of the effects on digester operation and the inherent variability in wastewater systems. The material of construction of the horn also affects process performance and longevity. Ultrasonic horns are typically constructed of aluminum, stainless steel, or titanium. Titanium is, by far, the most expensive material but exhibits the best acoustic properties and horn longevity, including resistance to pitting because of cavitation.

Exposure of WAS to ultrasonic energy is reported to improve digester operation through increased biogas production and additional solids destruction. Preprocessing of WAS before anaerobic digestion, the most common process configuration for technology, has been investigated for frequencies from 20 kHz, the lower range of ultrasonic frequencies, to as high as 3217 kHz (Tiehm et al., 2001). Onyeche et al. (2002) investigated the ultrasonic treatment of anaerobically stabilized sludge, but the results were variable. The reduction of pathogens with ultrasonic technologies is mixed. Some research has suggested a reduction (Chu et al., 2001); others showed no change (Laffite-Trouqué and Forster, 2002).

3.2.1.3 Process Description

The Ultrawaves technology is a patented process developed by Ultrawaves-Wasser & Umwelttechnologien GmbH, Hamburg, Germany, and is distributed in the United States as the Sonolyzer by EIMCO Water Technologies, Salt Lake City, Utah. The technology uses a serpentine reactor in which a 20-kHz ultrasonic probe is located at each turn, for a total of five probes. A single unit is capable of processing approximately 35 m^3/d (9250 gpd) of thickened WAS at concentrations from 5 to 7% total solids.

The process is typically installed to treat anywhere from 25 to 33% of the total WAS flow to the digester. Multiple units can be installed to meet the flow requirement of plants of varying size. Figure 9.10 is a photograph of the Ultrawaves unit.

Enhancing anaerobic digestion using the Ultrawaves technology has been reported to have relatively consistent results. Additional volatile solids destruction ranging from 15 to 30% and increased biogas production of 15 to 30% has been reported by facilities using the technology (Table 9.3).

3.2.2 *Potential Applications*

Ultrasonic technologies can be applied to enhance anaerobic digestion as described in the previous section. Process effects include changes in volatile solids destruction, biogas production, polymer use, and digester foaming. These effects are commonly reported for other ultrasonic technologies.

Another mode of operation for the technology is in the RAS line in the secondary treatment system. Disintegration of the RAS can be used in secondary systems to minimize solids and to generate an internal carbon source for denitrification processes.

3.2.3 *Process Evaluation*

3.2.3.1 *Levels of Contaminant Removal*

The Ultrawaves technology has been proven to increase volatile solids destruction and biogas production by 15 to 30%. In addition, the process has reduced digester

(a) (b)

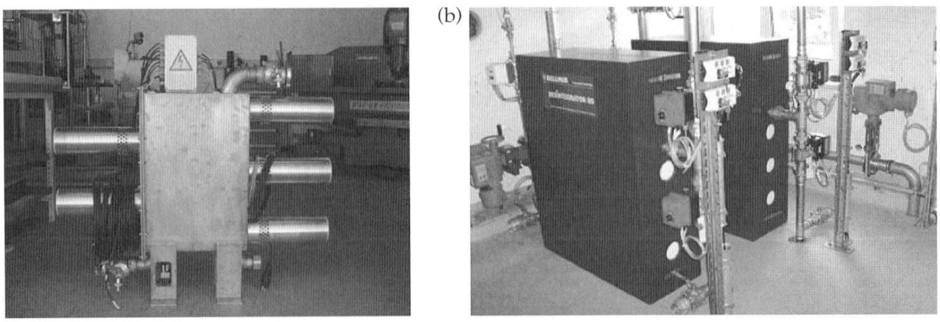

FIGURE 9.10 Photographs of (a) a single Ultrawaves unit and (b) installed units in acoustic housing.

Table 9.3 Summary of anaerobic digestion process changes using sonication to enhance performance for full-scale permanent installations (mgd × 3.785 = ML/d).

Wastewater treatment plant	Plant size (mgd)	Volatile solids destruction	Biogas production	Notes
Bamberg, Germany	12.15	+30	+30	
Meldorf, Germany	1.06	+25	+25	No foam or filamentous organisms present in the anaerobic sludge digester
Zeist, Netherlands	3.96	+25	+25	
Frederikshavn, Denmark	6.87	+25	+25	
Marselisborg-Arhus, Denmark	11.62	+15	+35	Polymer consumption decreased, 20%
Kleinsteinbach, Germany	2.11	+25	+25	
Bath, Holland	3.70	+15	+15	

foaming and resulted in a reduction in polymer usage in some cases. Additional process effects such as cake dryness and centrate or pressate strength and characteristics have not been reported. In cases where the technology was applied to secondary systems, additional solids destruction and reduced foaming and dryer cake (+2 percentage points) have been reported. Effects on aeration requirements have not been reported to date.

3.2.3.2 Status of Development

The Ultrawaves technology is a mature technology with multiple full-scale permanent installations in Europe since 2004. The capacity of the plants varies between 3800 and 88 000 m^3/d (1 and 23.2 mgd). Table 9.4 summarizes the installation size, length of operation, and operating mode (i.e., digester or secondary system). At the time of publication, no full-scale systems were operating in the United States in either mode.

3.2.3.3 Secondary Environmental Effects

Neither operation of ultrasonic technology before digestion nor as part of the secondary system resulted in reported adverse secondary environmental effects. Effects to consider are changes in centrate or pressate quality, especially nutrients, for

Table 9.4 Summary of facilities using ultrasonic WAS pretreatment to enhance anaerobic digestion (mgd × 3.785 = ML/d).

Wastewater treatment plant	Plant size (mgd)	Installation year
Waste activated sludge pretreatment before digestion		
Bamberg, Germany	12.15	2004
Meldorf, Germany	1.06	2004
Zeist, Netherlands	3.96	2005
Willem-Annapolder, Holland	2.91	2006
Frederikshavn, Denmark	6.87	2006
Marselisborg-Arhus, Denmark	11.62	2006
Kleinsteinbach, Germany	2.11	2006
Bath, Holland	3.70	2006
Nieuwgraaf, Holland	23.24	2006
Hennef, Germany	3.43	2006
Glarnerland, Switzerland	3.70	2007
Slupsk, Poland	13.21	2007
Return activated sludge treatment for secondary systems		
Bünde, Germany	2.11	2006
Leinetal, Heiligenstadt, Germany	~2.64	2003

anaerobic systems and an increase in effluent BOD for secondary systems, which do not completely degrade the material released.

3.2.4 *Limitations*

3.2.4.1 *Site/Matrix Considerations*

Ultrasonic units are relatively small when compared to other sludge minimization technologies. Another limitation is relatively small processing capacity per unit, which requires configuring multiple units in parallel for larger applications. This requires accurate flow splitting and measurement, which are relatively difficult in sludge streams. Furthermore, although the units require a minimum footprint, sound enclosures and cooling are recommended to protect operators and equipment.

Process efficiency is affected by solids concentrations. Typically, TWAS is treated by ultrasound because primary sludge is considered much more degradable.

Therefore, this process may not be considered if the digester feed consists of co-thickened WAS and primary sludge. Coordinating thickening process operation and ultrasound operation may increase the complexity of the feed system to the digester.

3.2.4.2 *Residue Treatment*

No additional residuals are generated from this system other than those typically generated during wastewater treatment.

3.2.4.3 *Process Risks*

Process risks associated with ultrasonics are not known at this time.

3.2.4.4 *Reliability*

The ultrasonic horns have a life expectancy of one to two years with replacement cost of $2,000 each and five horns per unit. The ultrasonic stacks have a life expectancy of 7 to 10 years.

3.2.5 *Technology Prognosis*

At the time of publication, there were no installations in the United States, but several in Europe and Southeast Asia. The technology has one of the highest degrees of implementation for side-stream sludge disintegration technologies. As process efficiency requirements and disposal costs increase, the technology may gain broader application.

3.3 Ozonation

3.3.1 *Process Identification and Description*

3.3.1.1 *Identification of Processes*

Ozone (O_3) is a chemical known for its oxidative power that is typically used for chemical oxidation or disinfection. This section discusses the Praxair's Lyso process, which uses ozone to oxidize and disintegrate RAS for further degradation in an aeration basin during secondary waste water treatment.

3.3.1.2 *Scientific Basis*

Ozone is a powerful oxidant that can be used for disinfection or chemical oxidation. When ozone reacts with a compound, it oxidizes the compound and is reduced to oxygen. Ozonation of WAS (or RAS) is intended to cause the lysis of cells within the floc matrix. Contact of the cell wall with ozone causes the wall and membrane to lose integrity, releasing the material from within the cell, which can then be degraded. The effect of ozone on sludge decomposition has been reported to change based on the chemical composition of the sludge. Scheminski et al. (2000) reported that the partial ozonation of digested sludge (doses up to 0.5 g O_3/g organic dry matter) resulted in 90% reduction in total protein; zero detected in the liquid phase; 63% release of polysaccharide to the liquid phase; and a 30% reduction in the lipid content of the sample.

Vergine et al. (2007) reported that partial treatment of a RAS stream (Lyso process) within a nutrient removal plant (nitrification/denitrification) resulted in 39% reduction in sludge mass for disposal, dryer cake (+3 percentage points), and significant reductions in foaming. The plant, which was treating textile effluent, had an observed reduction in the ammonia oxidation rate of 70% and nitrite oxidation rate of 35%. The author noted that the reduction in rate did not affect process efficiency of nitrogen removal in this case. Figure 9.11 is a photograph of the Lyso process installed at the Lariana Depur's Bulgarograsso WWTP, Como, Italy.

3.3.2 *Potential Applications*

Ozonated biosolids have been applied to both aerobic and anaerobic systems to increase the degradability of WAS. Full-scale implementation of technology has been limited to the Lyso process, which degrades ozonated solids in the aeration basin of secondary treatment systems (Figure 9.12).

The Lyso process has been successfully applied to activated sludge systems achieving reductions in waste sludge mass. The process uses 0.05 to 0.10 kg O_3/kg total solids to improve sludge degradability. (Ozonated solids should not be returned

FIGURE 9.11 Lyso process at Lariana Depur's Bulgarograsso wastewater treatment plant in Como, Italy.

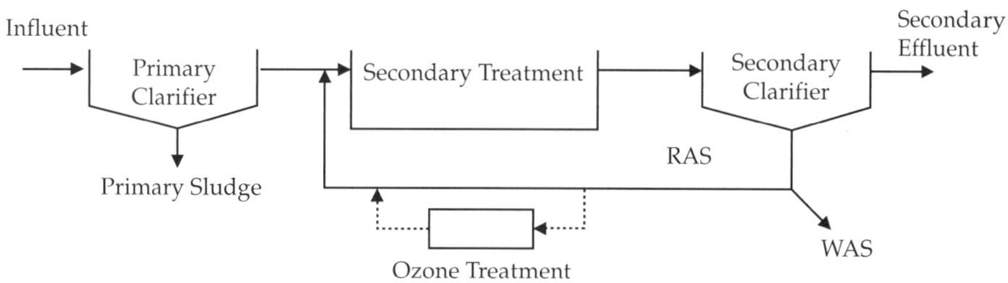

FIGURE 9.12 Flow diagram of Praxair's Lyso process (RAS = return activated sludge; WAS = waste activated sludge).

to anoxic zones for denitrification as the high oxygen levels may upset the process through a loss of anoxic conditions. Though not reported the same likely holds true for phosphorous removal systems as well.)

Anaerobic digestion of ozonated solids also needs to consider the effects of added oxygen in the feed to the digester. Care would need to be taken to prevent creation of explosive conditions in the digester headspace from significant additions of oxygen to the system or of a toxic environment for the strictly anaerobic methanogens.

3.3.3 *Process Evaluation*

3.3.3.1 *Levels of Contaminant Removal*

The single full-scale installation of the Lyso process has been shown to improve solids destruction and sludge cake dryness and did not appear to affect nitrogen removal efficiency, although rates were significantly reduced. In addition, when applied in the secondary treatment system, the process reduces undesirable filamentous organisms. Vergine et al. (2007) reported that both nocardioform and *Microthrix parvicella* were reduced with partial treatment of the RAS stream with a dose of 0.7 kg O_3/kg total solids removed.

Another effect that has not been quantified is the removal of color from the waste stream (Vergine et al., 2007). The ozone appeared to reduce pigments or dyes from the textile wastewater stream, which typically stains the water.

3.3.3.2 *Status of Development*

The Lyso process has been operating at the Lariana Depur's Bulgarograsso WWTP since 2006. At the time of publication, no other installations were reported to be operating anywhere else in the world. Although installations are limited, the technology appears to be viable at full-scale and to be a potential means of sludge reduction without anaerobic or aerobic digestion.

3.3.3.3 *Secondary Environmental Effects*

The Lyso process has not been reported to generate any byproducts of concern that would require additional or special handling. The process does produce sludge as part of the normal wastewater treatment process, but at reduced quantities. The potential for a negative effect on the ammonia-oxidizing bacteria and nitrite-oxidizing bacteria in terms of observed rate reductions should be weighed by facilities, which have strict nitrogen limits.

The degradation product of ozone is oxygen, which can be consumed. Unknown organic byproducts from partial oxidation have not been reported to date.

3.3.4 *Limitations*

3.3.4.1 *Site/Matrix Considerations*

The use of ozone to reduce solids at a facility has shown to be effective. The process does require that there is sufficient space and power available for the associated equipment at the facility. Process efficiency is affected by the concentration of solids in the liquid stream, and some process optimization may be required. It should also be noted that processing pure oxygen to generate ozone is more efficient than using air. Therefore, ozonation systems may be more suitable for WWTPs where high-purity oxygen is being used.

3.3.4.2 *Residue Treatment*

No residual materials is generated from the Lyso process other than those typically generated at a WWTP and no special handling has been reported.

3.3.4.3 *Process Risks*

The generation of ozone requires high voltages, which, for health and safety reasons, should be considered when designing the system. The documented negative effects on the metabolic rates of ammonia- and nitrite-oxidizing bacteria must be balanced with the solids mass reductions and permit requirements.

The effect of ozonation on the dewaterability of sludges has reported benefits and drawbacks. Vergine et al. (2007) reported an increase in cake solids concentration following ozonation of RAS at the Lariana Depur's Bulgarograsso WWTP. Scheminski et al. (2000), however, reported increased polymer demand based on the work of Muller et al. (1998) when digested sludge was exposed to ozone during anaerobic digestion. Both observations should be verified when conducting a cost/benefit analysis for such a system.

3.3.4.4 *Reliability*

At the time of publication, insufficient data was available to provide a reliability analysis for the Lyso process.

3.3.5 Technology Prognosis

Ozonation of solids will likely gain in application if disposal of biosolids becomes more difficult, and landfills no longer accept sludge products. Because the Lyso process is coupled with activated sludge, energy recovery from the sludge is minimized to reduce hauling and operating costs associated with reductions in mass. The sludge generated from the Lyso process is not suitable for direct land application because it does not meet the requirements for Class A or B biosolids in the United States. The system requires coupling with a more traditional stabilization technology to meet land application requirements and to treat primary sludge. If no primary sludge exists, such as soluble industrial wastewaters, then reduction in mass can significantly decrease the size of the ancillary stabilization processes.

3.4 OpenCEL Focused Pulsed Technology

3.4.1 Process Identification and Description

3.4.1.1 *Identification of Processes*

Pulsed electric field (PEF) technologies, based on well-established scientific principles, use different methods of discharging electrical pulses across a material to break apart organic materials such as cells. In this section, one such technology produced by OpenCEL, Glencoe, Illinois, is discussed.

3.4.1.2 *Scientific Basis*

Pulsed electric field technologies have an effect matter because of charged molecules in the material. When an electrical field is produced across electrodes, negatively charged particles or molecules will experience a force in one direction; whereas positively-charged particles or molecules will experience a force in the opposite direction. Because of these opposing forces, organic matter such as cells will experience a tearing action as some membrane-bound molecules will be forced in opposing directions. This causes the cell membranes of microbes such as bacteria, archaea, and larger eukaryotes to lyse and rupture. This causes the organic material inside of the microbes to be released into solution, increasing the soluble fraction of organics—material easily accessible to other microbes as a food source.

OpenCEL refers to a patented Focused Pulsed technology because it subjects biological wastes to focused, high-voltage pulses of electricity. This treatment lyses the cellular membranes and breaks down other complex organic materials, enabling improvements in subsequent digestion or processing. The technology underlying OpenCEL's process was developed and has been successfully used for pathogen destruction in food processing for nearly 40 years. Pulsed electric field technologies have been used in the past by microbiologists to reversibly lyse cells to manipulate the DNA of bacteria (Aly et al., 2001; Madigan et al., 2003; Xu and Xiao, 2006). By increasing the voltage further, cells can be irreversibly lysed, causing the desired effect of solubilizing organic material. The PEF process has also been used by the food industry in place of pasteurization because of its pathogen-killing ability.

3.4.1.3 *Process Description*

OpenCEL's technology requires a few separate components to work as shown in the schematic in Figure 9.13. The sludge being treated can be WAS, TWAS, primary sludge, or a mixture of WAS and primary sludge. The sludge is pulled off the main line, run through a grinder pump and into the Focused Pulsed treatment chamber, and returned to the main line. The pulses in the treatment chamber are produced by a transformer in the OpenCEL unit, which increases the voltage to produce an

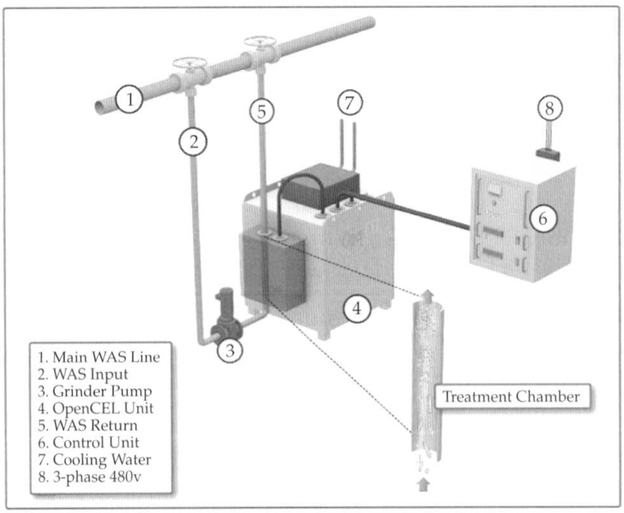

FIGURE 9.13 Components of focused pulsed technology (WAS = waste activated sludge).

electric field of 15 to 100 kV/cm. The pulses last for 2 to 15 μs, and occur with a frequency of 2 to 10 kHz.

The effectiveness of OpenCEL's process has been tested in laboratory-, pilot-, and full-scale operations. All data so far indicate that the process increases the soluble chemical oxygen demand (SCOD) of the sludge anywhere from 20 to 1000%, depending on the characteristics of the sludge (Salerno et al., 2008). (The more concentrated the sludge, the lower the percent increase; the higher the ratio of primary sludge to waste activated sludge, the lower the percent increase). This increase in soluble organic matter results in approximately 10% greater biosolids destruction in anaerobic digestion, and up to 60% more methane generation at full scale.

3.4.1.4 Status of Development

OpenCEL first tested their technology in a pilot-scale study at the Lancaster, Ohio, WWTP. It was used to treat 34% of TRAS flow in an aerobic digestion system. Because of the OpenCEL installation, dissolved organic carbon increased 73% and SCOD increased 35%; fecal coliform count was reduced 49% and biosolids production was reduced 24%.

OpenCEL has also commissioned laboratory-scale research on its applicability to anaerobic digestion. Results from this work showed that when Focused Pulsed-treated WAS was used, methane production doubled in anaerobic digesters versus digesters fed untreated WAS. This led to full-scale testing at the Mesa Northwest Water Reclamation Plant (NWWRP) in Mesa, Arizona, beginning in April 2007 (Rittmann et al., 2008). Mesa is a 37 850 m^3/d (10 mgd) plant, with a flow rate of TWAS/primary sludge mixture to the digesters of approximately 190 m^3/d (50 000 gpd) at approximately 5% solids. The anaerobic digesters operate at a approximately a 30-day HRT. Initially, despite only treating approximately 30% of the sludge volume entering the digesters, 20% more biogas had been produced, and the plant was generating 20% less biosolids requiring disposal than expected based on digester loading. In the beginning of 2008, biogas production rates had increased approximately 60% above the baseline rate, with the Focused Pulsed unit treating approximately 60% of the sludge volume entering the digesters. Between September 2007 and February 2008, a 10% increase in volatile solids reduction was reported, reducing the biosolids requiring disposal by more than 30%. In addition to these two effects, OpenCEL treatment also heats the thickened sludge to 10 °C (20° F) above the incoming temperature, which can significantly reduce the heat load on the plant.

Although these are promising results, Mesa is still the only full-scale installation of OpenCEL. Although OpenCEL's success does not need to be limited to this location, it has yet to be tested in other facilities.

3.4.1.5 *Design Data*

The only utility requirements for OpenCEL's unit are a 480V 3-phase electricity line and cooling water for the electronics. A grinder pump is required upstream of the unit to prevent clogging in the treatment chamber. Installation is relatively minor because two valves can simply be installed in the main sludge line to divert flow through a loop with the Focused Pulsed treatment chamber. A location for the unit should be selected that is:

- Near a main sludge line to minimize pressure drop through the loop;
- As close to the digester input as possible, to retain the greatest effect of the heating Focused Pulsed treatment causes;
- Enough space for two telephone booth-sized units (approximately 1 m × 1 m each) and necessary ancillaries (pump, piping, control panel, etc.); and
- Near utilities—a 480V electricity line and access to a cooling water line (which could be anything from potable water to the sludge being treated).

3.4.1.6 *Pre- and Post-Treatment*

The only pretreatment necessary for the OpenCEL system to function is a grinder pump installed in the WAS/primary sludge secondary loop. This is to ensure that the material flows through the treatment chamber without clogging the line. This is especially important when treating primary sludge because it can contain much larger solid particles than WAS. Because there are no chemicals added in the OpenCEL process, no other pre- or post-treatment is necessary.

3.4.1.7 *Environmental Effect*

No known negative environmental effects of the OpenCEL process have emerged. Although it is theoretically possible that some toxic radical or chemical derivative could be formed by the electrical discharge, there is no evidence of this. Based on the performance of the digesters at Mesa NWWRP, it is expected that there are no disruptive chemicals being created in the process.

3.4.1.8 *Operational Considerations*

The manufacturer states that the OpenCEL system is relatively easy to operate and control. The included PLC can be accessed remotely by the Internet, so that all operating conditions can be examined at any time. Mesa NWWRP staff often allows their OpenCEL unit to run continuously.

3.4.2 *Potential Applications*

OpenCEL's process is intended to increase the soluble organic matter in the sludge before anaerobic digestion to increase efficiency of the digestion process. Therefore, any material that has a significant amount of biological solids such as WAS, will be suitable for this technology. A sludge that has much of its chemical oxygen demand already in soluble form, such as primary sludge, will not be as dramatically affected by the Focused Pulsed process.

3.4.3 *Limitations*

3.4.3.1 *Site/Matrix Considerations*

The considerations for the site are minimal. As mentioned above, the OpenCEL system occupies the space of approximately two telephone booths (approximately 1 m × 1 m each) and a location for control panels and the PLC.

The Focused Pulsed process can handle thickened or unthickened sludge. It does require a grinder pump upstream to ensure there are no large items in the stream. At the Mesa NWWTP, the unit is treating a mixture of TWAS/primary sludge. The energy efficiency of the system may be improved, however, by only treating TWAS, which is more dramatically improved by this treatment. This occurs because primary sludge has more soluble organic material to start with; whereas TWAS contains more insoluble organic material before treatment.

3.4.3.2 *Residue Treatment*

There are no side streams from the OpenCEL process, so there is nothing to be treated.

3.4.3.3 *Process Risks*

OpenCEL has been proven on the laboratory-scale and at one full-scale installation for anaerobic digestion. It has not shown, however, that this success is transferable to other plants. Therefore, reproduction of the success of the process is a risk.

The system does use high voltage, which conveys inherent risk with that. The electricity is all contained and grounded within a lockable chamber, however, to reduce the possibility of accidental exposure.

3.4.3.4 *Reliability*

The OpenCEL system at Mesa NWWRP has been operating continuously since September 2007. Before that, the company had issues with the vibrations and physical stresses, which caused breaking parts in the treatment chamber. The affected parts were replaced with ones of a different material, and they have not had an issue since.

3.4.4 Technology Prognosis

OpenCEL's Focused Pulsed technology is promising for WWTPs that use anaerobic digestion and have the capability to reuse biogas. Once OpenCEL has proven itself with one or two more installations, it will be safe to recommend the technology in these situations. Without the capability to reuse the biogas, the economics of OpenCEL would be difficult to justify based on biosolids reduction alone. When the savings from biosolids reduction are combined with the potential energy savings, however, OpenCEL's technology is one that should be considered.

4.0 REFERENCES AND SUGGESTED READINGS

Adams, G. A.; Witherspoon, J.; Card, T.; Erdal, Z.; Forbes, B.; McEwen, D.; Geselbracht, J.; Glindemann, D.; Hargreaves, R.; Hentz, L.; Higgins, M.; Murthy, S. (2003) *Identifying and Controlling Odour in the Municipal Wastewater Environment Phase 2: Impacts of In-Plant Parameters on Biosolids Odour Quality.* Water Environment Research Foundation Report No 00-HHE-5T; Water Environment Research Foundation: Alexandria, Virginia.

Aly, R. E.; Joshi, R. P.; Stark, R. H.; Schoenbach, K. H.; Beebe, S. J. (2001) Repair Time of Bacteria after Pulsed Electric Field Application. *Proceedings of the IEEE International Conference on Plasma Science*, Las Vegas, Nevada, June 17–22; IEEE: New York.

Banaszak, J; Burrowes, P; Daigger, G; Enochs, M; Klein, A; Rittmann, B; Salerno, M; Schlegel, P. (2007) Focused-Pulsed Treatment of Waste Activated Sludge: Application to Waste Biosolids Reduction and Increased Methane Production. *Proceedings of the 80th Annual Water Environment Federation Technical Exposition and Conference* [CD-ROM]; San Diego, California, Oct 15–17; Water Environment Federation: Alexandria, Virginia.

Banaszak, J. E.; Burrowes, P.; Daigger, G.; Enochs, M. B.; Klein, A. M.; Lopez, R.; Rittmann, B. E.; Schlegel, P. R. (2008) Focused-Pulsed Treatment of Waste Activated Sludge: Application to Waste Biosolids Reduction and Increased Methane Production. *Proceedings of the 81th Annual Water Environment*

Federation Technical Exposition and Conference [CD-ROM]; Chicago, Illinois, Oct 18–22; Water Environment Federation: Alexandria, Virginia.

Chiu, Y. -C.; Chang, C. -N.; Lin, J. -G.; Huang, S. -J. (1997) Alkaline and Ultrasonic Pretreatment of Sludge Before Anaerobic Digestion. *Water Sci. Technol.*, **36** (11), 155–162.

Christi, Y. (2003) Sonobioreactors: Using Ultrasound for Enhanced Microbial Productivity. *Trends Biotechnol.*, **21** (2), 89–93.

Chu, C. P.; Chang, B.; Liao, G. S.; Jean, D. S.; Lee, D. J. (2001) Observations on Changes in Ultrasonically Treated Waste-Activated Sludge. *Water Res.*, **35** (4), 1038–1046.

Joyce, E.; Phull, S. S.; Lorimer, J. P.; Mason, T. J. (2003) The Development and Evaluation of Ultrasound for the Treatment of Bacterial Suspensions. A Study of Frequency, Power, and Sonication Time on Cultured Bacillus Species. *Ultrason. Sonochem.*, **10**, 315–318.

Kopplow, O.; Barjenbruch, M.; Heinz, V. (2004) Sludge Pretreatment with Pulsed Electric Fields. *Water Sci. Technol.*, **49** (10), 123–129.

Laffite-Trouqué, S.; Forster, C. F. (2002) The Use of Ultrasound and gamma-Irradiation as Pre-treatments for the Anaerobic Digestion of Waste Activated Sludge at Mesophilic and Thermophilic Temperatures. *Bioresour. Technol.*, **84**, 113–118.

Li, Y. -Y.; Noike, T. (1992) Upgrading of Anaerobic Digestion of Waste Activated Sludge by Thermal Pretreatment. *Water Sci. Technol.*, **26** (3–4), 857–866.

Madigan, M. T.; Martinko, J. M.; Parker, J. (2003) *Brock Biology of Microorganisms*. Prentice Hall: Upper Saddle River, New Jersey.

Mao, T.; Young, S. -Y.; Show, K. -Y.; Tay, J. -H.; Lee, D. J. (2004) A Comparison of Ultrasound Treatment on Primary and Secondary Sludges. *Water Sci. Technol.*, **50** (9), 91–97.

Mason, T. J.; Lorimer, J. P. (1988) *Sonochemistry: Theory, Applications and Uses of Ultrasound in Chemistry*, Ellis Horwood Series in Physical Chemistry. Ellis Horwood: Chichester, West Sussex, United Kingdom.

Muller, J.; Lehne, G.; Schwedes, J.; Battenberg, S.; Naveke, R.; Kopp, J.; Dichtl, N.; Scheminski, A.; Krull, R.; Hempel, D. C. (1998) Disintegration of Sewage

Sludges and Influence on Anaerobic Digestion, *Water Sci. Technol.*, **38** (8–9), 425–433.

Müller, J. A.; Winter, A.; Strünkmann, G. (2004) Investigation and Assessment of Sludge Pre-Treatment Processes. *Water Sci. Technol.*, **49** (10), 97–104.

Muller, C. D.; Abu-Orf, M. M.; Novak, J. T. (2007) The Application of Mechanical Shear in an Internal-Recycle for the Enhancement of Anaerobic Digestion. *Water Environ. Res.*, **79** (2), 297–304.

Nah, I. W.; Kang, Y. W.; Hwang, K. Y.; Song, W. K. (2000) Mechanical Pre-treatment of Waste Activated Sludge for Anaerobic Digestion Process. *Water Res.*, **34**, 2362–2368.

Neis, U.; Nickel, U. K.; Tiehm, A. (2000) Enhancement of Anaerobic Digestion by Ultrasonic Disintegration. *Water Sci. Technol.*, **42** (9), 73–80.

Novak, J. T.; Park, C.; Higgins, M. J.; Chen, Y. –C; Morton, R.; Gary, D.; Forbes. R.; Erdal, Z. (2007) WERF Odor Study Phase III: Impacts of the MicroSludge Process on Odor Causing Compounds. *Proceedings of the Water Environment Federation/American Water Works Association Joint Residuals and Biosolids Management Conference* [CD-ROM]; Water Environment Federation: Alexandria, Virginia.

Onyeche, T. I.; Schläfer, O.; Bormann, H.; Schröder, C.; Sievers, M. (2002) Ultrasonic Cell Disruption of Stabilised Sludge with Subsequent Anaerobic Digestion. *Ultrason.*, **40**, 31–35.

Portenlänger, G.; Heusinger, H. (1997) The Influence of Frequency on the Mechanical and Radical Effects for Ultrasonic Degradation of Dextranes. *Ultrason. Sonochem.* **4**, 127–130.

Rabinowitz, B.; Stephenson, R. (2005) Improvement of Anaerobic Digester Efficiency by Homogenization of Waste Activated Sludge. *Proceedings of the 78th Annual Water Environment Federation Technical Exposition and Conference* [CD-ROM]; Washington, D.C., Oct 30–Nov 2; Water Environment Federation: Alexandria, Virginia.

Rabinowitz, B.; Stephenson, R. (2006) Effect of MicroSludge on Anaerobic Digester Performance and Residuals Dewatering at LA County's JWPCP. *Proceedings of the 79th Annual Water Environment Federation Technical Exposition and Conference* [CD-ROM]; Dallas, Texas, Oct 21–25; Water Environment Federation: Alexandria, Virginia.

Raso, J.; Pagán, R.; Condón, S.; Sala, F. J. (1998) Influence of Temperature and Pressure on the Leathality of Ultrasound. *Appl. Environ. Microbiol.*, **64** (2), 465–471.

Rittmann, B. E.; Lee, H. -S.; Zhang, H.; Alder, J.; Banaszak, J. E.; Lopez, R. (2008) Full-Scale Application of Focused-Pulsed Pre-Treatment for Improving Biosolids Digestion and Conversion to Methane. *Water Sci. Technol.*, **58** (10), 1895–1902.

Roxburgh, R.; Sieger, R.; Johnson, B.; Rabinowitz, B.; Goodwin, S.; Crawford, G.; Diagger, G. (2006) Sludge Minimization Technologies-Doing More to Get Less. *Proceedings of the 79th Annual Water Environment Federation Technical Exposition and Conference* [CD-ROM]; Dallas, Texas, Oct 21–25; Water Environment Federation: Alexandria, Virginia.

Salerno, M. B.; Lee, H. -S; Parameswaran, P.; Rittmann, B. (2008) Using a Pulsed Electric Field as a Pretreatment for Improved Biosolids Digestion and Methanogenesis; *Proceedings of the 81th Annual Water Environment Federation Technical Exposition and Conference* [CD-ROM]; Chicago, Illinois, Oct 18–22; Water Environment Federation: Alexandria, Virginia.

Sanders, W. T. M.; Geerink, M.; Zeeman, G.; Lettinga, G. (2000) Anaeorbic Hydrolysis Kinetics of Particulate Substrates. *Water Sci. Technol.*, **41** (3), 17–24.

Scheminski, A; Krull, R.; Hempel, D. C. (2000) Oxidative Treatment of Digested Sewage Sludge with Ozone. *Water Sci. Technol.*, **42** (9), 151–158.

Scherba, G.; Weigel, R. M.; Brien, W. D. O., Jr., (1991) Quantitative Assessment of the Germicidal Efficacy of Ultrasonic Energy. *Appl. Environ. Microbiol.*, **57** (7), 2079–2084.

Stephenson, R.; Rabinowitz, B.; Laliberte, S.; Elson, P. (2005) Teaching an Old Digester New Tricks: Fullscale Demonstration of the MicroSludge Process to Liquify Municipal Waste Activated Sludge. *Proceedings of the Water Environment Federation Specialty Conference—19th Annual Residuals and Biosolids Conference*, Nashville, Tennessee, Apr 18; Water Environment Federation, Alexandria, Virginia.

Tiehm, A.; Nickel, K.; Zellhorn, M.; Neis, U. (2001) Ultrasonic Waste Activated Sludge Disintegration for Improving Anaerobic Stabalization. *Water Res.*, **35** (8), 2003–2009.

Tuziuti, T.; Yasui, K.; Sivakumar, M.; Iida, Y. (2005) Correlation Between Acoustic Cavitation Noise and Yield Enhancement of Sonochemical Reaction by Particle Addition. *J. Phys. Chem., A*, **109**, 4869–4872.

Vavilin, V. A.; Rytov, S. V.; Lokshina, L. Y. (1997) A Balance Between Hydrolysis and Methanogenesis During the Anaerobic Digestion of Organic Matter. *Microbiol.*, **66** (6), 712–717.

Vergine, P.; Menin, G.; Canziani, R.; Ficara, E.; Fabiyi, M.; Novak, R.; Sandon, A.; Bianchi, A.; Bergna, G. (2007) Partial Ozonation of Activated Sludge to Reduce Excess Sludge Production: Evaluation of Effects on Biomass Activity in a Full Scale Demonstration Test. *Proceedings of the International Water Association Specialty Conference: Moving Forward: Wastewater Biosolids Sustainability: Technical, Managerial and Public Synergy*, Moncton, New Brunswick, Canada, June 24; IWA Publishing: London, United Kingdom.

Weemmaes, M.; Grootaerd, H.; Simmoens, F.; Verstraete, W. (2000) Anaerobic Digestion of Ozonized Biosolids. *Water Res.*, **34** (8), 2330–2336.

Xu, F.; Xiao, D. (2006) Analysis of the Effect of Pulsed Electric Field on Cells with the RC Circuit Model and Experiment. *High Voltage Eng.*, **32** (6), 67–69.

Chapter 10

Anaerobic Digestion

(continued)

(*continued*)

1.0 INTRODUCTION

This chapter focuses on design of anaerobic digestion systems. It covers planning considerations, pre- and post-digestion, anaerobic digestion processes, digester design, odor control, digestion enhancement techniques, and gas management and includes case studies of anaerobic digestion facilities. Biogas management, including the design of gas piping, condensate control, pressure control, and the use of biogas for mixing, is also covered in this chapter. Methods to enhance digestion performance, such as improved feeding, heating, and mixing are discussed. Additionally, advanced digestion configurations, such as acid/gas and thermophilic digestion, are covered in the chapter. Fundamentals of co-digestion of domestic sludge with other organic wastes—such as fats, oils, and grease (FOG), restaurant waste, food processing waste, and the organic fraction of municipal solid waste—also are discussed as an important method to greatly increase biogas production.

2.0 PLANNING CONSIDERATIONS

Anaerobic digestion is a well-established solids stabilization process that is used extensively on a combination of primary and secondary solids generated as part of the wastewater treatment process. In a few circumstances, anaerobic digestion is suitable for smaller wastewater treatment plants (WWTPs). Typically, however, anaerobic digestion should be considered for sludge stabilization at WWTPs with primary sedimentation and average daily flows of approximately 11 000 to 19 000 m^3/d (3 to 5 mgd) or greater. In smaller facilities, especially facilities without a primary sedimentation process, aerobic digestion represents a typical approach, because it tends to be less complex from an operational standpoint and requires lower capital costs. Anaerobic digestion is part of a solids processing system aimed at reducing the mass of wastewater solids, stabilizing the remaining solids, producing and beneficially using biogas, and maintaining the integrity of the environment and water supply. It is affected by upstream processes, and it affects downstream processes. Therefore, anaerobic digesters must be designed with consideration for the entire solids processing system and its objectives.

When considering anaerobic digestion, the selection, design, and operation of the stabilization system will depend on the desired use of the biosolids and site-specific conditions. Chapter 2, *Considerations for Planning of Biosolids Management Projects*, provides information on planning and reviews U.S. Environmental Protection Agency (U.S. EPA) regulations relevant to digestion systems. Biosolids management planning and selection of the optimal stabilization method require an understanding of several key objectives. At least four main categories of objectives should be taken into account when planning the stabilization system: economic, environmental, operational, and social.

2.1 Economic

The capital and operating costs that contribute to the life-cycle cost of a digester must be considered for an economic sludge stabilization system. The cost of electricity and natural gas and system sustainability can significantly affect the selection of one stabilization system over another. In addition, it is important to consider whether a digestion process will consume energy or, conversely, if it will produce a useable energy source such as biogas. If a commercial product such as fertilizer is desired, then it is important to consider its production costs and potential benefits. Business case analyses and cost-benefit ratios may be helpful in determining the cost effectiveness of a stabilization system.

2.2 Environmental

Environmental considerations include quality of the digested sludge (referred to as biosolids when it can be beneficially used), the effect of plant operations on its surroundings, and the amount of biosolids produced. Anaerobic digestion provides a method for converting sludge to biosolids through the process of stabilization and pathogen reduction. Important quality parameters for the digested sludge include pathogen content, vector attraction, odor, and metals concentration as these all affect the potential for beneficial use. The amount of energy required to stabilize the solids is a consideration for sludge and its effect on the environment. In the case of anaerobic digestion, biogas can also be used to offset the amount of energy required by a plant. In addition, if a plant can decrease the amount of biosolids it produces, then less material needs to be processed and managed at the end of the digestion cycle. The quality of the biosolids and the applicable regulations (e.g., 40 CFR 503) are also a key consideration in planning decisions (U.S. EPA, 2003).

2.3 Operational

Several operational components should be considered during digester design, including the complexity of the system, ease of operation, redundancy requirements, and type of technology. This can lead to improved operating strategies at a facility and less maintenance of the system. Also important to consider is the potential for expansion of the digestion system to accommodate future demands. The effect of anaerobic digestion on the liquid stream needs to be considered. For example, anaerobic digestion will increase ammonia in recycle streams and energy requirements to process the ammonia recycle. Anaerobic digestion must be evaluated carefully as part of the overall treatment process at a WWTP to optimize the design and achieve the best overall solution.

2.4 Social/Political

Aesthetics and odors are two important social and political issues that should be considered. Depending upon the site location and visibility of the plant, the appearance of equipment and potential odors from plant operations can be an important consideration to the surrounding community. If the community is considered during planning and equipment siting, then public and other stakeholder acceptance may be more likely to occur. There is also social and political interest in the ability of anaerobic digestion to convert sludge into fertilizer and biogas.

3.0 ANAEROBIC DIGESTION THEORY

Anaerobic digestion is the biological process that stabilizes organic matter in the absence of oxygen. During this process, biodegradable organic matter is converted to water and biogas containing methane (CH_4) and carbon dioxide (CO_2). The biogas produced in the anaerobic digestion process is suitable for use as an energy source. Anaerobic digestion of municipal wastewater solids will reduce the quantity of organic matter for downstream processing and can produce a stabilized product meeting requirements of vector attraction and pathogen reduction. Digested sludge is often less odorous and attracts fewer vectors (such as rodents, flies, mosquitoes, or other organisms capable of transporting infectious agents) than raw material. Organic matter, nitrogen, and phosphorus in the digested sludge may allow for use as a soil amendment and fertilizer.

A simplified breakdown of the anaerobic digestion process is shown in Figure 10.1. The mechanism of anaerobic digestion involves solids hydrolysis, fermentation of complex organics to hydrogen and volatile fatty acids, and subsequent conversion of these simpler compounds to biogas. In hydrolysis, complex and particulate materials are transformed to simple and soluble material. The transformations include: lipids to fatty acids, polysaccharides to monosaccharides, proteins to amino acids, and nucleic acids to purines and pyrimidines. Fermentation, also referred to as acidogenesis, further breaks down products of hydrolysis into short-chain organic acids, such as propionic, butyric, and other fatty acids. Conversion of the organic acids to methane, or methanogenesis, is the final step of the process. The organisms responsible for methanogenesis, however, can only use a few substrates. In typical anaerobic digestion processes, acetic acid and hydrogen are used

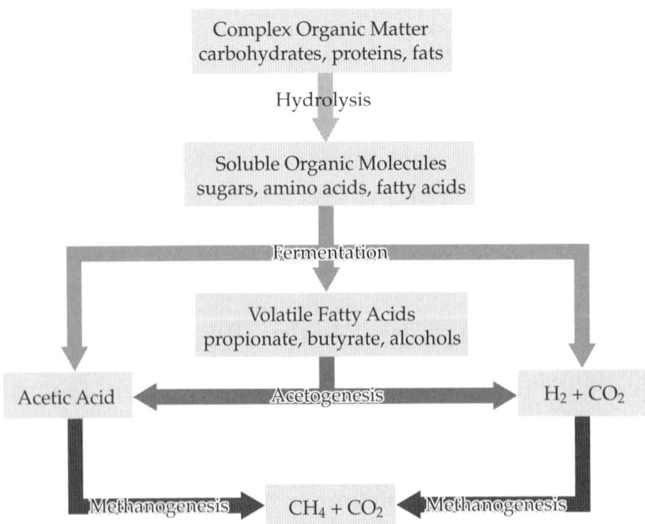

FIGURE 10.1 Simplified breakdown of the anaerobic digestion process (adapted from Pavlostathis and Giraldo-Gomez, 1991).

by the methanogenic organisms to produce biogas. Between the acidogenesis and methanogenesis step is a process called acetogenesis. In acetogenesis, the short-chain acids are converted to acetic acid and hydrogen, thus enabling methanogenesis to occur.

3.1 Digestion Loading Rate

Traditional methods for determining the required digestion volume include: (1) solids and hydraulic retention time (SRT and HRT); and (2) volatile solids loading rate. For digestion systems that do not practice recuperative thickening or decant supernatant from the tanks, the SRT and HRT are equal. Both require the simplest sizing considerations and have been used for many years. A complimentary method to HRT and SRT is to use the volatile solids loading rate. Loading rates for anaerobic digesters can also be based on chemical oxygen demand (COD) per unit volume. The COD loading rates typically are used for industrial digesters, particularly when high-strength liquid wastes are being digested. With variations in digester solids concentrations and feed material, more sophisticated digestion loading rates are needed. The section on co-digestion includes a discussion of the specific energy loading rate for sizing digestion facilities.

3.2 Solids and Hydraulic Retention Time

The most common factor used to determine the volume of any anaerobic process is the SRT and HRT; SRT will be used to represent both SRT and HRT for the remainder of this section. The SRT represents the average amount of time that sludge remains in the digester and is determined using the following equation:

$$SRT = \frac{V}{Q} \qquad 10.1$$

where
SRT = solids retention time (d);
V = volume of the digester tank (m^3); and
Q = sludge flowrate (m^3/d).

The SRT affects the treatment performance of anaerobic digesters and the solids destruction. Washout (the SRT below which organisms are removed faster than they can reproduce) for methanogens (organisms that convert organic acids into methane) occurs at SRTs less than approximately five days. A factor of safety should be applied to the design SRT to avoid washout from flow variations, process upsets, and effective volume reductions because of the accumulation of inerts in the reactor from changes in tank levels.

The design SRT should be maintained under all of the following conditions (U.S. EPA, 1979):

- Peak two-week hydraulic loading (i.e., corresponding to the 15-day minimum SRT). This can be estimated by combining poor thickener performance with the maximum plant loading expected during 15 consecutive days.
- Maximum scum and grit accumulation. Reduced tank volume resulting from scum and grit accumulation between digester cleanings.
- Variable liquid level. Variations in sludge feeding and withdrawal rates can create fluctuations in the digester liquid level, potentially affecting the operating SRT and HRT.
- Digester tank maintenance. Additional digester or storage volume should be supplied to allow for taking a tank offline for maintenance without causing a process upset.

The U.S. EPA's 40 CFR Part 503 regulations require that sludge is digested between a 15-day SRT at 35 to 55 °C and a 60-day SRT at 20 °C to meet its standards for "Processes to Significantly Reduce Pathogens" (U.S. EPA, 1993).

3.3 Volatile Solids Loading Rate

Volatile solids loading factors also are used to size digesters. Volatile solids loading rate refers to the mass of volatile solids added each day divided by the digester volume. A typical loading for sustained conditions for volatile suspended solids (VSS) (maximum week or maximum month) is 1.9 to 2.5 kg VSS/m^3·d (0.12 to 0.16 lb VSS/d/cu ft); a limiting value of 3.2 kg VSS/m^3·d (0.2 lb VSS/d/cu ft) is often used (WEF et al., 2009). *Recommended Standards for Wastewater Facilities* is conservative and recommends 1.3 kg VSS/m^3·d (0.08 lb VSS/d/cu ft) for digesters with intimate and effective mixing, and 0.65 kg VSS/m^3·d (0.04 lb VSS/d/cu ft) for systems where mixing is accomplished by circulation through a heat exchanger (GLUMRB, 2004). These volatile solids loading rates have been developed using a digester feed concentration of 5% solids. As thickening processes have been implemented, it is possible to increase the volatile solids loading rate and still maintain an acceptable SRT.

3.4 Volatile Solids Destruction

The degree of stabilization in a digester can be measured by the percent reduction in volatile solids. Volatile solids destruction in a high-rate complete-mix digester can be estimated using the following empirical equation (WEF et al., 2009):

$$V_d = 13.7 \ln(SRT_{des}) + 18.9 \qquad 10.2$$

where

V_d = volatile solids destruction (%) and

SRT_{des} = design solids retention time (day).

This method typically is used because untreated feed sludge is easily measured. This empirical equation provides a rough estimate of volatile solids destruction and does not account for variations in the sludge feed, digester mixing, or other digester operating conditions.

4.0 DIGESTION PRETREATMENT

Effective pretreatment is essential for the success of any anaerobic digestion process (conventional or high-performance). Specifically, provisions should be made for debris removal (i.e., fine screening at the WWTP headworks, sludge screening before digestion), grit removal, and thickening. Figure 10.2 illustrates a simplified process schematic showing thickening and storage before digestion and storage and dewatering after digestion. Note that sludge degritting precedes thickening and sludge screening follows thickening.

4.1 Sludge Characterization

The characteristics of the sludge fed to the digesters significantly affects the digestion process (refer to Chapter 4, *Solids Production and Characterization*, for a more detailed

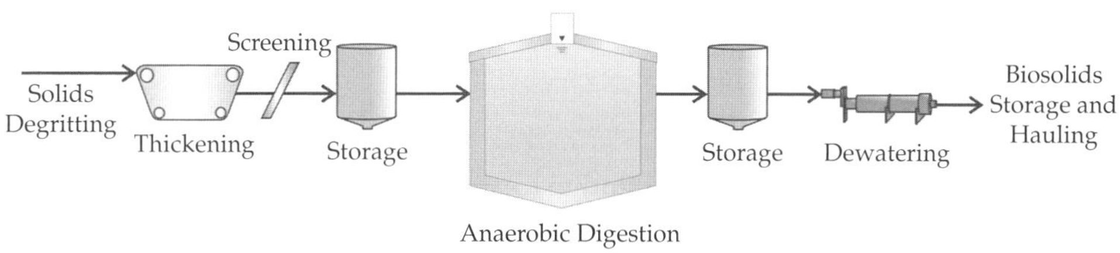

FIGURE 10.2 Simplified digestion process description.

discussion on solids processing). Because local population and industry vary, it is important to know the service area; expected chemical properties of the sludge; ratio of primary-to-secondary sludge (primary sludge digests more rapidly and completely than secondary sludge); ratio of volatile solids to total solids; biodegradability of the sludge; and amount of inerts in the sludge. Each of these characteristics affects the digestibility of the sludge and can influence the anaerobic digestion process selected. Characterization of supplemental feedstocks (including their energy content) such as fats, oils, and grease (FOG), food wastes, and high COD waste streams for co-digestion with municipal wastewater solids is also important for appropriate process selection and design.

4.2 Sludge Thickening

The required volume of digesters and amount of heating are inversely related to the concentration of solids in the feed. It is thus beneficial to thicken sludge to decrease the volume of digester tanks and heating needed. The upper limit for thickening is dictated by the digestion process and the type of mixing system in the digestion tanks.

4.3 Sludge Screening, Grinding, and Grit Removal

Influent screening and grit removal at the headworks of a WWTP are an important consideration during solids design planning. The spacing of the screens will determine the effectiveness of debris removal. Grinding solids ahead of pumping and processing protects equipment from plugging and blinding and makes debris in biosolids less identifiable. Removal of debris, including grit and fine screening (and sludge screening), is important for effective digestion. Debris that is allowed to accumulate will decrease the effective volume for digestion, produce excessive wear in pumps, clog heat exchangers, and blind mixing and pumping equipment with rags. Debris in sludge also will adversely affect the quality and marketability of biosolids products.

4.4 Sludge Pretreatment

Pretreatment processes can enhance sludge treatment and are found at some anaerobic digester facilities. These pretreatment methods are designed to enhance digestion or to pasteurize raw solids. Pasteurization pretreatment is aimed at reducing pathogens to produce a Class A biosolids product, as defined in 40 CFR Part 503. Lysing pretreatment is designed to physically alter the raw solids. Typically, lysing targets waste activated sludge (WAS) and releases intracellular biopolymers (e.g., proteins and sugars). Releasing these components can increase the biodegradability of raw solids, the rate at which the solids can be stabilized, and the production of biogas. Lysing also releases intracellular water, which can reduce viscosity, lower mixing requirements, and improve solids dewatering following digestion. Finally, when the waste activated solids contain filamentous organisms, lysing pretreatment destroys these organisms and can inhibit foaming. There are several techniques to enhance digestion via improvements to hydrolysis. A discussion on various hydrolysis techniques is included in the Section 7.0 *High-Performance Anaerobic Digestion* later in this chapter.

When implemented, sludge pasteurization, or prepasteurization, needs to be an integral part of the digestion process. Installing sludge pasteurization helps ensure the production of Class A biosolids following digestion. Pasteurization followed by digestion, as opposed to digestion followed by pasteurization, is necessary to avoid regrowth of bacteria (U.S. EPA, 1993). A batch pasteurization system uses four vessels, which allows for one vessel to be fed, another to be held at the required time and temperature, a third to be drained, and a fourth to provide a standby for redundancy. An alternative setup would be tanks in series with sequential withdrawal from the downstream unit followed by transfer from the preceding upstream unit to achieve the required detention time and to avoid short circuiting. Continuous flow through pasteurization systems are also available that rely on the length of pipe of a reactor to achieve the required time and temperature for pathogen reduction.

4.5 Sludge Blending

A raw sludge blending tank that receives thickened primary and secondary sludge can improve the digestion process by enabling continuous feeding to the digesters. A blend tank is particularly advantageous when thickened primary and secondary sludge needs to be pumped from multiple locations at a plant. The blending tank provides a wide spot in the line enabling the decoupling of the sludge flow to the digesters from the sludge flow from the primary and secondary clarifiers or sludge thickening processes. The blending tank can be vented to the atmosphere, vented to an odor control system, or connected to the digester gas system. Connecting the blending tank to the digester gas system eliminates the need for a separate odor control system and also reduces the risk of combustion. Having the gas space in the blending tank filled with digester gas without air, as opposed to a mixture of digester gas and air, reduces the potential for combustion.

The detention time of the blending tank should be relatively short (approximately four-to-eight hours of annual average sludge flow). Enough volume is needed to enable dampening of the sludge production and to equalize the sludge fed to the digestion system.

Blending tanks can be constructed in various shapes; both cylindrical and rectangular tanks are in operation. A blending tank is typically constructed out of concrete, with an emergency overflow and connections to a gas system that includes pressure/vacuum relief valves. Lining the blend tank to below the minimum sludge level and designing a steep bottom (greater than 20% slope) is preferred. Mixing of the blend tank is also desirable to prevent deposition. A blending tank may or may not be heated depending on the temperature of the feed sludge and the operating temperature of the digestion system. A blending tank that feeds a thermophilic digester is more likely to be heated.

A blending tank may reduce foam by premixing primary and secondary sludge before digester feeding. Foam is more pronounced during high loading and could be the result of mixing different sludges in the digester. Uniformly feeding sludge to the digester deters foam by reducing the peak feed flow rate. If a blending tank is not used, however, then thickened sludge can be fed directly to a digester with primary and secondary sludges combining in the pipes feeding the digesters.

Blending tanks are typically fed by previously thickened primary and secondary solids. An alternative approach is to co-thicken the primary/secondary solids and then feed the thickened combination into a "thickened sludge blending tank." This approach allows the equalization of flow. Further, co-thickening may result in a greater final solids concentration than blending individually thickened primary and secondary solids. The size of the blending tank should be adjusted to avoid development of anaerobic conditions and odors.

5.0 DIGESTER FACILITIES DESIGN

Anaerobic digesters have historically been designed to stabilize wastewater solids and to reduce both the pathogen content and the mass of solids. Anaerobic digesters can also be designed to perform specific functions, including producing biogas or Class A biosolids. Operational considerations for anaerobic digester design include the amount of sludge thickening and resulting solids concentration, how much grit and debris to remove, and what process configuration to use. For any digester configuration, it is important to uniformly feed the digester, provide adequate mixing, maintain a constant temperature, and ensure a microbial environment conducive to the growth of anaerobic bacteria. The following sections discuss key design features of anaerobic digester components needed to provide a system that achieves the operational objectives.

5.1 Process Configuration

Anaerobic digestion process configurations are classified based on the number of digestion stages (single, multiple stages of the same digestion phase) and the different phases (acid, gas, or thermophilic followed by mesophilic). High-performance configurations such as acid/gas phase digestion and thermophilic digestion are discussed later in this chapter. Note that throughout this chapter, stages will refer to multiple tanks in series of the same phase of digestion. For example, staged or extended thermophilic refers to multiple stages of thermophilic digesters. Phases will refer to a different step of digestion such as temperature-phased (thermophilic followed by mesophilic) or acid/gas (acid-phase followed by gas-phase digestion).

5.1.1 Two-Stage Digestion

In two-stage digestion, a high-rate digester is placed in series with a second tank. The first tank (primary tank) is used for digestion, and the second tank is typically unheated and unmixed. Two-stage digestion can be referred to as high-rate digestion followed by a digested sludge storage tank. Originally, the second stage was used for gravity thickening to increase the concentration of digested sludge. Supernatant was decanted off the top of the secondary digester. Two-stage processes typically were unsuccessful because digestion continued in the second stage (albeit at reduced rates). The biogas generated prevented settling and decanting, as anaerobically digested solids may not settle well. As a consequence, elevated concentrations of suspended solids were present in the supernatant. Several secondary digesters are either used simply for digested sludge storage or have been converted to high-rate digesters complete with mixing and heating.

Stages and phases can be confusing. Two-stage digestion (high-rate digestion followed by storage) should not be confused with multiple stages of high-rate digestion (e.g., thermophilic digestion in series), which is sometimes referred to as extended thermophilic.

5.1.2 *Single-Stage Digestion*

In a single-stage digestion system, the phases of anaerobic reactions take place in one tank. A simple schematic of single-stage anaerobic digestion is shown in Figure 10.3. For the purposes of this manual, a generic mixing system representing any of several mixing technologies (e.g., hydraulic, draft tube, or gas) is shown. For clarity, grinders and multiple-feed positions are omitted to make the schematic as simple as possible. A single boiler and hot water loop is shown for simplicity. Other heating system processes, such as steam injection, can be considered. Facilities with multiple digesters may also have a hot water system that receives heat from multiple boilers or combined heat-and-power equipment.

Single stage, low-rate digesters represent the oldest and most basic digester configuration, consisting of a cylindrical vessel that is typically unheated and without mixing devices. The biomass is exclusively mixed by the upward movement of gas through the solid matter, accumulating at the top of the tank. The biogas is drawn off for storage or as a renewable source of energy.

Most modern anaerobic digesters are single-stage, high-rate digesters. Single-stage, high-rate digesters are characterized by auxiliary heating and mixing and thickening and uniform feeding. The contents of the digester tank are mixed, and the

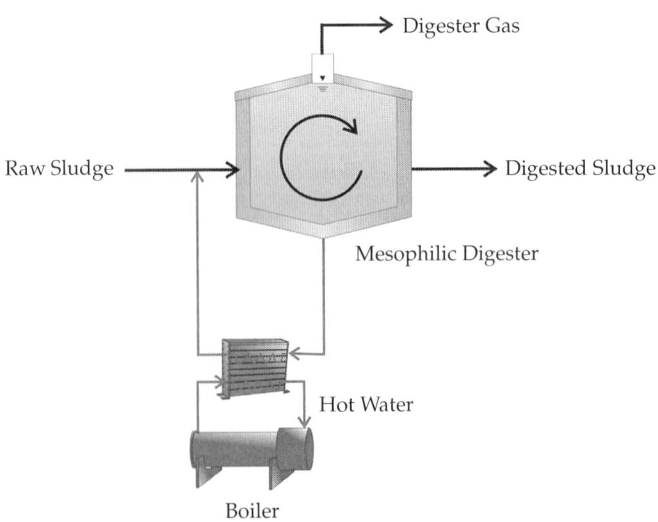

FIGURE 10.3 Single-stage mesophilic anaerobic digestion process configuration.

supernatant is not withdrawn separately. The required digester volume is smaller than the volume required for low-rate systems, and the process is more stable. Single-stage, high-rate digesters can be operated at mesophilic or thermophilic temperatures. Only high-rate digesters that are completely mixed and operate at mesophilic or thermophilic temperatures are discussed in this chapter.

5.2 Digester Tank Geometry

The physical shape and configuration of the digester affects its performance. Key design considerations include overall shape, type of cover, steepness of the bottom, and configuration of the gas dome, sludge feed, and sludge removal components. The sealing and lining of the digester is another important design consideration because it affects the maintenance and longevity of the digester.

The geometry of most anaerobic digesters is grouped into two categories that represent the shape of the reactor tank structure: cylindrical and egg-shaped. Cylindrical digesters can be classified based on the depth-to-diameter ratio. Tanks with ratios less than 1:1 are commonly referred to as "pancake" or "conventional American" digesters; tanks with ratios greater than 1:1 are referred to as "tall cylinder" or "German" digesters. Egg-shaped digesters are recognizable by the oval shape and height of the structure. *Design of Municipal Wastewater Treatment Plants* provides detailed information on cylindrical and egg-shaped digesters (WEF et al., 2009).

5.3 Covers and Appurtenances

Digester tanks require covers to collect gas produced, reduce fugitive odors, retain heat to maintain internal temperatures, support some types of mixing equipment, and to maintain anaerobic conditions in the tank. A liquid overflow system and a vacuum/pressure relief system equipped with a flame trap are among the typical appurtenances considered when designing digester covers. In addition to cover appurtenances, a large access hatch along the side of the tank will facilitate staff access, equipment removal, and tank cleaning. Additional information on covers and appurtenances can be found in *Design of Municipal Wastewater Treatment Plants* (WEF et al., 2009).

5.4 Sludge Feeding and Withdrawal

Digesters should be treated as biological reactors that perform best with consistent environmental conditions. Ideally, sludge should be fed to digesters at a constant rate. Fluctuations in the influent flow rate can upset the balance between acidogenesis and methanogenesis occurring in the digester. A uniform feed leads to more consistent operation with constant temperatures, greater volatile solids reduction, and more constant gas production.

Continuous feeding, however, is not always possible. Most plants institute a feed cycle of 5 or 10 minutes on a 30-minute to 2-hour cycle to maintain constant conditions in the reactor. For smaller plants staffed for only one shift, at least three feedings should be provided: one at the beginning, one in the middle, and one at the end of the day. To avoid short-circuiting and maintain low pathogen concentrations in the effluent sludge, digesters on an intermittent feed cycle can remove the digested sludge before feeding, compared to using the feed sludge to displace the waste sludge.

The influent sludge to the digesters—either primary sludge, secondary sludge, clarifier scum, or industrial waste sludges—can be blended and mixed in a blending tank, allowing for a homogenous sludge mixture to be fed into the tank. If the sludge feeding system does not include a blending tank, then each of the sludge feeding pumps will need to be sized to pump against the head of the digester operating level or the line pressure if a common feed pipe is used.

Raw sludge can be added into the digestion process by either feeding directly to the digester tank or by feeding into the digester sludge recirculation line just downstream of the sludge heating equipment. Raw sludge should be fed to a location in the tank with adequate mixing to ensure the incoming substrate comes into contact with established biomass. Additionally, raw sludge should be fed at a location that minimizes the chances of short-circuiting from digested sludge withdrawal.

Feeding raw sludge into the recirculation line will affect the performance of the recirculation pump but allows for heat transfer into the raw sludge. Additionally, blending raw sludge with the digester recirculation provides dilution and allows for some turbulence for mixing without excessive headloss.

All raw sludge feeding pipelines should be designed with provisions for pigging, cleaning, and flushing. Flushing and cleaning raw sludge pipelines is most effective when using steam; however, hot water or hot digested sludge can also provide some pipe cleaning and scouring action.

Digested sludge withdrawal design depends on the digester tank configuration and whether the tank has an overflow or pumped-out system. In the former, digested sludge will overflow into an overflow box or standpipe from the top of the tank and will either be pumped, or will flow by gravity, to postdigestion facilities. In a pumped withdrawal system, digested sludge is pumped typically from the digester cone sump to postdigestion facilities. Withdrawal from the bottom of the digester tank also removes accumulated grit; therefore, it is recommended that overflow systems be provided with provisions to pump out digested sludge from the cone. It is important to note that the efficiency of the grit removal and screening system in the headworks will have a significant effect on the amount of accumulated grit. Provisions for removing foam and scum from the surface of the digester should be provided. Withdrawing digested sludge from the surface of the digester, as in a submerged fixed cover, has proven to be an effective method for controlling foam and scum in digesters.

Digester tanks are infrequently taken down for planned or emergency maintenance or repairs. This activity requires removal or transfer of the digester sludge contents. In systems with multiple tanks, a common header connection between the withdrawal lines can transfer or equalize tank sludge concentrations. These common lines can be helpful in tank startup or seeding operations. Additionally, existing circulation or transfer pumps can be provided with piping and valves that allow the occasional transfer of sludge. The large digester volume and relatively slow pumping rates, however, can make this a lengthy process. Depending on the respective elevations of the digester cone and the potential withdrawal sludge pumps, the pumps net positive suction head (NPSH) could become a limiting factor by preventing the digested sludge pumps from draining the entire digester tank contents. Some digester systems have been designed with a future pipe connection to allow trailer-mounted solids handing pumps to quickly transfer sludge during these operations. Lastly, some digester systems have drain valves and lines that can allow sludge to flow by gravity back to the head of the plant. Care must be taken, however, to not overload the plant preliminary and primary treatment processes. Typically, sending digested sludge to the headworks of a plant is only done during emergency conditions.

5.5 Sludge Heating

Feed and temperature fluctuations can have detrimental effects on the digester operation. A temperature fluctuation of as little as 1 or 2 °C (1.8 or 3.6 °F) can result in digester foaming (National Biosolids Partnership, 2005). Maintaining a stable temperature can require a significant amount of heat. The greatest heat demand for maintaining digester temperature occurs during the heating of feed sludge. The heat loss from the digester tank represents a smaller heat demand but should be addressed in design. Digester tanks are often constructed to be partially sunk into the ground with the grade elevation at approximately halfway up the side wall. Soil above the groundwater level is an excellent insulator. Greater exposure to the soil, and more consistent temperature surroundings, will reduce the peak heat loss through the tank walls during cold winter months, particularly in colder climates. Tanks placed in areas with a high groundwater table will tend to have more heat loss through the digester floors and walls and will also require footing drains. If possible, it is desirable to design the digester so that the lowest point of the tank is above the groundwater.

Mesophilic digesters typically are designed to operate at or above 35 °C (95 °F). To maintain this temperature, the sludge must be heated enough to offset heat loss through digester roof, walls, bottom, and piping. An adequate supply of heat must be available from the heat source (e.g., boilers, combined heat-and-power equipment), and the heat exchangers or steam injectors must be properly sized to transfer the necessary heat. Design considerations for the heating system include maximum heat requirements, seasonal variations, and low heat demands, which can represent a significant portion of the operating time.

Numerous design guides, include WEF Manual of Practice No. 8, provide equations for determining the amount of heat required to maintain a digester operating temperature (WEF et al., 2009). Typically, the equations compute heat required for heating raw sludge and heat required to overcome heat losses through the walls, roof, and bottom of the tank.

One way to heat raw sludge is to blend it with heated digested sludge. By feeding the thicker raw sludge downstream of the sludge circulation heat exchanger, the heat exchanger is less likely to plug, and the circulation pump receives more consistent pressure. When it is necessary to heat raw sludge in a heat exchanger, it is important to maintain adequate sludge flows in the heat exchanger. Velocities can be maintained by the constant circulation of sludge through the heat exchanger.

Most plants use external heat exchangers to heat sludge from a hot water system. External heat exchangers include concentric tube, spiral, box, and water-bath heat exchangers. Some plants use a water jacket around a draft-tube mixer inside the digester; other plants inject steam. A few digesters are heated by combined boiler/heat exchanger equipment. In a combined boiler/heat exchanger, the boiler heats a water bath in which the tubes carrying sludge pass. A disadvantage of combined boiler/heat exchanger designs is that more sludge piping is required. Sludge is piped to the boiler instead of hot water. Concentric tube heat exchangers consist of a tube in a tube. Sludge flows in the tube and hot water flows in the opposite direction in the annual space between the two tubes. Spiral heat exchangers have pairs of passages wrapped to form spirals: one passage carries sludge; the other passage carries water flowing in the opposite direction. Box heat exchangers transfer heat between hot water and sludge as the fluids flow through box-shaped passages. In water-bath heat exchangers, sludge pipes pass through a heated water bath. Concentric tube and box heat exchangers are more resistant to plugging compared to spiral heat exchangers, which often require sludge screens or grinders.

Design and operation of the heat exchangers should consider effects of fouling, which will reduce the heat transfer efficiency of the exchanger. If the water temperature is too high, then sludge will cake on the surfaces of the heat exchanger. Keeping temperature below 68 °C (154 °F) should minimize caking. Also, to avoid caking and temperature shock to microorganisms, a reasonable temperature for sludge leaving a heat exchanger is 40 °C (104 °F). Because of the high viscosity and undigested constituents in raw sludge, heating digested sludge provides a more reliable method of heat transfer. The raw sludge being fed to the digester is blended with the heated digested sludge after the heat exchanger in the sludge circulation loop.

5.6 Digester Mixing

Mixing is one of the key requirements for achieving optimum performance in anaerobic digestion. Mixing distributes the digester feedstock to all parts of the digester, moves digestion byproducts throughout the bulk liquid, inhibits temperature and solids stratification, and creates homogeneous conditions in the digestion vessel. Inadequate mixing can result in localized unmixed volume, or "deadspots," in the digester where the digester feedstock is not distributed and where byproducts can accumulate. If the feedstock is not penetrating a deadspot, then both the effective reactor volume and SRT are reduced. As a result, the loading of the anaerobic digester places more stress on the remainder of the microorganisms.

Control of scum and grit is closely tied to the need for complete mixing in the digester. Foam, scum, and other floatables can stratify at the top of the reactor if mixing energy is insufficient for homogenization. Mixing energy, however, is not enough for surface management. The mixing energy must entrain the surface material with a downward force. This downward force can be achieved by contact with the digester cover or by pumped mixing. Adequate mixing is also needed to avoid deposition of grit at the bottom of the digester.

Wastewater solids contain a variety of materials that can impede processes and damage equipment. In anaerobic digesters, solids often contain sand and other high specific gravity materials that are difficult to suspend, presenting mixing challenges and causing damage to the mixing system. Additionally, wastewater solids contain plastics, rags, and other debris that can become caught in mixing and other digester components. Like grit, these rags and debris can have detrimental effects on the mixing equipment. Wastewater solids also can contain fats, oils, and greases with unique gravities that are much lower than the bulk liquid. This composition increases the problem of stratification in the digester. Finally, concentrated wastewater solids typically have fluid properties that change as the velocity of the fluid changes. At low velocities and laminar flows, solids have a high shear strength and viscosity that must be considered in the selection and design of the mixing system.

Effective mixing systems must account for all the materials that might be found in wastewater solids. The systems should be strong enough to prevent damage from grit and designed with sludge screening, chopping/macerating, or other methods to prevent plastics from becoming stuck on moving components or impellers.

Many different types of digester mixing systems are available with diverse characteristics and design criteria. Most designers work with the manufacturers of the mixing systems to specify the design criteria. The diverse characteristics of the mixing systems are not conducive to an overall criterion of power per unit volume. Some mixing systems provide adequate mixing with much less power than other mixing systems. Tracer studies, temperature measurements, and computational fluid dynamics (CFD) modeling are used to evaluate the effectiveness of a particular mixing technology.

Designers often specify requirements to demonstrate the effectiveness of a mixing system. This is typically achieved through tracer studies. One form is dye tracer studies, which can be used in clean water. To evaluate the effectiveness of a mixing technology on wastewater solids, the use of lithium tracer is also necessary. Care must be taken to measure the total lithium and not soluble lithium, which can be absorbed into the solids. In some instances, the designer will require a manufacturer to prove the effectiveness of their mixing system relative to a turnover time, where applicable, or will require a lithium tracer study to confirm tank mixing conditions. Temperature measurements at different depths and elevations also can be used to evaluate mixing effectiveness. A narrow temperature range from distributed sensors is an indication of effective mixing.

Computational fluid dynamics modeling is the use of mathematical equations to simulate the flow of fluids. Although available since the 1960s, this method only became practical for use by the wastewater industry with computer advancements made in the early 2000s. As a result, CFD modeling has seen increasing use as a tool to evaluate the characteristics of digester mixing systems. This method, however, should be used only as one means of comparing mixing systems. The accuracy of the CFD model in predicting flow patterns and other characteristics within digesters is dependent on several variables and assumptions regarding the sludge flow characteristics. Table 10.1 provides a description and a list of key features for the most common types of mixing used with anaerobic digesters.

5.7 Digester Cleaning

Over time, grit from sludge can settle onto the digester tank bottom and must be removed to restore the specified volume of the digester. Many different factors affect the rate of digester tank grit accumulation: adequacy of the grit removal facilities at the plant headworks, if installed; ability of digester mixing systems to keep grit suspended; method of sludge withdraw; and the digester tank shape. Digester tanks that pump out digested sludge from the bottom of the cone for postdigestion processing periodically waste some of the settled grit and often result in less grit accumulation.

TABLE 10.1 Description and key features of common anaerobic digester mixing systems.

Mixing system type	Description and key features
Floor mounted gas diffusers	External gas compressor, gas piping, stationary diffuser(s)
Gas mixing lance system	External gas compressor, gas piping, gas distribution valve(s)
Confined gas draft tubes	External gas compressor, gas piping, gas distribution valve(s), internal draft tubes
Bubble gun gas mixing system	External gas compressor, gas piping, gas distribution valve(s), internal draft tubes, gas bubble generator
Paddle mixing system	Internal rotating paddle(s), external drive motor
Mechanical draft tube (external)	Draft tubes mounted on tank perimeter, rotating mixer/propeller, external drive motor
Mechanical draft tubes (internal)	Draft tubes mounted inside tank, rotating mixer/propeller, external drive motor
Pumped/hydraulic mixing system	External recirculation pump(s), suction and discharge sludge piping, fixed/movable discharge nozzles
Linear motion mixing system	Internal mixing mechanism, external drive motor

Required digester cleaning frequency ranges from yearly to every 10 or more years, depending on the above factors.

It is critical that digester cleaning operations are considered during design. The digester should incorporate sufficient access points and structural supports to lower cleaning tools and equipment into the tank. Multiple access hatches should be located on the digester cover to allow equipment or personnel to be lowered into the tank. Confined space entry requirements will need to be taken into account. Some tanks have sidewall hatches installed for additional tank access. In some instances, side hatches are large enough to drive in a small walk-behind plow blade or other heavy machinery to help handle the grit. Such a large side hatch requires special structural design of the wall, however.

Cleaning procedures initially require pumping down sludge until the grit level is reached. At that point, a submersible solids handling pump is lowered down into the tank and high-pressure fire hoses are used to fluidize the remaining grit so that it is able to be pumped out of the digester tank. The removed material is pumped to a mobile belt filter press or other dewatering equipment and is hauled off as cake solids. Other cleaning operators elect to pump the grit and sludge back into the influent of the plant. The latter is only recommended if the headworks grit removal system is adequate to remove a large amount of grit and prevent carryover into the clarifiers. Some plants use their own staff for cleaning and others contract it out to specialized tank cleaning contractors, some of which have mobile dewatering equipment mounted on trailers.

Frequency and duration of digester cleaning activities affect the capacity and number of digester units required. Recycled residual solids and semi-liquid slurry generated by cleaning operations also have a significant effect on liquid and solids streams.

5.8 Biogas Management

A properly designed biogas management system is crucial to successful anaerobic digester operation. Biogas often is beneficially used in boilers for digester and building heat and for fuel in combined heat-and-power (CHP) systems. Biogas can be further processed and used as vehicle fuel in the form of compressed natural gas (CNG) or liquefied natural gas (LNG). Chapter 26 *Treatment and Use of Green Gas* provides further discussion of biogas management.

The biogas produced in the digesters must be safely conveyed to its use with provisions for dealing with both condensate removal and the variable production and demand rates that exist. The most important safety consideration is that biogas is explosive at relatively low concentrations, approximately 1 volume of gas in 15 volumes of ambient air. Any air and gas mixture could potentially reach a flame source and an explosion could occur. Appropriate biogas safety considerations should be made and code requirement should be met when designing biogas management systems.

Digester gas production depends upon a variety of factors, such as sludge characteristics, volatile solids destruction rate, sludge feeding method, and biological activity in the digester. Of particular importance is the destruction of volatile solids in the digester because the amount of biogas produced is in direct proportion to the percent of volatile solids destroyed during the process. Typical biogas production from anaerobically digested wastewater sludge varies between 0.8 and 1.1 m^3/kg volatile solids destroyed (13 and 18 cu ft biogas/lb VS destroyed) (WEF et al., 2009). Designers should consider biogas production, peaking factors, conveyance, safety equipment, waste gas burners, treatment for beneficial use, and storage and addition of ferric chloride for hydrogen sulfide biogas treatment.

5.9 Odor Control

As populations expand and move closer to WWTPs, odor control is becoming increasingly important. Anaerobic digesters can be a source of odors at the treatment plant. The best method for controlling anaerobic digestion odor is through containment. In areas that are sensitive to odors, fixed-cover digesters or other covers that provide a positive seal should be considered. Further, digester gas handling piping should be designed to minimize headloss conditions under instantaneous peaks during digester gas production. This will help prevent the digester gas pressure relief valves from opening and creating an odor emission.

6.0 PROCESS CONTROL MEASURING AND MONITORING OF ANAEROBIC DIGESTION

Anaerobic digesters are typically forgiving processes—total volume and long SRTs tend to keep the process stable. Sampling, monitoring, and control are required, however, to ensure that the facility is operated within applicable regulations and that performance is maximized to avoid unexpected upset. Recommended sampling and monitoring protocols for anaerobic digestion processes should include operating temperature, solids retention time, total and volatile solids, volatile fatty acids, alkalinity, ammonia, digester gas production, and digester gas composition. Additionally, monitoring of pathogens and pathogen indicators may be dependent upon the ultimate biosolids disposal practices.

Maintaining process stability is one of the most important operational considerations for anaerobic digestion. A successful anaerobic process depends on many different microorganisms; failure will result a highly odorous and unstable product. A failed digester is characterized by a drop in pH, reduced digester gas production, increase in volatile acids and the volatile acids-to-alkalinity ratio, and a decrease in methane content in the digester gas. Unfortunately, once the failure of the anaerobic process has progressed, it is difficult to rapidly correct because a depressed pH is toxic to the methanogenic bacteria. Further, facilities that rely on digester gas for heating may be affected if the fuel required to maintain operating temperatures is no longer available. A failed digester is colloquially called a "sour" or "stuck" digester. Operations staff should monitor several parameters to avoid anaerobic digester failure:

- Process SRT;
- Volatile solids loading rate (VSLR);
- Operating temperature;
- pH;
- Volatile fatty acids;
- Volatile acids to total alkalinity ratio; and
- Ammonia.

7.0 HIGH-PERFORMANCE ANAEROBIC DIGESTION

Several reasons exist for considering implementing advanced or high-performance digestion at WWTPs:

- Reducing the high cost of solids processing;
- Reducing pathogen concentrations to achieve higher quality biosolids;
- Reducing energy costs by producing more digester gas;
- Reducing the amount of biosolids requiring dewatering and trucking to a final destination; and
- Reducing reactor size or increasing its loading capacity for site-constrained facilities.

Improved operating characteristics of existing digesters sometimes are of greatest concern. Several changes can improve digestion performance: eliminating a foaming problem, improving the mixing, upgrading the digester gas system, providing additional heating, or enhancing the feeding and withdrawal system. If a WWTP needs more digestion capacity but has limited available space, then one option is to increase the capacity of existing digesters. This approach also can result in a faster return on the investment for the expansion through savings in capital or operating costs.

Typically, WWTPs are striving to reduce solids processing costs while being pressured to improve biosolids quality to satisfy users and the public. As a result, WWTPs are implementing productivity initiatives to achieve competitive performance, and high-performance anaerobic digestion is one option. The objectives of high-performance digestion include

- Greater bioenergy conversion;
- Improved biosolids quality;
- Greater energy conservation;
- Increased digester capacity,
- Enhanced operating characteristics;
- Faster return on investment; and
- Reduced solids mass.

8.0 ENHANCED

Hydrolysis, which is defined as the conversion of organic solids to soluble compounds, is often identified as a rate-limiting step in the anaerobic digestion process. Although this process occurs naturally, researchers have developed mechanisms to enhance hydrolysis. Predigestion processing of the digester feedstock typically will expedite the hydrolysis process and allow for increased organic loading to the anaerobic digester. Another benefit gained is an increase in volatile solids destruction in the digesters, which correlates to an increase in digester gas production. Finally, enhancing the hydrolysis step is also reported to decrease the viscosity of the biosolids, making them easier to pump, and to improve the dewatering characteristics of the digested product.

Several methods that focus on enhancing the hydrolysis step have been developed. WAS is difficult to digest because the large percentage of digestible material is enclosed within lignin-rich cell wall material. The intermolecular bonds in the cell wall are very stable and difficult to break. The primary objective of hydrolysis enhancement is to rupture the cell wall. Enhancing the hydrolysis step increases the amount of nutrient-rich cell material available for digestion, improves dewatering, and reduces the amount of biosolids to be disposed.

The methods for enhancing the hydrolysis step can be grouped into the following categories (see Chapter 9 for more details):

- Thermal hydrolysis;
- Biological hydrolysis;
- Mechanical hydrolysis; and
- Chemical hydrolysis.

8.1 Thermal

Thermal hydrolysis uses elevated temperatures and pressures to improve the digestibility of biosolids. Several companies have developed digestion process configurations that incorporate preprocessing of the digester feed by thermal hydrolysis.

Typically, digester feedstock is thickened to reduce the amount of water in the material. The material is then pumped through a macerator before entering the hydrolysis reactor tank. The thermal hydrolysis reaction occurs in a batch reactor tank operating at 160 to 170 °C (320 to 338 °F) for a 20 to 60 minute retention time, typically 30 minutes. The reactor tanks are operated under pressure, typically 500 to 800 kPa (70 to 120 psi). Although the processes of preheating, hydrolysis reaction, and depressurization can occur in one tank, an alternate process configuration is to have three tanks in series with a specific function occurring in each vessel. The addition of heat-and-energy recovery units is common through the process. Steam generated by boilers or heat captured from combined heat-and-power systems typically provide the heat for thermal hydrolysis processes. The use of digester gas as a fuel offsets the energy requirements of the thermal hydrolysis reaction.

Full-scale facilities have been constructed in more than 20 locations around the world with a service population equivalent of 12 million people (Panter and Auty, 2009). Examples of full-scale facilities include London, United Kingdom (see Figure 10.4); Aberdeen, United Kingdom; Dublin, United Kingdom; Manchester,

Figure 10.4 Thermal hydrolysis installation at Thames Water Utilities near London, England, United Kingdom.

United Kingdom; Palm Jumeirah, United Arab Emirates; Whitlingham, United Kingdom; Turku, Finland; and Hamar, Norway.

Design of Municipal Wastewater Treatment Plants describes ancillary equipment for installing a thermal hydrolysis system (WEF et al., 2009). Several design criteria should also be considered when installing a thermal hydrolysis process:

- Construction materials;
- Solids heating and recovery;
- Solids screening;
- Temperature and pressure monitoring and control;
- Odor management; and
- Sidestream treatment.

The advantages of thermal hydrolysis pretreatment include (Kepp et al., 2000)

- Decreased feed sludge viscosity;
- Increased process capacity;

- Improved dewaterability;
- Increased biogas production;
- High quality biosolids;
- Reasonable energy requirements; and
- Reduced foaming in digesters.

Thermal hydrolysis process gas is highly odorous and saturated with water vapor. Two treatment methods available for the process gas are:

- Passing the process gas through a condenser to remove water vapor. The condensed liquid is pumped to the anaerobic digester for treatment and the remaining gas is burned in a thermal oxidizer.
- Using the process gas to aerate the activated sludge tanks. The activated sludge acts as a biofilter to remove odors and other organics from the gas.

8.2 Biological

Biological hydrolysis, also referred to as enzymic hydrolysis, is an expansion of the acid phase of a digestion system into multiple tanks. The tanks are typically operated at or slightly above the mesophilic temperature range. The advantage of the multiple tanks is that specific reactions are allowed to occur in each tank. Typical SRTs for enzymic hydrolysis reactors (including all reactors) is two days at 42 °C (108 °F). The number of tanks range from one to six or more. A more detailed discussion of acid/gas digestion is included in a later section.

8.3 Mechanical

Mechanical hydrolysis, also referred to as disintegration, applies energy to break the cell walls. Although many technologies can be classified as using mechanical hydrolysis, the most typical are ultrasound, ball mills, and high-pressure or pulsed field homogenization. In each case, external energy is applied to make biosolids more available for digestion (Appels et al, 2008).

8.4 Chemical

Chemical hydrolysis involves the addition of chemicals to hydrolyze the cell wall and solubilize intracellular organics. Various methods have been used, but the most typical are acid and alkaline hydrolysis and oxidation techniques.

9.0 THERMOPHILIC ANAEROBIC DIGESTION

Thermophilic digestion occurs at temperatures 50 to 58 °C (122 to 136 °F). Design criteria and system performance for thermophilic digestion are different than those for mesophilic digestion. Thermophilic digestion can achieve higher volatile solids reduction per unit volume of digester than mesophilic digestion, as suggested by the Arrhenius relationship (Cooney and Wise, 1975). Although the Arrhenius relationship is applicable principally to chemical reactions, it also has relevance to biological reactions. The rule of thumb is that every 10 °C (18 °F) increase in temperature results in approximately a doubling in biological reaction rates. Although research has shown that biogas production decreases when operating outside the mesophilic (32 to 40 °C, or 90 to 105°F) and thermophilic (50 to 58 °C, or 122 to 136 °F) ranges (U.S. EPA, 1979). The microbial communities functioning at mesophilic temperatures are different from those that operate at thermophilic temperatures. Because the temperature is higher, the heating-energy needs are higher for thermophilic digestion than for mesophilic digestion.

Despite the greater heating requirements, thermophilic digestion has several advantages over mesophilic digestion:

- Improved pathogen destruction;
- More readily obtained Class-A sludge;
- Higher volatile solids loading capabilities;
- Faster volatile solids reduction and biogas production;
- Stable operations at lower SRTs; and
- Improved dewaterability.

In addition, thermophilic digestion has several disadvantages:

- Higher energy requirements;
- Poorer quality supernatant with significant quantities of dissolved solids;
- More odorous sludge; and
- Decreased process stability.

9.1 Configuration

This section includes an overview of single-stage and multiple-staged thermophilic digestion processes. Three basic multiple-staged methods using thermophilic digestion are discussed: single-stage digestion, staged thermophilic digestion, and temperature-phased digestion. Single-stage thermophilic digestion is similar in configuration to single-stage mesophilic digestion. Single-stage thermophilic digesters have had only limited application, and typically are coupled with mesophilic digestion in multiple-stage digestion processes. Staged thermophilic digestion includes a large reactor followed by one or more smaller reactors. This process tends to reduce pathogen short-circuiting and more readily produces Class A biosolids. The SRT for the first reactor is typically between 17 and 22 days; the additional reactor(s) can have SRTs of approximately two days each. Increased staging enhances digestion rates by simulating a plug-flow reactor arrangement. Temperature-phased anaerobic digestion (TPAD) incorporates both thermophilic and mesophilic digestion. Most configurations are operated as thermophilic-mesophilic, although mesophilic-thermophilic has also been used. In the thermophilic-mesophilic configuration, the thermophilic phase is operated at approximately 55 °C (131 °F) with a three-to-five day SRT, and the mesophilic phase is operated at 35 °C (95 °F) with an SRT of 10 days or greater.

9.2 Process Discussion

Because of the higher operating temperatures, thermophilic digestion provides faster pathogen inactivation than mesophilic digestion. The 40 CFR, Part 503 regulations that control land application of biosolids classify both mesophilic and thermophilic anaerobic digestion as a process to *significantly* reduce pathogens (PSRP or Class-B process), rather than classifying thermophilic digestion as a process to *further* reduce pathogens (PFRP or Class-A process) (U.S. EPA, 1993). In typical thermophilic operations, where sludge is fed almost continuously into the digester and biosolids are simultaneously discharged from the digester, the potential for pathogen short-circuiting exists. Therefore, continuous-flow thermophilic anaerobic digestion is not considered a process that further reduces pathogen counts, and the process, without batch adaptations, is considered a Class-B process (U.S. EPA, 1993).

The same regulations specify that any pathogen or pathogen-indicator destruction process must precede or be concurrent with the vector attraction reduction process (e.g., mesophilic or thermophilic anaerobic digestion), thus allaying the regrowth concerns posed by Clements (1982) and Keller (1982). For Class A anaerobic digestion systems, the requirement states that a 38% reduction in volatile solids must be met either during or after the pathogen destruction process. The regulations have also been interpreted that operation in batch or plug-flow mode is required of a Class A process. Thus, most Class A thermophilic digestion processes that are planned or in operation have been designed to satisfy U.S. EPA's batch requirement where every particle meets the time-temperature relationship.

In regulations, a thermophilic digestion process meets the pathogen or pathogen-indicator reduction requirements for Class A by maintaining the process temperature at or above 50 °C for a specific period of time in a batch operation (U.S. EPA, 1993). The amount of time is calculated using the formula below, eq 10.3, which can be applied for sludge with a solids content of less than 7% (U.S. EPA, 1993):

$$D = 50\,070\,00/10^{0.14T} \tag{10.3}$$

where
D = time (days) and
T = temperature (°C).

A graph of eq 10.3 is shown in Figure 10.5. Based on eq 10.3, the temperature of the biosolids must be maintained at 55 °C (131 °F) for a minimum of 24 hours or at 50°C (122 °F) for 120 hours. As the thermophilic digestion temperature increases, the batching time (and thus tank volume) required for pathogen or indicator destruction reduces considerably. Systems should be designed to operate in excess of compliance points between 55 and 56 °C (131 and 133°F) to coincide with a one-day interval and maintain a consistent daily operation. Operation at lower compliance temperatures requires additional volume and less uniform cycle times but might warrant consideration if available heating energy sources are scarce.

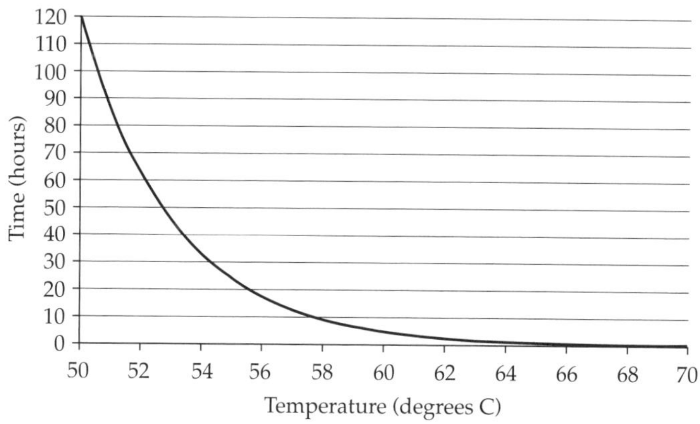

FIGURE 10.5 Time-temperature relationship for Class A biosolids from digestion with solids content less than 7%.

Equation 10.4 shows the time/temperature requirements for the disinfection of sludge higher solids contents of greater than 7% (U.S. EPA, 1993).

$$D = 131\,700\,000/10^{0.14T} \qquad 10.4$$

Based on the above equation, for solids content greater than 7%, the minimum batching time is 63 hours at 55 °C (131°F). In practice, however, it is unlikely that typical thermophilic digestion processes will be operated at such high solids concentrations because of the higher viscosity observed at these higher levels. The exception would be a thermophilic process coupled with a hydrolysis enhancement step. Higher viscosity results in a much greater energy requirement for pumping and mixing the thicker sludge.

It is also important to differentiate between the compliance temperature and normal operating temperatures in a thermophilic design. The compliance temperature is the temperature identified by eqs 10.3 and 10.4 and represents the minimum temperature at which a Class A systems must operate to meet the regulated conditions. The normal operating temperatures may need to be at least 1 °C (1.8 °F) higher than the compliance temperature or perhaps even 2 to 2.5 °C (3.6 to 4.5 °F) higher. The design operating temperature should be high enough to allow detection of any operational issues before the operating temperature drops below the compliance temperature.

Multistage thermophilic digestion, or two or more tanks in a series, has also been reported to show no observed *E. coli* regrowth following dewatering. This is something not observed in other anaerobic processes such as thermophilic or mesophilic (Chen et al., 2008).

Refer to Chapter 12 for more information on dewatering and to Chapter 15 for further discussion on pathogens in biosolids.

9.3 Temperature-Phased Anaerobic Digestion Process

Temperature-phased anaerobic digestion installations with thermophilic followed by mesophilic digestion (Figure 10.6) are still relatively uncommon compared to conventional mesophilic. An example of an operational TPAD installation is at the Orange Water and Sewer Authority (OWASA) plant in Chapel Hill, North Carolina. The OWASA configuration provides for Class A batch operation under Formula H.

Research done in Germany has identified the potential advantages of temperature-phased digestion. Since August 1993, anaerobic digesters in Cologne, Germany, were operated in a temperature-phased mode (Dichtl, 1997). In the United States, Han and Dague (1997) conducted laboratory studies documenting the advantages of

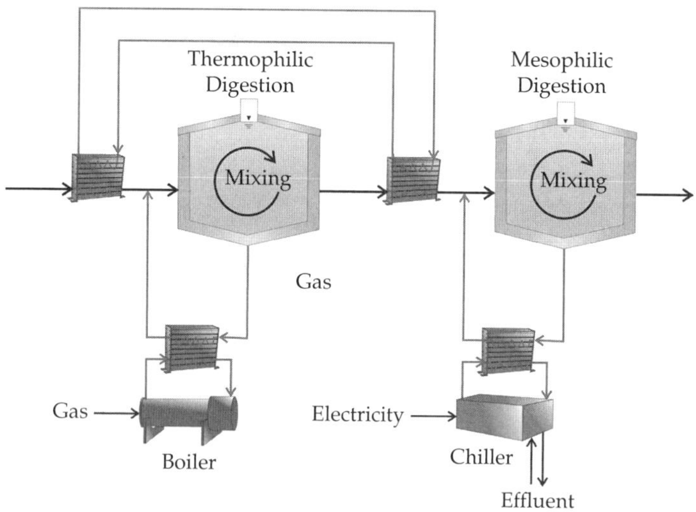

FIGURE 10.6 Temperature-phased anaerobic digestion process configuration.

temperature-phased anaerobic digestion. As of 2007, more than 20 full-scale TPAD plants had been constructed in North America (Le, 2007).

The TPAD process has a greater thermophilic digestion rate, which is typically four times faster than mesophilic. Additional benefits include enhanced performance during shock loads and higher volatile solids destruction efficiencies. The resulting solids are also reported to be easier to dewater than the product of mesophilic digestion (WEF Residuals and Biosolids Committee Bioenergy Technology Subcommittee, 2004).

The addition of a mesophilic phase also will reduce odorous compounds (specifically volatile fatty acids) that are not destroyed in the thermophilic phase and improve the stability of the digester. The tendency to produce foam, which is found in high-rate mesophilic digesters with mixed primary and WAS feedstock, is reportedly reduced in TPAD systems. Reduced coliform counts in the digested sludge and possible conformance with Class A pathogen criteria (via monitoring) under the 40 CFR Part 503 regulations are additional advantages of the process (Han et al., 1997).

Dague (1968) determined that the thermophilic stage could operate at 55 °C (131 °F) with a five-day detention time. The mesophilic phase was designed to operate at 35 °C (95 °F) with a 10-day detention time. Residence times other than those in Dague (1968) have been investigated for the thermophilic and mesophilic phases, and results have shown performance improvements at a variety of residence times for each stage. Few full-scale plants have operated the thermophilic phase at five days SRT or less, but work has indicated that this is successful (Bucher et al., 2001). The design criteria for TPAD encompass a range of values for residence times and temperatures because performance success has been demonstrated under various situations. The following observations encompass most of the experience from research and full-scale pilot testing:

- Thermophilic temperature range is 50 to 56 °C (122 to 133 °F).
- Thermophilic residence times are 4 to 10 days (longer SRTs are often observed at TPAD plants because larger-than-required tanks are often used for the thermophilic stage or early-year loads are less than design loads).
- Mesophilic temperature range is 35 to 40 °C (95 to 104 °F).
- Mesophilic residence times are 6 to 12 days (longer SRTs are often observed at TPAD plants because of tankage available or less than design loads).

The choice for specific design criteria to use at a certain location is a function of project objectives, sludge feed characteristics and variability, existing equipment, and space restraints. If wide variations in feedstock quantity and characteristics occur, then the designer may opt for a longer SRT (i.e., up to 10 days) in the first-stage thermophilic reactor. If an existing tank can be used for thermophilic digestion, but only has a four-to-five day SRT, then additional volume may be required in the mesophilic tank to provide a sufficient SRT and mitigate any variability in the performance of the first-stage thermophilic reactor. Minimum total SRTs of approximately 15 days are considered good design practice for peak 15-day or peak month loads. If the plant needs to ensure that Class B biosolids (per Part 503 regulations) are achieved, then a total 15-day minimum SRT is typically required. Another requirement for Part 503 rules is to ensure that digester temperatures in the mesophilic stage remain greater than 35 °C (95 °F), assuming the mesophilic SRT is needed to meet Class B requirements (U.S. EPA, 1993).

9.4 Design Criteria and Performance

Specific design criteria and normal operating ranges of thermophilic digestion are provided in Table 10.2.

Thermophilic digesters are more sensitive to changes in temperature than mesophilic digesters, and their temperatures can occur more rapidly if heating is interrupted. Two-stage digestion consisting of a mesophilic digester followed by a thermophilic digester has been used to solve this problem (Torpey et al., 1984). This form of two-stage digestion is rarely used, however, because it results in the more

TABLE 10.2 Typical operational and design criteria for mesophilic and thermophilic anaerobic digestion (SRT = solids retention time).

Criterion	Thermophilic	Mesophilic
Typical SRT	8–12 days	15–20 days
Minimum design SRT	4.5–6 days	12 days
Operating temperature	50–58 °C (122–136 °F)	35–39 °C (95 to 102 °F)
Feedstock concentration	4.5–6.5% total solids	3–6% total solids
Digester sludge concentration	2.5–4.5% total solids	1.5–4% total solids
Volatile solids (VS) loading	3.2–6.4 kg VS/m^3·d (0.2–0.4 lb VS/d/cu ft)	1.6–2.4 kg VS/m^3·d (0.1–0.15 lb VS/d/cu ft)
Volatile fatty acid concentrations (total as acetic acid)	400–1200 mg/L	<200 mg/L
pH	7.0–7.7	6.8–7.2

odorous thermophilic phase as the second and last phase before dewatering or product disposal.

In summary, thermophilic digestion can be more costly to construct and operate than mesophilic digestion. It is possible to offset these increased costs by using waste heat for thermophilic heating and from the resulting decreased dewatering costs and pathogen levels in biosolids (i.e., increased fertilizer value). A designer may wish to perform pilot tests with the actual feedstock before deciding to use thermophilic digestion. Thermophilic digestion offers additional process benefits relative to mesophilic processes, particularly greater gas production, solids reduction, and pathogen destruction.

9.5 Temperature-Phased Anaerobic Digestion Performance

Performance for TPAD systems has often been measured by improved volatile-solids reduction (VSR) or additional biogas production. Schafer et al. (2002) reported significant improvement in VSR at several plants with the use of the TPAD process, compared to the performance of a mesophilic system with similar SRT. Shafer et al. (2002) also review pilot or demonstration studies. Below is a summary of an example of pilot-testing results in King County, Washington, summarized as follows (Bucher et al., 2001).

- The system achieved 63.7 and 65.0% VSR when operated at thermophilic and mesophilic digestion temperatures, respectively, with a 20-day SRT.

- The TPAD systems with the same feedstock achieved an overall 73.3% VSR when operated at a 12–day thermophilic SRT, followed by a 12-day mesophilic SRT (12 T/12 M); 70.4% VSR when operated at a 16-day SRT (8 T/8 M); and a 69.1% VSR when operated at a 12-day SRT (4 T/8 M).

Direct comparisons of TPAD and mesophilic digester systems are difficult to quantify if the SRTs are different, however.

9.6 Physical Considerations

9.6.1 *Structural Considerations*

Thermophilic digesters are often created by converting mesophilic operations. Before process conversion, it is important to evaluate the ability of the existing infrastructure to accommodate increased thermal stresses imposed by higher operating temperatures. Structural evaluation is critical to avoid system failure. In particular, the following should be evaluated: tanks and vessels, covers, sludge and gas piping systems, and any lining or coating material used on the surfaces. Further, the adequacy of the current heat exchangers to handle and transfer the higher thermophilic temperatures should be evaluated. Insulation should be evaluated to verify that it can withstand thermophilic temperatures.

9.6.2 *Linings and Construction Materials*

A limited number of linings and coatings are compatible with thermophilic temperatures. Polyvinyl chloride liner materials should be avoided in high-temperature thermophilic applications, but high-density polyethylene liners can handle thermophilic temperatures.

9.6.3 *Sludge Feeding*

Although thermophilic digesters are considered to be more stable, it is still good practice to use constant-rate loading. Because thermophilic bacteria have a higher substrate use rate, thermophilic digesters can be loaded at a higher solids loading. They can also be loaded at a higher hydraulic rate and can typically withstand larger load variations than mesophilic digesters.

9.6.4 *Sludge Heating and Heat Recovery*

Because of the higher operating temperatures in thermophilic systems, heating efficiency and heat balances are important. The majority of the heat demand is attributed to heating the raw sludge feedstock before it is delivered to the digester. The increased ambient heat loss from pipes, tanks, and covers from higher operating temperatures increases the overall heating demand. These issues must be addressed in any digestion process that includes thermophilic temperatures. Three different options have been used to limit the effect of operation at these higher temperatures: increasing the feedstock solids concentration, providing heat recovery downstream of the thermophilic process, and using better insulation on the digester.

Many thermophilic systems thicken feedstock to between 5 and 7% solids to reduce the water content and, therefore, the overall heating load. Often, improved thickening is part of a thermophilic upgrade.

Applying heat recovery technologies to transfer heat from the discharge sludge to the feed sludge can also help supplement the thermophilic heat load. The most efficient mechanism for accomplishing this heat transfer would be to convey the heat directly from the hot sludge to the cold sludge. Such sludge-to-sludge heat recovery systems have proven problematic, however, because cold sludge, containing grease and fibrous materials, tends to blind heat exchangers much faster than digested sludge. Sludge-to-sludge heat recovery systems typically require considerable maintenance to keep clean and in service. In more recent designs, systems using sludge-to-water-to-sludge heat transfer have the option to sacrifice optimal efficiency for improved operability and maintainability. Such systems dedicate one heat exchanger for recovering the heat and transferring it to a closed, circulated water loop. The water loop is then used to heat the raw sludge in a second dedicated heat exchanger that is designed to handle the more problematic cold raw sludge.

In most cases, it is preferable to use a heat exchanger to transfer heat from the digested sludge to a closed water loop. Using this method reduces problems associated with pumping sludge through a heat exchanger and provides a source of low-grade heat that is easily used for other demands. Pumping raw sludge through a heat exchanger is difficult and energy-consuming because of the high viscosity of the sludge. Also, localized burning of the raw sludge within the heat exchanger can create sludge caking issues that require increased maintenance. Transferring the heat from warm sludge to a closed water loop eliminates problems with sludge caking, and the hot water loop can be used for digester heating, plant space heating, or hot-water maintenance demands.

9.7 Biogas Management

Gas production in thermophilic systems is similar to mesophilic systems with a few key differences. The biogas production rate in thermophilic systems can be significantly higher than mesophilic systems, especially if sludge of highly degradable feedstock such as food waste is added to the thermophilic digester. In this case, gas piping may need to be upsized due to the peak biogas production rates.

The relatively high temperatures in thermophilic systems results in biogas that contains significantly more water vapor than biogas from mesophilic systems. As the biogas moves through the gas piping, the gas is cooled and the water condenses. When existing digestion systems are converted from mesophilic to thermophilic operation, it is critical that these increases in condensate volume be addressed before conversion. Condensation typically is addressed by sloped gas piping to accumulation areas or evacuation systems. In addition, the frequency with which the evacuation systems need to be manually serviced or emptied also increases, encouraging the use of automated evacuation systems.

Additional design considerations for thermophilic and TPAD systems include modifications to the biogas handling system. Biogas can convey more water vapor with increasing temperatures; therefore, thermophilic digester gas contains more water than mesophilic digester gas. Also, thermophilic digesters convert more biosolids to biomethane, which can be beneficially used as an energy source. If a mesophilic digester is converted to a thermophilic digester, then the system must take into account the much larger biogas production rate as a result of the increased moisture content of thermophilic biogas and the higher biogas production rate from thermophilic digesters. The biogas piping will likely need to be reevaluated, and possibly replaced, to handle the greater biogas production.

9.8 Dewatering

Mechanical dewatering of thermophilically digested sludge reportedly achieves somewhat improved cake solids content. For instance, Los Angeles, California, reported an increase in centrifuged cake solids content of its digested sludge and a decrease in required chemical dose at its Hyperion WWTP when the plant operation changed from mesophilic to thermophilic digestion (Iranpour and Cox, 2007).

9.9 Odor Control and Issues from Thermophilic Digestion

Odors from thermophilically digested sludge can be significant, primarily because volatile acid concentrations are higher than in mesophilic digestion. Also, sulfur is more soluble at elevated thermophilic temperatures. The sulfur material is reduced to hydrogen-sulfide gas under anaerobic conditions, thereby increasing the odor level of the gas. For these reasons, floating cover digesters are used less often for thermophilic digestion, and more attention is devoted to preventing even accidental exposure of sludge or foam from thermophilic systems.

Dewatering of digested thermophilic sludge will also be somewhat more odorous than for mesophilic sludge. Therefore, self-contained dewatering technologies such as centrifuges are typically used. The centrate or filtrate from thermophilic sludge dewatering also tends to be odorous because of elevated volatile acid levels. Therefore, control of foul air within centrate/filtrate piping may be required.

The odor level from dewatered cake material following thermophilic digestion has not been a problem. Considerable research has been completed on cake odor level following dewatering. This work has discovered that cake odor level can climb significantly following high-shear centrifuge dewatering of digested sludge. (Murthy et al., 2003).

10.0 ACID/GAS PHASED DIGESTION

Acid/gas digestion (A/G), often called acid-phase digestion or two-phase digestion, is a staged digestion process. The purpose of A/G is to provide separate environments for the different groups of microorganisms involved in anaerobic digestion. The first stage of an A/G system is the acid phase, which is characterized by low HRT, acidic pH, and high volatile acids concentration. Volatile acid-producing organisms (acidogenic and acetogenic bacteria), sulfate-reducing bacteria, and hydrogen-consuming methanogens are capable of thriving under the conditions prevalent in the acid phase. The second stage is the gas phase, which is similar to a conventional anaerobic digester with near-neutral pH and low volatile acid concentrations. The environmental conditions in the gas phase favor the growth of volatile fatty acid consuming bacteria (propionate and butyrate degrading syntrophs) and methane-producing organisms. In theory, the low pH in the acid phase of the A/G process results in accelerated hydrolysis. The HRT in the gas phase is substantially less than the 15 to 25 days needed for conventional digesters because the A/G process can operate at higher loading rates. The level of volatile solids reduction and biogas production achieved in an A/G system is comparable to that of a conventional digestion system operated with a considerably larger HRT.

Acid/gas phased digestion involves separate reactors to optimize the acidogenic and methanogenic phases of anaerobic digestion. The first stage is operated at a pH of 6 or less and a short SRT, which selects for bacteria that hydrolyze particulate matter and produce high levels of volatile acids (>6000 mg/L). In the second phase, a neutral pH and a longer SRT select for methanogenic bacteria, which increase biogas production. Advantages of this process, compared to mesophilic digestion, include greater volatile solids reduction and better control of digester foaming. Either stage of A/G digestion can also be operated at mesophilic or thermophilic temperatures.

10.1 Configuration

Implementation of the A/G process has been performed using a variety of configurations. The most basic and prevalent configuration is the acid/gas mesophilic/mesophilic (AGMM) mode, in which the acid phase and the gas phase are operated at mesophilic temperatures, approximately 35 °C (95 °F). Other A/G digestion configurations include thermophilic/mesophilic (AGTM), mesophilic/thermophilic (AGMT), and multistage digestion. Multistage digestion combines A/G digestion and the TPAD process into a three-phase process that includes a mesophilic acid phase, a thermophilic gas phase, and mesophilic gas phase.

The acid phase can be staged to promote plug-flow conditions. Staging acid-phase digesters results in improved hydrolysis and prevention of short-circuiting. The enzymic hydrolysis process and the enhanced enzymic hydrolysis process are two proprietary A/G configurations that use plug-flow conditions in the acid phase. Both configurations include six small acid-phase digesters with a combined HRT of two to three days. Enhanced enzymatic hydrolysis process incorporates a batch thermophilic gas phase to achieve a Class A biosolids product.

10.2 Process

Most A/G facilities in North America are configured with a small, single-stage, acid-phase digester followed by two or more large mesophilic or thermophilic digesters. The acid phase is typically operated at mesophilic temperatures. The gas generated in the acid-phase digesters requires special attention to address gas quality and odor issues (see Section 10.9 for additional guidance). Several A/G facilities in North America operate with the AGMM configuration and do not meet Class A standards within the digesters. The A/G facilities that meet Class A standards are configured with a thermophilic acid phase and a thermophilic gas phase. Staged acid-phase digesters are standard throughout in Europe. Table 10.3 lists the key criteria for acid phase digestion.

TABLE 10.3 Key criteria for acid phase digestion.

Volatile solids (VS) loading rate	24–40 kg VS/m^3·d (1.5 to 2.5 lb VS/d/cu ft)
Solids retention time	1–3 days
pH	< 5.8
Volatile fatty acids concentration	> 5000 mg/L
Diameter/height aspect ratio	≤ 1

10.3 Acid/Gas Approaches

In the early 1990s, the Woodridge-Greene Valley WWTP in DuPage County, Illinois, implemented the first full-scale A/G process in North America. This A/G facility is configured as a multistage system where the third-stage digester provides both final stabilization and storage of digested sludge. The Woodridge-Greene Valley facility is often referred to as a two-stage process in publications that considered the third stage as merely a storage tank. Implementation of the A/G process resulted in increased digestion performance and decreased foaming in the digesters (Ghosh et al., 1995). Extensive research has shown that the Woodridge-Greene Valley facility produces Class A biosolids.

In the year 2000, the Inland Empire Utilities Authority (IEUA), Chino, California, implemented multistage digestion at the Regional Plant 1 (RP-1) (Kasirga et al., 2008). Figure 10.7 shows a process schematic of the IEUA A/G facility. Implementation of this process resulted in negligible digester performance improvements (i.e., volatile solids reduction). However, the A/G process provided other benefits, including increased dewatering performance, mitigation of digester foaming, and production of Class A biosolids. In 2001, IEUA RP-1 was certified as a Class A (Alternative 3) biosolids facility (U.S. EPA, 2006).

There are more than 30 A/G systems in final design, construction, or operation stage in North America and United Kingdom (Wilson et al., 2008).

10.4 Design Criteria and Performance

The A/G process can be much more heavily loaded than conventional digesters. Typical volatile solids loading rates for acid-phase digesters are 24 to 40 kg VS/m^3·d (1.5 to 2.5 lb/d/cu ft). These loading rates are an order of magnitude greater than conventional anaerobic digester systems. The key loading criterion in acid digesters

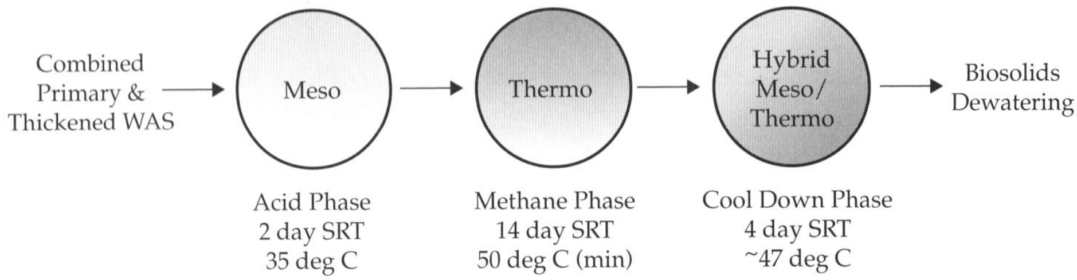

FIGURE 10.7 Inland Empire Utilities Authority, Chino, California, three-phase digestion.

is the SRT. Operational SRTs in an acid digester of more than three days can create foaming problems and should be avoided.

Pretreatment provided in the acid phase allows the gas-phase digesters to be operated at a higher loading rate than conventional anaerobic digesters. Gas-phase digesters can be successfully operated at SRTs as low as 12 days without affecting operational stability. Because of biosolids land application regulations, which require 15 days SRT for anaerobic digestion processes for a Class B, there is little incentive to design the gas-phase digester for less than 12 days SRT. Organic loading rates to the gas-phase digesters can be as high as 3.2 kg/m^3·d (0.20 lb/d/cu ft) volatile solids.

10.5 Physical Considerations

Biogas and sludge from the acid-phase digester will be different than conventional mesophilic digester products. Because the major processes in the acid phase include volatile acid generation and sulfate reduction, the biogas and the sludge in the acid-phase digester possesses a strong and unpleasant odor. The gas will contain H_2S, mercaptans, volatile acids, and other odorous compounds. Further, the gas contains little CH_4 and, thus, has a relatively low energy content. Another characteristic of acid digesters is that the sludge has a low pH, typically less than 5.8.

10.6 Sludge Feeding

The key operational criteria for an A/G system include both the volatile solids loading rate and the SRT in the acid phase. Maintaining acid phase operation with a volatile solids loading rate greater than 24 kg/m^3·d (1.5 lb/d/cu ft) and SRT of less than three days will keep the pH low and prevent foaming. Volatile solids loading rate and SRT control is typically accomplished through a variable volume acid-phase digester.

10.7 Sludge Heating

The A/G sludge heating can be conducted in a similar manner as conventional digesters. The acid-phase digester is heated to mesophilic or thermophilic temperatures.

10.8 Digester Mixing Considerations

Unlike conventional digestion, mixing during acid-phase digestion is critical because a plug-flow configuration is preferred. Some designers advocate acid-phase digester tanks with small diameter-to-height ratios and upflow configuration. Although this configuration may improve the process, it could lead to problems such as dead spaces in the digester or accumulation of grit and debris. The need for mixing should be evaluated on an individual basis.

10.9 Biogas Management

The A/G process produces two different biogas streams. Biogas from the gas-phase digester is similar to conventional digestion biogas, although its CH_4 content and energy value is typically higher. This gas can be used and handled in much the same way as regular digester gas.

The biogas from the acid-phase digester has a low energy content, is odorous, contains high concentrations of H_2S, and is more corrosive than typical digester gas. No standard method exists for handling this gas. Separate biogas burners for acid- and gas-phase digester gases should be considered. Alternatively, methods to blend the gas with the gas-phase biogas or to scrub the gas by injecting it into the gas-phase digester can be considered.

Characteristics of the biogas make it imperative that the biogas be contained. Fixed digester covers can be used to contain the digester gas in most operational situations. Odors, however, can still be problematic during equipment maintenance. Maintenance and cleaning scenarios should be addressed through well-designed operational protocols. Acid-phase digester sludge and gas are corrosive. A proper tank lining system is recommended for the acid-phase digester to protect the concrete. When steel tanks are used, selection of appropriate materials, such as stainless steel, or coatings should be considered.

10.10 Dewatering

Limited information is available on the effects of A/G on residuals dewatering. This limited information appears to show improved dewatering characteristics with higher solids concentration and lower polymer demand and higher solids capture.

10.11 Advantages and Disadvantages of Acid/Gas Phased Digestion

The A/G process is growing in use, and further research into A/G digestion improvements continues. Some of the purported advantages of A/G digestion include

- Increased volatile solids destruction;
- Smaller digestion facilities (although this may be limited by current regulations that require 15-day SRT to meet Class B land application criteria);
- Reduced mass of biosolids residuals for ultimate disposal or recycling;
- Foam control;
- Increased biogas production and increased stability in the gas digester; and
- Class A potential with a thermophilic process.

These advantages need to be weighed with the challenges of A/G digestion processes:

- Increased complexity of operating multiple digestion phases;
- Required maintenance of SRT in the acid digester, increasing control challenges;
- Special handling and consideration of the acid digestion biogas; and
- Significantly higher odor potential that must be addressed in design and maintenance procedures.

11.0 EXTENDED SOLIDS RETENTION TIME (RECUPERATIVE THICKENING)

The extended solids retention (ESR) process involves separating the hydraulic and solids retention times in an anaerobic digester by using sludge recycle thickening (see Figure 10.8). Thickening equipment separates the solids from the digested sludge. The thickened digested sludge is then blended with the incoming raw sludge, and

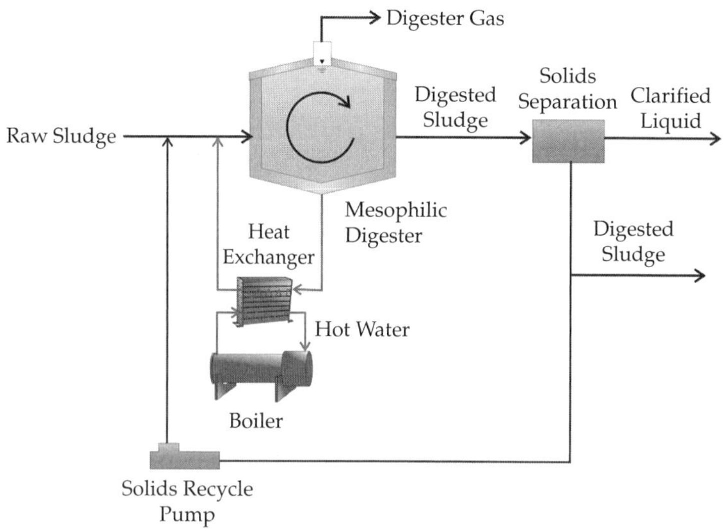

FIGURE 10.8 Extended solids retention digestion.

reintroduced back into the digester. This process allows for a longer SRT when compared to the HRT because the solids are separated from the liquid and returned. The benefit of this process is that active bacteria are returned back to the digester where they further decompose the organics, thereby increase methane formation. An additional benefit may be the elutriation of inhibitory metabolic byproducts during the thickening process, thus resulting in more complete digestion.

Various sludge thickening technologies can be used for solids separation of the recycle stream. When extended solids retention (sometimes referred to as recuperative thickening or the anaerobic contact process) was conceived over 40 years ago, gravity thickening was used for solids separation. Centrifugal and anoxic gas flotation technologies have been installed at full-scale (WEF Residuals and Biosolids Committee Bioenergy Technology Subcommittee, 2004). Other thickening technologies such as dissolved air flotation and gravity belt thickeners may also be applicable.

One concern with ESR has been whether anaerobic bacteria can survive the thickening action if they are exposed to oxygen and a lower temperature environment. This was found to be the case at the Washington Advanced Wastewater Treatment Plant, Spokane, Washington, where the methane-forming bacteria survived this temporary environmental change in dissolved air flotation thickening (Reynolds et al., 2001). This challenges the conventional thinking that the bacteria were overly sensitive to oxygen exposure and temperature change. King County, Washington, has tested ESR at their West Point Treatment Plant using centrifuges. They also conducted ESR pilot testing using anoxic gas flotation (AGF) with a 1900-L (500-gallon) digester at the King County South Treatment Plant in Renton, Washington. King County has adequate digester capacity and does not use ESR. Although it was not used for ESR, King County effectively seeded new digesters at the King County South Treatment Plant using dewatered digested sludge from its West Point Treatment Plant. The dewatered solids were hauled by truck between the plants and slurried into the new digesters. Startup was essentially immediate thus confirming the fact that the organisms are more tolerant of temperature variations and oxygen than previously believed. Additionally, Southwest Suburban Sewer District in Burien, Washington, operated a full-scale ESR using anoxic gas flotation (AGF) (Burke, 2001).

There are several advantages to using ESR: decreased volume of anaerobic digesters resulting in a lower life cycle cost; increased solids concentration, which can serve as a buffer to shock loadings; and an extended SRT, which allows for further organic conversion of solids to methane. An overall reduction in polymer use also can be achieved, and separating the SRT from the HRT provides greater flexibility in solids removal from the ESR digestion process. The potential for enhanced operating characteristics is high, especially if elutriation occurs in the thickening process. Increased solids concentration in the dewatering feed may result in improved dewatering results. In addition, traditional thickening equipment already onsite can be used for ESR, thus eliminating the cost of purchasing specialized treatment systems.

Despite its advantages, there are several disadvantages to the ESR process. These include the requirement for an additional thickening process, the capital and operating cost of the additional thickening equipment, and increased complexity. The net space savings from the reduced digester footprint are required for the thickening equipment. Before widespread implementation, more research is needed to understand the kinetics of recycling viable microorganisms in anaerobic digesters.

12.0 DUAL DIGESTION (AEROBIC/ANAEROBIC DIGESTION)

Dual digestion is a combination of autothermal thermophilic aerobic digestion (ATAD) and anaerobic mesophilic digestion. It combines the benefits of aerobic and anaerobic processes to produce Class A biosolids and biogas. This process requires two types of digesters in which the sludge is partially digested in the ATAD digesters and completed in the anaerobic digesters. Injecting air into the ATAD process sustains an exothermic reaction and maintains thermophilic temperatures (i.e., autothermal). The Tacoma, Washington, has operated a dual-digestion system successfully for more than a decade. Figure 10.9 provides a process flow diagram of the dual digestion process.

This process is readily capable of producing a Class A biosolids, and the biogas produced in the anaerobic process can offset the energy demand of the aerobic process.

13.0 CO-DIGESTION

Increasingly, anaerobic digesters are perceived as assets that do more than stabilize and process wastewater solids. Wastewater treatment plants with spare digestion capacity are now treating and stabilizing many organic wastes with co-digestion. Co-digestion refers to the digesting of domestic sludge with other organic wastes such FOG, restaurant waste, food processing waste, and the organic fraction of municipal solid waste.

Key drivers for co-digestion are maximizing the facility's assets, increasing biogas production, and generating a revenue stream from tipping fees associated with receiving organic wastes. Using the increased biogas for fueling boilers and combined heat and power (CHP) systems can significantly reduce a plant's energy costs. In addition, feeding excess waste to digesters is a way to keep it out of collection

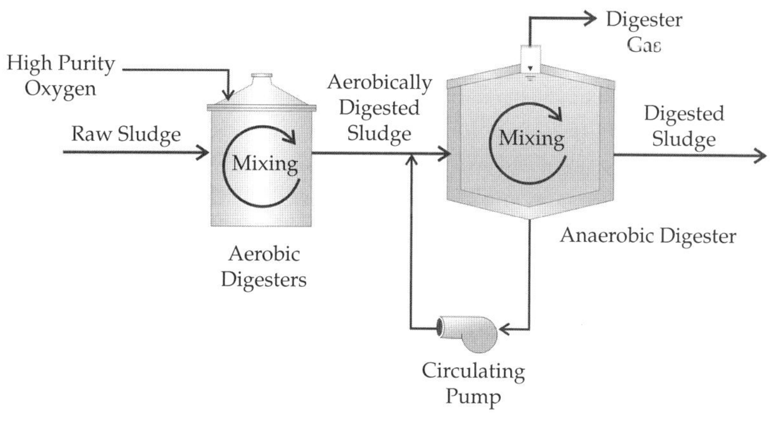

FIGURE 10.9 Dual digestion process.

systems and landfills, thereby decreasing material that might otherwise plug sewers and increase maintenance costs. Some WWTPs are collecting excess wastes and feeding it to their digesters at centralized FOG receiving facilities to accomplish these goals. The process is widely used in Europe and is being implemented more in North America.

Co-digestion with feedstocks that contain a high solids concentration of 15 to 20% are typically "dry" digested in plug-flow-type silo reactors. These processes were originally used in the solid waste industry. Feedstocks containing less than 10 to 15% solids are more easily pumped and mixed and are, therefore, more amenable to conventional "wet" digestion, which is discussed below.

There are economic, environmental, operational, social, and political benefits to co-digestion. Economically, the value of tipping fees for received organic waste, and the value of the additional biogas produced, cover the additional operation and maintenance costs and can help pay for the investment in co-digestion equipment. Environmentally, co-digestion will reduce greenhouse gas emissions by reducing the amount of methane that would otherwise be released during anaerobic decomposition in landfills, and by producing more biogas that can beneficially be used in lieu of nonrenewable fossil fuels. Operationally, co-digestion has the potential to improve the stability, biogas production, and VSR of the digesters and secondary treatment system. There are also social/political benefits to the nearby community who may value the conversion of waste products into a valuable renewable resource such as biogas.

13.1 Co-Digestion Loading Rates

For co-digestion of a mixture of organic matter, a more sophisticated loading rate beyond SRT and volatile solids is necessary. Volatile solids are not always the best process control parameter especially when high strength-low solids wastes are concerned. For example, waste cooking oil and grease and waste glycerin from biodiesel production are excellent co-digestion feedstocks. The volatile solids concentration of these feedstocks is low, however, and completely unrepresentative of the true digester loading.

Energy loading and biodegradable energy conversion are ultimately what is of interest in co-digestion process control. The energy content and the biodegradable energy conversion of the feedstock that is loaded to the digester are directly related to the amount of methane that is biochemically produced. The rate and extent of the conversion is related to the potential for and rate of biochemical hydrolysis. The specific energy loading rate (SELR) and the five-day biodegradable energy conversion (BEC_5) are parameters that are applicable to co-digestion loading rates and waste characteristics.

For co-digestion of a mixture of organic matter, research is underway (by the Water Environment Research Foundation, Alexandria, Virginia) to determine the SELR and BEC_5. For mesophilic digesters, the SELR appears to be 950 kJ/kg·d. The BEC_5 method provides a quick and effective metric to compare different organic wastes for co-digestion and to estimate the potential rate of increase in gas production given the addition of certain organic wastes to an anaerobic digester. It is determined by adding a known amount of a defined organic feedstock to an anaerobic digester and measuring the biogas produced after five days. The BEC_5 gives an indication of the digestibility of a particular feedstock that is helpful for estimating the stable digester loading rate.

13.2 Types and Sources of Co-Digestion Feedstocks

A variety of co-digestion feedstocks, also known as cosubstrates, have been tested at bench-, pilot-, and full-scale. These include aircraft deicing fluid, beet pulp, brewery wastes, cattle manure, cheese whey, chicken manure, clarifier skimmings, dissolved air floatation float, processed algae, fermentation wastes, FOG, food wastes, glycerin, grass clippings, grease trap pump-out, meat processing wastes, the organic fraction of the municipal solid waste stream, outdated beverages, restaurant wastes, thin stillage, various spent grains, vegetable wastes, and more (Totzke, 2009). Several researchers have performed literature reviews of cosubstrates (Alatriste-Mondragón et al., 2006; Erdal et al., 2005; Mata-Alvarez et al., 2000; Parry et al., 2009). Because there is no standard metric to gauge the digestion potential of different feedstocks, however, generalizing results from disparate experiments is difficult. Therefore, a site-specific analysis of potential cosubstrates is preferred when designing a co-digestion facility. Most of the cosubstrates listed above have been found to improve the anaerobic digestion process by increasing VSR or biogas yield, or both. A summary of two of the most common cosubstrates, FOG and food waste materials, is presented below.

13.3 Fats, Oils, and Grease

The most common organic waste cosubstrate is FOG, which is generated by myriad sources, including households, restaurants, hotels, commercial kitchens, bakeries, schools, prisons, and large food-preparation facilities. The solids concentration of

FOG can vary from less than 2% solids to more than 15%, and the volatile solids/total solids typically range from 94 to 97%.

To prevent blockages of the sanitary sewer system, FOG wastes typically are collected using waste drums, grease traps, and grease interceptors. Depending on the source of the waste, trucking of organic wastes and processing facilities should be considered. In many municipalities, dedicated vacuum pumper trucks are used to retrieve FOG from traps and interceptors and transport it back to the FOG receiving system. A FOG receiving system can be as simple as a connection to a digester to allow the truck to discharge directly to a heated receiving tank with settling tanks (rock traps) and a digested sludge circulation loop used to convey the FOG to the digester. Processing should include screening and grinding to remove stones, rags, metallic objects, and other inert material that could impair downstream operations, and may include FOG concentration.

13.4 Effect of FOG Addition on Digestion

Fats, oils, and grease are readily biodegradable and can provide both enhanced biogas production and increased VSR when added to anaerobic digestion systems. In fact, FOG is among the highest-rated cosubstrates in terms of methane production. In fact, FOG addition may enhance the digestion of other organic feedstocks (Schafer et al., 2007; Parry, 2009).

Digesters can remain stable at FOG loads as high as 30% of feedstock or volatile solids loading. At higher loadings of between 30 and 50%, digesters can become susceptible to instability during a peak FOG load (Schafer et al., 2007; WERF, 2009). One crucial characteristic for successful FOG digestion is good digester mixing, particularly at the surface. Without adequate surface mixing, FOG will tend to collect at the surface of the digester.

Literature regarding methane production rates from FOG co-digestion is limited and tends to differ in terms of units and FOG loading rate. Gas production potential from lipid-rich cosubstrates (i.e., FOG) is 1.4 m^3/kg (23 cu ft/lb) VSR, significantly higher than that expected from carbohydrate and protein-rich cosubstrates, which is 0.8 to 0.9 m^3/kg (13 to15 cu ft/lb) VSR (Erdal et al., 2005). For comparison, gas production from conventional mesophilic digestion fed municipal biosolids ranges from 0.8 to 1.1 m^3/kg (12 to 18 cu ft/lb) VSR. Also, because a relatively large percentage of the FOG waste is destroyed during digestion, little of the FOG feedstock remains in the digested product to be handled and disposed. (For more information on biogas production rates from FOG, see Alatriste-Mondragón et al., 2006; Kabouris et al., 2008; Kabouris et al., 2009; Parry et al., 2009; and Schafer et al., 2007.)

Fats, oils, and grease have been successfully fed to anaerobic digesters at several WWTPs including in Redwood City, California; East Bay, California; Millbrae, California; Riverside, California; Des Moines, Iowa; and Denver, Colorado.

Some WWTPs, such as East Bay Municipal Utility District (EBMUD), Oakland, California, have chosen to provide FOG collection as a service to its customers rather than enforce the local FOG disposal regulations. The utility includes the cost of collection in its sewer rates, and contracts with waste haulers to collect FOG. The utility realizes cost savings in decreased maintenance of their collection system. They also feed FOG to their digesters to increase biogas production or convert FOG to biodiesel for sale or use in fleet vehicles.

13.5 Food Wastes

Food waste is a broad category that includes wastes from industrial (e.g., fruit and vegetable processing, beverage industry), commercial (e.g., restaurant), and residential (e.g., the organic fraction of municipal solid waste) sources. Although most food wastes are easily digested and require little pretreatment, food wastes with large inert materials, such as animal parts and carcasses, may require detailed pretreatment and are not typically recommended for co-digestion. Other wastes from livestock, poultry, fish, and even fruit-juice processing may simply require chopping as part of the pretreatment scheme. Increased H_2S formation should also be expected from facilities that process alcoholic beverages (Erdal et al., 2005).

13.6 Advantages and Disadvantages of Co-Digestion

Co-digestion can improve the nutrient balance of the digester and help dilute many high-strength waste streams. Also, if a digester feed is too dry, cosubstrates can help supplement the moisture requirements of the digester. Easier access to handling of mixed wastes, the use of common access facilities, and economies of scale are also potential advantages of the process (Mata-Alvarez, 2000).

Both carbon and nitrogen are required in proper proportions to foster growth of microorganisms in the digester and to prevent process upset. If the carbon content is too high relative to nitrogen, then a nitrogen deficiency and process upset can occur as a result of propionate accumulation because of lower buffering capacity. Similarly, if the carbon content is too low relative to nitrogen, digester performance will likely suffer because of ammonia toxicity. The addition of certain cosubstrates to the digester can help balance nutrient deficiencies and improve performance. For example, adding a nitrogen-rich animal slurry (e.g., chicken manure) to a digester with a high C:N ratio can help achieve a favorable nutrient balance and reduce the potential toxic effects of free ammonia and hydrogen sulfide in the cosubstrate waste stream. Likewise, adding a carbon-rich cosubstrate (e.g., cattle manure) to a digester with a low C:N ratio can provide the necessary growth substrate and enhance microbial growth in the reactor. The optimum C:N ratio, measured based on biodegradable carbon and total nitrogen, in a typical digester ranges from 16:1 to 30:1, but more research is needed on this subject (Erdal et al., 2005, Ward et al., 2008). Wastewater solids, food waste, and mixed paper have C:N ratios of 10:1, 20:1, and 100:1, respectively. A site-specific analysis is preferred to determine the optimal C:N ratio for a combination of cosubstrates.

Potential disadvantages of co-digestion include construction of infrastructure required to receive and process additional waste, handling of additional wastes in dewatering systems, handling and treatment of additional gas, and the liquid-stream effects of increased loading on the digesters. Increased gas production and tipping fees, however, typically can compensate for the capital investment in co-digestion upgrades (Erdal et al., 2005).

14.0 FUTURE OF ANAEROBIC DIGESTION

Anaerobic digestion design is undergoing rapid change compared to the practices of previous decades. These changes are not only confined to process changes, pretreatment regimes, and changing feedstock, but perhaps more importantly, they may include increasingly sophisticated design and operation of anaerobic digesters. In the next decade, design and operation will likely move away from textbook parameters and "rules of thumb" that have been used to design and operate digesters for the last 50 years. Computing power and basic research have led to the development of sophisticated anaerobic digestion process models. Although these models are still not prevalent, and additional research is needed for calibration, design and operation of anaerobic digesters will factor in waste characterization and food-to-microorganism ratios that are common to other biological processes.

Historically, sludge characterization has been considerably less robust than the study of other wastewater streams and has been largely limited to total and volatile solids concentrations and masses. Liquid stream analyses have instead included carbon, nitrogen, and phosphorus speciation among many other factors. As modeled sludge systems and integrated liquid/solids plant-wide process models become more prevalent, a more comprehensive evaluation of sludge and biosolids characteristics will be needed to accurately predict process performance.

Anaerobic digestion is being viewed as an energy conversion process and not just as a sludge stabilization process. An estimated 340 MW of clean electricity can be generated if all 544 facilities greater than 20 ML/d (5 mgd) that use anaerobic digestion install CHP equipment (U.S. EPA, 2007). Co-digestion has the potential to increase the biogas production at WWTPs and convert enough organic matter into useable energy for self sufficiency. Eventually, anaerobic digestion facilities at WWTPs could be viewed more as energy conversion centers.

15.0 REFERENCES AND SUGGESTED READINGS

Alatriste-Mondragón, F.; Samar, P.; Cox, H. H. J; Ahring, B. K; Iranpour, R. (2006) Anaerobic Codigestion of Municipal, Farm, and Industrial Organic Wastes: A Survey of Recent Literature *Water Environ. Res.*, **78**, 607–636.

Aldin, S.; Elbeshbishy, E.; Nakhla, G.; Ray, M. (2009) Viability of Ultrasonication for Pre-Treatment of Biosolids. *Proceedings of the 82nd Annual Water Environment Federation Technical Exposition and Conference* [CD-ROM]; Orlando, Florida, Oct 10–14; Water Environment Federation: Alexandria, Virginia.

American Society of Heating, Refrigeration, and Air-Conditioning Engineers (2005) *2005 ASHRAD Handbook Fundamentals*, Inch-Pound ed.; American Society of Heating, Refrigeration, and Air-Conditioning Engineers, Inc.: Atlanta, Georgia.

Appels, L.; Baeyens, J.; Degréve, J.; Dewil, R. (2008) Principles and Potential of the Anaerobic Digestion of Waste-Activated Sludge. *Prog. Energy Combust. Sci.* **34**, 755–781.

Avallone, E. A.; Baumeister, T. (1996) *Marks' Standard Handbook for Mechanical Engineers*, 10th ed.; McGraw-Hill: New York.

Bird, R. B.; Stewart, W. E.; Lightfoot, E. N. (1960) *Transport Phenomena*. Wiley: New York.

Bucher, B.; Newman, G.; Moore, R.; Lepistö, R.; Stensel, H. D.; Ferguson, J. F. (2001) Pilot Plant Investigation of Thermophilic-Mesophilic Digestion for a Full-scale Retrofit. *Proceedings of the Water Environment Federation, American*

Water Works Association, Chesapeake Water Environment Association Joint Residuals and Biosolids Management Conference [CD-ROM], San Diego, California, Feb 21–24; Water Environment Federation: Alexandria, Virginia.

Burke, D. (2001) Producing Exceptional Quality Biosolids through Digestion Pasteurization and Redigestion. *Proceedings of the Water Environment Federation, American Water Works Association, Chesapeake Water Environment Association Joint Residuals and Biosolids Management Conference* [CD-ROM], San Diego, California, Feb 21–24; Water Environment Federation: Alexandria, Virginia.

Chen, Y.; Higgins, M. J.; Beghitol, S.; Murthy, S. N.; Hendrickson, D.; Schafer, P.; Farrell, J. (2008) Examination of *Salmonella* and Fecal Coliform Density After Dewatering of Anaerobically Digested Biosolids. *Proceedings of the 81th Annual Water Environment Federation Technical Exposition and Conference* [CD-ROM]; Chicago, Illinois, Oct 18–22; Water Environment Federation: Alexandria, Virginia.

Chiu, Y-C.; Chang, C-N.; Lin, J-G.; Huang, S-J. (1997) Alkaline and Ultrasonic Pretreatment of Sludge Before Anaerobic Digestion. *Water Sci. Technol.*, **36** (11), 155–162.

Chu, L.; Yan, S.; Xing, X-H.; Sun, X.; Jurcik, B. (2009) Progress and Perspectives of Sludge Ozonation as a Powerful Pretreatment Method for Minimization of Excess Sludge Production. *Water Res.*, **43** (7), 1811–1822.

Clements, R. P. L. (1982) Sludge Hygienization by Means of Pasteurization prior to Digestion. In *Disinfection of Sewage Sludge: Technical, Economic and Microbiological Aspects*, Bruce, A. M.; Havelaar, A. H.; L'Hermite, P. L.; Eds.; Springer-Verlag: Berlin, Germany.

Cooney, C. L.; Wise, D. L. (1975) Thermophilic Anaerobic Digestion of Solid Waste for Fuel Gas Production. *Biotechnol. Bioeng.* **17** (8), 1119–1135.

Dague, R. R. (1968) Application of Digestion Theory to Digester Control. *J. Water Pollut. Control Fed.* **40** (12), 2021–2032.

Dawson, M.; Leefe, S.; Harrison, D.; Cumiskey, A. (2004) *Modeling of Flows in Sludge Mixing Equipment—The Use of Physical Models and Computational Fluid Dynamics (CFD) as Design and Evaluation Tools*; University of Surrey Centre for Environmental and Health Engineering: Surrey, United Kingdom.

Dichtl, N. (1997) Thermophilic and Mesophilic (Two-Stage) Anaerobic Digestion. *J. Chart. Inst. Water Environ. Manage.*, **11** (2), 98–104.

Eastman, J. A.; Ferguson, J. F. (1981) Solubilization of Particulate Organic Carbon during the Acid Phase of Anaerobic Digestion. *J. Water Pollut. Control Fed.*, **53** (3), 352–366.

Erdal, U. G.; Soroushian, F.; Kitto, W.; Whitman, E. J. (2005) A Technology Review for Co-digestion of Food Waste with Biosolids and/or Manure. *Proceedings of the Water Environment Federation/American Water Works Association Joint Residuals and Biosolids Management Conference* [CD-ROM], Denver, Colorado, April 15–17; Water Environment Federation: Alexandria, Virginia.

Ferran, B; Huyard, A; Adamik, J. (2002) Two-Phase Anaerobic Digestion of Municipal Sewage Sludge Optimization of the Pathogen Destruction. *Proceedings of the 75th Annual Water Environment Federation Technical Exposition and Conference* [CD-ROM]; Chicago, Illinois, Sept 28–Oct 2; Water Environment Federation: Alexandria, Virginia.

Froud, C.; Schmitt, J.; Applegate, C. (2009) Anaerobic Digestion Enhancement Success Stories from Around the World. *Proceedings of the Water Environment Federation Residuals and Biosolids Conference 2009* [CD-ROM]; Water Environment Federation: Alexandria, Virginia.

Ghosh, S.; Buoy, K.; Dressel, L.; Miller, T.; Wilcox, G.; Loos. D. (1995) Pilot- and Full-Scale Two-Phase Anaerobic Digestion of Municipal Sludge. *Water Environ. Res.*, **67** (2), 206–214.

Great Lakes–Upper Mississippi River Board of State and Provincial Public Health and Environmental Managers (2004) *Recommended Standards for Wastewater Facilities— Policies for the Design, Review, and Approval of Plans and Specifications*, 2004 ed.; Health Research Inc., Health Education Services Division: Albany, New York.

Han, Y.; Dague, R. R. (1997) Laboratory Studies on the Temperature-Phased Anaerobic Digestion of Domestic Primary Sludge. *Water Environ. Res.*, **69** (6), 1139–1143.

Han, Y., Sung, S., Dague, R. R. (1997). Temperature-Phased Anaerobic Digestion of Wastewater Sludges. *Water Sci. Technol.*, **36** (6–7), 367–374.

Harrison, S. (1991) Bacterial Cell Disruption: A Key Unit Operation in the Recovery of Intracellular Products. *Biotechnol. Adv.*, **9**, 217–240.

Hernandez, G. L.; Redd, K. R.; Ingalla, L.; Wert, W. A.; Liu, A. M.; Haug, R. T. (2002) Hyperion Treatment Plant Advanced Digestion Pilot Program. *Proceedings of the 75th Annual Water Environment Federation Technical Exposition and Conference* [CD-ROM]; Chicago, Illinois, Sept 28–Oct 2; Water Environment Federation: Alexandria, Virginia.

Incropera, F. P.; DeWitt, D. P. (2002) *Introduction to Heat Transfer*, 4th ed.; Wiley & Sons: Hoboken, New Jersey.

Iranpour, R.; Cox, H. H. J. (2007) Evaluation of Thermophilic Anaerobic Digestion Processes for Full-Scale Class A Biosolids Disinfection at Hyperion Treatment Plant. *Biotechnol. Bioeng.*, **97**, 19–39.

Iranpour, R.; Cox, H. H. J.; Oh, S.; Fan, S.; Kearbey, R. J.; Abkian, V.; Haug, R. T. (2006) Thermophilic-Anaerobic Digestion to Produce Class A Biosolids: Initial Full-Scale Studies at Hyperion Treatment Plant. *Water Environ. Res.*, **78** (2), 170–180.

Iranpour, R.; Cox, H. H. J.; Fan, S.; Abkian, V.; Minamide, T.; Kearney, R. J.; Haug, R. T. (2006) Full-Scale Class A Biosolids Production by Two-Stage Continuous-Batch Thermophilic Anaerobic Digestion at the Hyperion Treatment Plant, Los Angeles, California. *Water Environ. Res.*, **78** (11), 2244–2252.

Kabouris, J. C.; Tezel, U.; Pavlostathis, S. G.; Engelmann, M.; Dulaney, J.; Gillette, R. A.; Todd, A. C. (2009) Methane Recovery from the Anaerobic Codigestion of Municipal Sludge and FOG. *Bioresour. Technol.*, **100,** 3701–3705.

Kabouris, J. C.; Tezel, U.; Pavlostathis, S. G.; Engelmann, M.; Todd, A. C.; Gillette, R. A. (2008) The Anaerobic Biodegradability of Municipal Sludge and Fat, Oil, and Grease at Mesophilic Conditions. *Water Environ. Res.*, **80** (3), 212–221.

Kathryn Scott, D.; Drew, R.; McHale, K.; Braccio, J. (2005) Connecticut's Approach to Minimizing Sanitary Sewer Overflows Through the Use of Centralized FOG Receiving Facilities. *Proceedings of the 78th Annual Water Environment Federation Technical Exposition and Conference* [CD-ROM]; Washington, D.C., Oct 30–Nov 2; Water Environment Federation: Alexandria, Virginia.

Kasirga, E.; Coopee, A.; Whitman, E.; Clifton. J. (2008) Innovative Anaerobic Digestion and Co-Digestion at Inland Empire Utilities Agency. *Proceedings of the 81th Annual Water Environment Federation Technical Exposition and Conference* [CD-ROM]; Chicago, Illinois, Oct 18–22; Water Environment Federation: Alexandria, Virginia.

Keller, U. (1982) Experiences and Development of Sludge Pasteurization in Alterhein. In *Disinfection of Sewage Sludge: Technical, Economic and Microbiological Aspects*, Bruce, A. M.; Havelaar, A. H.; L'Hermite, P. L.; Eds.; Springer-Verlag: Berlin, Germany.

Kepp, U.; Machenbach, I.; Weisz, N.; Solheim, O. E. (2000) Enhanced Stabilisation of Sewage Sludge Through Thermal Hydrolysis—Three Years Of Experience With Full Scale Plant. *Water Sci. Technol.*, **42** (9), 89–96.

Krugel, S.; Nemeth, L.; Peddie, C. (1998) Extending Thermophilic Anaerobic Digestion for Producing Class A Biosolids at the Greater Vancouver Regional District's Annacis Island Wastewater Treatment Plant. *Water Sci. Technol.*, **38** (8–9), 409–416.

Krugel, S.; Parrella, A.; Ellquist, K.; Hamel, K. (2006) Five Years of Successful Operation—A Report on North America's First New Temperature Phased Anaerobic Digestion System at the Western Lake Superior Sanitary District (WLSSD). *Proceedings of the 79th Annual Water Environment Federation Technical Exposition and Conference* [CD-ROM]; Dallas, Texas, Oct 21–25; Water Environment Federation: Alexandria, Virginia.

Le, M. S. (2007) Thermophilic and Enzymatic Treatments for Improved Digestion. *Proceedings of the International Water Association Leading-Edge Conference on Water and Wastewater Technologies*; IWA Publishing: London, United Kingdom.

Lewis, F. M.; Haug, R. T.; Poosti, A.; Redd, K. R. (2007) Energy Recovery from Digester Gas at Hyperion. *Proceedings of the Water Environment Federation/American Water Works Association Joint Residuals and Biosolids Management Conference* [CD-ROM]; Water Environment Federation: Alexandria, Virginia.

Lugowski, A.; Nakhla, G.; Patel, J.; Shular, K.; Singh, A. (2007) Biosolids Minimization Using Integrated Alkaline Sludge Hydrolysis and Biological Nutrient Removal Process. *Proceedings of the 80th Annual Water Environment Federation Technical Exposition and Conference* [CD-ROM]; San Diego, California, Oct 15–17; Water Environment Federation: Alexandria, Virginia.

Massart, N.; Bates, R.; Corning, B.; Neun, G. (2006) Design and Operational Considerations to Avoid Excessive Anaerobic Digester Foaming. *Proceedings of the 79th Annual Water Environment Federation Technical Exposition and Conference* [CD-ROM]; Dallas, Texas, Oct 21–25; Water Environment Federation: Alexandria, Virginia..

Mata-Alvarez, J.; Macé, S.; Llabrés, P. (2000) Anaerobic Digestion of Organic Solid Wastes. An Overview of Research Achievements and Perspectives. *Bioresour. Technol.*, **74**, 3–16.

McCarty, P. L. (1971) Energetics and Kinetics of Anaerobic Treatment. *Anaerobic Biol. Treat. Processes*, **June 1**, 91–107.

Müller, J. (2000) Disintegration as a Key Step in Sewage Sludge Treatment. *Water Sci. Technol.*, **41** (8), 123–130.

Murthy, S.; Higgins, M.; Chen, Y.; Covert, K.; Maas, N.; Toffey, W. (2003) The Impact of Dewatering Equipment on Odorant Production from Anaerobically Digested Biosolids. *Proceedings of the 76th Annual Water Environment Federation Technical Exposition and Conference* [CD-ROM]; Los Angeles, California, Oct 11–15; Water Environment Federation: Alexandria, Virginia.

National Biosolids Partnership (2005) *National Manual of Good Practice for Biosolids*; National Biosolids Partnership: Alexandria, Virginia.

Ohanian, E.; Haug, R. T.; Fan, S.; Voss, D. J.; Clark, J. H. (2005) Anaerobic Digester Evolution at the Los Angeles Hyperion Plant. *Proceedings of the Water Environment Federation Sustainability Conference Series* [CD-ROM]; Water Environment Federation: Alexandria, Virginia.

O'Rourke, J. T. (1968) Kinetics of Anaerobic Treatment at Reduced Temperatures. Ph.D. Dissertation, Stanford University, Palo Alto, California.

Owen, W. F. (1982) *Energy in Wastewater Treatment*. Prentice Hall, Inc.: Saddle River, New Jersey.

Panter, K.; Auty, D. (2009) Thermal Hydrolysis, Anaerobic Digestion and Dewatering of Sewage Sludge as a Best First Step in Sludge Strategy: Full Scale Examples in Large Projects in the UK And Strategic Study Including Cost and Carbon Footprint. *Proceedings of the International Water Association Sludge Conference*, Harbin, China; IWA Publishing: London, United Kingdom.

Parkin, G. F.; Owen, W. F. (1986) Fundamentals of Anaerobic Digestion of Wastewater Sludges. *J. Environ. Eng.*, **11**, 867–920.

Parry, D. L.; Vandenburgh, S.; Salerno, M. B.; Finger, R. (2009) A Review of the First Year for Biosolids Management Using a High Solids Anaerobic Digestion System; *Proceedings of the Water Environment Federation Residuals and Biosolids Conference* [CD-ROM]; Water Environment Federation: Alexandria, Virginia.

Parry, D.; Vandenburgh, S.; Salerno, M.; Finger, R. (2009) Codigestion of Organic Waste. Water Environment Federation Residuals and Biosolids Conference, Portland, Oregon, May 3–6; Water Environment Federation: Alexandria, Virginia

Pavlostathis, S. G.; Giraldo-Gomez, E. (1991) Kinetics of Anaerobic Treatment: A Critical Review. *Crit. Rev. Environ. Sci. Technol.*, **21** (5–6), 411–490.

Perry, R. H.; Green, D. W. (1997) *Perry's Chemical Engineers' Handbook*, 7th ed.; McGraw-Hill: New York.

Redner, J. A.; Hsi, R. P.; Esfandi, E. J.; Sydney, R.; Jones, R. M.; Won, D. (2002) *Evaluation of Protective Coatings for Concrete*. County Sanitation Districts of Los Angeles County: Whittier, California.

Reynolds, D. T.; Cannon, M.; Pelton, T. (2001) Preliminary Investigation of Recuperative Thickening for Anaerobic Digestion. *Proceedings of the 74rd Annual Water Environment Federation Technical Exposition and Conference* [CD-ROM]; Atlanta, Georgia, Oct 13–17; Water Environment Federation: Alexandria, Virginia.

Salerno, M. B.; Lee, H-S.; Parameswaran, P.; Rittman, B. E. (2008) Using a Pulsed Electric Field as a Pretreatment for Improved Biosolids Digestion and Methanogenesis. *Proceedings of the 81th Annual Water Environment Federation Technical Exposition and Conference* [CD-ROM]; Chicago, Illinois, Oct 18–22; Water Environment Federation: Alexandria, Virginia.

Schafer, P. L.; Farrell, J. B.; Newman, G.; Vandenburgh, S. (2002) Advanced Anaerobic Digestion Performance Comparisons. *Proceedings of the 75th Annual Water Environment Federation Technical Exposition and Conference* [CD-ROM]; Chicago, Illinois, Sept 28–Oct 2; Water Environment Federation: Alexandria, Virginia.

Schafer, P.; Trueblood, D.; Fonda, K.; Lekven, C. (2007) Grease Processing for Renewable Energy, Profit, Sustainability, and Environmental Enhancement. *Proceedings of the 80th Annual Water Environment Federation Technical Exposition and Conference* [CD-ROM]; San Diego, California, Oct 15–17; Water Environment Federation: Alexandria, Virginia.

Scrivener, K. L.; Cabiron, J.; Letourneux, R. (1999) High-Performance Concretes from Calcium Aluminate Cements. *Cement Concrete Res.*, **29**, 1215–1223.

Tanaka, S.; Kobayashi, T.; Kamiyama, K-I.; Signen Bildan, M. L. N. (1997) Effects of Thermochemical Pre-Treatment on the Anaerobic Digestion of Waste Activated Sludge. *Water Sci. Technol.*, **35** (8), 209–215.

Torpey, W. N.; Andrews, J.; Basilico, J. V. (1984) Effects of Multiple Digestion on Sludge. *J. Water Pollut. Control Fed.*, **56** (1), 62–68.

Totzke, D. (2009) Design Considerations: Tapping the Potential of Codigestion. *BioCycle*, 50 (6), 32.

Tyler, C.; Waitt, W.; White, N. (2002) The Dirty Dozen: Design, Construction, and Operational History of Deer Island's Egg-Shaped Digesters. *Proceedings of Water Environment Federation Annual Residuals and Biosolids Management Conference* [CD-ROM]; Water Environment Federation: Alexandria, Virginia.

U.S. Environmental Protection Agency (1979) *Process Design Manual Sludge Treatment Disposal*. U.S. Environmental Protection Agency: Washington, D.C.

U.S. Environmental Protection Agency (1993) Standards for the Use and Disposal of Sewage, Final Rule. *Code of Federal Regulations*, Part 503, Title 40, 1993; *Fed. Regist.*, **58**, 9248.

U.S. Environmental Protection Agency (1995) *Process Design Manual: Land Application of Sewage Sludge and Domestic Septage*. U.S. Environmental Protection Agency: Washington, D.C.

U.S. Environmental Protection Agency (2003) Standards for the Use or Disposal of Sewage Sludge. *Code of Federal Regulations*, Part 503, Title 40; *Fed. Regist.*, 58, 9248.

U.S. Environmental Protection Agency (2006) Biosolids Technology Fact Sheet: Multi-Stage Anaerobic Digestion. http://www.epa.gov/OWM/mtb/multi-stage.pdf (accessed July 2011).

U.S. Environmental Protection Agency (2007) *Opportunities and Benefits of Combined Heat and Power at Wastewater Treatment Facilities*; U.S. Environmental Protection Agency Combined Heat and Power Partnership: Washington, D.C.

Vandenburgh, S.; Mundhenke, L.; Strickland, J.; Campbell, K. (2008) Startup of a Thermophilic Digester in 30 Days. *Proceedings of Water Environment Federation Annual Residuals and Biosolids Management Conference* [CD-ROM]; Water Environment Federation: Alexandria, Virginia.

Ward, A. J.; Hobbs, P. J.; Holliman, P. J.; Jones, D. L. (2008) Optimisation of the Anaerobic Digestion of Agricultural Resources. *Bioresour. Technol.*, **99**, 7928–7940.

Water Environment Federation; American Society of Civil Engineers; Environmental and Water Resources Institute (2009) *Design of Municipal Wastewater Treatment Plants*, 5th ed.; WEF Manual of Practice No. 8; ASCE Manuals and Reports on Engineering Practice No. 76; McGraw-Hill: New York.

Water Environment Federation Residuals and Biosolids Committee Bioenergy Technology Subcommittee (2004) *High Performance Anaerobic Digestion*, Sieger, R.; Brady, P.; Donovan, J.; Shea, T.; Parry, D.; Bivins, J.; Buehler, V.; Coccaro, J.; Deshpande, S.; Forbes, R.; Finger, D.; Gillette, R.; Guild, J.; Igoe, J.; Lewis, M.; Richardson, J.; Panter, K.; Roberts, R.; Schafer, P.; Scission, J.; Shang, Y.; Sung, S.; Eds.; Water Environment Federation: Alexandria, Virginia.

Weemaes, M.; Grootaerd, H.; Simoens, F.; Verstraete, W. (2000) Anaerobic Digestion of Ozonized Biosolids. *Water Res.*, **34** (8), 2330–2336.

Werker, A. G.; Carlsson, M.; Morgan-Sagastume, F.; Le, M. S.; Harrison, D. (2007) Full-Scale Demonstration and Assessment of Enzymic Hydrolysis Pre-Treatment for Mesophilic Anaerobic Digestion of Municipal Wastewater Treatment Sludge. *Proceedings of the 80th Annual Water Environment Federation Technical Exposition and Conference* [CD-ROM]; San Diego, California, Oct 15–17; Water Environment Federation: Alexandria, Virginia..

Williams, T. O.; Borghesi, J.; Alexander, R.; Scharp, M.; Walters, L. (2007) Biosolids Markets—Class A or Class B? *Proceedings of the Water Environment Federation/American Water Works Association Joint Residuals and Biosolids Management Conference* [CD-ROM]; Water Environment Federation: Alexandria, Virginia.

Wilson, T.; Kilian, R.; Potts, L. (2008) Update on 2-phase AG Systems. *Proceedings of the Water Environment Federation Residuals and Biosolids Conference* [CD-ROM]; Water Environment Federation: Alexandria, Virginia.

Wilson, T. E.; Streicher, D. (2001) Full-Scale Application of AG Process: Update 2001. *Proceedings of the 74rd Annual Water Environment Federation Technical Exposition and Conference* [CD-ROM]; Atlanta, Georgia, Oct 13–17; Water Environment Federation: Alexandria, Virginia.

Winter, P.; Pearce, P.; Lee, K.; Ganidi, N. (2009) Aspects of Digestibility of Surplus Activated Sludge and Its Potential for Biogas Production. *Proceedings of the 14th European Biosolids and Organic Resources Conference and Exhibition*, Leeds, United Kingdom, Nov 9–11; Aqua Enviro: Wakefield, United Kingdom.

Chapter 11

Aerobic Digestion

(*continued*)

(continued)

1.0 INTRODUCTION TO AEROBIC DIGESTION

Aerobic digestion is a biosolids stabilization process in which biomass and other organic solids are converted into carbon dioxide, water, nitrite/nitrate, nitrogen gas, stabilized organic compounds, and energy in the presence of oxygen. This process extends the decomposition of solids to a point where the amount of volatile solids is sufficiently low to permit the waste sludge to be considered stable for land application.

Aerobic digestion typically is used to stabilize waste activated sludge (WAS), which may be thickened before, during, or after digestion. Aerobic digestion can be used to digest combinations of primary and secondary (biological) sludges, including trickling filter, biotowers, or rotating biological contactor (RBC) sloughings and waste biosolids from membrane bioreactors (MBR). This process is not recommended for digestion of pure primary sludge only because of severe oxygen requirements and additional biomass production.

Aerobic digestion is typically found in plants of less than 19 ML/d (5 mgd) because of its relatively lower capital cost and reduced system complexity compared to larger installations. At larger plants, anaerobic digestion is typically the process of choice because the potential for gas recovery and reduced energy consumption tends to favor anaerobic over aerobic systems. In recent years, however, aerobic digestion has been used in larger wastewater treatment plants (WWTPs), with capabilities up to 190 ML/d (50 mgd) (Metcalf and Eddy, Inc., 2003).

1.1 History of Aerobic Digestion

Studies of aerobic digestion of municipal wastewater sludge have been conducted since the early 1950s (Coackley, 1955; Coackley, 1958). By 1963, at least one major equipment supplier had approximately 130 installations in plants with flow from 0.4 to 0.04 ML/d (10 000 to 100 000 gpd) (Drier, 1963). By the late 1960s and early 1970s, consulting engineers across the United States were specifying aerobic digestion facilities for many of the plants they were designing.

Although the aerobic digestion process is a simple process that adequately stabilizes sludges, the disadvantages of the process led to additional research. One alternative, which was developed in the 1970s, was thermophilic aerobic digestion (Jewell and Kabrick, 1978; Matsch and Drnevich, 1977). Under appropriate operating conditions, this process was identified by the U.S. Environmental Protection Agency (U.S. EPA) as a process capable of producing Class A biosolids. Further research in the early 1990s, in response to the new performance requirements of the rules and regulations for beneficial reuse, led to the development of other alternatives such as aerobic/anoxic digestion and prethickening of solids (U.S. EPA, 1993).

Ongoing research and actual practice have provided additional capabilities to control aerobic digestion processes and to accurately predict the performance of the system. With these added capabilities and combined with the growing importance of controlling internal nutrient recycles in biological nutrient removal facilities, aerobic digestion has developed wider applicability. Two disadvantages of this include high energy requirements for aeration (with possible sustainability implications) and loss of the methane-based energy production from an anaerobic process.

1.2 Advantages and Disadvantages

Historically, two of the primary advantages of aerobic digestion, particularly for smaller plants with smaller operating staffs, have been reduced capital cost and the simplicity of operation. Many times, the equipment necessary for aerobic digestion (blowers or mechanical aerators) is similar to what is being used for secondary treatment, simplifying maintenance. Additional benefits of aerobic digesters are increased volatile solids reduction comparable to anaerobic digesters with shorter retention periods; reduced hazards for cleaning and repairing tasks; and no production of explosive digester gas.

In the past, the disadvantages associated with aerobic digestion included high energy costs, alkalinity depletion, poor pathogen reduction, poor mechanical dewatering characteristics, and poor standard oxygen uptake rates (SOURs). The aerobic digestion process initially was used in designs for new plants that typically treated WAS from treatment systems that did not contain a primary settling, waste activated or trickling filter sludge only, or mixtures of waste activated and trickling filter sludge.

The advantages and disadvantages of aerobic digestion are summarized in Table 11.1.

1.3 Microbiology

The process of aerobic digestion varies depending upon the individual mixtures of sludges and environmental conditions. In many ways, aerobic digestion is an extension of the activated sludge process, and the actual reactions will depend on the detention time in the digesters.

At retention times in the aerobic digestion process of 2 to 15 days, heterotrophic aerobic and facultative microorganisms will predominately use free or combined oxygen as an electron acceptor and obtain energy from the available biodegradable organic matter in the waste sludge. When the available food supply in the waste sludge is inadequate, the microorganisms will begin to lyse and consume their own

TABLE 11.1 Advantages and disadvantages of aerobic digestion (SRT = short solids retention time; ATAD = autothermal thermophilic aerobic digestion).

Advantages	Disadvantages
Lower capital costs for facilities with capacities less than 19 000 m^3/d (5 mgd)	Higher operating costs and other operational issues when treating primary sludge
Minimal nuisance odors (except for SRT ATAD processes)	
Best for waste activated sludge plants only	
Simple construction	Higher energy costs than other stabilization processes, especially traditional mesophilic anaerobic digestion
No dangers of explosions or suffocations	Limited pathogen reduction expect for ATAD
Simple operation	Lower cake solids except for ATAD
Weak sidestreams	Alkalinity depletion, except for short SRT ATAD processes

protoplasm to obtain energy for cell maintenance reactions. The products of aerobic digestion typically include carbon dioxide, water, and recalcitrant materials (i.e., polysaccharides, hemicellulose, and cellulose).

If additional aeration and reactor retention time is provided, then digestion may include nitrification, which oxidizes the remaining organics and includes autotrophic nitrification of ammonium ion to nitrites and nitrates. The result is the production of a mineral acid and a lowering of pH. Cyclic aeration or controlled airflow can promote denitrification, which partially restores the pH. If digestion is extended to more than 20 days, recalcitrant organics will also be hydrolyzed into available substrate (Kelly, 1990).

When multiple stages are used in the digestion process, each stage has the advantage of oxidizing additional organics, stabilizing the mass to greater and greater degrees of acceptability. Incompletely or partially oxidized organics that are odor precursors will eventually be oxidized sufficiently such that the final product can be used without additional blending or effort. The need for added stages should be considered based on cost and the required product use. Desired stability may go beyond the requirements for meeting Class A or B pathogen reduction under the U.S. EPA's Part 503 regulations.

At extended retention times, the digestion process will enter endogenous respiration or auto-oxidation, which is approximately described by the following equations (Metcalf and Eddy, Inc., 2003):

$$C_5H_7O_2N + 5O_2 => 4CO_2 + H_2O + NH_4HCO_3 \qquad 11.1$$

The term $C_5H_7O_2N$ in eq 11.1 is the "simple" stoichiometric formula for biomass or cellular material in an activated sludge system. The cellular material is oxidized to carbon dioxide, water, and ammonia. Approximately 75 to 80% of the cell material can be oxidized; the remainder is composed of inert material and poorly degradable organic compounds. Ammonia is released as shown in this equation, but it then combines with some of the carbon dioxide (CO_2) to produce a form of ammonium bicarbonate (NH_4HCO_3). The main observation from this reaction is that aerobic digestion produces both ammonia and alkalinity in the presence of oxygen in the initial phase.

If sufficient additional oxygen is provided, then the second reaction will occur as shown in eq 11.2. The nitrification of the released 1 mol of ammonia creates nitrate and 2 mol of acidity. The 2 mol of acidity are shown in the form of hydrogen (H^+), which can also be described as a loss of 2 mol of alkalinity.

$$NH_4^+ + 2O_2 => NO_3^- + 2H^+ + H_2O \qquad 11.2$$

Combining these two equations yields

$$C_5H_7O_2N + 7O_2 \Rightarrow 5CO_2 + 3H_2O + HNO_3 \qquad 11.3$$

Thus, in the first reaction, although the destruction of biomass produces 1 mol of alkalinity as ammonium bicarbonate, nitrification destroys 2 mol of alkalinity. Depending on the amount of alkalinity in the feed solids, it is easy to destroy the alkalinity and see a dramatic decrease in pH, to pH 5 or lower, which can adversely affect further digestion. In this case, the lower pH will also result in inhibition of the nitrifying organisms, resulting in only partial nitrification, as shown in eq 11.4:

$$2C_5H_7O_2N + 12O_2 \Rightarrow 10CO_2 + 5H_2O + NH_4^+ + NO_3^- \qquad 11.4$$

Aerobic digestion can recover the lost alkalinity biologically, through denitrification. In this case, the dissolved oxygen level in the digester is lowered to less than 1 mg/L, while maintaining digester mixing. The oxidation-reduction potential can be measured using standard redox probes (Wareham et al., 1993). The biomass then begins to use nitrate nitrogen as the electron acceptor, as shown in the following equation:

$$C_5H_7O_2N + 4NO_3^- + H_2O \Rightarrow NH_4^+ + 5HCO_3^- + 2N_2 \qquad 11.5$$

This equation assumes that adequate organic matter is available to sustain the denitrification reaction. If not, then an additional carbon source may be required.

These equations can then be combined to form

$$2C_5H_7O_2N + 11.5O_2 \Rightarrow 10CO_2 + N_2 + 7H_2O \qquad 11.6$$

The nitrification/denitrification cycle reduces the mass of nitrogen in the digested biosolids, which is important to land application and will be explored later in this chapter. The cycle also reduces the energy required for digestion. Denitrification conserves oxygen and can reduce aeration power requirements by up to 18% and recover up to 50% of the lost alkalinity, allowing for pH control without chemical addition.

2.0 CONVENTIONAL AEROBIC DIGESTION

2.1 Process Description

As noted above, in many ways, aerobic digestion was developed as an extension of the activated sludge process. This led to the initial development of aerobic digesters, both in semibatch and continuous-flow variants, operating in the range of 10° to 30 °C.

Originally, aerobic digestion was designed as a semibatch process, and this concept is still used in many facilities. Solids are pumped directly from the clarifiers into the aerobic digester. The time required for filling the digester depends on available tank volume, volume of waste sludge, precipitation, and evaporation. During the filling operation, sludge undergoing digestion is continually aerated. When the tank is full, aeration continues until the solids have been retained long enough to ensure that the solids are thoroughly stabilized. Aeration is then discontinued, and the stabilized solids are settled. Clarified liquid is decanted, and the thickened solids are removed at a concentration of between 2 and 4%. When a sufficient amount of stabilized sludge or supernatant has been removed, the cycle is repeated. Between cycles, it is customary to leave some stabilized sludge in the digester to provide the necessary microbial population for degrading the incoming wastewater solids. The aeration device does not need to operate for several days, provided no raw sludge is added (Folk, 1976).

A variant of the semibatch process has been developed. In this system, which the U.S. EPA (2003) refers to as "intermittent "operation, sludge is fed daily or at more frequent intervals to the digester. The residuals are aerated. Periodically, the sludge is allowed to thicken, typically by gravity in the tank, and the supernatant is removed. This cycle continues until the digester is full, or the necessary retention period has been met.

Many engineers have tried to make the semibatch process more continuous by installing stilling wells to act as clarifiers in part of the digester. This has not proven particularly effective (Folk, 1976; Paredes, 1976; Ritter, 1970).

The conventional, continuous aerobic digestion process closely resembles the completely mixed activated sludge process as shown in Figure 11.1. Solids are pumped directly from clarifiers into the aerobic digester. The aerator operates at a

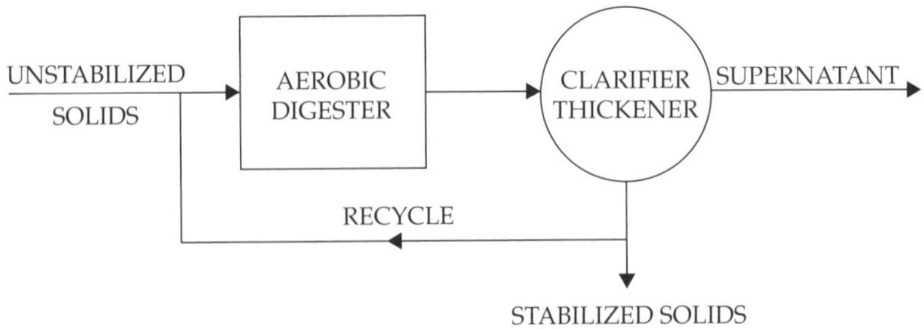

FIGURE 11.1 Process flow diagram for a conventional continuously operated aerobic digester (U.S. EPA, 1979).

fixed level, with the overflow going to a solids-liquid separator. Thickened and stabilized solids are either recycled back to the digestion tank or removed for further processing. Other reactor configurations include "true" batch systems using reactor tanks in parallel, and semibatch, continuous, and continuous-batch arrangements using two or more reactor tanks in series for each parallel train of digesters. Each condition provides advantages and disadvantages depending on the equipment used, facility requirements, and facility size. Figures 11.2a and b illustrate several process configurations currently in use.

In addition to conventional mesophilic aerobic digestion, several variations of the aerobic digestion process exist: (1) high-purity oxygen, (2) thermophilic, including micro-aerophilic systems, and (3) cryophilic aerobic digestion. Mesophilic conventional aerobic digestion is the most typically used process. As a result, the majority of this chapter addresses this process, including several modifications.

2.2 Process Design, Equipment, and Selection

2.2.1 *Types of Reactors*

Aerobic digesters come in many varying types and sizes. The design of a reactor should be based on local conditions and preferences wherever possible and should always consider reliability. Because of this, all but the smallest facilities should include a minimum of two aerobic digester basins. In some facilities, these can even be a single tank divided in half by a hydrostatic wall.

A variety of materials have been used to construct aerobic digesters. Concrete offers both corrosion resistance and flexibility in design. Concrete basins may be square, rectangular, or circular, and can be constructed to any depth. To maximize aeration efficiency, depths should be maintained at 6 m (20 ft) or greater whenever possible. Steel tanks can also be used, though care is required in the selection of interior and exterior coatings. Steel tanks typically are round, although other shapes are possible. Steel tanks can be welded or bolted. Various coatings are feasible for both concrete and steel.

2.2.2 *Single-Stage versus Multistage Operation*

Traditionally, aerobic digester design has used one or more digesters operated in parallel, typically in a semibatch fill/decant/draw operation. This is because of operational characteristics of both upstream and downstream processes. Many plants, especially smaller facilities, operate downstream process on a periodic basis. This may range from filling sludge drying beds on a daily or weekly basis,

2.3.5 *Jet Aeration*

Jet aeration devices provide many of the benefits of submerged mechanical aerators, but offer several other benefits. Venturi-pump design may provide sufficient aeration in smaller systems, and mixing, or multiple units may be used in larger facilities. The Venturi-pump design is a coupled pump and Venturi nozzle that provides recirculation of the tank contents and educts air through the reduced section of the Venturi to provide aeration and mixing of the tank contents. The pump can be provided with a variable speed control to adjust for temperature and oxidation requirements of the reactor.

Larger jet aeration systems use mixing pumps, discharging through a submerged discharge header with a series of nozzles to mix the tanks and inject air either from a Venturi eductor mechanism or separate process blowers. These nozzles can be arranged either lengthwise along a header or radially around a central hub. In either case, an upper line conveys air, and a larger header conveys the liquid (see Figure 11.4). The velocity of the liquid creates a Venturi effect that helps to draw air into the pumped liquid. Depending on oxygen requirements and tank depth, the Venturi effect may be sufficient so that a supporting blower is not required. More often, however, there is a blower that provides air. A mixing pump is used to provide the motive force for pumping the liquid. This combination offers the flexibility to mix with or without air. These devices typically are more easily installed than submerged turbines and have a high overall oxygen-transfer efficiency. Problems with plugging of the devices have occurred in the past where liquid flow paths have not been large enough to pass stringy solids typically found in aerobic digesters. Their use may promote floc shear and subsequent dewatering difficulties.

When compared to other diffused aeration technologies, jet aeration systems yield a higher alpha value. Based on typical design values, the alpha factor for jet aeration is

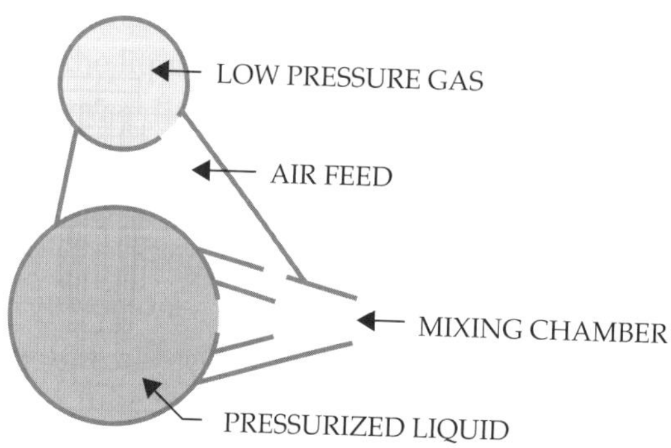

FIGURE 11.4 Cross-section of a jet-aeration nozzle.

approximately 28% higher than for coarse-bubble aerators. This high alpha value is because of the intense mixing provided by the high-velocity, turbulent jet plume with entrained air. Because of the high oxygen-transfer efficiency and mixing capabilities, jet aeration is often used in autothermal thermophilic aerobic digestion (ATAD) systems. This is an excellent application of the technology, with highly efficient oxygen transfer and mixing under very high (> 3.0 %) solids concentrations and reactor temperature (> 55 °C).

2.3.6 *Blowers*

Diffused aeration, jet aeration, and submerged turbine systems all require blowers to provide low-pressure compressed air. Sometimes air will be provided by the primary process blowers for the secondary treatment system. This is more common in smaller facilities (typically less than 0.2 m^3/s flow rate). The advantage of such a system is that the number of operating units is reduced, and maintenance costs may be lower. The disadvantage is the potential for differing operating conditions between the primary process basins, which typically are operated at a constant level, and the digesters, which may have variable levels. Therefore, the aerobic digesters will often be provided with separate, dedicated blowers. This allows both the secondary treatment aeration system and the digester aeration system to be optimized for the design conditions, reducing power costs. The capital cost may be higher, but the control issues are simplified.

In selecting a blower for aerobic digesters, the designer has the same choices available as for activated sludge system. These include centrifugal blowers, regenerative blowers, positive displacement blowers, and inlet guide vane/variable diffuser blowers. Each has advantages and disadvantages, and the designer needs to carefully consider the proposed operational conditions for the digester when selecting blower type. For example, centrifugal blowers operate much like centrifugal pumps, with an operating curve that results in a decrease in airflow as pressure increases. Centrifugal blowers, however, are typically more efficient than positive displacement units. Therefore, if the digester will be operated at varying levels, then varying back-pressure will result in variations in airflow unless control valves or variable-speed drives are used. In this case, a positive displacement blower may be more suitable. If the digesters will be operated at a relatively constant level, than the greater efficiency of a centrifugal blower would be more appropriate.

2.3.7 *Pumps*

The use of pumps for transferring sludge to and from aerobic digestion is discussed in Chapter 6. Pumps are not typically used within the actual digestion process unless jet aeration is used. In this case, it is important to consult with the manufacturer to determine the appropriate pump for the proposed application.

Several factors should be considered. First, the selected pump must be capable of meeting the flow and head requirements of the jet aeration system. Typically, these systems are a high-flow/low-head application, which require good solids-handling capabilities. Another concern is the temperature of the liquid being pumped. Many submersible pumps rely on the pumped liquid to provide motor cooling. In thermophilic digesters, the higher temperatures may adversely affect the ability to cool the motor. For these higher temperatures, it may be appropriate to use either conventional dry-pit pumps or submersible pumps that can be water-cooled from an external cooling source.

2.3.8 *Aeration and Mixing Requirements*

In most conventional aerobic digesters, if sufficient air is provided, then complete nitrification will occur. Hypothetically, eq 11.3 indicates that 1.98 kg of oxygen is required to oxidize 1 kg of cell mass. From pilot and full-scale studies, however, the kilogram of oxygen required to degrade one kilogram of volatile solids were found to be 1.74 to 2.07. For mesophilic systems, a design value of 2.0 kg of oxygen per kilogram of volatile solids is recommended.

In most aerobic digesters, the aeration equipment provides sufficient energy to sufficiently mix the digester. Typically, aeration rates of 1.2 to 2.4 m^3 air/m^3 volume per hour are provided to maintain mixing. When fine-bubble aeration or pure oxygen is used, however, the gas flow required to meet the oxygen demand in the digester may be insufficient to meet mixing requirements. Similarly, when sludge is thickened, airflow rates may be insufficient to maintain the solids in suspension. In some cases, therefore, supplemental mixing equipment may be provided. This has the added benefit of allowing anoxic/aerobic operation.

Mixing is required in an aerobic digester to keep solids in suspension and to bring deoxygenated liquid continuously to the aeration device. Whichever of these two requirements needs the most mixing energy controls the design. Typically, levels ranging from 20 to 40 kW/10^3 m^3 of tank volume are satisfactory. Designers should consult an experienced aeration equipment manufacturer for assistance in design.

2.4 Biosolids Dewatering Characteristics

Typically, aerobically digested biosolids do not dewater as well as anaerobically digested biosolids. Although reports of excellent operating systems have been published (Reynolds, 1967), other facilities have reported that mechanical dewatering of aerobically digested biosolids is much more difficult (Burton and Malina, 1964; Ganczarczyk and Hamoda, 1973). Several factors appear to cause this difficulty. For example, several investigations have found that the dewatering properties of aerobically digested sludge deteriorate with increasing sludge age (Bisogni and Laurence, 1971; Ganczarczyk and Hamoda, 1973; U.S. EPA, 1977). The makeup of the sludge is also important, with the concentration of divalent cations improving dewaterability (Murthy and Novak, 1999).

3.0 AUTOTHERMAL THERMOPHILIC AEROBIC DIGESTION

The history of the ATAD process is fully explored in a summary of research by Layden et al. (2007a) from its beginnings in the early 1960s to present day developments occurring largely in North America and Europe. Unlike conventional aerobic digestion, under certain operating conditions, U.S. EPA identifies ATAD systems as a PFRP that produces a final product that is low in pathogens. In this process, sludge is thickened to provide a digester feed solids concentration of between 4 and 8%. The heat liberated by the biological degradation of the organic solids and the energy added from mixing and aeration are sufficient to raise the liquid temperature in the digester. Heat recovery may be needed where higher latitudes keep sludge feed temperatures below 20 °C. Cooling is needed where dewatering or immediate use is required and where storage is not provided. Typically, the reactor tanks in the digesters are operated between 45 and 65 °C, although higher and lower temperatures are used. Table 11.2 summarizes the typical operating characteristics for these digesters.

Table 11.3 provides a list of advantages and disadvantages for ATAD (Jewell and Kabrick, 1978; Kelly, 2006; Kelly et al., 2009; Layden et al., 2007b; Matsch and Drnevich, 1977; Surucu et al., 1976).

Additional tankage to extend the hydraulic retention time (HRT) to more than 20 days provides the capacity for cooling, nitrification/denitrification, increased stability of the product, and opportunities for oxidizing incompletely oxidized organics (Kelly, 1990). Where space is not a constraint, extended HRT offers a potential improvement in stabilization.

3.1 Typical Design Considerations

Autothermal thermophilic aerobic digestion systems are available as design-build installations from at least two manufacturers in the United States. In addition, engineers have designed several ATAD systems (Kelly and Warren, 1997; Layden et al., 2007a and b).

TABLE 11.2 Typical operating parameters for autothermal thermophilic aerobic digestion reactor tanks.

Parameter	Value
Influent solids	40 to 80 g/L (4–8%)[a]
Volatile solids	≥ 30 to 70 g/L (3–7%)[b]
Hydraulic retention time (days)	10 to 15 days[c] and longer
Air supply (m^3/h air/m^3 digester volume)	0.5 to 4[d]
Oxidation reduction potential (mV)	−400 to +100 through reactors
Power density (W/m^3)	100 to 250[e]

[a] Solids concentrations less than 4% result in excessive energy needs for heating water and should be avoided; concentrations greater than 10 to 12% often exceed the mixing energy required to keep the solids in suspension; an exception is for the addition of fats, oils, and grease where the concentrations of solids can be much higher. Conversely, when adding high soluble organics, influent solids can be initially low as long as sufficient biomass is present for digestion.

[b] Volatile solids typically range from 70 to 85% of total solids.

[c] Short hydraulic retention times (HRT) of less than 10 days can only be sustained for short periods unless high-energy organics or high power densities are used; HRT and sludge retention time (SRT) is assumed to be equivalent.

[d] Air supply for micro-aerophilic and fermentation need not exceed 0.5 v/v·h (expressed as volume of air/hour per volume of tank); normal rates should vary from 0.5 to 1.5 v/ v·h and will depend on feed rates. Higher rates will result in heat loss through sensible cooling and requires added power density to compensate (Layden, 2007a and b; Kelly, 2009).

[e] Some facilities use more than 500 W/m^3 power densities and some reported as low as 80 W/m^3; higher values are typically used for startup or extreme conditions. Variable-speed operated equipment is often used to adjust for varying energy requirements.

The majority of installations in the United States have been based on the Fuchs process from Germany. Typical features of these systems include prethickening, a sludge holding tank before the ATAD process (where batch or semibatch systems are used), heat exchangers for preheating feed sludge, two or more enclosed and insulated reactors configured in series (unless batch operated), mixing/aeration and foam control equipment, and a final post-ATAD, or storage/post-thickening tank. Heat recovery may also be included in the final tank or from the product before storage or dewatering and final use.

3.1.1 *Types of Reactors*

Like most digestion processes, ATAD systems can exist as either single- or multiple-stage processes. Single-stage systems can achieve similar volatile solids reduction to

Table 11.3 Autothermal thermophilic aerobic digestion advantages and disadvantages.

Advantages	Disadvantages
• Meets Class A requirements for biosolids and produces a pasteurized sludge. • Rates of organic solids destruction are high, hence, smaller digester volume is needed. • Reactor vessels are fully enclosed to simplify odor control. • Requires no pretreatment of residuals feed. • Biosolids may be totally contained until they are pasteurized and stabilized. • No open tankage required. • Mechanical systems are simple and process is easy to operate, startup and shutdown. • Energy needs are less than other aerobic treatment systems. • Depending on design, product may dewater to 25%+ dry solids on belt filter press and 30%+ dry solids on a centrifuge. • Process may reuse existing digesters to save capital costs. • Process may be designed to specific plant needs and avoid proprietary system package systems. • Process has an excellent track record where proven equipment is used. • No boiler or gas handling combustion steps are needed. • 30 to 40% less oxygen requirement than mesophilic because few, if any, nitrifying bacteria exist in the temperature range. • First stages of the process can be used to recover volatile fatty acids for biological nutrient removal and struvite for phosphorus reuse (Kelly, et al, 2009).	• Product is odorous, and system requires complete emissions control and treatment. • Biosolids must be thickened to 5% dry solids before feed. • Product may need to be dewatered; polymer requirements are 2 to 3 times more per tonne than needed for mesophilically digested biosolids. • Product requires cooling to reduce odorous off-gassing and to reduce polymer needs for dewatering. • Sidestream may require treatment. • Foam control is necessary. • Some processes are proprietary. • Some system suppliers/designers are inexperienced, and inadequate systems have been installed. • Requires attention to erosion/corrosion control.

multiple-stage systems, but because of the possibility of short-circuiting are not considered a PRFP process. Unless batch operated, single-stage configurations have fewer disposal options and are limited to landfilling or restricted-use land application that does not require enhanced pathogen reduction. Alternatively, they can be operated to provide organic matter to a Class A compost operation. The multiple-stage or batch process is required by the U.S. EPA for systems where the sludge must meet Class A standards. Process configurations shown in Figures 11.2a, b, and d would meet U.S. EPA requirements. Configuration Figure 11.2c can meet acceptable level of risk for pathogen kill equivalent to U.S. EPA requirements by ensuring that multiple reactors in series are used to limit the probability of short-circuiting per Schafer (1994). This configuration, however, is not included in U.S. EPA's definition of a PFRP.

Typically, in a two-stage process, each of the tanks will be designed with an equal volume. In a batch mode as illustrated in Figure 11.2a, sufficient storage must be available before the reactors to ensure the batch process meets the mean cell residence time (MCRT) for Class A treatment. For example, if an MCRT of 10 days is required and only one batch reactor tank is in service, up to 10 days of storage must be provided. Conversely, if 10 batch reactor tanks are operated in parallel, one day of storage would be needed. In the Fuchs semibatch mode as illustrated in Figure 11.2b, a portion of sludge is withdrawn from the second or final stage and discharged to the holding tank. Sludge is then transferred from the first or prior stage to the second or last stage. Upon completion of this transfer, sludge is introduced into the first-stage digester. During the transfer, which takes approximately 1 hour, aeration equipment is stopped. This semibatch process minimizes the potential for short-circuiting of pathogens but requires 23 hours of storage of thickened feed sludge before each day of feeding. In a combination mode as illustrated Figure 11.2d, batch operation is maintained in the second or central reactor to meet PFRP requirements. Tank 1 operates in a continuous fill mode with varying volume and periodic withdrawals to tank 2; tank 3 can operate in a continuous draw mode with varying volume and periodic filling from tank 2. Transfers occur daily with recycle pumps or dedicated pumps to meet the time-temperature criteria for pathogen destruction. No additional tanks are needed for upstream storage. The simplest process configuration is illustrated in Figure 11.2c. This configuration allows continuous displacement of the tank contents. The final reactor tank of a multitank system could be used as a fill-and-draw tank where a continuous discharge is not needed. The use of this configuration is not defined as a Class A process.

Tanks may be either round or rectangular and either concrete or steel. They should be insulated to conserve heat. Original design values used a height/diameter ratio of 0.5

to 1.0 (U.S. EPA, 1990). This design was for tanks of 3 to < 5 m (10 to 16 ft) deep, because the vortex mixing and aeration systems used in early systems required a relatively shallow submergence for the vortex to educt the air. With the addition of recirculation pumps and jet aeration systems, this height/diameter ratio is not as critical because aeration and mixing can be designed for any tank size. In either case, the tanks must be covered and insulated, both for heat retention and odor control. Typically, the first stage will be operated at a temperature of approximately 50 ± 5 °C, and a pH of 7.2 or greater. The second stage will be operated at a temperature of 65 ± 5 °C, with a pH of 8.0. Operating temperatures in the reactors will vary depending on the operating strategy to be achieved.

3.1.2 Aeration Equipment

Multiple types of aeration devices are available for ATAD systems. The need for a high level of aeration efficiency reduces the choices available. Systems discussed previously that are suitable for ATAD systems include fine-bubble diffused aeration and jet aeration. For diffused aeration equipment, air requirements will depend on the efficiency of the fine-bubble aeration equipment but may be assumed to be 50 to 70%. For an 8-day HRT and a 43% COD reduction, and assuming a COD-to-VS ratio of 1.86/1, the volumetric airflow rate required is calculated at approximately 0.6 to 1.0 v/v·h. A higher airflow rate can be used but will increase heat loss, especially where the reactor operates at thermophilic temperatures because most of the heat will be lost through latent heat of evaporation. Course-bubble aeration should not be used for this reason. Oxygen is assumed to be 23.2% of air by weight and air density is 1.3 kg/m^3.

In addition to the previously discussed systems, several proprietary aeration systems are available that have been successful in ATAD systems. The most common systems are mechanical Venturi educator aerators, which develop an induced vortex action in the liquid and draw air from outside the ATAD vessel. The Venturi-pump mixer provides mixing and aeration by using recirculation pumps and Venturi educator aerators. These are now available from several manufacturers as either submersible or dry installed systems. Other manufacturers and designers use Venturi-pump jet aeration systems and Sinkair aerators (see Figure 11.5).

The Copa Sinkair aerator (Ovivo, Montreal, Canada) recirculates liquid by drawing liquid from the surface of the tank and discharging it vertically into the reactor. Air is added by Venturi effect. The Sinkair-aerator mixer has the advantage of included foam control but may be limited in depth for mixing and effective aeration (see Figure 11.6).

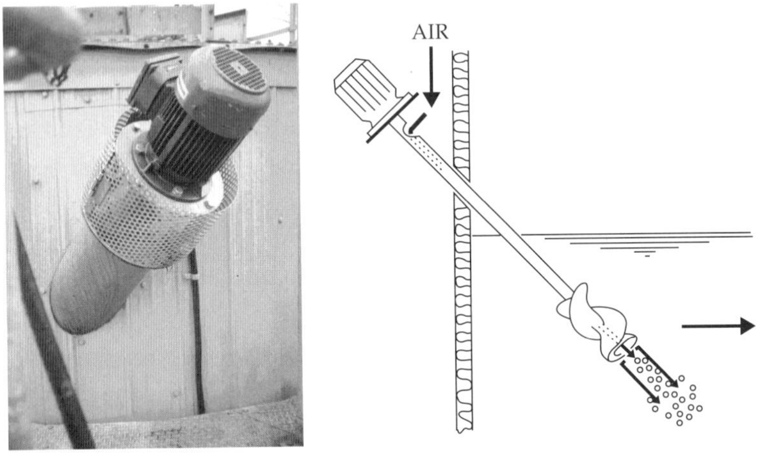

FIGURE 11.5 Fuchs self-aspirating aerator.

FIGURE 11.6 Sinkair aerator/mixer.

Mixing energy is needed to provide a maximum interaction between the thermophilic organisms and the substrate and to provide supplemental heat. A minimum shear gradient of 450 sec^{-1} for complete mix requires a power density of at least 100 W/m^3 of reactor volume at 50 to 60 °C.

The ATAD systems can generate large amounts of foam. Therefore, it is important to provide adequate foam control mechanisms in these systems. Several defense systems against excessive foam should be considered and are discussed by Kelly and Warren (1997) and Kelly and Wong (1997a, 1997b). The design of the reactor should include freeboard of approximately 1 m to manage foam levels of 100 to 400 mm. The tank's upper deck should be designed to contain foam, and, if a foam event occurs, then it should be allowed to discharge to a separate sump for retreatment. A foam-sensing device should be added to each tank to provide a signal to either slow or shutdown mixing and aeration equipment during a foam event. Foam suppressants can be used; typically, however, mechanical and hydraulic systems are used to reduce the foam or re-entrain it into the liquid.

3.1.3 *Use of High-Purity Oxygen*

One of the keys to maintaining the efficiency of the ATAD process is to minimize any heat losses. This is the reason that high-efficiency aeration is typically used in ATAD systems. As an alternative, HPO has also been used in ATAD systems (Matsch and Drnevich, 1977). In HPO systems, a lesser volume of gas flow is required to maintain the aerobic conditions because the concentration of oxygen is much higher than typical atmospheric concentrations. This translates into less stripping of heat, and a reduction in the volume of exhaust to be treated.

3.1.4 *Odor Control*

Odor control is a significant issue for ATAD facilities. This odor typically is characterized as a sharp, pungent ammonia or humus-like. Each stage of oxidation will exhibit different odor characteristics. Because of the high temperatures in the tanks and low HRT, nitrification is suppressed in an ATAD system, and ammonia is released during digestion and cannot be avoided. At the elevated pH typical in an ATAD reactor, ammonia will be stripped into the exhaust. Although many early facilities did not include facilities for odor control, design practice now calls for odor control on these facilities (Kelly, 2006; Kelly and Mavinic, 2003; U.S. EPA, 1990; Vesilind, 2003). Typically, these control systems treat the off-gas and will act as biostrippers; gas stream oxidizers that use UV, ozone, or ionization; wet and dry gas scrubbers; or biofilter systems. In some cases, recuperative thermal oxidizers have been used. Using a system with a low operating cost, such as a biofilter, before a wet or dry chemical scrubber can minimize operating costs and improve overall removal efficiencies. Treatment can also include the addition of oxidants to the liquid, such as iron salts or ferrite. If additional space is available, then extending the retention time as outlined below can help in reducing odors.

3.2 Sludge Feed Source and Thickening Requirements

As noted above, ATAD systems depend on using thickened sludge to achieve a sufficient energy density to be self-heating. Typically, sludge that has been thickened to 4 to 8% total suspended solids (TSS) with at least 2.5% mostly biodegradable volatile solids is sufficient to achieve autoheating with the addition of mixing energy of between 15 to 35 kWh/m^3 of treated sludge.

In Figure 11.2, configurations a and b would require a post-thickening tank to hold sludge between feed cycles to the ATAD system. Configuration of Figure 11.2b would require sufficient time for digestion and that all sludge feeding to the digester occurs within 0.5 to 1 hour. Therefore, a basin is required that can hold at least 23 hours worth of thickened sludge under normal operational conditions. This tank may be aerated. When aerated, it can be incorporated into the main ATAD flow train as shown in configurations shown in Figures 11.2c and d, which precludes the extra tanks.

3.3 Post Cooling and Holding Requirements

Digested biosolids exiting the second-stage digester typically will be greater than 60 °C. By cooling the biosolids to less than 35 °C (95 °F), chemical requirements for dewatering typically are lower and more controllable. Furthermore, vapor and odor emissions are significantly less, resulting in improved environmental conditions for operations. In addition to allowing cooling in the tank, it may be beneficial to use a heat exchanger to help further cool the sludge. Spiral heat exchangers can be used in a digested biosolids/water cooling loop to assist in cooling and water/sludge heating. Tube-in-tube direct sludge/biosolids or tube-in-vessel heat exchangers have also been used successfully (Kelly, 1996). Direct sludge/biosolids spiral heat exchangers should be avoided.

Holding may also be required to allow proper operation. In configuration Figure 11.2b, the semibatch operational mode, digested biosolids are withdrawn from the digester in 0.5 to 1 hour. Therefore, a holding tank is necessary to equalize this flow before dewatering. In configurations c and d of Figure 11.2, the final reactor is the holding tank and withdrawal of the digested sludge is undertaken to match the rate of the dewatering systems. The final tank can be fitted with piping for use as a heat exchanger to cool the final tank contents and add heat to the first tank.

suggested as the factors in the increased chemical requirements. In addition, SNDR provides an additional 10 to 15% total solids reduction (Scisson, 2006, 2009).

Figure 11.7 shows the soluble COD and ammonia concentrations across the ATAD system at Bowling Green, Ohio, which has ATAD reactors (shown as tank 1 and tank 2), SNDR (tank 3), and additional aerobic storage (tank 4). Data from this facility demonstrate a 65% reduction in soluble COD and ammonia across the SNDR system. The polymer dose for a high-solids centrifuge producing 35% cake solids is approximately 3.1 kg/dry ton active polymer with no metal salt addition (Scisson, 2009).

4.0 CRYOPHILIC AEROBIC DIGESTION

Cryophilic digestion, meaning aerobic digester operation at temperatures less than 15 °C (59 °F), may occur in WWTPs located in colder climates. Experience shows that bacterial metabolic rate is strongly correlated with temperature, and that digestion will be impaired as the temperature decreases (Mavinic and Koers, 1977, 1979, 1981). At 10 °C (50 °F), a minimum detention time of 55 to 60 days is necessary to obtain adequate (38 to 40%) volatile solids reduction. At temperatures of 5 °C (41 °F), even 80 days of detention time does not result in a stable sludge. Research for the U.S. Army indicated that at low temperatures, many of the basic kinetic models for predicting aerobic digester performance were inaccurate because of the changing nature of the

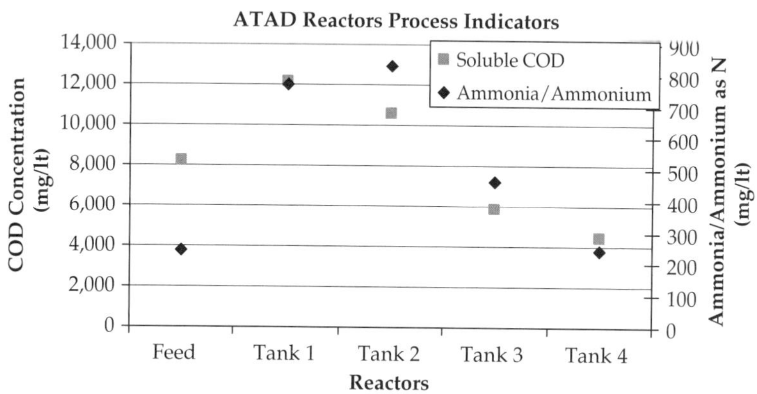

FIGURE 11.7 Soluble chemical oxygen demand (COD) and ammonia reduction through simultaneous nitrification/denitrification reactor and storage tanks, Bowling Green Wastewater Treatment Plant, Bowling Green, Ohio (Scisson, 2009).

microbial community (Schneiter et al., 1984). Under U.S. EPA guidelines, at temperatures below 10 °C (50 °F), it is not possible to meet Class B pathogen reduction requirements by the accepted time-and-temperature method. In these cases, WWTP staff will have to use alternate means to demonstrate adequate pathogen reduction. Cryophilic digestion is an extension of the conventional mesophilic aerobic digestion process that involves lower operating temperatures seasonally. The change in in situ microbial community, however, demands a more conservative design approach.

5.0 OTHER COMBINED PROCESSES

Many efforts have been made to augment processes by combining aerobic digestion system with other processes. Several of these combined processes are described below.

5.1 Cannibal Process

The Cannibal solids reduction process (Siemens, Warrendale, Pennsylvania) is a proprietary technology that incorporates extremely fine screening and degritting of the sludge, aerobic/anaerobic cycling, and return of biomass to the activated sludge treatment process. The fine screening removes larger nondegradable materials, and the degritting removes fine inert material that is nondigestible. The anaerobic/aerobic cycling and biomass return create a condition that allows the metabolic products of aerobic respiration to be anaerobically metabolized and vice versa. The result is a process that claims to reduce overall biomass yield to 0.3 kg biosolids/kg influent biochemical oxygen demand (BOD). The Cannibal process is covered in Chapter 9 on waste minimization.

5.2 Aerobic Digestion and Thermal Drying

Typically, dryers provide sufficient treatment to meet U.S. EPA's regulations for Class A solids without previous stabilization of the sludge. When sludge is not stabilized before drying, however, the dry material has the potential for odors if it is wetted. Stabilization of the sludge before drying can help to prevent odor issues in the field. Aerobically digested biosolids and WAS are dried with indirect or direct dryers at several locations in the United States.

5.3 Aerobic Thermophilic Pretreatment

Aerobic thermophilic pretreatment is an ATAD that is often aerated with pure oxygen, has an SRT of one to two days, and is followed by anaerobic digestion. This technology will be discussed in the chapter on anaerobic digestion.

5.4 Anaerobic/Aerobic Digestion (Theoretical)

Studies suggest that some solids in sludge are degraded only during anaerobic digestion and some only during aerobic digestion (Higgins and Novak, 1997; Novak et al., 2003). Therefore, dual digestion (anaerobic/aerobic) would be expected to provide volatile solids reduction beyond that achieved by either anaerobic or aerobic digestion alone. This is one of the operating principles also embodied in the Cannibal system.

As outlined in the anaerobic digestion chapter, there have been several installations that include an aerobic digestion step before the anaerobic digester. The reverse system, anaerobic pretreatment followed by aerobic digestion, is less common but has been studied by several researchers (Kumar et al, 2006; Ros and Zupančič, 2004; Subramanian, 2005; Tapana and Pagilla, 2000). These systems may be either mesophilic or thermophilic; the latter has been more widely documented. These studies have found that anaerobic/aerobic combination systems provide better volatile solids destruction and fecal coliform reduction and comparatively better dewatering properties compared to conventional mesophilic anaerobic digestion at comparable detention times.

Research by Kumar et al. (2006) compared dual anaerobic/aerobic digestion to single-stage anaerobic digestion using both mesophilic (35 °C) and thermophilic (55 °C) digestion. The results showed that dual digestion was able to achieve volatile solids reductions up to 65%. Aerobic digestion times of three, six, and nine days were studied, and increased volatile solids reduction was observed for all aerobic detention times. At aerobic digestion detention times greater than three days, the study also showed that more than 50% of total nitrogen and 80% of ammonia was removed from the anaerobic digester effluent.

Several studies of thermophilic anaerobic/aerobic digestion have been conducted (Ros and Zupančič, 2004). The goal of this work was to combine the benefits of anaerobic digestion, such as gas production, with the enhanced COD and volatile solids reduction of the aerobic digestion process. The study examined several combinations of anaerobic and aerobic digestion, ranging from three days anaerobic plus three days aerobic to 10 days in each zone. The best results occurred with a combination of 3 days anaerobic digestion plus 12 days aerobic digestion. This combination showed a volatile suspended solids (VSS) removal of 61.8% and COD removal of 57.4% in 15 days HRT. Testing in this study occurred in the thermophilic temperature range for both processes. Testing indicated that relatively little additional energy was required to maintain both stages in a thermophilic system compared to thermophilic anaerobic/mesophilic aerobic systems. The study found that any further stage after a first thermophilic stage requires only 2 to 3% additional heat.

6.0 PROCESS INSTRUMENTATION AND CONTROLS

Operational controls for aerobic digestion can vary from the simple to complex. As noted previously, time and temperature are key parameters to determine whether the process complies with Class B criteria. Therefore, it is important to be able to accurately determine solids retention time. At a minimum, the flow of biosolids from the final digester stage should be measured. This can be used to calculate digester detention time by dividing the digester volume by the discharge flow. Because of the solids concentration in sludge, magnetic flow meters will provide the most accurate flow data. To better quantify SRT, additional flow meters and solids concentration meters may be desirable. If the concentration of solids leaving the digester and the flow rate is known, then mass flow rate can be calculated and the actual solids detention time determined.

In addition to calculating detention time, it is necessary to measure temperature. If sludge is circulated, as in jet mixing, then a thermal well can be provided outside the digester for measurement of sludge temperature. Although a thermal well can be provided through the wall of the tank, it will not provide temperature readings across the entire tank. Otherwise, it may be necessary to manually measure the temperature at multiple points.

To ensure that adequate oxygen is available for the digestion process, it is also important to measure the dissolved oxygen in the digester. In smaller facilities, this can consist of periodic measurements with handheld devices. Other facilities may desire online dissolved oxygen monitoring and may use the measured value to control the operation of aeration equipment. Although measurement of dissolved oxygen can provide important information to the operator, dissolved oxygen probes can be maintenance-intensive, especially in high-solids environments.

As an alternative, standard oxidation/reduction potential (ORP) probes can be used. These probes can be used to control aeration equipment and to cycle the equipment to obtain anoxic and aerobic periods in the digester.

7.0 PROCESS PERFORMANCE AND CONTROL

7.1 Rules and Regulations

Rules and regulations governing disposal and reuse of aerobically digested biosolids vary from state to state within the United States and sometimes even within local jurisdictions. Differing regulations will apply in other countries (see Chapter 2).

Conventional aerobic digesters are classified as meeting Class B requirements. Under U.S. EPA regulations, aerobic digestion is defined as agitation of sludge with air or oxygen to maintain aerobic conditions for a specific MCRT (i.e., SRT) at a specific temperature. Values for the MCRT and temperature shall be between 40 days at 20 °C (68 °F) and 60 days at 15 °C (59 °F). Even for alternative systems, such as multistage thickened aerobic digesters, the time-temperature requirements must still be met to demonstrate Class B compliance. As noted previously, U.S. EPA guidance documents (2003) do allow for a reduction in residence time to 28 days at 20 °C for multistage or true batch systems. No provision exists, however, for a mesophilic aerobic digester to meet Class A requirements. Typically, if Class A residuals are required from a mesophilic aerobic digester, some type of ancillary process, such as sludge drying, will be required. On the other hand, under specified operating conditions, ATAD systems are defined by the U.S. EPA as a process capable of meeting Class A pathogen reduction requirements.

In reviewing U.S. EPA's vector attraction reduction (VAR) criteria, there are no criteria governing aerobic digestion directly. Typically, many facilities, especially those with primary sludge, can meet option 1, the 38% reduction in volatile solids across the digester. For facilities without primary sludge, however, and especially in facilities operating with long sludge ages, the remaining volatile solids may not be readily biodegradable, and it may be difficult to achieve this percentage reduction. If option 1 is not available, then compliance with the VAR criteria for solids leaving the facility can only be met using option 3, additional extended digestion, or option 4, the solids oxygen uptake rate (SOUR) test. Option 3 is suitable for all treated biosolids, but the extended time of the additional digestion test makes this test difficult to use for facilities that do not have a large amount of post-digestion storage. The SOUR test also will work for most facilities, but there are certain time-and-temperature limitations that must be recognized. These are discussed further below.

State regulations for the treatment of biosolids typically mirror federal guidelines. Some modifications may be required to meet specific concerns, however, and the designer should review the local biosolids rules.

7.2 Standards for the Use or Disposal of Biosolids

In addition to the requirements for pathogen reduction and vector attraction reduction control, there are also governing concentrations on certain pollutants found in biosolids. These concentrations determine the level of regulation and site control which is required for land application of these biosolids.

Controlling these pollutants in aerobically digested biosolids is essentially a function of controlling the concentrations at the headworks of the WWTP and may require implementation of an industrial pretreatment program. Each of these pollutants will partition to both the liquid effluent and the sludge. The proportion that will partition to the sludge varies and depends on the type of treatment process and the pollutant of concern. Once the pollutant is in the sludge stream, however, aerobic digestion typically will not significantly change the pollutant concentration.

8.0 PARAMETERS USED TO EVALUATE PERFORMANCE OF AEROBIC DIGESTERS

8.1 Standard Oxygen Uptake Rate

In an aerobic digester, the rate of oxygen use by the microorganisms depends on the rate of biological oxidation. As the biological activity decreases, the SOUR also decreases. A SOUR value of 1.5 mg O_2/h·g of total solids at 20 °C (68 °F) was selected by U.S. EPA to indicate that an aerobically digested sludge has been adequately reduced in vector attraction.

Because SOUR is a quick test and is independent of the initial value in the system or reduction of volatile solids in the upstream processes, it is becoming a more common testing procedure for most operators. Certain restrictions on the testing procedure must be observed, however. First, under the U.S. EPA guidelines (2003), the temperature of the sludge should not vary by more than 10 °C from the 20 °C value used for the standard method. If the temperature variance is less than 10°C, then the temperature correction process outlined in the U.S. EPA guidelines should be used. Also, by U.S. EPA methods, the maximum solids concentration in the digester is 2.0%. Dilution, even with distilled water, is not allowed.

During active aerobic digestion, a typical rate of oxygen uptake is from 3 to 10 mg O_2/h·g total solids. If primary sludge is added to the digester, then it may exhibit an uptake rate ranging from 20 to 40 mg O_2/h·g total solids. These values would be representative, as well, of the first stage of a multistage digester, and would decrease in subsequent stages. An oxygen uptake rate for a well-digested sludge from the aerobic digestion process ranges from 0.1 to 1.0 mg O_2/h·g total solids, well below the U.S. EPA standard of 1.5 mg O_2/h·g.

8.2 Pathogen Reduction

Pathogen reduction is one of the key regulatory requirements imposed by U.S. EPA. The U.S. EPA (2003) has indicated that, when operated in accordance with PSRP criteria, an aerobic digester will typically reduce biological densities by 2 $\log_{10}$ and viral pathogens by 1 log. Although the U.S. EPA definition of PSRP requires a minimum temperature of 15 °C (59 °F), pathogen reduction is possible at lower temperatures. A U.S. EPA study showed that at 28.6 °C, the daily reduction in *Escherichia coli* densities was 1.13 log; at 6.2 °C, it dropped to −0.23 log (Farrah et al, 1986). Studies in Ohio have demonstrated fecal coliform densities as low as 10 000/gram total solids at storage tank temperatures as low as 4 °C (Scisson, 1997). Actual fecal coliform counts tend to be plant-specific. The designer can provide for better results by

- Designing digesters for series operation and
- Covering tankage in cold climates.

8.3 Volatile Solids Reduction and Solids Reduction

Aerobic digestion results in the destruction of VSS. In addition, if a membrane thickening digestion option is selected, then some refractory solids may also be reduced. This occurs because both the organic and inorganic material in the slowly biodegradable suspended solids are solubilized and digested. Primary solids and WAS from a system with a short SRT will contain relatively high fractions of biodegradable material. Waste activated sludge from a long SRT, however, will contain a low fraction of biodegradable material and a high fraction of biomass debris (Grady et al., 1999).

Temperature affects the destruction of biodegradable suspended solids and can be characterized as a first-order reaction. The biodegradable suspended solids are typically active biomass, and the decay of this biomass is a first-order reaction. The decay coefficient is independent of the SRT at which the waste solids were produced because it is influenced strongly by the decay coefficient for heterotrophic bacteria, which is relatively constant.

One study evaluated two systems to determine the minimum SRT required to meet Class B requirements for both pathogens and volatile solids reduction while operating at minimum temperatures (Daigger et al., 1997). One system included two basins in series and the other, three basins in series. Table 11.5 shows the volatile solids reductions at different temperatures and different SRTs. The sludge used in this study had a very low digestible fraction. It was concluded that, even though the pathogen destruction was met in all the systems, an SRT of 29 days was required to exceed the required 38% volatile solids reduction. This study confirms that when sludge has a very low fraction of digestible organic content, it is difficult to meet the minimum requirements set by U.S. EPA.

TABLE 11.5 Volatile solids reduction at minimum operating temperatures and minimum solids retention time (SRT) (VS = volatile solids) (Daigger et al., 1999).

VS Reduction Data from Both Studies in all Basins						
°C	8–10°C	12°C	21°C	21°C	23°C	31°C
2 basins in series # of days	19.25	13.75	13.75	13.75	19.25	19.25
VS Removal	27	31%	31%	31%	28%	31%
3 basins in series # of days	29.25				29.25	29.25
VS Removal	28%				32%	40%

8.4 Solids Retention Time Times Temperature

The SRT is the total mass of biological sludge in the reactor divided by the mass of solids that are removed from the process on a daily average. The SRT significantly affects operation of aerobic digesters. Typically, increased SRT results in an increase in the degree of volatile solids reduction and can have other undesirable effects, such as a decrease in dewaterability.

The product of SRT times temperature (days × temperature, °C) curve can be used to design digester systems, considering the total days at a given temperature that the digester system will need to meet based on the quality of the source. The original U.S. EPA SRT times temperature curve was developed in the late 1970s based on pilot and full-scale data from operating digesters of that time (Mavinic and Koers, 1977). It has been updated by incorporating data from three full-scale installations and two extensive studies (Daigger et al., 1997).

8.5 Nitrogen Removal in Biosolids

The nitrogen contained in biosolids will vary with whether the digester is operated in continuous aeration or aerobic-anoxic mode. Annual design loading rates for land application of biosolids are typically limited by the nitrogen loading rate, although this may vary depending on state requirements. Compared to aerobic-anoxic systems, aerobic digestion with continuous aeration will have a lesser reduction in the total amount of nitrogen contained in the solids, limiting the application rate for the biosolids.

8.6 Supernatant Quality of Recycled Sidestreams

Supernatant quality from aerobic digesters will vary significantly, depending upon the digestion mode used (Table 11.6). Conventional aerobic digestion will tend to

TABLE 11.6 Characteristics of aerobic digester supernatant (COD = chemical oxygen demand; BOD = biochemical oxygen demand; ATAD = autothermal thermophilic aerobic digestion.

	National Technical Information Service, 1975[a]	Ahlberg and Boyko, 1972[b]	Stankewich, 1972[c]	ATAD supernatant[d]		
				Oxygen deprived	Oxygen satisfied	Excess oxygen
Turbidity, JTU	120	–	–			
NO–N, mg/L	40	–	30	0.3	0.65	9.05–9.70
Total Kjeldahl nitrogen, mg/L	115	2.9–1350	–	1035	1045	890–970
COD, mg/L	700	24–25 500	–	17 940	20 540	17 610– 18 300
Soluble COD, mg/L				406	3.1	4.0
PO-P, mg/L	70	2.1-930	35	77	46	51-56
Filtered P, mg/L		0.4-120				
BOD, mg/L	50	4-6350	2–5			
Filtered BOD, mg/L		3–280				
Suspended solids, mg/L	300	9–41 800	6.8			
Alkalinity, mg/L $CaCO_3$			150			
SO_4, mg/L			70			
Silica, mg/L			26			
pH	6.8	5.7–8.0	6.8			

[a]Average of seven months of data.

[b]Range taken from seven operating facilities.

[c]Average values.

[d]Boulanger, 1995.

return a supernatant stream high in nitrate nitrogen; anoxic/aerobic digestion will have a much lower nitrogen load. In ATAD systems, because only partial nitrification occurs, supernatant will have a different quality. The phosphorus loads of these streams will also vary.

9.0 DESIGN TECHNIQUES TO OPTIMIZE AEROBIC DIGESTION

Because of U.S. EPA's part 503 regulations and the need to develop more efficient treatment systems, new methods have been developed to optimize aerobic digestion. Some of the systems that have been found to be effective are discussed below.

9.1 Sludge Thickening

One way to improve aerobic digester performance is to thicken the sludge. This increases the solids residence time for a given reactor volume and provides other advantages, such as a more uniform temperature profile. Solids can potentially be thickened at one of four points in the aerobic digestion process:

1. Prethickening. Solids are thickened before entering the digester.
2. Settling and decanting the digester.
3. Recuperative thickening. Solids are withdrawn from the digester, thickened, and returned to the digester minus some water. The water is returned to the secondary treatment process. Depending on nutrient loadings and plant effluent limitations, the water may alternately be sent to a sidestream treatment process.
4. Post-thickening. This is where the biosolids are thickened for storage.

Advantages of prethickening include reduced volume requirements and heat loss and better pathogen reduction. Disadvantages are increased complexity compared to conventional systems and potential polymer requirements. In retrofit situations, it may be difficult to use existing equipment for aeration. Settling and decanting is limited in thickening abilities. Recuperative thickening has similar advantages to the prethickening process. In addition, because recuperative thickening can operate continuously, and sludge may be wasted intermittently to the digesters, it may be possible to use smaller equipment based on the lower continuous flow. Disadvantages include poor-quality filtrate unless polymer is used, potentially resulting in a higher load in recycle streams.

Because it occurs after the aerobic digestion process, and has no effect on treatment efficiency, post-thickening is not a way to optimize aerobic digestion, but rather a way to reduce liquid content before storage. Therefore, it will not be considered further in this section.

Several different types of thickeners are available, each of which is discussed in detail in Chapter 8. Therefore, these units are not described further in this chapter, except as specifically related to aerobic digestion. For use in aerobic digestion, the operational characteristics listed in Table 11.7 can be used.

9.1.1 *Prethickening Processes*

Prethickening can be used to increase the solids concentration of residuals entering the digester. Prethickening can be used for thickening of mixed liquor, for installations where the activated sludge solids retention time is controlled by wasting directly from the process. It can also be used for thickening clarifier underflows. Prethickening is most often done with mechanical thickeners such as gravity belt thickeners (GBT), rotary drum thickeners (RDT), disc thickeners, and centrifuges. Dissolved air flotation (DAF) units may also be used. Gravity thickeners may be used but typically will not achieve as high a solids concentration as a mechanical thickening device or DAF unit.

9.1.2 *Batch Operation and Decanting*

Thickening of aerobically digested biosolids by decanting (supernating) is as old as aerobic digestion and is still common today. Thickening by decanting is typically part of the "semibatch" mode of operation where the air is periodically turned off and the

TABLE 11.7 Characteristics of waste activated sludge thickeners for aerobic digesters.

Type	Point of use	Solids concentration, % total solids	Polymer dose
Gravity thickener	Pre-, post-, and recuperative	2–4	None
Gravity belt	Pre-, post-, and recuperative	4–13	2-6
Rotary drum	Pre-, post-, and recuperative	4–13	4-8
Centrifuge	Pre-, post-, and recuperative	6–12	0-4
Membrane bio-thickener	Recuperative	3–5	None

clear liquid is decanted. The air is then restarted. The process is repeated until the biosolids are sufficiently stabilized, and the biosolids no longer settle or the solids content becomes greater than the aeration capacity. The biosolids may then be land applied as a liquid, dewatered, or sent to liquid storage.

To allow sufficient sludge thickening and storage and flexibility in operations, the minimum decanting depth should be 1.67 m, or at least 33% of the total sidewater depth. For improved operation, the designer should allow the operator to decant at least 50% of the liquid volume in the tank, up to 2.5 m of decant depth.

The final solids concentration depends on local factors such as feed sludge composition and settleability of the solids. Where primary sludge is not present, unthickened WAS will enter the digester at 0.5 to1.0% total solids. Decanting will typically allow the solids to be concentrated to between 2.0 and 3.5% total solids. In the case of digesting mixed WAS and primary solids, the digested biosolids may concentrate to 6%.

Supernatant quality is site specific and highly variable. Variability depends upon the degree of digestion, source material, how long it takes to supernate the tank, type of decanter, and operator preference. If the air is off for a long period, then phosphorus will be released from the biosolids, increasing the concentration in the supernatant as discussed below.

9.1.2.1 Decanter Types

Typically used decanters for aerobic digesters include telescoping valves, fixed pipes through the wall, and floating and pumped decanters. Each of these has advantages and disadvantages.

Telescoping valves are the predominant decanter, especially for smaller facilities. They are relatively less expensive, but require more attention and take longer to decant the clear water than some of the alternatives. The entry area is relatively small in diameter, resulting in high velocities approaching the valve. This can lead to poor supernatant quality, which will draw in solids because of the high withdrawal velocity.

In older facilities, supernatant withdrawal typically was accomplished by multiple pipes passing through the walls of the digester. These pipes were at fixed vertical intervals (typically 0.3 to 0.6 m) from the high water level to the anticipated minimum level of decanting. Small sample lines may be tapped from the bottom of the pipe, allowing the operator to open each sample line and determine the lowest withdrawal port that is above the settled sludge. The operator may then open each pipe in sequence or may open the lowest pipe above the settled sludge. This system has the advantage of not requiring a great deal of operator attention after the initial sampling, and the withdrawal rate is relatively high. Depending upon the depth of water above the pipe and the size of pipe, however, the velocities may be excessively high, causing excess solids in the supernatant.

Floating sequencing batch reactor (SBR) decanters works well for aerobic digester and provide several advantages. Water is collected over a large area, reducing upflow currents and the potential for excess solids in the decant. It provides quick removal, because the decanter can be sized for flows up to several thousand liters per minute. Because the decanters typically draw-off below the water level, the potential for scum withdrawal is relatively low. Operator attention is minimal, as long as the low water level is above the settled sludge layer. The primary disadvantage is its additional cost compared to a telescoping valve.

Suspended or floating submersible pumps have been used as decanters. Although there are commercially manufactured decanters using pumps, these are often handmade devices, often lashed to an inner tube and set afloat on top of the digester. The pump discharges into a hose that runs to a nearby drain or channel.

9.1.2.2 Operational Issues

When decanting a digester, allowing an aerobic digester to settle adequately for supernatant withdrawal may require that the air be turned off for four hours and, sometimes, up to 24 hours or more. Especially at longer times, the sludge blanket tends to become anaerobic and release phosphorus to the liquid phase. Plants using BNR processes will have higher phosphorus releases than those using chemical precipitation. This is because the chemically precipitated phosphorus is bound in the solids. It may be necessary to design chemical precipitation facilities to add metal salts to the supernatant to prevent excessive phosphorus recycle.

Another issue is supernatant quality. Aerobic digesters operated at below 6.5 pH lose the ability to nitrify but not the ability to degrade proteins and make ammonia. In other cases, if the oxygen demand because of thickened sludge outstrips the aeration system's ability to transfer oxygen, then digesters will make ammonia as well. In these cases, supernatant streams can recycle ammonia at a concentration of 125 mg/L or more. If these high ammonia loads affect the plant's ability to meet effluent criteria, then the tanks will have to be supernated more often at a lesser volume or the supernatant will need to be directed to an empty tank and released over a longer time period.

9.1.3 *Continuous-Feed Operation with Separate Clarification*

As noted previously, many aerobic digesters are operated as continuously fed systems with separate clarification. This process closely resembles the activated sludge process. Solids are pumped directly from the clarifiers or reactor basin (for SBRs and membrane bioreactors, or MBRs) into the aerobic digester. The digester operates at a fixed level, with the overflow going to a solid-liquid separator. Thickened stabilized solids are removed for further processing. This process does not increase residence times in the digester but thickens the solids before the next processing step or disposal. Design of gravity thickeners is considered further in Chapter 8.

9.1.4 *Inloop (Recuperative) Gravity Thickening*

Recuperative thickening differs from continual-flow processes because sludge is drawn from the digester, thickened, and then returned to the digester. Installations as early as 1975 (Winter Haven, Florida) included the use of gravity thickeners set up to provide either continuous flow or recuperative thickening. Although this process has been possible for many years, its use has been limited because of issues with aerating thickened solids.

In older processes, the system consisted of a gravity thickener with proper piping to allow thickened sludge to be returned to the digester. Newer designs incorporate two main phases: an "inloop" phase and an isolation phase. Typically, the process includes four main basins: two digesters, one premix basin, and one gravity thickener. During the inloop phase, one of the two digesters, the premix basin, and the thickener operate in loop. Raw sludge and thickened sludge from the inloop phase digester are fed to the premix basin. The pre-mix basin overflows to a gravity thickener, which allows the sludge to settle in an anoxic environment. Thickened sludge is pumped from the thickener to the inloop digester. The inloop digester overflows back to the premix basin completing the cycle. The second digester is isolated, and no feed occurs during a stabilization period of 10 to 20 days. At the end of the isolation period, the "inphase" and isolation digester are exchanged.

The inloop phase reduces volatile solids and ammonia and increases the solids concentration. The anoxic environment of the inloop thickener allows denitrification, which helps to control pH in the digester system. The inloop digester provides volatile solids reduction. During the isolation phase, the digester's main function is to complete additional pathogen reduction to meet the sludge requirements. The process is considered a "modified batch process" because one digester is fed in short-batch intervals for an extended period of time—typically 10 to 20 days (equal to the length of the inloop phase)—before it goes to the isolation phase (10 to 20 days).

Typically, this process produces 2.5 to 3% solids. Because the digesters are operated at a 2.5 to 3% solids concentration, conventional mixing and aeration equipment can be used. Benefits include

- Improved pathogen destruction at reduced solids retention time;
- Increased solids concentration in gravity thickener;
- Continuous automated thickening of the aerobic digesters; and
- Enhanced pH and temperature control.

9.1.5 Inloop (Recuperative) Membrane Thickening

Membranes suitable for use in wastewater treatment created an additional option for thickening. Membranes allow the maintenance of a relatively high solids concentration while producing a higher quality supernatant.

The Enviroquip PAD-K process (Enviroquip Inc., Austin, Texas) operates similarly to the recuperative process discussed previously. Raw sludge is pumped into the premix (anoxic basin) and combined with sludge from the aerobic digester. The combined sludge mixture undergoes denitrification under anoxic conditions. Sludge is pumped from the anoxic basin, typically with an airlift pump, to the membrane thickener. Design parameters call for flow to be 300% of the design flow of the digester. In the membrane thickener tank (MBT), water (permeate) is withdrawn through the membrane plates using a small centrifugal pump. Thickened sludge overflows from the MBT into the aerobic digester where it is aerated and nitrified. Because of the alternation of anoxic/aerobic phases, less aeration is required than in a conventional aerobic digester. While one aerobic digester forms a recycle loop with the MBT and anoxic basin, the other digester operates in an isolated mode that allows for pathogen reduction. The "inloop" and isolated digesters are periodically exchanged.

The recuperative membrane thickening process provides several benefits:

- Digested sludge product exceeds Class B requirements;
- Improved pathogen destruction at shorter solids retention time;
- Smaller volume of digested sludge;
- Continuous and automatic thickening of solids without use of chemicals;

- Zero TSS and reduced phosphorus returned to main treatment process; and
- Enhanced pH and temperature control.

9.1.6 *Mechanical Prethickening*

The next logical step after gravity thickening is the use of mechanical thickeners to further increase sludge concentration to 4 to 8% solids. This process allows facilities to increase detention time to meet the requirements of U.S. EPA and state regulations without constructing additional storage volume. Increasing the solids concentrations may allow compliance with the detention time requirements, but it may also lead to other issues. Some aeration equipment may not perform adequately at high solids concentrations, with either inadequate oxygen transfer or inadequate mixing. Typically, the design should include bypass provisions around the thickener, allowing the mechanical thickener to operate at its optimum thickening point while blending sludge in the digester to meet the optimum concentration in the digester. The operator should have the ability to vary digester solids concentration as required to ensure that oxygen demand is met and to control temperature.

Sludge storage tanks may be beneficial to allow optimization of thickening operations. For example, if sludge thickening occurs only during one shift of plant operation, then a storage tank could hold sludge wasted during other operating periods. Typically, these tanks are cylindrical and may have either a flat or sloped bottom. Cylindrical tanks are preferable because of the elimination of potential "dead spots" in the corners of the tanks. The U.S. Army Corps of Engineers (1984) recommends a 4:1 floor slope and a minimum depth of 4.5 m (15 ft). These storage tanks can be either concrete or steel; corrosion may be an issue with steel tanks, however, which should be considered in evaluating materials.

Mixing is required to ensure that sludge transferred to downstream processes is homogeneous (Spinosa and Vesilind, 2001). Mixer manufacturers recommend mixing energies ranging from 10 to 12 kW/1 mil. L (40 to 50 hp/1 mil. gal) (Lottman, 2008). The key is to keep solids suspended without inducing excessive air into the residuals. Aeration of the sludge storage tank may also be required, especially if the sludge is unstabilized. In these cases, the sludge storage basin must be designed for oxygen requirements similar to aerobic digesters. On the other hand, if the material has been stabilized before storage, then oxygen requirements to maintain aerobic conditions will be significantly less than that required for aerobic digesters. Maintaining a minimum dissolved oxygen level of approximately 0.5 mg/L should be sufficient to maintain a basin free from anaerobic activity with adequate mixing. If aerobic conditions are not maintained, then nuisance odors develop and additional odor control may be required.

9.1.7 Basin Design and Configuration—Thickened Aerobic Digester

As with unthickened biosolids, the product of SRT and liquid temperature for thickened biosolids should be 600 to 1200 degree days, or 30 to 60 days at 20 °C (68 °F). This will help to achieve at least a 40% solids reduction. If liquid temperatures are lower, then a longer SRT will be required.

Thickened biosolids, especially from rotary drum or gravity belt thickeners, are viscous and need higher shear mixing than unthickened biosolids. To make best use of the mixing energy, tank dimensions should be reduced horizontally and increased vertically to create taller, narrower tanks that are easier to mix. Instead of the typical 4.5 to 5 m (14 to 15 ft) sidewater depth, aerobic digesters for thickened biosolids should be approximately 7 m (22 ft) high.

9.1.8 Staging Digesters

As with conventional aerobic digestion, operating thickened aerobic digesters in series can provide improved digestion. Series operation can reduce the SRT required by up to 30%. As noted previously, 70% of the air requirements will occur in the first stage. Given the thickened sludge and higher concentration of organic matter, process air requirements may exceed the mixing air requirements. The designer must take care to calculate process air requirements for each stage to ensure sufficient process air supply

To provide for the benefits of staged operation and flexible operation, a staged design should have three tanks. The staged digesters should have piping to allow them to be operated in parallel or for any one digester to be taken out of service. Operation in parallel can help during periods when the first stage is be overloaded or begins to autoheat. Similarly, it is advisable to remove basins from service periodically for cleaning and maintenance.

9.1.9 Air Requirements

When digesting unthickened WAS, the designer typically can assume that mixing air requirements are greater than process air requirements. When designing aeration systems for thickened WAS, the process air requirements may be greater than the mixing air requirements. This disparity is because the metabolic requirement for oxygen and the biosolids mass remain unchanged, but water has been removed, making the volume occupied by the biosolids smaller. Biosolids thickened from typical 1 to 4% total solids have the same mass oxygen demand but now occupy only 25% of the volume occupied at 1% total solids. Therefore, the air required per unit volume quadruples.

The total solids content and the biosolids viscosity also affect the air requirement. Alpha, the relative rate of oxygen transfer in wastewater compared to clean water, degrades with total solids content (Figure 11.8). As Figure 11.8 shows, alpha for fine-bubble diffusers may be 0.50 at a solids content of 1%, but will degrade to an alpha of less than 0.1 at 4% total solids.

This figure shows that both fine-bubble and coarse-bubble diffusers are adversely affected by digester biosolids concentration. Fine-bubble diffusers are affected to a greater degree than coarse-bubble diffusers

Another important design factor in the design of an aeration system is the thickened biosolids viscosity. Adding polymer for thickening further increases viscosity. Additional mixing energy is required to shear the viscous liquid and enable it to be circulated in the tank. Krampe and Krauth (2003) identify several studies demonstrating a relationship between viscosity and alpha factors. Figure 11.9 shows the relationship of viscosity and alpha factors at a shearing rate of 40 L/s (Krampe and Krauth, 2003). In some cases, over-thickening produces a thixotropic material that requires a significant amount of energy to reliquefy.

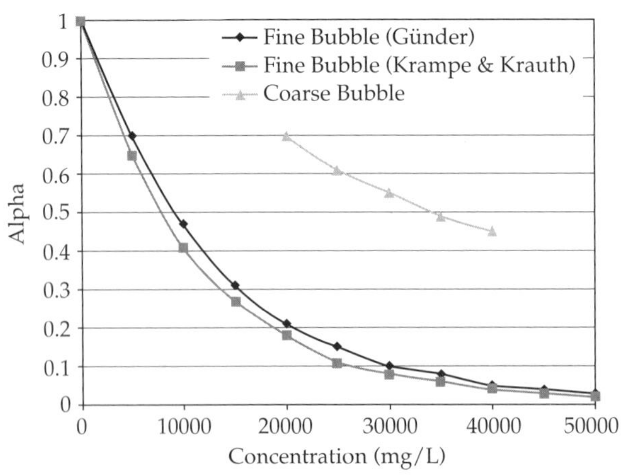

FIGURE 11.8 Offgas alpha testing—aeration diffusers (Schoenenberger et al., 2003; Stone and Livingston, 2008).

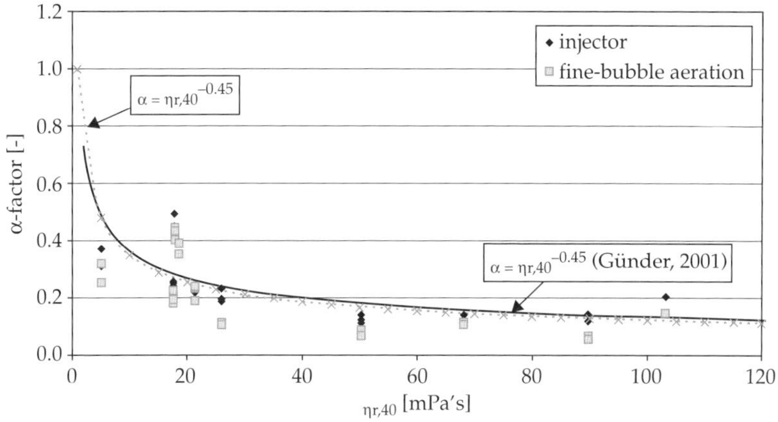

Figure 11.9 Relationship of viscosity and alpha factors (Krampe and Krauth, 2003).

Fabiyi and Novak (2008) confirmed an increase in viscosity with increasing solids and a corresponding decrease in alpha factors for fine bubble diffusers. Their study found that, at high concentrations, the fine bubbles tend to coalesce, forming large, vertically interconnected bubbles in the center of the diffuser plume. On the other hand, when an aeration system was used that combined mechanical mixing with air injection (Praxair In-situ Oxygenator), the alpha factor was much less affected by solids concentration.

This increase in viscosity affects the selection of aeration devices. Aeration systems that are suitable for use with thickened sludge include

- Shear tubes and point diffusers,
- Flexible-cap diffusers,
- Band diffusers, and
- Jet aeration.

Because of the increasing difficulty in transferring oxygen as solids concentrations increase, an optimal design point occurs where the additional value gained by decreasing the size of the digester basin is offset by the additional energy required for aeration. For each system, the tradeoff between energy costs, capital costs, aeration efficiency, and design needs must be evaluated. This will depend upon the type of aeration system selected. High-shear systems, such as shear tubes and jet aeration, will be able to operate economically at higher solids concentrations. For fine-bubble membrane aeration systems, manufacturer's data indicates that, over longer periods, concentrations of 15 000 mg/L or less are more cost efficient (Schoenenberger et al., 2003). This is consistent with studies by Krampe and Krauth (2003), which indicate that fine-bubble aeration is the most efficient system at solids concentrations of less than 18 000 mg/L.

9.2 Aerobic-Anoxic Operation

The aerobic sludge digestion process can be operated in either an alternating aerobic/anoxic mode in the same tank or in separate tanks to meet the requirements for aerobic nitrification and anoxic/anaerobic denitirification (Peddie and Mavinic, 1990). In this mode, a simple sequence of air on/off cycles can increase the removal efficiency for VSS and nitrogen (Klapwijk et al.,1998; Sakai et al., 1997). Aerobic-anoxic operation also offers significant energy savings and high oxygen-transfer efficiency during the initial period of aeration. Other advantages include partial recovery of alkalinity lost through nitrification, a stable pH, and slightly reduced final sludge volume (Daigger and Bailey, 2000; Jenkins and Mavinic, 1989b; Kim et al., 2000; Warner et al., 1985).

Nitrification is the biological oxidation of ammonia into nitrite followed by further oxidation of these nitrites into nitrate under aerobic conditions. Degradation of ammonia to nitrite is typically the rate-limiting step of nitrification. The process requires a significant amount of oxygen to produce a small amount of biomass and results in substantial destruction of alkalinity through the production of hydrogen ions as shown in eqs 11.2 to 11.4. One mole of alkalinity (as bicarbonate, HCO_3^-) is consumed in this reaction for each mole of ammonia-nitrogen oxidized to neutralize the acidity of the HNO_3 produced (Daigger and Bailey, 2000). The equilibrium pH of the reactor, therefore, will be dictated by the amount of alkalinity and CO_2 present in the system. If sufficient alkalinity is available in the feed solids, then an approximately neutral pH will be maintained. Alkalinity in the feed, however, is often insufficient to neutralize the acidity released, resulting in lowering pH.

Biological denitrification is because of the action of heterotrophic bacteria using available carbon sources under anoxic conditions. Oxygen will inhibit the activity of denitrifying enzymes. Nitrate and nitrite act as electron acceptors in the respiratory electron transport chain. Sufficient organic sources serve as carbon and electron donors and must be provided for proper denitrification. The denitrification process will produce alkalinity. If the sludge is nitrified and then fully denitrified during the anoxic cycle, then according to eq 11.6, little or no alkalinity is consumed. Therefore, a stable pH can be maintained in the aerobic-anoxic operation process.

Nitrogen removal from the alternating aerobic-anoxic operation can be controlled and optimized by monitoring the pH and oxidation-reduction potential (ORP) profiles. For a single digester system, the aerobic cycle can be controlled by an ammonia valley on the pH profile, which indicates the end of nitrification. Alternately, the end point of carbon oxidation, which is defined as the dissolved oxygen breakpoint, can be determined by monitoring the oxygen use rate and ORP in the process. The duration of aeration in the process can be optimized, significantly reducing costs for oxygenation (Casellas et al., 2006; Wareham, et al. 1994). The end of an anoxic cycle can be determined by the "nitrate knee" on the ORP profile, signifying the end of denitrification (Jenkins and Mavinic, 1989b; Ra et al., 1999; Wareham et al., 1993).

9.2.1 *Savings*

In the nitrification-denitrification reaction sequence, nitrogen oxides serve as electron acceptors in the absence of oxygen or at low oxygen tension. Aeration requirements and related energy needs are reduced (Oh and Silverstein, 1999). The energy requirements can be further reduced because energy for mixing does not need to be supplied during the anoxic cycle (Daigger and Bailey, 2000).

The final sludge quantities in an aerobic-anoxic process are typically lower than the conventional digestion process because substrate use rates are higher for nitrate versus oxygen respiration (McClintock, et al, 1988). The reduced quantity results in further cost savings from reduced sludge handling and disposal costs.

9.2.2 *Dissolved Oxygen and Oxygen Requirements*

Theoretically, less process oxygen is required for aerobic-anoxic operation. As shown in eqs 11.8 and 11.9, process oxygen requirements are reduced by 18% from 7 mol oxygen/1 mol of biomass destroyed in an aerobic digestion system to 5.75 mol oxygen/1 mol of biomass destroyed for the aerobic-anoxic operation (Daigger and Bailey, 2000). The oxygen transfer capacity of the first stage must be adequate to meet the process oxygen requirement.

Daigger and Bailey (2000) evaluated full-scale aerobic-anoxic digestion of activated sludge. When SOUR was less than 1.5 mg O_2/g VSS·h in the digestion process, high volatile solids reduction efficiencies in a range of 40 to 70% were obtained. This would also meet U.S. EPA Class B requirements.

9.3 Phosphorus Reduction in Biosolids and Biophosphorus

Phosphorus is released to the liquid phase in both anaerobic and aerobic digestion; however, the release is lower from aerobic processes than from anaerobic processes. If the conventional aerobic digestion process is operated in the aerobic-anoxic mode or under low dissolved oxygen conditions, then phosphorus release is minimized (Daigger et al., 2001). As shown on Figure 11.10, when sludge from a biological phosphorus removal facilities (Ozark WWTP, Ozark, Missouri) was digested under fully aerobic conditions, the phosphorus release in the liquid phase ranged from 100 to 120 mg/L (Ju et al., 2005). On the other hand, if the same sludge was digested under cyclic operation, it produced phosphorus in the range 70 to 90 mg/L after 500 hours of operation (see Figure 11.11).

The pH also affects phosphorus, ammonia, and nitrogen concentrations in the liquid phase. The work of Ju et al. (2005) showed that pH less than 6.0 should be avoided because it encourages dissolution of the inorganic metal phosphates. For the Ozark WWTP, a drop in the pH from 7.0 to 5.0 caused the supernatant concentration of phosphorus to increase by more than 50%, from 82 to 138 mg/L.

From an overall mass-balance prospective, the phosphorus entering the digester ends up either in the waste sludge or in the effluent. If the phosphorus is not allowed to move forward with the sludge, then it will be recycled back to the head of the plant. For plants that practice biological phosphorus removal, or where phosphate

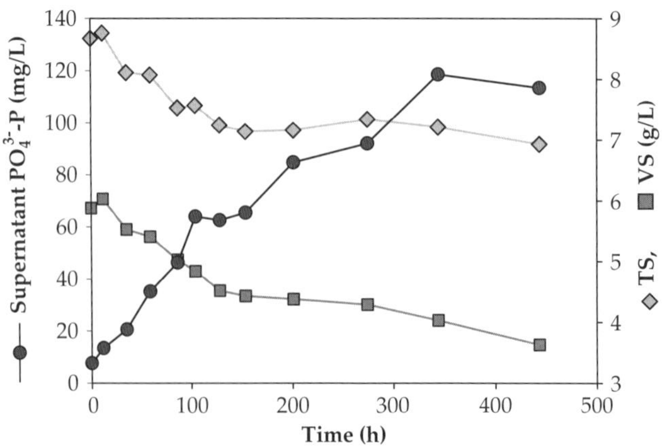

FIGURE 11.10 Time profiles observed with biological nutrient removal sludge from Ozark, Missouri, Wastewater Treatment Plant (VS = volatile solids; TS = total solids).

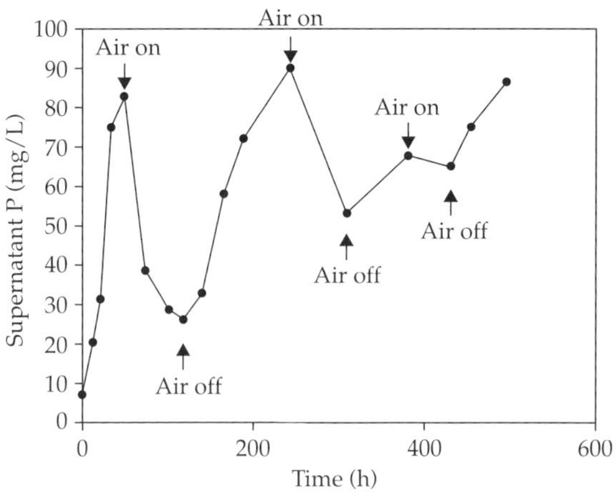

FIGURE 11.11 Phosphorus release in Ozark, Missouri, Wastewater Treatment Plant aerobic digester under cyclic aeration.

recycling may be a concern, two alternatives are available for managing the phosphorus release.

If there is no restriction on phosphorus for land application, then the main design consideration is to ensure that the sludge remains aerobic, especially if some intermediate step, such as prethickening or recuperative thickening is used. To prevent anaerobic conditions, the sludge should be wasted directly from the liquid sidestream to the prethickening device; or, if storage is required before prethickening, the detention time should be reduced to the minimum. Once in the digester, cyclical operation will help reduce nitrogen loadings while minimizing phosphorus release to the liquid phase. In this case, the phosphorus will be discharged as part of the sludge solids.

In some areas, there may be limitations on the amount of phosphorus in biosolids for land application. Several steps can be taken in this case:

- Assuming that phosphorus will be released during digestion and will be returned to the headworks.
- Controlling nitrification in first digester to encourage struvite production. This struvite can be separated and phosphorus removed from the biosolids stream. This may require use of an ATAD system to inhibit nitrification.

- Adding alum or ferric chloride to the supernatant and filtrate. Chemicals are only required for phosphorus removal. Because the supernatant contains few materials, chemicals for phosphorus removal are minimized. Phosphorus in the supernatant would then be fixed and can be removed separately.

10.0 REFERENCES

Ahlberg, N. R.; Boyko, B. I. (1972) Evaluation and Design of Aerobic Digesters. *J. Water Pollut. Control Fed.*, **44**, 634.

American Society of Civil Engineers (2000) *Conveyance of Residuals from Water and Wastewater Treatment.* American Society of Civil Engineers: Reston, Virginia.

Bisogni, J. J.; Laurence, A. W. (1971) Relationship Between Biological Solids Retention Time and Settling Characteristics of Activated Sludge. *Water Res.*, **5**, 753.

Boulanger, M. L. (1995) The Effect of Varying Air Supply Upon Supernatant Quality in Autoheated Thermophilic Aerobic Digesters Treating Waste Sludge from a Biological Phosphorus Removal Process. M.S. Thesis, University of British Columbia, Vancouver, British Columbia, Canada.

Burton, H. N.; Malina, J. F., Jr. (1964) Aerobic Stabilization of Primary Wastewater Sludge. *Proceedings of the 19th Purdue Industrial Waste Conference.* Purdue University: Lafayette, Indiana.

Casellas, M.; Dagot, C.; Baudu, M. (2006) Set Up and Assessment of a Control Strategy in a SBR in order to Enhance Nitrogen and Phosphorus Removal. *Proc. Biochem.*, **41** (9), 1994–2001.

Coackley, P. (1955) Research on Sewage Sludge Carried Out in the C.E. Department of University College London. *J. Inst. Sewage Purif.*, **59**.

Coackley, P. (1958) Laboratory Scale Filtration Experiments and Their Application to Sewage Sludge Dewatering. *Biol. Treat. Sewage. Ind. Waste.*, **2**, 287.

Daigger, G. T.; Bailey, E. (2000) Improving Aerobic Digestion by Prethickening, Staged Operation and Aerobic-Anoxic Operation: Four Full-Scale Demonstrations. *Water Environ. Res.*, **72** (3), 260–270.

Daigger, G. T.; Ju, L. K.; Stensel, D.; Bailey, E.; Porteous, J. (2001) Can 3% SS Digestion Meet New Challenges? *Aerobic Digestion Workshop, Vol. V;* Enviroquip, Inc.: Austin, Texas.

Daigger, G. T.; Yates, R.; Scisson, J.; Grotheer, T.; Hervol, H.; Bailey, E. (1997) The Challenge of Meeting Class B While Digesting Thicker Sludges, *Aerobic Digestion Workshop, Vol. I;* Enviroquip, Inc.: Austin, Texas.

Dirk, B.; Kelly, H. G.; Mavinic, D. S.; Frese, H. (2009) Impacts of Temperature and Hydraulic Residence Time on Odors Produced for Autothermal Thermophilic Aerobic Digestion; *Proceedings of the Water Environment Federation Residuals and Biosolids Conference 2009 [CD-ROM]; Portland, Oregon, May 3–6;* Water Environment Federation: Alexandria, Virginia.

Drier, D. E. (1963) Aerobic Digestion of Solids. *Proceedings 18th Purdue Industrial Waste Conference*. Purdue University: Lafayette, Indiana.

Fabiyi, M. E.; Novak, R. (2008) Evaluation of the Factors That Impact Successful Membrane Biological Reactor Operations at High Solids Concentration. *Proceedings of the 2008 Membrane Technology Conference;* Atlanta, Georgia; Jan 27–30; Water Environment Federation: Alexandria, Virginia.

Farrah, S.; Bitton, G.; and Zam, S. (1986) *Project Summary—Inactivation of Enteric Pathogens During Aerobic Digestion of Wastewater Sludge,* EPA/600/S2-86/O47; U.S. Environmental Protection Agency: Washington, D.C.

Folk, G. (1976) Aerobic Digestion of Waste Activated Sludge. *J. Water Pollut. Control Fed. Deeds and Data*. July.

Ganczarczyk, K. J.; Hamoda, M. F. (1973) *Aerobic Digestion of Organic Sludges Containing Inorganic Phosphorus Precipitates, Phase I,* Research Report #3; Canada-Ontario Agreement on Great Lakes Water Quality, Environment Canada: Gatineau, Quebec, Canada.

Grady, C. P. L.; Daigger, G. L.; Lim, H. C. (1999) *Biological Wastewater Treatment, 2nd ed.;* Marcel Dekker, Inc.: New York.

Günder, B. (2001) *The Membrane-Coupled Activated Sludge Process in Municipal Wastewater Treatment,* Technomic Publishing Co.: Lancaster, Pennsylvania

Hao, O. J.; Kim, M. H.; Al-Ghusain, I. A. (1991) Alternating Aerobic and Anoxic Digestion of Waste Activated Sludge. *J. Chem. Technol. Biotechnol.,* **52** (4), 457–472.

Higgins, M. J.; Novak, J. T. (1997) Characterization of Exocellular Protein and Its Role in Bioflocculation. *J. Environ. Eng.*, **123** (5), 479–485.

Jenkins, C. J.; Mavinic, D. S. (1989a) Anoxic-Aerobic Digestion of Waste Activated Sludge: Pt. 1, Solids Reduction and Digested Sludge Characteristics. *Environ. Technol. Lett.*, **10** (4), 355–370.

Jenkins, C. J.; Mavinic, D. S. (1989b) Anoxic-Aerobic Digestion of Waste Activated Sludge: Pt. II, Supernatant Characteristics, ORP Monitoring Results and Overall Rating System. *Environ. Technol. Lett.*, **10** (4), 371–384.

Jewell, W. J.; Kabrick, R. M. (1978) Autoheated Aerobic Thermophilic Digestion with Air Aeration. *Proceedings of the 51st Annual Water Pollution Control Federation Conference*, Anaheim, California, October 1978; Water Pollution Control Federation: Washington, D.C.

Ju, L. K.; Shah, H. K.; Porteous, J. (2005) Phosphorus Release in Aerobic Sludge Digestion. *Water Environ. Res.*, **77** (5), 553–559.

Kelly, H. G. (1990) *Demonstration of and Improved Digestion Process for Municipal Sludges*, KE405-8-6575/01-SE, UP-D8-010; Minister of Supply and Services Canada: Gatineau, Quebec, Canada.

Kelly, H. G. (1996) Autothermal Thermophilic Aerobic Digestion, Heat Exchange Budgets; *Proceedings of the Water Environment Federation Specialty Conference on Biosolids*, Denver, Colorado, July; Water Environment Federation: Alexandria, Virginia.

Kelly, H. G.; Warren, R. (1997) Autothermal Thermophilic Aerobic Digestion Design. *Proceedings of CSCE/ASCE Environmental Engineering Conference; Edmonton, Alberta, Canada, Jul 22–26;* American Society of Civil Engineers: Reston, Virginia.

Kelly, H. G.; Wong, V. (1997a) Stability in Autothermal Thermophilic Aerobic Digestion. *Proceedings of the Water Environment Federation and American Water Works Association Specialty Conference; Philadelphia, Pennsylvania, Aug*; Water Environment Federation: Alexandria, Virginia.

Kelly, H. G.; Wong, V. (1997b) Automation of Autothermal Thermophilic Aerobic Digestion. *Proceedings of the 2nd European Biosolids and Organic Residuals Conference; Wakefield, England, United Kingdom, November*; Aqua Enviro: Wakefield, England, United Kingdom.

Kelly, H. G.; Mavinic, D. S. (2003) Autothermal Thermophilic Aerobic Digestion Research, Application and Operational Experience. *Proceedings of the 76th Annual Water Environment Federation Technical Exposition and Conference* [CD-ROM]; Los Angeles, California, Oct 11–15; Water Environment Federation: Alexandria, Virginia.

Kelly, H. G. (2006) Emerging Processes in Biosolids Treatment, 2005. *J. Environ. Eng. Sci.*, **5** (3), 176–186.

Kelly, H. G. (2009) Personal communication.

Kelly, H. G.; Dirk, B.; Gibb, A. (2009) Carbon and Struvite Recovery from Centrate at a Biological Treatment Plant. *Proceedings of the Water Environment Federation Residuals and Biosolids Conference 2009 [CD-ROM]; Portland, Oregon, May 3–6;* Water Environment Federation: Alexandria, Virginia.

Kelly, H. G.; Dirk, B.; Gibb, A.; Frese, H. (2009) Autothermal Thermophilic Aerobic Digestion for Supplemental VFA in Enhanced Biological Treatment; *Proceedings 5th Canadian Residuals and Biosolids Conference*, Niagara Falls, Ontario, Canada, Sep 13–15; Water Environment Federation of Ontario: Niagara Falls, Ontario, Canada.

Kim, H.; McAvoy, T. J.; Anderson, J. S.; Hao, O. J. (2000) Control of an Alternating Aerobic-Anoxic Activated Sludge System Part 2: Optimization Using a Linearized Model. *Control Eng. Pract.*, **8** (3), 279–289.

Klapwijk, A.; Brouwer, H.; Vrolijk, E.; Kujawa, K. (1998) Control of Intermittently Aerated Nitrogen Removal Plants by Detection Endpoints of Nitrification and Denitrification Using Respirometry Only. *Water Res., G.B.*, **32** (5), 1700.

Krampe, J.; Krauth, K. (2003) Oxygen Transfer into Activated Sludge with High MLSS Concentration. *Water Sci. Technol.*, **47** (11), 297–303.

Kumar, N.; Novak, J. T.; Water, D. C. (2006) Sequential Anaerobic-Aerobic Digestion for Enhanced Volatile Solids Reduction and Nitrogen Removal. *Proceedings of the Water Environment Federation Residuals and Biosolids Management Conference 2006 [CD-ROM]*; Water Environment Federation: Alexandria, Virginia.

Layden, N. M.; Mavinic, D. S.; Kelly, H. G.; Moles, R.; Bartlett, J. (2007a) Autothermal Themophilic Aerobic Digestion (ATAD)—Part I: Review of Origins, Design, and Process Operation. *J. Environ. Eng. Sci.*, **6** (6), 665–678.

Layden, N. M.; Kelly, H. G.; Mavinic, D. S.; Moles, R.; Bartlett, J. (2007b) Autothermal Themophilic Aerobic Digestion (ATAD)—Part II: Review of Research and Full Scale Operating Experiences. *J. Environ. Eng. Sci.*, **6** (6), 679–690.

Lottman, S. (2008) Personal communication; Siemens: Berlin, Germany.

Matsch, L. C.; Drnevich, R. F. (1977) Autothermal Aerobic Digestion. *J. Water Pollut. Control Fed.*, **49** (2), 296–310.

Mavinic, D. S.; Koers, D. A. (1977) Aerobic Sludge Digestion at Low Temperatures. *Can. J. Civil Eng.*, **4**, 445.

Mavinic, D. S.; Koers, D. A. (1979) Performance and Kinetics of Low-Temperature Sludge Digestion. *J. Water Pollut. Control Fed.*, **51** (8), 2088–2097.

Mavinic, D. S.; Koers, D. A. (1981) Aerobic Sludge Digestion—A Suggested Design Procedure. *Can. J. Civil Eng.*, **8** (1), 9–15.

McClintock, S. A.; Sherrard, J. H.; Novak, J. T.; Randall, C. W. (1988) Nitrate versus Oxygen Respiration in the Activated Sludge Process. *J. Water Pollut. Control Fed.*, **60** (3), 342–350.

Metcalf and Eddy, Inc. (2003) *Wastewater Engineering: Treatment and Reuse*, 4th ed.; Tchobanoglous, G.; Burton, F. L.; Stensel, H. D.; Eds.; McGraw-Hill: New York.

Murthy, S. N.; Novak, J. T. (1999) Factors Affecting Floc Properties During Aerobic Digestion: Implications for Dewatering. *Water Environ. Res.*, **71** (2), 197–202.

Novak, J. T.; Sadler, M. E.; Murthy, S. N. (2003) *Mechanisms of Floc Destruction During Anaerobic and Aerobic Digestion and the Effect on Conditioning and Dewatering of Biosolids. Water Res.*, **37** (13), 3136–3144.

National Technical Information Service (1975) *Aerobic Stabilization of Waste Activated Sludge—An Experimental Investigation*, NTIS-PB-246-593/AS; National Technical Information Service: Alexandria, Virginia.

Oh, J.; Silverstein, J. (1999) Effect of Air On-Off Cycles on Activated-Sludge Denitrification. *Water Environ. Res.*, **71** (7), 1276–1282.

Paredes, M. (1976) Supernatant Decanting of Aerobically Digested Waste Activated Sludge. *J. Water Pollut. Control Fed.*, **13** (10), 4–5.

(continued)

1.0 INTRODUCTION

Dewatering is the removal of water from wastewater treatment plant (WWTP) solids to achieve a volume reduction and to produce a material for further processing or disposal. Dewatering is distinguished from thickening by the nature of the product produced. Sludge thickening concentrates the solids but still generates a product that behaves and flows like a liquid or fluid. Sludge dewatering produces a nonfluid sludge cake that behaves like a semisolid or solid material and is not free flowing. Reduction in volume can result in significant cost savings.

1.1 Process Evaluation

All dewatering processes generate two products: a sludge cake and a recycle liquid stream that carries the water removed from the cake and some residual solids. The recycle stream can be called supernatant, decant, under drainage, filtrate, centrate, evaporated moisture, or a combination of the above, depending on the specific dewatering process.

The performance of a dewatering process is measured by two primary parameters. Cake solids is a measure of solids content of the product and is expressed as the ratio of the weight of total solids in the cake divided by the total weight of the cake and is expressed as a percentage. Because the amount of dissolved solids is small compared to the suspended solids in dewatered cake, the total solids and total suspended solids are essentially the same. The percent solids typically can vary from approximately 15 to 50%. The second parameter is the percent capture, which represents the percentage of suspended solids in the feed to the dewatering process that are captured and remain in the sludge cake. The percent capture for dewatering processes typically exceeds 90% and can be as high as 99%. The solids that are not captured are contained in the recycle stream. By using these two measures, the designer can provide a solids balance for the dewatering device and determine the solids concentration in the recycle stream. In calculating percent capture from the suspended solids concentration in the cake and recycle, the affect of external wash streams and chemical addition needs to be considered.

A third parameter of importance is the bulk density of the cake, which represents the weight of the cake divided by its volume. The bulk density takes into account the air voids that can be present as the cake solids concentration increases and the product takes on soil-like properties. This parameter is useful in sizing cake conveyance and determining the number of trucks that may be required to remove the sludge cake. Some cake compaction and water release may occur as the cake is conveyed or loaded onto trucks. Alternatively, cake volume may actually increase slightly during holding because of further release of gas. In some cases, this can be significant.

Selection of a dewatering process needs to consider the nature of the sludge being processed, the additional sludge processing steps that will follow the dewatering step, and the sensitivity of subsequent sludge processing steps to cake solids concentration.

If incineration of the dewatered sludge is envisioned, then a drier sludge cake can result in significant savings in fuel costs. On the other hand, sludge processing alternatives such as alkaline stabilization may be unable to efficiently blend a dry sludge cake with added chemicals. The nature and constituents in the recycle and the method for treating the recycle stream also need to be taken into consideration in selecting the dewatering process.

1.2 Factors Affecting Performance

The ease with which different sludges dewater varies greatly. A waste activated sludge (WAS), for example, is difficult to dewater, whereas a well-digested primary sludge can be dewatered readily. Much of this variability results from the kinds of solids and the way in which the water is bonded to the solids. Much of the water in WAS is difficult to remove because it is attached to bacterial cells or tied up chemically in the cell structures.

Consequently, the larger the fraction of WAS in a blended sludge, the more difficult it will be to dewater. Several factors affect the dewaterability and performance of dewatering processes (Oerke, 1981):

- Solids concentration of the feed sludge;
- Hydraulic loading rate;
- Solids loading rate;
- Particle size of the solids in the sludge;
- Type sludge [primary, waste activated, biological nutrient removal (BNR), membrane, or blend];
- Preconditioning (chemical, thermal, or digestion);
- Temperature;
- Particle charge;
- Nature of the water present (free, interstitial, vicinal, or water of hydration);
- Presence of fibers or filamentous microorganisms;
- Evaporation rate, air temperature, and humidity (drying beds);
- Septicity;
- Compactability; and
- pH.

A higher solids concentration in the feed sludge has been shown to directly increase the cake solids that can be obtained for most mechanical dewatering processes because sludge that concentrates well in clarifiers and thickeners also concentrates well in dewatering equipment.

All mechanical methods for sludge dewatering benefit from some form of chemical conditioning. Chemical conditioning improves cake solids and capture efficiency. A detailed discussion of chemical and other forms of preconditioning is provided in Chapter 7. Typically, chemicals that are added include inorganic salts such as lime and ferric chloride and organic polymers.

The selection and design of a dewatering process can be based on performance observed at similar installations or by laboratory or full-scale testing. Because of the variability of sludge from plant to plant, the most accurate predictions of how a dewatering process will perform are best determined from full-scale pilot testing. Pilot testing will allow optimization of the chemical conditioning. Different types of polymers can be tested and the relationship between dosage, hydraulic and solids loading, capture efficiency, and cake solids can be ascertained. In many cases side-by-side pilot testing of different types of dewatering equipment from various manufacturers results in the selection of processes that minimize capital and operating costs while meeting the process objectives.

1.3 Methods of Dewatering

The methods for removing water from sludge use basic principles of natural drainage, evaporation, centrifugal force, or filtration to separate and remove water from sludge solids. These methods have not changed over the last twenty years, although some innovations have been made in the specifics of the equipment.

1.4 Selection of a Dewatering Process

Selection of the right dewatering process is a key element of WWTP design. The first step in the process is to establish the objectives of the dewatering step. This should include performance goals, such the minimum cake solids and capture efficiency, and operating requirements, such as the number of shifts equipment will be operated and redundancy desired. Procurement limitations and plant space requirements also need to be considered. Once these criteria have been established, a short list of technologies can be prepared that meet these requirements. In the case of a green field facility, the engineer will have to rely on vendor experience and site visits to treatment plants with similar sludges to be dewatered.

The next step is to contact manufacturers of equipment to find out if it has been used in a similar application. Field trips to operating installation will provide a good indication of how well the process will fit in a plant's current operations. Field trips will also help validate performance expectations.

Once the number of processes has been narrowed to two or three, full-scale pilot testing is advisable to confirm that the process will work with the sludge. Pilot testing also will allow the process to be optimized and to provide information about expected performance, chemical conditioning requirements, and power consumption. A report should be prepared that can be used to establish performance requirements for bid specifications. Finally, the detailed design of the dewatering equipment and ancillary facilities can be completed. The final design should take into account operating shifts and provide sufficient capacity and redundancy to be able to dewater sludge from seasonal loads and plant upsets.

The preceding discussion applies to the addition of dewatering capacity at an existing facility. For new facilities, solids dewatering operation at facilities with similar trains should be investigated.

1.5 Dewatering Costs

It is difficult to compare cost estimates from different dewatering studies because of differences in assumptions and sensitivity of cost comparisons to these assumptions. For example, if the price of land is high, then cost comparison will favor those dewatering technologies that have the smallest footprint. On the other hand, large sludge hauling costs and a high disposal charges will favor those technologies that produce the driest cake. Key parameters that will affect the relative cost of the dewatering technology include

- Cost of land, labor, chemicals, fuel, and power;
- Cost of subsequent processing and disposal of the dewatered sludge;
- Cost of the building to house the dewatering equipment;
- Number of spare units required;
- Degree of redundancy required for ancillary equipment (conveyors, chemical conditioning, etc.);
- Amount of clearance required between units;
- Number of operating shifts;
- Future growth and expansion requirements;
- Demolition costs to provide space for the equipment;
- Assumed peak processing rates;
- Life expectancy of the equipment;
- Interest rate and duration of bonds for financing;
- Need to keep other processes in operation during construction;

- Value of the dollar; and
- Year when the equipment will be installed.

A cost comparison should be done that is based on total lifecycle costs and includes both capital and operating and maintenance costs. The lifecycle-cost analysis can be used to test the sensitivity of the result to difference escalation factors of critical parameters such as power, fuel, chemicals, and labor.

In 1985, the U.S. Environmental Protection Agency (U.S. EPA; 1985) published *Handbook: Estimating Sludge Management Costs*. This handbook provided a series of curves for estimating capital and total annual dewatering costs, taking into account some of the above items. A spot check was made to check the validity of the drying-bed cost curve using *Engineering News-Record* (ENR) indices and comparing the cost to an independent detailed assessment for a Florida drying bed installation. The curves provided a cost that was approximately 25% lower than the detailed cost estimate. This suggests that the curves may still have some validity for relative comparisons. Hydromantis, Inc., has developed a commercial software product, CAPDET WORKS, which provides cost estimating for different dewatering technologies based on some of the early U.S. EPA models and updated equipment costs obtained from equipment manufacturers.

Although U.S. EPA curves and cost estimating software can provide order-of-magnitude cost comparisons, they should not be used for final cost estimates. Any detailed costing for equipment selection should be based on specific sizing of the dewatering technology for the particular sludge being dewatered, manufacturers' quotations, and labor and design practices applicable to the owner and operator of the dewatering equipment.

1.6 Quality Assurance/Quality Control

Although the dewatering process can have a significant effect on the quality of biosolids, it is important to have a closely managed and systematic quality assurance and quality control (QA/QC) process. This includes a system of checks and measures to maintain a high-quality product. If the solids construction or order characteristic of a product are not maintained within preset limits, then this will affect the ability to distribute or land apply the biosolids. In particular, some research has suggested that high-shear dewatering processes may produce a product that is more prone to odor generation and possible regrowth of pathogens. A discussion of the management of odors is provided in Chapter 20.

The liquid sidestream from the dewatering process can affect the performance of the wastewater treatment process and should also be monitored. Sidestream treatment is described in Chapter 21. Sidestreams may also be prone to struvite or vivianite formation, especially if the sludge being dewatered has undergone anaerobic digestion, and the WWTP is practicing biological phosphorus removal. A discussion of struvite and vivianite chemistry is provided in Chapter 4.

2.0 AIR DRYING

2.1 Introduction

Air drying is the oldest and simplest of the dewatering methods. Types of air-drying systems include sludge drying beds and lagoons. Both require a considerable amount of land. Air drying is typically used only by smaller facilities or those with access to a large amount of land. Air-drying systems take advantage of solar heat and mechanical drainage to separate water from solids. Another factor is local weather conditions. Drying beds in areas with high precipitation, frequent cloud cover, or low temperatures may not be practical.

2.2 Basic Components

Sludge drying beds are the most common air-drying method. Sludge drying beds have been used at more WWTPs than any other type of sludge dewatering system. Sludge drying beds typically consist of concrete sidewalls extending 0.3 to 1.0 m (1 to 3 ft) above a porous media composed of sand and gravel, although other media have also been used. Liquid sludge is applied on top of the porous media until a selected volume has been reached. The sludge is allowed to sit on the bed for days or weeks until the desired dryness has been achieved. Sludge may also sit on the bed until trucks are available to haul the material, thus providing onsite sludge storage.

An underdrain system under the porous media carries away liquid that drains through the media. The liquid or filtrate is typically is sent to the head of the plant for treatment. The filtrate drains through the media, and evaporation because of solar radiation further dries the material. It is important to consider rainfall if the beds are uncovered; sludge may not dry adequately during rainy periods, even in warm climates.

Sludge drying beds are a batch process. Liquid sludge is typically applied in less than an hour. The sludge is then left untouched, except for mechanically assisted systems, which will be discussed below. The dewatered sludge is typically removed by a front-end loader and placed in dump trucks for hauling to an application or disposal site. The drying bed is then ready for the cycle to be repeated.

Types of nonmechanical sludge drying beds include conventional sand drying beds, paved drying beds, and wedge-wire drying beds. Mechanical systems are proprietary and include a variety of configurations and components. Some mechanical systems that have been used in the past, such as vacuum-assisted drying beds, are no longer commercially available and are not included here. Sludge-drying lagoons are typically large, lined earthen pits. Liquid sludge is applied to the pit and allowed to set for an extended period of time. Lagoons typically do not use drainage as part of the dewatering mechanism, but rely on evaporation and time to dry the solids.

2.3 Conventional Drying Beds

2.3.1 Description

Sand drying beds are the most traditional and common type of air-drying technology. Figure 12.1 shows typical features of a conventional sludge drying bed. Drying beds are typically rectangular to allow convenient sludge removal with a front-end loader. Dimensions can be 4.5 to 18 m (15 to 60 ft) wide and 15 to 45 m (50 to 150 ft) long. Sidewalls are typically concrete, consisting of a vertical wall sitting on a small footer. Where multiple drying beds are required, common-wall construction can be used.

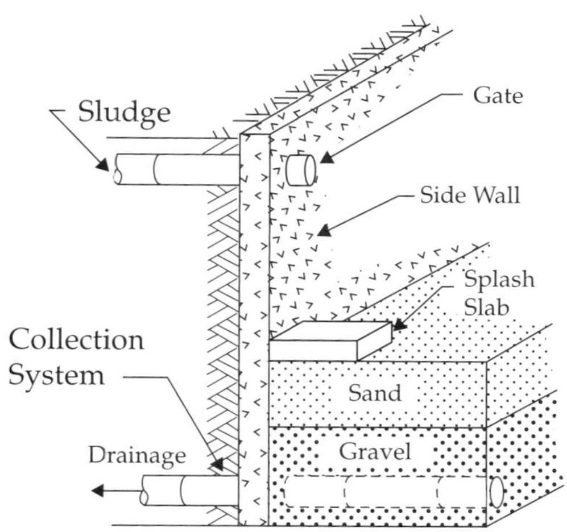

FIGURE 12.1 Conventional sand drying bed configuration (WEF et al., 2009).

The underdrain system typically consists of perforated plastic drain pipes that should be no smaller than 130 cm (4 in.) in diameter. The drain piping should be placed from 2.5 to 6 m (8 to 20 ft) apart and have a slope of no less than 1%. Often, the filtrate will ultimately need to be pumped to get it to the head of the plant. The drainage pipe should be placed in a gravel layer under a sand layer.

Drying beds are typically fed at one end of each bed through a valve or gate. A ramp needs to be provided on one end to allow a front-end loader to enter the bed for sludge removal.

Conventional sludge drying beds can be covered to extend the drying season in cold climates. Typical enclosures include glass, similar to green houses, to take advantage of sunlight to enhance evaporation. Covers also prevent rain from falling on the drying bed, enhancing evaporation and decreasing drying time even in warm weather. Covers can decrease the surface area required 25 to 35%.

Sludge drying beds are operated as a batch process, alternately being loaded and unloaded. Certain weather conditions will prevent sludge from being applied to the bed or removed. Thus, it is essential that sufficient storage for liquid sludge be provided to allow for times when the drying beds are not available.

2.3.2 *Principles of Operation*

Drying beds operate by draining free water through the media and through evaporation. Drying beds will not operate well during wet weather when rain can offset the amount of evaporation. Loading rates for sludge drying beds depend on type of sludge applied, concentration of sludge solids, climate, type of drying bed used, and whether drying beds are covered.

Loading rates are based on annual application rates because beds will be loaded and unloaded through many cycles that will vary with the amount of rain that falls in a period.

The drying time for a single application of liquid sludge is given by:

$$t_d = \frac{(y_o)(1 - s_o/s_f)(1 - D)}{(k_e)(E_v)} \qquad 12.1$$

where

t_d = dewatering time for a single application, month;

y_o = initial depth of applied sludge layer, cm;

s_o = initial dry solids concentration of sludge, %;

s_f = final dry solids concentration required for dewatered sludge, %;

D = fraction of water removed by drainage, % by decimal;
E_v = average pan evaporation, cm/month; and
k_e = reduction factor for sludge evaporation versus free water surface (0.6 typical).

The above equation is best applied monthly or seasonally because the pan evaporation rate will vary based on season of the year. In cold climates, sludge drying beds may not be available during parts of the winter.

The pan evaporation rate data is available for local areas on the Internet. The initial dry solids concentration should be based on information available for the given WWTP. Increased initial solids concentrations will decrease the drying time. Comparison of drying time and resultant area required can determine if prethickening of the sludge would be cost effective. The final dry solids concentration should be selected by the designer based on the use of the product. The sludge cake is considered "liftable," or relatively easy to remove, at 25 to 30% solids. The fraction of water removed by drainage can be 25% for anaerobically digested primary plus waste activated sludges to 75% for well-conditioned sludges. The fraction removed by drainage could be established by testing the specific sludge and the test will typically be complete in three to five days.

Equation 12.1 shows that the lower the initial depth of sludge applied, the lower the drying time will be. Lower initial depth, however, will proportionally increase the area required for the drying beds and the cost of facilities. A balance must be found between a reasonable drying time and a reasonable area. The initial depth is typically 20 to 30 cm (8 to 12 in.).

Calculation of total area required for sludge drying beds needs to consider drying time and the time it will take to remove the sludge cake from the beds. One complete cycle should include the time to apply the liquid sludge, drying time during the selected season, and time to remove sludge cake from the beds.

Drying time can be decreased by the addition of polymers for conditioning. Polymers tend to increase the amount of water that will drain initially, thus decreasing the amount of evaporation time required to reach a given solids concentration goal. Polymers, if used, should be applied to the full volume of liquid sludge applied to the bed. Blinding of the sand can result if excessive polymer is used.

Freezing can affect sludge drying beds. Freezing and thawing can enhance the rate of dewatering; prolonged freezing conditions, however, can deactivate sludge drying beds.

2.4 Paved Drying Beds

2.4.1 *Description*

Paving makes a good driving surface for sludge removal equipment. Paving can be concrete or asphalt and should have a slope toward the center of at least 1.5%. The characteristics of the sand, gravel, and underdrain are similar to conventional sand drying beds. One disadvantage of paved drying beds is that they tend to be more expensive than conventional beds and require more area because drainage will not be quite as good as conventional sludge drying beds. Other sizing criteria are similar to conventional sludge drying beds.

2.4.2 *Principles of Operation*

The principles of operation and calculation of drying time for paved drying beds are similar to conventional sludge drying beds. The main difference is that the allowance for drainage should be discounted moderately, increasing the overall drying time.

2.5 Wedge-Wire Drying Beds

2.5.1 *Description*

In wedge-wire drying beds, a wedge-wire screen replaces the porous media. Wedge-wire drying beds operate somewhat differently than conventional beds. Initially, clear water is applied to the underside of the wedge-wire screen, filling the enclosure to approximately 1 cm above the top of the screen. Liquid sludge mixed with polymer is then applied to the bed, initially with the drain valve closed. The sludge solids tend initially to float above the screen and settle down, forming a mat. The mat on the screen forms a barrier to the free movement of solids through the screen. When the bed is filled, the drain is initially partly opened and the filtrate is drained back to the WWTP. After approximately two hours, the valve is fully opened and the filtrate is allowed to drain freely. The valve is then left open while the sludge continues to drain for several days.

One advantage of the wedge-wire system is that there is no media to replace on a regular basis, as with conventional sand drying beds. In addition, the overall loading rate per year on a wedge-wire drying bed is greater than conventional drying beds.

2.5.2 *Principles of Operation*

Wedge-wire drying beds are used typically to achieve relatively rapid dewatering action but not high solids concentrations. Polymer is typically used with wedge-wire beds. Table 12.1 provides performance data reported for wedge-wire systems.

Table 12.1 Wedge-wire system performance data (WAS = waste activated sludge).

Sludge type	Feed solids, %	Final sludge solids concentration, %	Dewatering time
Primary	8.5	25.0	14 days
Digested primary and WAS	3.0	10.0	12 days
Fresh WAS	1.1	9.9	8 days
Thickened WAS	2.5	8.1	41 hours

Sludge removed from wedge-wire drying beds with concentrations similar to those in Table 12.1 may be difficult to handle and to use in land application systems.

2.6 Quick Dry Filter Beds

2.6.1 Description

One manufacturer (F. D. Deskins Co., Inc., Alexandria, Indiana) provides a proprietary sludge drying system consisting of a sand layer over specialized drainage panels. Physically, the drying beds look similar to conventional beds. The key to this process is the upstream polymer activation system. Sizing a Quick Dry system is similar to the sizing of conventional sludge drying beds, but should be confirmed through the manufacturer. One advantage of this system is that the sludge turnover time is typically one to three weeks instead of two to three months. This turnover time results in typical solids loading rates of 5 to 10 times conventional sand drying beds. This system also requires significantly less land than conventional sand drying beds. This system can also be used as part of a U.S. EPA-approved Class A biosolids process. The drying beds would be used to dewater the biosolids to approximately 40% solids. The dewatered sludge would be removed and placed in windrows and turned each day. The manufacturer claims that, in normal weather, liquid biosolids can be converted to dried, Class A biosolids in approximately 21 days.

2.6.2 Principles of Operation

Sludge is applied to the Quick Dry system similar to a conventional drying bed. Sludge must be conditioned with polymer before application. The underdrain system is designed to provide a siphoning effect that provides some amount of vacuum assistance to the drying process. The WWTP water system is used to apply plant water to the underdrain system, which creates the siphoning effect. The combination of the polymer conditioning and siphoning allows the sludge to dewater more rapidly than a conventional drying bed, according to the manufacturer.

2.7 Mechanically Assisted Solar Drying System

One manufacturer (Parkson Corp., Pasadena, California) provides a solar drying system consisting of a series of drying beds covered by a translucent, climatically controlled chamber. Sensors monitor the atmosphere in the chamber and control air louvers and ventilation fans to optimize drying conditions. The monitors also control a mobile, electric "mole" that plows the sludge cake during the drying cycle up to 13 times per day. The manufacturer claims that systems can be designed to produce dried sludge solids concentrations of 50 to 90%. The advantages of this system are protection from rain, elevated temperatures because of the "greenhouse effect," potentially drier cake solids, and shorter drying time. This system is proprietary, and information on sizing and design should be obtained from the manufacturer.

2.8 Drying Lagoons

2.8.1 Description

Sludge drying lagoons are similar to drying beds, except the liquid sludge is placed in the lagoons at depths at least three to four times greater than in a drying bed. Typically, sludge is allowed to dewater and dry to the design concentration over a period of one to three years or more. Sludge should be stabilized before being applied to a lagoon to minimize odors. Lagoons can be a source of odors as sludge goes through periods of wet and dry cycles.

Sludge drying lagoons require a considerable amount of available land but little operator attention. Evaporation is the most important factor in dewatering sludge in a lagoon.

Sludge drying lagoons consist of retaining walls that are typically earthen dikes 0.7 to 1.5 m (2 to 4 ft) high. The earthen dikes typically form rectangular lagoons. Design considerations include climate, subsoil permeability, depth and area, drainage, and dike slope.

2.8.2 Principles of Operation

A typical operating cycle for a lagoon is as follows:

- Stabilized liquid sludge is pumped into a lagoon over a period of up to several months.

- The supernatant from the liquid sludge is decanted periodically.
- The sludge is allowed to sit in the lagoon for months or years. A solids content of 20 to 40% can be achieved in 3 to 12 months, depending on climate.
- The sludge cake is removed by mechanical equipment, such as a front-end loader.
- The cycle is ready to be repeated.

3.0 BELT PRESSES

3.1 Principles of Operation

Belt filter presses continuously dewater sludge using two or three moving belts and a series of rollers. Sludge is typically conditioned before being applied to the belt filter press. Water is removed from the sludge through the belt or filter fabric through a combination of gravity drainage and compression. After dewatering, the sludge cake is typically discharged to a conveyance device or hopper.

Belt filter press machines evolved from the application of papermaking machine technology to the dewatering of municipal wastewater sludge. The belt filter press was introduced to North America in the 1960s as a lower-energy alternative to centrifuges and vacuum-filter equipment. Belt filter presses are used throughout the United States and are available from more than a dozen manufacturers. The availability of polymers has allowed belt filters to work on wastewater sludges as an alternate to rotary vacuum filters, recessed-plate filters, and drying beds.

3.1.1 Components of a Belt Press

Many designs of belt filtration processes are available, but all incorporate these basic zones: polymer conditioning, gravity drainage, low-pressure squeezing, and high-pressure squeezing. Figure 12.2 shows a schematic of a typical belt filter press. Advanced designs provide a larger filtration area, additional rollers, and variable belt speeds that can increase cake solids concentration by several percentage points.

3.1.2 Gravity Drainage Zones

Sludge is dewatered in the belt filter press process through three stages: chemical conditioning, gravity drainage to a nonfluid consistency, and compaction in a pressure and shear zone.

The dewatering operation of the belt filter press begins when the polymer-flocculated solids enter the gravity drainage zone. This section has a continuous

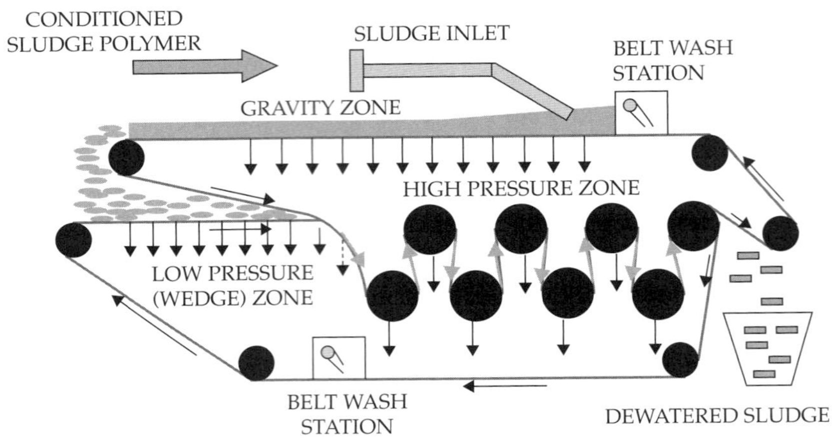

FIGURE 12.2 Belt filter press schematic (courtesy of Ashbrook-Simon-Hartley, Houston, Texas).

porous belt that provides a large surface area through which free water drains by gravity. A distribution system, which typically is provided, evenly applies the mixture onto the gravity feed belt. Filtrate from the gravity zone is collected and piped to a drain system. The gravity drainage zone is a flat or slightly inclined belt that is unique to each manufacturer. The gravity zone is also equipped with plows that move the sludge back and forth to facilitate water drainage.

Effectiveness of the gravity drainage zone is a function of the sludge type, chemical conditioning, belt fabric, and length of time in the gravity zone. Two-belt machines have a continuous flow path for the sludge because the belt on which gravity drainage occurs carries the material directly into the pressure zone. Three-belt machines have a single-belt gravity drainage zone that is similar to a gravity belt thickener but mounted above, and discharging to, a two-belt pressure zone. This configuration provides for a larger gravity zone that would likely benefit the dewatering of more dilute sludges. Equipment manufacturers select the length of the gravity drainage zone based on the inlet solids concentration and the relative drainage rate of the sludge/polymer mixture. Gravity-zone lengths range from 2 to 4 m (6.6 to 13 ft).

3.1.3 *Pressure Zones*

Following the gravity drainage section, the sludge enters a low-pressure zone often called the wedge compression section, where the belts converge and increase pressure on the draining material. The wedge compression section can be configured to permit liquid to escape through both belts.

Following the wedge compression section, the sludge enters a dewatering section of increasing pressure, where the two belts pass sequentially over a series of rollers of decreasing diameter. The roller diameters and belt tensioning system exert increasingly greater pressure as liquid is removed from the sludge, providing efficient dewatering and solids capture.

3.2 Design Criteria

The main design elements of belt filter presses include hydraulic and solids throughput capacity, sludge and polymer feed, belt washing, filtrate and dewatered-cake conveyance, equipment access and layout, and odor control. The characteristics of the feed sludge and use of the dewatered cake must be considered.

Belt press performance data indicate significant variations in the dewaterability of different types of sludges or biosolids. Although the press is capable of producing a dewatered cake that is 18 to 25% solids for typical combined primary and secondary sludges, many plants produce a 15 to 18% solids cake when dewatering anaerobically digested material. The solids capture rate, or total solids recovery, including washwater solids, ranges from 85 to 95%.

In recent years, new press designs have been developed (with additional rollers or separate gravity drainage or gravity-thickening deck) to produce higher dewatered-cake solids concentrations. Although these presses have provided significant dewatering improvement for some applications, for others it has been only marginal.

The best method for evaluating belt filter press performance on a specific material is to dewater the sludge using a pilot-scale test unit. Several belt press manufacturers have mobile trailer-mounted pilot units that can be rented for testing. Most pilot units are small production machines and perform comparably to larger models. Data that should be collected from the pilot test include hydraulic- and solids-loading rates, polymer type and usage, percent solids in the feed sludge and the dewatered cake, and percent capture of solids (total solids recovery).

As an alternative to a full-scale pilot test that many manufacturers offer is in-house testing equipment that can be used to predict belt press performance. A sample of the material to be dewatered can be sent to the manufacturer who will provide design criteria based on their tests. These tests have been shown to provide adequate projections of belt filter press performance for most smaller applications.

Whenever possible, polymer dosages and feed rates should be specifically optimized for characteristics at the facility. Specific resistance is used to determine filtration characteristics and optimum coagulation requirements of the sludge. A pilot- or bench-scale filter press test unit is used to conduct these tests. These tests typically indicate that material with a higher WAS proportion requires higher polymer doses.

When evaluating the performance of the belt filter press, like that of any other dewatering process, the quantity and quality of the filtrate and belt washwater and their effects on the wastewater treatment system should be considered. Typically, the recycle flow biochemical oxygen demand (BOD) varies from 150 to 300 mg/L, and total suspended solids (TSS) vary from 600 to 1100 mg/L.

3.2.1 *Structural Elements*

Structural elements within belt presses that must be considered include

- Frame,
- Rollers,
- Belt,
- Bearings,
- Belt drives/tension/tracking system,
- Press enclosures,
- Belt wash boxes, and
- Plows.

Refer to *Design of Municipal Wastewater Treatment Plants* (Water Environment Federation et al., 2009) for detailed information concerning the structural elements within belt presses.

3.2.2 *Hydraulic Loading Rate*

The throughput capacity of a belt filter press is the primary design criterion for sizing a belt press dewatering facility. Throughput capacity is considered to be either hydraulically limited or solids limited, depending on the feed solids concentration. Belt filter presses for a given unit of width have a maximum liquid or solids loading capability that can be attained only with the correct conditioning.

Typical hydraulic loading rates for a belt filter press range from 3 to 4 L/s per meter of belt width (15 to 22 gpm per foot of belt width). The hydraulic loading limit is typically 6 to 9 L/s per meter of belt width (30 to 45 gpm per foot).

3.2.3 *Solids Loading Rate*

Solids characteristics, origin, and degree of stabilization each have a significant effect on belt press loading and obtainable dewatering performance. Dilute sludge (0.5 to 1.0% total solids) requires more gravity drainage and polymer and a longer dewatering time than more concentrated material. Most manufacturers indicate that solids throughput, and possibly the percentage of cake dryness, increase as the percentage of solids in the feed increases. In addition, the type of process that is used for stabilization has a direct effect on the maximum solids content achievable. Although manufacturers estimates differ regarding the amount of dewatering that can be achieved, they typically agree that anaerobic digestion will yield a product more easily dewatered than will aerobic digestion. Typically, the lower the volatile content of the digested sludge, the higher the cake solids will be. In addition, as the primary to secondary sludge ratio increases, dewatering becomes easier and the cake solids concentration increases. Other variables that affect the maximum solids loading rate include: degree of stabilization, amount and type of fiber in the sludge, shear strength of the sludge, type of chemical conditioning, belt type, and maximum pressure applied to the sludge.

Typical dry solids applications rate range from 150 to 300 kg/m·h (100 to 200 lb/hr/ft). Typical maximum solids loading rate is 450 kg/m·h (300 lb/hr/ft). Recommended loading limits vary per manufacturer and should be verified when performing sizing calculations and comparing units.

3.2.4 *Number/Pattern of Rollers*

Each manufacturer of belt presses has a different configuration of rollers and most produce multiple models or variations with different roller configurations. Typically, belt presses are provided with seven or eight rollers.

Following the gravity drainage section, sludge enters a low-pressure section often called the wedge compression section, where the belts converge and increase pressure on the draining material. The wedge compression section can be configured to permit liquid to escape through both belts.

Following the wedge compression section, the sludge enters a dewatering section of increasingly greater pressure, where the two belts pass sequentially over a series of rollers of decreasing diameter. The roller diameters and belt tensioning system exert increasing pressure as liquid is removed from the sludge, providing efficient dewatering and solids capture.

3.2.5 High-Pressure Zones

Several of the belt filter press manufacturers provide the option to increase the number of rollers, expanding the pressure section of the press. This higher-pressure zone can be provided with the original equipment and, in some cases, can be added on to an existing machine. The number of additional rollers varies by manufacturer and application. The additional rollers increase the pressure and shear on the sludge and the potential for extrusion of sludge from between the belts. In some applications, the addition of the higher-pressure zone has not resulted in drier cake. The additional rollers increase the length of the machine.

3.2.6 Belt Speed

Compressive and shear forces are exerted as the sludge passes between the belts while winding its way through the belt press. Belt speed directly relates to solids detention time for passage through the various press sections, the dryness of cake solids, and throughput. Belt speed should be adjustable at the belt press control panel.

3.2.7 Belt Tracking

The belt tracking system maintains proper belt alignment by keeping the belts centered on the rollers. It uses sensing arms connected to a limit switch to sense movement in the belt position. A continuously adjustable roller senses the shift and automatically adjusts the belt to compensate. This roller is connected to a pneumatic, hydraulic, or electrically operated response system. An automatic, continuous modulating control must be an integral part of the system.

3.2.8 Belt Tension

Belt tension can be adjusted on the machine and is maintained and controlled either pneumatically, mechanically, or hydraulically during operation. Increasing belt tension will increase belt dewatering pressure. Several manufacturers offer separate control systems for the upper and lower belts so that the belt tension of each can be adjusted independently. An automatic adjustment system, similar to that described for the tracking system, is necessary. A pressure gage or similar device to indicate belt tension is recommended. The belt tensioning system should be able to accommodate at least a 3% increase in belt length. The system should adjust to maintain the desired belt tension as the belt stretches under normal use and wear. Belt life decreases as belt tension increases.

Systems with exposed gearing present a safety hazard; those that are not continuously running will impose a jarring action on belts. Retrofit of continuous tensioning and tracking systems deserves consideration, although this is typically difficult and costly. For new facilities, specifications should require continuously operating systems with ready access for easy maintenance and properly covered gearing to minimize potential safety hazards.

3.2.9 Belt Washing

The belt cleaning system includes a discharge blade and belt-wash system. The discharge or scrapper blade is often referred to as a "doctor blade." It is typically a knife edge constructed of ultrahigh-molecular-weight plastic. The doctor blade, typically located at the outlet end of the high-pressure section, scrapes or peels dewatered solids from the press belt into the cake disposal or conveyance system. A belt-tensioning system provides the ability to adjust the pressure exerted by the scraper blade against the belt and the angle at which the blade hits the belt. Components of the blade-tensioning-system parts should be made of corrosion-resistant material (for example, polycarbonate) and should be inspected frequently. Worn blades or poor blade adjustment reduces belt life and causes deterioration of the belt seam. Doctor blades are considered a wear item and should be removable for easy replacement.

Belt presses require a washing system consisting of piping, nozzles, drip pans, and spray containment shields to clean the belt that was in contact with sludge after discharge of the cake and before returning to the pressing zones. A belt-wash station typically is provided for each belt. The belt-wash pipe and nozzle, housed in either stainless-steel or fiber-glass enclosures, provide a high-pressure water spray to cleanse the belt of any dried or residual sludge, grease, polymer, or other material that binds. Self-cleaning nozzles are suggested; however, most manufacturers provide a manual cleaning feature that includes a hand-wheel-operated brush device internally mounted in the nozzle header pipe. Spray piping and nozzles should be adequately braced and of sufficient pressure rating to withstand changes caused by sudden valve closures.

Drainage from the gravity dewatering section, the pressure dewatering section, and each belt-wash area must be provided for the collection and transport of filtrate and washwater.

Drainage pans, shield, and piping should be designed to confine spray and splashed liquids and should discharge to a sump or floor drainage system directly below the unit. Drainage connections should be self-venting to prevent overflow. Drainage capacity must be sufficient to allow washdown of the unit. If possible, the drainage piping should be hard piped to the floor drainage system to minimize turbulence, thereby reducing odor generation.

When the drainage system is sized, filtrate and washwater flows must be included. One 2-m belt press, for example, can discharge 450 to 950 L/min (120 to 250 gpm) of drainage flow (filtrate and washwater). This recycle flow typically is routed back to the headworks or primary clarifiers. Combined filtrate and washwater flow TSS concentrations typically range from 400 to 800 mg/L.

3.2.10 Washwater Requirements

A reasonably clean washwater supply is needed to ensure adequate belt cleaning, especially when dewatering secondary WAS and scum that tends to rapidly clog the belt. This water supply, which amounts to 50 to 100% of the flowrate of the sludge to the machine, typically is pressurized up to 700 kPa (100 psi); sometimes, a booster pump is needed. Belt washwater can be potable water, secondary effluent, or even recycled filtrate water, although a clean supply is preferable.

3.3 Performance

Many variables affect the performance of a belt filter press including type and mix of sludge, sludge characteristics, method and type of chemical conditioning, pressure developed in the low-and high-pressure sections, and belt speed.

3.3.1 Cake Solids

The type of sludge to be dewatered significantly affects dewatering capabilities of the belt filter press. A dilute sludge (less than 1.0% total solids) will require a greater amount of gravity drainage and dewatering time than a more concentrated sludge. Typically, the dryness of the dewatered cake increases as the percent solids of the incoming feed sludge increases. In addition, the type of process that is used to stabilize the sludge affects the maximum solids content achievable. Anaerobically digested sludge typically will dewater better than aerobically digested sludge. Dewaterability and resulting cake solids concentration increases as the primary-to-secondary-sludge ratio increases. Table 12.2 provides typical performance data for various types of sludges.

3.3.2 Loading Rate/Belt Width

Hydraulic and solids loading rates depend on the type and concentration of the feed solids. Table 12.2 presents ranges of loading rates for various types of sludges.

3.3.3 Polymer Requirements

Chemical conditioning is important in the preparation of feed sludge for dewatering on a belt filter press. A polymer solution feed system is provided in the conditioning stage. The addition of polymer to the sludge, with proper mixing and sludge interaction time, will lead to the formation of large heavy agglomerated solids that maximize gravity water release. Table 12.2 presents typical polymer addition rates for various types of feed sludges.

Table 12.2 Typical data for various types of sludges dewatered on belt presses (WAS = waste activated sludge).

Type of wastewater sludge	Total feed solids, %	Hydraulic loading, L/s·m	Solids loading, kg/h·m	Polymer, g/kg	Total cake solids, %
Raw primary	3–10	150–190	450–725	1–5	28–44
Raw WAS	0.5–4	150–190	800–1000	1–10	20–25
Raw primary and WAS	3–6	150–190	360–725	1–10	15–35
Anaerobically digested primary	3–7	150–190	360–550	1–5	25–36
Anaerobically digested WAS	3–4	150–190	450–725	2–10	12–22
Anaerobically digested primary and WAS	2–6	150–190	225–450	2–8	18–44
Aerobically digested without primary	1–3	100–170	90–225	2–8	12–20
Oxygen activated WAS	1–3			4–10	15–23
Thermally conditioned primary and WAS	4–8	130–200	450–800	0	25–50

3.3.4 *Power Requirements*

Belt presses typically provide the lowest energy consumption per liter of sludge dewatered of the mechanical dewatering devices available. The largest electrical loads are associated with feed pumps, belt drives, washwater system, cake conveyance, and ventilation systems.

4.0 RECESSED PLATE PRESSURE FILTERS

The main advantage of a filter-press system is that it produces cakes that are drier than those produced by other dewatering alternatives. If cake solids content must be greater than 35%, then filter presses can be a cost-effective dewatering option. Filter presses can be adapted for a wide range of solids characteristics and have acceptable mechanical reliability, comparable energy requirements to vacuum-filter dewatering systems, and high filtrate quality that lowers recycle stream treatment requirements.

The main disadvantages of filter presses are their high capital cost, the substantial quantities of conditioning chemicals or precoat materials that are required, periodic adherence of cake to the filter medium that requires manual removal, and relatively high operation and maintenance costs.

Facilities England in the United Kingdom were using filter presses for wastewater solids dewatering with and without chemical pretreatment during the mid-1800s; however, before the mid-1900s, few cities in the United States were using pressure filter presses. From its inception through the 1960s, the essential mechanical features of the pressure filter presses remained virtually unchanged.

Improvements in mechanization and automation, such as automatic plate shifting, cake discharge, and washing, have reduced labor requirement. Moreover, the capacity range of the filter press has substantially increased, and fewer presses are required for larger facilities, enhancing cost effectiveness.

In terms of both capital and operation and maintenance costs, the pressure filter press system remains more expensive than other dewatering alternatives; however, when disposal requirements dictate drier cakes, pressure filter presses have often been proven cost effective because of the lower disposal costs (however, the added conditioning chemical mass should be considered). Moreover, many landfills have adopted more stringent criteria for moisture content of solids cakes in the United States; consequently, cake solids contents exceeding 35% are typically required. Typically, other dewatering devices cannot reliably achieve cake solids contents greater than 35% on a routine basis like the pressure filter press. They also have also been shown to be cost effective when solids cake must be incinerated. Typically, the increased dry solids content of the filter press cake (which increases the proportion of volatile matter to water content) enables autogenous combustion in incinerators, thus reducing the need for other fossil fuels such as natural gas or fuel oil.

4.1 Principles of Operation

Pressure filtration separates suspended solids from a liquid slurry using a positive pressure differential as the driving force. Recessed-chamber filter presses are operated as a batch process. Solids pumping to the filter press under pressure ranging from 700 to 2100 kPa (100 to 300 psig) force the liquid through a filter medium, leaving a concentrated solids cake trapped between the filter cloths that cover the recessed plates. The filtrate drains into internal conduits and collects at the end of the press for discharge. Finally, the plates separate and the cake drops downward via gravity, thus completing the press cycle.

On units of 1600 kPa (225 psig) pressure and higher, pressures typically are expressed in terms of atmospheres, or bars. A machine rated at 1600 kPa (225 psig) could also be called a "15-bar" unit; a pressure filter press rated at 2100 kPa (300 psig) is called a "20-bar" filter press.

Filter press dewatering is a combination of a constant-rate and constant-pressure process because of the boundary conditions set by the type of equipment used and the complex and often unpredictable interrelationships of the process variables. In this process, the beginning of the cycle uses a constant filtration rate up to the maximum pumping head available and then switches over to constant pressure filtration until the rate diminishes to a predetermined low level.

The typical filtration cycle is characterized by temporal variations in flowrate, pressure, and solids loading. Design capacities and controls further define these relationships by limiting the maximum pressure for system operation and maximum and minimum flowrates based on the design limitations of the feed system selected. The portion of the overall cycle where wastewater solids are actually being applied to the filter is termed the "form cycle." At the onset of feed solids to the filter press, resistance to filtration remains relatively low and constant until enough solids collect on the media to cause the press chambers to fill. This portion of the cycle is characterized by high and constant feed rates and relatively low pressure.

As solids accumulate and resistance to filtration builds, the flowrate declines and pressure increases. Solids will accumulate at a relatively high but steadily declining rate until the formed cake experiences a significant change in porosity that will severely restrict the amount of flow discharged as filtrate. Thereafter, because of the increased resistance to flow through the cake, pressure will continue to increase, but at a declining rate; similarly, flowrate will continue to decrease, but at a continually decreasing rate. The pressure of the system will continue to increase until the setpoint pressure is reached, and, thereafter, the system pressure will remain relatively constant while the flowrate continues to decline. Although pressure remains constant until the end of the cycle, cake porosity will continue to permit flow to pass, and solids will continue to accumulate in proportion to the flowrate, provided the particle concentration remains constant. Pressure filtration is affected by process factors such as particle size, specific gravity, and particle concentration.

The effect of particle size on filtration can best be illustrated using several examples discussed by Thomas (1971). If particles are of equal size, then the resulting cake will be loosely packed and relatively unstable (like a stack of marbles), particularly if the cycle incurred a large pressure drop. If the particles are relatively flat or platelike in nature, then the resulting cake will be a relatively impervious envelope characterized by a highly fluid center with correspondingly high moisture content. Municipal wastewater solids have a wide variety of particle sizes and shapes that can assist in keeping an open matrix of particles, thus promoting free filtration. Ideally, with the wide range in particle sizes, the voids between larger particles can be filled with smaller particles creating adequate flow channels between the individual particles and through the physical structure of the particles to promote and maintain free filtration. For most wastewater solids, the particle range to the filter press must be altered through chemical conditioning. Providing an open matrix to filter biological solids is difficult to accomplish. Biological solids are gelatinous in nature and result in relatively small void space to pass filtrate, even when well conditioned with inorganic chemicals. To provide adequate flow channels, higher mixtures of biological solids require increased quantities of lime.

Particles with a wide range of specific gravities pumped to the filter in relatively low concentrations can cause settlement to occur in the lower chambers of the press and result in poor cake formation and unbalanced cake pressures in the press. As solids concentration increases, the viscous drag created among the particles inhibits the coarser solids from settling out, unless there is a significant difference in specific gravity. This effect becomes less important when the feed consists of finely sized particles.

Particle concentration in the solids has a significant effect on filtration cycle time. Higher solids concentration in the feed will directly increase yield and correspondingly decrease cycle time.

4.1.1 Types of Filter Presses

Two types of filter presses are typically used to dewater municipal wastewater solids: fixed-volume, recessed-chamber filter press, which is the most common; and variable-volume, recessed-chamber filter press, also referred to as the diaphragm filter press. *Design of Municipal Wastewater Treatment Plants* provides more detailed information (Water Environment Federation et al., 2009).

4.1.2 Operating Cycle

The pressure filter feed system includes precoat, rapid fill, pressurization, and cake removal. Filter press feed systems must deliver conditioned, flocculated feed solids to filter presses under varying flow and pressure requirements.

4.2 Design Criteria

Performance criteria and process design conditions are critical to a pressure filter press. Design elements include operating pressure, number of plates, feed method, type of feed system, layout and access, type of press, mechanical features, and safety. Considerations associated with each of these design elements are discussed below.

As previously discussed, filter presses are either fixed or variable volume. Both types can be reliable when proper attention is given to operation and maintenance. The main operational difficulty encountered in pressure filter installations is inconsistent clean separation of the cake from the filter media. This problem may indicate the need for filter media wash or increased chemical conditioner dosages.

The main mechanical components of filter press equipment include the structural frame, filter press plates, diaphragms, filter cloths, and plate shifters. Ancillary equipment and controls include the feed system, operating pressure, washing system, cake removal, chemical conditioning, operating cycle, layout and access, and safety. Various options are available for each component. In many cases, individual manufacturers provide only one option for a particular component. A designer should carefully evaluate all requirements of a contract that may either specify or exclude a specific option. Refer to *Design of Municipal Wastewater Treatment Plants* for more detailed information (Water Environment Federation et al., 2009).

4.3 Performance

Performance will vary based on the following

- Cake solids,
- Cycle time,
- Chemical requirements, and
- Power requirements.

5.0 SCREW PRESSES

5.1 Principles of Operation

Two major types of screw presses are used in municipal dewatering applications: horizontal and inclined. The major elements of a screw press dewatering system are the sludge feed pump, polymer makeup and feed system, polymer injection and mixing device (injection ring and mixing valve), flocculation vessel with mixer, sludge inlet headbox or pipe, screw drive mechanism, shafted screw enclosed within a screen, a rectangular or circular cross-section enclosure compartment, and an outlet for the dewatered cake. Some horizontal screw press systems include a rotary screen thickener before the screw press. This screen is part of the combined dewatering and pasteurization process and may be used to reduce the hydraulic load to the screw press for certain feed sludge characteristics in conventional applications.

A screw press is a simple, slow-moving device that achieves continuous dewatering. Polymer is combined with sludge in flocculation vessels upstream of the screw press to enhance solids dewatering characteristics. Screw presses dewater sludge first by gravity drainage at the inlet section of the screw and then by squeezing free water out of the sludge as it is conveyed to the discharge end of the screw under gradually increasing pressure and friction. The increased pressure is generated by progressively reducing the available cross-sectional area for the sludge. The released water is allowed to escape through perforated screens surrounding the screw while the solids are retained inside the press. The liquid forced out through the screens is collected and conveyed from the press, and the dewatered solids are dropped through the discharge outlet of the screw at the end of the press. Screw speed and configuration and screen size and orientation can be tailored for each dewatering application.

Sludge is combined with polymer and pumped into the flocculation vessel. After flocculation, the sludge is transferred to the screw press unit. In the horizontal screw press configuration, sludge is fed by gravity from the flocculation tank to the screw press headbox. If a rotary screen thickener is used, then sludge flows from the flocculation tank to the rotary screen thickener and then to the screw press headbox. Sludge flows from the headbox to the inlet of the screw press. In the inclined configuration, sludge is pumped through the polymer injection and mixing device followed by the flocculation chamber to the inlet pipe to the screw press. The polymer injection-and-mixing device is designed to provide intense mixing of sludge and polymer, which determines the strength of the floc that is formed. The flocculation tank is designed to provide the reaction time needed to create the appropriate floc size.

A slowly rotating screw conveys the flocculated sludge through a wedge-wire drum. Reducing the pitch of the flights, increasing the center shaft diameter, and reducing the flight diameter gradually increases sludge compaction. The inclined screw press includes a pneumatic (or manually) adjusted dewatering cone at the discharge end of the screw press. The position of the dewatering cone of the screw press can be adjusted to provide an opening between 0.95 and 1.9 cm (0.375 and 0.75 in.) and used to regulate sludge pressure, which, in turn, provides a balance between increasing cake dryness and associated decrease in pressate clarity (solids capture). The pressure of the sludge in the inclined screw press is typically between 34 and 152 kPa (5 and 22 psi), with a maximum of 276 kPa (40 psi), depending on the type of unit and sludge characteristics. In the horizontal screw press, the screen is cleaned from the inside by the rotating screw flight with a nominal clearance of 0.5 mm (0.2 in.). In the inclined screw press, the wedge-wire screen is cleaned from the inside by brushes fitted on the screw flights to prevent sludge from attaching to the inner surface of the wedge wire drum. A spray wash system is used to clean the screen from the outside.

A screw press is also used in a process that combines dewatering and pasteurization. In this patented process, lime is added to the sludge to raise the pH to 12. The lime-treated sludge is then flocculated with polymer and fed to the screw press. Steam is used to heat the screw press to achieve pathogen reduction requirements for producing Class A biosolids. The generated biosolids typically have 30 to 50% solids concentration. The process has increased odor control requirements because of the increased temperature and pH. Selection of the combined dewatering and pasteurization process for meeting future Class A requirements may be required at the start of a project because different screw press design characteristics are required. Significant cost is required to upgrade a dewatering screw press to a dewatering-pasteurization screw press. The first Class A process installations started up in 2003, with minimal previous application in the United States.

Screw press cake solids concentrations vary widely depending on polymer use, sludge characteristics, and dewatering application. The higher cake solids concentrations are typically achieved in applications with primary sludge or with the combined dewatering and pasteurization process. Table 12.3 provides a summary of screw press dewatering performance data for various facilities. The table includes information on facility process and solids type, screw press type, feed and cake solids concentrations, solids capture, polymer dose, operating schedule, labor requirement, and manual cleaning schedule. The cake concentration varies significantly among facilities. Typical screw press cake solids concentrations are 15 to 28% for secondary waste activated sludge and 13 to 40% for a blend of primary and secondary sludge. Digested sludge cake solids concentration is typically higher than undigested sludge, but requires higher polymer dosage. The combined dewatering/pasteurization screw press results in higher cake concentrations (up to 50% solids) because of the added lime. Cake dewatering performance for sludge with poor dewatering properties such as, for example, in treatment facilities without primary clarifiers, can be significantly lower compared to alternative dewatering processes (Kabouris et al., 2005).

Table 12.3 Full-scale screw press data (WWTP = wastewater treatment plant; SRT = solids retention time; SBR = sequencing batch reactor; DAF = dissolved air flotation; BOD = biochemical oxygen demand).

Facility	Facility information/ process[a]	Solids type	Screw press type	Average feed solids, %	Average cake solids, %	Solids capture, %	Typical feed rate L/min	Through-put per dress, kg/d	Polymer dose, g/ kg dry solids	Operating schedule	Labor required	Manual cleaning schedule
1[b]	3 ML/d WWTP: OD (SRT = 18–21 d), SC, SH (SRT = 3–7 d)	Secondary sludge	Horizontal	4–6	20–45		57–114		11.5	24/7	8 hours/ day	
2	7.6 ML/d WWTP: OD (SRT= 20 d), SC, SH (SRT = 20 d)	Secondary sludge	Horizontal	2–3	15–20		106			16 hours/ day	1–2 hours per week	
3[b]	1.1 ML/d WWTP	Secondary sludge	Horizontal	1.5	30		14–16			24 hours/ day, 5 days/week	40–50 hours/ week	Once per week
4[c]		Secondary sludge	Horizontal	1.2	16		13			70–100 hours/ week		2–3 times per week
5[c]		Secondary sludge	Horizontal	1.2	16		13			150 hours/ week		Once a day
6	3.8 ML/d WWTP: AS (SRT = 24 d), SC, SH (SRT = 7–10 d)	Secondary sludge	Inclined	0.07	N/A[d]		38			8 hours/ day, 5 days/week		
7	17 ML/d WWTP: OD (SRT = 2–3 d), SC	Secondary sludge	Inclined	0.5–1	12–17	92	Up to 280		7.5–10	24/7	2 hours/ day	Once per week
8	7.7 ML/d WWTP: OD, SC, SH	Secondary sludge	Inclined	3.5–4	18–20		Up to 150		7.5	8 hours/ day, 7 days/week		
9	17.8 ML/d WWTP: PC, RBC, SC	Combined primary/ secondary sludge	Inclined	2.5	18–30		Up to 150	0.0033	9	2 days/ week, 6–8 hours per day		Twice per month

10	17 ML/d WWTP: RBC, SC, SH (SRT = 2–4 d)	Combined primary/ secondary sludge (40%/60%)	Inclined	1.1–1.2	30–40		300–340	0.0066		8 hours Monday, 2–4 hours Tues–Fri		Once a month
11[3]	0.9 ML/d WWTP: PC, TF, SC, AD	Aerobically digested secondary sludge	Horizontal	1.2	18		26–38			24/7	8 hours/ day	Once per week
12	16 ML/d WWTP: PC, AB, SC	Primary/ secondary (60%/40%) digested sludge combination	Horizontal	3–4	> 18	95	190–378	0.0055	9–11	5 hours/ day, 5 days/week		Once every 6 months
13	31 ML/d WWTP: BOD removal (SRT = 20 d), SH (SRT = 1 d)	Aerobically digested secondary sludge	Horizontal	1	15–19	90	95–397	0.0066	8.2	16 hours/ day M–F, 8 hours/day Sat–Sun	Minimal	Once per week
14	9.9 ML/d WWTP: OD (SRT = 10 d), SC, AD (SRT = 1.5 d)	Aerobically digested secondary sludge	Inclined	0.9	12–14		170			7 hours/ day, 5 days/week	Operator checks every 1–2 hours	Once per week
15	1.0 ML/d WWTP: AB, SBR, AD (SRT = 40 d)	Aerobically digested secondary sludge	Inclined	N/A	22–25		30–34			8 hours/ day		Once per week
16	6.1 ML/d WWTP	Aerobically digested secondary sludge	Horizontal	0.75–1.0	21–28		22–64			8–12 hours/day, 5 days/ week	1–2 checks by operator per hour	Once per month
17[3]	83 ML/d	Anaerobically/ aerobically digested primary/ secondary sludge	Horizontal	3.2	17–21		114					

(*continued*)

Table 12.3 Full-scale screw press data (WWTP = wastewater treatment plant; SRT = solids retention time; SBR = sequencing batch reactor; DAF = dissolved air flotation; BOD = biochemical oxygen demand). (*continued*)

Facility	Facility information/ process[a]	Solids type	Screw press type	Average feed solids, %	Average cake solids, %	Solids capture, %	Typical feed rate L/min	Throughput per dress, kg/d	Polymer dose, g/kg dry solids	Operating schedule	Labor required	Manual cleaning schedule
18	23 ML/d WWTP: PC, OD (SRT = 20 d), SC, AND (SRT = 30–60 d)	Anaerobically digested primary/ secondary sludge	Inclined	3–3.5	13–16		90–227			Midnight to 8–9 pm (3 hours downtime/ day)	4–6 hours/ day	
19	3.7 ML/d WWTP: PC, AB (SRT = 3.5 d), SC, CCT, DAF, AND (SRT = 41 d)	Anaerobically digested primary/ secondary (55%/45%) sludge	Horizontal	1	20–24	High	26–30			4 days/ week, 10 hours/day		Once per month
20[c]		Anaerobically digested primary/ secondary sludge	Horizontal	2–4	25		156–312					
21[3]		Anaerobically digested primary/ secondary sludge	Horizontal	2	22		163					
22	7.6 ML/d WWTP: PC, AB (SRT = 15 d), RBC, SC, SF, UV, AD (SRT = 20 d)	Aerobically digested primary/ secondary sludge	Inclined	2	19–24		106–170			24/7	1 hour/ day	

[a] PC = primary clarifiers, OD = oxidation ditch, AS = activated sludge, TF = trickling filter, SC = secondary clarifiers, AD = aerobic digestion, AND = anaerobic digestion, SH = sludge holding, DAF = dissolved air flotation, CCT = chlorine contact tank, SF = sand filter, SRT = solids retention time of liquid or solids treatment process.

[b] Class A installations with heat/lime process.

[c] Information included for these facilities was provided by the screw press manufacturer.

[d] Facility had only been in service for two weeks at time of survey. Cake solids percentage had not yet been tested.

Screw presses run continuously at low speeds and do not require close operator supervision; therefore, they are easy to maintain and have low power consumption. Typical manual cleaning schedule is once per week to once every 30 days.

5.2 Process Design Conditions and Criteria

The screw press dewatering process is determined by sludge characteristics, proper chemical conditioning, and hydraulic and solid load affecting the detention time/speed of the screw. It is essential to select the proper conditioning agent/polymer. The flocculation should agglomerate solids to large and strong flocs, which release as much water as possible. The released water drains in the thickening section of the screw press by gravity. Screw press performance improves with less water to be removed in the dewatering zone.

5.2.1 *Mechanical Features*

Significant variations exist in screw design among the manufacturers, and detailed information is proprietary. Primary features include the screw, screen, and flocculation system. Stainless steel (either 304 or 316L) is typically used for all wetted screw surfaces. *Design of Municipal Wastewater Treatment Plants* provides more detailed information (Water Environment Federation et al., 2009).

5.2.2 *Structural Elements and Building Requirements*

Structural elements for screw presses include a concrete base/pad or elevated support system for horizontal screw press installations, screw press support system for inclined screw presses, flocculation vessel, screw press, and support for the rotary screen thickener, if present. Depending on screw press installation configuration, a stair and landing system or movable ladder may be required to gain access to all parts of the screw press.

Because of the fully enclosed design for odor containment and low operating noise levels, it is possible to install screw presses outdoors in mild climates. For indoor applications, adequate room must be provided around the perimeter of the press to accommodate maintenance and manual washdown. Adequate overhead space and provisions for lifting must be considered for maintenance. Either fixed lifting equipment can be used, such as a crane rail centered above the press or mobile equipment. If mobile equipment will be used, then access to the equipment needs to be considered.

Because screw presses are fully enclosed, ventilation requirements are minimal. If odors are a concern, then screw press covers can be fitted with ventilation connections. For the connection to the screw press cover, flexible ductwork that can be removed by the operator without the use of tools is recommended. A flex hose slipped over a pipe stub held in place with a hose clamp with a thumbscrew is one example. Typical ventilation airflow rate requirements range from 340 to 680 m^3/h (200 to 400 cfm) for direct connection to the screw press. The screw press housing, the discharge of the screw press, or sludge cake conveyor system can be attached to a ventilation system for odor control. Several design and layout considerations exist for screw presses:

- Installing a curb around the screw press area to protect other areas from spills and washdown water.
- Providing access to all parts of screw press via platforms or movable stepladders.
- Including a hoist or crane for installation, removal, or repair of screw press components.
- Providing plant water connection for automatic wash down system.
- Installing connections to plant odor control system if odor ducts from screw press are desired. The dewatered sludge conveying system should also be connected to the odor control system.
- Including adequate space between screw press units for maintenance and manual cleaning operations.
- Providing sludge storage tanks from which sludge is fed to the press with sufficient mixing to ensure near-constant feed solids concentration during the dewatering process.
- Blending primary and secondary sludge in an upstream storage tank; mixing of sludge in the pipe feeding the dewatering system is not recommended.

A typical screw press dewatering system is presented in Figure 12.3.

5.3 Performance

5.3.1 *Hydraulic and Solids Loading Rates*

In addition to solids loading, sludge type, and desired discharge dryness, the hydraulic loading rate should be considered in screw press sizing. Higher hydraulic loading grates typically require large-diameter presses, coarser screens, or both. Sludge conditioning affect hydraulic loading rate because the maximum capacity of the screw press can be attained only with the optimal polymer type and application. Optimum operation occurs when the flocculated sludge has sufficient time to fully gravity drain before being conveyed beyond the inlet end of the screw press. Typical horizontal screw press hydraulic loading rates range from 3.8 to 2081 L/min (1 to 550 gpm),

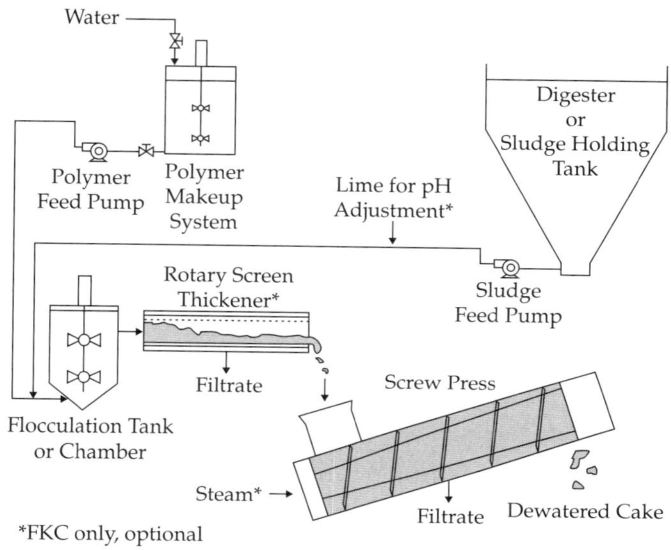

FIGURE 12.3 Schematic of screw press dewatering system (WEF et al., 2009).

depending on model. Typical inclined screw press hydraulic loading rates are 18.9 to 227 L/min (5 to 60 gpm), depending on the model.

The solids loading rate for screw press dewatering varies depending on sludge characteristics, screw size, and rotational speed. Horizontal screw press solids loading rate capacity ranges from 0.91 to 703 kg/h (2 to 1550 lb/hr), depending on model. Typical inclined screw press solids loading rates are often 22.7 to 295 kg/h (50 to 650 lb/hr), depending on model. Although the inclined screw press design is mainly controlled by solids loading, hydraulic loading is an important factor if the sludge is low in solids.

Screw presses can operate with a wide range of load rates simply with a change in screw speed. The design involves balancing increased solids/hydraulic loading rate via increasing screw speed and higher cake solids concentration, which improves with decreasing screw speed. Operation of the screw press is determined by the sludge flow (hydraulic/solids loading) and the speed of the auger (retention time of conditioned sludge in dewatering system). Cake solids will improve with decreasing auger speed, provided sludge flow is constant. The speed can be decreased until the point at which incoming flow exceeds the amount of water that drains by gravity in the lower part of the screw press. At which point, sludge will start backing up.

If the screw speed is constant, then the screw press can be operated with variable sludge feed flowrate. Because cake solids improve with increasing sludge flow, the flowrate can be increased to the point that incoming flow exceeds the amount of water drained by gravity in the initial part of the screw press and sludge starts to back up. Flowrate is then decreased to keep the system in a steady state.

The screw press will operate at its maximum capacity as long as the drainage of water is not limited by high-solids load, and the volume of the auger is filled properly with solids to ensure maximum pressure in the dewatering zone. The loading is highly determined by the correct conditioning; therefore, it is not possible to predict the values of the maximum loading rate. This value needs to be determined onsite (e.g., by pilot tests or during start up) and will vary from plant to plant.

5.3.2 *Units Redundancy*

Unlike many other dewatering systems, screw presses typically are designed for continuous operation, which ranges from five to seven days per week. Because of relatively infrequent maintenance requirements, screw press systems typically are designed without a redundant (backup) unit. In single-press installation, the plant must have facilities to store the sludge for several days in case the press is out of operation. Multiple screw presses may be necessary for larger facilities to increase system reliability. Multiple screw presses also allow for operating a smaller number of screw presses during extended periods of lower solids production.

5.3.3 *Rotation Speed*

Typical rotation speeds range from 0.1 to 2.0 revolutions per minute (rpm) for horizontal screw presses and from 0.5 to 2.0 rpm for inclined screw presses. An increase in screw rotation speed typically increases production capacity but decreases cake solids concentration. For example, in pilot testing at a municipal WWTP, cake solids concentration decreased from 16.6 to 14.2% when screw speed was increased from 0.3 to 0.5 rpm. The associated solids loading rate increased from 5.5 to 6.6 dry kg/h (12.2 to 14.5 dry lb/hr). In another test at the same facility, cake solids concentration increased from 14.4 to 17% when screw speed was decreased from 0.5 to 0.3 rpm. In a full-scale application, increasing rotational speed from 1 rpm to 1.25 rpm, decreased cake concentration from 23% to 20% (Atherton et al., 2005).

5.3.4 *Ancillary Equipment and Controls*

Ancillary equipment needed for screw press dewatering includes a sludge feed pump, polymer makeup and feed system, flocculation vessel, and dewatering control system. The screw press manufacturer may provide all of these components as a package system.

The dewatering operational control system is provided by the manufacturer and includes an interface to monitor and adjust the screw press. The main control panel operates the complete dewatering system including the sludge feed pump, polymer system, flocculation vessel, rotary screen thickener (if applicable), and screw press. The control system is equipped with operating and warning lights for various monitoring points, audio alarms, emergency shutoff devices, and display panel. Screw press dewatering systems are typically operated in an automatic mode.

The performance of the screw press (and any other dewatering system) relies on a consistent flow to the screw press. Therefore, positive displacement or progressive cavity pumps are highly recommended because their performance is not affected by sludge consistency and changing water levels in storage tanks. The progressive cavity pumps need to be designed properly to minimize wear and tear caused by grit and other abrasive material. The polymers used for dewatering with screw presses are not specifically designed for this application and are widely available from different suppliers. The type of polymer needs to be determined by jar testing.

Although there are no specific requirements for the polymer makeup system, it needs to be sized properly for the maximum solid load. Controls of the system should allow automatic operation and pace the polymer feed rate with the sludge flow. The highest efficiency of the polymer and, therefore, the lowest polymer consumption can be attained by using polymer systems with an aging tank.

The system always needs some operator attention, especially if the solid concentration of the sludge is not consistent. Control systems can only adjust the polymer dosing rate with changing sludge flow (hydraulic load); any change in the solids loading because of concentration changes need to be adjusted manually by the operator. Therefore, it is essential for the operating reliability that the solid concentration is constant (proper mixing in storage tank). Continuous solids content monitoring is a possible requirement. Sensors often require a lot of maintenance and operator attention. Therefore, it may be easier to readjust the settings of the dewatering system than to maintain the solid content sensor. Adjustment requires the change of the solids concentration parameter at the operator interface: solid content of raw sludge.

For the horizontal press, sludge pump speed is typically automatically adjusted based on headbox level analog input. The primary control for the sludge pump is maintaining a constant headbox level.

5.3.5 *Chemical Conditioning*

Polymer addition assists with particle flocculation and increases the degree of dewatering and solids capture rate. Jar and pilot testing can be used to estimate the type and quantity of polymer necessary for each application, which may vary significantly depending on sludge characteristics. Polymer consumption is affected by multiple parameters such as grit content of the sludge, presence or absence of primary clarifiers, type of biological treatment, and type and duration of sludge digestion. Polymer doses for screw press systems can range from 3 to 17.5 g active polymer/kg dry solids (6 to 35 lb active polymer/ton dry solids), with a typical range of 6 to 10 g active polymer/kg dry solids (12 to 20 lb active polymer/ton dry solids). Lime is also required for the horizontal screw press dewatering/pasteurization process. Typical lime dosage ranges between 100 to 400 g/kg dry solids (200 and 800 lb/ton dry solids).

Typically, an increase in polymer dose causes an increase in cake solids concentration, although other factors, such as screw speed, can affect performance. For example, in pilot testing at a WWTP, cake solids concentration increased from 16 to 17.5% when polymer dose was increased from 10.5 to 24.5 g active polymer/kg dry solids (21 to 49 lb/ton dry solids). In another pilot test, cake solids concentration increased from 14 to 15.5% when polymer dose was increased from 10 to 14 g active polymer/kg dry solids (20 to 28 lb/ton dry solids).

Polymer type and consumption also largely determines the solids capture rate. Solids capture is also affected by efficiency of polymer injection and mixing, design of the screen (size of opening), and pressure in the dewatering zone. For example, during a pilot test of WAS dewatering using the polymer used by the plant's belt filter presses, relatively weak floc was formed, resulting in a capture rate of 83 to 95.7% and cake of 12.2 to 15% solids. When cross-linked emulsion polymer was used to generate stronger floc, capture exceeded 95% and cake concentration increased to 14 to 18% solids.

5.3.6 *Energy Requirements*

Screw presses have relatively low power requirements. Screw press motor horsepower ranges from 0.67 to 10 kW (0.5 to 7.5 hp) for horizontal screw press installations and from 0.67 to 2.7 kW (0.5 to 2 hp) for inclined screw press installations, depending on screw press size. Installations also require a sludge feed pump and polymer system with varying power requirements depending on system size. The flocculation tank mixer motor is typically 2 kW (1.5 hp) or less.

5.3.7 *Washwater and Pressate*

Pressate quantity is a factor of the inlet flow and consistency and washwater flow. No screw presses use continuous washwater and most have automatic intermittent showers. Typically, horizontal press automated wash cycles occur for one minute each hour. The resulting washwater flow volume averages 2 to 5% of sludge feed rate. Depending on the size of the screw press, the instantaneous flowrate can range from 76 to 454 L/min (20 to 120 gpm). The presses require manual washdown at a frequency ranging from once per week to once per month. Typical volume required for manual washdown is an average of 0.2 to 0.5% of sludge feed rate.

The inclined screw press automatic spray wash system is controlled by a timer and runs intermittently, typically for one revolution (approximately 60 seconds) every 10 to 15 minutes. Wash cycles at some installations, however, operate only every 30 minutes. Typical inclined press washwater volume demand is 49 to 163 L (13 to 43 gal) per wash cycle per screw press, depending on size. The actual flowrate is 79 to 132 L/min (21 to 35 gpm). Total washwater volume is 4 to 9% of the processed sludge volume on average, with a maximum of 15%. Water demand depends on performance of the sludge flocculation. Manual spray-down cleaning of the flocculation vessel, screw, and screen is also done as needed depending on installation conditions and dewatering schedule.

Limited data is available documenting full-scale solids capture for screw presses under various field conditions. Screw press solids capture is typically 85 to 97%. Capture of up to 99% has been reported in certain applications; however, a maximum of 95% capture is typical in most installations. The quality of the pressate is typically clear when WAS is dewatered. It is slightly cloudy when primary sludge is present or digested sludge is processed. The particles are typically small, and the solid content varies depending of the screw press operation status: low solid content during dewatering mode and higher solid content during the wash cycle.

6.0 ROTARY PRESSES

6.1 Principles of Operation

Rotary press and rotary fan press dewatering technology relies on gravity, friction, and pressure differential to dewater sludge. Figure 12.4 provides the principal components of a rotary press. Sludge is dosed with polymer and fed into a

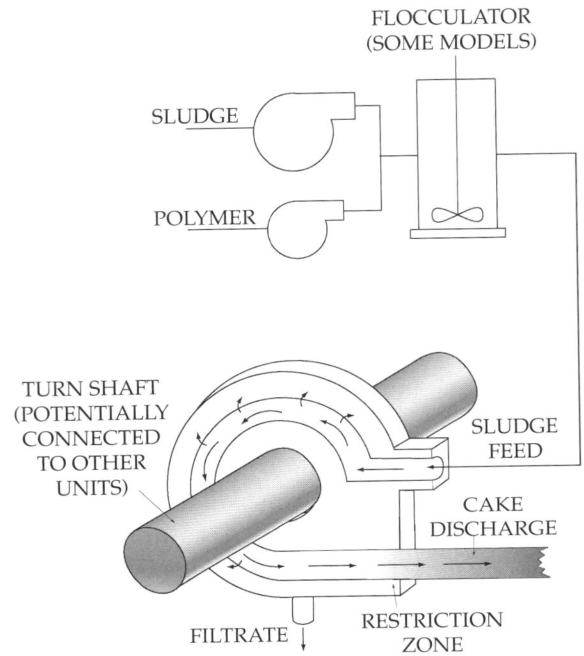

Figure 12.4 Rotary press process schematic (WEF et al., 2009).

channel bound by screens on each side. The channel curves with the circumference of the unit, making a 180º turn from inlet to outlet. Free water passes through the screens, which move in continuous but slow, concentric motion. The motion of the screens creates a "gripping" effect toward the end of the channel where cake accumulates against the outlet gate, and the motion of the screens squeezes out additional water. The cake is continuously released through the pressure-controlled outlet.

Figure 12.3 shows a process flow schematic of a typical screw press dewatering system installation. The major elements of a rotary press are the polymer feed and mixing system, parallel filtering screens, a circular channel between the screens, the rotation shaft, and a pressure-controlled outlet.

Entry zones of rotary presses and rotary fan presses function much like the gravity phase of belt press dewatering. Free water "falls" through the filtering screen pores and is collected in a filtrate channel. Pressure builds gradually as the solids travel toward the machine outlet. Because the outlet controls the pressure at which cake can be released, cake solids accumulate against the outlet and are further dewatered via friction from the continuous motion of the screens. In the rotary press, the friction generated between the screens and the cake plug translates into mechanical pressure that deflects the cake away from the center and forces it sideways against the restricted outlet. In the rotary fan press, frictional force also is generated in the outlet zone to dewater the sludge, but mechanical pressure is not generated at the same magnitude. In both designs, water is released through friction and is collected in the filtrate channel along with water released by gravity in the entry zone.

A key feature of rotary press and rotary fan press dewatering technology is the slow rotational speed of the moving unit. Typical installations use speeds of 1 to 3 rpm. This provides the advantages of low vibration, low shear, and low noise.

6.2 Design Criteria

Hydraulic loading rate, solids loading rate, operational control, and capture efficiency should be considered.

6.3 Performance

Performance is determined by requirements for chemical conditioning, energy, and washwater.

7.0 CENTRIFUGES

7.1 Principles of Operation

Centrifuges used in wastewater treatment use sedimentation to separate wastewater feed into a lighter phase, the centrate, and a heavier phase, the cake. The process is similar to an efficient clarifier or thickener. Because the centrifuge rotates at high speed, it creates significant acceleration, which drives the separation. Traditionally the acceleration is referred to as "g-force." The acceleration of the Earth's gravity is 9.81 m/s^2 (32 ft/s^2), also referred to as "one g," which is used by centrifuge manufacturers. Most centrifuges in service operate between 153 to 306 m/s^2 (1500 to 3000 g). In practical terms, if a centrifuge operating at 153 m/s^2 (3000 g) had a 0.45 kg (1 lb) imbalance on the bowl wall, then the centrifuge would shake with a 1260 kg (3000 lb) force. The relationship between rotational velocity and force (as a multiple of g)

is calculated as: relationship between speed and g-force, or acceleration by gravity, is calculated as:

$$\text{Rotational force} = k \times rpm^2 \times \text{diameter} \qquad 12.2$$

where

rpm = rotational rate, rpm, and

k = 0.000 014 2 when the diameter is expressed in inches or 0.000 000 56 when the diameter is in mm.

The "best" operating g-force varies, and higher is not always better. Higher g-forces do require better materials and manufacturing techniques and, therefore, more money to build. As a result, the g-force at which a centrifuge can meet its vibration and noise specification is an excellent measure of the quality of the mechanical design.

In operation, the feed material passes down a stationary feed tube that is an open pipe. The feed travels axially and spills into the feed zone, which is the center portion of the conveyor. The conveyor (also called the scroll) rotates within a few revolutions per minute of bowl speed. The liquid flows along the surface of the pool (also called the pond), and leaves the centrifuge over the dams at the centrate end. The liquid residence time from the point of leaving the feed tube to leaving the centrifuge is approximately one to two seconds. All of the energy imparted to the feed sludge, on the order of 2 kW/m3·h (0.6 hp/gpm) takes place in a fraction of a second in the feed zone and in the pool area outside the feed zone. As the solids flow from the feed area to the dams, the heavier material settles outward to the bowl wall. Liquid that is trapped between the compacting solids is displaced and migrates upward to the pool surface (see Figure 12.5). The adjacent illustrations show the water rising through less compacted particles toward the pool surface. By contrast, in filtration,

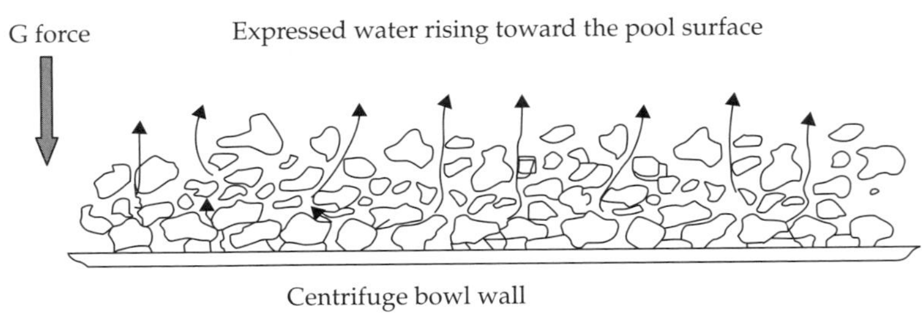

FIGURE 12.5 Water removal in centrifuge.

the water must pass through the densest solids on the filter media. As a result, filtration devices are more limited by thin feed material than centrifuges. The conveyor turns at a slightly different speed than the bowl, and this differential rpm (also called delta rpm or relative speed) conveys the solids to the conical end of the centrifuge (also called the solids end or small end) where it leaves the centrifuge. As with all screw conveyors, the conveyed solids exert a viscous drag on the screw. The conveyor motor measures this drag as amperage, and the controls convert it to torque.

7.1.1 *Major Elements*

Elements of a solid bowl centrifuge consist of the bowl shell, scroll, frame, and scroll drive. The bowl shell consists of the front hub (centrate end), rear hub (small end), conical bowl extension, and cylindrical bowl. Within the bowl is a conveyor operating at a somewhat different speed. Whether the conveyor turns faster than the bowl or slower than the bowl does not make a difference. Presently, all centrifuges offered to the environmental industry are manufactured of stainless steel. It is difficult to maintain critical tolerances necessary for higher speed operation with carbon steel. The structural stainless pieces are manufactured from centrifugal castings, static castings, and rolled and welded assemblies as chosen by the manufacturer. Centrifugal castings are better because this process produces castings with more uniform properties; they are preferred for use in higher speed centrifuges. At lower speeds, or in lower stress areas, manufacturers sometimes use static casting and rolled and welded steels because they are lower in cost. In operation, the vibration caused by an imbalance increases as the speed increases. The amount of imbalance is a result of quality of the design and manufacturing. Issues may include parts that are not quite on the center line, pieces that do not mate up square, and metals that stretch under load. In most cases, the balance can be trued up by adding or removing weight at strategic points. When metal bends under stress, however, it is a big problem for balancing, because the imbalance changes with speed and load. If the conveyor hub is not stiff enough, then it will whip, creating imbalance there. Centrifuges that balance well in air may develop an imbalance when water enters the bowl. Centrifuges will have greater vibration in operation at the plant compared with at the manufacturer's factory. Manufacturing quality is best defined by the noise and vibration level when running at the nameplate speed and full of water. A low-speed design cannot be balanced to run smoothly at a higher speed. Design engineers and owners expect the nameplate speed to be the operating speed at which the centrifuge can meet the factory vibration standards and can run continuously. Unfortunately, some manufacturers create their own definitions and, if challenged, will admit that the centrifuge nameplate speed is not an operating speed but a stress level. It may walk across the floor from vibration at the nameplate speed but will not spontaneously disassemble. The maximum operating speed (without vibration alarm shutting everything down) is lower, and the normal operating speed lower even lower.

The purpose of the frame is to hold support covers, motors, guards, and to keep the main bearings in alignment. The ideal frame would be infinitely rigid and weigh nothing. This frame would allow the centrifuge to oscillate (vibrate) with minimum dynamic load on the bearings and maintain precise bearing alignment, thus increasing bearing life. It also would require precision manufacturing to minimize the amount of movement of the centrifuge. As a rule of thumb, to be sufficiently rigid, the frame needs to be about the same weight as the rotating assembly. Frames are made out of box channel, cast iron, and reinforced concrete. Owners value lower noise and vibration, and manufacture value lower manufacturing costs. All nominal objectives can be met by loading frames with 100 to 200% dead weight, which results in a smooth-running centrifuge. Manufacturing costs go down with looser manufacturing and design standards. Massive frames reduce vibrations, which may conceal high rotor imbalances with resulting high dynamic loading that shortens bearing life. Prospective purchasers should be aware of the total static weight/total dynamic weight ratio:

$$\text{Static/Dynamic Weight ratio} = \frac{\text{Total static weight on the four main vibration isolators}}{\text{Weight of the rotor filled with water}} \qquad 12.3$$

The static loading on each isolator is typically shown on the installation drawing supplied by the manufacturer. The sheet will include the maximum weight of the rotating assembly. Assuming the total weight is 15 000 kg (33 000 lb) and the rotor is 4500 kg (10 000 lb).

$$\text{Static/dynamic weight ratio} = 15\,000\ \text{kg}/4500\ \text{kg} = 3.3\ \text{S/DW} \qquad 12.4$$

S/DW $\leq$ 2 is excellent,
S/DW = 2–3 is fair, bearing life will be shortened, and
S/DW = 3–4 is poor.

Typically, smaller centrifuges are built to lower standards than larger ones, and higher S/DW are more acceptable if the centrifuge will only operate a few days per week.

Torque generated by the conveyor moving the solids out of the centrifuge is proportional to the sludge viscosity, which, in turn, is proportional to cake dryness. In dewatering applications, the torque is huge, so all centrifuges have a torque reducer of some sort, typically with a reduction ratio between 60 and 250:1. Three reducers are typically used: two- and three-stage planetary gear boxes; one-, two, and three-stage cyclo gear boxes; and hydraulic motors. It is nearly impossible to judge the quality of the torque reducers or to determine a realistic continuous rating without disassembling and studying the engineering design. Instantaneous torque ratings are not useful because gearbox life is a function of cycles and loading. The cycles are the differential rpm, which is low, and the loading is the torque, which is typically high. A quality torque reducer should last 20 000 to 40 000 hours without mechanical problems or rebuild; the best designs last 50 000 to 100 000 hours or more. Asking reference installations about their experience and what the typical torque load is *as a percentage of the allowable load* is the only practical way to estimate repair costs. It is impossible to know, however, if the new device is exactly same as the one that was used in the past. The gearboxes, in practice, have a fixed ratio, typically between 90 and 200:1. One centrifuge manufacturer has a variable gear ratio, which changes, depending on pulley and belt combinations installed. The benefit of this design is unclear, because the pulleys are rarely changed. Regardless of the hardware, all of these mechanical reducers behave the same. Changing the speed of the scroll drive motor results in a change in the differential rpm. The formula for this is

$$\text{Differential, rpm} = (\text{Blow speed} - \text{Gearbox input speed})/\text{Gearbox ratio} \qquad 12.5$$

For example

$$\text{Bowl speed} = 2800 \text{ rpm}$$
$$\text{Gearbox input speed} = 2650 \text{ rpm}$$
$$\text{Gear ratio} = 95{:}1$$
$$= (2800 \text{ rpm} - 2650 \text{ rpm})/95 = 1.58 \text{ rpm}$$

The difference between the speed of the bowl, and the conveyor would be 1.58 rpm. The gearbox input speed is the speed of the conveyor backdrive motor. By varying the speed of this motor, the operator can change the differential rpm. The larger the gear ratio, the smaller the size of the backdrive motor but unfortunately, the smaller the maximum differential rpm. The maximum differential speed is important when the centrifuge is near plugging, and should be 12 to 15 rpm. The solids create a viscous drag on the scroll that is proportional to the cake dryness. In electronic drives, the torque is measured by the amperage on the backdrive motor. The Viscotherm is a hydraulic system consisting of two hydraulic pumps: (1) a stationary pump that pumps oil under high pressure to the rotating assembly; and (2) a Rotodiff pump that is connected between the bowl and the conveyor. In this system, the differential rpm is proportional to the volume of oil, and torque is proportional to the oil pressure.

Energy efficiency of the various designs depends on details of the centrifuge itself, the motors, the drive, and the load. A power calculation consists of the windage and friction to rotate the centrifuge, the acceleration of the feed to the tangential velocity of the pool, the energy consumed by scrolling the solids, and the efficiency of the electronics and hydraulics. The larger number is the power to accelerate the feed material up to bowl speed. Since 2000, most centrifuges operate with a smaller solids discharge and pond radius than older centrifuges. Reducing the discharge radius reduces the tangential discharge velocity and, therefore, reduces the power. Power consumed by the various backdrives and gear must be added to the total power requirements. Planetary gears are more energy efficient than cyclo gears, which in turn are more efficient than hydraulic devices. As with the torque ratings of the drives, there is no practical way to confirm the energy claims of a manufacturer; however, for the most part, the energy differences are minimal.

7.1.2 Factors That Affect Process Efficiency

The dewatering process occurs between the feed pipe where feed material enters the centrifuge and the centrate/solids hubs. The four steps that must take place are:

1. Acceleration of the feed material up to bowl speed;
2. Separation of the solids from the moving liquid layer;
3. Compaction of the solids; and
4. Removal of the solids from the centrifuge.

Certain techniques are thought to enhance each of these steps. The purpose of the feed zone is to accelerate the incoming feed up to bowl speed and distribute the flow around the circumference of the bowl. For a 750-mm (30-in.) diameter centrifuge, the feed zone is where most of the power is transferred to the incoming liquid. Either the conveyor or the pool area must accelerate the feed to a tangential speed of 200 to 250 km/h (120 to 150 mph). The more that this energy transfer can take place in the conveyor hub, the less disturbance and turbulence in the pool where separation is occurring. The simplest feed zones consist of two or more large openings in the conveyor hub, which results in much of the feed spilling into the pool and relatively little acceleration. As a result, the pool of liquid/sludge in the centrifuge must accelerate the feed material, creating significant turbulence. The better process designs are ones that do as much of the acceleration of the feed in the conveyor hub as possible, making use of an otherwise useless area. When feed leaves the conveyor hub through nozzles, it has a large radial velocity that stirs up the settled cake under the feed zone. For much the same reason that clarifiers have a baffle to absorb incoming velocity, some centrifuge designs have a deflector that absorbs the radial velocity, making the overall process more efficient.

7.1.3 Flow Path

Clarification takes place between the feed zone and the centrate dams. One of the designs to do this was a simple helical path. The feed spilled into the space between the flights and flowed in a spiral like a nut on a bolt down the length of the bowl. On the one hand, feed distribution was unimportant, the width of the flow path on the larger centrifuges was approximately 300 by 2290 mm (12 by 90 in.). This creates a long, narrow flow path with high velocities, which is not effective for clarification. Helical centrifuges that produce "dry solids" are prone to bridging solids between the flights that creates a dam and causes feed to suddenly back up and divert out the cake end. Designs where the liquid flowed in one direction, and the cake traveled in the other is called countercurrent flow. For a time there was also a variation of the helical flow called the cocurrent design, where the feed entered at the centrate end of the centrifuge, and flowed toward the solids discharge end in the same direction as the solids transport. Approximately midway up the centrifuge, the centrate turned 180° and flowed down channels welded to the conveyor, past a rotary seal immersed in sludge, over the dams, and out of the centrifuge. Maintenance problems unique to this design resulted in the elimination of cocurrent scrolls in modern centrifuges. At around the same time as the cocurrent design was developed, axial flow was introduced.

Axial flow centrifuges have flights in the path from the feed zone to dams mounted on spokes or, in less costly designs, openings cut in the flights to reduce the flow velocity and enhance solids separation. This flow path is two to three times longer than it is wide. As a consequence, changes in feed rate do not change the differential head pressure (ΔH) as much as with the helical design, allowing these centrifuges to handle a wider variation in feed rates without changing the dams. As the differential rpm decreases to obtain drier cake, the space between the flights tends to bridge. In helical centrifuges, the blockage may cause the liquid level to rise, and cause a washout. With axial flow, bridges may form, but the feed easily flows around the blockage. As a result, axial flow is the most common type of centrifuge. Axial vanes were offered at one point, but, in the end, the expense to build them exceeded the value to the owner and are no longer in production.

7.1.4 *Solids Concentration Mechanism*

In sedimentation, it is the net weight of the sludge particles, one bearing on the other, that compresses the pile of sludge and expresses water to the surface of the pool. Within limits, the higher the pile of solids and the higher the *g*-force, the higher the resulting cake concentration. In centrifuges, efforts to increase the solids depth resulted in larger bowl diameters and smaller diameter pools. A second problem was that the cross-sectional cylindrical area became much larger than the cross-sectional beach area. This resulted in a higher volume of sludge being scrolled in the cylindrical section to the beach, causing large recycle of settled sludge and inefficient operation that limited cake dryness. The solution was to raise the liquid level 15 to 20 mm higher than the solids discharge radius. This is variously called a negative pond or a "super pool." Although the negative pond has a slightly increased pond depth and volume, its principle effect is to increase the difference in radius between the pond and the solids discharge, called the differential head pressure (ΔH). The ΔH extrudes much of the sludge out of the centrifuge, allowing the operator to control the separation by making small changes to the differential rpm staying within the range of 1 to 2 rpm. The low differential rpm significantly reduces the internal recycle, making the process more efficient. The centrifuge controls need to be precise enough to allow the operator to set the differential rpm to ±0.05 rpm.

Some centrifuge manufacturers reduce the solids discharge diameter to make the pond deeper without moving the dam openings. This limits the differential head pressure, and, therefore, the cake dryness the centrifuge can achieve. Limiting cake dryness limits torque, allowing smaller torque reducers and thinner flight material without bending flights. It is easier to limit cake dryness by limiting how negative the pond can be than to limit the allowed torque. A good centrifuge design should have the capability of a 25-mm negative pond, and the largest diameter of the conveyor hub should be at least 50 mm smaller than the solids discharge diameter.

One unfortunate consequence of the negative ponds is that all dewatering centrifuges achieving dry cake discharge feed material (i.e., "sloppy cake") out of the cake end during startup, approximately 5 to 15 minutes, which must be considered during installation. Two approaches can be used to address this issue: (1) using a diverter gate to send the startup sloppy cake to the drain; (2) placing an inclined screw conveyor under the centrifuge. During startup, the screw reverses to the drain and, during normal operation, conveys the sludge upward to a transverse conveyor.

7.1.5 *High-Solids Dewatering*

Older centrifuges had analog controls and no great effort to ensure accuracy. High-solids centrifuges need precise control of the differential rpm within ±0.05 rpm and good torque control. The auto torque should hold torque to ±3 percentage points of setpoint, with no more than a 0.5 rpm range in the differential rpm.

Because the net weight of the solids compresses the cake, the deeper the pool (pond radius–bowl radius) becomes and the more centrifugal pressure compressing the cake. As a result, the centrifuge must have a very deep pond. When comparing narrow centrifuges, the deeper the pond the better. Large centrifuges have an advantage over smaller centrifuges because they nearly always have deeper ponds.

Pond use is important. To completely fill the centrifuge volume with sludge and to minimize internal recycle, the differential rpm should be a low 0.5 to 2 rpm, regardless of throughput. This requires a negative head of approximately 15 to 25 mm or so. Very dry cake has a very high viscosity, which in turn requires a large torque capacity.

7.1.6 *Bowl/Scroll Geometry*

Scale-up issues are important to bowl/scroll geometry. The single most important dimension is pool depth. As with sedimentation, the deeper the pool, the better the cake dryness. The pool depth of large centrifuges is typically deeper than that of smaller centrifuges. Hence, for two centrifuges with equal volume, a long, shallow pool centrifuge is not equal to a shorter deep pool centrifuge. The length-to-diameter (L/D) ratio is another standard. Economics results in long bowls with a length approximately four times the diameter. The L/D is a design parameter that influences centrifuge throughput capacity and should be included in specifications.

7.1.7 *Bowl Speed*

Most centrifuges in wastewater actually operate between 245 and 285 m/s^2 (2400 and 2800 g). Nevertheless, 306 m/s^2 (3000 g) has emerged as a quality standard. As operating speed increases, conveyor hubs have to be stiffer, better materials used, and components machined to tighter tolerances. All this adds to cost. If the owner reduces the speed of a 306 m/s^2 (3000 g) centrifuge to 255 m/s^2 (2500 g), then vibration, noise, power, and erosion will decrease, and bearing life will be extended compared with a centrifuge designed for 255 m/s^2 (2500 g). In specifying centrifuges, it is important to insist that the vibration, noise, and process guarantee be done at the nameplate g-force. Otherwise, the owner may discover that the 306 m/s^2 (3000 g) claimed by the manufacturer is not an operating speed, but a "design stress speed." A lower speed, or "maximum operating speed," is the speed that triggers the vibration alarm and the "normal running speed" is even lower.

7.1.8 *Bowl/Scroll Differential Speed*

The difference between the bowl speed and the conveyor speed is typically between 0.5 and 6 rpm for dewatering centrifuges. At times, the centrifuge needs to operate at a high differential rpm of 15 or greater to clear solids out of the bowl. If a centrifuge is running at 2600 rpm with a differential speed of 1.5 rpm, then the conveyor will be rotating at 2598.5 or 2601.5 rpm, depending on whether the centrifuge has a conveyor that turns slower or faster than the bowl. There is no credible evidence to show that there is any advantage to the conveyor turning faster or slower than the bowl. It is important that the controls hold the differential rpm to ±0.05 rpm.

7.1.9 *Pool Depth*

Centrifuges work by the process of sedimentation. Solids are classified by settling rate, and the pressing of one particle on those below it expresses moisture from between the particles. The pressure is a function of g-force and the depth of the solids blanket. Because of their shallower pool depths, smaller centrifuges tend not to produce a cake as dry as larger centrifuges. It may seem unclear why a small change in pool depth often makes a significant difference in performance. The reason, however, is tied to the necessary geometry of the bowl. For successful separation, the solids have to separate from the liquid moving through the centrifuge. The solids, once settled out, have to scroll out of the centrifuge; the pond affects this scrolling rate. When removing the solids, the conveyor scrolls them up the beach and out of the centrifuge. As long as the solids are submerged, buoyancy helps lift them up the beach. If the pond is lower than the top of the beach, then solids would emerge from the pool, loose the benefit of buoyancy, and slide back down the beach. If the pool is much less than 3 mm (0.125 in.) below the solids discharge, little or no wastewater sludge will be transferred out of the solids end of the centrifuge. If the pool level is higher (smaller radius) than the solids discharge radius, then water will spill out during startup; eventually, the seal will reform and dewatering will resume. As the ponds are set higher, they exert hydraulic head pressure, which extrudes some of the solids. As a result, the solids range from the centrifuge is actually

$$\text{Cake rate } \alpha \, \Delta\text{rpm} + \Delta H \qquad 12.6$$

where

Cake rate = rate of cake leaving the centrifuge, which is proportional to Δrpm,
Δ rpm = difference in speed between the scroll and the conveyor, and
ΔH = difference in height between the pond radius and the solids discharge radius.

All things equal, as pond depth increases, the differential rpm will decrease. It is important that the "dry solids" centrifuge be designed to operate with a ΔH of at least 19 mm (0.75 in.). Another design consideration, is how the plate dams affect cresting. Water flowing over a weir has a crest height proportional to flow. As the flowrate increases, the pool height increases. Therefore, the ΔH is proportional to the flowrate. Good designs minimize this by using a ring dam that has a long weir length. The next best design is a plate with a radius cut that matches the curvature of the pond. Round holes and flat plates do not result in good performance.

7.2 Design Criteria

7.2.1 Structural Elements

Virtually all centrifuges now offered have all wetted parts made out of stainless steel. Major structural parts are best made from centrifugally cast stainless steel. Centrifugal casting produces a more uniform product with no soft spots or slag. As a result, it is stronger than the same size piece made by static casting or by rolling and welding. To some extent, the weakness introduced by static casting can be minimized by using more metal, which makes the piece thicker. Rolling and welding bowls is not typical because the strength of the weld is suspect. Erosion protection is important. Areas of the bowl shell that need protection are the bowl under the feed zone and at the base of the beach. Solids with a significant amount of grit sliding on the bowl can cause serious wear. Most manufacturers now tack weld strips directly to the bowl to lock in a static heel to protect the bowl underneath. Inserting a ribbed liner is a more elegant and expensive procedure that minimizes welding on the bowl. The area where the solids leave the bowl is another wear point. The hubs and bowl together form a structural unit called the bowl shell. The opening where the solids leave the centrifuge is subject to severe abrasion, and failure can be structural. Typically tungsten carbide is the best for wear protection, but its brittleness is sometimes a problem. More resilient but less abrasion-resistant materials can have a longer life. Likewise, the area in the casing where the solids leave the bowl has a linear velocity greater than 240 km/h (150 mph). Urethane and rubber liners provide, by far, the best service (as long at the temperatures are less than approximately 55 °C (130 °F), followed by plasma applied steel, and, worst of all, stainless steel. Liners should extend below the frame to protect the flex connectors. All centrifuges act as blowers, sucking in air near the centerline and blowing it out of the solids and liquid chutes. Inadequate venting can result in solids filling the center casing and building up to the point that they are rubbing on the bowl, causing serious damage. The design engineer should follow exactly the manufacturers' venting requirements. Structural elements include bowls, scrolls and motors and drives.

Lastly, the choice of centrifuge will affect facility design and will require additional structural support compared with units such as belt filter presses.

7.2.2 *Dynamic Loads*

Imbalance of the rotor is often designed-in by the manufacturer. The dynamic load on the bearings is a significant contributor to bearing life. The imbalance in a centrifuge results in a dynamic load applied to the bearings and transferred to the frame and through the vibration isolators to the structure the centrifuge sits on. A large centrifuge with a static weight of 9100 kg (20 000 lb) operating at 2600 rpm with a 0.5 kg (1.1 lb) imbalance exerts a 43 Hz dynamic load of 1392 kg (3066 lb). The dynamic load is proportional to the vibration and the harmonics of the centrifuge and its suspension system. All manufacturers supply an installation sheet that details these loads and necessary hook heights and clearances. The dynamic loads are typically in the vertical and horizontal plane, with minimal axial dynamic loading.

7.2.3 *Vibration/Noise Control*

All centrifuges vibrate and make noise. To the extent the manufacturer has done a good job with the design, both the noise and the vibration are minimized. In the best practice, the weight of the rotor is approximately half of the total weight on the isolators. A well-manufactured centrifuge can achieve 1 mil. vibration rms. It is expensive to manufacture to these standards. As a result, excess weight typically is added to the frame to suppress vibrations and the rotor is 25 to 30% of the weight on the isolators. For this reason, operating *g*-force specifications should be exact to ensure that the centrifuge can meet its noise and vibration requirements. Evaluation of centrifuges should consider the rotor-to-total weight ratio as a quality measure of the centrifuges offered. Factory noise and vibration tests are typically conducted with a dry centrifuge far from the WWTP and do not include plant staff. A more realistic place to measure noise and vibration is during the performance testing. Manufacturers argue that they do not control the environment or the structure, which will affect noise and vibration; although this is true, the issue that remains is how the centrifuge will operate at the WWTP. A reasonable compromise is that the peak vibration the centrifuge should reach while operating online during its performance test should be at or below 150% of the factory vibration level. Older centrifuges had a simple vibration switch that was a safety device to shut the centrifuge down in the event of a severe mechanical problem. Vibration meters tell operators how much imbalance is occurring and whether it is getting worse or better. This instrumentation is inexpensive and should always be specified.

Even if the installation met the 85 dB standard when new, the noise will typically increase over time. Sound deadening covers and sound absorbent panels on walls and ceilings are cumbersome and expensive. The best solution is for operators to minimize time spent in the same room, or even on the same floor, as the centrifuges, other than for a walk through once or twice per shift. The operator control room should typically be on the floor below the centrifuge with the centrate, cake and process sampling and instrumentation.

7.2.4 *Hydraulic Loading Rates*

With all dewatering devices, as hydraulic loading increases, performance decreases. Centrifuges throughput can vary from 76 to 6800 L/min (20 to 1800 gpm) each. Ideally, picking a single hydraulic rate and optimize the centrifuge for that rate is the best approach. The wider the variation in feed rates expected, the more the centrifuge will have to be de-adjusted to accomplish the rate variation. Centrifuges typically are less limited by hydraulic loading than solids loading primarily because the extra polymer that is added for a drier cake in rapidly settling floc. Within the centrifuge, feed material flows in a thin, moving layer on top of the sludge blanket below. The solids only need to settle a fraction a few millimeters to enter the sludge blanket. A second reason centrifuge are not as hydraulically limited as filtration devices is that the extracted water flows from the densest cake on the bowl wall up through less dense cake to the surface or the pond. Filtration devices have to push the water through the dense cake on the filter media.

7.2.5 *Solids Loading Rates*

The sludge blanket inside the centrifuge is more or less fixed in volume—as solids loading rate increases, solids residence time decreases, resulting in wetter cake. This can be offset somewhat by increased polymer or higher *g*-force to a point. Four performance parameters occur with centrifuges: cake dryness, polymer dosage, feed rate, and centrate quality. The operator can improve one parameter at the expense of another. For example, reducing the feed rate will increase the solids residence time, which results in a drier cake, lower polymer, or cleaner centrate. One limitation often overlooked is feed concentration. For the centrifuge to dewater wastewater sludge, the polymer has to be well mixed with the sludge. As the feed is concentrated, its viscosity increases, and it becomes harder to mix the polymer with the feed sludge. Most wastewater centrifuges operate on feed material between 0.75 and 4.0% solids concentration; WWTPs have minimal experience outside of that range.

7.2.6 *Bowl Speed*

Bowl speed is more important for quality than process. As mentioned earlier, as the speed increases, the manufacturer is forced into redesigns to strengthen bowls, stiffen conveyors, improve hard surfacing, and tighten manufacturing tolerances. Centrifuges that can operate continuously at high *g*-forces are better designed than ones that cannot. They are also more expensive. The prospective owner must choose between price and performance.

7.2.7 *Scroll/Bowl Differential Controls*

Originally, centrifuges only operated in speed control. The operator changed the speed of the backdrive motor by turning a knob. The best they could set the speed to was ±50 rpm, which might be ± 0.5 differential rpm. Dry-solids centrifuges require an order of magnitude tighter control to ±0.05 differential rpm. When torques increase, they can approach the limits of the centrifuge, so torque control is essential. All controls should be able to run in differential control, and hold the differential rpm to ±0.05 rpm. All controls should have a torque control option where the operator enters in a torque setpoint, and the controls hold that setpoint within ±3 percentage points. At a minimum, the controls should be simple to operate; the operator enters a load setpoint, and the centrifuge operates at that point.

7.2.8 *Pool Depth*

As with all sedimentation devices, the deeper the pond, the greater the force will be compressing the solids. The pool depth is the depth of the liquid relative to the bowl of the centrifuge. Because larger centrifuges intrinsically have deeper ponds than smaller ones, they typically will achieve drier cake. Most centrifuge manufacturers have reduced the solids discharge diameter to make the pond deeper and to reduce the power consumption. Some manufacturers have moved the dam openings inward by a similar amount. This limits the differential head pressure and, therefore, the cake dryness the centrifuge can achieve. Limiting cake dryness limits torque, allowing smaller torque reducers and thinner flight material without bending flights.

7.2.9 *Effect of Chemical Conditioning*

All dewatering centrifuges and filters require polymer conditioning of the sludge to achieve reasonable results. Polymers were invented before centrifuges could be used in wastewater applications. As polymer dosage increases, initially nothing happens. At some point, the polymer flocs are strong enough to clarify the water and allow the resulting sludge to scroll out of the centrifuge. Adding polymer results in better compaction of the sludge and a drier cake. To maximize the benefits, polymer dose cannot continue to be increased; rather, the operator also needs to reduce the differential revolutions per minute and may have to increase the pond setting and change the polymer addition point. Four performance parameters affecting process economics are related to chemical conditioning:

- Cake solids,
- Polymer dose,
- Feed rate, and
- Centrate quality.

Again, tradeoffs exist for each. Adding more polymer improves the performance, which can result in drier cake, higher feed rate, or cleaner centrate. Of course, other operational changes may also be necessary, or a different polymer may be required. As the polymer dosage increase in belt filter presses, eventually the belts will become clogged with a polymer slime, and dewatering performance decreases. Centrifuges have no belts to plug, so they typically can continue to derive benefit from higher polymer dosages.

Inorganic chemicals are sometimes used with centrifuges, including ferric chloride and alum. Ferric chloride improves dewatering, but rarely enough to justify its cost. Plant operators using ferric chloride in liquid treatment processes for phosphorus removal or for odor control have noticed the sludge becomes easier to dewater in proportion to the ferric dosage. Conversely, alum added to the wet end results in poorer performance, presumably because of the associated bound water.

Struvite ($MgNH_4PO_4$) is sometimes a problem on anaerobically digested sludge. It forms in the centrate, rarely in or on the centrifuge itself, and only when feed is greater than ph 7. Adding ferric chloride has two affects: it removes some of the phosphate and reduces the pH. Adding waste acid will also help, and may be a less costly option. Several processes are available that recover struvite from wastewater.

The addition of hydrogen peroxide and potassium permanganate will have minimal effect on centrifuge dewatering. Adding lime slurry to feed sludge before dewatering may offer some process benefit to belt filters because this may reduce specific resistance to filtration. This may, however, cause problems in the centrifuge because lime settles and scrolls out as a wet slurry, resulting in wetter cake.

8.0 REFERENCES AND SUGGESTED READINGS

Ashbrook-Simon-Hartley (undated) *Electroacoustical Dewatering Press: Squeezing More Moisture from Dewatered Cake Solids Hasn't Been Easy. Until Now,* brochure; Ashbrook-Simon-Hartley: Houston, Texas.

Atherton, P. A.; Steen, R.; Stetson, G.; McGovern, T.; Smith, D. (2005) Innovative Biosolids Dewatering System Proved a Successful Part of the Upgrade to the Old Town, Maine Water Pollution Control Facility. *Proceedings of the 78th Annual Water Environment Federation Technical Exposition and Conference* [CD-ROM]; Washington, D.C., Oct 30–Nov 2; Water Environment Federation: Alexandria, Virginia.

Banks, L.; Davis, S. (1983) Desiccation and Treatment of Sewage Sludge and Chemical Slimes with the Aid of Higher Plants. *Proceedings of the 15th National Conference on Municipal and Industrial Sludge Utilization and Disposal*, Atlantic City, New Jersy, Apr 6–8; Hazardous Materials Control Research Institute: Silver Spring, Maryland.

Baxter, J. C.; Martin, W. J. (1982) Air Drying Liquid Anaerobically Digested Sludge in Earthen Drying Basins. *J. Water Pollut. Control Fed.*, **54** (1), 16–26.

Campbell, H. W.; LeClair, B. P. (1979) Sludge Dewatering Alternatives for Waste Activated Sludges from Phosphorus Removal Facilities. *J. Water Pollut. Control Fed.*, **51** (5), 991–998.

Clutten, T. (2003) Effective, Economic and Easy Dewatering of Sludges. *Filtr. Sep.*, **40** (3), 26–27.

Coackley, P.; Allos, R. (1962) The Drying Characteristics of Some Sewage Sludges. *J. Proc., Inst. Sew. Purif.* (G.B.), **Part 6**, 557.

Condie, D.; Miller, S. (2001) *Improved Filtration of Sewage Sludges Using Electrodewatering Final Report of Pilot Scale Trials*, CSIRO Investigation Report; Queensland Department of Local Government and Planning: Queensland, Australia.

Crosswell, S.; Young, T.; Benner, K. (2004) Performance Testing of Rotary Press Dewatering Unit Under Varying Sludge Feed Conditions; *Proceedings of the 77th Annual Water Environment Federation Technical Exposition and Conference* [CD-ROM]; New Orleans, Louisiana, Oct 2–6; Water Environment Federation: Alexandria, Virginia.

Day, P.; Giles, P. (2002) Innovative Belt Filter Press Takes the Hard Work out of Sludge Dewatering. *Filtr. Sep.*, **39** (8), 18–20.

Decker, W.; Sieger, R. B.; Lamont-Black, J. (2006) A New, Real Technology for Solids Dewatering; *Proceedings of the 79th Annual Water Environment Federation Technical Exposition and Conference* [CD-ROM]; Dallas, Texas, Oct 21–25; Water Environment Federation: Alexandria, Virginia.

Fuji Electric (1989) Electro-Osmotic Dewatering System Producing a Low Water: Electro-Osmotic Beltpress. Eur. Pat. EP0578182A2.

Genter, A. L. (1934) Adsorption and Flocculation as Applied to Sewage Sludges. *Sew. Works J.*, **6,** 689.

Hagstrom, L. G.; Mignone, N.A. (1978) What to Consider in Basket Centrifuge Design. *Water Wastes Eng./Ind.*, **15** (13), 58–63.

Hamel, K.; Higgins, M.; Yen-Chih, C.; Sudhir, M.; Travis, M.; Barben, E. J.; Livadaros, A.; Mass, N. (2005) Part I of Field Research: Impact of Centrifuge Torque and Polymer Dose on Dewatering Performance from Anaerobically Digested Biosolids. *Proceedings of the Water Environment Federation Residuals and Biosolids Management Specialty Conference;* Water Environment Federation: Alexandria, Virginia.

Hansen, B. E.; Garrison, W. E.; Smith, D. L. (1980) Start-Up Problems of Sludge Dewatering Facility. *J. Water Pollut. Control Fed.*, **52** (10), 2558–2570.

Haseltine, T. R. (1951) Measurement of Sludge Drying Bed Performance. *Sew. Ind. Wastes*, **23,** 1065.

Imhoff, K.; Fair, G. M. (1940) *Sewage Treatment*. John Wiley and Sons, Inc.: New York.

Jeffrey, E. A. (1959) Laboratory Study of Dewatering Rates for Digested Sludge in Lagoons. *Proceedings of the 14th Industrial Waste Conference*; Purdue University: West Lafayette, Indiana.

Jeffrey, E. A. (1960) Dewatering Rates for Digested Sludge in Lagoons. *J. Water Pollut. Control Fed.*, **32,** 1153.

Jeong Su Environmental Engineering Co. Ltd. (1994) *Belt Press: EPD—Wring Jack*, brochure (in Korean); Jeong Su Environmental Engineering Co. Ltd.: Korea.

Kabouris, J. C.; Gillette, R. A.; Jones, T. T.; Bates, B. R. (2005) Evaluation of Belt Filter Presses, Centrifuges, and Screw Presses for Dewatering Digested Activated Sludge at St. Petersburg's Water Reclamation Facilities. *Proceedings of the 78th Annual Water Environment Federation Technical Exhibition and Conference* [CD-ROM]; Washington, D.C., Oct 29–Nov 2; Water Environment Federation: Alexandria, Virginia.

Koch, C. M.; Chao, A; Semon, J. (1980) Belt Filter Press Dewatering of Wastewater Sludge. *J. Environ. Eng.*, **114** (5), 997.

Koch, C. M.; McKinney, D. E.; Fagerstrom, A. A.; Palmer, E. W. (1989) Comparison of Centrifuge Performance on Oxygen Activated Sludge; *Proceedings of the American Society of Civil Engineers Environmental Engineering Specialty Conference*; Austin, Texas; American Society of Civil Engineers: Reston, Virginia.

Kondoh, S.; Hiraoka, M. (1990) Commercialization of Pressurized Electroosmotic Dehydrator (PED). *Water Sci. Technol.*, **22** (12), 259–268.

Langeloh, T.; Bott, R. (1996) High-Pressure Dewatering of Sludges with the Sico-W.A.P.—Results and Automation of the Process. *Proceedings of the American Filtration and Separations Society Annual Technical Conference 96*; Pennsylvania, April 22–24; American Filtration and Separations Society: Richfield, Minnesota.

Lecey, R. W. (1980) Polymers Peak at Precise Dosages. *Water Wastes Eng./Ind.*, **17** (3), 39–41.

Lightfoot, D.; Raghavan, G. S. V. (1995) Combined Fields Dewatering of Seaweed with a Roller Press. *Appl. Eng. Agric.*, **11** (2), 291–195.

Lockhart, N. C. (1986a) Electrodewatering of Fine Suspensions. In *Advances in Solid Liquid Separation*; Muraldihara, H. S., Ed.; Battelle Press: Columbus, Ohio.

Lockhart, N. C. (1986b) New Technique—Electrical Dewatering of Tailings and Fine Suspensions. *Minfo*, **11**, 52–56.

Metcalf and Eddy, Inc. (2003) Dewatering in *Wastewater Engineering: Treatment and Reuse,* 4th ed.; Trobanoglous, G.; Burton, F. L.; Stensel, H. D., Eds.; McGraw-Hill: New York.

Miller, S. A.; Yan, Y. D.; Firth, B. A.; Jameson, G. J. (2005) *Innovations in Dewatering Sludges*, Project no. 02-CTS-3; Water Environment Research Foundation: Alexandria, Virginia.

Novak, J. T.; Haugan, B. E. (1979) Chemical Conditioning of Activated Sludge. *J. Environ. Eng.*, **105** (5), 993–1008.

O'Donnell, C.; Keith, F., Jr. (1972) Centrifugal Dewatering of Anaerobic Waste Sludges. *J. Water Pollut. Control Fed.*, **44,** 2162.

Oerke, D. W. (1981) Fundamental Factors Influencing the Dewatering of Wastewater Solids. M.S. Essay, Marquette University, Milwaukee, Wisconsin.

Perry, R. H.; Chilton, C. H. (1963) *Chemical Engineer's Handbook,* 4th Ed., McGraw-Hill: New York.

Purchas, D. B. (1977) *Solid/Liquid Separation Equipment Scale-Up*. Uplands Press Ltd.: Fulton, New York.

Quon, J. E.; Johnson, G. E. (1966) Drainage Characteristics of Digested Sludge. *J. Sanit. Eng. Div., Proc. Am. Soc. Civ. Eng.*, **92,** 4762.

Raats, M. H. M.; van Diemen, A. J. G.; Laven, J.; Stein, H. N. (2002) Full Scale Electrokinetic Dewatering of Waste Sludge. *Colloids Surf., A.*, **210** (2–3), 231–241.

Randall, C. W. (1969) Are Paved Drying Beds Effective for Dewatering Digested Sludge? *Water Sew. Works*, **116,** 373.

Randall, C. W.; Koch, C. T. (1969) Dewatering Characteristics of Aerobically Digested Sludge. *J. Water Pollut. Control Fed.*, **41,** 215.

Reimers, R. S.; Little, M. D.; Klein, N. A.. (1981) *Parasites in Southern Sludges and Disinfection by Standard Sludge Treatment*, Project Summary; U.S. Environmental Protection Agency: Washington, D.C.

Rolan, A. T. (1980) Determination of Design Loading for Sand Drying Beds. *J. N.C. Sec. Water Pollut. Control Assoc.*, **L5,** 25.

Sharman, L. (1967) Polyelectrolyte Conditioning of Sludge. *Water Wastes Eng./ Ind.*, **4,** 50.

Snyman, H. G.; Forssman, P.; Kafaar, A.; Smollen, M. (2000) Feasibility of Electro-Osmotic Belt Filter Dewatering Technology at Pilot Scale. *Water Sci. Technol.*, **41** (8), 137–144.

Tenney, M. W.; Stumm, W. (1965) Chemical Flocculation of Microorganisms in Biological Waste Treatment. *J. Water Pollut. Control Fed.*, **37,** 1370.

Thomas, C. M. (1971) The Use of Filter Presses for the Dewatering of Sludges. *J. Water Pollut. Control Fed.*, **43,** 93.

U.S. Environmental Protection Agency (1977) *Handling and Disposal of Sludges from Combined Sewer Overflow Treatment Phase III, Treatability Studies*, EPA-600/2-77-053C; U.S. Environmental Protection Agency, Environmental Protection Technology Services: Cincinnati, Ohio.

U.S. Environmental Protection Agency (1979) *Process Design Manual for Sludge Treatment and Disposal*, EPA-625/1-79-011; U.S. Environmental Protection Agency, Municipal Environment Research Laboratory, Office of Research and Development: Cincinnati, Ohio.

U.S. Environmental Protection Agency (1985) *Handbook: Estimating Sludge Management Costs;* EPA-625685010; U.S. Environmental Protection Agency: Washington, D.C.

U.S. Environmental Protection Agency (1987) *Design Manual Dewatering Municipal Sludges*, EPA/625/1-87/014; U.S. Environmental Protection Agency: Cincinnati, Ohio.

U.S. Environmental Protection Agency (2000a) *Biosolids Technology Fact Sheet; Belt Press Filter*, EPA 832-F-00- 057; U.S. Environmental Protection Agency, Office of Water: Washington, D.C.

U.S. Environmental Protection Agency (2000b) *Biosolids Technology Fact Sheet; Centrifuge Thickening and Dewatering,* EPA 832-F-00-053; U.S. Environmental Protection Agency, Office of Water: Washington, D.C.

U.S. Environmental Protection Agency (2000c) *Biosolids Technology Fact Sheet; Recessed-Plate Filter Press,* EPA 832-F-00-058; U.S. Environmental Protection Agency, Office of Water: Washington, D.C.

Vesilind, P. A. (1974a) Scale-Up of Solid Bowl Centrifuge Performance. *J. Environ. Eng.,* **100,** 479.

Vesilind, P. A. (1974b) *Treatment and Disposal of Wastewater Sludges.* Ann Arbor Science Publishers, Inc.: Ann Arbor, Michigan.

Vesilind, P. A. (1979) *Treatment and Disposal of Wastewater Sludges,* rev. ed.; Ann Arbor Science Publishers, Inc.: Ann Arbor, Michigan.

Walski, T. M. (1976) Mathematical Model Simplifies Design of Sludge Drying Beds. *Water Sew. Works,* **123,** 64.

Water Environment Federation (1994) *Safety and Health in Wastewater Systems.* Manual of Practice No. SM-1; Water Environment Federation: Alexandria, Virginia.

Water Environment Federation; American Society of Civil Engineers; Environmental and Water Resources Institute (2009) *Design of Municipal Wastewater Treatment Plants,* 5th ed.; WEF Manual of Practice No. 8; ASCE Manuals and Reports on Engineering Practice No. 76; McGraw-Hill: New York.

Water Pollution Control Federation (1983) *Sludge Dewatering,* Manual of Practice No. 20; Water Pollution Control Federation: Washington, D.C.

Water Pollution Control Federation (1987) *Operation and Maintenance of Sludge Dewatering Systems,* Manual of Practice No. OM-8; Water Pollution Control Federation: Washington, D.C.

Zablatzky, H. R.; Peterson, S. A. (1968) Anaerobic Digestion Failures. *J. Water Pollut. Control Fed.,* **40,** 581.

Zacharias, D. R.; Pietila, K. A. (1977) Full-Scale Study of Sludge Processing and Land Disposal Utilizing Centrifugation for Dewatering. *Proceedings of the 50th Annual Meeting of Central States Water Pollution Control Association., Inc.,* Milwaukee, Wisconsin.

Chapter 13

Composting

(continued)

1.0 INTRODUCTION

Composting, as defined in this chapter, is a controlled, aerobic, exothermic, microbial process that converts organic waste materials to a more stable form of a humus-like substance that can be beneficially used as a soil conditioner. During the composting process, exothermic heat will raise the temperature of the composting mass through mesophilic to thermophilic processes.

1.1 Composting Objectives

The four main objectives of sludge or biosolids composting as defined in this chapter are to

1. Kill disease-causing organisms (reduce pathogens);
2. Further stabilize biosolids by decomposing odor-producing compounds;
3. Dry the biosolids; and
4. Produce a stable, manageable, and marketable product.

The primary objective of any composting process should be to kill disease-causing organisms (Willson et al., 1980). The aerobic biological process of composting generates sufficient heat to raise the temperature in the composting mass into the thermophilic temperature range. The composting process must be controlled so that high enough temperatures are maintained in the mass for long enough to destroy pathogens.

A second objective of composting is to further stabilize the biosolids through microbial decomposition, which reduces odor production by simply degrading biosolids rather than completely oxidizing them.

The third objective of composting is to dry the biosolids. Before composting begins, moisture content of a properly prepared feedstock (sludge or biosolids cake and bulking agent) is approximately 60%. Heat generated during the composting process helps moisture to evaporate from the composting mass, as air introduced to the mass transports moisture from the compost pile to the atmosphere. Composting must be controlled so that drying allows the finished product to be properly screened, stored, and used. The composting process is inhibited when the moisture content is lower than approximately 35%.

The fourth objective of composting is to produce a product that is stable and mature that can be conveniently stored, managed, and used in different ways. To meet this objective, pathogen reduction, drying, aging, and finishing of the compost are necessary.

1.2 Historical Overview of Biosolids Composting

Historically, composting of organic wastes has been more of an art than a science. Windrows were the most typical way to compost at small compost facilities. Approximately 50 years ago, several mechanical composting systems were introduced in Europe. The aerated static-pile method was introduced in the mid-1970s by the U.S. Department of Agriculture at the Agricultural Research Center in Beltsville, Maryland.

Many advances made in composting since the 1970s are based on or are refinements of this earlier work. In the late 1980s and early 1990s, growth in the number of biosolids composting facilities and in populations living near compost plants resulted in many cases where odors from the process created tension between facility owners/ operators and nearby residents. This occurred at both open and enclosed facilities. As a result, many facilities were closed. The engineering community responded to this issue by designing composting systems with improved odor control and treatment systems so that facilities could be built and operated as good neighbors. Today, several large composting facilities (several hundred tons of biosolids cake processed per day such as Davenport, Iowa; Inland Empire, California; South Kern Industrial Center in Taft, California; Burlington County, New Jersey) operate near neighbors without odor issues plaguing their operations. Aerated static-pile composting systems are used by more than 90 municipalities in the United States; within-vessel systems of various configurations are operated as approximately 36 municipalities. Windrow composting facilities are used primarily in the more arid climates of the southwestern United States where precipitation rates are low enough to not severely affect operations.

2.0 FACILITY DEVELOPMENT

Developing a composting facility takes careful consideration of the materials to be composted, site constraints, neighboring property users, climate conditions, final use, seasonal demands for the product, and other issues. This section describes the necessary steps to ensure successful development of compost facilities.

2.1 Planning and Feasibility

As noted in Chapter 3, public acceptance, available outlets for the product, and, to a lesser degree, costs are the driving factors affecting the development decisions for a composting facility. Facility location, technology selection, and degree of enclosure and odor treatment will affect these factors and are explored in depth in this chapter.

2.2 Locating a Facility

The old real estate adage of "location, location, location," definitely applies to composting facilities. The following should be considered when selecting a location:

- Availability of land;
- Proximity of neighbors;

- Surrounding land use;
- Distance to raw materials and markets;
- Transportation routes of raw materials and finished product to and from the facility;
- Site configuration;
- Availability of utilities such as water, wastewater, and electricity;
- Microclimate; and
- Surrounding topography.

2.3 Transition from Planning to Design

Capacity, level of technology, and cost must be factored in when moving from planning to design stage. A balance exists between these three factors, and owners must decide which of these factors are most important. Typically, capacity is the first factor that will determine design. Most owners will not have the luxury of an open-ended budget. Therefore, priorities regarding technology and amenities at the facility will need to be established, and appropriate choices will need to be made. Wastewater solids quantity and planning horizon will need to be determined, including how much solids and what type and over what period of planning. Site master planning should be considered if staged development of a facility is anticipated.

Owner preference will affect level of technology. Owners who are planning to build a composting facility should visit and tour facilities with various technologies and capacities in similar environments. A wealth of knowledge can be gained through this process that will be needed to help make informed decisions regarding level of controls and type of technology. Both good features and non-desirable ones can be identified at this stage in the planning process.

Capital costing typically is included in the early stages of design. A sufficient project definition is needed to adequately estimate project costs. This is typically not possible until well into the preliminary design stage. Establishment of a capital budget and comparison of life-cycle costs is needed to justify further development of a compost project as compared to other solids management alternatives being considered.

Preliminary studies are typically considered at this point in project planning or in the early stage of preliminary engineering design. Pilot studies using the wastewater treatment plant (WWTP) solids and available bulking agents typically are conducted to demonstrate the viability of composting, to allow operators to become familiar with the process, and to generate small quantities of compost that can be analyzed and used for preliminary market analysis. Bulking agent surveys may be needed at this stage in planning to determine cost and availability of suitable bulking agents for the composting process. It is highly recommended that a compost market analysis be performed to determine potential demand for a product. This effort is more important for larger facilities because of larger volume of product that will be produced.

2.4 Preliminary Engineering and Design Development

During preliminary engineering or basis of design for composting facilities, many factors must be considered to adequately site, size, and layout a suitable facility to meet project requirements. These components are discussed below.

- **Solids quantity and characteristics**. The type of solids will affect many aspects of design. Undigested solids, for example, will have much more available energy than digested biosolids. This can lead to greater potential for odor development, but also will allow for the use of less active bulking agent material such as dried, cured wood waste instead of fresh ground yard waste. Further, the type of solids and the equipment used for dewatering at the WWTP will affect the solids content of the cake. Drier cake will require less bulking agent than a wet cake. For example, a 16% total solids cake will require twice as much bulking agent as a 24% total solids cake, effectively doubling the volume of bulking agent required and the mixture to be composted. In addition, because the density does not change significantly, the wetter the cake the greater the volume of cake per dry megagram. These factors have the greatest affect on the design for composting facilities. Some WWTPs produce wetter cake seasonally, so the lowest sustained monthly average solids content cake should be used as the basis of design target.
- **Bulking agent type and availability**. The available bulking agent will affect facility development and should be determined early in the design process to ensure sufficient quantities are available to meet facility demands. Storage of bulking agent should also be considered to ensure a consistent supply to the operation year round. Further, in areas where precipitation is significant, covered or enclosed bulking agent storage is recommended to prevent overwetting of the bulking agent supply. Facilities without storage may be forced

to purchase bulking agents that are dry enough during wet seasons of the year when an otherwise perfectly suitable source of bulking agent cannot be used. This is an important consideration in the design and layout of facilities in the eastern and northwestern regions of the United States.

- **Siting and infrastructure development**. Land requirements, site characteristics, and proximity to neighbors must be addressed early in the design process to determine suitability of the compost process type, the ability of the selected process to fit onto the available land parcel, and the level of odor control needed. If neighbors are located within 1.6 km (1 mi.) of small facilities (less than 10 wet Mg/d) or within 4.8 km (3 mi.) of larger facilities, then odor modeling is recommended to determine the level of odor controls needed for the planned technology. Comparison of the odor effects of various compost technology choices or odor containment and treatment choices should be done to effectively engineer the level of control needed. In addition, the proximity to the point of generation of the solids, availability of suitable roads, and the availability of utilities such as power, water, and wastewater collection must be accounted for in selecting an appropriate site for a chosen technology.
- **Materials balance development**. The development of a materials balance is crucial for engineering a composting facility, and is required in the early stages of design. Quantities and characteristics of the solids to be composted and the bulking agents to be used are needed to determine daily and annual quantities of material that must be handled. The characteristics of various materials often change significantly each season, which should be considered in the development of materials balances. This information is necessary to determine spatial needs, equipment sizing, power loads, and other important basic information for developing a site design. Operating schedules should be considered as well, such as if the facility will operate five or seven days per week.
- **Process layout**. Process layout options can be compared in the project definition stage of design once design criteria have been established. Factors such as the number of days per week solids will be received and processed and expected generation and management needs at the WWTP must be considered. Other items that must be defined include the amount of solids, bulking agent, and final product storage; mixing, composting, screening, and curing detention times; and pile sizes and equipment, which should be sized to ensure needs will be met. Materials handling needs and quantities should be

defined at this stage, and odor control technology and sizing criteria should also be established.

- **Process controls and staffing support facilities.** The level of process controls and automation and the degree of enclosures and odor controls should be defined and the type of support facilities for operator support offices and maintenance facilities must be considered and appropriately planned for at this stage in design development.
- **Cost and budget development.** Both capital and operating costs should be considered as early in the design process as possible to establish expected capital costs for developing a facility and to plan for appropriate staffing. This effort can be done in the planning stages as well but should be revalidated once a concept level of design is determined to ensure appropriate budgeting.

2.5 Final Design

Once the above basic features for a planned composting facility are determined, the design concept must be sufficiently defined so that detailed process and facility design is possible. At the final design stage, sizing criteria and technology type should be defined and a preliminary layout established, which is typically "frozen." At this stage in the design process, detailed equipment selection is made, materials of construction are determined, the level of enclosure, and building types are developed. The purpose at this stage is to develop a set of contract documents (including equipment and technical specifications and project drawings) sufficient to submit to permitting agencies for review and approval and, ultimately, for owner acceptance to allow for bidding to contractors to build the facility.

3.0 FACILITY DESIGN

3.1 Process Selection

Process selection must consider multiple variables simultaneously to achieve the desired objectives. In virtually every case, key objectives are to eliminate disease causing organisms in the biosolids, to reduce biosolids odors, and to produce a marketable product. The technology chosen to perform these objectives will be influenced by site size and constraints, the proximity of neighbors, local weather conditions, bulking agent availability, degree of process controls desired, and ultimately, capital and operating cost effects. The following section describes many of the process variables in composting systems that are available for use and the pros and cons of each. This section also presents a multi-attribute utility analysis tool that will aid the practitioner in making choices regarding the type of process that is most appropriate based on priorities most important to a given utility or situation.

3.1.1 *Biosolids Composting Process Alternatives*

Although composting is a naturally occurring biological process, the degree of control imposed on a system can vary from the simple method of periodically turning a pile or windrow to the more involved enclosed or within-vessel system with mechanical agitation and forced aeration.

Over the years, several composting methods have evolved that offer many benefits such as accelerating a naturally occurring biological process; providing for process control over variables such as moisture, carbon, nitrogen, and oxygen; containing odors and particulates; reducing land area requirements; reliably producing consistent product quality; and integrating aesthetically pleasing facilities into local and regional sites.

Three categories of composting technologies are available: aerated static pile (ASP); within-vessel; and windrow. Of these three, ASP is the most commonly used. According to a survey done by *BioCycle*, of the 265 biosolids composting facilities in the United States, approximately 45% are ASP; the rest are divided between windrow and within-vessel facilities (Beecher and Goldstein, 2010).

Table 13.1 presents a comparison of the advantages and disadvantages of five composting technologies based on consideration of physical facilities, processing aspects, and operation and maintenance. No one technology is appropriate for every situation. Climate, site considerations, operational concerns, sensitivity to odors, and other factors will influence the choice of systems.

3.1.2 *Aerated Static Pile Systems*

The aerated static-pile method of composting was developed in Beltsville, Maryland, in the 1970s by the U.S. Department of Agriculture (USDA), which has been referred to as the Beltsville Method. The ASP method is highly flexible and thus is the most typically used biosolids composting method in the United States. As the name suggests, this method involves aerating piled feedstock material. The compost mixture is constructed into an approximately 2- to 4-m (6-to 12-ft) deep pile over an aeration floor, plenum, or perforated piping. The entire pile is covered with an insulating

TABLE 13.1 Composting technology comparison.

Composting technology	Advantages	Disadvantages
Aerated static pile	• Adaptability to various bulking agents • Flexibility to handle changes in loading • Simple equipment requirements (lower maintenance) • Good odor control possible	• High labor requirements • Large area requirements • High energy requirement
Within-vessel agitated bay	• Automated movement of material • Lower initial solids content required; lower volume of material processed • Adaptability to various bulking agents • Regular agitation re-homogenizes mixture and allows for moisture addition during processing • Good odor control possible	• Low flexibility with respect to volume of material processed • Potential for high dust generation • Positive aeration increases worker exposure • High maintenance • Process is highly corrosive to structures from hot, humid gasses released during agitations
Within-vessel vertical plug flow	• Smaller footprint • Automated operation • Low worker exposure	• Low flexibility with respect to volume of material processed • No redundancy • Only two facilities in United States • High maintenance • Supplemental curing outside of reactor is needed
Within-vessel horizontal plug flow	• Completely enclosed with low air-change requirements • Low worker exposure • Adaptability to various bulking agents • Automated operation	• Low flexibility with respect to volume of material processed • No redundancy • Only two operating facilities in United States • High maintenance • Supplemental curing outside of reactor is needed
Windrow	• Flexibility to handle volume changes • Low maintenance • Adaptability to various bulking agents • Simple, low-tech operation	• Large area requirements • Low ability to control odors • Dust potential • Severe effect of precipitation on open facilities

blanket of wood chips or unscreened finished compost (150 mm to 300 mm [6 in. to 12 in.] deep) to ensure that all parts of the mixture meet temperature standards for pathogen kill and vector attraction reduction. Figure 13.1 provides a typical cross section and layout of an aerated static pile. In small operations, individual piles may be constructed. In large operations, a continuous or extended pile is divided into sections representing each day's contribution. The continuous-pile construction is referred to as extended aerated static pile (EASP). The EASP allows for construction of a new pile against the shoulder of the previous pile, which extends the pile length and minimizes the footprint. In a like manner, piles are broken down once regulatory criteria such as process to further reduce pathogens (PFRP) temperatures [three days above 55 °C (131 °F)] and vector attraction reduction (VAR) temperatures (14 days above 40°C, 104°F) are completed. Piles breakdown in near-vertical walls that can cross two piles of material. Space must be provided to allow room for pile building and breakdown in facility design.

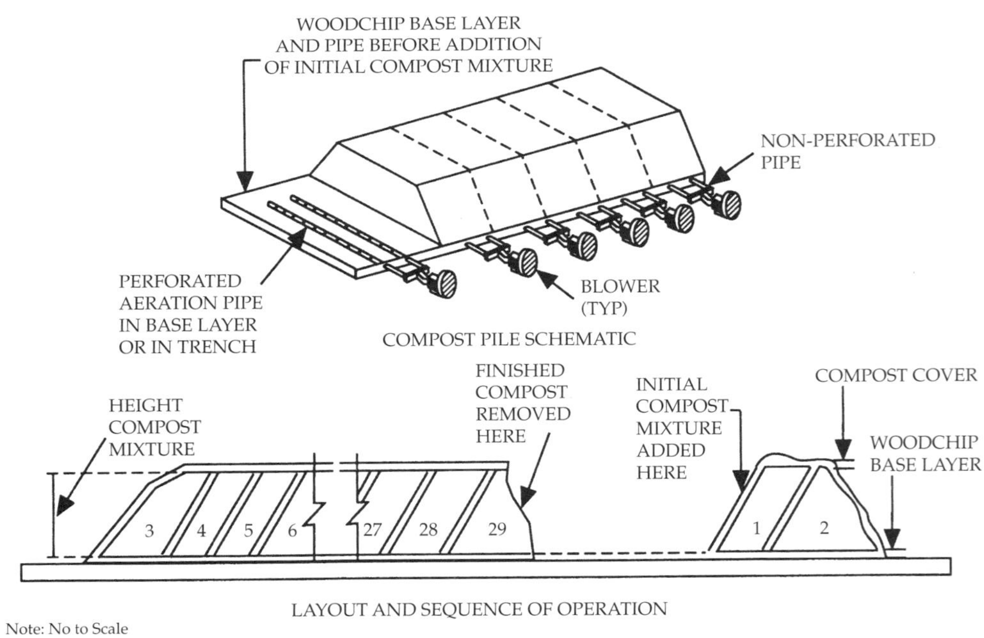

FIGURE 13.1 Extended aerated static pile (WEF et al., 2009).

When using the aerated static-pile method, the mixture remains in the pile for the active composting period, typically from 21 to 28 days. Following this, the piles are broken down, and the material is either moved to a curing area or screened and then moved to a curing area. It is critical that the compost solids content be at least 55% for screening. In some facilities, an intensive drying step with a higher aeration rate than active composting precedes screening. Alternatively, screening can also follow curing. Curing before screening reduces the amount of area needed for curing and minimizes the amount of bulking agent that is degraded (increases the amount of bulking agent which can be recycled). Compost typically remains in curing for a minimum of 30 days to further stabilize the material. This may be longer if the curing process is not aerated.

Aerated static-pile composting was originally developed for outdoor sites. Many facilities have been constructed, however, that are either partially or fully enclosed. The level of enclosure is typically dictated by the required level of odor capture and control. Local environmental conditions such as extremes of temperature or rainfall may dictate at least partial enclosure to facilitate operations.

3.1.3 Within-Vessel Composting Systems

Although a wide variety of within-vessel composting systems have been developed, only a few systems been installed at multiple installations; most systems are found at only one or two facilities. Typically, within-vessel systems incorporate some type of automated material movement combined with aeration. The detention time for within-vessel systems varies from approximately 14 to 21 days depending on system supplier recommendations, regulatory requirements, and costs. Detention time should be based on desired product characteristics, especially stability, and should take into account the detention time in all process phases. Further stabilization of 30 to 60 days typically is required after the active composting phase to achieve desired product stability. Four types of within-vessel composting systems are available:

1. Vertical plug flow reactors (no longer being built) (Figure 13.2);
2. Horizontal plug flow reactors (no longer being built) (Figure 13.3);
3. Agitated bay systems (Figure 13.4); and
4. Hybrid systems.

Vertical plug-flow reactors are vessels constructed of steel, concrete, or reinforced fiber-glass panels. The mix of dewatered cake, bulking agent, and recycle is loaded in the top of the reactor. The composting mix is aerated but not agitated or mixed and

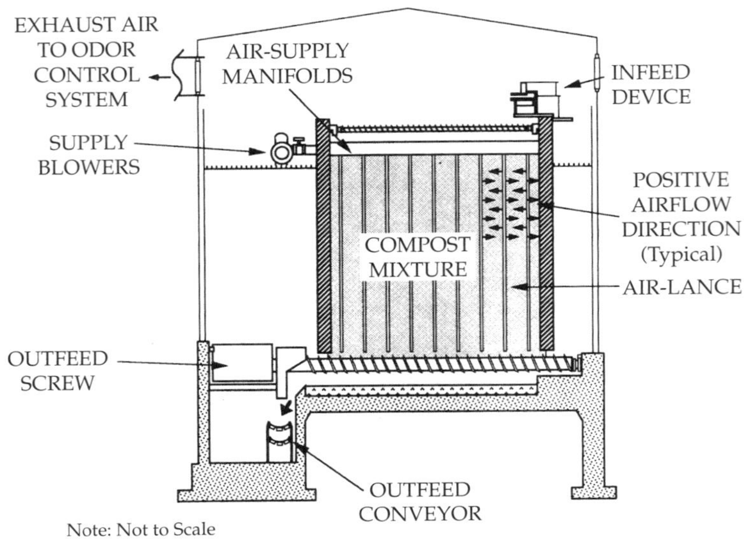

Figure 13.2 Schematic for a vertical plug flow reactor (WEF et al., 2009).

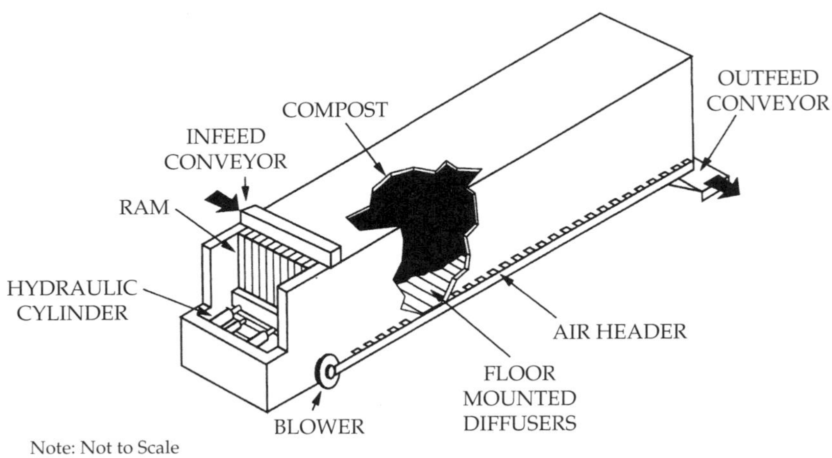

Figure 13.3 Schematic of a horizontal plug flow reactor (WEF et al., 2009).

FIGURE 13.4 Photograph of an agitated bay.

moves as a plug to the bottom of the reactor. From there it is removed by a traveling auger discharge device.

Horizontal plug-flow reactors are similar to vertical reactors in that the contents are not mixed within the reactor. The compost is moved laterally through the reactor by the action of a hydraulic ram.

Agitated bay reactors, however, do not move the compost as an unmixed mass through the reactor. Instead, mechanical devices periodically agitate and convey the composting material down the length of the reactor. Physically, the reactors are open-topped bays with air supplied with blowers and piping systems from the bottom. A variety of methods are used to transfer compost from the reactors. In concept, these systems are similar to aerated windrows. The repeated agitation of these systems tends to break apart bulking agents, thereby reducing the amount of agent that can be recovered during screening for recycling.

The most typically used within-vessel system is the horizontal agitated-bay reactor illustrated in Figure 13.4. These reactors are rectangular, aerated from the bottom with independently programmable aeration zones, and enclosed in a building. Fresh mix is loaded into the front end by a loader. The agitation device is completely automatic and it typically makes one pass through the reactor each day. The composting material is dug out and redeposited approximately 4 m (11 ft) behind the machine. Eventually, the composting material is moved through the entire length of the reactor.

3.1.4 *Windrow Composting*

In windrow composting, the composting mix is formed into long parallel windrows with a trapezoidal or triangular cross section. The material is then periodically turned with a front-end loader or a dedicated windrow-turning machine. The purpose of turning is to expose the material to air, release moisture, and loosen and fluff the material to facilitate air movement through the windrow. Turning also is needed to mix the cooler, outside material into the hot core of the pile to expose all material to high enough temperatures for uniform pathogen kill. Further, anaerobic zones occur in the center of windrows, and the agitation is intended to provide needed oxygen to the core of the pile. In practice, during the active phase of composting, oxygen is consumed rapidly after turnings in the core of the pile creating anoxic or anaerobic zones. Windrow composting is performed at open outdoor sites or covered sites. Windrow systems require a significant amount of space compared to other composting technologies primarily because of pile geometry and the required allowance between piles and for maneuvering a windrow-turning machine.

3.1.5 *Hybrid Processes*

Hybrid composting technologies combine elements of more than one of the three technologies described above. One such system is an aerated windrow system. This method incorporates turned windrow process with aeration similar to ASP. The windrows are constructed over air channels to protect aeration piping from the turning equipment. As with ASP, the air can be forced up through the windrow or pulled down into the channel. Aerated windrows are also periodically turned. The combination of forced aeration with physical turning makes the most of the composting rate and moisture release. The physical turning exposes new interfaces between particles in the mass. Tunnel composting systems can be hybrids as well. In a tunnel system, the compost mixture is stacked in a rectangular-shaped closed vessel that is just high enough to allow for the material to be piled and removed. Some tunnel systems are ASP systems with less air movement because the space is unoccupied except for pile building and teardown. Other tunnels have conveyor systems that distribute material in the tunnel and remove it after active composting is complete. Such tunnels incorporate elements of ASP and within-vessel technologies.

3.1.6 *Membrane Covered Pile Process*

One emerging technology that has been used for composting yard and food wastes at facilities in Europe, and more recently in North America, is the membrane covered process. This technology is beginning to be used to compost biosolids as well. The technology includes an aeration pad with two aeration trenches below each compost pile. Compost piles are approximately 50 m long × 8 m long × 3 m high (164 ft × 26 ft × 10 ft) are used to process the mixture initially for 28 days. Each pile is covered with a proprietary membrane fiber cover that allows oxygen and carbon dioxide to pass through and which captures moisture and odors beneath the cover. These odorous compounds are reportedly degraded by the composting biology. A single aeration fan is provided for each pile that is turned on and off by a computer based on pile temperature and oxygen-content levels measured by sensor probes in the pile.

After the initial 28 days of primary composting, the cover will be removed and the pile will be moved by front-end loader to the secondary compost pad. The cover then is replaced for another 14 days while the pile continues to be aerated. After 14 days in the secondary composting stage, the cover is removed and the pile is aerated for an additional 14 days.

3.1.7 *Composting Process Selection Criteria*

Design is the process of balancing a vast array of interest and goals. Selection of a composting technology falls into this same category. Each set of circumstances requires a different solution. Factors considered important in selecting a composting technology include

- Physical facilities,
- Process considerations, and
- Operation and maintenance.

Although cost is always a significant factor in facility development, available space and proximity of neighbors to the facility typically have the greatest influence on overall design. The size and shape of available land may influence the technology selection. The proximity of neighbors will influence the level of enclosure required.

As with any industrial process, the proximity and type of raw materials available influence the technology decision. For example, an aerated static pile process can require greater amounts of bulking agent that is coarser than either an aerated turned windrow or an agitated bed process. Adding aeration reduces the amount of bulking agent required because the initial moisture content can be lower than that of a turned windrow-only operation. Frequent agitation of an agitated bay process allows for finer bulking agent and less overall volume because the initial solids content of the mixture can be as low as 38%.

Although product quality is important, any of the composting technologies can produce a marketable product if properly used. Post-processing such as blending the compost with other materials or nutrients typically is not affected by the initial composting process.

3.2 Mass Balances and Process Steps

Composting facilities are primarily materials and air-handling operations. Therefore, the first step in the design process must be to estimate the amount of material that will be present in each step of the operation at any given time. This must include allowances for peaks in raw material (biosolids) production. The tool used to establish the material flows is the mass or materials balance (see Section 5.0 Design Example for Aerated Static Pile for an example of a materials balance).

Critical quantities in each step of the process are the wet or dry weight, the moisture content, and the bulk density of the material. Bulk density is a critical value. Whenever possible, field measurements should be made to establish target bulk densities. This is not always possible, however, and past experience or data from other facilities may need to be used.

The amount of bulking agent recaptured after screening depends on type of bulking agent used. Coarse wood chips will yield a higher percentage return; ground wood and yard waste, which has a high percentage of fine material, have a significantly lower return rate. The return percentages shown in the design example are typical for the material types listed. An important part of the materials balance is to balance the amount of recycled bulking agent to be used in the initial mix with the amount that is produced. Simultaneously, the total bulking agent amount must be balanced against the biosolids so that the target initial-mix solids content is reached. An iterative process is used to determine the amount of amendment needed to hit the target initial-mix solids content. This is accomplished by approximating the recycle and new bulking agent amounts and by allowing the solids content and the eventual recycled bulking agent to be calculated. The bulking agent can be adjusted until the total recycle at the end of the process equals the input plus any recycle used for cover and base.

One of the significant process differences between the various technology types is the required target initial-mix solids content. A range of target initial-mix solids contents for each technology exists. Achieving these targets is essential for successful operation of the composting facility and for meeting regulatory pathogen kill requirements. The following is a summary of the targeted mix solids contents for the three major technology types:

1. **Aerated static pile.** An initial-mix solids content of 40 to 45% is targeted. Wetter than this will cause heat energy to be lost to evaporation, slowing the process. Drier material will run low on moisture content during the process, which will slow the biological process.
2. **Within-vessel.** For automated loading tunnel and vertical plug flow systems, the initial-mix solids content should be in the same range as ASP. For agitated bay systems, the initial-mix can range from 38 to 40%. The frequent agitation and forced aeration cause material in these systems to dry out much faster than other systems. Experience has shown that in an agitated bay system, as much as 2% moisture can be lost during one agitation.
3. **Turned windrow.** The target initial-mix solids content is approximately 45%. Wetter mixes will not have enough porosity to allow for convective airflow in the piles. In wet climates, the initial mixture should go in to the windrow slightly drier.

3.3 Mass Balance Approach to Composting

The materials balance provides detailed information needed for sizing each operation in a composting facility. The material balance provides a summary of all operations in a single day. Each processing step must be sized based on the full residence time for that process. In addition, allowance must be made for intermediate actions for each step. For example, the active composting process in an ASP operation requires 21 days residence time plus one to two days allowance for removing and constructing piles.

The materials balance can be developed for any time frame. For instance, in some small facilities, biosolids deliveries may occur two or three times per week. In this instance, a materials balance may be established fortweekly operation.

3.4 Biosolids Characteristics

The important characteristics of biosolids going into a composting operation are solids content, metal content, and ability to support biological activity. Typical biosolids from a WWTP are biologically active enough to reach pathogen-kill temperatures when blended with most bulking agents. Even digested biosolids are typically biologically active enough. Caution must be taken when adding inert materials into the biosolids. Large quantities of inert material can dilute the biosolids to a point at which temperature increases in the mixture being composted is inhibited.

Solids content has a significant effect on the composting process. The wetter the biosolids, then larger the amount of bulking agent required to reach feedstock target solids contents. More bulking agent means greater volume and, thus, more space requirements. At the other extreme, if the biosolids are too dry, then not enough bulking agent will be added and the mix will have insufficient porosity. Insufficient bulking agent can also result in a low carbon-to-nitrogen ratio that could inhibit temperature generation and increase nitrogen losses and ammonia-based odors.

Biosolids are composed of uniform small particles. Because polymer is typically added in the dewatering process, the cake produced has a gelatin-like consistency. This does not affect biological activity, but is important for materials handling equipment. The effect of consistency is discussed in detail in the materials handling section of this chapter.

3.5 Amendments and Bulking Agents

All composting technologies require mixing sufficient quantities of bulking agent with dewatered wastewater solids to adjust the initial solids content and provide porosity. In addition, bulking agents provide supplemental carbon to adjust the carbon-to-nitrogen ratio and energy balance. Table 13.2 lists some typically used bulking agents and their characteristics.

Although yard debris can be used as a bulking agent, grass clippings and substantially green yard waste are unsuitable because of their high water and nitrogen content and lack of porosity. If grass clippings and substantially green yard waste are composted, then the pile will require supplemental bulking agent. The bulking agent-to-biosolids ratio depends on the characteristics of agent available and desired solids content. For biosolids in the 18 to 24% solids range, a blend of woody yard debris at 55 to 65% solids will require a bulking agent-to-biosolids ratio of approximately 3 or 4 to 1 by volume to attain a 40% solids mix. For a 45% solids blend mix, 5 or 6 to 1 by volume may be needed. For 38% solids cake, blend ratios as low as 2.5 to 1 may be used. At less than 2.5 to 1, the mix is likely to have insufficient porosity, and the decomposition process will be slowed.

TABLE 13.2 Characteristics of bulking agents.

Bulking agent	Characteristics
Wood chips (25 to 50 mm or 1 to 2 in.)	• Must typically be purchased • High recovery rate in screening (60 to 80%) • Good source of supplemental carbon
Chipped yard or land clearing debris	• May be available as waste material • Low recovery rate in screening (40 to 60%) because of higher percentage of fines • Green waste fraction can add moisture and nitrogen, requiring more wood chips • Can be a source of supplemental carbon
Ground waste lumber	• May be available as waste material • May be poor source of supplemental carbon if old and extremely dry because more volatile forms of carbon missing • Contaminants such as lead (paint) and arsenic (old pressure-treated lumber) could be present and need to be screened upon receipt
Leaves	• Insufficient porosity to be used alone • Rapidly available source of supplemental carbon • Available as waste material • Not recovered by screening; adds to compost volume
Saw dust	• Insufficient porosity to be used alone • Rapidly available source of supplemental carbon • Must be purchased; typically expensive • Not recovered by screening; adds to compost volume
Shredded paper	• Insufficient porosity to be used alone • Rapidly available source of supplemental carbon • Available as waste material • Not recovered by screening; adds to compost volume

3.6 Energy Balances

Heat is generated by conversion of organic carbon to carbon dioxide-and-water vapor. The fuel is provided by the rapidly degraded volatile fraction of the solids supplemented by the bulking agent. Heat is primarily removed by evaporative cooling through aeration and agitation. To a lesser extent, heat is lost at the pile surface.

The process temperature will not rise if the rate of heat loss exceeds the rate of heat generation. Haug (1980) provides a detailed discussion of the energy balance and concludes with the following relationship:

$$W = \text{Weight of water evaporated}/\text{Loss of volatile solids} \qquad 13.1$$

If W is less than 8 to 10, sufficient energy should be available for heating and evaporation. If W exceeds 10, then the mix will remain cool and wet. This generalization is based on heat of vaporization and does not consider the effect of ambient conditions on evaporation and surface cooling.

3.7 Product Generation and Characteristics

Although the U.S. Environmental Protection Agency Part 503 (1993) regulations have stipulations for producing Class A or B biosolids compost, the time in process required to achieve Class B is not less than Class A; thus, Class B is seldom the goal of any biosolids composting operation. Assuming Class A requirements are met, other characteristics may affect marketability of the product. These include particle size, solids content, odor, appearance, nutrient content, and physical contaminants. Particle size is dictated by the screening operation. A coarse product has large amounts of bulking agent retained and may be desirable in erosion control or for landfill cover but not in top dressing or soil blending products. Solids content affects the cost of shipping the material because water in the product adds weight. In addition, material handling and aesthetics are also affected. Wet material is sticky and harder to handle. It also has a less desirable appearance, looking more like mud than top soil. Odor relates to stability. Well-stabilized compost will not have any residual biosolids odor but rather a soil-like odor. Stability is discussed at greater length in the product marketing section of this chapter.

Appearance typically is related to solids content and color. Color typically is consistent in biosolids, however, additions of other materials into the compost mix can cause variations. For example, compost made from a combination of biosolids and MSW often is a significantly lighter color than biosolids compost. Material with solids content less than 45 to 50% will be sticky and less friable. Nutrient content, including macronutrients nitrogen, phosphorous, and potassium, is important for users. Although compost has the greatest benefit for its soil amendment properties, a higher nutrient content will increase its value. Physical contaminants such as glass, plastic bags, and metals will decrease the value of the compost. Plastic bags can often be found in yard waste/brush used as a bulking agent if it is not carefully controlled. Other contaminants should be prevented or removed from compost before marketing.

3.8 Composting Process Steps

Regardless of the composting technology selected for a facility, the basis processing steps are the same: material receiving; mixing; active composting; screening; curing; and product storage and shipping.

Figures 13.5, 13.6, and 13.7 provide typical process flow diagrams for ASP, agitated bay, and windrow composting operations, respectively. The details governing the design of each of these processes steps are examined below.

3.8.1 Materials Receiving/Handling

Raw materials for the composting process are the biosolids and the bulking agents. Just as with any manufacturing process, it is vital to keep the raw materials separated and properly stored to maintain their properties. Proper storage allows the materials to be mixed in the proper proportions to maintain optimum process performance and product quality.

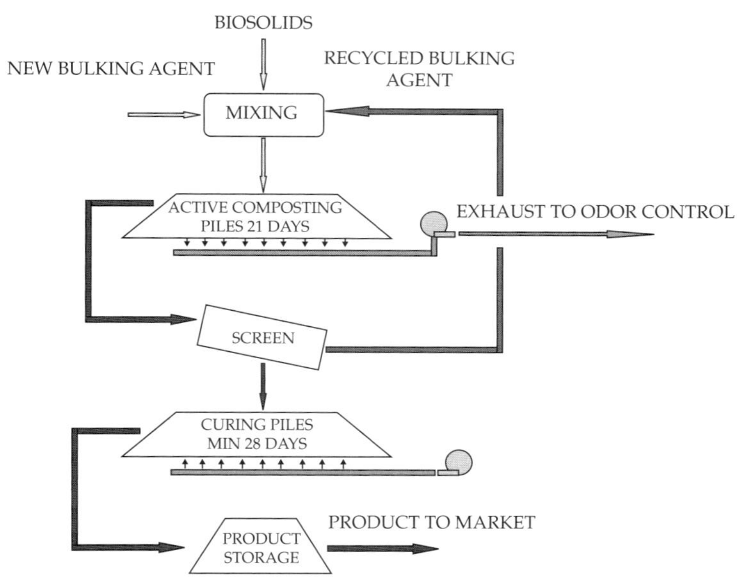

FIGURE 13.5 Typical aerated static pile process flow.

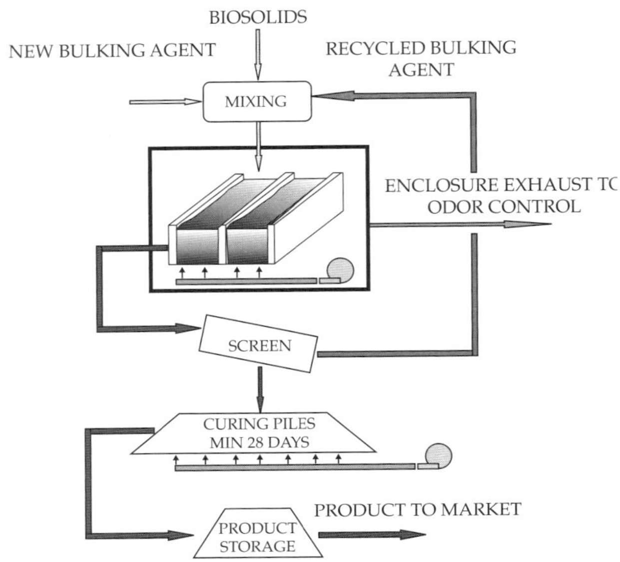

FIGURE 13.6 Typical agitated bay process flow.

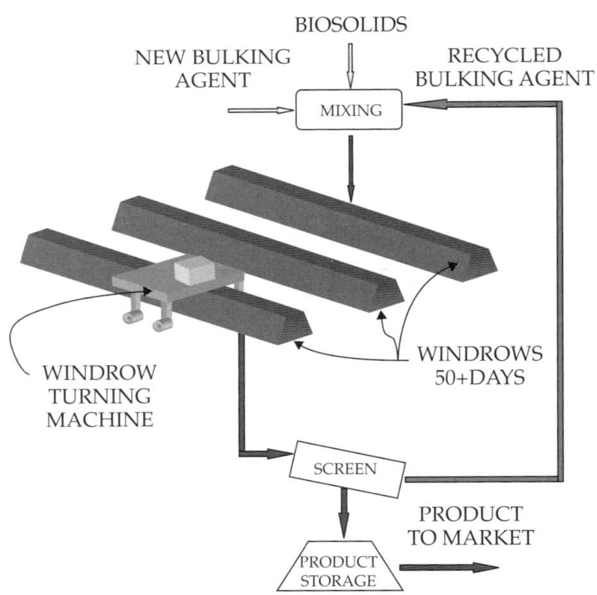

FIGURE 13.7 Typical turned windrow process flow.

The amount of required storage and the methods of handling the material dictate the type of materials receiving. A basic rule for any feedstock material is that if it is not to be blended and added into the composting process immediately, then it must at least be stored under a roof. This will keep rain and sun from significantly altering the solids contents of the materials before blending.

Ideally, at least 15 to 30 days worth of storage should be available for bulking agent. The actual storage capacity is dependent on availability. If bulking agent is only available seasonally or if prolonged heavy periods of precipitation are normal, then more storage is required. As noted above, the process should be designed to use the recycled bulking agent as it is produced. In actual practice, however, minor upsets will occur, and the exact amount of recycled bulking agent used will vary. Therefore, a minimum of approximately 15 days of recycled bulking agent storage should be available. The amount of biosolids storage is dependent on the delivery schedule. Most facilities have storage for approximately one day of usage of biosolids. Because of odor generation, it is undesirable to have more than one or two days of biosolids storage onsite. State regulations should also be considered to ensure the amount of storage provided is appropriate.

Material typically is moved around a composting facility either by front-end loader (FEL) or by conveyor. Regardless of automation, FELs moves at least some portion of material at all compost facilities. Bulking agents, biosolids, and finished compost have relatively low densities and, thus, FELs can be outfitted with large-volume, light-material buckets. In addition, rollout or pushout buckets are also highly advantageous. These allow greater vertical and horizontal reach of the FEL. Care must be taken to not use the same bucket to handle fresh feedstock or blended material for material that has met the pathogen-kill temperature requirements. If the same bucket must be used, then it must first be cleaned with a power washer and a bleach-and-water mix to avoid cross-contamination of the finished product.

Several types of conveyors are available. Typically, however, belt and screw conveyors are used. Belt conveyors must have shallow incline angles or material will tumble or flow backwards. The incline angle depends on the material being conveyed; typically, the maximum is 15 to 20 deg. The drier bulking agent material can only accommodate a 15-deg belt conveyor rise; biosolids can handle up to 20 deg.

Screw conveyors allow steeper inclines but require more maintenance. Biosolids typically contain grit and can be abrasive and cause wear on the screws. Bucket elevators can be used in some areas of a composting facility but care must be taken in the design. The enclosure must curve and match the path of the buckets. It is important to avoid void spaces where spilled material can collect creating vector and odor nuisance conditions. This material will dry out if left in place for long periods and can be a fire hazard because of sparks or heat generated by the conveyor mechanism. In addition, bucket elevators should be avoided for raw biosolids. This material is wet and sticky and will rapidly accumulate in the bucket.

Regardless of the type of conveyor used, it is important to follow three basic material conveyance rules: (1) Careful attention must be paid to transition-point design, including providing adequate belt or screw scraping and a method to catch and collect the scraped off material; (2) Nonperpendicular conveyor transition angles should be avoided when possible to prevent spillage; and (3) If the material is dry, then an enclosure or water mist should be provided to control fugitive dust.

In addition to conveyors, live bottom hoppers are used for material surge storage and to load conveyors, mixing, and screening equipment. Although the different materials at a composting facility have different properties, design requirements for live bottom hoppers are similar for all of them. All hoppers must have steep sides. Biosolids are wet and sticky and will cling to the side walls. Although bulking agents may be dry, they are coarse and can interlock and bridge easily. Any unscreened compost also will bridge and can have an angle of repose as much as 90°.

3.8.2 *Mixing*

The initial blending of the raw ingredients is crucial to successful production of the final product. In composting, proper mixing is required to produce the desired solids content and mix porosity. A good mixture consists of bulking agent particles uniformly coated with solids containing no large balls (> 12 cm, or 5 in.) of dewatered WWTP solids; large sludge balls indicate poor mixing. Three basic forms of mixing are used at composting facilities, and several vendors provide various types of equipment. Below is a brief description of these mixing methods:

- **Front-end loaders.** An FEL portions feedstock in discrete piles and "tosses" (similar to tossing a salad) the material several times until blended. The mix quality typically is poor, and mixing is time consuming. Front-end loaders are best for small facilities (up to 40 Mg wet biosolids cake per day) and as a backup to a primary means of mixing.
- **Batch mixers.** Batch mixers are hoppers that are either stationary or truck/trailer mounted and are equipped with internal paddles or augers that mix the

material. A short side-mounted conveyor with a slide gate is opened to discharge the blended batch. Batch mixers are equipped with internal scales and a weight display to aid in material portioning. Mixers typically are loaded by FLEs, but can also be loaded by a conveyor from a live bottom hopper. Batch mixers are best for small- to medium-sized (up to 100 Mg wet biosolids cake per day) facilities.

- **Continuous mixers.** Pug mills and plow mixers are examples of continuous mixers. Feedstock material is loaded into separate live bottom hoppers. The variable-speed augers in the hoppers allow for metering the correct portions of the individual feedstocks. Weigh elements in the hoppers or on the discharge conveyors are used to meter the material. The material is conveyed to the mixers, and conveyors discharge the blended material into or near the composting piles or vessels. Continuous mixers are the most automated and complex systems, and are appropriate for only medium (40 to 100 Mg wet biosolids cake per day) and large (> 100 Mg wet biosolids cake per day) facilities where labor offsets justify the increased capital cost.

Regardless of the mixing technology selected, the mixture must have uniform porosity and all particles of cake should be in close contact with the bulking agent. Dewatered biosolids of 18 to 25% solids should be mixed with a bulking agent so that each wood chip or other bulking agent particle is coated with a thin layer of solids. Dewatered cake of 30 to 35% solids will break into clumps, which must be uniformly small and mixed with the bulking agent. Large clumps and balls will become anaerobic and will contribute to excessive odors. If mixing is not uniform, then zones with a disproportionate amount of bulking agent will divert the flow of air, causing other zones to become anaerobic.

3.8.3 Active Composting

It typically takes at least 50 days to produce completely stabilized compost. The level of process control required to produce the product decreases with time. The highest level of control is needed in the first 14 to 21 days. It is during this time that the material is most biologically active and requires the most moisture and heat removal. This stage is known as "active composting." It is during active composting that regulatory requirements for pathogen kill and vector attraction reduction is typically met.

Three key variables are controlled during the active composting phase: pile oxygen content, temperature, and solids content. In a turned windrow system, all of these variables are controlled by the initial porosity of the mix and the frequency of the turning operation. In an ASP facility, the mix porosity and aeration system provide the control. For within-vessel systems, although mix porosity and aeration system are the primary control, material agitation also has an effect. Solids content is the most affected by agitation of a within-vessel system.

In turned windrow systems, little distinction exists between active composting and the curing phase of the operation. Because porosity is vital to air movement into the pile in all phases of the process, the only thing that changes is the frequency of turning; the material does not need to be moved from active composting to curing. In ASP and within-vessel operations, the active composting stage should be designed for a minimum of 15 days. Because vector attraction reduction requirements take 14 days, it makes sense to keep the material in the most controlled environment for at least 15 days. Typically, 21 days is preferred because the material is still generating significant heat and odor up to that point, depending on the technology used.

3.8.4 Aeration Systems and Controls

In the United States, each state is responsible for regulating biosolids reuse within their borders according to Federal government guidelines that must be met. These are U.S. EPA (1993) Title 40 Part 503 Regulations (referred to as the "503 Regulations"). Current 503 regulations require pathogen kill temperatures greater than 55 °C (131 °F) for three or more days for aerated static pile and within-vessel systems, or 14 days for windrow systems with five turnings during the 14-day period. In addition, 14 days with an average temperature of 45 °C (113 °F) with a minimum of 40 °C (104 °F) is required for 503 vector attraction reduction.

In addition to regulatory requirements, it is important to control temperatures within certain ranges to optimize the decomposition process. The optimum temperature range for volatile solids destruction is 45 to 50 °C (113 to 122 °F). It is also desirable to prevent material temperatures from climbing too high. Pile temperatures in excess of 70 °C (158 °F) inhibit the biological decomposition process, which can result in immature compost with significant amounts of energy and heat remaining that can generate malodors during curing. Also, if high temperatures persist for periods longer than several weeks, the potential of spontaneous combustion can occur in dry material (> 80% solids).

In turned windrow operations, temperature and oxygen are controlled by porosity and frequency of turning. Initial porosity is controlled by thorough blending of the feedstock and having the proper bulking agent-to-biosolids mix ratio. Once the windrows are in place, both temperature and oxygen content are controlled by turning of the windrow. Turning incorporates oxygen and releases heat and moisture. Although turning releases heat, pile temperature will spike shortly after turning because of redistribution of feedstock and infusion of oxygen. These temperature spikes are typically last a few hours.

In ASP and within-vessel composting systems, forced aeration is used to supply oxygen and to maintain aerobic conditions within the material, control temperatures, and remove moisture. In the first one to two days of composting, a slow increase in airflow typically kick-starts the process and causes pile temperatures to rise quickly. Over-aeration at the beginning of the process can inhibit temperature increases and must be controlled carefully to match system demands. This is especially true of low-energy biosolids (digested) or when wet, nonbiologically active amendment is used. Throughout the rest of the process, as the rate of airflow is increased in a forced aeration system, pile temperature decreases, and the rate of water vapor removal increases. As with a turned windrow system, agitation releases heat and water vapor.

Experience has shown that an aeration rate approaching 300 m^3/dry Mg·h (10 000 cu ft/h/dry ton of wastewater solids) may be needed to maintain pile temperatures below 60 °C (140 °F) during peak activity. This aeration capacity may be impractical in large systems because of the high horsepower associated with these blowers and their effect on electrical charges for demand and connected horsepower. Practical aeration capacities range from 90 to 140 m^3/dry Mg·h (3000 to 5000 cu ft/hr/dry ton of wastewater solids). These rates have been widely used in design of existing composting facilities. Aeration in this range will control temperatures through most of the composting period and will provide adequate moisture-removal capacity. Higher aeration rates are possible, but require a greater energy use, larger and more closely spaced piping or trenches, and larger odor collection and treatment systems, if provided. If highly reactive bulking materials are used, such as ground-up leaves, then mass of bulking agent and dewatered cake should also be considered in the sizing of aeration capacity.

Forced aeration of compost can be either pushed up through the compost material (positive aeration) or pulled down through the material (negative aeration). Positive-mode aeration generally requires less energy to move the same volume of air than negative aeration. In positive aeration, when outside supply air is used, the air is cooler and contains less moisture, which requires less volume for the same weight of air. Negative aeration (especially continuous) provides a significant advantage for enclosed and worker-occupied operations. Most odors and moisture from the material are captured and prevented from entering the air above the pile where greater airflows are required to capture and treat emissions. One disadvantage to negative aeration is the accumulation of condensate in ductwork and blowers. When using negative-mode aeration, it is vital to provide ample opportunity for duct and blower drainage. Drainage must be supplied at all low points in the collection duct and at the bottom of the fan housing. All of these drains must be equipped with traps to prevent drawing in unwanted air from the drain piping. In addition, all of the condensate that is collected must be treated.

Aeration can be either intermittent or continuous. If the aeration is intermittent, then it must not be in the off mode for more than 15 minutes. Microbial oxygen demand during composting can reduce the available oxygen in the air to as low as 3 to 5% in as little as 15 minutes, thus causing anaerobic conditions and further contributing to malodor production. If the aeration is continuous, then fans must be equipped with variable frequency drives to allow for adjustments in airflow for pile temperatures and solids content.

Aeration demands for controlling pile temperature will vary through the composting process. At the start of composting, the challenge is to bring the material temperature up to the thermophilic range as quickly as possible. As the process continues, the challenge changes to removing excess heat to maintain the process in the thermophilic temperature range. Near the end of the process, it is important to dry the material without removing too much heat or over-drying.

Water vapor is removed during composting, increasing the solids content of the mixture from approximately 40 to 55% or higher. Proper drying near the end of the active compost stage is critical to the screening operation because screens do not perform well on compost with less than 55% solids.

Most within-vessel systems use positive aeration for a variety of reasons that depend on the actual system. For example, in tunnel systems, negative aeration provides no advantage because there is minimal headspace above the piles and no workers in the vessel during active composting. In ASP operation, negative aeration is popular because the odors and moisture are captured directly. Many ASP operations are configured to allow either negative or positive aeration. This allows odor and moisture capture during the periods when decomposition is occurring at the greatest rate. During the later stages of composting before screening, the aeration mode can be changed to positive, and the rate is increased. This accelerates material drying in preparation for the screening process.

Figures 13.8 through 13.11 show several different air floor configurations that can be used in ASP. Agitated bay systems have perforated pipe embedded in a gravel plenum. Regardless of the type of air floor or plenum used, it is vital that it be designed to deliver air evenly over the entire length of the pile. Three methods are used to accomplish this:

1. Provide progressively more air outlets along the pipe or trench to offset friction headloss along the pipe or trench is offset because of reduced velocity loss through the outlets.
2. Change the cross-section of the pipe or trench to provide a constant air velocity along the entire length.
3. A combination of the two.

The spacing of pipes or trenches typically ranges from 1 to 2 m (3 to 6 ft). The spacing selection is dependent on the size of the distribution piping or trench—the larger the spacing, the larger the distribution elements. If the spacing is too great, then anaerobic zones will develop in the lower portion of the piles between the pipes or trenches. Air will always seek the path of least resistance through the pile. Within-vessel systems may use continuous plenums or gravel floors with permanent piping. Regardless of the plenum type used, regular cleaning is required. In the gravel plenum of many within-vessel systems, a hard pan develops on the top surface of the gravel. If this is not routinely removed, then it will block and redirect the airflow. In ASP systems, equipment moving over the outlets to place and remove material can block the air outlet from pipes or trenches. These must be cleaned after every one or

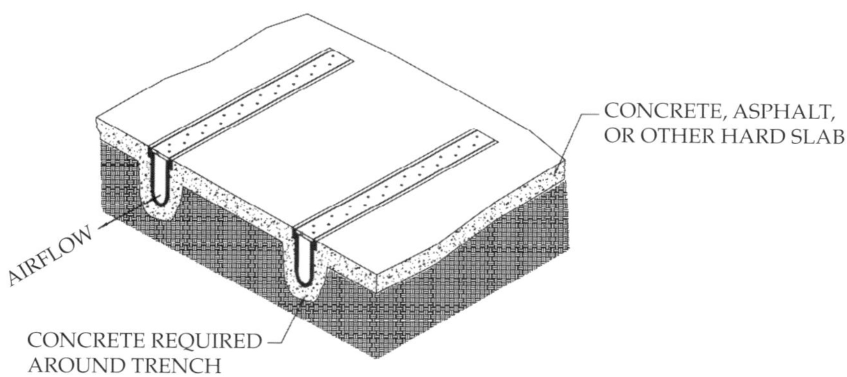

Figure 13.8 Aeration trench in floor (WEF et al., 2009).

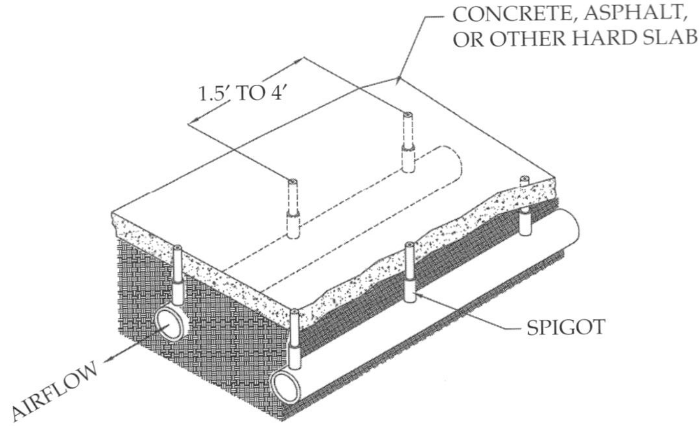

FIGURE 13.9 Embedded pipe and spigot aeration floor (ft × 0.3048 = m) (WEF et al., 2009).

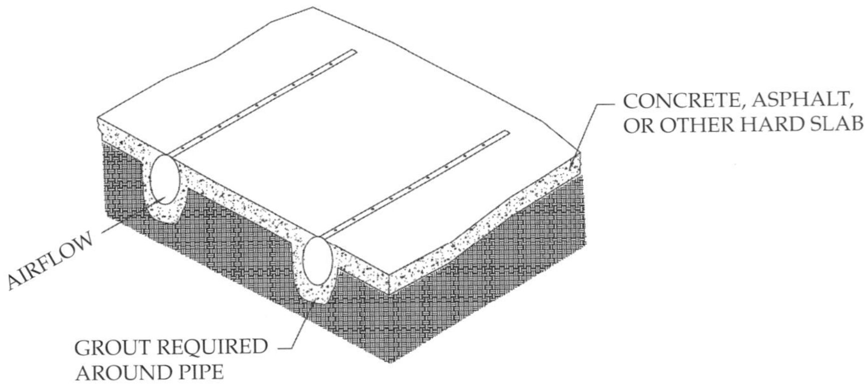

FIGURE 13.10 Embedded pipe aeration floor (WEF et al., 2009).

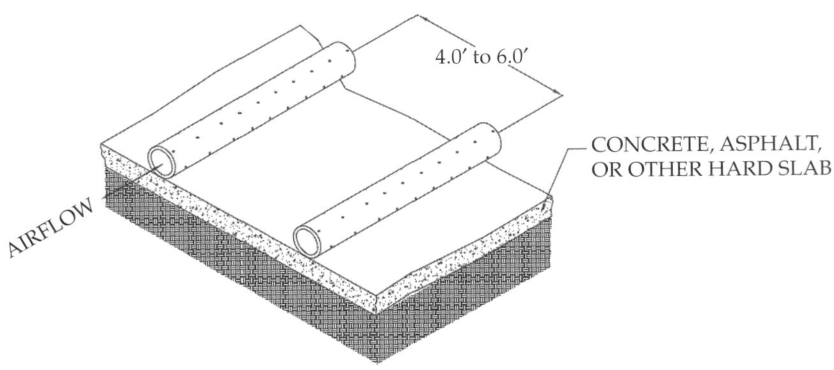

FIGURE 13.11 Pipe on slab aeration floor (ft × 0.3048 = m) (WEF et al., 2009).

limiting the number of aeration fans that are running by using a computerized control system that examines aeration needs for each pile and makes continuous adjustments.

In cold climates, heating intake air into a compost building can be a significant energy consumer. For enclosed and partially enclosed facilities, building walls and roofs create a large surface area. At least one facility uses solar panels mounted on the facility walls to heat the air entering the building. Large roof areas also provide an opportunity for solar power, but the cost of the structure will increase to accommodate the additional roof load. These economics must be carefully examined. Because the buildings are typically large, small changes in the structure can result in significant increases in overall cost.

3.11.3 *Contingency Planning*

Two main areas of composting operations require contingency planning. First is planning for loss of power. As noted above, composting can be an energy-intensive operation, and the cost of providing backup generation can be high. It should not be assumed that all systems require backup power. If outages are typically short (less than eight hours), then backup power for compost aeration may not be needed. Composting is a steady-state process during which changes taking place slowly. The most significant consequence of not aerating material for a few hours is a significant increase in odor generation and piles that become anaerobic. Although this is not desirable, it can be tolerated for isolated power outages. If power can be expected to be lost for a week or more, then there is danger of spontaneous combustion occurring in the piles. In this case, aeration may need to provide a minimal level of power backup.

Power in a facility must be prioritized to best determine the level of backup power generation needed. For example, a facility located in an area with close neighbors who are sensitive to odor may want to list the odor collection fans and systems as the highest priority. The facility would provide enough backup power to run aeration blowers at minimal levels, for example, running only 1 minute in 15 minutes and allowing only a few fans to operate at any one time.

The second area requiring contingency planning is for the complete or partial shutdown of operations. Many states require a contingency plan for treating biosolids as part of the facility permitting. Severe storms, fires, and natural disasters can destroy or significantly damage a facility. It is, therefore, important to have an alternative treatment or disposal method available for such instances.

4.0 COMPOST MARKETING AND USE

Compost is a soil conditioner that possesses the following characteristics:

- Increases water-holding capacity for sandy soils;
- Increases aeration and drainage for clay soil;
- Provides slow-release nutrients of nitrogen (1 to 2%), phosphorus (2%), and potassium (zero to 1%);
- Provides essential plant micronutrients; and
- Increases nutrient holding capacity of soil.

In the composting process, most nitrogen is converted to ammonia and volatilized. Compost contains less nitrogen than the input feed cake.

Compost use is regulated by each state with minimum standards set the U.S. EPA. Federal regulations specify high quality and ceiling limits for inorganics and minimum quality requirements for pathogen and vector-attraction reduction (U.S. EPA, 1993).

States can have more stringent regulations but must petition U.S. EPA to have less stringent rules. Several state regulations establish classes of compost based on inorganic concentrations. The compost of the highest (best) class may be distributed to the public. In some states, compost must be labeled to indicate whether it is suitable for crops grown for direct human consumption. Lower classes of compost have more restrictions and may require site-based permitting for use.

Marketing requirements vary according to the intended use and the demand for compost. Following are several market requirements:

- Stability or maturity
  - Low carbon-dioxide generation or oxygen depletion per gram of compost volatile solids;
  - Minimal odors;
  - Minimal phytotoxic effects because of incomplete decomposition; and
  - Minimal nitrogen depletion because of unstabilized carbon (bulking agents).
- Moisture content
  - Prevention of excessive moisture, which adds to transportation costs and may generate odor;

  - Avoidance of insufficient moisture, which will cause dust; and
  - Achievement of ideal moisture range, which is between 35 and 40%.
- Texture and particle size
  - Coarse compost production for land reclamation and erosion control applications; and
  - Fine compost production, as required for topsoil blending and turf top dressing.
- Salinity or conductivity
  - Prevention of excessive salinity, which will inhibit sprouting of seeds; and
  - Avoidance of ferric-chloride conditioning, which can cause salinity.

Uses in which compost serves as a seed-germination medium are particularly sensitive to phytotoxicity and salinity. Top-dressing applications are sensitive to particle size. For landfill cover, land reclamation, and erosion control, coarser compost may be preferred because it is more resistant to erosion. Retail or homeowner applications are particularly sensitive to texture and residual odor. Many applications are sensitive to nitrogen depletion. Frequently, fertilizer is applied with the compost to ensure adequate nitrogen.

It is critical that a marketing plan for compost is developed long before production is started. Without a good marketing plan, the potential exists for material to build up at the facility, which can ultimately affect its throughput.

5.0 DESIGN EXAMPLE FOR AERATED STATIC PILE

One design example is provided below for an ASP for a facility treating 20 Mg/d dry biosolids. The following conditions apply:

- Seven day per week operation;
- Yard waste supplemented by ground wood waste as needed for bulking agents;
- Biosolids storage, mixing, active composting, and screening operations are fully enclosed;
- 30 days storage of new bulking agent under cover;

- One day storage of biosolids;
- Seven days storage of recycled bulking agent under cover;
- 21 days minimum retention time in active composting;
- 28 days minimum retention time in aerated curing; and
- 90 days outdoor storage of finished product.

In this exercise the various areas of the facility will be sized. Overall size of areas depends on expected vehicle types and topography of the site. The first step is to develop a materials balance for the facility. Table 13.4 provides the ASP materials balance. A target solids content of 40% has been used to develop the materials balance.

Total recycled bulking agent produced is 232 m^3 (304 cu yd) calculated as follows

$$\begin{aligned}&\text{Total recycled bulking agent used in the process}\\&\quad= \text{Screened recycled input} + \text{Base (recycled bulking agent)}\\&\quad= 204\ m^3 + 28\ m^3 = 232\ m^3\end{aligned} \qquad (13.2)$$

For area sizing, the following areas are constructed with concrete walls on three sides: biosolids and all bulking agent storage areas, active composting, and curing.

For most FELs, the maximum height, H, will be approximately 3 to 4 m (10 to 12 ft). For design purposes a maximum of 3 m (10 ft) should be used. The following equation represents the total volume for a given area. The equation is manipulated to determine the desired value. For example, a site may be narrow and the allowable length may be limited. Width should be at least 4.6 m (15 ft) for each day's worth of material. This is wide enough for a FEL to dig out the material while leaving the surrounding piles intact. More area is needed if aboveground aeration pipes are used to allow space for pipe removal between compost cycles.

$$\text{Pile volume per day} \times \text{Number of days} = H \times (L - H/2) \times W \qquad (13.3)$$

where
H = height,
L = length, and
W = width.

For ground yard waste bulking agent storage area with 30 days storage area with an assumed L of 30 m (100 ft) and H of 3 m (10 ft):

$$W = 146\ m^3/d \times 30\ d/[3\ m \times (30\ m - 3\ m/2)] = 51.2\ m$$

Table 13.4 Aerated static pile materials balance.

Material	Volume (m^3)	Wet Weight (Mg)	Dry Weight (Mg)	Volatile Solids (Mg)	Bulk Density (kg/m^3)	Solids Content (%)	Volatile Solids (%)
Biosolids	128.0	100.0	20.0	13.0	781	20.0	65.0
Yard waste (processed)	145.9	42.7	23.5	16.5	293	55.0	70.0
Wood waste	44.4	10.8	6.5	6.2	244	60.0	95.0
Screened recycled bulking agent	223.5	75.8	41.7	38.8	339	55.0	93.0
Unscreened recycle	0.0	0.0	0.0	0.0	381	55.0	72.5
Mixture	516.9	229.4	91.7	74.4	444	40.0	81.1
Base (recycled bulking agent)	25.8	8.8	4.8	4.5	339	55.0	93.0
Cover (unscreened)	62.0	23.6	13.0	9.4	381	55.0	72.5
Composting losses		99.8	7.4	7.4			
Cover (unscreened)	62.0	23.6	13.0	9.4	381	55.0	72.5
Screen feed	425.4	162	89.1	71.5	381	55.0	80.2
Recycled bulking agent	247.0	83.8	46.1	42.9	339	55.0	93.0
Curing	177.9	78.1	43.0	28.6	439	55.0	66.5
Curing losses		2.6	1.4	1.43			
Compost to storage	112.3	49.4	41.6	27.1	439	55.0	65.3

Assumptions:

Recovery by screening

- Yard waste — 50% by volume
- Wood waste — 70% by volume
- Recycled bulking agent — 50% by volume
- Pile base — 95% by volume

Processing losses

- Losses during composting — 10% of volatile solids
- Losses during curing — 5% of volatile solids

For the biosolids storage area with an assumed *W* of 9.1 m (30 ft) and an *H* of 0.9 m (3 ft):

$$L/d = 128\ \text{m}^3/\text{d} \times 1/(9\ \text{m} \times 0.9\ \text{m}) + 0.9/2 = 16\ \text{m}$$

Biosolids have a gelatinous consistency and a relatively high bulk density. These factors prevent the material from stacking. Biosolids with solids contents between 18 and 24% will only stack to approximately 0.9 to 1 m high (3 to 4 ft). Wetter biosolids will flow and not stack to more than 0.3 m (1 ft).

In sizing the active compost for most composting facilities, a single bay or pile will hold a single day's worth of material. In small facilities, however, a bay may be sized for approximately two or more days. The number of bays is driven by the desired retention time; 21 days is the typical minimum with two additional bays provided to allow for a bay to be torn down and another to be constructed without reducing retention time. In ASP facilities, the active composting area is constructed with bays on either side of a center aisle with no physical obstruction between bays. The bay designation is for keeping track of material and aeration. In most facilities, aeration of the compost in each bay is controlled separately.

Below is the sizing of the active compost hall using an assumed *W* of 6.1 m (20 ft) and a depth of mix of 2.3 m (8 ft). Although the overall pile depth will be 3 m (10 ft), an allowance of 0.3 m (1 ft) for a base plenum layer and 0.3 m (1 ft) for a cover layer must be made. The minimum bay *W* should be 4.6 m (15 ft). This allows a FEL to pile and remove a single days worth of material.

$$L/\text{bay} = 517\ /(2 \times 6) + 3/2 = 38\ \text{m}$$

The overall composting hall width must include allowance for the piles, the center aisle, and a separate area for the aeration blowers. The blowers are most often housed behind the piles. The minimum allowance for the blower gallery will depend on the size of the blowers and ductwork and access to the area. If the only access to the blower gallery is at the ends of the compost building, then the gallery must be made wide enough to move the blowers without being dismantled. If access doors can be available closer to the blowers, then the hall can be narrower. For this example, a 4.6-m (15-ft) wide gallery is assumed.

The center aisle must be a minimum of 9.1 m (30 ft) wide to allow for movement of FELs to construct and teardown piles. If the operation plan is to load trucks directly from the active compost pile and to provide curing at another location, then the center aisle should be increased to a minimum of 14 to 15 m wide (45 to 50 ft).

$$\text{Hall width} = [2 \times (\text{Pile width} + \text{Blower gallery width})] + \text{Center aisle width} + (4 \times \text{Wall thickness})$$

$$\text{Hall width} = [2 \times (38.3\ \text{m} + 54.6\ \text{m})] + 9\ \text{m} + (4 \times 0.3\ \text{m}) = 95.5\ \text{m} \tag{13.3}$$

$$\text{Compost hall length} = (\text{Pile width} \times \text{Number of pile per side of hall}) + (2 \times \text{Wall thickness})$$

$$\text{Compost hall length} = (6\ \text{m} \times 12) + (2 \times 0.3\ \text{m}) = 73.8\ \text{m} \tag{13.4}$$

In most ASP facilities, compost aeration is provided individually for each day's worth of material (one bay). Aeration typically can be provided by a single blower for each bay. This configuration provides the greatest flexibility and the least interruption to operations if a blower goes out of service. In small facilities, one continuously running blower can serve several bays, and the individual piles can use butterfly valves to control the airflow. For this example, a single blower will be used for each day's worth of material. The aeration rates for each blower are calculated below based on the rates defined in section 13.3.9.4.

$$\text{Required aeration} = 140\ (\text{m}^3/\text{h})/\text{dry Mg} \times 20\ \text{dry Mg biosolids} = 2800\ \text{m}^3/\text{h}\ (1648\ \text{cu ft/min}) \tag{13.5}$$

6.0 CASE STUDY FOR AERATED STATIC PILE

In 1995, Davenport, Iowa, began operating an enclosed ASP composting a facility. The facility processed 28 Mg/d dry biosolids (140 Mg/d wet) using ground yard waste as a bulking agent. The facility is located along the Mississippi River adjacent to the city's WWTP. There is a 2- to-3 m (8- to 10-ft) high berm surrounding the facility to protect against flooding. The facility as originally constructed consisted of the following:

- Fully enclosed mixing and active composting areas;
- Enclosed product bagging area;
- Partially enclosed screening and curing area;
- Outdoor yard-waste processing operation;
- Outdoor product storage pad; and
- Two open biofilters for odor control.

Biosolids from the WWTP are tipped into two 53-m^3 (70- cu yd) live bottom hoppers. Recycle and new bulking agent is blended together with an FEL and then tipped into a 15-m^3 (20-cu yd) live bottom hopper. The material is metered using weighing elements and transported by conveyors to one of two pugmill mixers. The blended feedstock is dropped into the composting hall by conveyor where an FEL builds composting piles. The residence time in active composting is a minimum of 21 days.

After active composting, the material is screened and the compost is placed in aerated curing piles for 28 days. The facility has successfully marketed the product under the brand name *Earth Cycle*. The compost is sold in both bulk and bagged form. In addition, the facility rents compost-spreading equipment and sells mulch from the yard waste processing operation. Capital costs for the original facility constructed in 1995 was $7.41 million. In addition, moving stock at the compost facility consists of two 8-m^3 (10-cu yd) capacity FELs and two 4-m^3 (5-cu yd) capacity FELs and a yard waste grinder. This equipment was purchased for $1.2 million. Since 1995, approximately $2 million in improvements have been made at the facility. These improvements include

- Recoating the composting hall;
- Replace aeration floor trenches;
- Replace the variable frequency drive controls system;
- Replaced aeration fan housings; and
- Replacement of biofilter laterals.

The building is a prefabricated metal structure that was painted and insulated. Major structure members were not insulated. Between 2006 and 2007, the composting hall was recoated with the Stayflex coating, which consists of several inches of polyurethane foam insulation covered with a polyester hard outer shell.

The aeration trenches of the original aeration floor had stainless-steel cover plates, but the frame angles embedded in the concrete floor were made of galvanized steel, which corroded over time and have been gradually replaced by facility staff. With improvements in technology, the facility has replaced the variable-frequency drives (VFDs) and the computerized controls. The new VFDs are more energy efficient.

The compost aeration system as originally designed was capable of operating in either positive or negative mode. Operation has always been in negative mode, however, and the city has replaced piping and valves to only allow negative aeration, which provides better odor control. The original aeration fan housings were coated steel. Over time, these have corroded and have been gradually replaced.

The aeration piping laterals in the bottom of the biofilters have been upgraded from the original corrugated high-density polyethylene pipe to PVC pipe, including cleanouts at the end of the filters. This has allowed for easier maintenance of the biofilter cells.

The annual operating cost in 2004 was estimated at $1.431 million. Revenues from product sales that year amounted to $512,000. Therefore, the net operation and maintenance cost for the facility in 2004 was $919,000. Based on an annual throughput of 25 369 Mg of biosolids, the net operating cost per wet megagram of biosolids processed was $36.23.

7.0 REFERENCES AND SUGGESTED READINGS

Beecher, N.; Goldstein, N. (2010) Biosolids Composting in the United States: 2010 Update. *BioCycle*, **51** (12), 35–41.

Braccio, J.; Dievert, D.; O'Brien, D.; Kuter, G. (2008) Modern Approaches to Enhance Biosolids Composting. *Proceedings of the Water Environment Federation Residuals and Biosolids Conference* [CD-ROM], Mar 30–Apr 2, Philadelphia, Pennsylvania; Water Environment Federation: Alexandria, Virginia.

Epstein, E. E. (1997). *The Science of Composting*; CRC Press: Boca Raton, Florida.

Haug, R. T. (1980) *Compost Engineering*; Ann Arbor Science Publishers: Ann Arbor, Michigan.

Higgins, A. J.; Chen, S.; Singley, M. E. (1982) Airflow Resistance in Sewage Sludge Composting Systems. *Trans. Am. Soc. Agric. Eng.*, **25** (4), 1010–1014.

Murray, C. M.; Thompson, J. L. (1986) Strategies for Aerated Static Pile Systems; *BioCycle*, **6**.

National Fire Protection Association (2008) *NFPA 820: Standard for Fire Protection in Wastewater Treatment and Collection Facilities*; National Fire Protection Association: Quincy, Massachusetts.

North East Biosolids and Residuals Association (2007) *National Biosolids Regulation, Quality, End Use and Disposal Survey*; North East Biosolids and Residuals Association: Tamworth, New Hampshire.

U.S. Environmental Protection Agency (1993) Standards for the Use and Disposal of Sewage, Final Rule. *Code of Federal Regulations*, Part 503, Title 40, 1993; *Fed. Regist.*, **58**, 9248.

Verschueren, K. (2001) *Handbook of Environmental Data on Organic Chemicals*, 4th ed.; Wiley and Sons: New York.

Water Environment Federation; American Society of Civil Engineers; Environmental and Water Resources Institute (2009) *Design of Municipal Wastewater Treatment Plants*, 5th ed., WEF Manual of Practice No. 8; ASCE Manual and Report on Engineering Practice No. 76; McGraw-Hill: New York.

Willson, G. J.; Parr, J.; Epstein, E.; Marsh, P.; Chaney, R. (1980) *Manual for Composting Sewage Sludge by the Beltsville Aerated-Pile Method*, EPA-600/8-80-022; U.S. Environmental Protection Agency: Washington, D.C.

Chapter 14

Alkaline Treatment

(continued)

1.0 INTRODUCTION

Alkaline treatment is viewed by many in the wastewater treatment industry as a low cost, easy way to meet the requirements of the federal, state, and local sludge and biosolids management rules for use of biosolids in agriculture, horticulture, and reclamation. Although alkaline treatment can effectively disinfect and stabilize wastewater solids when applied correctly, it is not a simple process and must be properly designed and operated. If it is not well designed and operated, then the process typically does not comply with pathogen-reduction requirements and becomes a source of public complaints, particularly regarding obnoxious odors.

In this chapter, the terms "lime" and "alkaline" are used somewhat interchangeably. Lime is an alkaline earth material, and when mixed with water, produces a pH of 12.45 at 25 °C. At this high pH, all pathogenic bacteria, viruses, and fungi are destroyed (Berg et al., 1968; Farrell et al., 1974; U.S. EPA, 1975). This work led to the design criteria that the U.S. Environmental Protection Agency (1979) included in its *Process Design Manual for Sludge Treatment and Disposal* that recommended: (1) sludge should be treated in the liquid state, and b) sludge pH should be raised to12.454 (as corrected to 25 °C) by lime addition and maintained above 12.45 for 30 minutes (which keeps pH >12 for two hours).

The objective was to maintain the sludge greater than a pH of 12 for 2 hours to ensure destruction of most pathogenic organisms and to provide enough residual alkalinity so that the pH would not drop less than 11 for several days, allowing sufficient time for land application and incorporation of the biosolids, thus avoiding the possibility of putrefaction.

Good mixing of lime and sludge in the presence of adequate moisture is critical to ensure uniform distribution of calcium and pH throughout the sludge mass, which temporarily stabilizes it. This action temporarily ensures control of putrefaction and reduction of bacteria and viruses. No direct reduction of organic matter occurs in lime stabilization (Class B, Alternative 2). This has two important effects. First, as noted above, lime addition does not permanently make sludge chemically and biologically stable; if the pH drops below 10.5, then biological decomposition will resume, producing noxious odors. Second, the quantity of residuals for use or disposal is not reduced as it is by biological stabilization methods. On the contrary, the mass of dry sludge is increased by the added lime and by the chemical precipitates that derive from this addition.

2.0 ALKALINE MATERIALS

Alkaline materials, alkaline processes, and alkaline byproducts refer to materials and products that have an alkaline pH (by definition a pH > 7, but typically pH 8 to 14). Suitable alkaline treatment agents are all lime-based materials. Lime itself is found in two forms: calcium oxide (CaO) and calcium hydroxide [$Ca(OH)_2$]. Calcium oxide is the result of heating limestone [calcium carbonate ($CaCO_3$)] to a sufficiently high temperature [approximately 1090 °C (2,000 °F)] to drive off carbon dioxide (CO_2), resulting in CaO. This process is called calcination or burning. When CaO, or quicklime, is mixed with water, it forms a fine white powder, $Ca(OH)_2$, or hydrated lime, and gives off considerable heat [63 965 kJ/kg·mol (27 500 Btu/lb·mol)], called the heat of hydration. The heat generated when CaO is reacted with water can be used to disinfect residuals.

The types and dosages of the alkaline chemicals to be added are critical to the success of the alkaline-stabilization process. The quality of the chemicals—lime, cement-kiln dust (CKD), Portland cement, and lime-kiln dust (LKD)—should be consistent. Different types or sources of additives result in different product texture and granularity. Lime is available from many sources, ranging from a high-calcium lime obtained from oyster or clam shells to a relatively low-calcium dolomitic lime. Major considerations in selecting the type of chemical to be used include economics, availability, desired mixing, and desired product characteristics. Table 14.1 provides typical alkaline chemical characteristics.

TABLE 14.1 Characteristics of common alkaline materials [adapted from Lewis, C. V.; Gutschick, K. A. (1980) *Lime in Municipal Sludge Processing*; National Lime Association, Washington, D.C., with permission.]

Common name formula	Available forms	Appearance and properties	Bulk density	Commercial strength	Referenced standards
Quicklime/CaO	Crushed; ground; pulverized (virtually all passes 20 sieve)	White (light gray, tan) lumps to powder; unstable, caustic irritant; slakes to hydroxide slurry evolving heat 1.1 MJ/kg (490 Btu/lb); reacts with Co_2 in air to form $CaCo_3$; saturation solubility approximately pH 12.5	880 to 1200 kg/m^3 (55 to 75 lb/cu ft); specific gravity 3.2 to 3.4	70 to 96% CaO (below 88% can be poor quality)	ASTM C911[a] and PFHWA/RD-82/167
Hydrated lime/$Ca(OH)_2$	Powder (at least 75% passes 200 sieve)	White, 200 to 400 mesh; powder free of lumps; caustic dusty irritant; absorbs H_2O and CO_2 from air to form $Ca(HCO_3)_2$; saturation solubility approximately pH 12.4	400 to 640 kg/m^3 (25 to 40 lb/cu ft); specific gravity 2.3 to 2.4	$Ca(OH)_2$–82 to 98%; CaO–62 to 74% (standard 70%)	ASTM C51–95, ASTM C911[a] and FHW/ARD-82/167
Lime kiln dust	Powder (at least 75% passes 100 sieve; at least 50% passes 200 sieve)	Light gray to tan; properties variable, depending on kiln and operating conditions	50 to 75 lb/cu ft	Variable	ASTM C911[a] and FHW/ARD-82/167
Cement kiln dust	Powder (at least 75% passes 100 sieve; at least 50% passes 200 sieve)	Light gray to brown; properties variable, depending on kiln and operating conditions	50 to 75 lb/cu ft[b]	Variable	ASTM C911[a] and FHW/ARD-82/167
Portland cement	Powder; Type I cement is typically used	Light gray	NA[c]	NA	ASTM C:50
Fly ash	Powder (at least 70% passes 200 sieve)	Variable	Variable	Variable	ASTM C593

[a] ASTM C911–96 primarily addresses the following water treatment: softening, silica removal, color removal, and clarification. An ASTM standard. "Lime for Municipal Treatment of Biosolids," has been drafted

[b] lb/cu ft x 16.02 = kg/m^3.

[c] Not applicable.

2.1 Materials Used in Treating Sludge

Alkali (sodium and potassium)-based hydroxides must not be used to raise the pH of the wastewater solids to achieve regulatory requirements. The chemical compounds formed by reacting alkali with wastewater solids can cause high total dissolved solids levels, especially sodium and potassium, in effluent and groundwater. At high enough concentrations, these salts can be toxic to most plant life. Only lime-based alkaline earth materials should be used because they form insoluble compounds, protecting groundwater, and the environment. As long as the application rate of lime does not exceed the buffering capacity of the soil to which it is applied, the pH will not rise to more than approximately a pH of 7 or the target pH. Rates that exceed the buffering capacity of the particular soil and limit micronutrient availability for target crops should not be applied.

2.2 Typical Chemical Reactions

When mixed with water, hydrated lime separates into the calcium (Ca^{+2}) ion and the hydroxide (OH^{-1}) ions. The OH^{-1} concentration is responsible for the high pH that lime imparts to a sludge solids/lime/water mixture. A saturated solution of limewater attains a pH of 12.45 at 25 °C. When mixed with water, quicklime reacts to form hydrated lime and generates heat because this is an exothermic (heat generating) chemical reaction. If the operator needs more heat, then he or she should use more quicklime or supplemental heat. If a higher pH is needed, then more lime should be added.

2.3 Type of Material to Use

Industrial wastes and coproducts such as LKD, CKD, and fly ash are suitable to use only if they contain sufficient free lime for practical use and are reactive with water. Only free lime (not chemically bound with other compounds) can react with water to liberate heat and form the OH^{-1} needed to give the high pH for stabilization. Many wastes and byproducts are not reactive with water because of the environment in which they were formed. These products typically contain small particles, have been subjected to intense heat (often greater than fusion point), and have a coating of insoluble calcium sulfate (from absorbing sulfur). They typically have low-reactive rates with water and may be nonreactive materials. The insoluble sulfate coating and intense heating seals the pores in the material, slowing the absorption of water into it. The rate of absorption of water into the material determines how quickly the hydration reaction occurs, forming the desired products—$Ca(OH)_2$ and heat. Thus, large quantities of these materials must be added to the sludge mixture to provide sufficient free lime to accomplish stabilization. Typically, insufficient free lime is available to adequately increase pH and maintain it, thus accomplishing the desired reactions. The sludge, therefore, does not achieve stabilization, resulting in regrowth of bacteria and odors. Using low-reactive lime creates the same problem. Several conditions can cause low reactivity with water:

- **Hard burning.** In the production of lime (CaO), limestone ($CaCO_3$) is heated to the dissociation temperature to release carbon dioxide (CO_2), forming CaO, sometimes referred to as quicklime. As the carbon dioxide escapes from the limestone, tiny pores are left in the resulting quicklime. If heating is continued or the temperature is excessively high, then these pores begin to close. The size of the pores determines the rate that water can enter and react with the lime. If water cannot enter the lime particle, then a reaction does not occur to form the hydrate (particles containing water molecules).
- **Water-insoluble coating.** When quicklime is exposed to sulfates or fluorides, an insoluble coating is formed on the particle. This impedes the absorption of water, slows the hydration reaction, and wastes the inside part of the coated quicklime particle.
- **Air slaking.** If lime-based alkaline materials are exposed to the atmosphere and moist environments, then they will absorb moisture and carbon dioxide from the air and chemically revert to their original chemical composition, limestone ($CaCO_3$). Limestone is slowly reactive with water and does not create heat or high pH.

2.3.1 *Available Lime Index*

The evaluation of alkaline materials and their free or available lime content is presented in *Standard Test Methods for Chemical Analysis of Limestone, Quicklime, and Hydrated Lime* (ASTM International, 2006); and in a derived standard, *Quicklime and Hydrated Lime* (American Water Works Association, 1993). Use of these standards with *Specifications for Quicklime, Hydrated lime, and Limestone for Chemical Purposes* (2011)will allow an operator to evaluate the suitability of an alkaline material for the process being used. Results are expressed as available CaO or available $Ca(OH)_2$ (ASTM International, 2006).

2.3.2 *Reactivity Rates and Byproducts*

The evaluation reactivity of a lime with water is measured by its slaking rate (as defined in AWWA specification B202-93, Sec. 5.4). Small-pore limes react slowly, requiring 20 to 30 minutes to fully react with water, forming the hydrate but with a

2.4.3 *Measurement of pH using a Flat-Surface pH Electrode*

The efficacy of lime treatment is confirmed by demonstrating that operational conditions for pH and temperature have been met. The current method for measuring pH, which involves inserting a pH probe into a solids/water slurry, does not address mixing uniformity. The addition of water will hydrate any unreacted lime, and the vigorous mixing required to create the slurry homogenizes the sludge, potentially masking poor mixing (North, 2003). Complete mixing of lime and sludge is essential for eliminating low pH regions that can remain biologically active and lead to pH decay and odor generation. Well-mixed lime-treated wastewater solids will have a uniform distribution of calcium and OH ions and, thus, pH. The pH will consistently be approximately greater than 12. A flat surface pH electrode has been shown to be effective for monitoring mixing (see Figure 14.1).

Flat surface pH electrodes allow for pH measurement of semisolids without the addition of water. Studies have evaluated the use of a flat surface pH electrode to measure soil pH without creating a slurry or adding additional moisture (Adamchuck et al., 2002; Breitenbeck and Bremmer, 1984). Because additional water is not required, flat surface pH electrodes can be used to measure precise

FIGURE 14.1 Flat surface pH probe.

regions of sludge to demonstrate whether the entire sludge mass is demonstrating high pH. North (2003) used a flat-surface pH electrode on lime-treated wastewater solids from Blue Plains Advanced Wastewater Treatment Plant (AWTP) operated by DC Water, Washington, D.C. (formerly District of Columbia Water and Sewer Authority). North (2003) concluded that although it had limited ability to quantify lime incorporation, the flat-surface pH electrode was superior to the slurry method for evaluating pH variability. The Pennsylvania Department of Environmental Protection (PA DEP) and Pennsylvania State University, Harrisburg, Pennsylvania, further developed use of a flat-surface pH electrode to evaluate lime mixing. Localized pH values measured with the flat-surface pH electrode reflected the total Ca content of the biosolids. Regions with low Ca content had lower pH values; regions with high Ca contents had higher pH values. The flat-surface pH electrode identified low pH regions of lime treated sludge, which were masked by the slurry method.

2.5 Lime-Based Alkaline Material Dose Rate

To determine proper lime dose rates, "jar" tests must be performed in the laboratory on samples using the same sludge-to-alkaline reagent ratio as used in the full-scale process. The full-scale operation typically requires an additional 1% over the amount of lime determined by the jar test to compensate for less efficient mixing in the field. A viable base dosage must be established for the specific alkaline material to be used. Conducting a jar test is discussed in detail below.

1. Representative samples of the liquid sludge or sludge cake must be obtained by taking several samples from throughout the mass of materials. A 4-L (1 gal) sample is sufficient.
2. Sample should be mixed thoroughly.
3. Sample should be dried to 100% solids (typically done in a drying over at greater than 100 °C).
4. Sample should be divided into at least five equal-weight samples.
5. Weight of each dry sample should be measured to the nearest gram.
6. Each sample should be placed into individual 250-mL beakers to which 100 mL of distilled water is then added to make a slurry.
7. Fresh hydrated lime (determined to be fresh by measuring the pH in distilled water at 25 °C; pH must be 12.4) should be added in increasing amounts to

each sample. For example, 5% is added to the first sample; 10% to the second sample; 15% to the third sample.

8. Beaker is covered and the mixture is stirred using a magnetic stirrer or swirled to ensure complete mixing of the lime, water, and sludge. This should be repeated every 15 minutes for an hour. Cover should be removed to allow off gassing of NH_3.
9. After the NH_3 has been driven off, a high-alkalinity pH probe on an accurate pH meter (standardized to a slurry of hydrated lime at 25 °C and at a pH of 12.45) is used to measure the pH of each sample, which is then recorded. The pH test probe should be immersed into the supernatant solution only after the slurry has settled. The settled material should not be touched because it will contain unreacted lime and will always have a pH reading of 12.4.
10. If none of the sample reaches a pH of 12.4, indicating a saturated solution of $Ca(OH)_2$, more CaO should be added to the samples in increments of 5% and the samples should be retested after mixing. If NH_3 can be detected, then additional time must be allowed for the NH_3 to fully escape before reading the slurry pH.
11. The sample that maintains a pH of 12.4 with the least amount of lime for 24 hours with no NH_3 smell contains the amount of lime needed to treat the wastewater solids and meet the requirements of many regulatory agencies for Class B disinfection and vector-attraction reduction.
12. Results of the amount of lime needed to stabilize the sludge samples should be reported in grams of lime per gram of cake (pounds of lime per pound of cake), or grams of lime per liter (pounds of lime per gallon) of liquid sludge.
13. Actual doses should include an additional percentage (at least 1%) of lime to ensure stabilization in the field. This can be determined by checking pH of the biosolids after one to two weeks. The pH should be in the range of 11 to 12 at that time.

The above procedure will yield a lime-stabilized product but not necessarily one in which all pathogens are less than analytical detection limit. To ensure complete disinfection, a pasteurization step typically is applied. This can be done by adding sufficient CaO to the sludge to elevate the temperature to > 70 °C for > 30 minutes.

3.0 DESIGNING TREATMENT FACILITIES AND SYSTEMS

This section includes process descriptions and design considerations for several alkaline stabilization technologies.

3.1 Process Descriptions

Several alkaline stabilization technologies are available. Each system has advantages and disadvantages; therefore, it is important to evaluate and select a process that will produce the desired product on a site-specific basis. The following subsections describe several alkaline stabilization processes including dry-lime stabilization and advanced technologies and the considerations for design of alkaline treatment facilities and systems.

3.1.1 Dry-Lime (Post-Lime) Stabilization

Dry- or post-lime stabilization involves the addition of dry quicklime or hydrated lime to dewatered cake. Dry lime stabilization at wastewater treatment plants (WWTPs) has been practiced since the 1960s (Stone et. al., 1992). The lime typically is mixed with the dewatered cake material using a pug mill, plow blender, paddle mixer, ribbon blender, screw conveyor, or similar device. Thorough mixing of dry lime with a dewatered solids material can be energy-intensive, and it is difficult to ensure complete mixing without verification by constant sample testing. Figure 14.2 shows a process schematic for a typical dry-lime stabilization system with pneumatic lime conveyance.

Either quicklime, hydrated lime, or other dry alkaline materials can be used for dry-lime stabilization, although the use of hydrated lime typically is limited to smaller installations.

Quicklime is less expensive and easier to handle than hydrated lime. Additionally, the heat of hydrolysis released in slaking the quicklime when it is added to dewatered cake can enhance pathogen destruction. The addition of sufficient dry alkaline material can achieve either Class B or Class A requirements.

The proper design of lime-storage and transfer facilities is critical. Clogging of lime-transfer facilities maybe occur in areas with humid weather. Fine lime particles can generate significant operational and maintenance problems with dust collection and filter systems. The combination of humidity/moisture and fine lime particles may cause problems with dry (post-lime) alkaline stabilization that use pneumatic

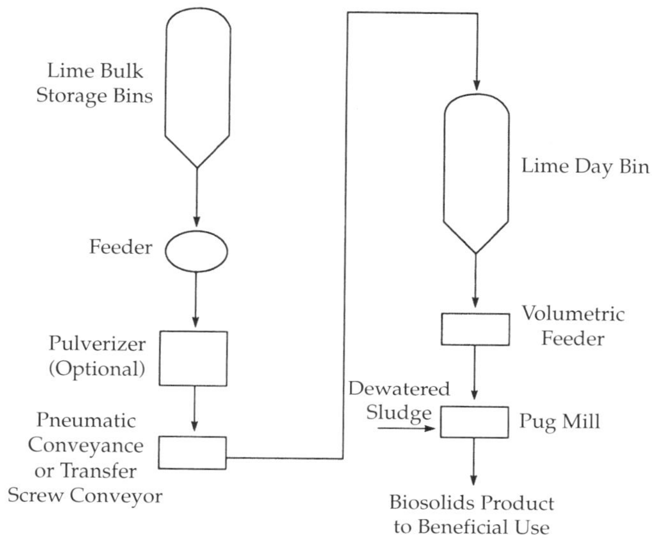

FIGURE 14.2 Typical dry-lime stabilization system process schematic (Oerke and Rogowski, 1990).

lime transfer systems unless the moisture is removed from the transfer air. Lime and sludge/lime transfer and conveyance facilities should be designed under negative pressure to minimize dust issues and other problems. For more detailed information, see design considerations and criteria sections to follow.

3.1.2 *Advanced Stabilization Technologies*

Advanced alkaline stabilization technologies typically involve a high alkaline chemical dose and significant retention time to achieve Class A time-and-temperature stabilization requirements. Typical advantages and disadvantages of advanced alkaline stabilization are shown in Table 14.2 (Oerke, 1996). Several municipalities are using materials other than lime in alkaline stabilization technologies that have been developed in the last 30 years. Most of the technologies that rely on additives, such CKD, LKD, Portland cement, or fly ash, are modifications of traditional dry-lime stabilization. The most common modifications include addition of other chemicals, a higher chemical dose (depending on the chemical), and supplemental drying. These processes alter the characteristics of the feed material and, depending on the process, increase stability, decrease odor, reduce pathogens, and provide product enhancement.

TABLE 14.2 Typical advantages and disadvantages of generic advanced alkaline stabilization processes.

Advantages	Disadvantages
Meets Class A stabilization requirements	High annual cost
	High chemical use
Multiple product markets	Extensive odor-control systems required to treat ammonia and other off gases
Typically lower capital cost when compared with other Class A stabilization processes	Dewatering facilities required
Proven with more than 200 installations in the U.S.	Some proprietary processes; annual patent fee could be required
Easy to operate, start up, and shut down	Worker safety concerns with dust from alkaline chemical and ammonia off gas
Metals concentrations in biosolids are diluted	Increase on total solids/chemical mass to transport
Product has value as a *** agent	Product not appropriate for alkaline soils
Property stabilized product is easy to handle and can be stored in smaller storage facilities	

Many of the processes are proprietary and are available through private firms. The following descriptions are not intended to cover all available technologies or to rank one process as superior to another. Rather, they are intended to present the scope of some processes available to municipalities. A summary of more detailed case study planning, design, and operational considerations on advanced alkaline stabilization processes is available elsewhere (Engineering-Science, Inc., and Black and Veatch, 1991; Stone, et. al, 1991).

Pasteurization processes use the exothermic reaction of quicklime with water to achieve process temperatures of more than 70 °C. This temperature is maintained for more than 30 minutes as required by federal regulations for add-on pasteurization to meet Class A criteria. This pasteurization reaction must be carried out under carefully controlled and monitored mixing-and-temperature conditions to ensure uniform treatment and inactivation of pathogens by heat generated during the reaction.

The process can produce a soil-like material that is nonviscous and, therefore, not subject to liquefaction under mechanical stress. Varying the process additives and mix ratios produces a range of biosolids-derived materials suitable for use as daily, intermediate, and final landfill cover or land reclamation material (Sloan, 1992). A process schematic for a typical pasteurization process is shown in Figure 14.3. In one variation of this process, pasteurization is carried out in a heated and insulated vessel reactor where temperatures are maintained at 70 °C or higher for a minimum of 30 minutes.

A chemical stabilization/fixation technology process typically involves the addition of silicate materials to dewatered cake. The addition of these materials causes cementitious reactions and produces, after drying, a soil-like material of approximately 35 to 50% solids concentration. The only use of this soil-like product to date has been as landfill cover material. In many cases, the treated material is further dried at the landfill for two to three days in small windrows. Class A, or the process to further reduce pathogens (PFRP), equivalency has not yet been proven for this process (Oerke and Rogowski, 1990; Reimers et al., 1981). Figure 14.4 shows a typical chemical stabilization process. Another process combines advanced alkaline stabilization with accelerated drying. The U.S. EPA has approved two alternative versions of this technology to meet Class A requirements. These alternatives involve the addition

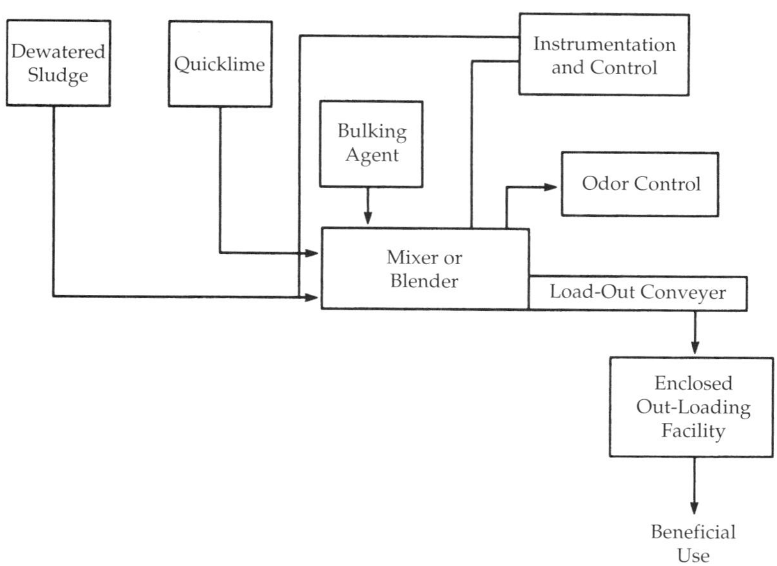

FIGURE 14.3 Typical pasteurization-like process schematic (WEF et al., 2009).

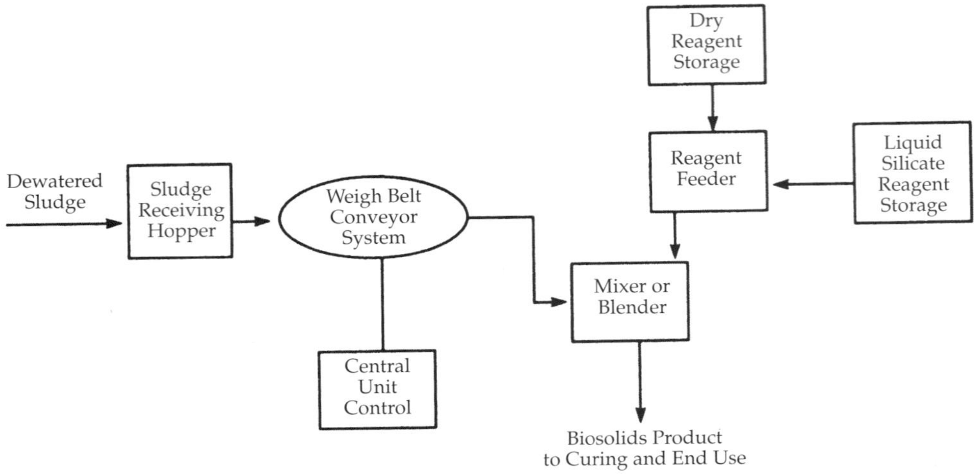

FIGURE 14.4 Typical chemical stabilization process (WEF et al., 2009).

of quicklime, CKD, or other alkaline admixtures and other subsequent processes to create stresses on the pathogens with pH, temperature, ammonia, salts, and dryness (Burnham et al., 1992). One of the alternatives requires an elevated temperature in which the material is contained for a minimum of 12 hours at 52 to 62 °C to allow the heat generated by the chemical reaction to further reduce pathogens. The second alternative involves alkaline addition followed by mechanical drying in windrows to achieve a 50 to 60% solids product. A process schematic of a typical system is shown in Figure 14.5. The Class A product is used predominantly as an agricultural liming agent, soil conditioner, or landfill cover or as a component of blended topsoil.

3.1.3 Process Variations

Several alternative approaches or modifications to the basic alkaline stabilization process have been developed. Some of these alternatives have evolved from other processes used in a WWTP; others are associated only with alkaline stabilization.

An example of using other processes to accomplish alkaline stabilization is combining lime-treated (lime used for phosphorus removal) primary solids with raw secondary solids (Paulsrud and Eikum, 1975). Another example is using existing digesters or other available plant tankage to thicken alkaline-stabilized wastewater solids before dewatering and disposal (Farrell et al., 1974).

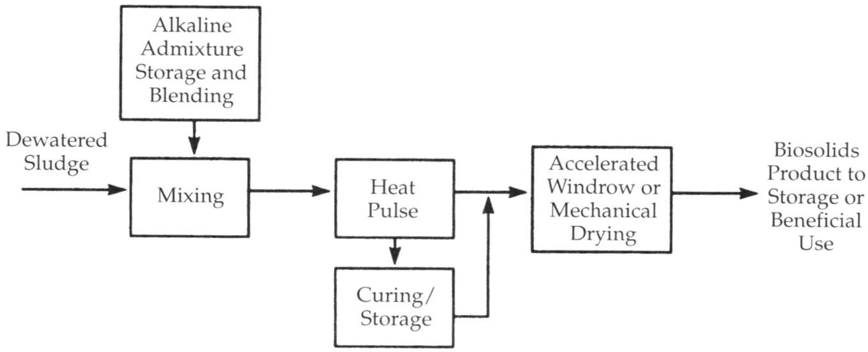

FIGURE 14.5 Typical alkaline stabilization with subsequent drying process schematic (WEF et al., 2009).

Another alternative strictly associated with the alkaline stabilization process is use of two mixing vessels, one for initially raising the pH to more than 12 and the other for providing adequate contact time combined with excess lime addition to maintain the pH within the desired range (Counts and Shuckrow, 1975).

Other, nonproprietary processes can also be used to produce Class B alkaline-stabilized sludge. For instance, Waukegan, Illinois, mixes fly ash and dewatered cake at ratios of 2.0:1 to 2.5:1. The resultant material is structurally stable and is used to "build" a biosolids-only monofill, as opposed to buying and importing fill material (Byers and Jensen, 1990).

3.2 Design Considerations and Criteria

This section summarizes design considerations for alkaline stabilization processes based on research and the experiences of pilot- and full-scale facilities using the process. Because of the interdependency between product quality and process design, the importance of defining process and product goals cannot be overemphasized. Several design criteria should be evaluated before implementing an alkaline stabilization process (Oerke, 1999).

Typical advanced alkaline stabilization design criteria are shown in Table 14.3, which vary from site to site (Fergen, 1991). Typical system design criteria include

- Feed sources and characteristics (quantity, type, quality, and solids concentration);
- Contact time, pH, and temperature;

TABLE 14.3 Design criteria depend on the technology used and the desired product.

Item	Descriptions or equipment	Parameter	Units	Range of values: Minimum	Range of values: Maximum	Selected design value
Materials	Sludge	Solids	percent	20	30	25
		Density	lb/cu ft[a]	45	55	50
	Alkaline bulking chemical	Solids	percent	90	98	95
		Density	lb/cu ft[a]	50	65	65
	Lime	Solids	percent	90	96	95
		Density	lb/cu ft[a]	55	60	60
	Stabilized products	Solids	percent	55	65	60
		Density	lb/cu ft[a]	65	75	75
Curing	Technology	Windrow				
	Detention time	Average	Days	3	7	5
		Peak	Days	3	7	4
	Temperatures		°C	—	—	52 for 12 hr
	Pile dimensions	Bottom width	ft[b]	6	14	10
		Mix height	ft[b]	2	3	3
		Top width	ft[b]	4	8	6
		Area/unit length	Sq ft/ft[c]	10	33	24
		Pile spacing	ft[b]	—	—	5
	File turning		Lb/d[d]	—	—	1 (typical)
Odor control	Building air	Number of stages	Number	1	3	1
		Air changes	Number/hr	6	15	12

(continued)

Table 14.3 Design criteria depend on the technology used and the desired product (*Continued*).

Item	Descriptions or equipment	Parameter	Units	Range of values Minimum	Maximum	Selected design value
	Product storage	Number of stages	Number	1	3	1
Storages	Sludge	Days of storage	Days	0	1	1
	Chemicals	Days of storage	Days	5	30	5
	Product	Days of storage	Days	80	180	60

[a] Item R × 1602 – kg/m^3
[b] ft × D.304 8 – m
[c] sq ft/ft × D.304 8 – m^2/en
[d] lb/d × D453.6 = kg/d.

- Alkaline chemical types and dosage rate;
- Solids concentration of the alkaline stabilized product;
- Energy requirements;
- Process optimization in alkaline treatment;
- Process monitoring and control;
- Mixing requirements;
- Economic considerations; and
- Solids cake and dry lime product handling.

3.2.1 *Feed Sources and Characteristics*

The amount, sources, and composition of the feed to be treated determine sizing of the alkaline stabilization facility. Variable performance of thickening or dewatering equipment is an important consideration when sizing equipment because poor dewatering performance significantly increases facility sizing. Solids concentration of the dewatered cake to be stabilized by the alkaline process affects the chemical dosage and facility sizing. Equipment capacities must be sized to accommodate the volume of feed to be processed. For example, processing a "wet" cake (10 to 15%

solids) requires larger equipment and more alkaline chemical processing than a drier cake (20 to 25% solids).

The nutrient content of the feed affects the characteristics of the product. The agronomic benefit of an alkaline-stabilized residuals product depends on the amount of plant nutrients contained in the biosolids and the need for a liming agent at the application site. Alkaline stabilization may be advantageous for untreated solids with relatively high metal concentrations because of dilution (on a dry-weight basis) of metals by the alkaline additives and immobilization of some trace metals.

The type of solids should also be considered when evaluating process design. After being anaerobically digested, biosolids contain 5 to 8 times the concentration of ammonia-nitrogen, which is converted to a gaseous form and released at the elevated pH required for the alkaline stabilization process. Consequently, the potential for odors from ammonia and other nitrogen compounds, such as amines, is increased when anaerobically digested sludge (biosolids) are treated by alkaline stabilization. The use of alkaline-unstable polymers can also contribute to the formation of odorous methyl amines. As with all solids-processing alternatives, odor control facilities typically are required at alkaline stabilization facilities located near residences or sensitive commercial areas.

3.2.2 *Contact Time, pH, and Temperature*

Contact time and pH are related because the pH must be maintained at the required level for an adequate time to destroy pathogens. The chemical added must provide enough residual alkalinity to maintain a high pH until used or discarded. High pH prevents growth or reactivation of odor-producing and pathogenic organisms.

The drop in pH (referred to as pH decay) occurs in a specific sequence. Atmospheric carbon dioxide or acid rain (which forms a weak acid when dissolved in water) is absorbed and gradually consumes the mixture's residual alkalinity, decreasing the pH. Eventually, the pH drops to less than 11.0, allowing bacterial growth to resume. Renewed production of organic acids and carbon dioxide then causes pH decay to continue (similar to the reactions within the anaerobic digestion process).

For the pre-lime stabilization process, if an inadequate lime dose is used and settled, thickened, partially stabilized solids are allowed to become septic in process settling basins/storage tanks, higher levels of reduced sulfides will form. These reduced sulfides will volatilize and produce odors. For the post-lime stabilization process, higher levels of reduced sulfides will also form in partially stabilized dewatered solids material that has an inadequate lime dose. These odors will be emitted later during transport, storage, and application of the partially stabilized solids material.

The pH typically drops during the stabilization process and, therefore, should be raised to more than pH 12 and maintained at that value. Biosolids do not have to be contained within a contact vessel as long as the pH can be monitored to ensure that it remains at the desired value for the desired period of time.

3.2.3 *Chemical Types and Dosage Rate*

The types and dosages of the alkaline chemicals to be added are critical to the success of the stabilization process. The quality of the chemicals (lime, CKD, Portland cement, and LKD) should be consistent. Different types or sources of additives result in different product texture and granularity. Lime is available from numerous sources, ranging from a high-calcium lime obtained from oyster or clam shells to a relatively low-calcium dolomitic lime. Considerations in selecting the type of chemical to be used include economics, availability, desired mixing, and desired product characteristics. Table 14.3 lists typical alkaline chemical characteristics.

Some alkaline reagents, such as CKD or LKD, are considered byproducts from other industries, and it is important to ensure that this material does not introduce contaminants or additional pollutants that can jeopardize the quality of the product. Cement kiln dust from hazardous waste kilns should be avoided. A quality assurance/quality control program with frequent sampling and analysis should be developed to ensure consistent quality of the biosolids product.

Because quality of the alkaline additives may affect product quality, adequate monitoring and proper management are important to ensure consistency. Because the characteristics of a byproduct, such as LKD, CKD, or fly ash, can vary from one location to the next, consistent vendor quality-control procedures are essential. The material from a single kiln or furnace will remain fairly consistent, provided that operating conditions and feedstock characteristics do not drastically change. More importantly, pilot- or bench-scale testing should be performed to determine how variations in the alkaline additives will affect product quality and how the process and chemical dosages should be adjusted to compensate for these variations.

There are two predominant types of lime available: CaO and $Ca(OH)_2$. Slaked lime in a liquid slurry (carbide lime) is also available. Carbide lime is a byproduct of the manufacture of welding-grade acetylene from calcium carbide. The application principles of this material are the same as those for $Ca(OH)_2$ in slurry form; therefore, carbide lime is not specifically discussed. The decision of which type of lime to use should be based on economics and materials-handling characteristics (such as alkaline material particle size). Production and transportation costs of $Ca(OH)_2$ are approximately 30% more than for CaO. For liquid operations, however, $Ca(OH)_2$ may require less equipment at the facility because it has already been hydrated (slaked). Typically, $Ca(OH)_2$ is economical for use at small facilities; however, when WWTP capacity exceeds $Ca(OH)_2$ demand of 2.5 to 3.5 Mg/d (3 to 4 ton/d), quicklime should be considered.

Aside from the cost, the primary advantage of $Ca(OH)_2$ compared with CaO is the elimination of slaking equipment at the facility. If CaO is added directly to dewatered cake (dry lime stabilization), then slaking is not necessary; however, the additional handling precautions that must be addressed because of the exothermic reaction of CaO and water should be evaluated. Another advantage of dry-lime stabilization is the elimination of lime sidestreams and the related abrasion and scaling of piping and mechanical equipment.

Several disadvantages of dry-lime use occur. When treating solids with dry lime, transferring lime, and maintaining equipment, lime dust could be a significant health hazard because of its effects on the eyes and respiratory systems. The dry-lime material should be kept sealed in conveyance systems under negative pressure. The conveyance system seal must be maintained. The characteristics of the dry-lime material will vary depending on the lime kiln mine location and actual lime-processing procedures (i.e., temperature, retention time, gradation before kiln). Lime gradation may also vary significantly from source to source, which may affect postlime transfer and dust-control equipment.

The required dosages of the specific chemicals will depend on the raw feed type (primary, waste activated sludge, trickling filter, or septage) and quality and chemical composition (including organic content) of the feed and liquid, the feed solids concentration, the desired product, and the type and quality of the alkaline material.

Lime type and grade can greatly affect mixing uniformity. The basic reactions between CaO and $Ca(OH)_2$ and sludge are hydration and hydrolysis, both of which require interaction with free water in the sludge. The CaO is hydrolyzed to $Ca(OH)_2$ by combining with water, and the $Ca(OH)_2$ dissolves in water to raise pH. Both CaO and $Ca(OH)_2$ are relatively insoluble in water and, because dewatered sludge does not contain much free water, intimate and immediate contact between the CaO or $Ca(OH)_2$ and sludge is critical. For Class A processes, which use CaO's reaction with water in sludge to generate heat, solubility of the formed $Ca\ (OH)_2$ decreases as temperature increases. For these reasons, the use of a CaO product with a smaller particle size is recommended to improve mixing efficiency. The use of an onsite lime grinder, however, may cause significant maintenance and dust issues.

Economic considerations will also guide these decisions. Finer-grade CaO is more expensive than pebble lime but less effective, particularly with high-solids sludge. Hydrated lime is more expensive than CaO. If byproducts are used instead of lime, then extra care must be taken to account for their variable CaO content to achieve the desired results.

Chemical composition of the feed also determines the required chemical dose. This composition depends on the type of solids and the process used (for example, chemical coagulation) for treatment.

The minimum dose rate is determined by process parameters. Typically, lower minimum doses of 15 to 30% on a dry-weight basis as calcium hydroxide are required for effective dry-lime versus liquid-lime stabilization. Figure 14.6 shows a range of adequate dry-lime Class B stabilization doses of 13 to 40% as calcium hydroxide on a dry-weight basis is required to achieve pH 12 for two hours and pH 11.5 for the next 22 hours. A 13% dose is the minimum CaO required, and does not guarantee that odors will be prevented. A higher dose of 20 to 30% (on a dry-weight basis) is

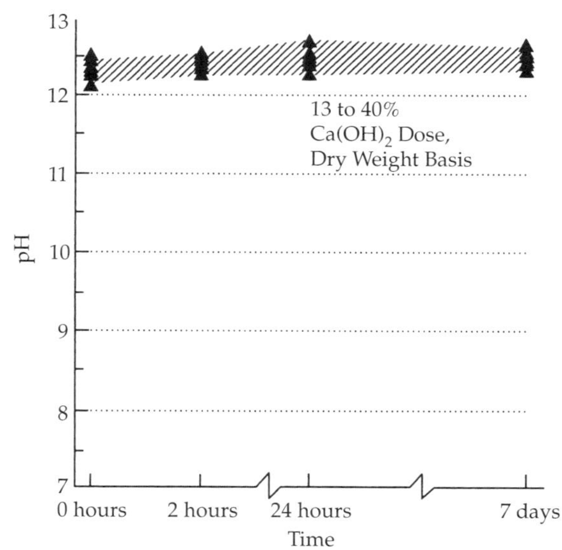

FIGURE 14.6 Example of pH decay following dry-lime stabilization of raw primary/waste activated dewatered cake (Westphal and Christensen, 1983).

recommended for adequate stabilization. This amount should be increased based on bench-scale testing, which is always necessary, if the sludge is particularly odorous (raw sludge, sludge with high concentrations of fats, oil, and grease) or the treated material will be stored for extensive periods (Christensen, 1987).

Class A processes that use the pasteurization criteria of 70 °C for 30 minutes require a range of CaO or its equivalent of 60 to 70% on a dry-weight basis (12 to 15% wet-weight basis). The EnVessel Pasteurization Process (EVP) process by RDP Technologies, Inc. (Norristown, Pennsylvania), uses 40%; the Schwing, Inc., (Somerset, Wisconsin) Bioset process uses approximately 75% CaO on a dry-weight basis and gains supplemental heat from electricity (RDP) and sulfamic acid (Bioset). Generic CaO addition (Biofix process by Bio Gro Systems, Inc., Annapolis, Maryland) requires approximately 100 to 150% lime dose on a dry weight basis to achieve the pasteurization criteria. The N-Viro Advanced Alkaline Stabilization with Subsequent Accelerated Drying (Toronto, Ontario, Canada) process uses approximately 60% CaO (or equivalent amount of CKD or LKD) on a dry-weight basis to achieve 52 °C for 12 hours. Additional byproduct alkaline materials (CKD or LKD or fly ash) are added to achieve a final solids concentration of 55 to 65%, the amount added depending on whether air drying or heated-air drying is used.

Figure 14.7 shows the theoretical dry lime stabilization dose for both Class B and Class A stabilization. The lower line shows maximum pH requirements, and the upper line shows Class A temperature requirements. The lower line of Figure 14.7 is based on a quicklime dose requirement of 25% on a dry-weight basis. Although the quicklime requirement for Class B stabilization theoretically increases with increased solids, the quicklime requirement for Class A stabilization decreases with increased solids. This is because of the predominance of the heating requirements of water for Class A versus the pH effects of Class B (Lue-Hing et al., 1992).

The following assumptions were used for the Class A temperature requirements in Figure 14.7:

- Feed input temperature equals 20 °C (68 °F);
- 100% of CaO reacts with water in the feed to produce heat [1140 kJ/kg (490 Btu/lb) of quicklime];
- Quicklime is 100% CaO (typically, this value is 90%);
- The specific heat of the feed is 0.58 kJ/kg (0.25 Btu/lb); and
- No heat loss from the feed to the air or the equipment occurs.

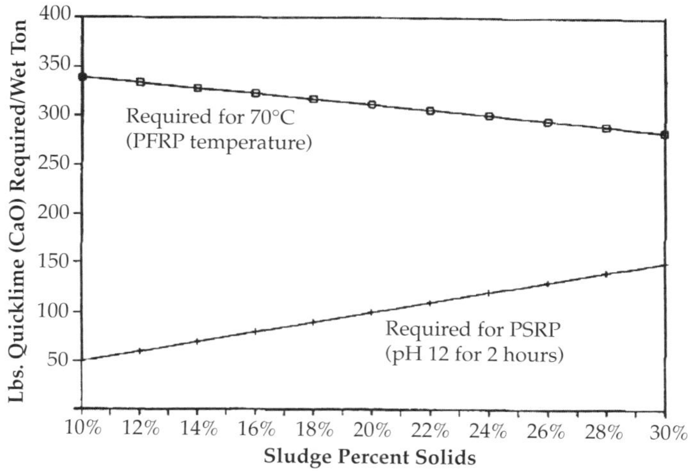

FIGURE 14.7 Theoretical lime requirements for process to significantly reduce pathogens (PSRP) and process to further reduce pathogens (PFRP) at varying sludge solids concentrations (Lue-Hing et al., Eds., 1992).

In practice, these conditions rarely exist; therefore, the amount of quicklime used to achieve Class A can be up to 50% greater than the Class A temperature line (top) in Figure 14.7. A drier, more easily crumbled product can be achieved by increasing the quicklime addition up to as much as twice the theoretical value shown.

Chemical dosages for advanced alkaline stabilization technologies vary by process, types of chemical, and product requirements. A typical material balance for an advanced alkaline stabilization facility assuming a 65% chemical dose on a wet-weight basis is shown in Table 14.4. Lime dose expressed on a wet-weight basis is 0.25 times greater than a dose expressed on a dry-weight basis for a dewatered cake material with a 25% solids concentration. For example, the chemical dose in Table 14.4 of 65% on a wet-weight basis is equal to approximately 245% on a dry-weight basis. Material balances should be used to size alkaline stabilization facilities and determine initial and product characteristics.

The data presented in Tables 14.3 and 14.4 can be used for preliminary design of dry-lime stabilization facilities; however, the required dosage should be determined on a case-by-case basis because of the many factors that affect this dosage (Farrell et al., 1974). To prevent pH decay and the associated regrowth of organisms, the lime dose may have to be increased beyond that necessary for stabilization (Ramirez and

TABLE 14.4 Typical materials balance for advanced alkaline stabilization (Fergen, 1991).

		Solids balance				
Process	**Items**	**Solids contents, %**	**Volume, ca ft[b]**	**Total weight, ton**	**Dry weight, ton**	**Bulk density, lb/cu ft[c]**
Mixing	Sludge ***	25.0	6.400	160.0	40.0	50.0
	Chemicals	95.0	3.186	103.5	98.4	65.0
	Initial mix	52.5	9.586	263.5	138.4	55.0
Windrow	Initial mix	52.5	9.586	263.5	138.4	55.0
	Evaporation loss	—	3.437	32.9	—	—
	Product	60.0	6.149	230.6	138.4	75.0

Chemical dose:

Wet weight basis 65%

Dry weight basis 245%

[a] For peak conflict ***

[b] cu ft × 28.32 – L.

[c] lb/cu ft × 16.02 – kg/m[2]

Malina, 1980). The exact dosage for any particular feed can be estimated through laboratory testing. Additional information may be found in Section 2.0.

3.2.4 *Solids Concentration of Stabilized Product*

The solids concentration of the alkaline-stabilized product is an important design consideration for materials-handling purposes. Minimum solids concentration of the alkaline stabilized cake (biosolids) material may be required to meet regulatory requirements such as those for landfilling or extended storage. Solids concentration (dryness) and granularity affect the materials-handling characteristics for trucking, application/disposal equipment, and marketing of the product.

The solids concentration of the initial alkaline-stabilized product also affects any supplemental drying step in advanced alkaline-stabilization processes. Through the addition of the alkaline additive, chemical reactions occur that increase the apparent solids concentration of the mix. This increase in solids is caused by the cumulative consequence of the addition of solids, chemical binding, and evaporation of the water in the feed. The alkaline material, particularly quicklime, produces a fast reaction resulting in a temperature increase within a matter of minutes. Thorough mixing of the feed and alkaline material is important to achieve the target solids concentration and pathogen destruction and to reduce residual odors (such as ammonia) in the product.

By using a high chemical dose, it is possible to achieve the desired solids concentration, thereby reducing or eliminating the need for supplemental drying. Excessive chemical costs may make this practice prohibitively expensive. It is possible to add other bulking materials to increase product dryness and improve handling characteristics, but not to increase the chemical dosage. Potential bulking materials include fly ash, wood ash, sawdust, sand, and soil. In addition, the quality of the product may be affected. Mechanical mixing in a windrow operation enhances drying, blends the material, and releases trapped ammonia and other volatile gases created during dewatering operations, resulting in a more homogeneous product. The final design should reflect the best balance between the chemical dose and the amount of subsequent drying required.

3.2.5 *Energy Requirements*

During the liquid-lime stabilization process, most energy is used for mixing the liquid sludge and lime slurry. For dry alkaline stabilization, mixing energy requirements are minimal (10 to 20 HP per mixer) and vary with throughput, chemical dose, and mixer type.

Air ventilation and scrubbing equipment for ammonia and odor control can also be substantial energy users.

3.2.6 *Pilot-Scale Testing*

Because quality and consistency of feed vary from site to site, qualitative and quantitative analyses must be performed to determine specific chemical dosages and process design parameters. Pilot-scale testing should be used to determine the optimum chemical dosages and mixer performance. Pilot-scale testing also allows a WWTP to evaluate different operating procedures and products.

Bench- and pilot-scale testing should be conducted before implementation of an alkaline-stabilization process. Four primary areas to be evaluated include:

1. Process requirements, including alkaline material types, sources, size gradation, and dosages;

2. Equipment, including energy requirements to ensure thorough blending of sludge and lime materials;
3. Product quality, including desired product solids concentration and granularity; and
4. Generation and control of odors.

Process concerns include chemical types and dosages, feed and product solids concentration, and other process steps, such as supplemental heating and drying, as required. The chemical dose must be determined to meet the pH, solids concentration, heat rise, and product requirements. The chemical dosages can be estimated in bench-scale tests using carefully measured volumes of feed mixed with varying dosages of chemicals. All pilot-scale testing should include generation of mass balance calculations to ensure consistency between chemical dosages and product solids concentration. Whenever possible, full-scale pilot equipment should be used to assess actual chemical dosage rates and mixing performance.

It is extremely important that testing conditions be controlled to simulate field conditions to the greatest extent possible. For example, during winter, a different chemical dose or a modification of the chemical formula may be required to achieve the desired product. In large-scale windrow drying operations, carbon dioxide mixed into the product can lower the pH. Therefore, the dosage of alkaline material may have to be increased to maintain the proper pH. Samples should be cured in the same type of closed or open containers that will be used in the full-scale system.

The initial solids concentration of the alkaline stabilized (biosolids) product should be tested for compatibility with the proposed drying technique. It may also be useful to investigate various chemical dosages in different drying/curing configurations. The chemical dose can have a significant effect on the rate of drying and the corresponding drying-area requirements. Pilot-scale testing of equipment can be considered to evaluate equipment requirements. The goal of the testing is to determine the equipment, energy, and chemical requirements for producing the desired product compatible with use or next processing step. It was found that use of an inappropriate paddle configuration or operating speed on a pug mill feed chemical mixer at a dry-lime stabilization facility resulted in an undesirable material. Proper mixing is necessary not only to achieve desired product characteristics, but to ensure thorough blending of the alkaline additive. Excessive energy in mixing can result in a nongranular mass that is difficult to handle.

Several other process parameters that should be considered during pilot- or bench-scale testing: emission of odors from the process and product; levels of plant nutrients, metals, and organic chemicals; and compatibility of the alkaline material dose with the dewatering polymer. Some polymers may deteriorate in high-alkaline conditions and exhibit strong trimethylamine "dead fish" odors (Jacobs and Silver, 1990). Various dosages of alkaline material should be tested with different polymers to determine the effects of odor release and physical characteristics of the products (such as compaction and granularity).

The final item that should be evaluated by pilot-scale testing is the alkaline-stabilized (biosolids) product. Pilot- and bench-scale testing provides excellent opportunities to investigate product quality and marketability before beginning full-scale production. It is helpful to invite prospective users to observe pilot-scale tests or small-scale demonstration programs to encourage interest in the product. Physical characteristics, such as solids concentration, pH decay, leachability, permeability, or unconfined compressive strength, should be evaluated if the product will be used for landfill or slope stabilization. The quality of the product should also be tested to provide the data and documentation required for regulatory approval (WEF, 1995).

3.2.7 *Process Optimization in Alkaline Treatment*

Facilities using lime for sludge stabilization can experience inconsistencies in odors and fecal destruction if not properly processed. The staff at Blue Plains AWTP has used lime stabilization to achieve Class B biosolids standards for many years. Although always achieving the Class B standard of < 2 mil. colony-forming units (CFU)/g for fecal coliforms, the results were inconsistent. One week, the results would be approximately 10 000 CFU/g; the next week 1 million; the next 100 000. In addition, empirical evidence gathered in the field revealed that the odors were also inconsistent. Field inspectors described odors during these incidents as rotten eggs or rotting cabbage (methyl mercaptan, dimethyl sulfide, dimethyl disulfide); rancid pungent meat (volatile fatty acids); and fecal matter (indole, skatole). All of these odors are products of anaerobic microbial activity, indicating less-than-optimum microbial inactivation. Staff suspected these inconsistencies in odor and fecal coliform destruction were related and that poor mixing might be responsible. Research described below was undertaken to determine if this was the case. Results indicated that poor mixing of sludge and lime materials and the use of coarse lime with a large particle size was the cause. Researchers found a means to measure mixing efficiency. Better mixing of the raw sludge and lime materials was implemented with the use of lime with a smaller particle size. DC Water biosolids is now consistently below 1000 CFU fecal coliforms with low odors.

3.2.8 *Process Monitoring and Control*

It is critical that feed and alkaline materials be monitored frequently so that process adjustments can be made to achieve adequate stabilization and a consistent product. The effects of incomplete stabilization are not readily apparent and may not be seen at a WWTP; therefore, proper process control is important. Absence of odors and acceptable dewatering characteristics, however, are not good indicators of adequate stabilization.

Monitored characteristics include total solids, pH, and temperature of both the feed and the product. For Class A (PFRP) products, fecal coliform indicator organisms must be monitored at the frequency specified in 40 CFR Part 503.16 (U.S. EPA, 1993). Metals must be monitored if the product is to be used for agricultural purposes. Toxic characteristics leaching procedure tests and a paint filter test may be necessary if the product is going to be landfilled. Quality-control data may be required for regulatory approval. The method and frequency depend on regulatory requirements. In some cases, odor characterization and emission monitoring may also be required.

Chemical dosage adjustments may be made in response to manual measurements of temperature and pH and visual inspection of the feed–chemical mixture. Some automatic process control can be incorporated if desired. Weighing and volumetric systems should be calibrated every month. For instance, thermocouples can be used to measure the heat pulse if an enclosed vessel is used. The chemical feed rate can be controlled by pacing it with the incoming-feed flowrate or dewatered cake using a weigh belt or similar means. The use of a rapid feed sludge microwave drying system before and after sludge/lime mixing has been successful in quickly analyzing the feed solids concentration and pacing and optimizing lime addition.

Programmable logic control of the chemical feed system aids in producing a consistent product. A typical system may include an electronic chemical metering system linked to feed-weight belt scales. Chemical feed is automatically controlled based on the weight of the feed solids. With this type of system, reagents are added in proportion to the amount of dewatered cake. Special care should be taken to maintain calibration and operation of the weigh scales to ensure that the appropriate dosage of alkaline materials is being added. An automated system also decreases the number of personnel needed to operate the process. Pacing the lime addition to the feed-sludge flow (4- to 20-mA signal from a magnetic meter) to the upstream dewatering equipment can also provide an accurate way to pace and optimize lime addition. The measurement of feed sludge-flow with a magnetic flow meter typically is more accurate and requires less maintenance than using a weigh-belt scale.

Sensors, particularly pH electrodes (both laboratory and automatic process control units), must be properly cleaned, calibrated, and maintained. Special pH electrodes are necessary for routine measurements greater than pH 10. The pH must be carefully monitored to ensure it is maintained high enough for a sufficient period. Portable pH pen probes are acceptable for process monitoring. Microbiological examination for indicator organisms (fecal coliforms and fecal streptococci) should be regularly performed by a qualified laboratory.

Measurement of pH has been discussed above. Calibration standard solution should be selected for the alkaline range, and pH electrodes should be suited to an alkaline environment. Results should be recorded in bench log and spreadsheet form for regulatory compliance.

Temperature can be measured manually by a variety of analog or digital probes. In-vessel Class A systems have built-in temperature sensors. Temperature logs should be kept for regulatory compliance.

Solids can be determined by oven or microwave drying. Temperatures should be kept below 105 °C to avoid loss of volatile solids.

3.2.9 *Mixing Requirements*

This section includes a discussion of mixing requirements and issues for dry-lime solids stabilization. Mixing performance, as already noted in Section 3.2.7, can significantly affect product quality. Inadequate mixing or low dose rate can result in loss of pH control in storage and formation of putrescent odors. Over-mixing can destroy the granular structure of the product. Nonuniform mixing will produce variable calcium carbonate equivalence (CCE) of the product and impair its acceptance for use as a lime-treated product for land application.

3.2.10 *Dewatered Cake/Chemical Mixing for Dry-Lime Stabilization*

Mixing or blending dewatered cake and the alkaline material is the most critical component of dry-alkaline stabilization. The goal is to provide intimate contact between the cake and chemical, resulting in a complete pH adjustment throughout the cake-chemical mixture. The lack of adequate cake-chemical mixing has led to incomplete stabilization, odors, and dust problems at several facilities in the United States (Oerke and Stone, 1991; WEF, 1995).

The mixing step typically is accomplished using a mechanical mixer, such as a pug mill or plow blender. Typical cake–lime pug mill and plow blender mixers are shown in Figures 14.8 and 14.9, respectively. Dewatered cake and chemicals are added together at the "head" of the mixer. Both batch and continuous mixers are available. The proportioning of the alkaline material dosage rate to the rate at which cake is introduced is important.

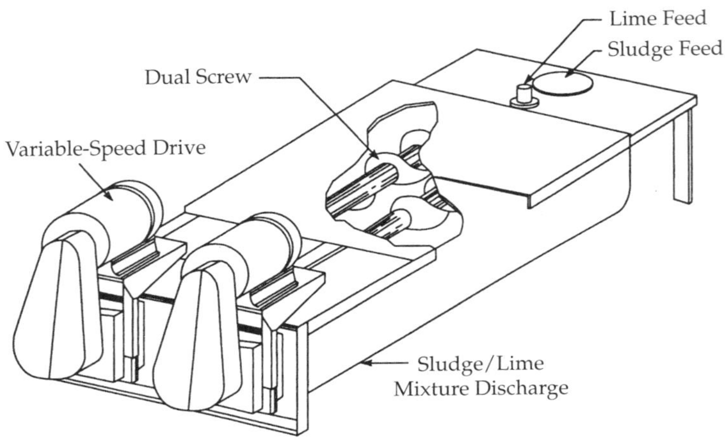

FIGURE 14.8 Typical dual-screw pug mill mixer (WEF et al., 2009).

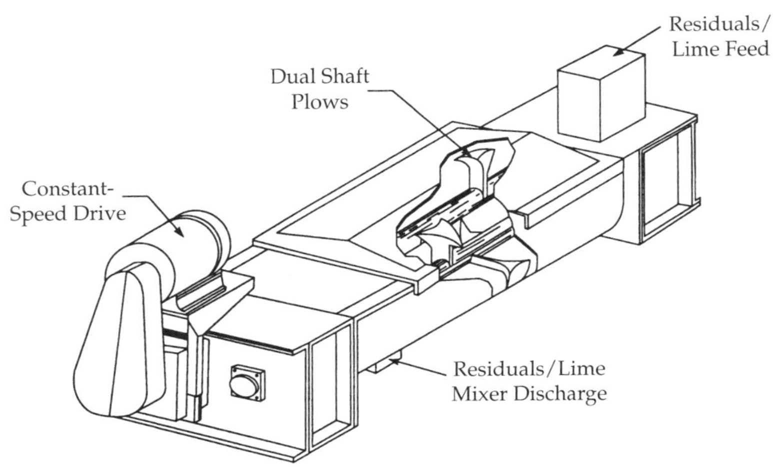

FIGURE 14.9 Typical plow blender mixer (WEF et al., 2009).

Mixing dewatered solids with dry lime is an inexact science if mixing equipment is not properly evaluated in a pilot-scale system. Techniques have been developed that can compare mixers and evaluate proper mixing settings. Because pH is a regulatory permit parameter, most facilities use pH as the only evaluation criteria. This is a typical mistake that has led to poor mixing at many facilities. Lime distribution in the final sludge/lime product must be evaluated. This distribution pattern is critical to ensure a proper mix. At least ten 0.5 g samples should be taken from the final biosolids/lime mixed product and evaluated for calcium levels. The variance can be used to evaluate if proper lime distribution/mixing and optimization of the equipment has occurred.

Several variables affect the mixing process and, consequently, the characteristics of the biosolids product. Mixer selection typically is based on past experience and trial-and-error testing. Many mixer manufacturers have mobile pilot full-scale testing equipment available. Whenever possible, this equipment should be used in the evaluation and selection of the proper and most effective mixing equipment.

Mixing characteristics of a dewatered cake vary with solids concentration, polymer used for conditioning before dewatering, chemical type and dosage, temperature, mixing intensity, available mixer equipment surface area per volume exposed to the cake, and mixing retention time. When selecting a mixer, consideration should also be given to minimum and maximum cake production, hours of operation, and other operating conditions. To account for these variable mixing characteristics and conditions, mixers can be equipped with variable-speed drives, adjustable mixer paddle configurations, weir plates, and other options to adjust mixing intensity and retention time (Figure 14.10) (Christy, 1992).

Physical characteristics of the alkaline product vary according to the mixing parameters. The physical consistency of a biosolids product can range from sticky and plastic to granular and dusty. A granular product is most desirable for farmers and other users because it provides for more uniform spreading at agricultural land application and other sites. The goal of the mixing step is to produce a product compatible with the use or next processing step. Characteristics of the alkaline biosolids product may continue to change for up to several days after mixing because of ongoing chemical reactions, temperature, and other parameters (Oerke, 1989).

3.2.11 Economic Considerations

Economics is another important factor when selecting a solids management option. Evaluating the costs of an alkaline stabilization process should include total cost

FIGURE 14.10 Photo of pug mill mixer with adjustable mixer paddle configuration and variable-speed drivers to allow adjustment of mixing intensity and retention time.

over its useful life, using present worth, equivalent annual cost, or similar measures. Capital cost for alkaline stabilization typically is less than for other sludge stabilization processes that can produce a Class A product. The cost of disposal or beneficial use of the product can be significant for hauling and land application and must be included in the cost analysis (Jacobs et al., 1992). In addition, the costs of a privatized option, if the preferred procurement method, should be compared to alternative solids management options that are to be owned and operated by the municipality. A capital, annual, and present worth cost comparison of Class A alkaline stabilization versus Class A composting and Class A thermal drying is included in an analysis by the National Lime Association (1996).

Annual operation and maintenance costs include operating labor, chemical costs, fuel, utilities, maintenance, and transportation of chemicals to the facility and the product to sites. Other annual costs may include public education and relations; marketing of biosolids, and soil; agronomic testing; and analyses. An owner should exercise caution in examining the annual costs at other facilities. The cost often includes several site-specific factors such as the local labor cost, distance of the chemical supply from the facility, and power. Moreover, minimum biosolids production amounts specified in the contract also affects total annual costs. When alkaline-stabilization technologies are operated under private contracts, it is important that the contract accurately reflects actual biosolids production.

Many site-specific factors influence costs and make economic evaluations and comparisons difficult. These include physical layout, feed type and characteristics, product use, local regulations, local product market availability, and seasonal effects of local climate. Additionally, at some alkaline-stabilization facilities, existing equipment has been retrofitted and used in the process.

Flexibility (adaptability) of the alkaline-stabilization process and the use of existing facilities should be considered when evaluating solids management options. Although it may not always be possible to use or retrofit existing equipment, cost savings can be achieved if existing equipment is used as part of the process train.

3.3 Solids Cake and Dry-Lime Stabilized Product (Biosolids) Handling

Dewatered cake and dry-lime stabilized wastewater solids product handling equipment consists of belt-and-screw conveyors and pumps. Belt conveyors typically are used on horizontal or gently sloping planes. Belt conveyor problems, such as minor spillage, slippage, and frequent bearing maintenance, have occurred at many facilities. Screw conveyors are also used to transfer dewatered cake to the alkaline stabilization process mixer or storage hopper. Vertical and inclined screw conveyors that are operated at a relatively high speed and high-pressure dewatered cake pumps can physically "condition" the cake. This can make it difficult and sometimes impossible to homogeneously mix the cake with a dry alkaline chemical. Some screw conveyors have a tendency to roll the cake–chemical mixture into "balls." Pumps can compact the dewatered cake into a long tube that requires breaking at the mixing step. The rolled balls and compacted cake, which are either desirable or undesirable depending on the final objective, can be especially critical when dealing with the product (Oerke and Stone, 1991).

Although alkaline stabilization processes are relatively simple, a regular inspection and maintenance program is essential. The conveying system and other moving parts must be closely monitored for wear. If only one conveyor is used to feed the alkaline stabilization process, then it is important that the equipment be routinely inspected, maintained, and calibrated. Conveying-system downtime can delay or halt the stabilization process. If multiple process trains are used, then bypasses and crossovers should be provided to avoid excessive downtime. Design of the alkaline-stabilization system should minimize length of conveyance from the dewatering equipment to the mixer. Likewise, the design for storage should also minimize conveyance length.

Redundant storage and processing equipment trains should be seriously considered to allow routine maintenance and calibration and operational flexibility without downtime. Although they are not as desirable as dual operating systems, temporary portable units can be placed in operation in a matter of hours or days, if necessary. Storage hoppers/bunkers between the dewatering system and the alkaline stabilization system may be used to dampen variations in dewatering system output and to allow each process to independently operate.

4.0 DESIGN EXAMPLE

This design example is for production of Class A biosolids by alkaline treatment of wastewater residuals. The treatment system is to handle solids produced from a generic 38 ML/d (10 mgd) secondary activated sludge WWTP.

4.1 Design Example Process Assumptions

The following process assumptions were made for the generic 38-ML/d (10-mgd) WWTP:

1. Primary and waste activated sludge (WAS) liquid stream treatment;
2. Primary-to-WAS ratio of 60:40;
3. Total combined solids production of 240 g/m^3 (2000 lb/mil. gal) of WWTP capacity;
4. Influent biochemical oxygen demand (BOD) of 280 mg/L;
5. 0.85 kg solids/kg (0.85 lb solids/lb) of BOD removed;
6. Primary solids thickening in primary clarifiers to 5% solids;
7. WAS thickening in secondary clarifiers to 1% solids;
8. WAS storage/blending facilities;
9. WAS thickening in gravity belt thickeners to 5% solids;
10. Storage/blending facilities for combined thickened primary and WAS;
11. Belt filter press dewatering of raw solids to 20% total solids;
12. Class A lime stabilization;
13. Odor control facilities; and
14. 90-day covered onsite product storage.

4.2 Product Considerations

It is important to "start with the end in mind" when designing alkaline-stabilization facilities and deciding what type of final material will be produced. The designer should ask the following questions:

- What is the proposed product use?
- Does the application site for the alkaline stabilized residuals have public access?
- What is the desired class of biosolids product?
- What type of alkaline material is to be used?
- Is a relatively high dose of alkaline material needed?
- Is the use of additives that reduce the product pH and alkaline properties needed?

The answers to these questions are project and site specific and will have a direct effect on the design of the alkaline-stabilized residuals treatment system. Typical alkaline-stabilized product considerations are included in Table 14.5.

4.3 Process and Equipment Descriptions

The EVP and Biofix (lime only) processes were chosen to be used as design examples for a Class A lime stabilization process. Use of these two processes as design examples in this chapter is not intended to serve as an endorsement or to rank them as superior to other processes.

Facilities associated with the EVP process include dry chemical (quicklime) storage silos, pneumatic chemical transfer systems, day bins, chemical feeders, screw and belt sludge transfer conveyors, pug mill mixers, and a pasteurization vessel. In the EVP process, dewatered solids are preheated in an insulated, electrically heated screw conveyor called a Thermofeeder. Before being mixed with dry quicklime and transferred to a heated and insulated pug mill mixer, called a Thermoblender, the solids/lime mixture is conveyed to a heated and insulated pasteurization vessel reactor where it is retained for a minimum of 30 minutes at a temperature of at least 70 °C to meet the Class A requirements. The resultant solids concentration of the lime-stabilized material for the RDP EVP process should be approximately 30% solids if the dewatered cake material is approximately 20 to 22% solids. The quicklime dose for the RDP EVP process (30 to 40% dry weight) is significantly lower than the Biofix

TABLE 14.5 Typical alkaline stabilized biosolids-product considerations (LKD = lime-kiln dust; CKD = cement-kiln dust).

Proposed product use	Does application site have public access?	Desired class of biosolids product	What type of alkaline material to be used*	Is a relatively high dose of alkaline material needed?	Use additives that reduce the product pH/alkaline properties?
Solids conditioning for dewatering	NA	NA	Hydrated lime	No	NA
Land application of liquid biosolids	No	B	Hydrated lime	No	No
Land application of liquid biosolids	Yes, (i.e. turf farm)	A	Hydrated lime	Yes	Yes, to provide a more neutral pH product
Land application of dewatered cake	No	B	Quicklime, LKD, CKD	No	Maybe, if alkaline soils present
Land application of dewatered cake	Yes	A	Quicklime, LKD, CKD	Yes	Maybe, if alkaline soils present
Land reclamation	No	B	Quicklime, LKD, CKD	No	No
Land reclamation	Yes	A	Quicklime, LKD, CKD	Yes	Yes
Landfill disposal	No	NA	Quicklime, LKD, CKD	No	No
Landfill daily cover	No	B	Quicklime, LKD, CKD	Yes	Yes, structural properties
Landfill final cover	Yes	A	Quicklime, LKD, CKD	Yes	Yes
Manufactured organic topsoil	Yes	A	Quicklime, LKD, CKD	Yes	Yes
Bulk fill application (slope stabilization)	Yes	A	Quicklime, LKD, CKD	Yes	Yes, structural properties
Horticulture (nurseries and sod farms)	Yes	A	Quicklime, LKD, CKD	No	Yes, to provide a more neutral pH product

*Use of LKD or CKD is acceptable if metal concentrations are low and below U.S. EPA 503 Table 3 limits (1993).

(lime only) process because it uses electrical or other means of supplemental heat to raise the solids/lime mixture temperature.

The Biofix (lime only) process uses similar equipment to mix the solids with dry quicklime as the EVP process but does not use heat tracing to transfer electrical or other heat to the solids/lime mixture in the feed screw conveyor, pug mill mixers or pasteurization vessel. Instead, the Biofix process uses only the heat generated from the exothermic reaction of quicklime with the water in the solids (1141 kJ/kg, or 491 Btu/lb of quicklime) to meet Class A temperature requirements. A significant amount of quicklime (100 to 150% dry weight) is required for the Biofix (lime only) versus EVP to meet Class A stabilization criteria. The resultant solids concentration of a lime-stabilized material for the Biofix (lime only) process should be approximately 40 to 50% assuming a 110 to 120% lime dose on a dry-weight basis and a dewatered solids concentration of approximately 20 to 22%.

For either process, dry quicklime would be delivered in pneumatic trucks that would transfer the chemical into storage silos with a truck-mounted blower. Traditionally, lime chemical storage should include provisions to meet a minimum of seven days of demand, with a two- to three-week supply being optimal. Railcars can also be used for transport of quicklime. In this design example, two weeks of quicklime storage was assumed to be provided by silos with dust collection and vibrating activators to minimize clogging. The absolute minimum lime storage capacity recommended is 250% of the bulk chemical shipment volume, depending on the distance between the chemical supplier and the user. Therefore, a minimum 54.4-Mg (60-ton) lime silo is typically recommended equal to approximately 2.5 times the size of 18.1- to 22.7-Mg (20- to 25-ton) lime delivery trucks. A typical arrangement with two silos is shown in Figure 14.11.

Dry alkaline materials can be conveyed mechanically with a screw conveyor if the distance from the bulk storage silo to the chemical addition point is short. Dry alkaline materials can also be conveyed pneumatically, either under pressure or under a vacuum. Each type of pneumatic conveyance has its benefits. Dust problems are reduced under a vacuum because any leaks are into the system. The pressure systems can move larger volumes of material. Pneumatic conveying air should be predried to reduce hydration and other moisture-related problems; however, pneumatic transfer systems may have problems maintaining homogeneous chemical bulk densities if a variety of alkaline materials are used. A typical pneumatic lime conveyance system is shown in Figure 14.12. The system was designed to allow for the

FIGURE 14.11 A typical arrangement with two silos with truck unloading and dust collection facilities and a pneumatic conveyance system.

FIGURE 14.12 A typical arrangement of a lime pneumatic conveyance system to transfer lime from two silos to two day bins.

top of day bins to penetrate the roof to minimize potential dust issues inside the lime stabilization building.

If a positive pneumatic system is used to transfer dry quicklime from a silo to a day bin, then rotary airlock feeders should be used. Day bins could be located adjacent to pug mill mixers to ease quicklime chemical transfer. Day bins should be sized to hold more than one day of chemical usage and should be filled approximately once per operating day. The pneumatic conveying system should have a cross-pipe connection so that either lime silo can transfer material to either one of two day bins. Volumetric feeders should meter the chemical flow from the day bins to the pug mills. A day bin and volumetric screw feeder is shown in Figure 14.13. Flexible piping with tight seals is typically provided to connect all lime-stabilization equipment to minimize dust (National Lime Association, 1996; Oerke, 1996).

Mechanical equipment should be provided to withdraw foul air and to ventilate the storage/blending tanks, gravity belt thickeners, belt filter presses, pug mills, conveyors, and pasteurization vessels to odor control facilities.

FIGURE 14.13 Typical arrangement of a day bin and volumetric screw feeder to transfer lime from the day bin to the pug mill mixer.

A typical EVP Class A alkaline stabilization process schematic that includes solids storage, thickening, and dewatering unit processes is shown in Figure 14.14.

Design criteria for the alkaline stabilization facilities used in these design examples are shown in Table 14.6. (See National Lime Association, 1996, and Oerke, 1992, 1996, and 1999, for more detailed typical design criteria and a description of characteristics that affect lime dose.) Note that 160 Mg (three 55-Mg silos) of lime storage (two weeks) is required for the Biofix process compared with 55 Mg for the EVP process because the lime dose is approximately 3 times higher.

5.0 DC WATER CASE STUDY

The Blue Plains AWTP operated by DC Water is one of the largest WWTPs in the United States using postlime stabilization (North, 2003). Blue Plains treats the wastewater of more than 70% of the Washington D.C. area and areas in Virginia and Maryland. The Blue Plains WWTP has a capacity of 1400 ML/d (370 mgd). The region that it serves is mainly residential because there are few industries in the Washington

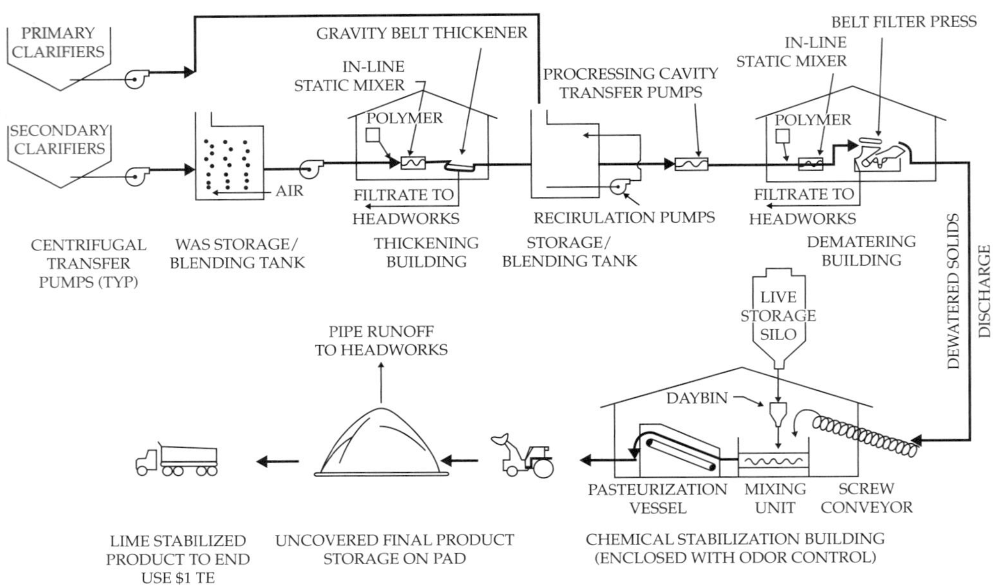

FIGURE 14.14 Lime stabilization process schematic (National Lime Association, 1996).

TABLE 14.6 Class A alkaline stabilization design criteria for generic 38 ML/d (10 mgd) wastewater treatment plant design example (DS = dry solids; NA = not applicable).

Item	EnVessel Pasteurization Process (EVP) design criteria	Biofix (lime only) design criteria
PROCESS		
Operation		
Days per week	5	5
Hours per day	7	7
Wastewater feed solids (raw dewatered cake)		
Dry solids	20% DS (average)	20% DS (average)
Density	1120 kg/m^3 (70 lb/cu ft)	1120 kg/m^3 (70 lb/cu ft)
Capacity	9072 kg/d (20 000 lb DS/d)	9072 kg/d (20 000 lb DS/d)
Alkaline material		
Type	Quicklime	Quicklime
Dose as CaOH2	0.4 [kg lime ($CaOH_2$)/kg DS or lb lime ($CaOH_2$)/lb DS on dry-weight basis]	1.1 [kg lime ($CaOH_2$)/kg DS or lb lime ($CaOH_2$)/lb DS on dry-weight basis]
Dose as CaO	0.36 [kg lime (CaO)/kg DS or lb lime (CaO)/lb DS on dry-weight basis]	1.0 [kg lime (CaO)/kg DS or lb lime (CaO)/lb DS on dry-weight basis]
Density	880 kg/ m^3 (55 lb/cu ft)	880 kg/m^3 (55 lb/cu ft)
Quicklime purity	90% CaO	90% CaO
Wastewater solids/lime mixture		
Dry solids	30% DS (average)	40% DS (average)
Density	928 kg/m^3 (58 lb/cu ft)	880 kg/m^3 (55 lb/cu ft)
Volume	53.8 m^3/d (1900 cu ft/d)	62.3 m^3/d (2200 cu ft/d)
pH	12.5 units (initial)	12.5 units (initial)
Wastewater solids/lime reaction		
Detention time	2 hours	2 hours
Temperature	70 °C for 30 minutes	70 °C for 30 minutes
pH	More than 12.0 for 2 hours	More than 12.0 for 2 hours
	More than 11.5 for 24 hours	More than 11.5 for 24 hours

(*continued*)

TABLE 14.6 Class A alkaline stabilization design criteria for generic 38 ML/d (10 mgd) wastewater treatment plant design example (DS = dry solids; NA = not applicable) (*continued*).

Item	EnVessel Pasteurization Process (EVP) design criteria	Biofix (lime only) design criteria
EQUIPMENT		
Solids/lime Mixers		
Type	Pug mill	Plow blender
Quantity	2	2
Capacity	10.5 m^3/h (370 cu ft/h)	12.5 m^3/h (440 cu ft/h)
Detention time (minimum)	30 seconds (variable)	30 seconds (variable)
Drive	Variable speed	Variable speed
Motor horsepower	7.5 kW (10 hp) each	7.5 kW (10 hp) each
Pasteurization vessels		
Type	Belt	Concrete bunker
Quantity	2	2
Capacity	283 m^3/h (370 cu yd/h) each	92 m^3/h (120 cu yd/h) each
Temperature	70 °C	70 °C
Detention time	30 minutes	30 minutes
Onsite product storage		
Type	Uncovered concrete storage pad	Uncovered concrete storage pad
Number	1	1
Capacity	4741 m^3 (6200 cu yd)	5658 m^3 (7400 cu yd)
Dimensions	40 m × 30 m × 3 m deep (130 ft × 130 ft × 10 ft deep)	44 m × 44 m × 3 m deep (145 ft × 145 ft × 10 ft deep)
Lime Silos		
Lime delivery	Bulk pneumatic truck	Bulk pneumatic truck
Type	Free-standing cylindrical steel silo with pneumatic loading and bag house	Free-standing cylindrical steel silo with pneumatic loading and bag house
Quantity	1	3
Storage capacity, each silo	61.8 m^3 or 54.4 Mg (2180 cu ft (60 tons) each	61.8 m^3 or 54.4 Mg (2,180 cu ft (60 tons) each

(*continued*)

TABLE 14.6 (*Continued*).

Item	EnVessel Pasteurization Process (EVP) design criteria	Biofix (lime only) design criteria
	2 weeks at design capacity	2 weeks at design capacity
Unloader	Cone vibrator with slide gate to screw conveyor	Cone vibrator with slide gate to screw conveyor
Pneumatic conveyance system		
Type	Positive displacement blower	Positive displacement blower
Number	2	2
Transfer capacity	4.5 Mg/h (5 tons/h)	4.5 Mg/h (5 tons/h)
Feeder, type	Rotary airlock	Rotary airlock
Lime feed day bins		
Type	Cylindrical steel bins	Cylindrical steel bins
Quantity	2	2
Capacity	3.6 Mg or 4 tons (1 day storage)	9.8 Mg or 11 tons (1 day storage)
Storage volume	4.3 m^3 (150 cu ft)	11.3 m^3 (400 cu ft)
Lime feeders		
Type	Volumetric screw	Volumetric screw
Number	2	2
Capacity	0.85 m^3/h (30 cu ft/h)	1.7 m^3/h (60 cu ft/h)
Front-end loaders		
Type	NA	Diesel-powered, four-wheel drive, articulating chassis
Quantity		1
Bucket capacity		4.5 m^3 (6 cu yd)
Additional equipment		Enclosed air-conditioned cab with activated charcoal filter for ammonia removal
Transport trucks		
Type	Diesel-powered, two-wheel drive, chassis-mounted dump bed	Diesel-powered, two-wheel drive, chassis-mounted dump bed
Quantity	1	1
Bed capacity	15.3 m^3 (20 cu yd)	15.3 (20 cu yd)

D.C. area. The facility generates approximately 408 240 wet Mg (450 000 wet tons) of biosolids per year, of which approximately 10% is land applied in Maryland and 90% in Virginia. All of the land-applied sludge is lime treated to Class B standards. The Class B lime treatment process is summarized in Figure 14.15.

Historically, wastewater solids at Blue Plains were treated partly by anaerobic digesters and partly by lime stabilization. Lime treatment began to play a larger role as the digesters aged and were taken out of service. Currently, lime treatment is the sole treatment for sludge at Blue Plains. Blue Plains required additional capacity for dewatering (only seven centrifuges were available) and sludge treatment. Therefore K-F Environmental Technologies, Inc., Lafayette, New Jersey, was contracted to dewater and lime-treat a portion of the sludge produced by Blue Plains; K-F processes the sludge onsite.

The K-F and Blue Plains' solids handling facilities receive the same wastewater solids and use similar treatment methods, although there are some differences. K-F uses a combination of belt presses and centrifuges for dewatering. Belt conveyors are then used to transfer the dewatered material to Willowtech plow blender mixers that are used to mix CaO into the material. The Blue Plains' solids handling facility dewaters sludge with centrifuges. The dewatered sludge is conveyed by screw conveyors to a plow blender that is used to mix CaO into the sludge. The stabilized biosolids from K-F and Blue Plains are then combined in storage bunkers and later transported to land application sites. The biosolids from both sources pass

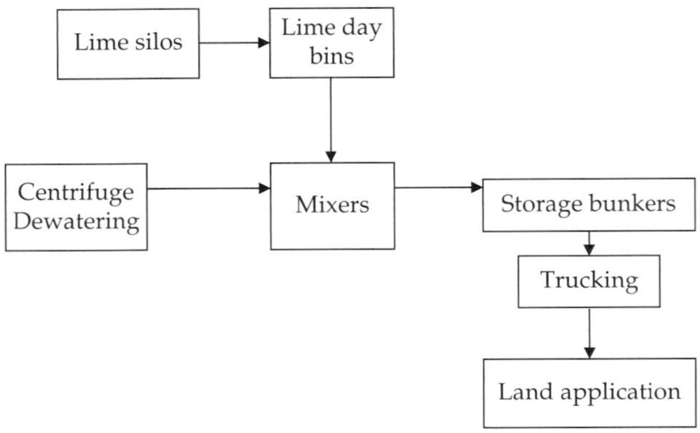

FIGURE 14.15 DC Water, Washington, D.C., Class B lime treatment process diagram.

current regulations for Class B sludge disinfection and vector-attraction reduction (VAR) by achieving the minimum pH levels as required by disinfection Class B Alternative 2 Option 5 and 503.33(b)(6) of the U.S. EPA 40 CFR Part 503 regulations (1993).

The lime-treated sludge at the Blue Plains is a good candidate for land application because of the low amounts of industrial waste in the service area and the constant need by farmers for lime addition to their naturally occurring acidic agricultural soils in the Mid-Atlantic area. As expected, Blue Plains' wastewater has low concentrations of heavy metals and organic chemicals. Therefore, for pollutant limits, Blue Plains' biosolids is of high quality. In terms of biological stability, however, there is room for improvement.

Although K-F and Blue Plains treated the same sludge and had similar treatment processes, the biosolids products they produced were not of equal quality. The biosolids produced by K-F had much better lime incorporation based on visual assessment and, correspondingly, less odor and consistently lower fecal coliform counts. The Blue Plains' biosolids appeared to be poorly mixed. As a result of poor mixing, the biosolids contained many "chunks" of unincorporated lime, which caused the biosolids to have an appearance referred to by treatment personnel as being like that of "chocolate chip cookie dough." Consistent with the visual appraisal, Blue Plains' wastewater solids had inconsistent treatment, more odors, and higher fecal coliform counts than the K-F biosolids. The Blue Plains staff conducted a study of lime treatment at the plant (North, 2003).

Based on the research conducted at Blue Plains, several modifications were made to the process trains. The existing mixers were replaced with Loedige mixers. The Loedige mixers have variable-speed drives and adjustable weirs so that mixing energy and retention time in the mixer could be varied. The Loedige mixers include multiple lime feed locations so that the solids cake can mix with the dry lime quickly and with more efficiency. The coarse-pebble lime previously used in the DC Water operation was changed to lime fines (> 90% material passing 100 mesh). These steps have significantly improved consistency of the final product and have reduced fecal coliform levels. Staff at DC Water is considering modifications to lime conveyance to better handle the finer material being used. Another modification being considered is installation of a sludge-cake hopper between the centrifuges and the mixers. This would allow better synchronization between sludge feeding to the mixers and lime addition. A chronic problem with the system, which would require a major upgrade to solve, is the long stretch of

FIGURE 14.16 Earthen storage lagoon for large quantities of DC Water, Washington, D.C., lime-treated wastewater solids.

screw conveyors from the centrifuges to the mixers and from the mixers to the storage bunkers. The extensive continuous use of vertical screw conveyors operated at relatively high speeds before and after liming results in a material with a dense, toothpaste consistency.

Because of the size of the Blue Plains' land application program, field storage is an important logistical issue and was compromised in the past because of odor issues. Lime-treated biosolids is trucked to agricultural areas in Virginia and placed in storage facilities before application. The type of storage depends on the length of time material will be held in storage (U.S. EPA, 2000). Figure 14.16 shows a storage lagoon used to hold large quantities of biosolids that will then be distributed to land applications sites in the area.

6.0 REFERENCES AND SUGGESTED READINGS

Adamchuck, V. I.; Morgan, M. T.; Ess, D. R. (2002) System and Method for Automated Measurement of Soil pH. U.S. Patent 6,356,830, March 12, 2002.

American Water Works Association (1993) *Quicklime and Hydrated Lime,* AWWA B202-93; American Water Works Association: Denver, Colorado.

ASTM International (2006) *Standard Test Methods for Chemical Analysis of Limestone, Quicklime, and Hydrated Lime*, ASTM C-25-06; ASTM International: West Conshohocken, Pennsylvania.

ASTM International (2011) *Specifications for Quicklime, Hydrated Lime, and Limestone for Chemical Purposes*, ASTM C-911; ASTM International: West Conshohocken, Pennsylvania.

Berg, G.; Dean, R. B.; Dahling, D. R. (1968) Removal of Poliovirus 1 from Secondary Effluents by Lime Flocculation and Rapid Sand Filtration. *J. Am. Water Works Assoc.*, **60**, 193–198.

Breitenbeck, G.A.; Bremner, J. M. (1984) Use of a Flat-Surface Combination pH Electrode for Measurement of Soil pH. *Comm. Soil Sci. Plant Anal.*, **15** (2), 87–98.

Burnham, J. C., et al. (1992) *Use of Kiln Dust with Quicklime for Effective Municipal Sludge Treatment with Pasteurization and Stabilization with the N-Viro Soil Process*; Stand. Tech. Publication 1135; ASTM International: West Conshohocken, Pennsylvania.

Byers, H. W.; Jensen, B. (1990) Stabilizing Sludge with Fly Ash-Sludge. Paper presented at Department of Engineering Professional Development, University of Wisconsin, Madison, Wisconsin.

Christensen, G. L. (1987) Lime Stabilization of Wastewater Sludge. In *Lime for Environmental Uses*; Gutschick, K. A., Ed.; ASTM International: West Conshohocken, Pennsylvania.

Christy, R. W. (1992) Process and Mechanical Design Considerations for Sludge/Lime Mixing. *Proceedings of the Water Environment Federation Specialty Conference*, Portland, Oregon; Water Environment Federation: Alexandria, Virginia.

Counts, C. A.; Shuckrow, A. J. (1975) *Lime Stabilized Sludge: Its Stability and Effect on Agricultural Land*, EPA-670/2-75-012; Battelle Memorial Institute: Richland, Washington.

Engineering Science, Inc.; Black and Veatch (1991) *Technology Evaluation Report: Alkaline Stabilization of Sewage Sludge*; Report prepared for U.S. EPA, Contract No. 68-C8-0022, Work Assignment No. 01-08; U.S. Environmental Protection Agency: Washington, D.C.

Farrell, J. B., et al. (1974) Lime Stabilization of Primary Sludge. *J. Water Pollut. Control Fed.*, **46**, 113.

Fergen, R. E. (1991) Stabilization and Disinfection of Dewatered Municipal Wastewater Sludge with Alkaline Addition. *Proceedings of the American Water Works Association/Water Pollution Control Federation Joint Residuals Management Conference*, Durham, North Carolina; Water Pollution Control Federation: Washington, D.C.

Jacobs, A.; Silver, M. (1990) *Sludge Management at the Middlesex County Utilities Authority. Water. Sci. Technol.*, **22,** 93.

Jacobs, A., et al. (1992) Odor Emissions and Control at the World's Largest Chemical Fixation Facility. *Proceedings of the 6th Annual Water Environment Federation Residuals Management Conference,* Portland, Oregon; Water Environment Federation: Alexandria, Virginia.

Lewis, C. J.; Gutschick, K. A. (1980) *Lime in Municipal Sludge Processing*. National Lime Association: Washington, D.C.

Lue-Hing, C.; Kuchenrither, R. D. (1992) *Municipal Sewage Sludge Management: Processing, Utilization and Disposal*; Technomic Publishing Co., Inc.: Lancaster, Pennsylvania.

National Lime Association (1996) *Biosolids Stabilization: Which Class a Stabilization is Most Economical?* Bulletin 334; National Lime Association: Arlington, Virginia.

North, J. (2003) Using Calcium to Measure Mixing and Mixer Performance. M.S. Thesis, University of Maryland, College Park, Maryland.

O'Connor, G. A. (1996) *Understanding Fate, Transport, Bioavailability, and Cycling of Metals in Land-Applied Biosolids*; Water Environment Research Foundation: Alexandria, Virginia.

Oerke, D. W. (1989) The Role of Lime Stabilization Processes in Wastewater Sludge Processing and Disposal; *Proceedings of the 1989 National Lime Association's Effective Use of Dry Lime for Sewage Sludge Stabilization Seminar*; Jun 7–8; National Lime Association: Washington, D.C.

Oerke, D. W. (1992) Alkaline Stabilization. In *Municipal Sludge: Management, Utilization, and Disposal Options*; Lue-Hing, C.; Kuchenrither, R. D., Eds.; Technomic Publishing Co., Inc.: Lancaster, Pennsylvania.

Oerke, D. W. (1996) Which Class a Biosolids Stabilization Process is the Most Economical: Lime Stabilization, Composting or Thermal Drying? *Proceedings of the RMWEA/RMAWWA Annual Conference*, Steamboat Springs, Colorado, Sep 8–11; Rocky Mountain Water Environment Association: Denver, Colorado.

Oerke, D. W. (1999) Alkaline Stabilization of Biosolids Can Save Money, Space. *WaterWorld*, 14–16.

Oerke, D. W.; Rogowski, S. M. (1990) Economic Comparison of Chemical and Biological Sludge Stabilization Processes; *Proceedings of the American Water Works Association/Water Pollution Control Federation Joint Residuals Management Conference*, New Orleans, Louisiana, Dec 2–5; Water Pollution Control Federation: Washington, D.C.

Oerke, D. W.; Stone, L. A. (1991) Detailed Case Study Evaluation of Alkaline Stabilization Processes; *Proceedings of the American Water Works Association/Water Pollution Control Federation Joint Residuals Management Conference*, Durham, North Carolina, Aug 12; Water Pollution Control Federation: Washington, D.C.

Paulsrud, B.; Eikum, A. S. (1975) Lime Stabilization of Sewage Sludge. *Water Res.*, 9, 297.

Ramirez, A.; Malina, J. (1980) Chemicals Disinfect Sludge. *Water Sew. Works*, **127** (4), 52.

Reimers, R. S.; Little, M.; Englande, A.; Leftwich, D.; Bowman, D. (1981) *Parasites in Southern Sludge and Disinfection by Standard Sludge Treatment*, EPA-600/2-81-166; U.S. Environmental Protection Agency: Washington, D.C.

Sloan, D. (1992) Design and Process Considerations for Advanced Alkaline Stabilization with Subsequent Accelerated Drying Facilities. *Proceedings of the 5th Annual International Conference on Alkaline Pasteurization Stabilization*; Somerset: New Jersey.

Stone, L. A., et al. (1991) Detailed Case Study Evaluation of Alkaline Stabilization Processes. *Proceedings of the American Water Works Association/Water Pollution Control Federation Joint Residuals Management Conference*, Durham, North Carolina, Aug 12; Water Pollution Control Federation: Washington, D.C.

Stone, L.A., et al. (1992) The Historical Development of Alkaline Stabilization. *Proceedings of the Water Environment Federation Specialty Conference*, Portland, Oregon; Water Environment Federation: Alexandria, Virginia.

U.S. Environmental Protection Agency (1975) *Lime Stabilized Sludge: Its Stability and Effect on Agricultural Land*, EPA-670/2-75-012; U.S. Environmental Protection Agency: Washington, D.C.

U.S. Environmental Protection Agency (1979) *Process Design Manual for Sludge Treatment and Disposal*, EPA 625/1-79-011; U.S. Environmental Protection Agency: Washington, D.C.

U.S. Environmental Protection Agency (1993) Standards for the Use and Disposal of Sewage, Final Rule. *Code of Federal Regulations*, Part 503, Title 40, 1993; *Fed. Regist.*, **58**, 9248.

U.S. Environmental Protection Agency (2000) *Guide to Field Storage of Biosolids and Other Organic By-Products Used in Agriculture and for Soil Resource Management;* EPA-832-B/00-007; U.S. EPA OW-OWM; U.S. Environmental Protection Agency: Washington, D.C.

U.S. Environmental Protection Agency (2003) *Environmental Regulations and Technology: Control of Pathogens and Vector Attraction in Sewage Sludge*, revised July 2003; EPA/625/R-92/013; U.S. Environmental Protection Agency: Cincinnati, OH.

U.S. Environmental Protection Agency (2011) Text Methods. 9000 Series Methods, SW-846, Methods 9040B, and 9045D http://www.epa.gov/epawaste/hazard/testmethods/sw846/online/9_series.htm (accessed July 2011).

Water Environment Federation (1995) *Wastewater Residuals Stabilization*, Manual of Practice No. FD-9; Water Environment Federation: Alexandria, Virginia.

Water Environment Federation; American Society of Civil Engineers; Environmental & Water Resources Institute (2009) *Design of Municipal Wastewater Treatment Plants*, 5th ed.; WEF Manual of Practice No. 8; ASCE Manuals and Reports on Engineering Practice No. 76; McGraw-Hill: New York.

Westphal, A.; Christensen, G. L. (1983) *Lime Stabilization: Effectiveness of Two Process Modifications. J. Water Pollut. Control Fed.*, **55,** 1381.

Chapter 15

Disinfection and Stabilization Processes

(*continued*)

1.0 INTRODUCTION

In 2004, 16 583 wastewater treatment plants (WWTPs) in the United States were processing 1474.6 m^3/s (33.657 bil. gpd) of municipal wastewater. This wastewater contains feces and urine from both humans and animals, which may carry disease-causing organisms or pathogens. Pathogenic microorganisms from humans may enter a community's wastewater from sites such as hospitals and nursing homes. Humans come into contact with pathogens and, thus, potential disease, through direct contact with wastewater, after eating food or drinking water contaminated with pathogens, or through contact with human, animal, or insect carriers. Animal wastes may enter wastewater from farms, meat packing and processing facilities, and from rats and other animals or vectors found in or around wastewater or collection systems.

During the course of typical treatment, the microorganisms in wastewater are reduced in number, becoming concentrated in wastewater sludge. Table 15.1 shows the primary pathogens of concern in municipal wastewater and sewage sludge and associated disease/symptoms (U.S. EPA, 2003). The pathogens listed can cause typhoid, gastroenteritis, cholera, hepatitis A, polio, giardiasis, hook worms, cryptosporidiosis, and amebiasis. Table 15.2 gives typical pathogen levels in unstabilized sludges (Smith et al., 2005; U.S. EPA, 1981). Frequently, the pathogen load to municipal WWTPs is low because of the high health status of the human population; therefore, actual numbers in sludge are low and are further reduced or eliminated by treatment.

But what constitutes an infectious dose of a pathogenic organism? The infectious dose depends on the organism and, particularly for bacteria and viruses, can vary widely. Table 15.3 gives examples of reported infective dose data, showing values as low as 100 for various species of *Salmonella*, less than 1 plaque-forming unit (PFU) for poliovirus, and 1 egg for helminths.

Infective dose data coupled with lifespan of pathogens on soil and plants provided in Table 15.4 form the basis for regulations in the United States (U.S. EPA, 1993). Regulations state that sewage sludge must be adequately disinfected, have its vector attractiveness reduced, and in some instances be applied with site access and harvesting restrictions that prevent the biosolids from being a risk to public health and the environment.

TABLE 15.1 Pathogens present in raw domestic sludge.

Pathogen class	Examples	Disease/symptom
Bacteria	*Shigella sp.*	Bacillary dysentery
	Salmonella sp.	Salmonellosis (gastroenteritis)
	Salmonella typhi	Typhoid fever
	Vibrio cholerae	Cholera
	Enteropathogenic-*Escherichia coli*	A variety of gastroenteric diseases
	Yersinia sp.	Yersiniosos (gastroenteritis)
	Campylobacter jejuni	Campylobacteriosis (gastroenteritis)
	Listeria monocytogenes	Listeriosis (sepsis, meningitis)
Viruses	Hepatitis A	Infectious hepatitis
	Norwalk virus	Acute gastroenteritis
	Rotaviruses	Acute gastroenteritis
	Polioviruses	Poliomyelitis
	Coxsackie viruses	"Flu-like" symptoms
	Echoviruses	"Flu-like" symptoms
Protozoa	*Entamoeba histolytica*	Amebiasis (amoebic dysentery)
	Giardia lamblia	Giardiasis (gastroenteritis)
	Cryptosporidium sp.	Crytosporidiosis (gastroenteritis)
	Balantidium coli	Balantidiasis (gastroenteritis)
Helminths	*Ascaris sp.*	Ascariasis (roundworm infection)
	Taenia sp.	Taeniasis (tapeworm infection)
	Necator americanus	Ancylostomiasis (hookworm infection)
	Trichuris trichuria	Trichuriasis (whipworm infection)

TABLE 15.2 Typical pathogen levels in sludges and their changes over time (PFU = plaque-forming units).

Organism	Before 1982	Present
Fecal coliforms number/g dry sludge	2.0×10^7	2.0×10^7
Salmonella sp. number/g dry sludge	4.1×10^2	5.7×10^0
Enteric viruses PFU/g dry sludge	4.0×10^2	1.0×10^0
Helminth ova number/g dry sludge	< 1 to 50	< 1

TABLE 15.3 Reported infective dose data (HID = human infective dose; TCID50 = tissue culture infectious dose for 50% response; PFU = plaque-forming units; FFU = focus-forming units; NR = not reported).

Organism	Infective dose	Range	Reference
Bacteria			
Clostridium perfringens	10^6	10^6–10^{10}	U.S. EPA, 1985
Escherichia coli	10^4	10^4–10^{10}	U.S. EPA, 1985; Keswick, 1984
Salmonella (various species)	10^2	10^2–10^{10}	U.S. EPA, 1985
Shigella dysenteriae	10 to 10^2	10–10^9	U.S. EPA, 1985; Keswick, 1984; Levine, 1973
Shigella flexneri	10^2	10^2–10^9	U.S. EPA, 1985
Streptococcus faecalis	10^9	10^9–10^{10}	U.S. EPA, 1985
Vibrio cholerae	10^3	10^3–10^{11}	U.S. EPA, 1985; Keswick, 1984
Viruses			
Echovirus 12	HID50 919 PFU HID1 17 PFU, estimated	17–919 PFU	U.S. EPA, 1985
Poliovirus	1 TCID50, <1 PFU	4 × 10^7 TCID50 for infants 0.2 to 5.5 × 10^6 PFU for infants	U.S. EPA, 1985
Rotavirus	HID50 10 FFU HID25 1 FFU, estimated	0.9 to 9 × 10^4 FFU	Ward et al., 1986
Parasites			
Entamoeba coli	1–10 cysts	1–10 cysts	U.S. EPA, 1985
Cryptosporidium	10 cysts	10–100 cysts	Casemore, 1991
Giardia lamblia	1 cyst estimated	NR	U.S. EPA, 1985
Helminths	1 egg	NR	U.S. EPA, 1985

TABLE 15.4 Survival times of pathogens on soil and plants.

	Soil		Plants	
Pathogen	**Absolute maximum**	**Common maximum**	**Absolute maximum**	**Common maximum**
Bacteria	1 year	2 months	6 months	1 month
Viruses	6 months	3 months	2 months	1 month
Protozoa	10 days	2 days	5 days	2 days
Helminths	7 years	2 years	5 months	1 month

The U.S. Environmental Protection Agency's (U.S. EPA's) Part 503 Rule (1993) includes alternatives for achieving a reduction in the number of pathogenic microorganisms present and divides them into Class A or Class B. The intent of using Class A pathogen-reduction processes—like heat drying, composting, and pasteurization—is to reduce the level of pathogenic organisms in the sludge to *below detectable levels* consistent with available standard methods. When Class B processes like aerobic digestion, anaerobic digestion, are used, however, it is recognized that the level of pathogenic organisms is *only significantly* reduced by approximately one order of magnitude. Pathogens may still be present. As such additional protective measures must be exercised to minimize the risk to public health. These measures include restricting the public's access to land application sites, controlling animal grazing, and preventing crop harvesting for a period of time (U.S. EPA, 1993).

Similarly, Part 503 identifies options for controlling the vector attractiveness of residuals, which are discussed in Section 4.0 of this chapter. The most frequently used methods for demonstrating vector-attraction reduction (VAR) are

- Showing > 38% volatile solids reduction by a biological treatment process like anaerobic or aerobic digestion;
- Showing the specific oxygen uptake rate (SOUR) for aerobically digested sludge at 20 °C (68 °F) is < 1.5 mg oxygen/g·h total sewage sludge solids;
- Demonstrating during composting of sludge that the average temperature of the mixture is greater than 45 °C (113 °F) for at least 14 days and always greater than 40 °C (104 °F); and

- Adding sufficient alkali to raise the pH to at least 12 at 25 °C (77 °F), and maintaining a pH > 12 for 2 hours and a pH > 11.5 for 22 more hours.

The intent of these requirements is to indicate removal of all putrescible (biodegradable) organic material from sludge so that it will no longer be attractive to vectors. Possible vectors for sewage sludge pathogens include insects, rodents, and birds. It is recognized, however, that although the above requirements meet the *letter* of the law, they do not necessarily mean all putrescible material is gone, vectors will not be attracted, or odors will not be present. Odors are a particular problem because they can lead to public concern.

Four approaches or barriers are frequently applied to control infectious disease causing organisms:

1. Application of a disinfection process, such as time and temperature, to reduce the number of pathogenic and indicator organisms;
2. Application of a solids-destruction process, such as aerobic or anaerobic digestion with the goal of reducing the biodegradable material and thus the food supply for organisms remaining in the sludge and for vectors that could carry disease-causing organisms;
3. Application of a chemical or physical factors such as lime or desiccation that retards organism growth or renders the material unattractive to vectors temporarily or permanently; and
4. Allowing time following the application of material to land for the natural attenuation of remaining organisms and organics in sludge and before public contact occurs.

2.0 CLASS A DISINFECTION PROCESSES

According to a recent survey, approximately 23% of sludge produced in the United States meets Class A criteria (North East Biosolids and Residuals Association, 2007). The requirements for sludge to be Class A for pathogens can be met with any one of six alternatives. Every alternative, however, requires that the density of *Salmonella* species be reduced to less than 3 MPN (most probable number) per 4 grams of dry sludge solids, or the fecal coliforms must be reduced to less than 1000 MPN per gram of dry sludge solids. Because numbers of fecal coliforms in the untreated sludge have historically approached 10^8 per gram of sludge by dry weight, it is expected that there will be at least a 5 log reduction following treatment or < 10^3. The four most typically used alternatives are described below.

2.1 Class A—Alternative 1

Time and temperature can be described with the following equations

$$D = 131\,700\,000 / 10^{0.14t} \quad 15.1$$

$$D = 50\,070\,000 / 10^{0.14t} \quad 15.2$$

where

D = time in days and t = temperature, °C.

The first equation, eq 15.1, applies when total solids are ≥ 7%; $t \geq 50$ °C, and time is ≥ 20 minutes. If the sludge particles are small and are heated by warmed gases or an immiscible liquid, however, then the minimum time is 15 seconds. This equation further applies when the solids are < 7%, $t > 50$ °C and time is ≥ 15 seconds to < 30 minutes. Equation 15.2 applies for total solids < 7 %; $t \geq 50$ °C and time ≥ 30 minutes. These requirements were established based on U.S. Food and Drug Administration requirements for eggnog, data from German sources, and other data collected during composting experiments in the United States (U.S. EPA, 1992). Many facilities consider Alternative 1 without realizing that it was derived from experience with fluids such as milk. In fluids, it is not difficult to ensure that all particles meet the requirements for time and temperature. As a result, when facility personnel submit a plan to the permitting authority, they are shocked when asked for a monitoring plan that will ensure that all parts of the sludge will meet the appropriate time and temperature requirements. Facilities considering this alternative to achieve Class A using thermophilic anaerobic digester, heat dryer, or some other process should carefully consider plans for temperature monitoring. This includes where temperature sensors might be placed and when time measurements will be made to collect data that successfully demonstrates the conditions of Alternative 1 have been met. It is typically easiest to demonstrate these conditions with batch or plug flow reactors rather than continuous flow reactors. Tracers may be helpful in showing time requirements are met. It is critical that these sludge treatment processes are completed without short-circuiting occurring.

2.2 Class A—Alternative 2

This alternative is based on disinfection research done by N-Viro Energy Systems, Ltd., in the late 1980s (U.S. EPA, 2003). It is essentially the same as the process to further reduce pathogens (PFRP) equivalency recommendation given to N-Viro under Class A, Alternative 6.

2.3 Class A—Alternatives 3 and 4

These alternatives are not discussed in detail here. They depend on a quality assurance and quality controlled monitoring plan to determine the presence or absence of enteric viruses and helminth ova. This is only useful when significant numbers of enteric viruses and helminth ova are present in the raw sludge (this is seldom; see Table 15.2). Monitoring can be performed to measure the effectiveness of the treatment process (Smith, et al., 2005). Otherwise, it is meaningless because the presence or absence of enteric viruses and helminth ova says nothing about the presence or absence of other pathogens. Several states have removed these two alternatives from their regulations. More states and the federal government are either expected to do the same thing or significantly change how they are applied including a requirement that they be used with some sort of treatment process.

2.4 Class A—Alternative 5

The sludge is treated by one of the PFRP shown in Table 15.5 and discussed here.

2.4.1 Composting

In this process, it is critical that all parts of the sludge pass through a zone where they can be held for no less than three days at 55 °C and then be removed without contamination. Composting is discussed in detail in Chapter 13.

2.4.2 Heat Drying

This process is discussed in detail in Chapter 16. When heat drying processes deviate from the description given in Table 15.5, they will frequently need to qualify as Class A by Alternative 1. As such, the state or regional permitting authority will want facility personnel to demonstrate that the time and temperature requirements are achieved. Where this is not possible, authorities may not accept the product as meeting Class A criteria.

TABLE 15.5 Processes to further reduce pathogens (PFRP).

Process	Description
Composting	Using either within-vessel or static aerated pile composting method, the temperature of sludge is maintained at 55 °C (131 °F) or greater for three consecutive days. Using the windrow composting method, the temperature of sludge is maintained at 55 °C (131 °F) or greater for 15 consecutive days or more during which there shall be a minimum of five turnings of the windrow.
Heat drying	Sludge is dried by direct or indirect contact with hot gases to reduce moisture content to 10% or lower. Either temperature of the sludge particles exceeds 80 °C (176 °F) or the wet bulb temperature of the gas in contact with the sludge as it leaves the dryer exceeds 80 °C (176 °F).
Heat treatment	Liquid sludge is heated to a 180 °C (356 °F) or greater for 30 minutes.
Thermophilic aerobic digestion	Liquid sewage sludge is agitated with air or oxygen to maintain aerobic conditions and the mean cell residence time (i.e., the solids retention time) of the sludge is 10 days at 55 °C (131 °F) to 60 °C (140 °F).
Beta ray irradiation	Sludge is irradiated with beta rays from an electron accelerator at dosages of at least 1.0 megarad at room temperature (20 °C [68 °F]).
Gamma ray irradiation	Sludge is irradiated with gamma rays from certain isotopes, such as Cobalt 60 and Cesium 137, at dosages of at least 1.0 megarad at room temperature (20 °C [68 °F]).
Pasteurization	Temperature of sludge is maintained at 70 °C (158 °F) or greater for 30 minutes or longer.

The facility or dryer manufacturer will need ways to measure the temperature within the material being dried at different locations in the dryer and have a way to measure the time of passage of the material through the dryer.

Equation 15.1 applies to sludges with 7% or higher solids content that are small particles heated by contact with either warmed gases or an immiscible liquid. For these sludges, the equation is applied to times not less than 15 seconds and temperatures not less than 50 °C. Two examples of sludges to which this requirement applies are: (1) Sludge cake that is mixed with previously dried solids to make the entire mass a mixture of separate particles and is then dried by contact with a hot-gas stream in a rotary drier; and (2) Sewage sludge dried in a multiple-effect evaporator system in which the sewage sludge particles are suspended in a hot oil that is heated by indirect heat transfer with condensing steam.

2.4.3 *Heat Treatment*

In the typical process arrangement in which sludge is either being conditioned for improved dewaterability or digestibility, sludge is heated by countercurrent heat exchange from approximately 20 to 170 °C or even higher and discharged to a large vessel. Steam is added to bring the temperature up to 180 °C or more; after approximately 30 minutes residence time, the sludge leaves and is cooled from 180 °C to approximately 30 °C by countercurrent heat exchange with the incoming sludge. The storage vessel is designed to avoid the possibility of sludge bypassing rapidly from inlet to outlet because short thermal treatment of a portion of the sludge would produce poor conditioning. These conditions are extreme, essentially sterilizing the sludge. Zimpro, Cambi, and other manufacturers offer processes capable of achieving these conditions (U.S. EPA, 2006d).

2.4.4 *Thermophilic Aerobic Digestion*

Aerobic digestion is discussed in detail in Chapter 11.

2.4.5 *Beta Ray and Gamma Ray Irradiation*

These processes are seldom used and the reader is referred to other literature for more information (U.S. EPA, 2003).

2.4.6 *Pasteurization*

Pasteurization of a sludge slurry typically requires holding the temperature of the sludge at 70 °C or above for at least 30 minutes. Although milk pasteurization is a good example, its temperature requirements are 66 °C for 30 minutes. Either liquid sludge or sludge cake with approximately 80% moisture can be pasteurized. In fact, by following the conditions for using eq 15.1 or 15.2 depending on the solids concentration under Alternative 1, pasteurization can also be accomplished by maintaining a temperature of 65 °C for one hour. Sludge can be brought up to temperature by indirect heat exchange or by steam injection. After a holding period at the desired operating temperature, sludge may be cooled by heat exchange or by transfer to another vessel where natural convection, evaporation, and radiation decrease the temperature.

The U.S. EPA has indicated that to produce a Class A biosolids that meets the requirements in Alternative 1 or Alternative 5 under 40 CFR Part 503, every particle should be exposed to a minimum temperature for a minimum time. Design engineers should avoid using completely mixed systems or systems with potential for back-mixing or short-circuiting as pre-pasteurization tanks. Most vendor-supplied systems are batch tanks; currently, only one vendor supplies a plug flow tank for pre-pasteurization. Design engineers should consult U.S. EPA staff or other pertinent regulators before using a nonbatch system.

Batch pre-pasteurization systems are operated in a fill/hold/draw mode, with several batch vessels used to perform each cycle if continuous operation is desired. The vessels should be well mixed to ensure the monitored temperature reflects the entire contents (i.e., every solids particle meets time and temperature requirements). If the downstream anaerobic digestion process, for example, uses an intermittent feed cycle and upstream storage is adequate, then the system needs fewer than three batch vessels for the required fill, hold, and draw cycles.

2.4.6.1 *Ancillary Equipment for Pasteurization*

Design engineers should consider three important ancillary features when installing a pasteurization process:

1. Solids heating and cooling;
2. Solids screening; and
3. Temperature monitoring and control.

Because the temperature of pasteurization systems typically is maintained at 70 °C, solids must be heated and then cooled. Heat exchangers are typically used for heating and cooling solids. If desired, design engineers could include a heat recovery step to use heat from cooling biosolids to preheat raw solids. The recovery step will require a substantial amount of heat-exchange capacity. If heat exchangers are used, then solids may need to be screened before pasteurization—even if fine screens are used in the plant headworks. Screening also helps produce a more aesthetically pleasing product.

Good temperature monitoring and control are required to maintain Class A compliance. It is critical to have a well-automated system to ensure pasteurization and to prevent downstream contamination, which would take months to remedy. The system should prevent unpasteurized solids from passing through, and either waste or recirculate material that did not meet the time and temperature requirements. If necessary, then standby equipment should be included to maintain time and temperature, because compromising these parameters could contaminate the downstream anaerobic digestion process.

2.4.6.2 *Performance*

The pasteurization process can meet the *Salmonella* criteria in Part 503. Ward et al. (1999) showed that pasteurized solids resisted regrowth even after they were seeded with *Salmonella*; instead, the organisms died off. Higgins et al. (2008) also observed that pasteurization effectively destroyed *Salmonella*, although fecal coliform regrew (suggesting that it may be necessary to measure the actual pathogen rather than the indicator to ensure Part 503 compliance).

In summary, pasteurization is an effective method for destroying pathogens in solids. Used throughout the world, it is one of the more prevalent solids disinfection processes and typically is associated with anaerobic digestion. Several U.S. facilities use this process, including a WWTP with a capacity of 2.4 m^3/s (54 mgd) operated by the Alexandria Sanitation Authority in Virginia (see Figure 15.1).

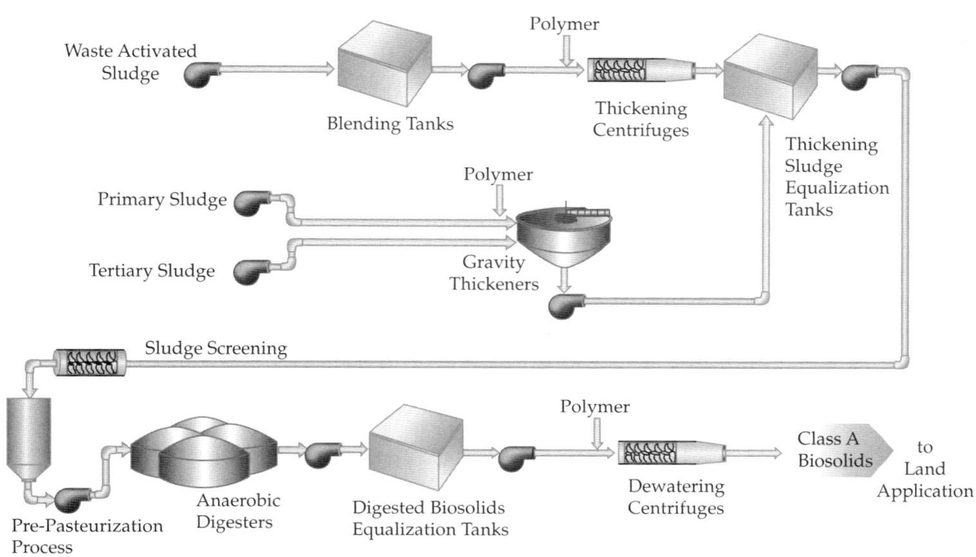

FIGURE 15.1 Process flow diagram of Alexandria Sanitation Authority in Virginia, including screening and pasteurization vessels (courtesy of Alexandria Sanitation Authority, Alexandria, Virginia).

2.4.6.3 Alternative Approach to Achieve Pasteurization

Achieving the same results by mixing a powder like quicklime (or similar reagent), with semisolid material, sludge, is not easy. Mixing has to be thorough to ensure that all parts of the sludge comes in contact with the alkaline material and the temperature increases to 70 °C or greater and is held there for at least 30 minutes. Properly placed temperature sensors are needed to ensure 70 °C is achieved throughout and for the prescribed time. Some minimal level of moisture is necessary to ensure that thorough mixing can occur. These systems need to be engineered and carefully monitored to ensure proper operation. Alkaline treatment of sludge is discussed in detail in Chapter 15. Several proprietary processes exist for achieving pasteurization and include Ecotherm, Burch Biowave, RDP's EnVessel Pasteurization , and Bioset among others (U.S. EPA, 2006d).

2.5 Class A—Alternative 6

Sludge is treated by a process equivalent to a PFRP. To date, 13 technologies have received equivalency under Class A Alternative 6. These PFRP equivalent processes are listed on the U.S. EPA Pathogen Equivalency Committee's website, http://www.epa.gov/nrmrl/pec/index.html. Detailed information on what is needed to obtain a PFRP equivalency recommendation can be found below.

3.0 CLASS B DISINFECTION PROCESSES

Requirements for a sludge to be Class B for pathogens are met by using one of three alternatives described below and combined with access and harvesting restrictions. The US 40 CFR Part 503 Rule applies access and harvesting restrictions because Class B sludges can still contain pathogenic organisms. These restrictions allow for any remaining pathogens in the sludge to die before the public comes into contact with the treated sludge or crops for human consumption are harvested.

3.1 Class B—Alternative 1

The geometric mean fecal coliform density (either MPN or CFU per gram of dry sludge solids) of seven samples shall be less than 2 000 000 (2×10^6). Because levels of fecal coliforms in untreated sludge often approach 10^8, a 2 log reduction is expected. This alternative is problematic because many sludges meet this requirement before any kind of stabilization. It is essential when this alternative is used that the sludge has first gone through some kind of stabilization/treatment method.

3.2 Class B—Alternative 2

The sludge is treated by a process to significantly reduce pathogens (PSRP), see Table 15.6.

Anaerobic digestion, aerobic digestion, composting, and lime stabilization processes are discussed in detail in Chapters 10, 11, 13, and 15.

3.2.1 Air Drying

From the design/operational point of view, liquid sludge is air-dried by applying it to outdoor, under-drained sand beds, or paved or unpaved basins, to depths of approximately 230 mm (9 in.), and allowed to drain and dry. It is removed for disposal when it reaches a solids content that allows removal by shovels or front-end loaders. Drying for three months is needed for significant reduction of pathogens. Temperatures should be greater than 0 °C for at least two of these months. Low

TABLE 15.6 Processes to significantly reduce pathogens.

Process	Description
Aerobic digestion	Sludge is agitated with air or oxygen to maintain aerobic conditions for a specific mean cell residence time (i.e., solids retention time) at a specific temperature. Values for the mean cell residence time and temperature shall be between 40 days at 20 °C (68 °F) and 60 days at 15 °C (59 °F).
Air drying	Sludge is dried on sand beds or in paved or unpaved basins for a minimum of 3 months. During 2 of the 3 months, the ambient average daily temperature is greater than 0 °C (32 °F).
Anaerobic digestion	Sludge is treated in the absence of air for a specific mean cell residence time (i.e., solids retention time) at a specific temperature. Values for the mean cell residence time and temperature shall be between 15 days at 35 to 55 °C (131 °F) and 60 days at 20 °C (68 °F).
Composting	Using within-vessel, static aerated pile, or windrow composting methods, the temperature of the sludge is raised to 40 °C (104 °F) or greater and remains there for five days. For four hours during the five-day period, the temperature in the compost pile exceeds 55 °C (131 °F).
Lime stabilization	Sufficient lime is added to the sludge to raise the pH 12 for at least two hours of contact.

temperatures retard microbial metabolism of available substrate. Therefore for the greater part of the drying period, the temperature must be greater than 0 °C.

Drying of sludge is a batch operation, so the time of exposure to the process can be well controlled. Efficiency is time and climate dependent. Sludge on underdrained sand beds dries much faster than sludge dried in basins (because drainage can take place on sand beds). In hot dry weather, sludge dried on sand beds can be "liftable" in two weeks.

3.3 Class B—Alternative 3

Sludge is treated by a process equivalent to a PSRP. Only one of these processes needs to be mentioned here: alkaline addition to achieve lime stabilization, which was developed by N-Viro Energy Systems, Ltd., of Toledo, Ohio. The process description reads, "Use of cement kiln dust and lime kiln dust (instead of lime) to treat sludge by raising the pH. Sufficient lime or kiln dust is added to sludge to produce a pH of 12 for at least 12 hours of contact" (U.S. EPA, 2003).

4.0 PROCESSES TO SIGNIFICANTLY REDUCE PATHOGENS AND PROCESSES TO FURTHER REDUCE PATHOGENS

Table 15.7 shows U.S. EPA's numeric requirements for demonstrating equivalency of an innovative or alternative technology to a PSRP or PFRP. In addition, U.S. EPA must be able to understand how the proposed technology disinfects and how key

TABLE 15.7 Process equivalency criteria (PSRP = process to significantly reduce pathogens; PFRP = process to further reduce pathogens; MPN = most probable number; TS = total solids; dw = dry weight; CFU = colony-forming units).

PSRP equivalency	PFRP equivalency
> 2 log reduction of fecal coliform bacteria	(1) > 3 log reduction of total enteric viruses, and (2) > 2 log reduction of viable helminth (*Ascaris* spp.) ova, and (3) > 3 log reduction of fecal coliform bacteria
Final product contains < 2 000 000 MPN or CFU fecal coliform/g TS dw	Final product contains < 1000 MPN/g TS dw fecal coliform bacteria or < 3 *Salmonella* spp./4 g TS dw; plus < 1 PFU/4 g TS dw of total enteric viruses and < 1 viable helminth ova/ 4 g TS dw

process control parameters work. To demonstrate adequate pathogen destruction, the untreated sludge must contain adequate numbers of organisms. For example, to demonstrate PFRP equivalency, the untreated sludge needs to contain at least 1000 PFU of enteric viruses per 4 grams total solids on a dry-weight basis (1000 PFU enteric viruses/4 g TS dw); and 100 viable helminth (*Ascaris* spp.) ova per 4 g total solids on a dry-weight basis (100 viable helminth ova/4 g TS dw). If the untreated sludge does not naturally contain these density levels, then the applicant must augment (spike) the sludge with the appropriate microbe to achieve these levels.

Work by several investigators has shown the danger of relying on only one or two stressors to disinfect sludge (Meckes and Rhodes, 2004; Reimers, et al. 2005). Similarly, it is difficult to rely on the reduction of a few organisms to demonstrate adequate disinfection for protection of public health. Every organism tolerates specific stressors to varying degrees. Further, the degree of tolerance among organisms is not uniform between stressors. Therefore, it has become increasingly apparent that if disinfection technologies are to provide adequate protection then they must impose multiple stressors that consistently result in effective destruction of organisms including bacteria, viruses, worms, and protozoa.

The decay of pathogen density with temperature is known to follow first-order kinetics with time. Stern and Farrell (1977) stored digested primary sludge at temperatures of 4 °C and 20 °C for up to 24 weeks. At 4 °C and in six months, fecal coliform densities dropped almost 5 logs, *Salmonellae* dropped 2 logs, and enteroviruses fell approximately 0.5 log. At 20 °C densities of these microorganisms fell much more rapidly. Fecal coliform densities dropped 3 logs in eight weeks, and *Salmonellae* and enteroviruses fell more than 2 logs. In testing at the University of North Carolina, it was found that the ova of *Ascaris* spp., considered by many investigators to be the most difficult organism to inactivate, could be inactivated at 53 °C in 30 minutes, although large numbers of fecal coliforms continued to exist (Aitken et al., 2005). Elevating the pH above 12 and keeping it elevated for two hours is the basis of the lime-stabilization process (U.S. EPA, 2003). This process significantly reduces the density of viruses and pathogenic bacteria. At this elevated pH, ammonia is liberated from sludge and effectively kills microorganisms (Cramer, et al. 1983; Ward, 1978; Ward and Ashley, 1976, 1977, 1978). Ammonia as an uncharged molecule is capable of permeating microbial cellular membranes and, therefore, enhancing pathogen kills. Desiccation alone requires drying to 95% solids to inactivate helminth eggs (Reimers et al., 1981). Irradiation (gamma or electron beam) directly damages the deoxyribonucleic acid (DNA) of living organisms, inducing cross-linkages and other changes that make an organism unable to grow or reproduce. When these rays interact with water molecules in an organism, they generate transient free radicals that can cause additional indirect damage to DNA.

Table 15.7 shows the numeric requirements for demonstrating process equivalency as a PSRP or PFRP process. Adequate supporting documentation must be supplied by technology developers or those individuals or entities that want to use a PSRP- or PFRP-equivalent process. The technology supplier/developer typically collects and assembles supporting documentation. Potential applicants are strongly advised to contact their state or regional biosolids coordinator or the U.S. EPA's Pathogen Equivalency Committee (PEC) directly at the start of a project. Getting the PEC involved as early as possible reduces the risk of collecting inadequate or inappropriate data that may waste efforts. For this reason, data in support of an application for equivalency beyond proof-of-concept level should not be collected before consulting with the PEC.

The following sections describe the information that must be included as part of the documentation in support of an application for a PSRP- or PFRP-equivalency recommendation. Detailed information on the entire equivalency recommendation process can be found on the PEC's website at http://www.epa.gov/nrmrl/pec/index.html.

4.1 Detailed Process Description

The process used to treat sludge must be described clearly so that the PEC fully understands how the technology is to be applied and under what operating conditions it will provide adequate treatment. For example, the type of sludge and other materials used in the process should be described. Specifications for these materials should be provided as appropriate.

The process should be broken down into key steps and graphically displayed in a quantified flow diagram of the wastewater and sludge treatment processes. The application should precisely define which steps constitute the beginning and end of treatment. The earliest point at which sludge treatment can be defined as beginning is the point at which the sludge is collected from the wastewater treatment process. Sufficient information should be provided for a mass balance calculation (i.e., actual or relative volumetric flows and solids concentration in and out of all streams, additive rates for bulking agents or other additives). A description of process parameters should be provided for each step, which includes typical ranges and mean values where appropriate. Parameters that should be detailed will depend on the type of process and could include any processes that affect pathogen reduction or process reliability as provided below.

- Sludge characteristics
  - Total and volatile solids content of sludge before and after treatment;
  - Proportion and type of additives (dilutents) in sludge;
  - Chemical characteristics (as they affect pathogen survival/destruction, e.g., pH);
  - Types of sludge (unstabilized versus stabilized, primary versus secondary, etc.);
  - Wastewater treatment process performance data (as they affect sludge type, sludge age, etc.);
  - Quantity of treated sludge;
  - Sludge age; and
  - Sludge detention (retention) time.
- Process characteristics
  - Scale of the system (e.g., reactor size, flowrate);
  - Sludge feed process (e.g., batch versus continuous flow);
  - Organic loading rate (e.g., kg volatile solids/$m^3 \cdot d$);
  - Operating temperatures (including maximum, minimum, and mean values);
  - Operating pressures if greater than ambient;
  - Type of chemical additives and loading rates;
  - Mixing (include number of mixers and horsepower ratings; if gas mixing is used, a complete description of how it is applied);
  - Oxygen/air requirements for biological systems;
  - Duration and type of storage; and
  - Odor control equipment.
- Climate (at the location of the sludge sampling/testing)
  - Ambient temperature (average and range) over the course of sampling;

- Ambient seasonal temperatures (averages and ranges);
- Precipitation (seasonal averages and the average over sampling period);
- Humidity (both seasonal averages and the average over sampling period);
- Solar radiation;
- Wind speed, direction, and fetch, if applicable; and
- Possible affects parameters may have on the efficiency or reliability of the process, including supporting data if necessary.

4.2 Supporting Literature and Data for a Process

Supporting literature for specific processes may be available. For example, if the process uses a combination of treatment techniques that are known to affect disinfection, then citations from peer reviewed publications may be submitted in support of its potential effectiveness. Also, information regarding trials of the process used on other waste streams may be of value in determining if the process could be used to effectively disinfect sludges. Whenever possible, supporting documentation should be provided from unbiased sources. Additionally, bench- or pilot-scale data that demonstrates the process' ability to disinfect sludge may be collected and submitted as proof of concept.

4.3 Preparation and Implementation of a Quality Assurance Project Plan

A quality assurance project plan (QAPP) must be developed before initiating testing so that the desired quality in sample collection, laboratory analysis, data validation and reporting, and documentation and record-keeping is achieved and maintained. A QAPP is a written document that provides a blueprint for the entire project and each specific task to ensure that the project produces reliable data that can be used to meet the project's objectives and goals. The QAPP should be reviewed and approved by the PEC before implementation. The information that should be contained in a QAPP is discussed below.

Most of the information included on the QAPP is essential to the final application. Once sampling and analysis is completed, only minor updating to the information is required, as noted, to convert sections of the QAPP into sections of the application itself.

4.3.1 *Developing Process-Specific Performance Data*

Some thought must go into the experimental approach and sampling procedures. In addition to the results from sampling, the PEC will evaluate the overall design including representativeness of the samples and adequacy of the sampling techniques. The following key points of the experimental approach (using diagrams and tables where applicable) should be described:

- Approach and the test conditions for each experimental phase;
- All details of the experimental design and sampling strategy;
- If spiking is necessary, then detailed description of the spiking method (see section 4.3.4);
- Location of sample collection; sampling points should correspond to the beginning and end of the treatment process;
- Frequency of sampling/monitoring events and the number of samples for each sample type and location, including quality control and reserve samples;
- All measurements or parameters of interest for each sample—analytical (e.g., chemical, microbiological, assays); physical (e.g., temperature); and process (e.g., flowrate); and
- Evidence demonstrating that the design/strategy was appropriate to support process equivalency (i.e., a description of the statistical method or scientific rationale used to select sample sites, number of sampling events, number of replicates per sample, and sample timing).

The experimental design should include sampling under all possible operating conditions, particularly least favorable, such as wintertime.

4.3.2 *Data Quality Objectives*

The quality assurance objectives for each analyte/organism and for the project as a whole should be discussed. The objectives include accuracy (matrix spikes); precision (agreement of replicate analyses); detection limits; completeness for critical measurements; and any split sampling events or project-specific quality assurance criteria. Objectives should also include maximum holding times or the time that elapses between sample collection and analysis. Holding time requirements can be found in the analytical procedures for each microbiological contaminant.

In the final application, this section should include a comparison of the collected data to predetermined quantitative acceptance criteria for each specified quality assurance objective. If at any time the objectives were not met, then the discussion should also include a description of the corrective actions performed. Tabularized summaries are encouraged.

The quality control measures to be taken for each analyte/organism and for the project as a whole (e.g., negative and positive controls, field blanks, laboratory blanks, etc.) should be discussed. This section in the final application should include a comparison of the collected data to predetermined quantitative acceptance criteria for each quality control measure. If at any time the quantitative acceptance criteria were not met, then the discussion should also include a description of the corrective actions performed. Tabularized summaries are encouraged.

Qualifications of the intramural or extramural laboratory staff should be described. Their experience in analyzing sludge samples for the organisms/analytes included in the study should be presented.

4.3.3 *Sampling Technique*

Several following key points of the sampling procedures to be employed should be described (including tables where applicable):

- General sampling monitoring procedures,
- Sampling methods to be used (i.e., grab versus composite); if composite, then how individual samples will be taken and how many samples will be compiled;
- Sampling equipment to be used, if any, and the measures taken to ensure cross-contamination between samples is avoided;
- Sample sizes and sub-sampling procedures, if any;
- Sample containers to be used (type, cleaning, and pretreatment, if any);
- Sample handling, preservation, packaging, and transportation or shipping procedures;
- Sample holding-time requirements;
- Representativeness of the sampling procedure:
    - Steps that will be taken to ensure samples are representative of sludge as it enters and exits the disinfection process; for example, whether processing time was included;

- Schematic of the pile that shows where samples were taken if taken from a pile;
- Timing of sample collection related to important process parameters that could affect results (e.g., turning over, beginning of drying).

4.3.4 *Need for and Implementation of Spiking*

As the criteria in Table 15.8 shows, the ability of a process to significantly or further reduce the number of specific microbial indicators is required if that process is to be recommended as a PSRP or PFRP process. Some sludges, however, may not contain a sufficient number of enteric viruses (≥ 1000/ 4 g TS dw) or viable helminth ova (≥ 100/ 4 g TS dw) to demonstrate effectiveness. In such cases, it is necessary to add sufficient numbers of the organisms of interest to the sludge before processing. When such augmentation (or spiking) is required, then it is important to describe how the organisms are to be obtained and how they will be added and mixed with the sludge. A representative amount of the spiked sludge must be sampled following incorporation, but before it is processed through the disinfection system for proof that the anticipated organism levels were achieved. The spiking description should include:

- Source of spiking stock (it is recommended that *Ascaris* eggs for this purpose be obtained from fecal discharges of humans or pigs because they will have developed maximum hardiness);
- Any techniques to be used in preparation and storage of the spike;
- Method of adding the spike to the process feed (include when and where with respect to the treatment process, amounts to add, target concentrations, method of ensuring uniform incorporation, etc.).

Table 15.8 Stability assessment methods.

Process	Monitoring methods
Composting	CO_2 respiration, O_2 uptake
Heat drying	Moisture content
Alkaline treatment	pH; pH change with storage; moisture; ammonia evolution; temperature
Aerobic digestion	Specific oxygen uptake rate; volatile solids reduction, additional volatile solids reduction
Anaerobic digestion	Gas production; volatile solids reduction, additional volatile solids reduction

4.3.5 *Analytical Methods*

The analytical methods to be used must be identified; standard or U.S. EPA-approved methods should be identified by name and number. Any nonstandard methods should be described in detail including standard operating procedures for implementation. Methods preferred by the PEC are published in *Environmental Regulations and Technology: Control of Pathogens and Vector Attraction in Sewage Sludge* (U.S. EPA, 2003). In the final application, any deviation from predetermined methods must be reported and discussed. Acceptable analytical methods for microbial indicators listed in Table 15.8 and related methods include those listed below.

Fecal coliform (either U.S. EPA method is preferred):

- **U.S.** EPA Method 1680: Fecal Coliforms in Sewage Sludge (Biosolids) by Multiple-Tube Fermentation using Laurel-Tryptose Broth (LTB) and EC Medium (U.S. EPA, 2006a);
- U.S. EPA Method 1681: Fecal Coliforms in Sewage Sludge (Biosolids) by Multiple-Tube Fermentation using A-1 Medium (U.S. EPA, 2006b); and
- *Standard Methods for the Examination of Water and Wastewater,* 9221 E: Multiple-Tube Fermentation Technique Fecal Coliform Procedure, Used in Conjunction with Control of Pathogens and Vector Attraction in Sewage Sludge, Appendix F: Sample Preparation Fecal Coliforms in Sewage Sludge (Biosolids) by Multiple-Tube Fermentation using A-1 Medium (American Public Health Association et al., 2005).

Salmonella spp.:

- U.S. EPA Method 1682: *Salmonella* in Sewage Sludge (Biosolids) by Modified Semisolid Rappaport-Vassiliadis (MSRV) Medium (U.S. EPA, 2006c) and
- Analytical Method for *Salmonella* sp. Bacteria Used in Conjunction with Control of Pathogens and Vector Attraction in Sewage Sludge, Appendix F: Sample Preparation for Fecal Coliform Tests and *Salmonella* sp. Analysis (U.S. EPA, 2003).

Enteric viruses:

- Appendix H: Method for the Recovery and Assay of Total Culturable Viruses from Sludge (U.S. EPA, 2003).

Viable helminth ova:

- *Control of Pathogens and Vector Attraction in Sewage Sludge,* Appendix I: Test Method for Detecting, Enumerating, and Determining the Viability of *Ascaris* Ova in Sludge (U.S. EPA, 2003).

Percent total and volatile solids:

- U.S. EPA Method 1684: Total, Fixed, and Volatile Solids in Water, Solids, and Biosolids (U.S. EPA, 2001);
- *Standard Methods for the Examination of Water and Wastewater,* Methods 2540 B & E: Total Solids Dried at 103–105 °C and Fixed and Volatile Solids Ignited at 500 °C or Method 2540 G: Total, Fixed, and Volatile Solids in Solid and Semisolid Samples (APHA et al., 2005).

Specific methods for the analysis of optional and surrogate indicator organisms are not mandatory unlike the organisms listed above that are specified for microbial compliance monitoring of biosolids. Acceptable protocols for analysis of microbial indicator organisms and specific pathogenic microorganisms have been developed for water, wastewater, soils, foods, and other matrices. These protocols have been incorporated into compendiums of industry standard assays or are defined as U.S. EPA-approved methods. Specifically, the references listed below identify methods which may be useful for analysis of biosolids.

- *Standard Methods for the Examination of Water and Wastewater* (American Public Health Association et al., 2005).
- *Compendium of Methods for the Microbiological Examination of Foods* (American Public Health Association, 2001).
- *Bacteriological Analytical Manual* (U.S. Food and Drug Administration, 1998).
- EPA Microbiology Home Page http://www.epa.gov/nerlcwww/

Caution should be used when selecting and using methods for alternate or surrogate indicator organisms because none of these methods have been subject to multi-laboratory validation studies for sludge or biosolids.

4.4 Analytical Results

Following approval and implementation of the QAPP, the final application must provide a discussion of the results. Summaries of the analytical results in table or graphic form should be provided to support discussions and interpretations. All tables and graphs should be appropriately labeled with headings (tables) or captions (figures). Each table or graph should be presented with a detailed description and adequate interpretation of the data within the report text. Examples of results that should be presented in detail in this section are presented below.

Monitored characteristics—results of any monitored process parameters, including

- Means or averages of process parameters;
- Ranges and stability/variability in the data collected;
- Comparison of observed data to expected data;
- Any trends observed (e.g., diurnal temperature swings); and
- Any upsets and what was done to correct the situation.

Physical characteristics of sludge used in testing the treatment process, before and after treatment, including

- Sludge type (unstabilized versus stabilized, primary versus secondary, domestic versus industrial, activated sludge treatment process, etc.);
- Sludge age;
- Sludge quantity as compared to sample size;
- Chemical characteristics of the untreated sludge (as they affect pathogen survival/destruction);
- Possibility of the formation of harmful byproducts;
- Temperature of sludge before and after treatment (both at time of sampling and analysis);
- Moisture/total solids content of sludge before and after treatment;
- Total suspended solids of sludge before and after treatment;
- Volatile solids content of sludge before and after treatment; and
- Other characteristics that may affect the efficiency of the treatment process.

Microbiological characteristics—results of all microbiological analyses including

- Distribution of both influent and effluent sludge samples and the variability experienced (minimum, maximum, variance);

- Any exceedances in the target effluent concentrations (number and percentage of samples); and
- Log reductions (mean, minimum, maximum, variance).

All results must be presented in the appropriate units—fecal coliform as number per gram of total solids (dry weight); *Salmonella* spp., enteric viruses, and viable helminth ova as number per 4 grams of total solids (dry weight).

4.5 Conclusions

The following should be included in the final application in a conclusions section:

- Type (PSRP or PFRP) and scope (site-specific or national) of equivalency being sought;
- Summary of all supporting information from the literature on why the process qualifies for equivalency;
- Summary of microbiological data that supports consistent achievement of the target log reductions and effluent concentrations;
- Review of particular aspects of the process that contribute to pathogen reduction and summary of the monitoring data that support expectations of consistent operation of the process;
- Discussion of why the effectiveness of the process is expected to be independent of the location of operation if national equivalency is being sought.

4.6 Use of Statistics

Microbiological data is typically log-normally distributed, and a basic assumption for the common average is that the underlying population is normal. Thus, all averages of microbiological data should be geometric averages (also known as a geometric or log mean).

- Microbial data is known for its relatively high variability. Statistics, in particular confidence intervals, should be used for comparison of the microbial results to the target log reductions. The confidence interval around a mean is a common statistic found in spreadsheet tools and statistical software and described in basic statistics texts. Similar to the arithmetic mean, confidence intervals assume the underlying population is normal. Therefore, confidence intervals around a mean should only be used for log-transformed microbiological data.

- In a statistical analysis, zeros do not provide useful data. If an assay results in a zero, then the applicant must determine the detection limit for the assay and report less than that number. In cases where the data is below the detection limit, then it is typical to include the point in the analysis as one-half the detection limit (e.g., < 5 = 2.5).

5.0 REQUIREMENTS FOR REDUCING VECTOR-ATTRACTION REDUCTION

Sludges that are subject to significant biodegradation and can be a food source will attract vectors such as insects, rodents, and birds. If vectors come in contact with sludges and then humans, pathogens remaining in the sludge and/or with the vectors can become a public health risk because of the increased risk of transmission. [Furthermore, sludge decomposition can generate odors that most people find objectionable. When these odors are derived from a biosolids recycling or storage facility, attention by the public will result in investigation by regulators, public health officers, and political officials. Even when no reasonable public health risk exists, the probability of the facility being the subject of a lawsuit or shutdown because of odors is high. Thus, reducing the attractiveness of sludges to vectors not only reduces public health risk because of potential for disease transmission, but also minimizes the release of objectionable odors and the probability of legal action.

The U.S. EPA's Part 503 Rule provides specific requirements for sludges that are stored, recycled, or disposed of in the environment to avoid environmental and public health problems. These requirements are based on reducing the attractiveness of sludge to vectors through biological stabilization, adjusting physical or chemical properties to minimize pathogen growth or survival, or isolation of the sludge from contact by vectors or the public. The Part 503 Rule contains 11 options for demonstrating reduced vector attraction for sludges (U.S. EPA, 1993).

Of these 11 options for achieving VAR, only biological stabilization irreversibly reduces attractiveness of sludge to vectors. Chemical and physical stabilization reduces attractiveness temporarily; and isolation within the soil reduces vector and public access to the sludge. Biological stabilization, however, will continue within the soil ecosystem.

At this time, equivalent options that satisfy the requirements of VAR do not exist. The U.S. EPA encourages submission of additional measures that may be equivalent to one or more of the eight options for biological stabilization or chemical or physical stabilization.

5.1 Biological Stabilization and Vector-Attraction Reduction

The U.S. EPA's regulations cause engineers to think of stabilization in relation to VAR, pathogen reduction (PFRP or PSRP), and regrowth. From a regulatory and public health standpoint, this is a logical connection. From an engineering design and process-control perspective, however, stabilization is a biological aerobic process to convert some percentage of biodegradable (volatile) solids (BVS) into carbon dioxide and water or an anaerobic process that converts BVS into methane and carbon dioxide. In the case of high-temperature oxidation (i.e., thermal conversion), all volatile solids are converted (destroyed) into carbon dioxide and water, leaving only an inert but oxidized ash. Suitable methods to directly quantify stabilization are not available from U.S. EPA. Nonetheless, VAR can be at least partially accomplished by using one of the following methods:

- Biological processes that break down BVS, reducing the available food nutrients for microbial activities and the odor producing potential of the solids;
- Chemical or physical conditions that stop microbial activity (in some cases only temporarily); or
- Physical barriers between vectors and volatile solids in the sludge.

Stabilization and a reduction in vector attraction are not necessarily the same. The term "stability" typically is used to describe a biochemical quality of sludge and its resistance to biochemical change. Although the term is associated with VAR, stability is not regulated by U.S. EPA. For sludge, stability is typically defined as the point at which food for rapid microbial activity is no longer available. In other words, rapid biochemical change in the material is significantly diminished, typically leading to reduced odor potential.

Stabilization, ideally, should be a nonreversible process, as are all natural processes according to the second law of thermodynamics. During stabilization, biosolids are degraded into a more chemically and biologically inert material. Certain processes such as composting and thermal destruction are nonreversible within the geological timescale. In composting, BVS are oxidized microbially into carbon dioxide and water. Combustion also irreversibly converts volatile solids into carbon dioxide and water. Once such processes have taken place, the process is not reversible.

Protocols for quantifying stability have not been published; however, it is clear by odor production or vector attraction when sludge is not well-stabilized. Nonetheless, several processes, such as composting and digestion, are known to irreversibly increase the overall stability of biosolids because of conversion of BVS into gaseous products such as water, carbon dioxide, and methane. Table 15.8 identifies some of the methods used for assessing stability for a given treatment process.

An example of an existing technology is discussed below. The technology, the "Chicago process," is a series of conventional, low-tech biosolids unit processes that was documented through a three-year Water Environment Research Foundation study to achieve both PFRP and VAR (Farrell et al., 2004). It is a non-reversible process.

In contrast, reversible processes, such as heat drying and alkaline treatment, may only temporarily reduce VAR. Once moisture content has increased (i.e., heat-treated biosolids) or the alkaline pH has diminished (i.e., alkaline-treated biosolids) and microorganisms will grow and the material will again have the potential for biochemical change and attraction of vectors because of the availability of food or generation of odors. Microbial regrowth does not, however, necessarily indicate either the presence or growth of pathogens.

5.2 Vector-Attraction Reduction

The VAR options are discussed in the 503 Rule (U.S. EPA, 1993). The key point is to understand that VAR and stability are not synonymous—achieving VAR does not mean stability has been achieved. In contrast, achieving stability will achieve VAR. Options 1 through 5 for VAR depend on the action of microorganisms and their ability to reduce biodegradable organic material in the sludge. Although the concept is good, it is essential that the biodegradable organic matter be reduced to a low enough level that once the sludge is used in a beneficial manner, like land application, little or no further degradation is possible. Unfortunately, conditions like time and temperature are not always optimum for this to occur under the conditions of the Part 503 Rule. For example, it may be possible to achieve 65% volatile solids reduction during either anaerobic or aerobic digestion, but the Rule says it is only necessary to achieve 38% to meet the VAR requirement.

For processes that can use option 5 (e.g., composting), 14 days typically is not enough time for adequate biostabilization. Some states have required additional time, typically where active composting plus curing total 40 to 50 days.

In contrast, heat drying removes water and, through time, temperature, and desiccation, kills living organisms. Biological and chemical oxidation is minimal. In contrast to compost, heat-dried material consists of dehydrated soup. The product, only temporarily stabilized by drying, is reversible. With wetting, the condition for biodegradability (i.e., available food) of the product increases. Invasion of microbes will initiate biological decomposition, and stabilization will continue along the path to inertness.

5.3 Types of Stabilization Processes

Nonreversible processes are considered truly stabilized and include

- Thermal reduction processes that reduce volatile solids to ash;
- Pressure, heat, and oxidation processes (for example, Zimpro or Cambi), which need to be run in a complete oxidation mode so that all organic matter is destroyed and a highly degradable broth is not created;
- Biological stabilization processes (for example, the Chicago process, composting, and vermicomposting); and
- Incineration or formation of glass (such as the Minergy process).

Biological stabilization processes can demonstrate a reasonable extent of volatile-solids reduction and movement of material along the stabilization curve toward inertness.

Pathogen reduction during biological stabilization (e.g., composting or air drying), is a separate but, typically, simultaneous step. The influence of heat, time, pH, ammonia, UV light, desiccation, scarcity of food, and microbial competition on pathogen reduction are reasonably well understood. On the other hand, the biological basis for pathogen and vector attraction reduction and during vermicomposting is not clear. Some evidence suggests the process may reduce pathogens by processes other than heat or pH. Time, desiccation, microbial activity, and microbial competition appear to be important.

Reversible processes may require subsequent biological stabilization or operational standards to isolate the material from the public or to prevent physical changes in the material (such as rewetting of the product after drying or decrease in pH). Examples of processes that achieve pathogen reduction but not biological/permanent stabilization include

- Disinfection processes based on changes in pH or addition of chemical agents that are designed primarily to kill pathogens;

- Heat-drying processes that evaporate water and temporarily stabilize the material until it is rewetted and reinoculated by microbes; and
- Ionizing radiation, which disrupts DNA and RNA of bacteria, viruses, and other pathogens.

5.4 Stabilization as a Nonreversible Process

The second law of thermodynamics holds that entropy of an isolated system always increases. Energy once lost cannot be recaptured and used to reverse the process. Following this law, during stabilization, biosolids are degraded into a more biochemically inert material.

Biodegradation is a continuous process that will continue as long as the environmental conditions allow for the survival of some bacterial species. The biodegradation rate of organic matter is determined by environmental conditions: temperature, microbial population, nutrient balance, moisture content, and presence of toxic materials. For example, the time and temperature requirement to achieve VAR, as expressed in 40 CFR Part 503, option 5, assumes that the environmental conditions are suitable to support a biodegradation rate typical of well-composted material. The goal of the VAR requirement is to ensure that such a biodegradation rate occurs for a minimum of 14 days. This minimum rate is operationally defined as maintaining a minimum temperature and an average temperature. Thus, the intent of VAR option 5 is to ensure that a certain *quantity* of stabilization or volatile-solids reduction has occurred. Quantity of stabilization is a function of biodegradation rate multiplied by the time during which biodegradation has occurred:

$$\text{Quantity of stabilization} = f\,(\text{biodegradation rate} \times \text{Time}) \tag{15.3}$$

It follows that if the biodegradation rate is decreased, then a proportional increase in time will result in the same quantity of stabilization.

In addition, biodegradation is continuous, although the process may slow when the temperature drops below the regulated temperature. If the VAR requirement is met by 14 consecutive days, then the quantity of stabilization could be written as:

$$\text{Quantity of stabilization} = \int_0^{14\text{ days}} (\partial s/\partial T)\, dt \tag{15.4}$$

Or, more typically,

$$\text{Quantity of stabilization} = \int_0^{t} (\partial s/\partial T)\, dt \tag{15.5}$$

where
$\partial s/\partial T$ = the rate of stabilization with respect to temperature and t = time during which the stabilization process has occurred.

Graphically, the quantity of stabilization is the area under the time–temperature curve as shown in Figure 15.2. It follows that the quantity of stabilization (i.e., quantity of volatile solids destroyed) is more accurately related to the area under the complete time–temperature curve than to number of consecutive days at the regulated time and temperature. As shown in Figure 15.2a, the quantity of stabilization (i.e., the area under the curve) may be greater if the days satisfying the VAR rule are not

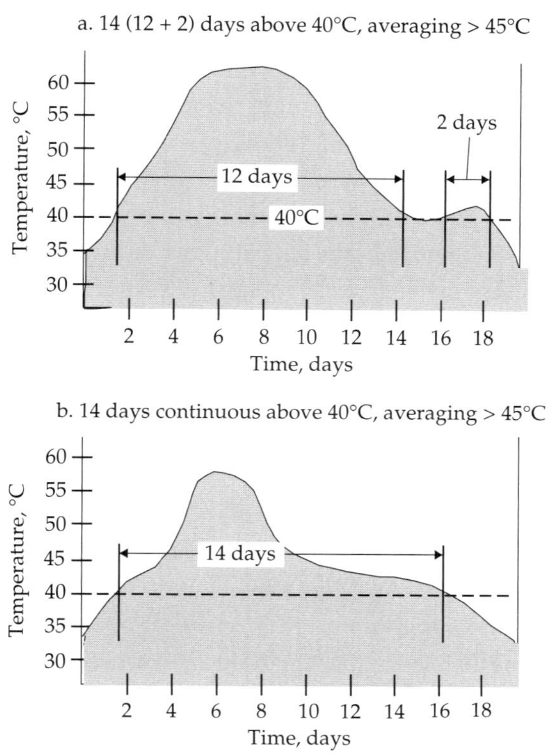

FIGURE 15.2 Quantity of stabilization = f (time × temperature), the area under the curve.

continuous because the degradation process continues, even when the temperatures are lower than the regulatory requirement.

5.5 Reversible "Stabilization" Processes

In contrast with biological stabilization, processes such as microwave treatment, electron irradiation, or chemical (alkaline or acid) treatment can kill pathogens but only have a temporary or insignificant effect on biological stabilization.

For example, the MagnaGro (Magna Management Inc., Carson City, Nevada) process kills pathogens and disinfects with a fumigant, but does not biologically stabilize the solids. The N-Viro process disinfects because of the effects of pH, temperature, and ammonia. The Neutralizer process uses ClO_2 as a primary disinfectant and nitrous acid as a secondary disinfectant. Burch BioWave (Burch BioWave, Inc., Fredericktown, Ohio) uses microwaves to heat the material and disrupt cell membranes. The Eco-Therm process (Ashbrook Simon-Hartley, Houston, Texas) heats the material to satisfy pasteurization temperature and time requirements but does not appreciably move the material along the stability continuum.

Such disinfection processes—whether operating by heat, electrons, fungicides, microwaves, or high pH—were designed primarily to produce a pathogen-free product and meet 40 CFR Part 503 requirements for pathogen reduction by creating an adverse environment for microorganisms to exist and grow; VAR must then still be satisfied.

5.6 Alternative Approaches to Vector-Reduction Attraction and Stabilization

For decades, before promulgation of 40 CFR Part 257 in 1979, digested sludge was either lagooned, air-dried on outdoor drying beds, or stockpiled for lengthy periods to facilitate further stabilization and recycling (U.S. EPA, 1979). Stockpiled sludge was located for easy access to local gardeners and landscapers. Casual evidence suggests that this sludge was not considered by users to be highly infectious, although the potential for and quite possibly instances of infection and illness certainly existed. To better safeguard public health, pathogen destruction and related data were gathered and reviewed (Farrell et al. 2004). Part 257 and 503 standards were based on such data. The rules eliminated unregulated distribution of sludge from practices now considered to inadequately destroy pathogens. Nonetheless, experience suggested such a regulated series of treatment and stabilization processes (including lagooning, air drying, and long-term storage) had the potential to effectively reduce pathogens and stabilize the sludge, rendering the biosolids product safe for use. One city developed a research program to quantitatively assess the potential for both pathogen and vector-attraction reduction using a series of low-tech unit processes. Although no single process may have been adequate, combining several in series took advantage of anaerobic/aerobic processes, desiccation and ultraviolet light, and microbial competition to achieve the requirements of PFRP and VAR.

5.6.1 *Metropolitan Water Reclamation District of Greater Chicago*

For many years, the Metropolitan Water Reclamation District of Greater Chicago (MWRD) practiced a series of conventional, low-tech sludge treatment processes that included anaerobic digestion, lagooning, and air drying on outside drying beds. A three-year, comprehensive, Water Environment Research Foundation (WERF) study of the stabilization and pathogen kill achieved suggested that the Chicago process train would consistently achieve production of a Class A biosolids product (Farrell et al., 2004). This practice would enable beneficial use of the Class A product in agriculture and other uses. Table 15.9 shows the operating parameters for the Chicago operation. The process train and characteristics of the product were submitted to the U.S. EPA's PEC and Region V Office for review. The conclusion drawn from the WERF study, with some limitations and additional requirements, was that all batches of the final air-dried product complied with Class A and VAR criteria, as specified by 40 CFR Part 503 regulations.

5.7 Process Descriptions and Pathogen Reduction

The process assessment described in part in the WERF publication by Farrell et al. (2004) is shown in Figure 15.3. The MWRD treatment train serves as a model for

TABLE 15.9 Codified operating parameters of biosolids processing at the Metropolitan Water Reclamation District of Greater Chicago.

Process step	Parameter	Criteria (codified)
Anaerobic digestion	Detention time	≥ 20 days
	Digestion temperature	35+2 °C (95 ±3.6 °F)
Lagoon aging	Holding time	≥18 months
Air-drying	Application depth	≤ 46 cm (18 in.)
	Agitation frequency	Average ≥ 3 times/week
	Solids content	≥ 60%

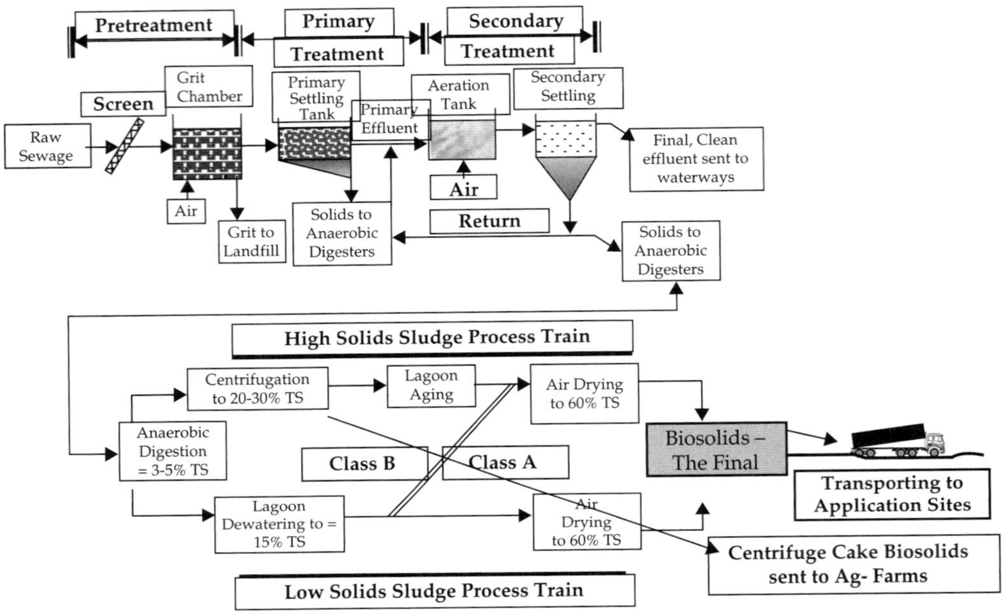

FIGURE 15.3 Wastewater treatment and Biosolids processing train at the Metropolitan Water Reclamation District of Greater Chicago.

combined lagoon-storage and air-drying systems. The MWRD's lagoons operate with a minimum of 18 months of unfed storage time. The air-drying systems that follow are operated during the warm-weather drying season to produce a greater than 60% solids product within four to six weeks. Windrows are turned three to four times a week. Treatment times for other coupled systems may be shorter in warmer climates and longer in colder climates. The operating conditions of the MWRD solids processing trains are summarized in Table 15.9. Treatment processes and probable pathogen reduction mechanisms are described in the following sections.

5.7.1 *Storage Lagoons*

Anaerobically digested biosolids in the form of centrifuge cake (20 to 30% solids) or liquid (3 to 6% solids) are placed in the storage lagoons either continuously or intermittently. The solid materials settle to the lagoon bottom, and supernatant is typically returned to the WWTP. After the lagoons fill with biosolids, they are left unfed for at least 18 months, allowing for further biosolids decomposition and pathogen reduction. Mechanisms that may be used to reduce pathogens of concern in lagoon storage systems are a continuation of catabolic activity of the digestion process, thermal effects, suppression by nonpathogenic bacteria, and chemical action by ammonia or organic acids.

5.7.2 *Air Drying*

The lagoon-aged biosolids are dredged from the lagoons and placed on asphalt-paved drying pads for air-drying operations. The rate of air drying is facilitated by mechanical-agitation using horizontal augers or rototillers. When the biosolids have dried sufficiently (solids content reaches 30 to 35%), the mechanical-agitation process mounds the biosolids into windrows approximately 30 cm (1 ft) high. Water evaporates from the windrows. The windrow-turning machine uses a front-mounted horizontal auger to accelerate drying by introducing dry air to the windrow, releasing moist air, and replacing dry particles on or near the surface with moist particles from the interior. Turning also provides oxygen that oxidizes remaining biodegradable material, sometimes generating enough heat to elevate windrow temperature. Higher-than-ambient temperatures speed the process and accelerate pathogen reduction. Desiccation may also play a part in pathogen reduction. The presence of nonpathogenic bacteria and chemical action can further suppress pathogen activity.

Alternatively, liquid biosolids can be dried on porous media, typically sand beds. Liquid biosolids are applied to the beds at depths up to 30 cm (1 ft). The beds drain and dry over a period of one to several months, depending on ambient conditions. Typically, biosolids are left on sand beds until their solids content is approximately 30 to 50%. Then the material can be readily removed. In humid climates, additional drying is likely to be slow. If a Class A product is desired, then biosolids may need to be removed from the bed and further dried in windrows. In dry, hot climates, however, biosolids can remain on the beds, shrinking to a thin layer of high solids content. High pathogen-kill levels can be reached by heating this thin layer completely through to high temperatures on a sunny day, particularly if the biosolids rest on a dark-colored surface.

5.7.3 *Cake Storage*

Cake storage is similar to air drying on impervious pads; except the windrows tend to be larger and windrow-turning is minimal. The cake is stored until pathogen densities have dropped to Class A levels. There is no intent to produce a dried product, although some drying may occur. Pathogen-destruction mechanisms are the same as for windrow drying, except desiccation plays a minor role.

5.8 Process to Further Reduce Pathogens Equivalency: Class A Alternative 6

Class A Alternative 6, as discussed in Section 2.5, allows processes that have been determined to be equivalent to a PFRP to be used to meet Class A requirements. Pathogen-reduction data collected over the years by the Chicago MWRD showed that the process train used routinely and exclusively at two of its WWTPs could produce a Class A product. If this process train was approved by the U.S. EPA PEC or Region V Office, then MWRD could eliminate or reduce analyses for enteroviruses and viable helminth eggs and still ensure safe distribution of a Class A biosolids.

5.8.1 Metropolitan Water Reclamation District of Greater Chicago Low-Tech Class A Protocol

The Chicago MWRD applied for PFRP equivalency for sludge-treatment trains at its Calumet and Stickney WWTPs. Virus and helminth densities at these plants are very low, even in raw (untreated) sludge. Table 15.10 provides example pathogen die-off data.

The PEC requested that MWRD conduct large-scale tests to demonstrate 3-log reductions of enteroviruses and 2-log reductions of helminth ova. This demonstration would require spiking the sludge stream with these pathogens. The MWRD, however, felt it was impractical and possibly dangerous to spike viruses and helminth ova into a large-scale test. Instead, it used a statistical approach to estimate helminth and virus inactivation within its sludge-treatment trains in the case of a surge in density of these particular pathogens (Tata et al., 2000). Eventually, MWRD was awarded site-specific equivalency for its Calumet and Stickney plants by the U.S. EPA's Region V Office. This approval reduced (but did not eliminate) virus and helminth monitoring. Specifically, the MWRD was required to monitor for enteric virus and helminth ova monthly for the first year after receiving certification and once every other month after that, as long as it operates its sludge treatment processes according to the process and operating protocols specified in its application (Traub, 2002). This is the only known U.S. EPA approval for PFRP equivalency for sludge treatment trains involving these low-technology processes.

TABLE 15.10 Reduction of pathogens and pathogen indicators in Calumet and Stickney sludge treatment systems (HSSPT = high-solids sludge processing train; LSSPT = low-solids sludge processing train; PFU = plaque-forming units; MPN = most probable number) (Tata et al., 2000).

		Pathogen density			Log_{10} reductions		
System	Organism	Digester Draw	Lagoon draw (codified[a])	Air-dried final product (codified)	Lagoon	Air-drying system	Overall reduction
Calumet HSSPT	Fecal coliforms MPN/g DS[b,c]	6.78×10^4	105	5.58	2.81	1.27	4.08
	Salmonella sp. MPN/4 g DS[b] Virus,	74.6	0.828	0.419	1.95	0.3	1.87
	PFU/4 g DS[d]	0.428	0.00	0.00	–	–	–
	Viable helminth ova number/4 g DS[d]	0.286	7.44×10^{-3}	1.54×10^{-3}	1.58	0.68	2.26
Calumet LSSPT	Fecal coliforms MPN/g DS *Salmonella*	6.78×10^4	68.5	17.1	2.99	0.60	3.59
	sp. MPN/4 g DS	74.6	2.9	0.673	1.41	0.63	2.05
	Virus PFU/4 g DS	0.428	0.00	0.00	–	–	–
	Viable helminth ova number/4 g DS	0.286	0.625	3.75×10^{-3}	0.66	1.22	1.88
Stickney HSSPT	Fecal coliforms MPN/g DS *Salmonella*	2.02×10^5	188	18.3	3.03	1.01	4.04
	sp. MPN/4 g DS	25.7	0.278	0.506	1.96	(0.26)	1.70
	Virus PFU/4 g DS	0.121	0.00	0.00	–	–	–
	Viable helminth ova number/4 g DS	0.379	0.784	0.0191	(0.31)	1.61	1.30
Stickney LSSPT	Fecal coliforms MPN/g DS *Salmonella*	2.02×10^5	151	25.6	3.13	0.77	3.90
	sp. MPN/4 g DS	25.7	0.533	0.532	1.68	0.00	1.68
	Virus PFU/4 g DS	0.121	0.00	0.00	–	–	–
	Viable helminth ova number/4 g DS	0.379	0.0394	0.0132	0.58	0.48	1.46

[a] Codified means operating at conditions as stringent or more stringent than specified by the operating protocol.

[b] Geometric mean.

[c] DS = dry solids.

[d] Arithmetic mean.

5.8.2 *Low Technology Class A Disinfection and Vector-Attraction Reduction or Stabilization Process Guidelines*

The product from an ideal, low-technology, sludge treatment process, or combination of processes can satisfy Class A for pathogens and VAR by fulfilling the following five requirements (Farrell et al., 2004):

1. Processes that precede the low-technology process(es) must significantly reduce pathogen densities and the bacterial food supply. They must also develop and sustain a healthy population of nonpathogenic bacteria.
2. The treatment process(es) must be of sufficient intensity and duration to ensure requisite pathogen kill. Helminth ova are typically more resistant to treatment than other regulated organisms (fecal coliform, salmonellae, and viruses). Providing the conditions needed to inactivate helminth ova should ensure satisfaction of Class A criteria for these other organisms.
3. The overall treatment train must satisfy VAR needs.
4. Bacterial regrowth must be prevented.
5. Treatment must be uniform to ensure that all biosolids have been properly treated.

Guidelines for individual processes, separate and combined, are summarized below. For more details, refer to Farrell et al. (2004).

5.8.3 *Process Description*

5.8.3.1 *Lagoon Storage*

Lagoon storage begins with anaerobic digestion (at least 15 days solids retention time with mesophilic temperatures) or aerobic digestion. Sludge is stored, unfed, at 25 °C for 16 months or 25 months at 4°C.

5.8.3.2 *Air Drying or Cake Storage*

After anaerobic digestion (at least 15 days of solids retention time with mesophilic temperatures) or aerobic digestion, biosolids moisture content is reduced to less than 5% for desiccation-oriented air drying systems. For time- and temperature-dependent systems, biosolids should be air dried for at least 250 days in warm climates and at least 350 days in cold climates while greater than 40% moisture is maintained with periodic turning to ensure uniform treatment.

5.8.3.3 *Combined Lagoon-Storage and Air-Drying Systems*

Lagoon storage is typically used in combination with air drying for Class A applications (see Figures 15.4 and 15.5). Because organisms exit the lagoons in a debilitated

FIGURE 15.4 Metropolitan Water Reclamation District of Greater Chicago lagoons and drying beds.

FIGURE 15.5 Air drying on paved bed with lagoon aged biosolids that are then agitated three to four times per week using horizontal auger to achieve greater than 60% solids.

state, they are vulnerable to the environment of the air-drying process. This sequential combination is probably more powerful than lagoon storage or air drying alone, even when temperature conditions are the same. Therefore, guidelines for sequential lagoon-storage and air-drying systems are different from guidelines presented previously, which were for individual systems. The process is preceded by anaerobic digestion (at least 15 days solids retention time with mesophilic temperatures) or aerobic digestion. Unfed lagoon storage requires a minimum of 18 months, followed by air drying with periodic turning to produce a product (60% solids) within 9 to 10 weeks. Figure 15.5 provides the results of air drying on paved beds with lagoon-aged biosolids that are agitated three to four times per week using a horizontal auger to achieve > 60% solids.

Table 15.11 shows the volatile solids in processing and volatile solids reduction in Calumet and Stickney Water Reclamation Plant's air-dried final product during 2005 to 2009.

5.8.4 *Costs of the Chicago Process*

Costs reported by Bergman and Rohloff (1996) in Farrell et al. (2004) for Chicago's biosolids-treatment operations from1992 through 1995 are presented in Table 15.12.

Table 15.11 Volatile solids in processing and volatile solids reduction in the Metropolitan Water Reclamation District of Chicago's Calumet and Stickney plants' air-dried final product, 2005–2009.

Water reclamation plant		Digester feed	Digester draw	Air-dried final product (codified)	VSR[a]
		Volatile solids (%)			%
Calumet					
	Minimum	52.0	43.6	18.9	20.7b
	Mean	64.9	52.0	36.3	66.9
	Maximum	75.0	60.1	55.3	88.4
Stickney					
	Minimum	52.0	46.3	14.2	40.0
	Mean	61.4	51.9	33.2	61.7
	Maximum	71.6	58.6	47.0	88.4

[a]Volatile solids reduction calculated using the Van Kleeck equation.

[b]Vector attraction reduction demonstrated by additional anaerobic digestion test according to Option 2 of section 503.33(b)

TABLE 15.12 Average costs for combined lagooning/air-drying processes at the Metropolitan Water Reclamation District of Greater Chicago, 1992–1995 (HSSPT = high-solids sludge processing train; LSSPT = low-solids sludge processing train; DS = dry solids).

	Average Cost \$/ton DS[a]	
Operation	**LSSPT**	**HSSPT**
Lagoon dewatering	11.08	0.44
Lagoon outloading	14.09	10.00
Transport to drying	21.39	31.97
Air drying	28.57	32.73
Loading for transport to disposal	3.09	3.24
Transport to disposal	14.75	5.14
Total	92.99	85.83

[a] \$/Megagram DS = 1.10 × \$/ton DS.

6.0 REFERENCES AND SUGGESTED READINGS

Aitken, M. D.; Sobsey, M. D.; Blauth, K. E.; Shehee, M.; Crunk, P. L.; Walters, G. W. (2005) Inactivation of *Ascaris suum* and Poliovirus in Biosolids under Thermophilic Anaerobic Digestion Conditions. *Environ. Sci. Technol.*, **39** (15), 5804–5809.

American Public Health Association; American Water Works Association; Water Environment Federation (2005) *Standard Methods for the Examination of Water and Wastewater*, 21st ed.; American Public Health Association: Washington D.C.; American Water Works Association: Denver, Colorado; Water Environment Federation: Alexandria, Virginia.

American Public Health Association (2001) *Compendium of Methods for the Microbiological Examination of Foods*, 4th ed.; American Public Health Association: Washington D.C.

Bergman, W. A.; Rohloff, G. M. (1996) *Complete Costs for Biosolids Processing at the Metropolitan Water Reclamation District of Greater Chicago*; Water Environment Federation: Alexandria, Virginia.

Casemore, D. P. (1991) The Epidemiology of Human Cryptosporidiosis and the Water Route of Infection. *Water Sci. Technol.*, **24** (2), 157–164.

Cramer, W. N.; Burge, W. D.; Kawata, K. (1983) Kinetics of Virus Inactivation by Ammonia. *Appl. Environ. Microbiol.*, **45** (3), 760–765.

Farrell, J.; Merrill, D. T.; Shafer, P. L. (2004) *Producing Class A Biosolids with Low-Cost, Low-Technology Treatment Processes*; WERF project no. 95-REM-2; Water Environment Research Foundation: Alexandria, Virginia.

Higgins, M. J.; Chen, Y.; Hendrickson, D. H. (2008) *Evaluation of Bacterial Pathogen and Indicator Densities after Dewatering of Anaerobically Digested Biosolids: Phase II and III*, WERF project no. 04-CTS-3T; Water Environment Research Foundation, Alexandria, Virginia.

Keswick, B. H. (1984) Sources of Groundwater Pollution. In *Groundwater Pollution Microbiology*, Bitton, G.; Gerba, G., Eds.; Wiley and Sons: New York.

Levine, M. M.; DuPont, H. L.; Formal, S. B. (1973) Pathogenesis of *Shigella dysenteriae* (Shiga) Dysentery. *J. Infect. Dis.*, **127** (3), 261–270.

Meckes, M. C.; Rhodes, E. R. (2004) Evaluation of Bacteriological Indicators of Disinfection for Alkaline Treated Biosolids. *J. Environ. Eng. Sci.*, **3** (4), 231–236.

North East Biosolids and Residuals Association (2007) *A National Biosolids Regulation, Quality, End Use, and Disposal Survey—Final Report*; North East Biosolids and Residuals Association: Tamworth, New Hampshire.

Reimers, R. S.; Little, M. D.; Englande, A. J.; Leftwich, D. B.; Bowman, D. D.; Wilkinson, R. F. (1981) *Parasites in Southern Sludges and Disinfection by Standard Sludge Treatment*, EPA600152-81-166 Project Summary NTIS Report No.PB82-102344; U.S. Environmental Protection Agency: Washington, D.C.

Reimers, R. S.; Jeng, H. W.; Bowman, D. D.; Oleszkiewicz, J. A.; Meckes, M. C.; Fitzmorris, K. B. (2005) Development of Surrogates to Monitor Pathogens in Biosolids, *Proceedings of the Water Environment Federation/America Water Works Association Joint Residuals and Biosolids Management Conference*, Nashville, Tennessee, Apr 17–19; Water Environment Federation: Alexandria, Virginia.

Smith, J. E., Jr.; Millner, P. D.; Jakubowski, W.; Goldstein, N. (2005) Contemporary Perspectives on Infectious Disease Agents in Sewage Sludge and Manure. *Compost Sci. Util.*, Special Ed., 1–249.

Stern, G.; Farrell. J. B. (1977) Sludge Disinfection Techniques, *Proceedings of the National Conference on Composting of Municipal Residues and Sludges*, Washington, D.C.; Information Transfer Inc.. Rockville, Maryland.

Tata, P.; Lue-Hing, C.; Knafl, G. J. (2000) Statistical Evaluation of Pathogen Inactivation for a Conventional Low-Cost Technology Class A Biosolids Process. *Water Environ. Res.*, **72** (4), 423–431.

Traub, J. L. (2002) Letter from Jo Lynn Traub, Director, Water Division, U.S. Environmental Protection Agency, Chicago, Illinois, to Jack Farnan, General Superintendent, Metropolitan Water Reclamation District of Greater Chicago, Chicago, Illinois, June 20.

U.S. Environmental Protection Agency (1979) Part 257—Criteria for Classification off Solid Waste Disposal Facilities and Practices. *Code of Federal Regulations*, Part 257, Title 40, 1979.

U.S. Environmental Protection Agency (1981) *Density Levels of Pathogenic Organisms in Municipal Wastewater Sludge: A Literature Review*, EPA/600/2-81-170; U.S. Environmental Protection Agency, Municipal Environmental Research Laboratory: Cincinnati, Ohio.

U.S. Environmental Protection Agency (1985) *Health Effects of Land Application of Municipal Sludg*e, EPA/600/185/015; U.S. Environmental Protection Agency, Health Effects Research Laboratory: Research Triangle Park, North Carolina.

U.S. Environmental Protection Agency (1992) *Technical Support Document for Reduction of Pathogens and Vector Attraction in Sewage Sludge*, EPA/822/R-93-004; U.S. Environmental Protection Agency: Washington, D.C.

U.S. Environmental Protection Agency (1993) Standards for the Use and Disposal of Sewage, Final Rule. *Code of Federal Regulations*, Part 503, Title 40, 1993; *Fed. Regist.*, **58**, 9248.

U.S. Environmental Protection Agency (2001) *METHOD 1684: Total, Fixed, and Volatile Solids in Water, Solids, and Biosolids*, EPA-821/R-01-015; U.S. Environmental Protection Agency: Washington, D.C.

U.S. Environmental Protection Agency (2003) *Environmental Regulations and Technology: Control of Pathogens and Vector Attraction in Sewage Sludge*, EPA /625/R-92/013; U.S. Environmental Protection Agency: Washington, D.C.

U.S. Environmental Protection Agency (2006a) *Method 1680: Fecal Coliforms in Sewage Sludge (Biosolids) by Multiple-Tube Fermentation using Lauryl Tryptose Broth (LTB) and EC Medium*, EPA-821-R/06-012; U.S. Environmental Protection Agency: Washington, D.C.

U.S. Environmental Protection Agency (2006b) *Method 1681: Fecal Coliforms in Sewage Sludge (Biosolids) by Multiple-Tube Fermentation using A-1 Medium*, EPA/821-R/06/013; U.S. Environmental Protection Agency: Washington, D.C.

U.S. Environmental Protection Agency (2006c) *Method 1682: Salmonella in Sewage Sludge (Biosolids) by Modified Semisolid Rappaport-Vassiliadis (MSRV) Medium*, EPA/821/R-06/014; U.S. Environmental Protection Agency: Washington, D.C.

U.S. Environmental Protection Agency (2006d) *Emerging Technologies for Biosolids Management*. EPA 832-R-06-005; U.S. Environmental Protection Agency: Washington, D.C.

U.S. Environmental Protection Agency (2011) EPA Microbiology Home Page. http://www.epa.gov/nerlcwww/ (accessed July 2011).

U.S. Food and Drug Administration (1998) *Bacteriological Analytical Manual*, 8th ed.; U.S. Food and Drug Administration: Washington D.C.

Ward, A.; Stensel, H. D.; Ferguson, J. F.; Ma, G.; Hummel, S. (1999) Preventing Growth of Pathogens in Pasteurized Digester Solids. *Water Environ. Res.*, **71** (2), 176–183.

Ward, R. L. (1978) Mechanism of Poliovirus Inactivation by Ammonia. *J. Virol.*, **26**, 299–305.

Ward, R. L.; Ashley, C. S. (1976) Inactivation of Poliovirus in Digested Sludge. *Appl. Environ. Microbiol.*, **31**, 921–930.

Ward, R. L.; Ashley C. S. (1977) Identification of the Virucidal Agent in Wastewater Sludge. *Appl. Environ. Microbiol.*, **33**, 860–864.

Ward, R. L.; Ashley C. S. (1978) Comparative Effects of Ammonia and Related Compounds on Poliovirus." *Appl. Environ. Microbiol.*, **36**, 198–200.

Ward, R. L.; Berstein, D. I.; Young, E. C.; Sherwood, J. R.; Knowlton, D. R.; Schiff, G. M. (1986) Human Rotavirus Studies in Volunteers: Determination of Infectious Dose and Serological Response to Infection. *J. Infect. Dis.*, **154** (5), 871–880.

Chapter 16

Thermal Drying of Biosolids

(continued)

(continued)

1.0 INTRODUCTION

1.1 Background

Thermal drying involves using heat to evaporate water from biosolids. It reduces the moisture content to a level below that achievable by conventional mechanical dewatering methods. Thermal drying has been practiced since the 1920s when the Milwaukee Metropolitan Sewerage District (MMSD), Milwaukee, Wisconsin, began the production and sale of Milorganite, its thermally dried biosolids fertilizer. Until the early 1990s, however, only a few municipalities across the country had implemented the technology. Since then, a steady increase in the number of biosolids drying facilities has occurred, with more than 50 facilities operating or under construction by 2011. The throughput of the facilities ranges from less than 1 to more than 100 dry Mg/d. This growth has been driven by several factors, including improvements in the technology, more stringent regulations, and social factors affecting land application and landfilling of biosolids.

The heat-dried product typically has a dry solids content of approximately 90 to 95% (or 5 to 10% water content). Thermally dried biosolids products can be used as fertilizer, soil conditioner, or biofuel. The suitability of the product for these outlets typically is affected by the source of the solids, pretreatment processes, the type of drying systems used, and local conditions.

1.2 Advantages and Disadvantages of Thermal Drying

Thermal drying has several advantages compared with other solids processing methods:

- Production of a Class A product;
- Well-proven technology;
- Wide-range of system capacities;
- Volume and mass reduction;
- Nutrient availability;
- Multiple outlets for product;
- Thermal value enhancement; and
- Public acceptance.

Disadvantages include

- System complexity;
- High cost;
- Safety concerns because of the potential for dust explosions, product overheating, fires, and auto-oxidation of biological material if rewetted (which can lead to self-heating and smoldering fires);
- Requirement for air permits; and
- Product marketability.

1.3 40 CFR Part 503 Requirements

Thermal drying is listed as a process to further reduce pathogens (PFRP) in 40 CFR Part 503 (U.S. Environmental Protection Agency, 1993). The PFRP standards call for the biosolids to be dried by direct or indirect contact with hot gases to reduce their moisture content to 10% or less, and the temperature of either the biosolids or the wet bulb temperature of the gas in contact with the biosolids to exceed 80 °C (176 °F) as the biosolids leave the dryer. The high temperatures used in thermal drying typically meet both of the above criteria.

The vector-attraction reduction criteria require stabilized biosolids to be dried to at least 75% solids and biosolids with an unstabilized fraction to at least 90% solids. Drying typically increases the solids concentration to more than 90%, which meets the vector-attraction reduction criteria.

1.4 Sustainability

The ability of thermal drying to remain economically and environmentally viable will depend on the demands placed on it by regulatory agencies and the public and on economic conditions.

Regulatory agencies are continually scrutinizing emissions of pollutants and odorous air from thermal-drying plants and imposing increasingly tighter criteria on new facilities. Federal and state statutes also regulate the quality of the product. For a heat-dried product to be applied to land as an "exceptional quality" product, it must meet stringent quality parameters, which include pathogen-density reduction (Class A), vector-attraction reduction, and low metals concentrations. Operating experience at thermal-drying facilities has shown that the dryers operating at higher temperatures can consistently meet the pathogen and vector-attraction reduction criteria for Class A compliance. Some newer dryers that operate at lower temperatures, however, may not be able to demonstrate Class A compliance by definition and may need a monitoring plan in place to demonstrate Class A compliance. Thermal drying does not affect the mass loading of metals in biosolids, which are typically dependent on influent wastewater characteristics.

The public perceives thermal drying as an environmentally acceptable technology for solids processing. Odors, which were the primary concern from thermal drying facilities in the past, have been addressed through design improvements incorporated by system manufacturers. Thermal dryers also now incorporate recirculation of exhaust gases and use of regenerative thermal oxidizers or other means to treat exhaust gas from the dryers.

One significant challenge with thermal drying is its high energy demands, particularly high fuel use. Thermal drying uses considerably more fuel than other solids processing methods. Value and acceptability, however, is significantly higher for a heat-dried product than for a dewatered product. Municipalities generating a heat-dried product typically have had little difficulty in marketing the product. Thus, energy demands and associated costs of thermal drying have been acceptable because the municipality is assured that the final product can be safely used and, in most cases, will generate some income.

2.0 DRYING THEORY AND PRACTICE

2.1 Drying Process Fundamentals

The drying process depends on the internal mechanism of liquid flow and the external mechanism of evaporation. During the drying process, a temperature gradient develops from the heated surface inward, causing moisture to migrate from within the wet sludge to its surface by the mechanisms of diffusion, capillary flow, and internal pressures generated by shrinkage during drying. The transfer of heat from the medium to the wet sludge raises the temperature of the sludge and evaporates water at the solid surface. External conditions affecting the transfer of heat or the drying process include temperature, humidity, rate and direction of gas flow, exposed surface area, physical form of sludge, agitation, detention time, and the method of support used during drying. Understanding these external conditions and their effects is necessary when investigating the drying characteristics of sludge, choosing the correct dryer, and determining the optimal operating conditions. Although the internal and external mechanisms occur simultaneously, either may limit the drying rate.

2.1.1 Stages in Thermal Drying

The three general stages of thermal drying: include warm up, constant rate, and falling rate (Figure 16.1).vDuring the warm-up stage, sludge temperature and drying rates increase to steady-state conditions of the constant-rate stage. The warm-up stage is typically short and results in little drying. During this stage, sensible heat is transferred to the solids, and the contained moisture and solids are heated up from the inlet condition to the process condition. The rate of evaporation increases dramatically during this period, with primarily free moisture being removed.

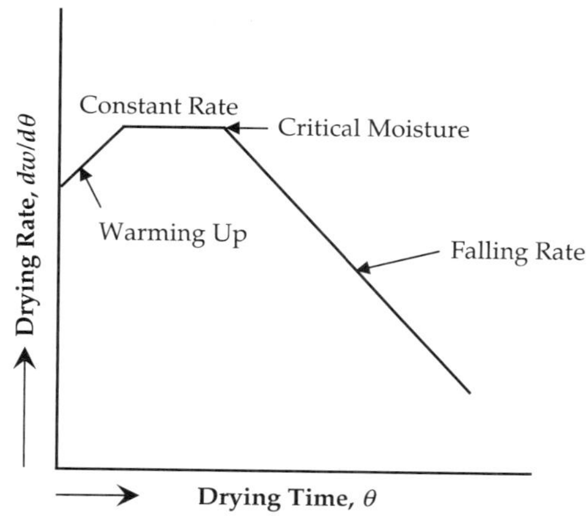

FIGURE 16.1 Stages in thermal drying (WEF et al., 2009).

During the constant-rate stage, interior moisture replaces external moisture as it evaporates from the saturated surface of the sludge. The transfer of heat to the evaporating surface controls the drying rate. This stage typically is the longest and results in most of the drying. The drying rate, essentially independent of the internal mechanism of liquid flow, depends on three external factors: (1) heat- or mass-transfer coefficient, (2) area exposed to the drying medium, and (2) temperature and humidity differences between the drying medium and the wet surface of the sludge. As long as unbound surface moisture is present, the solid is heated only to the wet bulb temperature of the gas. The solids may, therefore, be dried with hot gases and not attain elevated temperatures. A gradual and relatively small rise in product temperature occurs during this period.

During the falling-rate stage, external moisture evaporates faster than it can be replaced by internal moisture. As a result, the exposed surface is no longer saturated and the drying rate decreases. The temperature of the solid-gas interface rises because the latent heat cannot be transferred from the solids to the gas phase as rapidly as sensible heat is received from the heating medium. The average moisture content in the solids at the transition point between the constant-rate and falling-rate stages is called *critical moisture*.

2.1.2 *Structural Changes to Biosolids during Thermal Drying*

Biosolids undergo structural changes as they are dried from 15 to 30% dryness (dewatered solids) to more than 90% dryness. Dewatered solids, up to approximately

40% dryness, remain fluid, and their structure and handling characteristics remain stable. As the water content is reduced, biosolids pass through a "sticky" or "plastic" phase, between 40 and 60% dryness. During the plastic phase, viscosity of the biosolids increases and large lumps form that can adhere to the drying equipment. Biosolids handling in the plastic phase can cause other problems, such as lower drying capacity, wear of the dryer equipment, and high energy consumption. Although some dryers are designed to handle solids through the sticky phase, others incorporate back-mixing of dried product with wet dewatered material to eliminate the sticky phase upstream from the dryers. Between 60 and 85% dryness, biosolids start forming granules. The solids concentration of the thermally dried product is typically higher than 90%, and the volume of biosolids is reduced by approximately 80% through evaporation of water. Figure 16.2 depicts the changes in physical characteristics and volume of biosolids as a function of the dry solids concentration.

2.2 Dryer Sizing Criteria

2.2.1 *Evaporative Capacity*

The first step in sizing a drying system is to establish the amount of water to be evaporated in a given time. Solids concentration of the wet feed solids, the desired product solids concentration, and the operating time are key parameters for calculating the desired evaporative capacity. Evaporation capacity is typically expressed in terms of kilograms

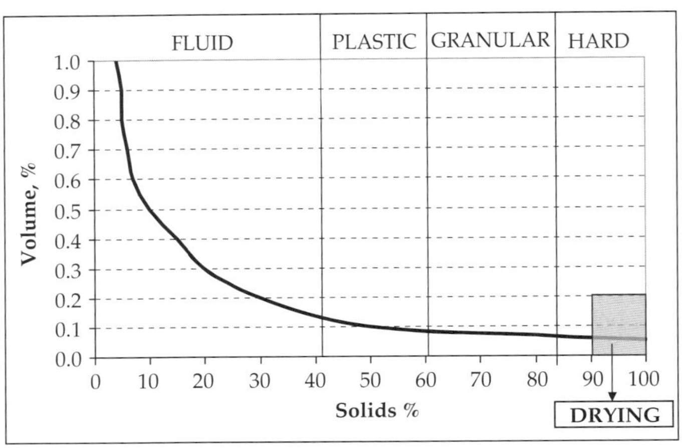

FIGURE 16.2 Changes in physical characteristics of biosolids.

of water per hour (pounds of water per hour) to be evaporated. Manufacturers may establish different evaporative capacities for units based on the source of the solids (e.g., undigested, digested combined primary and secondary, etc.) because of the different heat transfer coefficients and materials handling characteristics of the material.

Performance of the mechanical dewatering process has a significant effect on dryer capacity. It is more cost-effective to remove water mechanically, to the extent possible, than thermally.

2.2.2 *Operating Schedule*

The acceptable operating schedule is a key parameter in determining the size of the dryer. A dryer has a fixed evaporative capacity. Consequently, larger quantities of solids can be processed by operating the dryer for longer periods. For example, a dryer designed to operate 24 hours per day will require only one-third the evaporative capacity to process the same daily quantity of solids as a dryer designed to operate 8 hours per day. Because of their high cost, most drying systems are sized to run on a continuous basis, such as five days per week, 24 hours per day (or approximately 100 to 105 hours per week to fit typical work schedules). This allows time for preventive maintenance during the shutdown period. Another advantage of running on a more continuous basis is less wear on the equipment. Frequent heating and cooling cycles will increase metal fatigue and the rate of component failure.

The dryer feed hopper typically can accommodate enough dewatered solids for a few hours of operation. Consequently, it is not always necessary to operate the dewatering units in tandem with the dryer system. Further, because initial sizing of equipment is typically based on a projected future solids quantity, at least at the outset, the dryer may be used only for a fraction of the possible 168-hour week.

2.2.3 *Residence Time*

Residence time in a dryer varies depending on the type of drying system and the manufacturer's design. Residence time must be determined by the manufacturer based on the heat transfer rate and the system capacity. Therefore, the temperature of the system does not control residence time within the dryer. For example, a rotary drum dryer and a belt dryer could have similar residence times although the operating temperature of the rotary drum dryer is significantly higher than the belt dryer's. Some manufacturers are able to use the residence time in the dryer to demonstrate compliance with pathogen-reduction criteria; others use alternative methods to show compliance.

2.2.4 *Operating Temperature*

The rate of heat transfer during the drying process is directly related to the temperature of the drying medium (hot gases, oil, or steam). For similar types of systems, those operating at higher temperatures will typically dry faster than those operating at lower temperatures, which can be a factor in determining the size of the dryer needed for a given application. Factors such as area of the exposed wetted surface and the unit's heat transfer coefficient also affect the heat transfer rate. Therefore, operating temperatures alone cannot be used as a direct indicator of unit size.

2.2.5 *Storage*

Storage, an important and often-overlooked component of dryer systems, typically precedes and follows thermal drying. Sufficient predrying storage is needed upstream from the dryer system to allow for shutdowns and to attenuate variations in dewatered solids production; postdrying storage should be able to accommodate seasonal fluctuations in market demand and product hauling schedules. Because of their high cost, careful consideration must be given to determining the optimum storage required. Rather than providing excess storage, alternative disposal methods, such as landfilling, may be used when market demand for the product is low. In addition to costs, there are safety concerns associated with storing dried product for long periods.

2.3 Types of Dryers

Dryers are classified according to the method used to transfer heat to wet solids. The predominant methods of heat transfer used in biosolids drying systems are convection (direct dryers) and conduction (indirect dryers). Although most systems benefit to some extent from multiple methods of heat transfer, typically one method is more dominant than the others.

In direct-drying systems, the heat transfer medium, typically hot gas, is in direct contact with the wet solids. In indirect-drying systems, the evaporative heat is transferred from the heat transfer medium to the solids through a material separating the solids from the medium. Typically, the heat transfer medium is hot oil or steam and the separating surface is a steel plate.

2.4 Dryer Systems in Use

The following sections provide basic process descriptions for established dryer systems for biosolids applications. The systems are further categorized according to the predominant method of heat transfer used.

2.4.1 *Direct-Drying Systems*

Direct dryers that have been successfully used for drying municipal wastewater solids include the rotary drum dryer and the belt dryer.

2.4.1.1 *Rotary Drum Dryer*

Rotary drum dryers have been used to process biosolids since their first application in Milwaukee in the 1920s. Although system components differ with the manufacturer, there are common elements. For example, all comprise long, cylindrical steel drums that rotate on bearings. Some are designed with concentric drums to allow the product to make multiple passes through the drum (triple pass); others consist of a single drum in which the product makes a single pass. Hot gas is used to dry and help move the product through the system. In 2010, rotary drum dryers were in use or under construction at more than 20 facilities in the United States and at more than 75 facilities in Europe.

A mixer blends dewatered solids with recycled thermally dried product, creating granules and reducing the moisture content of the feed material typically to less than 35%. Feed granules continuously enter the upper end of the drum dryer, with hot furnace gas ranging from 400 to 650 °C (750 to 1200 °F). The granules and hot gas are conveyed (typically cocurrently) to the discharge end of the dryer. During conveyance, axial flights along the slowly rotating interior wall of the dryer pick up and cascade the granules through the dryer. This creates a thin sheet of falling granules that directly contact the hot gases and dry rapidly.

In most systems, approximately 80% of the gas is recycled to the furnace. Recycling the gas enhances the thermal efficiency of the system and helps maintain low oxygen concentrations within the drying loop. The importance of maintaining an inert atmosphere in the dryer is addressed in the section on Drying Safety. Recycling the gas also reduces the volume of exhaust gas to be treated. Exhaust gas exits the dryer at 93 to 105 °C (200 to 220°F), and is conveyed to emissions control equipment for particulates removal and odor control. Figure 16.3 illustrates typical flow schematic of a rotary drum system.

2.4.1.2 *Belt Dryer*

The belt dryer has been in use in Europe for drying wood wastes and other homogeneous materials for decades. As of 2010, more than 30 systems were in use in Europe, with only one operating system and one under construction in the United States. Configuration of the belt dryer depends on the manufacturer, but there are some similarities.

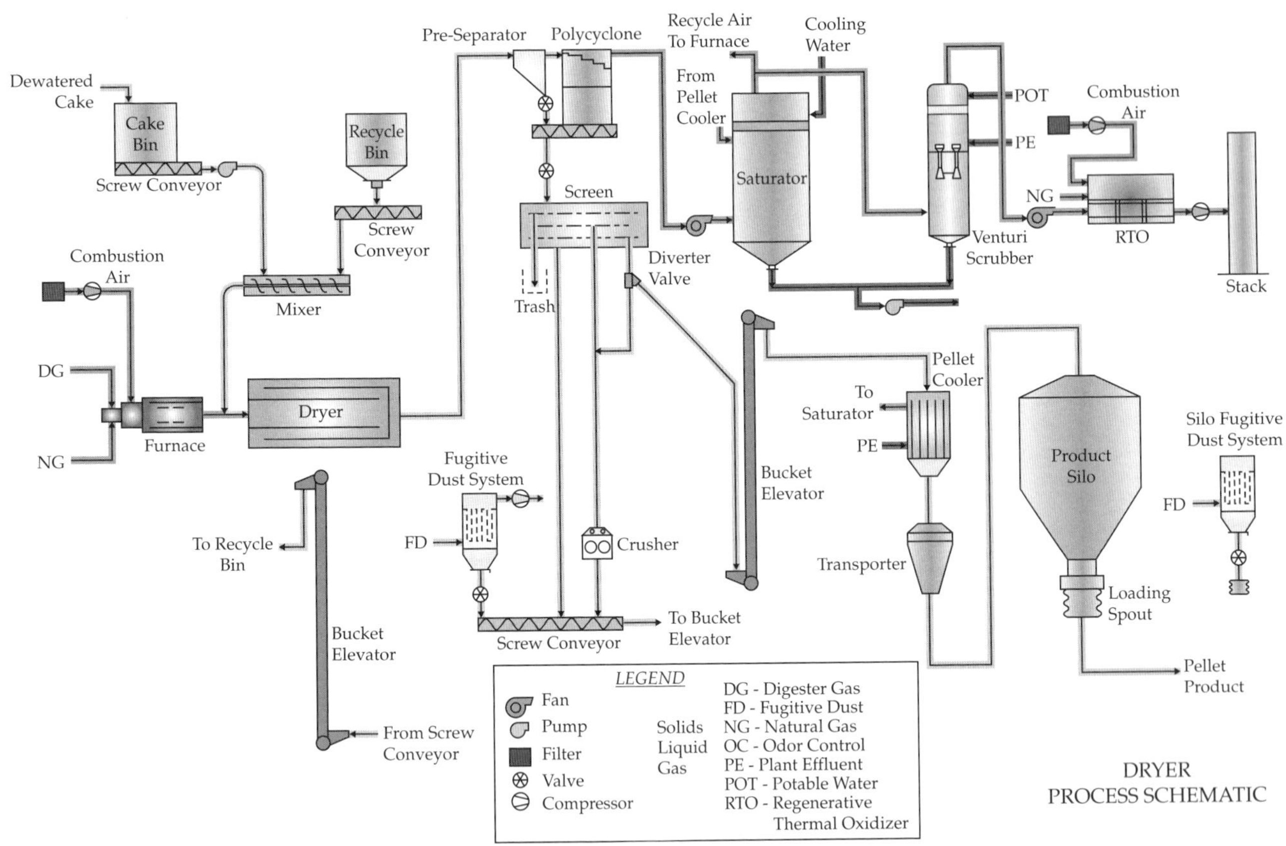

FIGURE 16.3 Flow schematic of a rotary drum dryer.

Belt dryers operate at relatively low temperatures that range from approximately 130 to 177 °C (265 to 350 ° F). Sludge is placed on a porous conveyor belt, and the heated gas is blown or drawn through the layer of solids on the belt. Belts are made of either steel mesh or synthetic material (similar to the filter media used in a belt filter press). One system uses two belts, and the product makes two passes through the system; the others use a single pass on one belt.

Because of the low operating temperature, the belt dryer uses larger volumes of drying air. The method of management of this air differs among manufacturers. Typically, a portion of gas is removed from the process air loop by a fan and is passed through a condenser to remove moisture from the system. One supplier diverts a portion of this scrubbed process gas to mix with combustion air, and the majority is returned to the dryer. Others recommend that the scrubbed gas be treated before discharge for odor control using a biofilter or chemical scrubber. A significant portion of the process gas will be returned to the combustion process.

Product from the belt dryer is typically less uniform in size compared to product from some of the more complex systems. This is because of limited materials handling downstream from the dryer; additional product screening can be included to improve uniformity. Although product characteristics for the same types of dryers are similar, attention should be given to the differences between suppliers. For example, one belt drying system uses backmixing, and another extrudes material directly onto the belt, which produces different product characteristics.

Alternative energy sources can be used for belt dryers because of the low temperatures of the drying gas. This allows for a wide range of energy recovery options. Figure 16.4 illustrates a typical flow schematic of a belt dryer system.

2.4.2 *Indirect-Drying Systems*

Indirect dryers that have been successfully used for drying municipal sludge include the fluidized-bed dryer, tray dryer, paddle dryer, and rotary chamber dryer.

2.4.2.1 *Fluidized Bed Dryer*

Fluid bed drying of biosolids began in Europe in the early 1990s; today, more than 30 facilities are using this technology. In the United States, this dryer was first used in the late 1990s; there are two units in operation.

The main component of the fluidized bed dryer is a stationary, vertical chamber that is divided into three zones. The first zone is the wind box, or gas plenum, which distributes the hot fluidizing gas (typically air) evenly through the bed of granules. The gas is forced or induced through the system by a blower arrangement in a closed-loop

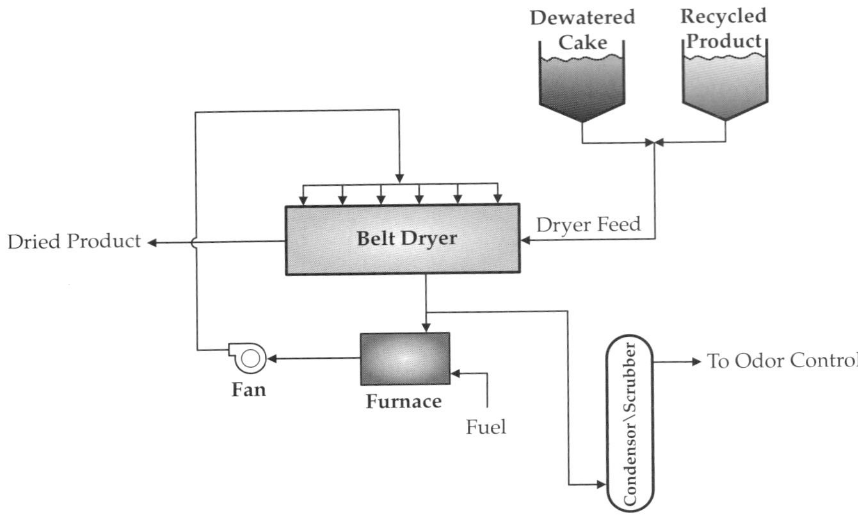

FIGURE 16.4 Flow schematic of a belt dryer system.

system. The second zone consists of a heat exchanger within the fluidizing zone of granules. A heat transfer medium, typically steam or hot oil, is circulated through the heat-exchanger tubes. The fluidizing gas keeps the granules in suspension, helps transfer heat from the exchanger tubes to the granules, and removes moisture from the system. The third zone is the hood where the fluidizing gas exits the unit and the bulk of the solids are separated from the gas. The exhaust gas is then processed for removal of particulates and moisture using equipment such as a cyclone separator and a condenser. The temperature of the fluidizing gas in the dryer is approximately 85 °C (185 °F).

The method of introducing dewatered solids into the dryer varies with the manufacturer. Some systems pump the solids into the dryer, where a rotating device cuts the dewatered cake into small, irregular pieces that fall into the fluid bed. Fluidizing gas promotes intermixing between the gas and solids, which leads to mass and heat transfer. Agitation in the fluidized bed smoothes the solids fragments to a relatively uniform shape. The dried product exits the dryer by overflowing an adjustable weir equipped with a rotary air lock.

The low operating temperature of the drying process makes fluid bed systems well suited for applications where waste heat is available from other processes. Systems in Europe have been operated using waste steam from garbage incineration plants. Figure 16.5 presents a typical flow schematic of a fluidized bed drying process.

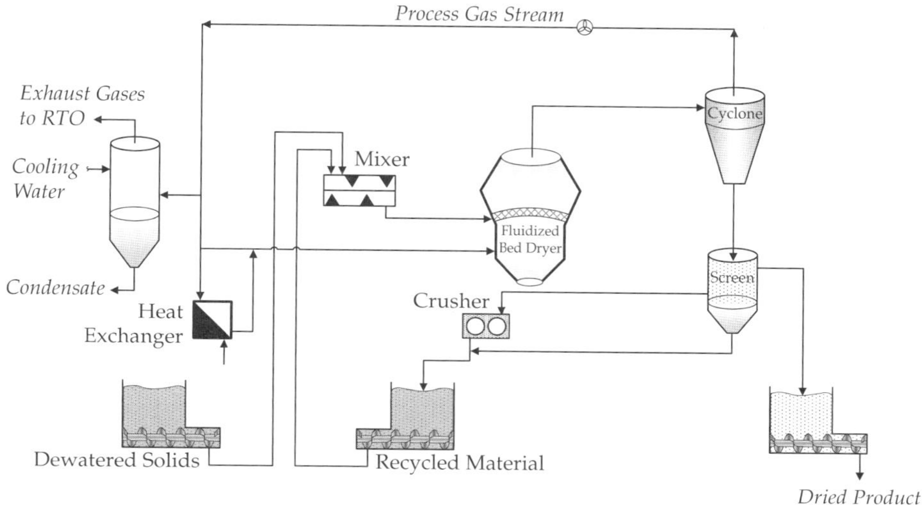

FIGURE 16.5 Schematic of fluid-bed drying system.

2.4.2.2 *Paddle Dryers*

Paddle dryers have been used in industrial applications for many decades, but were first used for drying biosolids in Japan in the late 1980s and in the United States starting in the mid-1990s. Approximately 10 paddle drying facilities are operating or under construction in the United States.

The paddle dryer consists of a stationary horizontal jacketed vessel containing two intermeshing, counter-rotating agitator shafts with paddles. The rotors and paddles are hollow, allowing the heat-transfer medium to circulate through the hollow core. The heat-transfer medium also circulates through the jacketed vessel. Heat is transferred to the biosolids by conduction through the surfaces of the paddles. One of the manufacturers also provides stationary agitator ploughs or breaker bars between the rotating agitators to achieve better mixing and to prevent cake buildup on the agitator surface.

Dewatered biosolids are typically fed directly to the dryer by a positive displacement pump or a screw conveyor. Back-mixing of dewatered biosolids with dried product is typically not done with this type of dryer. The resulting dried granules are of irregular shape and vary in size. The agitation caused by the paddles can increase the concentration of fines in the product. Screening or the addition of conditioners can produce a more uniformly sized product. Dried product leaves the dryer at an average temperature of 105 to 110 °C (220 to 230 °F), depending on the dry-solids content and the operating temperature. The product is cooled to approximately 40 °C (104 °F) using a cooling screw conveyor. The shaft and trough of the screw conveyor is jacketed and cooled by a separate cooling-water system. The carrier gas is exhausted from the dryer and is conveyed to a condenser. Noncondensable gas is routed to odor control units.

The paddle systems can be either continuous feed or batch type. In continuous-feed systems, the product is conveyed through the dryer by the action of the paddles or discs and by volume displacement. Some systems include a weir at the discharge end of the dryer to help submerge the heat-transfer surface in the material being dried. One manufacturer's system is placed at a slight incline, so that the product moves through the machine by gravity. Figure 16.6 presents a typical flow schematic of a paddle dryer system.

2.4.2.3 Rotary Screw Dryer

Rotary screw dryers use an indirect method of drying and operate in batch mode. Controlled amounts of dewatered solids are fed into the insulated dehydration chamber, where they are heated by a slowly turning rotor, which reverses its direction

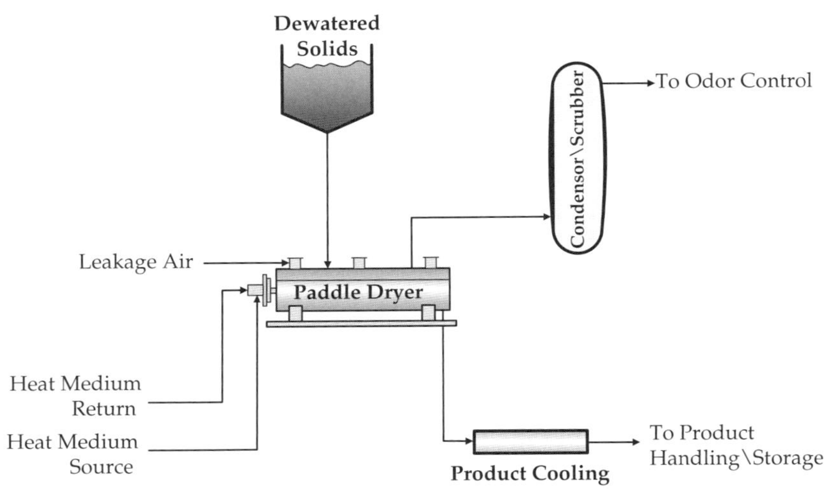

FIGURE 16.6 Schematic of paddle dryer system.

of rotation every few minutes. The rotor is heated by a thermal fluid that circulates through the rotor and the hollow sections of the dehydration chamber shell. The heat exchanger is fired with a single burner that can burn natural gas, fuel oil, propane, or methane.

The dewatered solids are dried at controlled time and temperature to the desired dryness. At the preset temperature, the dried product is discharged by a small auger to a surge bin. In the surge bin, the product is allowed to cool before being conveyed to load-out or to storage silos. The surge bin is equipped with a screen gate to control the size of the finished product. Larger particles are recycled to the cake hopper to be mixed with the dewatered cake.

Exhaust gas containing evaporated moisture is collected in a duct at the top of the dehydration chamber and conveyed by an exhaust blower to a wet scrubber/condenser. Condensed moisture is returned to the headworks for treatment. Gas is vented to a biofilter or the aeration basins for odor removal. Figure 16.7 is a schematic showing the primary components of a rotary screw dryer.

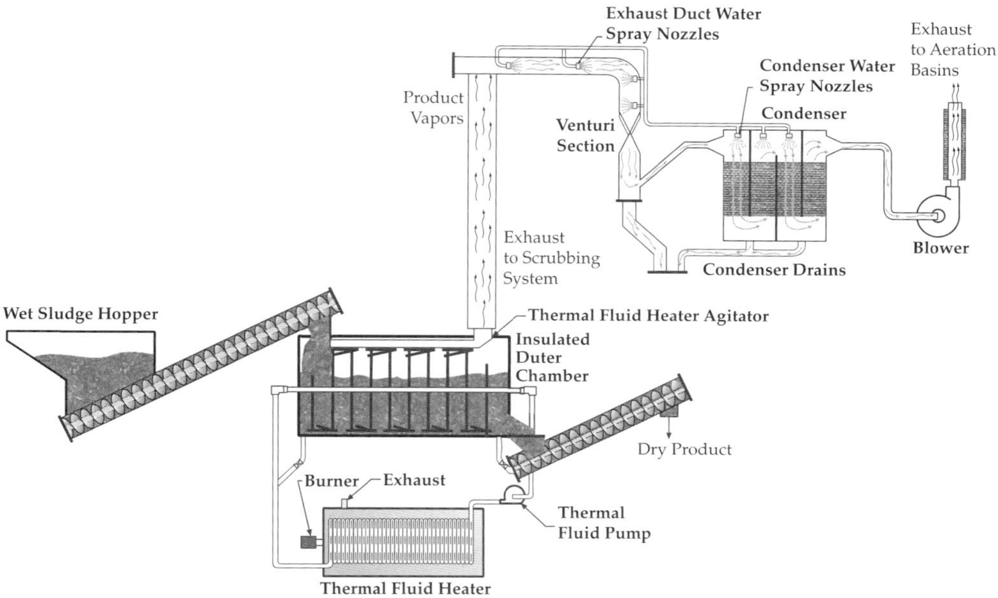

FIGURE 16.7 Flow schematic of a rotary screw dryer.

2.4.2.4 *Tray Dryers*

The tray dryer is essentially a vertical disc dryer. In addition to the difference in configuration, dewatered solids are mixed with recycled dried product to form granules before they are fed to the dryer. The blended granules are fed through a top inlet in the vertical multistage dryer. The dryer has a central shaft with attached rotating arms that move the granules from one heated stationary tray to another in a rotating motion until they exit at the bottom as a dried product. The product is conveyed by a bucket elevator to a separation hopper, where is screened, and oversized material is crushed and returned to the mixer along with fines and a portion of properly sized pellets. The recycled product is then recoated with dewatered cake and sent back through the dryer. A portion of the properly sized pellets downstream from the screen is not recycled, but is cooled in a product cooler and then pneumatically conveyed to storage.

The dryer is maintained under negative pressure by a vapor fan that draws the evaporated moisture, noncondensable gas, and a small amount of bleed air through a contact condenser. In the condenser, moisture is removed from the gas stream, and the noncondensable gas is sent to the oil heater for thermal treatment of odors. A closed-loop thermal oil system provides heat for the dryer.

Tray dryers have seen limited growth in North America over the last decade. The Back River Wastewater Treatment Plant (WWTP) in Baltimore, Maryland, has used tray dryers since 1995. New tray drying systems started operation at the Stickney plant in Chicago, Illinois, and at the Ashbridges Bay Treatment Plant in Toronto, Canada, in 2009. Figure 16.8 presents a flow schematic of a tray dryer system.

2.5 Dryer Selection Considerations

2.5.1 *Product Quality and Use*

The expected final use of the product should be carefully considered when selecting a dryer system. Some uses, such as high-end fertilizer production, will require better structural strength and uniform product size than general uses, such as for soil blending, bulk application, or biofuel. Production rates and facility location must also be considered.

Physical characteristics of the product vary with the system and the manufacturer. Systems that incorporate multiple materials handling components generally produce a higher-quality product. For example, backmixing systems can help produce a more durable product that does not disintegrate easily when handled. Screening can be

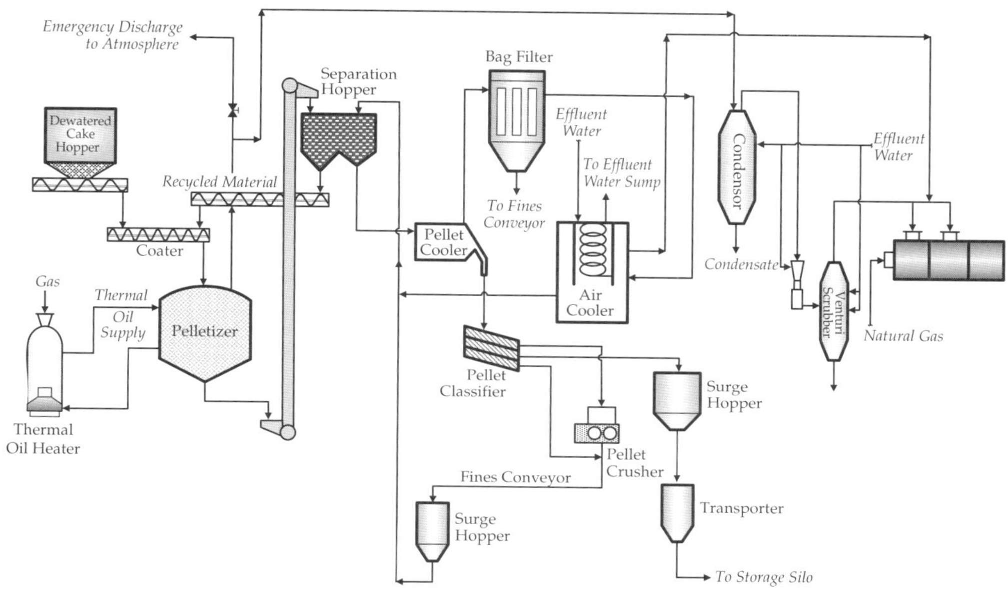

FIGURE 16.8 Flow schematic of a tray dryer.

used to achieve uniform product size. Conditioning agents (typically food-grade oils) can be used to agglomerate particles and reduce dust. Dried product quality considerations are described in detail in Section 3.0.

2.5.2 *Processing Train Unit Capacity*

Capacity requirements and the number of processing trains required to supply the initial and ultimate processing needs should be considered in selection. Systems should be sized to provide an optimal balance between capital cost, redundancy, and site space constraints. Careful consideration to turn down capacity should be given to systems being sized to accommodate significant increases over time. Dryers typically have limited turn down capability, possibly requiring the use of multiple trains to meet future processing needs or shorter initial operating cycles to optimize throughput. Throughput optimization will also provide better thermal efficiency for the systems.

It is typically more cost-effective to install a single large-capacity processing train than several smaller trains. Reasons for this include the economy of scale associated with equipment, smaller building space, and fewer ancillary support systems. A single train, however, may not meet the need for redundancy. Many facilities in the United States have installed single processing trains to reduce capital costs and rely on dewatering and alternate disposal methods, such as landfilling, for periods when the system is out of service.

Typically, direct-drying systems (rotary drum and fluid bed) have larger capacities than indirect systems (paddle, disk, and rotary chamber). The rotary drum and fluid bed systems can be sized to process between 2000 and 11 000 kg H_2O/h. Tray dryers also are considered to be large-capacity systems. Indirect systems can be less than 500 kg H_2O/h.

2.5.3 *Labor Requirements*

Labor requirements depend on the type of system and its design. Some systems (e.g., rotary drum dryers), require higher operator and maintenance staff skills and continuous supervision compared with others. Other systems have been designed to operate with minimal supervision in automatic mode.

2.5.4 *Utility Requirements*

Utility requirements include electricity, fuel, and water.

2.5.5 *Sidestreams*

Odorous liquid sidestreams produced by condensation of water vapor from dryers contain both organic liquids and ammonia. Wet scrubbers and other ancillary equipment generate other sidestreams, which also have to be considered.

2.5.6 *Emissions and Odor Control*

Drying equipment and product handling and storage areas should be enclosed and vented to air pollution control equipment. Cyclone separators, wet scrubbers, baghouses, or a combination of these can be used to remove particulates from the air stream. Thermal oxidation is the most typically used odor control method for thermal dryers, as discussed further in Section 5.0.

3.0 HEAT-DRIED BIOSOLIDS PRODUCT

Heat-drying of municipal biosolids results in a product that, in most cases, meets U.S. EPA and most state regulations for unlimited distribution as a soil conditioner or fertilizer. This aspect of heat-drying makes it attractive to WWTPs. Examples include MMSD, which has produced and sold Milorganite since approximately 1926, and

Houston, Texas, which has produced and distributed its product, Hou-actinite, since the 1950s. Both have successfully developed markets for their dried biosolids, and the revenues from sales of Milorganite and Hou-actinite are an important part of their budgets. These communities, however, have invested many years in developing and serving their markets. A new facility is unlikely to fully emulate the marketing programs of Milwaukee and Houston for several years after starting up the drying process.

The potential for marketing heat-dried biosolids remains among the most intriguing aspects of thermal-drying technology. It is recommended that WWTPs assess local and regional market opportunities for the product before investing the capital necessary to construct a biosolids drying facility. The assessment should focus both on capacity of outlets to accept the product and on the qualities and characteristics of the product that will meet the requirements of prospective customers. Some of these issues are discussed below.

3.1 Product Quality Requirements

To raise interest in heat-dried biosolids, and to subsequently distribute and market the product, it is necessary to define the product characteristics that may be of interest to potential users. Conducting market assessments for defining the required product characteristics can be difficult without having actual product to show prospective customers. For estimating physical properties, it is recommended that WWTP managers who are considering drying obtain samples of dried biosolids from systems identical or similar systems to those being considered; ideally, the reference samples should reflect similar type of feed biosolids. The chemical characteristics of the heat-dried biosolids can be projected by extrapolation of data from analyses of the dewatered biosolids stream to be dried.

3.1.1 *Physical Characteristics*

3.1.1.1 Particle Size

Particle size is a measure of the size of the particles of dried biosolids after completion of processing. Many biosolids drying processes will incorporate screening of the final product, which is as an important process step. The screening step enables the producer to specify the size of the particles by selecting the screen size. Preferred particle size might vary from market to market; for example, users seeking a standard-grade fertilizer might prefer a product with a particle size ranging from 2 to 4 mm in diameter; whereas, a golf course seeking to use the dried biosolids on tees and greens might prefer a finer product that has a 0.3 to 0.8 mm diameter.

Particle size is an important consideration for several reasons. If the dried biosolids are going to be blended with other fertilizer ingredients, then using a product that is a similar size as the other ingredients will reduce the potential for segregation of ingredients after blending. To an extent, particle size also will affect the rate at which dried biosolids might release nutrients, with a smaller particle releasing nutrients quicker than a larger particle.

3.1.1.2 *Durability (or Hardness)*

The durability or resistance to degradation of the heat dried particles is important because the product undergoes further handling following drying (conveyance, dumping, and moving with front-end loaders). A particle that is poorly resistant to degradation will easily deteriorate into finer particles and dust.

3.1.1.3 *Dust*

The most important physical parameter used to assess whether a dried biosolids product is well-suited for a particular use is dust content. Dried biosolids dust can adversely affect a marketing program and a dusty product tends to be poorly received by users.

One aspect of upstream wastewater treatment that influences the amount of dust in the dried biosolids is the extent to which grit, fine fibers, and hair are removed. These extraneous materials make the dried biosolids more brittle and likely to break into fine particles during handling and transfer. If the input stream contains a significant fraction of raw primary solids or if the plant's headworks do not do an adequate job of removing grit and fibers, then the product will contain extraneous materials and can easily fracture. Therefore, drying of undigested solids is likely to result in a dustier product than biosolids that have been well digested before dewatering and drying.

3.1.1.4 *Odor*

Most heat-dried biosolids have a musty, earthy odor. Anecdotally, dried biosolids produced from well-digested solids streams tend to be perceived as less odorous than those produced from raw solids. In most applications, odor is not the limiting characteristic for beneficial use. Occasionally, comments about product odor are heard after application and subsequent rewetting after rainfall or irrigation. As in any other situation, if heat-dried biosolids are surface-applied at a high rate without being incorporated into the soil, some odors are likely. The intensity of the odor will depend on the application rate, but in most cases, the odor should dissipate over time. If such applications are planned, then sites should be selected where odor is less likely to be noticed.

3.1.1.5 *Bulk Density*

Bulk density is a measure of the mass of a product per unit volume. The bulk density of dried biosolids typically ranges from 560 to 880 kg/m^3 (35 to 55 lb/cu ft). A product with relatively low bulk density will fill up the volume capacity of a transport unit before the allowable weight limit is reached, which will result in higher unit costs for delivery.

3.1.2 *Chemical Characteristics*

3.1.2.1 *Nutrients*

The value of heat-dried biosolids to users is typically associated with the plant-food content of the biosolids. The nutrients of interest can be classified as primary nutrients (nitrogen, phosphorus, and potassium); secondary nutrients (calcium, magnesium, and sulfur); and micronutrients (boron, chlorine, copper, iron, manganese, molybdenum, and zinc).

Most states in which heat-dried biosolids are marketed or distributed as fertilizer require product labeling and registration, including a guarantee of primary plant nutrients in the form of the percent of nitrogen, phosphoric acid, and potash (the three numbers commonly seen on a fertilizer bag, for example 10-10-10); and guarantees of secondary or micronutrients. These guarantees become regulatory standards, and failure of a product to contain the promised amount of nutrients can subject the generators to fines, stop-sales orders, or both.

3.1.2.2 *Salt Index*

The concentration of dissolved ions in an extract of water held within the soil matrix is referred to as soluble salts. In some areas, soils may accumulate excessive concentrations of soluble salts, especially during the summer, as a result of increased transpiration, insufficient leaching of salts from the soil solution, and over-fertilization. Extremely high salt concentrations may inhibit water uptake and lead to chlorosis, scorching of foliage, premature dropping of leaves, dieback, and inhibition of growth, frequently referred to as "burning." The salt index (SI) is the measure of a fertilizer's ability to cause these effects in plants rather than the percentage of salt content.

3.1.2.3 *Heating Value*

Heating value is an important parameter for biosolids are used as a fuel in industrial processes, such as cement manufacturing. Heating value is used to determine the fuel characteristics of biosolids and is a function of the organic content in the solids. Typical heating values for dried biosolids are expected to range from 12 000 to 16 000 kJ/kg (5000 to 7000 Btu/lb). The heating value of coal is approximately 28 000 to 30 000 kJ/kg (12 000 to 13 000 Btu/lb); woody biomass has a heating value of approximately 19 000 kJ/kg (8000 Btu/lb).

3.2 Product Uses

Heat-drying has become a broadly accepted method of biosolids management in North America, used by more than two dozen communities. Communities operating biosolids drying systems include large systems such as Boston, Massachusetts; New York; Houston, Texas; and Milwaukee, Wisconsin; and smaller municipalities such as Hagerstown, Maryland, and Newport, Tennessee. These systems are producing "exceptional quality" material suitable for use as a soil conditioner, fertilizer, or fuel. It has been projected that by 2008 approximately 270 000 to 360 000 dry Mg (300 000 to 400 000 dry tons) of biosolids would be heat-dried and distributed through the United States and Canada.

With the generation of this quantity of heat-dried biosolids, a growing trend in the industry is to focus on local or regional markets for the management of the product. Because many of these markets may not be familiar with the use of heat-dried biosolids, however, development of local market demand may require several years before revenue-generating demand approaches supply. Available outlets for heat-dried biosolids are discussed below.

3.2.1 Bulk Distribution

The first, and typically least difficult, market is bulk distribution of the product as agricultural fertilizer. In this high-volume outlet, the product can be routinely collected from the processing facility and incorporated into beneficial use programs. This outlet may provide the processor with a buffer that can reliably handle material until other markets can be developed or serve as a "widespot" in place of some storage.

In most instances, little to no revenue is generated from this outlet. In fact, typically a net cost is associated with the transport and management of the product that must be borne by the processor, at least until the local markets recognize the value of the dried biosolids.

In a bulk-application market, the dried biosolids must be managed in the same manner as any other biosolids product. Applications at agronomic rates, setbacks from water, and other management practices should be followed. In some states, bulk land application of any "exceptional quality" product requires site-specific approval or authorization.

3.2.2 *Blended Fertilizers*

Blends of dry fertilizer components are often developed to produce a specific analysis to meet the requirements of one or more cropping systems or to respond to economic conditions. The ratio of ingredients is manipulated to produce a mixture that offers as close to the desired analysis as possible. Because it may not be possible to prepare the fertilizer with the desired nutrient ratio by blending only chemical fertilizer components, blenders typically use inert filler that contributes no plant nutrients to the mixture. Some typical filler materials include granular limestone, corn cob, sand, and clay. Dried biosolids may be used to replace some or all of these fillers. Incorporating heat-dried biosolids into the fertilizer mixture would not only satisfy the needs for filler, but would replace the inorganics in the fillers with organic materials from biosolids and contribute some nitrogen, reducing the amount of nitrogen to be purchased.

Bulk blenders typically have strict limits in terms of product sizing, dust generation, and other physical characteristics for any product incorporated into a fertilizer blend. The value of dried biosolids as a constituent of blended fertilizer is a function of the cost of the materials replaced by the biosolids.

3.2.3 *Specialty Fertilizer*

Any product that is not classified agricultural is typically referred to as a specialty fertilizer and may include products that are

- Sold at wholesale and retail for turf-grass management (including home lawn and garden fertilizers);
- Applied to "specialty" agricultural products (such as vegetables and citrus fruits); or
- Included in management of nursery, greenhouse, or house plant stocks.

Specialty fertilizers can be standalone products, such as Milorganite, or may be blends of fertilizer products. In most regions, such markets may be relatively small, but the value of the product can be substantially higher than in other markets.

3.2.4 *Nontraditional Outlets —Fuel*

With increasing cost of fossil fuels, cement manufacturers and other industries are considering the use of alternative fuels such as biomass to provide the energy for their operation. The cement industry is an especially attractive outlet for biosolids because the ash (including most of the regulated pollutants in the biosolids) is incorporated into the cement product. Therefore, the pollutants are not released to the environment. This outlet is used by Lehigh Cement in Maryland, which is using dried biosolids from Boston and Baltimore as a fuel.

Cement manufacturers are not likely to be interested in the fertilizer components of the dried biosolids, but may be interested in other chemical constituents that may affect the kiln's air permits. Each kiln may have its own limits based on modeling of emissions. It will be important to users to know the concentration of some of these elements to project the viability of using the dried biosolids as a fuel under their air pollution control permits.

Cement manufacturers typically will have to make significant capital investments before they can accept dried biosolids. The required improvements might include modifications to air and other permits, engineering and construction of material handling systems, including product storage, and any additional air pollution or process control systems required by changes to permits.

When a new biosolids drying facility begins to market its product, it will be competing with other products already accepted and in use. A key to creating demand for the new product is to demonstrate comparable or superior performance to the products already in use. Biosolids managers should plan to support several demonstrations or research programs with local agencies or universities to better promote the product.

Users of heat-dried biosolids will expect a reliable, consistent supply of product that meets their specific needs. As with any fertilizer product, market demands will fluctuate. The manner in which the drying facility stores its product will determine which of the several outlets can reasonably be serviced.

3.3 Valuing Heat-Dried Biosolids

Assuming that the physical characteristics and other attributes of heat-dried biosolids are acceptable to the marketplace, the value of the product is typically comparable to other commercial fertilizers in the region. In 2008, the cost of nutrients in available chemical fertilizers was approximately $0.50 per 0.5 kg (1 lb) for nitrogen and $0.30 per 0.5 kg (1 lb) for phosphorus (as phosphoric acid). These values vary with location and time of year; nitrogen values fluctuate with the price of natural gas.

Based on this level of pricing for competing types of fertilizer, a dried biosolids product of 4% total nitrogen and 2% total phosphorus could be expected to have a fertilizer equivalent value of approximately $67.50 per 0.9 Mg (1 ton). Dried biosolids would not necessarily bring this price, however, because they contain a slow-release form of nutrients. As a result, all of the nitrogen and phosphorus might not be available to the user in the year of application. A lower price is also often used as a lever to gain entry into the marketplace. New producers are encouraged to avoid "giving away" the product to the extent possible, as even a minimal price imparts value and allows future increases in price once performance has been demonstrated.

3.4 Marketing Heat-Dried Biosolids

Dried biosolids can be marketed in several ways, which may influence the value of the product. Producers who rely on outside brokers or marketing agents typically realize lower revenues per megagram of product than those who do their own marketing, because most brokers or outside contractors are not motivated to seek outlets that may offer the highest price for the material. The marketing entities typically are asked to move product expeditiously from the plant, rather than maximize its value.

An internal marketing program is not without its costs (including training, packaging, outreach, etc.). These costs may or may not be specifically allocated to product marketing, which makes it difficult to fully project net revenues from product distribution. Moreover, most internal sales programs begin to recognize revenue after several years of market development, which typically is not factored into product value.

Regardless of how the product is marketed, however, dried-biosolids sales can generate some level of revenue for a WWTP. Whether all facilities contemplating biosolids drying and marketing will enjoy these levels of success is a function of local market conditions, competing products, and product characteristics.

4.0 ENERGY USE AND RECOVERY

4.1 Energy Requirements for Drying

Removing water from biosolids by thermal drying is energy-intensive and represents the most significant operating expense for any facility. The heat needed to evaporate approximately 0.5 kg of water in a thermal drying system is far greater than the 2257 kJ/kg (970 Btu/lb) heat for vaporization of water. Heat is required for the sensible heat needed to raise the temperature of the solids, the water in the biosolids, and the air used for combustion and product and transport, and for shell losses from the furnace and the dryer by convection and radiation. Ancillary equipment such as thermal oxidizers, which are often used to treat off-gases from the dryers, can also increase the consumption of thermal energy. The overall thermal energy requirement for drying ranges from about 3250 to 3950 kJ/kg (1400 to 1700 Btu/lb) of water evaporated. The high energy requirement is also one of the most significant drawbacks of the process.

4.2 Energy Sources for Dryers

The most typically used fuel for biosolids drying is natural gas. Where mesophilic digestion is used for solids stabilization, digester gas can be a cost-effective option for drying. Landfill gas can also be used, if a landfill is within a reasonable distance from the drying facility.

Anaerobic digestion and heat drying can form a synergistic pair. Digestion can provide digester gas for the drying process and, in return, waste heat recovered from drying can provide process heat for digester heating. When digester gas is used as the primary fuel, natural gas should also be available for use during maintenance of digester gas systems, and to supplement the heating capacity for processing peak loads. Because of variability in digester gas production, the ability to store and adjust use rates to match availability is highly recommended.

Both digester and landfill gas are composed primarily of methane and carbon dioxide, but can also contain impurities such as sediments, hydrogen sulfide, and siloxanes. The gas will also be saturated with moisture at the operating temperature of its source facility. The quality requirements for digester or landfill gas as an alternative drying fuel are low. Hydrogen-sulfide concentrations and moisture in the gas, however, should be reduced to avoid condensation and formation of corrosive acidic solutions that can damage equipment. Moisture in the gas can also cause problems with valves and the burner nozzles. If the gas contains high concentrations of siloxanes, then it is advisable to include a siloxane treatment step to reduce the maintenance requirements on thermal oxidizers used for off-gas treatment in drying.

Other fuel options available for facilities that practice thermal drying include heat-dried biosolids product, wood waste, or refuse. The heated air from the combustion of these fuels can be fed through an air-to-air heat exchanger to raise the temperature of the drying air. Typically, dried biosolids can be used to provide some or even all of the energy needed for drying. The use of these fuels has to be accompanied by appropriate air pollution control and off-gas treatment systems. Air pollution control agencies typically permit equipment on a "worst-case" scenario (i.e., permitting on the fuel that has the most pollution potential when multiple fuels are available).

Additionally, energy requirements can be reduced by recirculating or recovering heat from dryer exhaust gas for preheating combustion air. Accordingly, the net purchased heat can be lower than the total heat required to dry biosolids. The following sections discuss the heat recovery equipment available for drying systems and potential uses for recovered energy.

4.3 Heat-Recovery Equipment

To reduce the energy use by a thermal dryer and to improve operating efficiency, it is important to recover as much waste heat as possible. Heat recovery is accomplished using a variety of equipment. The equipment may be as simple as a duct transferring heated gas from one point to another or it may be as complex as a multifuel burner designed to combust a continuously variable fuel supply. This section describes some types of heat-recovery equipment used with drying systems.

4.3.1 Furnace Extensions

A furnace extension is an add-on to a furnace that provides an inlet for recovered gas from another source (Figure 16.9). With proper design, a furnace extension effectively

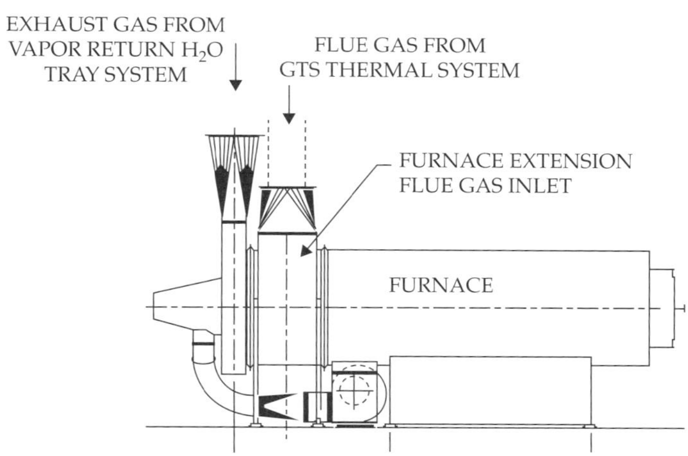

FIGURE 16.9 Direct dryer furnace with extension for waste heat recovery.

mixes recovered gas with combustion products from the burner. Furnace extensions make it possible for heat to be recovered from other sources or from the drying system itself.

Heat recovery for the biosolids dryers is most easily achieved by recirculating exhaust gas back into the dryer furnace to recover any residual heating value remaining in the exhaust. Recirculation is limited by the requirement that mass outputs from the system equal mass inputs. Therefore, the mass flow of gas leaving the system (gas that is not recirculated) must be equivalent to or greater than the combined mass flow of the various gases entering the system. Gases entering the system typically include products of combustion, excess combustion air, water evaporated from the biosolids, and infiltration air. In some drying systems, products of combustion (and excess combustion air) are not fed to the dryer.

Typical drying systems use condensers to remove water from the dryer exhaust stream to make the exhaust suitable for recirculation. The condenser serves the added function of removing particulates from the dryer exhaust stream. Blowdown from the condenser wastes some heat. The temperature of the blowdown is too low to recover any meaningful amount of heat for the drying operation, but condenser blowdown heat has been recovered in some installations for heating anaerobic digesters. In direct rotary drum dryers, recirculation rates are limited to approximately 70 to 80% of the total air flow. Recirculation serves the added benefit of reducing oxygen in the dryer gas to levels below that required to support a deflagration.

A furnace extension may also be used to recover heat from a source external to the drying system. For example, if a boiler system is used for heating an anaerobic digester, combustion gases from the boiler can be recovered in the furnace extension. One notable installation at the Jones Island Biosolids Drying Facility in Milwaukee uses turbine exhaust for 100% of the heat needed for biosolids drying using a rotary drum system. Turbine exhaust is ideal for this use because of its temperature (700 to 900 °F) and its reduced oxygen content.

4.3.2 *Heat Exchangers*

An alternative to direct-heat recovery is heat exchange from one medium to another. Heat exchangers offer a mechanism for noncontact heat transfer from a high-temperature fluid to a low-temperature fluid (Figure 16.10). Heat exchangers typically are used to transfer heat from a liquid to a liquid. For example, heat can be transferred from a high-temperature fluid such as hot oil (used in an indirect dryer) to water (used

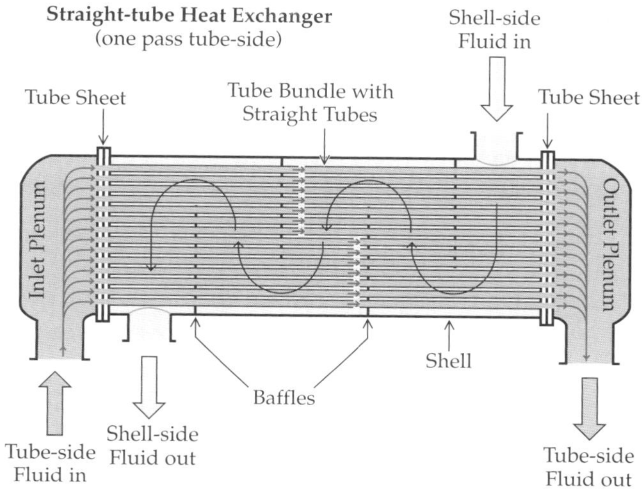

FIGURE 16.10 Example of a heat exchanger.

in a boiler system). The other approach is to transfer heat from a gas to a liquid or vice versa.

Some drying systems (e.g., Kruger belt dryer) use air-to-air heat exchangers to recover heat. These systems typically pass exhaust gas from the dryer through the hot side of the heat exchanger to the cold side to preheat incoming dryer air. This approach improves overall system efficiency by lowering the energy requirement for heating the dryer air. Another use of air-to-air heat exchangers is to preheat the burner combustion air with dryer exhaust air or with waste heat from another source.

4.3.3 *Dried Biosolids*

With the rapid increase in the cost of energy in recent years, making beneficial use of dried biosolids for fuel has become an increasingly attractive option. As fuel, dried biosolids can provide some, or even all, of the energy needed for initially drying biosolids. In this case, dried biosolids are fed to a moving grate-type combustor with an afterburner to control volatile organic compounds (VOC). The resulting heated gas is then fed through an air-to-air heat exchanger to raise the temperature of the dryer air. The cooled exhaust gas from the combustor can often be discharged to the atmosphere without further treatment, although particulate removal may be required.

Combustor ash typically is disposed of offsite as nonhazardous material unless concentrations of regulated metals exceed allowable limits.

4.3.4 *Multifuel Burner*

In some cases, an inexpensive or free fuel, such as biogas from landfills or anaerobic digesters, is available for use in the drying system. The methane content of digester or landfill gas is typically lower than that of natural gas. Landfill gas contains approximately 50 to 55% methane, and anaerobic digester gas contains 60 to 70%. Consequently, a burner designed to operate only on natural gas will not operate effectively with these alternative fuels. A multifuel burner that can fire on either natural gas or biogas, or a blend of both, can take advantage of all available fuel sources. Working with a burner supplier will enable the drying-system engineer to identify the best burner configuration for a specific situation.

4.3.5 *Multiple-Burner System*

Instead of blending gases of different methane contents, it may be more efficient to install two separate burners. Biogas can be supplied directly to a burner configured specifically for biogas in conjunction with a separate natural gas burner. If biogas supplies a fraction of the total energy demand by the drying system, then a small biogas burner supports its effective use. One convenient type of burner is a duct burner, which fits inside a recirculation or gas feed duct and heats the gases flowing through it (Figure 16.11). The duct burner is a simple and economical way to use available biogas in a drying system.

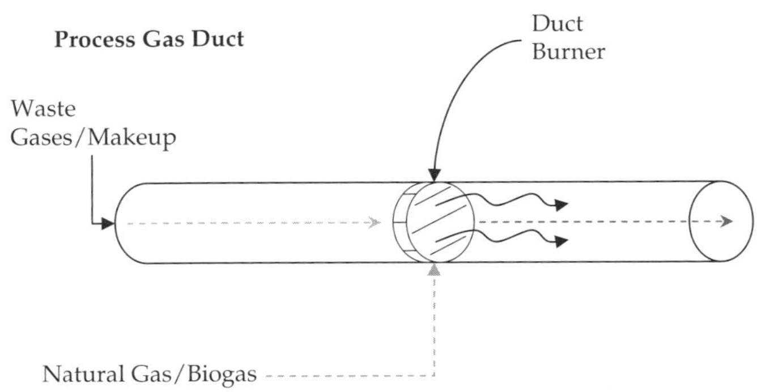

FIGURE 16.11 Burner inside process gas duct.

4.4 Potential Uses for Recovered Energy

Waste heat from a drying process comes in the form of exhaust gas, cooling water return streams, scrubber blowdown streams, and spent heating fluids. If recovered, then waste heat has several potential uses. The following sections describe some of the more common uses for heat recovered from the drying process.

4.4.1 *Process Heating*

The heat contained in the dryer off-gas offers a high-grade energy source usable in other unit processes. The exhaust gas typically passes through a condenser-scrubber for removal of the water evaporated from the solids being dried. In the process of cooling the off-gases to condense the water vapor, the cooling water is heated and combines with the water vapor to produce hot water of approximately 50 to 65 °C (122 to149 ° F). Although typically too low for building heating, the heated water is ideal for heating anaerobic digesters using spiral or tube-in-tube heat exchangers. These typically are sized for lower-temperature water to prevent fouling. Scrubber water is not used directly as a source of hot water for heat exchangers, but passes through a high-efficiency plate-type heat exchanger that recovers the heat and uses it to heat clean, recirculated water. Figure 16.12 illustrates the heat recovery scheme.

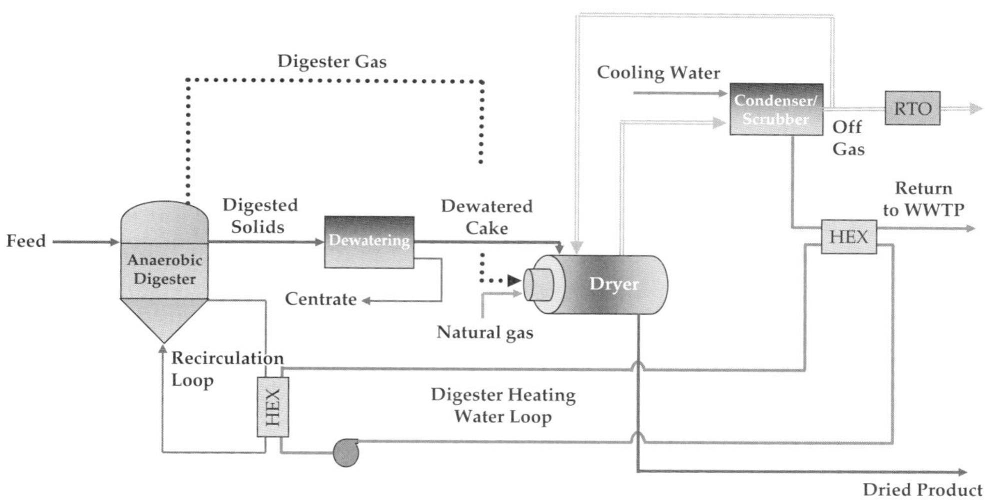

FIGURE 16.12 Recovered heat for digester heating.

4.4.2 *Product Energy Content*

Because dried biosolids contain organic compounds, they have an inherent energy value. The manual, *Sludge Incineration: Thermal Destruction of Residues*, reported a heating value of 23 700 kJ/kg of volatile solids for biosolids (WEF, 1992). Once water is removed by drying, biosolids offer a net positive heat value for combustion. As shown in the manual (1992), the energy content of wet cake is related to the solids concentration according to the following equation:

$$\Delta H_{\text{wet cake}} = (\%TS/100) \times (\%VS/100) \times (HHV) \qquad 16.1$$

where

$\Delta H_{\text{wet cake}}$ = gross heating value on wet cake basis, kJ/kg;
$\%TS$ = total solids content, %;
$\%VS$ = volatile solids concentration in total solids, %; and
HHV = high-heating value of volatile solids, kJ/kg.

Some heat content of the biosolids is lost through evaporation of water during combustion. Thus, an efficiency factor (η) accounts for energy consumed by water (WEF, 1992).

$$\eta = 1 - 270\,000 \times (100 - \%\text{TS})/(\%\text{TS} \times \%\text{VS} \times \text{HHV}) \qquad 16.2$$

As a result, the net heating value (ΔH_{net}) on a wet cake basis is illustrated as follows (WEF, 1992):

$$\Delta H_{\text{net}} = \Delta H_{\text{wet cake}} \times \eta \qquad 16.3$$

When the net heating value of wet cake exceeds approximately 4600 kJ/kg (2000 Btu/lb), biosolids burn without an auxiliary fuel. Consequently, once the total solids concentration approaches 30%, biosolids combust autogenously. Dried biosolids have a much higher solids content than needed to burn autogenously, which means excess heat is available from biosolids combustion. Consequently, dried biosolids have a fuel potential similar to that of low-grade coal.

5.0 OFF-GAS TREATMENT

As with most biosolids treatment processes, odor control for heat drying should be given adequate consideration to provide a suitable working environment and to avoid public complaints. This chapter provides an overview of odor control considerations for heat drying, including odor sources and characteristics, odor control strategies and treatment equipment, and design issues.

5.1 Sources and Characteristics of Odors from Heat Drying

Both direct and indirect dryers include an area where dewatered biosolids are heated to temperatures ranging from 260 to 482 °C (500 to 900 °F). It is hypothesized that the drying process produces odors in at least two ways: first by volatilizing odorous compounds in the solids, and second as a result of the elevated temperatures causing reactions that create a burnt odor.

Thermal drying incorporates several distinct odor sources, such as process gas produced by drying, ventilation air from material handling, and ventilation air from storage. Each source has unique odor characteristics, which will determine the appropriate treatment options.

5.1.1 *Exhaust Gas from Dryers*

During drying, when water is converted to vapor by the heating process and removed from solids, odorous compounds will be produced through volatilization and heat reactions. These odorous compounds will be entrained in the water vapor and the process gas stream that must be continuously removed from the drying area.

5.1.2 *Dried Product Material Handling Equipment*

In the material handling equipment, dried product will be at elevated temperature under saturated conditions. It is thought that odors are produced by compounds emanating from the material and from the limited amount of process gases that are removed along with the material.

5.1.3 *Dried Product Storage Areas*

Odors produced in storage areas emanate from the product itself. Because the product is often cooled and ventilated before storage, much less odor typically is produced in storage areas than in material handling areas.

5.1.4 *Dryer Rooms*

Drying systems are typically enclosed, and processing areas are maintained at negative pressure to prevent escape of gas and odors. Measurements taken in March 2001 from the dryer room of a Florida drying facility indicated an odor detection threshold concentration of 150 and an odor recognition threshold concentration of 90. These relatively low readings indicate that the room ventilation air would not be a significant source of odors.

Table 16.1 summarizes the odor data from different areas of the drying processes based on testing performed at a limited number of facilities. As such, the results were influenced by site-specific solids characteristics and the types of drying processes used. The information on the odor emissions is quantified according to areas of the drying process where they originated, which in turn should provide useful information regarding the relative production of odors.

TABLE 16.1 Odor characteristics from various drying sources (Hoener et al., 2007).

Sample	Odor detection threshold units	H_2S concentration, ppm	Total volatile organic compounds, mg/Nm^3
Dryer process gas[a]			
1	190993	0.2	180.98
2	416803	3.4	163.06
3	367374	1.8	65.97
4		0.25	
Average	325057	1.4	136.67
Dried product material handling			
1	16412	9.8	26.12
2	14803	13.4	3.85
3	11907	21.0	11.26
4		16.8	5.38
Average	14376	15.3	11.65
Dried product storage[b]			
Storage area ventilation	400	0.017	
Ambient conditions at storage area	80	0	

[a] From a rotary drum drying system in the British Isles (May 2006).

[b] From the product storage silo of a rotary drum drying system at the Morris Forman Wastewater Treatment Plant, Louisville, Kentucky (May 2006).

Odor threshold concentration is a measure of the number of times an odor concentration must be diluted with an equal volume of air before the odor is undetectable. Odor threshold concentration indicates that the process gases from the dryer equipment are highly odorous. The odor threshold concentrations of scavenging air from the material handling equipment are approximately 10 times lower than from the dryer area; this scavenging air, however, is still a significant odor source. Surprisingly, the H_2S concentration in the scavenging air from the material handling equipment is higher than from the dryer area. Given the counterintuitive nature of the relative H_2S concentration, independent confirmation of this result is needed to ensure that this reading was not an anomaly. Scavenging air from storage areas is a much smaller source of odors than the dryer areas or the material handling equipment.

To put these concentrations in perspective, odor concentration detection thresholds for the most odorous areas, the interior of a thickened waste activated sludge (TWAS) tank and solids blending tank, at the Morris Forman WWTP in Louisville, Kentucky, were 11 000 and 7100, respectively.

5.2 Odor Control Strategies and Treatment Systems

Odor control treatment systems for thermal drying include thermal destruction and discharge to aeration basins. Selection of a suitable system should be based on the volume and characteristics of the gas to be treated and on the applicable air pollution emission requirements.

5.2.1 *Thermal Destruction*

With thermal destruction, odorous compounds are combusted to create compounds with no or greatly reduced odors. For this treatment to be effective, complete combustion must occur, which occurs only if the treated gas is heated to a sufficiently high temperature for an adequate length of time. Thermal destruction for drying systems has been accomplished either by using thermal oxidizers or by incinerating the process gas in equipment such as furnaces, boilers, or oil heaters.

5.2.1.1 Thermal Oxidizers

All thermal oxidizers include a burner and combustion chamber. They are effective for odor control because they can be designed for optimum destruction conditions of temperature, typically 815 °C (1500 ° F), and retention time, typically 0.75 to 1 second. Thermal oxidizers are classified as regular, recuperative, or regenerative:

- Regular thermal oxidizers do not include heat recovery and, therefore, require relatively high heat input for the amount of gas treated. They are less expensive than other types of oxidizers.
- Recuperative thermal oxidizers include a gas-to-gas heat exchanger to recover heat from treated gas exiting the combustion chamber to preheat the incoming gas. They typically recover 50 to 60% of the heat from the treated gas and use only 40 to 50% the amount of fuel used by regular thermal oxidizers. Their capital cost is higher than that of a regular thermal oxidizer, but lower than the cost of a regenerative thermal oxidizer.
- Regenerative thermal oxidizers consist of multiple heat exchanger canisters, which contain ceramic media for storing heat and a gas distribution manifold. The heat recovery process operates in cycles. While one canister is recovering heat from the gas leaving the combustion chamber, another canister is preheating the gas entering the combustion chamber. Periodically, the function of the canisters will be reversed. Regenerative thermal oxidizers can be designed to have high thermal efficiency—it is common that over 95% of the heat needed can be supplied by the heat recovery system. Because regenerative thermal oxidizers are more complex than regular or recuperative thermal oxidizers, their capital costs are higher and they have greater maintenance requirements.

Regular and recuperative thermal oxidizers are less efficient than regenerative oxidizers and are best suited for treating small gas flows from small dryers. Smaller systems will have relatively small fuel cost, which may not justify the high capital cost and complexity of a regenerative thermal oxidizer. For larger dryers, a regenerative thermal oxidizer should be used to treat process gas from the dryer area and material handling areas.

5.2.1.2 Combustion Equipment

At some facilities, the combustion equipment for the drying process has been used to treat odorous gas to avoid the costs of separate pollution control equipment. Process gas is directed to the combustion equipment as part of the required combustion air. The disadvantage of this practice is that the equipment is designed to produce heat, not to destroy odors. Most of this equipment does not have sufficient combustion area or volume to provide adequate retention time at elevated temperatures. The odor complaints at some facilities that use furnaces and boilers for odor treatment indicate that this treatment option does not always adequately reduce odors.

5.2.2 *Discharge to Aeration Basins*

Small modular dryers with minimal material handling equipment may produce relatively low volumes of process gas to be treated for odor removal. At several facilities, the process gas has been directed to diffusers located in aeration basins to be scrubbed using aeration-basin liquid.

5.3 Design Considerations

This section addresses design considerations for regenerative thermal oxidizers, which are the most prevalent means of treating process gas streams from larger thermal drying systems.

5.3.1 *Number of Canisters*

Regenerative thermal oxidizers include heat exchanger canisters filled with ceramic media to absorb and store heat and to transfer it to the process gas to reduce fuel consumption. The gas will cycle between canisters, alternately accepting heat from the process gas treated in the combustion chamber and preheating gas flowing to the combustion chamber. Two, three, or more canisters may be included. Units with two canisters have the least capital cost, but they cannot treat the entire process gas stream for odors. When the intake canister begins to serve as the exhaust canister, a portion of the gas in the bottom of the canister will flow directly into the discharge manifold, bypassing the combustion chamber. Although the percentage of the untreated gas is typically 5 to 10%, this will reduce overall destruction performance. In addition, because the untreated gas will be discharged undiluted during the switchover, the chance for development of detectable odors will increase. Providing regenerative thermal oxidizers with at least three canisters will ensure that the entire process gas stream is treated. With this arrangement, a third canister is used during the switchover to accept the exhaust gas, while the contents of the canister that had been serving as the intake canister will be purged to the combustion chamber.

5.3.2 *Types of Media*

The two main types of media typically used in regenerative thermal oxidizers at drying facilities are random and structured. Random media consists of ceramic saddles arranged in no particular order that have high thermal efficiency but a relatively high pressure drop. Structured media consist of ceramic blocks with small, parallel passages, similar to those in honeycomb, that also has high thermal efficiency but a relatively low pressure drop. There are differing opinions as to which type of media requires less maintenance: structured media may be subject to more frequent blockages, but the blockage area is typically limited to the top two layers, which can be readily cleaned or replaced (when cleaning is no longer effective). Random media may not plug as frequently, but when cleaning is no longer effective, it must be completely replaced. Experience in Europe suggests that oversizing the regenerative thermal oxidizer and using random media is an effective method to minimize maintenance.

Recently, one Florida facility experimented with an "egg-crate" type media in an effort to reduce maintenance. The media is laminated, extruded in sheets, and glued together to form blocks. The initial results are promising—the need for cleaning has decreased from every 1.5 to 2 months to every 3 to 5 months. Although concern has been expressed about the durability of the glued blocks, after 1.5 years of operation no problems have been experienced.

5.3.3 *Media Fouling*

Over time, particulates will build up in the ceramic media and can increase the pressure drop through the media. When the pressure drop becomes greater than the design capacity of the regenerative thermal oxidizer fan, the media must be cleaned or replaced. Media typically can be cleaned two to three times before it must be replaced. Information from five facilities indicated a broad range of cleaning frequency: from every six weeks to once a year. The cleaning process takes two or three workers three to four hours, requiring an overall shutdown of two to three days to allow time for time for the media to cool before cleaning and to be heated after cleaning. Cleaning consists of spraying the media with water and, in some cases, either vacuuming or replacing the media at the top of the canisters.

6.0 OPERATION AND MAINTENANCE

The operation and maintenance of dryer systems can be complex because of the amount of main process and ancillary equipment. A set of standard operating procedures (SOP) that covers prestartup, startup, normal operation, shutdown, troubleshooting, and preventive maintenance activities, should be established for each dryer system. The SOP will be site-specific and will depend on dryer type, solids characteristics, and other variables. Procedures outlined in this section, however, provide guidelines for safe operation and maintenance of dryer systems.

6.1 Prestartup

Prestartup includes activities the operator must perform before startup of the drying equipment. These tasks include, but are not limited, to the following:

- Reviewing the operations log and ensuring all alarms and process issues noted in the last log entry have been addressed;
- Ensuring that hatches, doors, and ports are closed and sealed as required for operation and bolts or latches are secure;
- Placing all valves in the appropriate operating position;
- Checking the associated motor control center and verifying that equipment has not been locked out by the maintenance staff;
- Inspecting bins or piping for plugging and clearing debris from the area;
- If applicable, ensuring utilities such as water, compressed air, and natural gas are available to the dryer system;
- Ensuring that all equipment is in the correct mode of operation for control from the supervisory control and data acquisition (SCADA) system;
- Clearing all alarms from control panels and from the operator interface terminal (OIT).

6.2 Startup

Startup is the period from when an automatic start sequence is initiated until steady-state operation is achieved. Startup of a dryer system should be initiated only by qualified personnel. Dryer systems are typically started in automatic start mode, which implies a software program will start individual subsystems in a controlled manner to avoid electrical overload or damage to equipment. The dryer operator should monitor the system during startup for alarm conditions or other issues that may arise.

Once the dryer is online and at operating temperature, the operator should check for correct pressures and sufficient water and airflow rates, and should listen for unusual noise or vibrations from the equipment. Once the operator is satisfied that the dryer is operating as intended, feed to the dryer system can be initiated. Biosolids should be fed into the dryer in a controlled manner to avoid overloading the system or causing large variations in process parameters such as inlet and outlet temperatures and pressure. The operator should monitor the feed solids concentration and dryness of the product at least every 10 to 15 minutes until the dryer reaches steady-state conditions and normal operating procedures can be followed.

6.3 Normal Operation

Normal operation is reached when steady-state operating conditions have been established following startup. Normal operation will involve periodic adjustments to the system to maintain steady-state conditions. Adjustments should be made in small increments, one at a time to avoid large variations in the process. The operator should wait at least 10 minutes between process adjustments to see the effect of an adjustment and to minimize swings in process parameters. Typical tasks during normal operation include

- Monitoring feed and product solids concentration once each hour or more frequently as recommended by the system supplier.
- If applicable, coordinating with the dewatering system operator to ensure availability of consistent dewatered cake solids to the dryer system.
- Monitoring process system parameters including pressures, temperatures, and flow of air, biosolids, water, and natural or digester gas at the OIT and verifying that they are in the correct control range.
- Using the OIT to trend process parameters to ensure that the process is operating within defined limits and that it maintains steady-state conditions.
- Filling out plant data log sheets and log books and filing the documentation in the appropriate location for use by the supervisor and next shift personnel.
- Taking representative samples for laboratory analysis as required by the plant SOP. Poor material characteristics, whether solids concentrations outside acceptable ranges or too much fiber in the feed, can cause dryer malfunction. Solids concentrations and material characteristics have to be carefully controlled at several locations in the drying system to avoid malfunctions.

Parameters that require close monitoring include solids concentrations of the dewatered cake, dryer feed mix, and dried product. The location of sampling ports should be a critical consideration during system design.

6.4 Shutdown

Shutdown is initiated by the operator at the OIT using an automatic stop sequence, which will stop dryer subsystems in a controlled manner. Sufficient time is needed for the shutdown sequence to allow dried material to clear the system. The operator should monitor pressures, temperatures, and flow rates during shutdown to verify that all rotating equipment has come to a complete stop and should complete all operations logs as required.

6.5 Troubleshooting

Troubleshooting of the dryer process begins by understanding normal operating limits for monitored parameters. If the operator observes a parameter deviating from the defined operating limits, then he or she should take appropriate corrective action based on the information presented at the OIT.

The staff at each plant should develop their own troubleshooting guide by forming an operations team to address various scenarios and how each should be handled. The best troubleshooting guides are in tabular format and should contain the problem, its causes, and solutions. Entries in the table should be concise to allow the operator to quickly identify the solution for each problem.

Establishment of the troubleshooting procedures starts with the vendor-supplied operation and maintenance manuals; although typically related to individual equipment components, the manuals are a good starting point. Additional troubleshooting guidance can be obtained from users of similar dryer systems.

6.6 Preventive Maintenance

Preventive maintenance minimizes unwanted situations or emergency shutdowns of equipment. It should be scheduled for times when most maintenance personnel are available to do the work. The duration of preventive maintenance depends on system complexity and availability of personnel.

The type of dryer will determine preventive maintenance requirements. The system vendor should supply a comprehensive list of preventive maintenance tasks for each piece of dryer equipment. Some of the most common preventive maintenance tasks include, but are not limited, to the following:

- Checking and adjusting drive belts;
- Checking, adjusting, and greasing bearings;
- Checking lubrication levels;
- Checking for unusual or excessive wear at typical wear points such as rollers, belts, seals, fan blades, mixer blades, pump stators, and other points;
- Checking and replacing air and oil filters as needed;
- Checking, cleaning, and calibrating instrumentation such as dust monitors, level indicators, flame sensors, carbon-dioxide sensors, and dissolved oxygen sensors.
- Checking the motor control center for tripped or failed circuits.

Preventive maintenance is crucial to safe and stable operation of the dryer system. It instills confidence in the operating staff and will ensure consistent product quality.

7.0 DRYING SAFETY

This section discusses safety practices at a drying facility. It is not intended to be all-inclusive but to identify basic needs and sources of information for specific applications. Referenced codes, industry standards, and other sources presented here should be consulted for details that may apply to a specific case. Also, it is suggested a hazardous operations study be performed during design of the dryer system to allow for implementation of appropriate engineering controls during detailed design.

7.1 Hazards Associated with Thermal Drying of Biosolids

Dried biosolids are a combustible material that will burn in the presence of air and an ignition source. In addition to the typical hazards associated with combustible materials, heat drying of biosolids can pose some unique hazards. These include overheating because of auto-oxidation of dried material and unstable operating conditions caused by fluctuations in the moisture content of the dewatered feed solids. Fire hazards exist in the drying area where dried material can come in contact with ignition sources from the heating process. Fire hazards also exist in the material handling and storage areas, particularly where material remains for extended periods under conditions that allow the auto-oxidation process to generate significant amounts of heat.

Combustible dust is produced by handling of the dried product. Dust can accumulate if large amounts of it are produced from solids that are too dry and dust is not properly removed from equipment. Combustible dust can be an explosion hazard if it is suspended in air in sufficient concentrations near an ignition source. Explosion hazards can develop throughout the drying facilities where combustible dust has built up. These areas include where dried material is produced and handled and areas where large volumes of dust could accumulate, such as recycle bins and storage silos.

Dried biosolids contain biological material that can undergo auto-oxidation if it is rewetted by condensation in storage bins or if it contains too much moisture after drying. The auto-oxidation process generates heat that, if not dissipated, could produce smoldering and, if left unattended, develop into an uncontrolled fire. Smoldering materials and fire can trigger ignition leading to the explosion of nearby combustible dust. The smoldering material can produce carbon monoxide, which is a combustible gas; although opinions are divided whether explosive levels would ever be produced in a drying system.

Treatment plants have only limited control over the constituents of the influent wastewater and in the solids processed for ultimate disposal. Excessive fiber or grease in influent can cause problems when solids are heat dried. Large amounts of fiber in the feed can clog air filters, screens, and product coolers, and can lead to dust production. Grease can volatize during drying and produce combustible vapors. Deposits of grease inside equipment can lead to blockages and formation of combustible material.

7.2 Safety Provisions for Dryers

Ensuring the safety of drying systems requires both prevention and mitigation.

7.2.1 *Prevention and Mitigation Systems*

Some of the more important prevention systems include

- Inert atmosphere to ensure the oxygen content stays below the lower explosive limit;
- Water deluge for drenching material in the drying area if monitoring indicates incipient fire conditions;
- Temperature control to ensure that temperatures do not rise to unsafe levels and material does not become too dry, which would generate excessive dust; and to activate other preventive systems in case of elevated temperatures;

- Product cooling to reduce the potential for auto-oxidation;
- Ventilation to keep dust concentrations below the lower explosive limit and to remove moisture that could lead to buildup of material inside equipment; and
- Spark detection to detect glowing or burning material at key transition areas, such as the discharge of the dryer or the gas product separator, and to activate deluge systems if necessary.

Some of the more important mitigation systems for biosolids dryers include

- Venting for the product storage areas of drying systems to minimize hazards from deflagration; and
- Isolation to ensure that a deflagration does not spread to other areas where large amounts of dried material are stored such as recycle bins or storage silos, or areas of high concentrations of dust, such as baghouses.

7.2.2 *Monitoring Systems*

Adequate instrumentation is important for providing operators with sufficient information about the state of dryer operation and for alerting them to upset conditions. Some of the critical components of the instrumentation system include

- Temperature monitoring;
- Oxygen and carbon dioxide monitoring; and
- Pressure monitoring.

7.3 Regulations, Codes, and Standards

Many national, state, and local regulations, codes, and standards are applicable to the design and operation of drying facilities. Some regulations and codes will be enforceable based on their applicability for a particular facility, considering the operating entity and location; others will serve as guidelines. In all cases, the underlying safety principles should be considered and incorporated in the design and operation of drying facilities. The following sections include a listing of some of the regulations, codes, and standards that are applicable to drying facilities. The most recent version of all national, state, and local regulations, codes, and standards should be consulted for design of drying facilities.

7.3.1 *Occupational Safety and Health Standards*

The Occupational Safety and Health Act established general national Occupational Safety and Health Standards (29 CFR 1910) (U.S. Department of Labor, 1970). These standards apply directly to WWTPs, and it is typical practice for facilities to meet these standards. Some of the items of particular application to drying facilities include

- Walking work surfaces;
- Means of egress;
- Occupational health and environmental control (ventilation and noise exposure);
- Personal protective equipment;
- Permit requirements for confined spaces;
- Control of hazardous energy (lock out/tag out);
- Fire protection;
- Machinery and machine guarding; and
- Electrical requirements.

7.3.2 *Building, Fire, and Mechanical Codes*

Local codes typically include safety requirements that are applicable to drying facilities. Code requirements will be location-specific, based on the codes adopted for the area. The International Building Code and International Fire Code are International Code Council model code regulations that are applicable to drying facilities.

7.3.3. *National Fire Protection Association*

The National Fire Protection Association (NFPA) maintains codes and standards that can apply to different aspects of fire and explosion safety for municipal wastewater solids drying equipment. Relevant codes are as follows:

- *NFPA 30: Flammable and Combustible Liquids Code* (National Fire Protection Association, 2008a);
- *NFPA 31: Standards for the Installation of Oil-Burning Equipment* (National Fire Protection Association, 2011a);
- *NFPA 54: National Fuel Gas Code* (National Fire Protection Association, 2009);

- *NFPA 68: Guide for Venting of Deflagrations* (National Fire Protection Association, 2007);
- *NFPA 69: Standard on Explosion Prevention Systems* (National Fire Protection Association, 2008b);
- *NFPA 86: Standard for Ovens and Furnaces* (National Fire Protection Association, 2011b);
- *NFPA 654: Standard for the Prevention of Fire and Dust Explosions from the Manufacturing, Processing, and Handling of Combustible Particulate Solids* (National Fire Protection Association, 2006); and
- *NFPA 820: Fire Protection in Wastewater Treatment and Collection Facilities* (National Fire Protection Association, (2008c).

7.3.4 *Insurance and Other Industry Standards*

In addition to regulatory codes and standards listed above, insurance and other industry standards, such as burner safety regulations, can also apply to drying systems. It is important to determine the facility's insurance carrier and requirements to ensure that design meets these specifications. In many cases, insurance standards can be more stringent than local codes or NFPA regulations, and should be considered, especially when specific hazards are present.

7.4 Personnel Training and Safety

Drying systems consist of many different types of equipment and subsystems; operation of each will affect interrelated subsystems. Therefore, adequate training of dryer operators, especially during initial startup of the system, is essential. Training materials and information from initial training and lessons learned by experience should be passed on to new operators.

Drying system safety depends on a strong plant-wide program that creates a culture emphasizing safety in all decisions and procedures. Included in such a program are established security roles for control systems and lockout/tagout procedures. Security roles allow only qualified personnel to modify safety controls. This will ensure that alarm conditions and interlocks are not modified without adequate review and understanding of potential effects. Adequate lockout/tagout procedures are critical for drying systems because many equipment safety interlocks are dependent on the electronic control system.

The unique hazards associated with biosolids drying cannot be prevented solely by providing protection systems. They can be minimized or eliminated only by proper operation of the drying system, which underscores the importance of sound operating practices.

8.0 AIR PERMITTING

This section discusses the issues related to air permitting for biosolids dryers. The Clean Air Act Amendments of 1990 (CAAA) define stationary sources of air emissions as "any building, structure, facility or installation which emits any air pollutant" (42 U.S.C. § 7401-7671q; 40 CFR 50–88). Air construction permits are required before construction and operation of stationary sources and modification to existing stationary sources. The CAAA also defines modification as, "Any physical change in, or change in the method of operation of, a stationary source which increases the amount of air pollution emitted by such source or which results in the emissions of any air pollutant not previously emitted" (42 U.S.C. § 7401-7671q; 40 CFR 50–88). Direct and indirect dryers are stationary sources, and combustion of fuels of any form in these stationary sources will result in air emissions. Consequently, air construction permits are required before implementing sludge dryer projects.

The primary goal for facilities contemplating the installation of a biosolids dryer is to obtain an air construction permit with minimal requirements and least number of permitting hurdles. Air construction permits are classified as minor and major source construction permits, depending upon the extent of change in emissions from either a new installation or modifications to an existing installation. Typically, minor source air construction permits have fewer permitting hurdles to overcome and regulatory requirements to meet. Major source permits are implemented under the federal CAAA New Source Review (NSR) provisions under two programs: the Prevention of Significant Deterioration (PSD) program outlined in 40 CFR 52.21; and the Nonattainment NSR (NNSR) program outlined in 40 CFR 51 and 52 (42 U.S.C. § 7401-7671q; 40 CFR 50–88). The PSD regulations apply to major stationary sources and major modifications at major existing sources undergoing construction in areas designated as attainment or unclassifiable. Air quality in a given area is designated as being in attainment for a pollutant if the monitored concentrations of that pollutant in the ambient air are less than the applicable National Ambient Air Quality Standards [NAAQS] or is designated as unclassifiable if sufficient monitoring data are not available to make an attainment decision. A given area is classified as nonattainment for a pollutant if the monitored concentrations of that pollutant in the ambient air in the area are above the NAAQS. The PSD regulations are designed to ensure that air quality in attainment areas does not significantly deteriorate or exceed the NAAQS while providing a margin for future industrial and commercial growth. NNSR applies in regions that are classified as nonattainment for a particular pollutant. Biosolids dryers are not listed as belonging to any of the named source categories. A given area is classified as nonattainment for a pollutant if the monitored concentrations of that pollutant in the ambient air in the area are greater than the NAAQS.

8.1 Typical Air Permitting Process

The air permitting process is initiated by the owner or operator (referred to as the "applicant") of the proposed biosolids drying facility. The air permitting process is summarized in Figure 16.13.

FIGURE 16.13 Air permitting process flow chart.

8.1.1 Establishing the Engineering Design

Establishing the preliminary/conceptual design for the project is an important first step. The engineering design typically identifies the drying technology, air emission sources, air-quality control equipment, and type of fuels (primary and supplemental) combusted for the given project site based on a set of design criteria. The emission sources at a new drying facility can be grouped into two main categories:

1. Biosolids dryer—direct or indirect system, including post-combustion off-gas treatment;
2. Dried product conveyance, handling, storage, and transportation.

8.1.2 Estimating Potential to Emit

The CAAA regulates six criteria air pollutants and 187 hazardous air pollutants (HAPs). The six criteria air pollutants are nitrogen dioxide (NO_2); sulfur dioxide (SO_2); carbon monoxide (CO); particulate matter (PM_{10} and $PM_{2.5}$); ozone (volatile organic compounds or VOCs are considered surrogates for ozone and are regulated); and lead (Pb). In addition to these pollutants, some states and local agencies also regulate state-specific air toxics.

The EPA defines potential to emit as, "[T]he maximum capacity of a stationary source to emit under its physical and operational design. Any physical or operational limitation on the source to emit an air pollutant, including air pollution control equipment and restrictions on hours of operation, or on the type or amount of material combusted, stored, or processed, shall be treated as part of its design if the limitation is enforceable by the (EPA) Administrator" (42 U.S.C. § 7401-7671q; 40 CFR 50–88). Potential emissions of regulated pollutants must be estimated as part of the air construction permit application. The major or minor source permit applicability for a dryer project typically depends on the anticipated annual emission rate in tons per year from the proposed project (for greenfield stand-alone facility only) and the major or minor source classification of the current site (for modification).

8.1.2.1 Criteria Pollutants

The potential emissions of NO_x, CO, PM/PM_{10}/$PM_{2.5}$, VOCs and SO_x from a proposed dryer project can be based on different sources of emission data, which include emissions and performance guarantees provided by the vendors of dryer and emissions control equipment, emission factors from U.S. EPA's *Factors Emissions Factors & AP 42, Compilation of Air Pollutant Emission Factors*, which are based on stack tests conducted at similar facilities elsewhere, and other engineering calculations. If digester gas is used, then gas must be analyzed to determine the concentrations of sulfur and siloxanes in it.

8.1.2.2 Air Toxics and Hazardous Air Pollutants

In some states and local jurisdictions, air toxics regulations will require the applicant to estimate emissions of specifically identified pollutants, some of which may also be classified as federal HAPs. Biosolids dryer facilities typically emit trace amounts of metallic particulates such as arsenic, cadmium, mercury, and nickel; trace amounts of volatile organic HAPs; and other pollutants such as hydrogen sulfide and ammonia. Emissions of metallic particulates can occur from both the dryer and the materials handling operations. The amounts of metallic particulates in such emissions typically are determined by multiplying the expected metal concentrations in the biosolids and the suspended PM and PM_{10} emission rates for the dryers and the solids storage and load-out, respectively. Metal concentration data for biosolids are obtained by solids analysis.

Emissions of mercury from direct-fired sludge dryers are regulated by the National Emission Standards for Hazardous Air Pollutants (NESHAP). The mercury NESHAP (40 CFR part 61, subpart E) *National Emission Standard for Mercury*, limits the emissions of mercury from sludge drying plants (that use direct-fired dryers only) to less than 3200 g (7.1 lb) of mercury per 24-hour period (42 U.S.C. § 7401-7671q; 40 CFR 50–88). The mercury NESHAP also requires affected facilities to conduct stack testing or sludge sampling, and relevant recordkeeping and reporting that are triggered if mercury emissions exceed 1580 g (3.5 lb) per 24-hour period.

8.1.2.3 Odor

Most state and local agencies have nuisance odor regulations that limit odor emissions beyond the property line. Odor emissions can be estimated based on design input and outlet loading rates to the odor control units. Hydrogen sulfide is widely accepted as a surrogate for odor emissions.

8.2 Development of Air Permitting Strategy

Strong interdependence exists between air permitting and engineering design for biosolids drying projects. Several iterative steps may be needed before an optimal process design is developed that would result in an installation that can pass the air permitting process. Optimal project characterization is an important first step in the air permitting process because it can affect several project considerations and should include

- Facility location;
- Facility size;
- Facility layout;
- Emission rate;
- Fuels;
- Stack parameters;
- Type of fuel pretreatment or add-on emission control technology to be selected; and
- Operational flexibility.

8.3 Project Schedule

An air construction permit is needed before starting construction of a biosolids dryer project. As part of the project schedule and planning, it is important to budget time for

- Iterations to be performed during the preliminary or conceptual design when the project is being optimally characterized;
- Preparation of air construction permit application documents; and
- Agency review.

Approval of minor source permit reviews typically takes three to six months after a complete application has been submitted. Agency review and approval of major source review can take from 12 to 24 months. It is, therefore, important to include an air permitting schedule into the overall project schedule.

9.0 REFERENCES

Clean Air Act. *U.S. Code*, Section 50–88, Title 40, 1963.

Clean Air Act. *U.S. Code*, Section 7401–7671, Title 42, 1990.

Hoener, W.; Santha, H.; Bates, R.; Taylor, R. (2007) Odor Control Considerations for Biosolids Heat Drying; *Proceedings of the Water Environment Federation/American Water Works Association Joint Residuals and Biosolids Management Conference*, Denver, Colorado, April; Water Environment Federation: Alexandria, Virginia.

National Fire Protection Association (2006) *NFPA 654: Standard for the Prevention of Fire and Dust Explosions from the Manufacturing, Processing, and Handling of Combustible Particulate Solids*, 2006 ed.; National Fire Protection Association: Quincy, Massachusetts.

National Fire Protection Association (2007) *NFPA 68: Guide for Venting of Deflagrations*, 2007 ed.; National Fire Protection Association: Quincy, Massachusetts.

National Fire Protection Association (2008a) *NFPA 30: Flammable and Combustible Liquids Code*, 2008 ed.; National Fire Protection Association: Quincy, Massachusetts.

National Fire Protection Association (2008b) *NFPA 69: Standard on Explosion Prevention Systems*, 2008 ed.; National Fire Protection Association: Quincy, Massachusetts.

National Fire Protection Association (2008c) *NFPA 820: Fire Protection in Wastewater Treatment and Collection Facilities*, 2008 ed.; National Fire Protection Association: Quincy, Massachusetts.

National Fire Protection Association (2009) *NFPA 54: National Fuel Gas Code*, 2009 ed.; National Fire Protection Association: Quincy, Massachusetts.

National Fire Protection Association (2011a) *NFPA 31: Standards for the Installation of Oil-Burning Equipment*, 2011 ed.; National Fire Protection Association: Quincy, Massachusetts.

National Fire Protection Association (2011b) *NFPA 86: Standard for Ovens and Furnaces*, 2011 ed.; National Fire Protection Association: Quincy, Massachusetts.

U.S. Department of Labor (1970) Occupational Safety and Health Standards. *Code of Federal Regulations*, Title 29, Chapter XVII, Part 1910; *Fed. Regist.* **39**, 125.

U.S. Environmental Protection Agency (1993) Standards for the Use and Disposal of Sewage, Final Rule. *Code of Federal Regulations*, Part 503, Title 40, 1993; *Fed. Regist.*, **58**, 9248.

U.S. Environmental Protection Agency (2011) Technology Transfer Network Clearinghouse for Inventories & Emissions; *Factors Emissions Factors & AP 42, Compilation of Air Pollutant Emission Factors* http://www.epa.gov/ttn/chief/ap42/ (accessed August 2011).

Water Environment Federation (1992) *Sludge Incineration: Thermal Destruction of Residues,* Manual of Practice No. FD-19; Water Environment Federation: Washington, D.C.

Water Environment Federation; American Society of Civil Engineers; Environmental and Water Resources Institute (2009) *Design of Municipal Wastewater Treatment Plants*, 5th ed.; WEF Manual of Practice No. 8; ASCE Manual and Report on Engineering Practice No. 76; McGraw-Hill: New York.

Chapter 17

Thermal Oxidation

(continued)

1.0 INTRODUCTION

Thermal oxidation (referred to as incineration) has been used to manage municipal wastewater solids since the early 1930s. Approximately 17 to 22% of the wastewater solids generated in the United States are currently being managed via thermal oxidation. Thermal oxidation is a viable, environmentally sound, and cost-effective wastewater solids management option that is used by wastewater treatment plants (WWTPs) throughout the United States, Canada, Europe, and Asia.

Thermal oxidation has several potential advantages over other wastewater solids management options:

- **Detoxification.** Degradation of potentially toxic organic chemical compounds, pharmaceuticals, and personal care products that may otherwise create adverse environmental effects.
- **Pathogen control.** Destruction of all pathogens contained in the wastewater solids.

- **Maximum volume reduction.** Evaporation of water and combustion of organic matter results in an 85 to 95% reduction in volume and weight.
- **Energy recovery.** Beneficial reuse of excess heat in the hot exhaust gases significantly reduces the use of auxiliary fuels (e.g., natural gas, fuel oil). In some cases, steam for building comfort heat and unit processes is produced from the excess heat by using waste-heat boilers. In other cases, electricity is generated by sending the steam generated in the waste-heat boilers to steam turbine generators. All of these beneficial reuse options reduce the overall cost of incineration.
- **Small footprint.** Incineration has lower land requirements compared to several other biosolids management options.
- **Storage requirements.** Significant reduction in both onsite and offsite storage requirements for the dewatered solids and the final product.
- **Final product.** Production of inert nontoxic/nonhazardous ash that can be beneficially reused.
- **Reduced truck traffic.** Substantial reduction in truck traffic, resulting in low hauling costs, land application fees, and tipping fees.

On the other hand, several challenges are associated with thermal oxidation:

- **Initial capital costs.** Initial capital cost for thermal oxidation is frequently greater than for the other management options.
- **Skilled personnel.** Experienced and licensed operators are required for thermal oxidation and energy recovery.
- **Permitting requirement.** Obtaining required permits for new incinerators can be more complex than obtaining required permits for other management options.
- **Misconceptions.** Several misconceptions concerning the incineration of wastewater solids exist including the belief that incineration is not a viable, cost-effective, or environmentally friendly method for managing wastewater solids.

This chapter focuses on thermal oxidation of wastewater solids and associated energy recovery methodologies. For more detailed information on thermal oxidation of wastewater solids, including combustion theory and fundamentals; air pollution control devices; ash handling and management; ancillary systems; and case studies, refer to the Water Environment Federation's *Wastewater Solids Incineration Systems* (2009).

2.0 THERMAL OXIDATION AND COMBUSTION TECHNOLOGY

Thermal oxidation is the rapid exothermic oxidation of combustible elements in fuel. At WWTPs, incineration refers to complete combustion of the volatile content of wastewater solids and results in the production of a nonhazardous, nontoxic, inert ash.

Incineration is a three-step process involving evaporation of water, combustion of the volatile content of wastewater solids, and cooling of the ash. Although presented in simplified form, incineration is a complex process involving thermal and chemical reactions that occur at varying times, temperatures, and locations in the incinerator.

The volatile solids fraction of the wastewater solids is the fuel that ultimately reacts with oxygen and releases heat. The feature that makes the combustion of dewatered wastewater solids different from the more common fuels is its high moisture content (typically 70 to 75%).

2.1 Composition of Wastewater Solids

It is important to develop a basic understanding of the composition of wastewater solids. From a thermodynamic perspective, the dewatered cake fed to the incinerator is composed of three key components: moisture, volatile solids, and inserts (ash).

In an incinerator, moisture changes from liquid to vapor during the evaporation process; there is no change in chemical composition. Ash is typically chemically inert and does not undergo any chemical reactions in a furnace. The volatile (combustible) fraction, however, reacts with the oxygen in the air, undergoes changes in composition, and liberates heat.

An effective study of the combustion process depends on a detailed knowledge of specific fuel characteristics of the wastewater solids fed to an incinerator. Tests conducted on wastewater solids include

- Proximate analysis—moisture, volatile matter, fixed carbon, ash;
- Ultimate analysis—moisture, ash, carbon, hydrogen, nitrogen, oxygen, chlorine (sometimes), sulfur, heating value;
- Ash ultimate analysis—up to 14 major constituents; and
- Ash fusion temperature—performed in oxidizing and reducing atmospheres.

The heat content of the wastewater solids [kilojoules (kJ) or British thermal unit (Btu)] can vary greatly from WWTP to WWTP because of materials discharged to the plant, treatment methods used, and techniques used to process wastewater solids before incineration. Values typically range from 20 900 to 23 200 kJ/kg combustibles (9000 to 10 000 Btu/lb combustibles).

2.2 Combustion Technology

The first multiple hearth wastewater solids incinerator (MHI) was built in 1935 in Dearborn, Michigan. From that time until the late 1980s, MHIs were the thermal oxidation method of choice at WWTPs. Presently, 167 MHIs are in service in the United States, and numerous units are located around the world. Figure 17.1 presents a cross-section of a typical MHI.

In the late 1980s, fluidized bed incinerators (FBI) became the preferred thermal oxidation technology for combustion of wastewater solids, primarily because of their

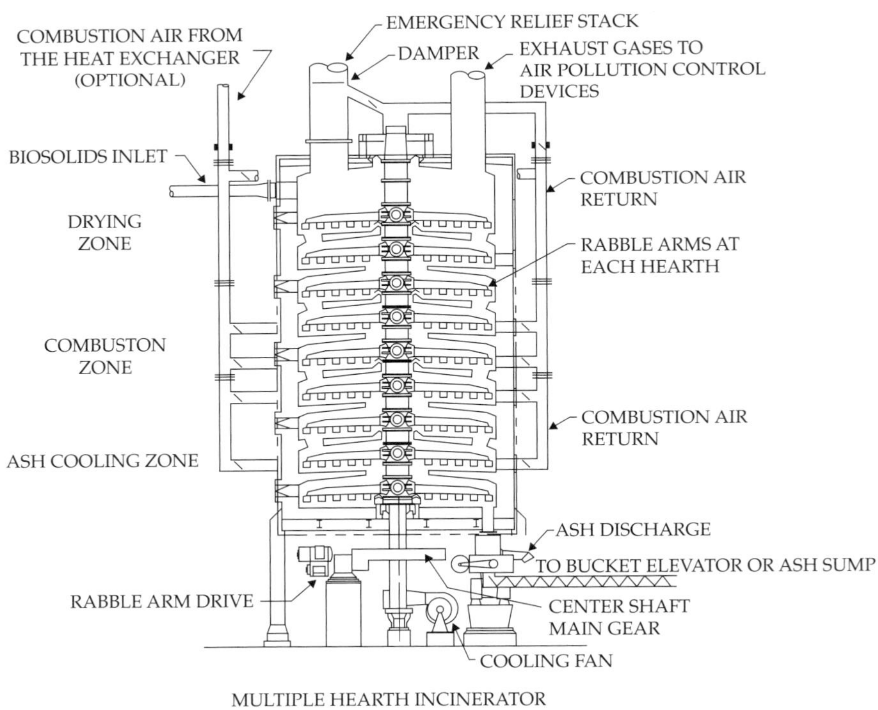

FIGURE 17.1 Multiple hearth incinerator cross-section.

lower emissions and lower operation and maintenance costs. Presently, 63 FBIs are in service in the United States, and significantly more are operating in Canada, Europe, and Asia. Figure 17.2 presents a typical cross-section of an FBI.

Although all new wastewater solid incinerators will be FBIs, a significant number of multiple hearth incinerators will likely be in service for many years to come.

2.2.1 *Fluidized Bed Incinerators*

To design an FBI that is as efficient as possible with minimum emissions, the system must operate without formation of scale in the exhaust system or clinkers in the sand bed at minimum auxiliary fuel consumption. Although scaling and clinkers depend on characteristics of the feed material, emissions of pollutants and auxiliary fuel consumption are directly linked to the fundamental parameters of design.

To determine equipment size and characteristics, it is necessary to determine required airflow to the incinerator, flue gas quantities, supplementary fuel

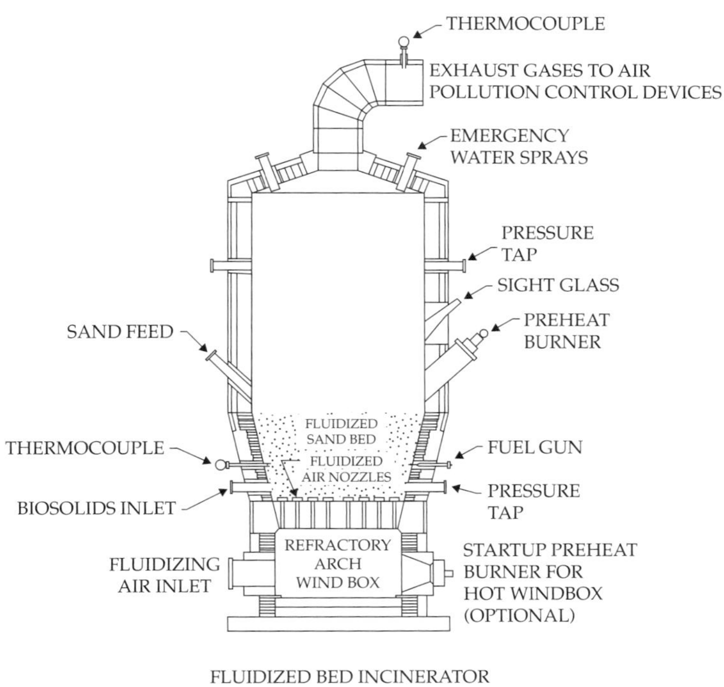

FIGURE 17.2 Fluidized bed incinerator cross-section.

requirements, and cooling water requirements. A mass balance and a heat balance must be prepared, and system exhaust characteristics must be determined. Several fundamental parameters are described in the sections below. Table 17.1 contains typical design parameters.

In designing an FBI, the selection of bed material is critical, and solid particle size has a direct effect on the quality of fluidization. For incineration of wastewater solids, sandlike material with a median size of 550 μm (30 mesh) is typically used. The bed fluidizing gas velocity is typically 0.64 to 0.76 m/s (2.1 to 2.5 ft/sec); although some FBIs are designed and operated with velocities as great as 1.52 to 1.83 m/s (5 to 6 ft/sec) because of higher solid contents in the feed.

Once the ultimate analysis of the feed and feed rates is known, the combustion air can be calculated from the oxidation reactions with an excess of 40% greater than stoichiometry. Designing the system based on an excess air that is too low presents the risk of having incomplete combustion; having excess air greater than 40% excess air is not efficient and can generate higher emission of nitrogen oxides.

For wastewater solids, a combustion temperature of 843 °C (1550 °F) is typically used, with a minimum freeboard gas residence time of 6.5 seconds. Although there are no regulatory temperature and residence time requirements for incinerators in the United States, the previous figure is conservative compared with European regulatory requirement of a freeboard temperature of 850 °C (1562 °F) for a minimum of two seconds.

One of the most important criteria in designing the incineration system is to minimize fuel consumption. Supplementary fuel consumption is calculated from the heat and mass balances and depends on two factors: (1) heat content of the feed material

Table 17.1 Fundamental parameters of design.

Parameter	Values
Bed material size range, μm (U.S. mesh)	1680–177 (10–80)
U_o gas velocity in bed, m/s (ft/sec)	0.75–1 (2.5–3)
U_f gas velocity in freeboard, m/s (ft/sec)	0.76–0.64 (2.5–2.1)
Excess air, over stoichiometry	40%
Freeboard temperature, °C (°F)	843 (1550)
Residence time in freeboard, sec	6.5 minimum
Combustion air temperature	Depends on percent total solids

(specifically its solid content or percent dry solids); and (2) heat content of combustion air (or its temperature).

Solids content of the feed depends on the dewatering equipment and on the quantity of polymer used as a dewatering aid. Heat content of combustion air depends on how intense heat recovery is in the heat exchanger. Typically, a heat exchanger can recover up to 40% of flue gas enthalpy to preheat the combustion air to approximately 675 °C (1250 °F).

Figure 17.3 presents the theoretical curve of supplementary fuel consumption. The calculation was based on a combustion gas temperature of 843 °C (1550 °F) and a throughput capacity of 454 kg (1000 lb) of dry solids per hour. The feed material is typical of product from a WWTP with 75% volatile solids and a high heating value of 23 260 kJ/kg (10 000 Btu/lb) of volatile solids. Fuel consumption decreases with increases in percent dry solids or combustion air temperatures.

The greater the solids content and combustion air temperature, the lower the auxiliary fuel requirement. With a wind box temperature of 648 °C (1200 °F), feed material is burned autogenously, or thermally self-supporting, at a solid content of 27%.

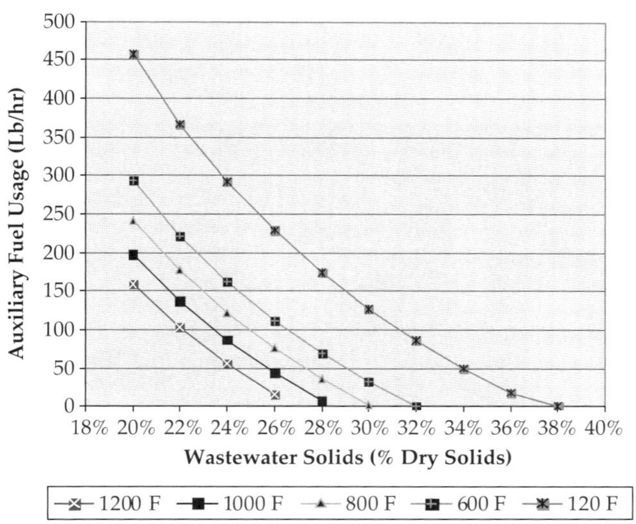

FIGURE 17.3 Auxiliary fuel consumption versus wastewater solids content as a function of windbox preheat air temperatures (kg/h = 2.205 lb/hr; °C = (°F- 32) × 0.555).

To reduce emissions of nitrogen oxides (NOx) and polymer consumption, the system is typically designed based on autogenous combustion at maximum air temperature and minimum dry solids.

2.2.1.1 Description of Major Components

The FBI system typically is composed of the incinerator with subsystems for conveying and feeding the wastewater solids and sand; recovering heat; handling ash; and controlling air pollution. Although FBI equipment typically is located indoors, it can be located outdoors as in Puerto Nuevo, Puerto Rico (Figure 17.4).

The Puerto Nuevo WWTP's FBI system is composed of a hot wind box fluidized bed, a heat exchanger to preheat combustion air to approximately 675 °C (1250 °F), a quench section followed by a cooling tray and multiple Venturi scrubber, a wet electrostatic precipitator, and an exhaust stack. The storage and feed system is composed of a live-bottom bin and piston pumps for the feed. Figure 17.5 provides a detailed cross-section of the Puerto Nuevo WWTP's FBI.

Incineration, heat recovery, and air pollution control systems, however, can differ from project to project. The FBI could be of either the hot or cold wind box types. The heat recovery system could be composed of heat exchangers for preheating the combustion and for plume suppression or a waste-heat boiler for steam generation, or both. A dry-ash system or a wet-ash handling system can be used. Although typical

FIGURE 17.4 Fluidized bed incinerator in Puerto Nuevo, Puerto Rico (courtesy of Infilco Degremont, Inc.).

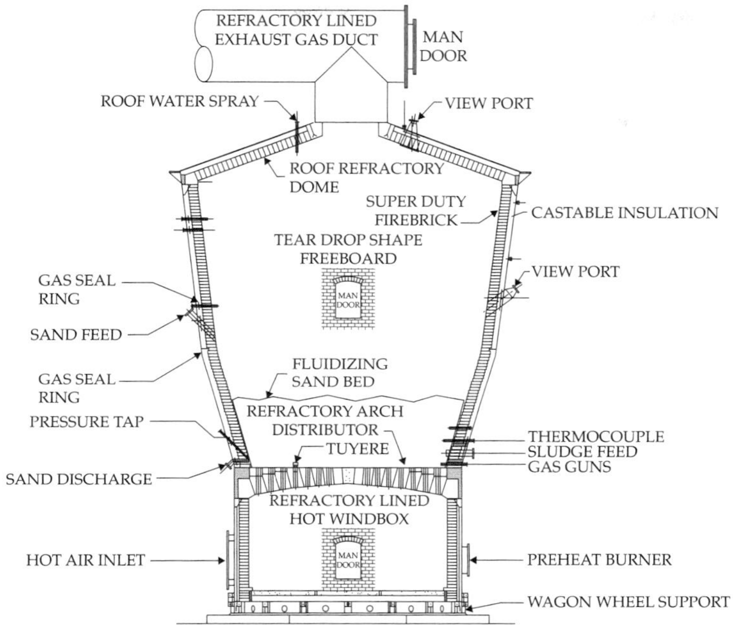

FIGURE 17.5 Cross-section of the Puerto Nuevo, Puerto Rico, wastewater treatment plant fluidized bed incinerator (courtesy of Infilco Degremont, Inc.).

air pollution control trains consists of Venturi/impingement tray scrubbing systems, other systems can include baghouse filters, wet electrostatic precipitators, and activated carbon polishing systems.

Figure 17.6 shows a process flow diagram of the Puerto Nuevo FBI system using a wet-ash system with a hot wind box and heat recovery by heat exchangers. At the Puerto Nuevo WWTP, wastewater solids are dewatered using belt filter presses, and the dewatered cake is pumped via piston pumps to the incinerator. Other facilities use centrifuges to dewater wastewater solids and transport the dewatered cake to the incinerator using belt conveyors. Auxiliary fuel (either natural gas or No. 2 fuel oil) is used to heat the incinerator up to operating temperature.

The freeboard portion of the incinerator operates at a design temperature of approximately 843 °C (1550 °F). The incinerator offers an expanded freeboard to allow deceleration of the larger particles to minimize sand carry-over and maximize carbon burnout.

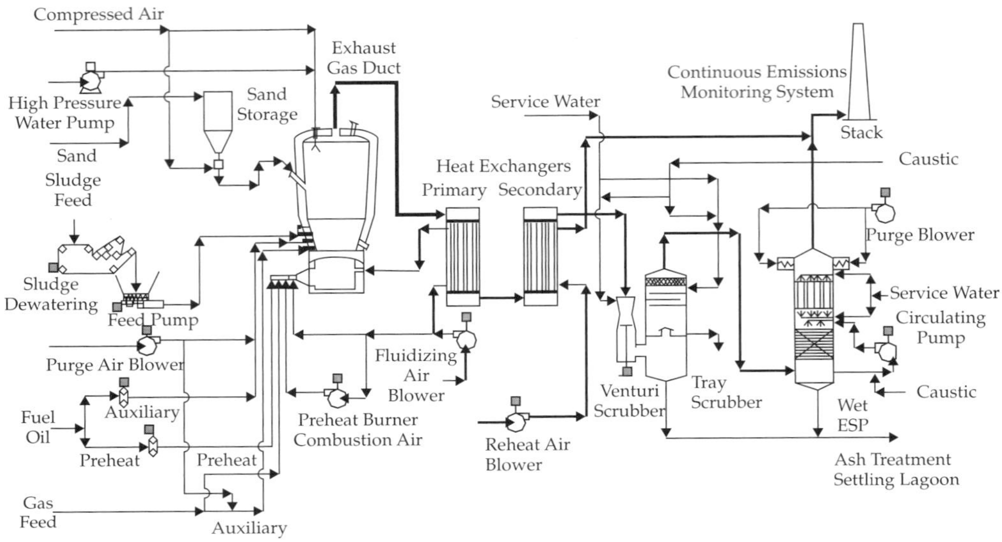

FIGURE 17.6 Puerto Nuevo, Puerto Rico, wastewater treatment plant process flow diagram (wet-ash handling system) (courtesy of Infilco Degremont, Inc.).

Additional sand may be fed to the incinerator pneumatically during operation, if required. The hot wind box furnace in this example is equipped with a refractory arch that supports the sand bed and evenly distributes the air. To minimize auxiliary fuel use, fluidizing air is preheated to approximately 675 °C (1250 °F) in an external tube and shell heat exchanger using the exhaust flue gas of the reactor as a heat source.

The air pollution control system includes a Venturi scrubber followed by a tray tower. Plume suppression air is preheated in a secondary heat exchanger using the exhaust flue gas from the primary heat exchanger.

Ash and fine-sand particles carried in flue gas are removed in the high-pressure drop Venturi scrubber, which produces an ash-slurry. The slurry is either pumped or flows by gravity to an outdoor ash settling lagoon system for dewatering.

In the wet-ash system, acid gases such as SO_2 and HC1 are removed by water in the Venturi scrubber and cooling tray. These gases are soluble in water, which means that as much as 95% of the acids can be removed by effluent plant water alone.

Because of its simplicity and the availability of effluent water and space, the wet-ash system is used at more than 90% WWTPs in North America.

In the dry-ash system, the flue gas temperature has to be in the range of 150 to 205 °C (300 to 400 °F) before hot incinerator exhaust gases enter a bag filter (or a dry electrostatic precipitator). A waste-heat boiler or an economizer can be installed between the fluidized bed and the air pollution control system or between the heat exchanger and the air pollution control system to generate steam or hot water. Figure 17.7 contains a process schematic with a dry-ash handling system. Four systems compose an FBI:

- **Wind box.** The lower section of the FBI is the wind box, which is a distribution chamber for fluidizing air and a combustion chamber for the preheat burner. The hot wind box is a refractory-lined plenum in which the hot combustion air is received. The wall of the wind box has openings for fluidizing air supply, a preheat burner, observation port, and instrument ports. Hot and cold wind boxes are discussed in more detail later in this section.
- **Bed support and air distributor.** A roof (arch) that is typically constructed of refractory, separates the wind box from the next compartment of the FBI. Metal alloy was used to construct three roofs that are in service at the Metropolitan WWTP, St. Paul, Minnesota. The refractory arch construction is self-supporting

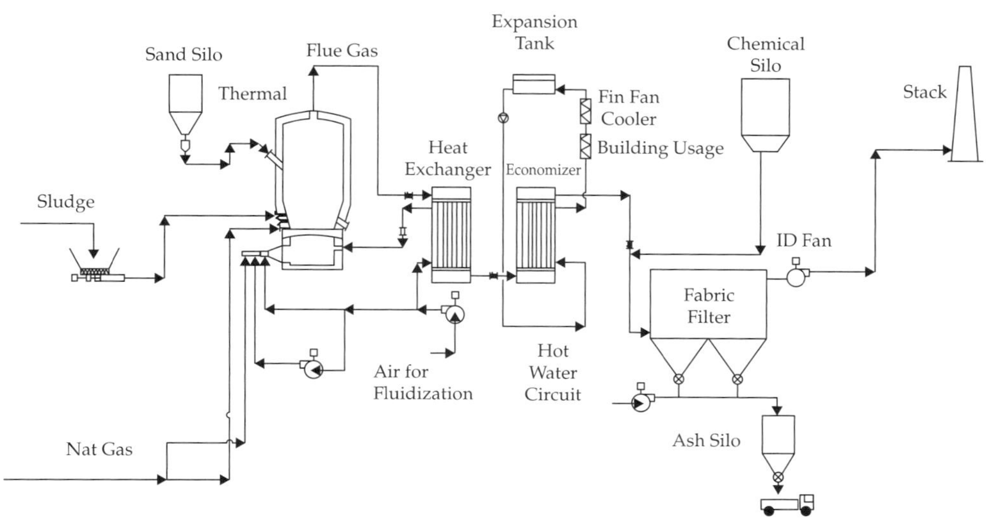

FIGURE 17.7 Process flow diagram (dry-ash handling system) (courtesy of Infilco Degremont, Inc.).

because of the special shape of the refractory elements used, and is typically referred to as the "dome." It supports the weight of the bed material when the bed is not fluidized and is a distributor plate for the fluidizing air. To allow passage of air without the back drainage of sand, the dome has several air nozzles typically referred to as "tuyeres." These specially shaped to prevent sand drainage, to provide uniform air distribution, and to withstand operating temperatures. The refractory arch distributor and the refractory-lined wind box are designed for a wind box temperature of approximately 980 °C (1800 °F). In a hot wind box with refractory arch distributor, the combustion air is typically preheated to approximately 675 °C (1250 °F).

- **Bed.** The section immediately above the distributor that contains the fluidized mass of sand is called the "bed" or combustion zone, which has no physical upper boundary. The air from the distributor causes the bed of sand to fluidize. The height of the bed depends on the amount of sand it contains. For incineration of wastewater solids, an expanded bed of approximately 1.5 m (5 ft) is typically used. The sidewalls slope outward from the bottom to the top of the bed to ensure water vapor expansion and to keep the gas velocity within acceptable limits. The walls are also equipped with nozzles and ports for wastewater solids and auxiliary fuel injection and for various instruments.
- **Freeboard.** The space above the bed is called the freeboard or disengagement zone. It acts as a retention chamber for combustion gases to ensure a sufficient residence time at the temperature required and as a separation chamber for the bed material particles to disengage from the gas. To ensure complete combustion of any escaping volatile hydrocarbons (from the bed), the freeboard must be sized to provide approximately 6.5 seconds of gas residence-time minimum. It is typically 4.6 m (15 ft) high. The shape of the freeboard could be cylindrical straight or conical teardrop. The cylindrical straight shape typically is designed based on a gas velocity of 0.76 m/s (2.5 ft/sec). In the conical teardrop shape, the freeboard is expanded laterally along its height to maximize residence time and to reduce gas velocity further. The gas velocity at the top of the teardrop shaped freeboard is 0.64 m/s (2.1 ft/sec). The exhaust gas duct is installed in the center of the roof dome to minimize gas bypassing and dead zone and to maximize residence time in the freeboard. The gradient of decreasing gas velocity in the freeboard and a lower gas velocity at the exhaust minimize sand losses.

2.2.1.1.1 Wind Box Design

The FBI is equipped with either a hot or cold wind box (also known as a warm wind box). The hot wind box is typically provided to incinerate wastewater solids that have low heat values and require intensive air preheating to minimize auxiliary fuel consumption. The cold wind box furnace is typically used to burn heat-treated solids, solids that have been dried, primary wastewater solids, scum and grease, or other materials such as wood chips or sawdust that can be burned autogenously without heat recovery or with moderate heat recovery.

2.2.1.1.2 Hot Wind Box

The hot wind box fluid bed is designed for maximum temperature exposure of approximately 650 to 980 °C (1200 to 1800 °F) for most applications. It is used when the wind box air temperature is greater than approximately 400 °C (750 °F). The unit is a vertical steel shell made of carbon steel. The inside lining is made of refractory and insulating brick. The refractory lining is necessary because of the inside temperature of approximately 980 °C (1800 °F).

2.2.1.1.3 Cold (Warm) Wind Box

A cold wind box (or warm wind box) incinerator is used for feed material that can be incinerated without heat recovery (or with moderate heat recovery). Wind box air temperature is typically limited to less than 400 °C (750 °F). As with the hot wind box, the cold (or warm) is also composed of four sections, and the designs are similar, except as follows:

- The wind box is not refractory-lined and the bed support and air distributor can be a metal alloy plate. To sustain the high temperature of the bed, the top of the plate is typically refractory lined.
- The preheat burner is installed in the freeboard, angled downward to heat the top of the fluidized sand bed during startup.

2.2.1.2 Description of Subsystems

As previously shown in the process flow diagrams in Figs. 17.6 and 17.7, the incineration system can be divided into subsystems, which are described in the next several sections.

2.2.1.2.1 Air System

Air supplied to the incineration system could be divided in three categories: fluidizing air, purge air, and atomizing air.

Combustion air (same as fluidizing air) is typically supplied by a multistage centrifugal blower. The system shown in 17.6 operates under positive pressure and is referred to as a push-type system. Complete gas tightness is essential because the pressure in the system is greater than atmospheric, and the temperature of the air and gases is high. This type of system is typical in North America, where heat recovery is aimed at autogenous operation and plume suppression and wet-ash types of air pollution are typical.

In contrast, the system shown in Figure 17.7 operates at both positive and negative pressure, with the zero point typically at or near the incinerator outlet. This type of system is referred to as a push-pull type system. In this case, an induced draft fan is required to overcome the pressure drop across the baghouse, waste heat boiler, wet electrostatic precipitator, etc.

In a hot or warm wind box incinerator, the air is preheated in a shell and tube heat exchanger, which uses the reactor off-gases as the hot medium. Heat recovery by combustion air preheating reduces supplemental fuel and increases capacity of the plant. In a cold wind box, combustion air from the centrifugal air blower is supplied directly to the metal wind box.

The fluidizing air blower is designed with several stages to provide the required outlet pressure. The main control of the incineration process capacity is regulated by the quantity of fluidizing air entering the wind box. Airflow is controlled by a damper on the blower inlet side and measured by a flow meter.

Purge air is used in all ports to the incinerator and for all expansion joints and pressure taps in the ductwork to keep them cool and free of sand and ash. Purge air can be either high or low pressure. For example, high-pressure air is supplied from a compressed system for roof spray nozzles and pressure ports to keep them cool and free of deposits. Purge air from the fluidizing air blower can be used in annular sleeves of all ports, including site ports, bed oil guns, sand inlet/outlet, sludge inlet, and roof sprays.

Atomizing air at approximately 48 to 55 kPa (7 to 8 psig) is required when fuel oil is selected as auxiliary fuel. Atomizing air can be provided by an injection-purge air blower.

2.2.1.2.2 Feed System

Solids from the wastewater process are dewatered first. Dewatering can be achieved in a plate filter press, a belt filter press, or a centrifuge decanter. A belt press typically can dewater undigested wastewater solids to approximately 25 to 30% dry solids.

High-performance centrifuges typically can dewater to approximately 28 to 35% dry solids. Polymer is used in both the belt press and centrifuge decanter as a dewatering aid.

Continuous, even transport and distribution of the dewatered cake to the incinerator is an important to ensure stable and economical operation. Past practice was to use screw extrusion feeders for dry cake and progressing cavity pumps for wet cake. Today, hydraulic piston pumps are used to convey cake from dewatering equipment to the furnace. The piston pump is preferable to others because of its flexibility and insensitivity to feed quality. Recent developments have been made in the design of progressing cavity pumps for pumping cake.

Inbed feeding is used for incineration of wastewater solids. The feeding location, at 1.2 m (4 ft) under the bed surface or 30 cm (1 ft) above the distributor, ensures that maximum possible retention time of feed cake particles is obtained before it reaches the bed surface where carryover may occur. It is important that the feed cake release its maximum energy to the sand bed to counteract the quenching effect of water evaporation.

Depending on the diameter of the furnace, two or four feed injection points are included to ensure the cake is evenly distributed throughout the bed.

2.2.1.2.3 Sand System

Sand-like material is used as bed media. The furnace is typically filled with sand to a static height of 0.9 m (3 ft). When the fluid bed is in operation, the bed material will expand, because of fluidizing air, to a height of approximately 1.5 m (5 ft). With time, the sand gets abraded, and makeup is required. Makeup sand can be pneumatically fed into the furnace during normal operation. The feed system is typically of the dense-phase type pneumatic conveying.

Bed removal systems have been provided with some existing systems to cool the sand. Removing sand from a hot bed is unsafe. As a safety precaution, sand should be removed from the bed when the furnace has been cooled and the sand bed is at approximately 38 °C (100 °F).

Hydrodynamics of the fluidized layer depends on size and density of the media because the incinerator is sized based on gas flowrate. The media should have a bulk density of approximately 1600 kg/m^3 (100 lb/cu ft). The sand must be angular, dry, and free of sodium and potassium. It must not grind into fines at an operating temperature of 870 °C (1600 °F) or fuse at 980 °C (1800 °F). Two types of fine bed media can be used: silica or olivine sands. Although silica sand is lower in unit cost, abrasion of the silica is higher than with olivine sand.

Abrasion results in sand particle size being reduced and in the sand exiting the furnace with the flue gases, typically referred to as elutriation. This lost sand must be replaced. Depending on the quality of the feed, the selection of media needs to be carefully planned. Olivine sand is not desirable in two particular cases. First, if the feed contains grit, then bed material can build up, and excess material will need to be removed, resulting in unnecessary shutdowns. Second, if the feed contains high alkali metals, then accumulation of these elements on the olivine bed can lead to low melting eutectics.

2.2.1.2.4 Auxiliary Fuel System

Auxiliary fuel is used at startup in a preheat burner or by direct fuel injection into the fluidized sand bed via fuel guns. Typically either natural gas or No. 2 fuel oil is used. Other fuels including coal, saw dust, and digester gas can be used as supplementary fuel. The only stipulation is that the fuel must be suitable for feeding in a reliable and controllable manner.

2.2.1.2.5 Preheat Burner

When starting up from cold temperatures, a proper burner system must preheat the fluid bed to the temperature at which it is possible to inject fuel to the bed itself. It is important that proper ignition takes place. In a hot wind box, a standard natural gas or No. 2 fuel oil burner located in the sidewall provides preheating. The fluidizing air from the heat exchanger (preheater) is heated by mixing with the hot gases from the burner. The resulting hot air fluidizes the sand bed. Heat from the air is then intimately transferred to the fluidizing sand. In a cold or warm wind box where the preheat burner is installed in the freeboard, the heat transfer from the gas to the bed is less efficient. Therefore, fuel consumption is greater and length of time to startup is higher in a cold wind box than in a hot wind box.

The air supply to the preheat burner is taken from the outlet side of the fluidizing air blower. The air is then pressurized further by the combustion air blower to ensure a burner supply pressure that is approximately 14 kPa (2 psig) greater than the pressure inside the wind box.

2.2.1.2.6 Bed Fuel Injection

Once the fluidized bed is sufficiently hot to support instantaneous ignition of fuel, the preheat burner can be shut down; preheating is continued by injection of auxiliary fuel directly to the fluidized bed. This mode of operation is much simpler than with the preheat burner and is typically preferred to such an extent that a bed once heated will be kept hot for an extended period by reheating. Another reason for keeping the bed hot at all times is the extension of the refractory life.

Oil injection takes place through oil guns located at the periphery of the bed approximately 0.3 m (1 ft) above the distributor. To prevent fouling of the injector guns, the oil is mixed with purging air at the supply end of the gun and blown into the bed in the form of a coarse mist. Natural-gas injection takes place through gas guns located at the distributor level to homogenously distribute the gas throughout the sand bed; and the gas guns vary in length.

It is important to maintain airflow on the oil or gas guns whenever they are inserted to the bed. Therefore, the air supply line is equipped with a flow indicator.

2.2.1.2.7 Water System

To protect the heat exchanger from excessive temperatures, the fluidized bed is equipped with water quench spray nozzles. The nozzles are installed through the roof of the furnace above the freeboard. The nozzles operate at a high gauge pressure (approximately 2100 kPa or 300 psig) to create a fine water mist that will evaporate and quench the exhaust gas as fast as possible. Evaporation takes place close to the freeboard exhaust gas duct, limiting the cooling effect of flue gas leaving the incinerator. The nozzles are used in sequence, depending on the temperature of the gas at the heat exchanger inlet.

The spray nozzles have small orifices and swirl grooves. To maintain clear internal passages and provide some cooling when it is not spraying, the nozzles are purged with air from the compressed instrument air system at a gauge pressure of approximately 410 kPa (60 psig). Both air and water supply include check valves to prevent backflow of one medium into the pipes of the other.

The water supply includes the water pump, pressure regulator, filter, and relief valve.

2.2.1.2.8 Ductwork and Expansion Joints

All hot gas and air ducts are furnished with internal refractory lining and thermal insulation and expansion joints. The refractory lining must resist abrasion because the hot gas contains abrasive ash and elutriated sand. Internal thermal insulation is selected and sized for a typical steel temperature of 93 °C (200 °F). Expansion joints are constructed of high-temperature metal. The expansion joints are refractory lined and constructed of metallic bellows allow for duct and equipment movement caused by temperature fluctuations.

Purge air is provided to the bellows to prevent sand and fly ash from passing through and restricting movement of the joint.

2.2.1.2.9 Bed Temperature

Thermocouples are placed in the sidewall of the bed to measure the bed temperature. The operating range for bed temperature is typically from approximately 675 to 790 °C (1250 to 1450 °F) during stable operation. The temperature of the fluidized sand bed must be high enough to ensure that ignition of the cake and fuel takes place. A sufficient part of the cake combustion must occur in the bed itself to provide liberated heat for water evaporation and to reduce use of supplementary fuel.

The bed temperature must also be kept sufficiently high to ensure that the freeboard temperature is greater than the required minimum of 843 °C (1550 °F). This temperature ensures complete combustion of hydrocarbons. Alternatively, the bed temperature should not be kept too high to minimize fuel consumption and to respect the upper limit to the heat exchanger gas inlet temperature of 870 °C (1600 °F). Interlocks are installed to stop cake feed if the bed temperature drops below the ignition temperature of wastewater solids and fuel.

2.2.1.2.10 Oxygen

An analyzer installed in the hot gas duct, which connects the heat recovery system and the air pollution control system, measures oxygen. Oxygen is recorded in real time and displayed in the operator console. To ensure complete combustion, it is necessary to have excess oxygen available. Design excess air of the FBI is 40%, which corresponds to approximately 3.5 to 4.0% oxygen (by volume on a wet basis) in the incinerators exhaust gases.

To ensure good fluidization, fluidizing airflow rate must be maintained constant at design levels. Excess air is, therefore, a function of the total amount of combustible material (dewatered cake and fuel) being fed to the combustor.

If the oxygen content is high, then combustion air is not being efficiently used for combustion, which can mean that capacity of the unit is less than optimal. If oxygen concentration in the reactor off-gas decreases to less than 2% (by volume wet basis) for two minutes, then an alarm and interlock will be activated and cake feeding will be interrupted.

2.2.1.3 Advantages of Fluidized Bed Technology

As outlined above, FBI is the most efficient thermal technique for wastewater solids disposal. The technique offers several advantages over other techniques, including

- **Flexibility, suitable for intermittent operation.** The inventory of hot bed solids acts as a thermal reservoir, thereby causing only a small change in temperature on shutdown. This situation permits quick startup following daily or weekend shutdown.

- **Flexibility, suitable for feed variability.** Within a defined size range, FBIs can easily handle feeds with varying chemical properties, moisture, and volatile contents, particularly for short periods of time. This characteristic is primarily a result of the large quantity of heat stored in the sand bed and because the feed represents approximately 1% of total bed material.
- **Ease of control and automation.** The rapid mixing of solids caused by the turbulence in a fluid bed provides uniformity of bed temperature. Temperature control is reduced to one point of control, which is the average of all measurements.
- **Lower auxiliary fuel use.** Significantly less fuel is required, primarily because of relatively low excess air requirements (only 40% excess over stoichiometry compared with typically 100% excess over stoichiometry in other thermal technologies) and possibility of heat recovery from flue gas and preheating of combustion air to approximately 650 °C (1200 °F).
- **Reduced maintenance costs.** No moving parts are exposed to the combustion section. Absence of thermal shock from the thermal reservoir of the bed solids, which results in slow temperature changes and leads to longer refractory life and lower maintenance cost.
- **High efficiency of combustion.** High turbulences of the sand bed increase the contact surface of cake particles and oxygen, increasing combustion efficiency. Combustion is instantaneous, with low emissions of carbon monoxide (CO) and total hydrocarbons (THC).
- **Low NOx emissions.** Because of low excess air and good distribution of temperature throughout the bed, NOx emissions are reduced.

2.2.2 *Multiple Hearth Incinerators*

The MHI is durable, relatively easy to operate, and can handle wide fluctuations in feed quality and loading rates. The MHI is designed for continuous operation. Startup fuel requirements and the extended time needed to increase the temperature of hearths and internal equipment from a cold condition typically preclude intermittent operations. The MHI is a series of 6 to 14 vertically oriented, cylindrically shaped, refractory hearths stacked one on top of the other. They are available with diameters ranging from 1.4 to 8.8 m (4.5 ft to 29 ft). Figure 17.1 shows a cross-section of a typical MHI. More details can be found in *Wastewater Solids Incineration Systems* (WEF, 2009).

Since the early 1990s, several MHIs have been upgraded to meet U.S. EPA's 40 CFR Part 503 Regulations or have been replaced with FBIs.

As previously discussed, it is unlikely that any new MHIs will be constructed at WWTPs because of the new fluidized bed technology. However, many currently operating MHIs have significant useful life remaining.

3.0 ENERGY RECOVERY AND REUSE

Energy recovery and reuse is an integral part of a modern wastewater solids incineration system. For MHIs, the dewatered wastewater solids fed to the incinerator is not sufficiently dry enough to be autogenous (i.e., thermally self-supporting without the use of auxiliary fuel). Therefore, it is necessary to provide additional energy to through use of auxiliary fuel (i.e., natural gas, fuel oil, etc.) or thermal energy recovered from the incinerator's exhaust gases to properly sustain the combustion process. Energy in the incinerator exhaust gases represent a valuable energy resource which, when effectively recovered, will result in reduced operating costs and a reduction in greenhouse gas emissions. Heat recovery from the hot incinerator exhaust gases also can reduce the volume flowing through air pollution control devices.

A careful technical and economic evaluation must be conducted, however, before embarking on an energy recovery and reuse program. Although a heat recovery system does require capital investment and routine operation and maintenance effort and cost, these systems often can offset capital and operation and maintenance costs.

Recovered energy can greatly reduce auxiliary fuel requirements and greenhouse gas emissions. Moreover, recovered energy can be used to serve other portions of the plant, through generation of steam, hot water, or heated thermal fluid, or, taken a step further, for generation of electrical power. Whenever there is a "market" for the recovered energy in the WWTP, there is an opportunity for savings. Taking advantage of the energy properties of the feed solids is what makes incineration with energy recovery a true form of beneficial use.

Waste heat recovery can take many different forms and can be classified as primary recovery or secondary recovery, based on the ultimate use of the recovered energy. Primary recovery refers to energy recovery that is used specifically to improve performance of the combustion process, by reducing or eliminating the need for auxiliary fuel. This can include using recovered energy to preheat combustion air to the incinerator, to thermally dewater wastewater solids, or to dry incoming feed cake to the incinerator.

Secondary recovery uses heat in the flue gases for beneficial purposes outside of the combustion envelope. Heat recovery could be in the form of producing steam for building comfort heating or for a plant process, generating electricity, and reducing the steam plume from the incinerator's exhaust stack.

For building comfort heating, an economizer may be added as a source of hot water for space heating or inplant processes. Steam generated in a waste-heat boiler can be used throughout the WWTP, sold to neighboring users, used for production of electricity in a cogeneration system, or used directly in steam turbine drives. The exhaust flue gases can also be used, directly or indirectly, to preheat flue gases downstream of the scrubbing system before being exhausted to the atmosphere to provide plume suppression.

Heat transfer technology also can be classified as either direct or indirect. For example, a direct process would include preheating or drying feed cake by direct contact with flue gases. This approach has gained limited acceptance as a means of preheating or thermally dewatering the feed cake.

Indirect-heat transfer refers to a process in which a physical barrier separates the heat source from the material or flowing stream being heated. An example would be incinerator flue gases being used to preheat combustion air in a heat exchanger. This is referred to as recuperation and is the most common and economical approach. In an FBI, the fluidizing air is typically preheated to 200 to 650 °C (400 to 1200 °F). In an MHI, the utility of combustion air preheat is practically limited to a maximum of approximately 200 °C (400 °F) because the nature of the process and equipment precludes operation with high combustion air temperatures. In some variations of the MHI process, wet scrubber exhaust is reheated to 760 to 870 °C (1400 to 1600°F) in a secondary combustion unit. Here the energy in the incinerator exhaust could be used to preheat the scrubber exhaust flue gas, thus reducing auxiliary fuel requirements for secondary combustion. In this case, such recovery allows high-temperature treatment with lower energy costs.

Regardless of the form of heat recovery, the goal should be to economize plant operation, including fuel efficiency and operating and maintenance costs. Energy recovery should first be directed to the primary objective: efficient and effective combustion of wastewater solids. The remaining recoverable energy can then be used for secondary purposes.

The source of recoverable energy is the hot incinerator exhaust flue gas. As a resource, these flue gases contain most of the heat energy input to the system, including the heating value of the feed cake and auxiliary fuel and the sensible heat of the combustion air.

Although many options are available for heat recovery, the most effective method is recuperation/air preheating. Options for secondary heat recovery include heat exchangers for plume suppression, economizers, waste heat boilers, or thermal oil heaters. Optimization of heat recovery options is essential to the efficient energy operation of a plant and makes incineration with energy recovery a bona fide form of beneficial use. See *Wastewater Solids Incineration Systems* for more details (WEF, 2009).

4.0 AIR EMISSIONS CONTROLS AND MONITORING

4.1 Introduction

Control of air emissions has become an increasingly important and crucial part of the design and operation of incineration facilities because of the numerous and more stringent regulations imposed by federal, state, and regional air pollution control boards. This section describes the formation of pollutant emissions, factors affecting their generation and release, technologies, and operating modes to control emissions, advantages and disadvantages of various air pollution control (APC) systems, and devices used to monitor emissions.

4.2 Pollutants

Three main types of pollutants are found: solid, liquid, and gaseous. Solid are particulate matter and metals. Gaseous pollutants result from the oxidation and volatilization of sulfur, nitrogen, and chlorine in the feed cake and from incomplete combustion of hydrocarbons and other organic compounds. The gaseous pollutants of primary concern are the acid gases: sulfur dioxide (SO_2); hydrogen chloride or hydrochloric acid (HCl); carbon monoxide (CO); volatile organic compounds (VOCs); polycyclic organic matter (POM); and oxides of nitrogen (NOx).

4.3 Control Devices

A variety of control devices can be used to reduce emission. Each has its own specific characteristics, efficiency, and effectiveness in controlling one or more types of pollutants. See *Wastewater Solids Incineration Systems* for more details (WEF, 2009).

5.0 EMISSION REGULATIONS AND PERMITTING

Since the early 1970s, the Clean Water Act (33 U.S.C. § 1251) and the Clean Air Act (40 CFR 50–88) have served as the foundation of the U.S. EPA's regulatory authority. The scopes of both programs have been amended over the years to continue to improve the environment and define the nation's regulatory agenda.

The Clean Water Act (CWA) has resulted in tremendous improvements in the quality of the wastewater effluent from WWTPs and in the quality of the water in receiving waterways. The CWA Amendments of 1987 required U.S. EPA to establish standards for the use or disposal of wastewater solids, including incineration, and were incorporated into Title 40 Part 503 of the *Code of Federal Regulations* (40 CFR 503) in 1993.

Where incineration is used to combust wastewater residuals, the requirements of the Clean Air Act (CAA) must also be considered. The Part 503 regulations recognize that the CAA includes requirements for the combustion of wastewater solids. For wastewater solids that are thermally oxidized, the requirements of both Acts must be satisfied.

The CAA has significantly reduced emissions to the atmosphere and improved air quality in urban areas. The CAA not only established emission limits for individual emission units, it also considered other emission sources at the facility and the attainment status of the ambient air in the region that the facility is located. Construction of a new facility cannot commence until all of these considerations have been addressed.

With the passage of the CAA Amendments of 1990 (42 U.S.C. § 7401–7671); many new programs were implemented including emission limits for hazardous air pollutants from specific source groups and Title V Operating Permit program to document and track emissions on a plant-wide and continuing basis.

The critical issue related to permitting thermal oxidation units under the CAA is that approval from the reviewing agency must be received before construction commences. The air permit review is a two-step process in which a preconstruction permit must be applied for and approval received before construction of the facility can commence.

Although complying with the requirements of the Clean Air Act can be achieved, other factors will influence the specific emission limits that may apply to a facility. These factors include attainment status of the facility site, local meteorology or terrain features, and state regulatory requirements. Once the facility is constructed, a performance test is conducted to demonstrate that the preconstruction limits have been met and an operating permit application must be submitted and approval issued for the facility to continue operating. Thus, careful consideration of air permitting requirements for an incineration system must be made at the early planning and design phases of a project.

This section reviews air quality permitting requirements and emission limitations that apply to wastewater solids incineration systems. The goal is to provide WWTP managers, operators, engineers, and planners with an understanding of the requirements for this source group. The information is organized by project phase, as is relevant to know when to do something as it is to know what to do. The project phases consist of planning, implementation, construction, startup, and operation.

Online resources for environmental regulations include U.S. EPA (http://www.epa.gov) and the Government Printing Office (GPO), which provides access to the *Code of Federal Regulations* (http://www.gpoaccess.gov/cfr/index.html). State environmental regulations can be found online for most state and local agencies. Some states caution that the online regulations are provided for informational purposes only, and the published version is the official document.

5.1 Project Planning and Pre-Permitting

The nature of the regulatory review and emission limitations depend on the type of facility, the magnitude of the potential emissions, and attainment status for the area where the plant is located. If the facility is located in an area that is not in attainment of the National Ambient Air Quality Standards (NAAQS), then stringent air pollution control requirements and emissions limits will apply (CAA, Title I, Part D, Plan Requirements for Nonattainment Areas, as amended). A facility located in an area meeting the NAAQS, may still be subject to a comprehensive permit review (CAA, Title I, Part C, Prevention of Significant Deterioration of Air Quality, as amended), but the resulting emission limits may be less stringent than a facility located in a nonattainment area.

Air permitting should begin in the project planning stage when detailed information about the proposed facility is still being developed. Information that can be obtained at the planning stage includes

- Attainment status of criteria pollutants for the proposed facility's location;
- Classification of the facility as a major or minor air emissions source considering ownership of the adjacent facilities and industrial classification; and
- Potential emission rates of the new facility or net change in emissions from an existing facility.

5.1.1 *Attainment Status*

Air quality standards in the United States are mandated by the CAA and its amendments. The U.S. EPA Office of Air Quality Planning and Standards has set NAAQS for six principal pollutants, which are called "criteria" pollutants (40 CFR 50):

- Carbon monoxide (CO);
- Sulfur dioxide (SO_2);
- Nitrogen dioxide (NO_2);
- Ozone (O_3);
- Categories of particulate matter, including particulate matter less than 10 μm in size (PM_{10}) and particulate matter less than 2.5 μm in size ($PM_{2.5}$); and
- Lead (Pb).

The U.S. EPA has identified two types of standards for these pollutants:

1. Primary ambient air quality standards, which define levels of air quality necessary to protect public health with an adequate margin of safety; and
2. Secondary standards, which define levels needed to protect the public welfare from any known or anticipated adverse effects of a pollutant.

Such standards are subject to revision, and additional primary and secondary standards may be promulgated as the U.S. EPA deems necessary to protect public health and welfare.

Geographic areas in which the NAAQS for all criteria pollutants are met are called "attainment areas"; areas in which one or more standards are violated are called "nonattainment areas." A nonattainment area must develop and implement a plan to meet and maintain CAA standards. When a nonattaining region again meets the standard, the area can be redesignated as a "maintenance area." A maintenance area is a geographic region redesignated by the U.S. EPA from nonattainment to attainment as a result of monitored attainment of the standard and U.S. EPA approval of a plan to maintain air quality standards for at least a 10-year period. This determination is made on a pollutant-specific basis, so an area can be nonattainment for ozone and in attainment for the other criteria pollutants. Because emissions of nitrogen oxides (NOx) and volatile organic compounds (VOCs) can lead to the formation of ozone, regions designated as being nonattainment for ozone also have more restrictive limits for NOx and VOCs. For example, an area may be classified as a serious nonattainment area for ozone resulting in lower emission threshold limits for NOx and VOC but be classified as attainment for CO, SO_2, and particulate matter.

The official listing of attainment status designations is found in 40 CFR 81 Subpart C—Section 107 Attainment Status Designations. This subpart of the federal regulations lists attainment areas by state and air quality control region. These can be accessed through the online resources described above. The U.S. EPA also operates a web page call the Green Book that lists nonattainment areas for criteria pollutants (http://www.epa.gov/oar/oaqps/greenbk). This Web site offers a variety of ways (by state, county, or pollutant) to search area attainment designations.

5.1.2 *Facility Classification*

Three factors determine a facility's classification: common owner or operator, adjacent facilities, and the same industrial classification. For a municipal facility, the operation of a WWTP may be under the direction of a public works director who may also oversee other municipal facilities.

A regional authority may have responsibilities only for the WWTP operations even if the facility serves people in multiple communities. An operator of a privately operated plant would be responsible only for the operations at that plant. It is possible for the solids-handling facility to be privatized and operated by a private company even if the rest of the WWTP is run by a municipal or regional authority. In this case, the owner/operator of the solids-processing facility would have responsibilities only for emissions associated with that process.

If a regional authority is responsible for more than one WWTP and associated solids processing, then treatment plants would be considered separate facilities as long as they are not adjacent to each other. A public roadway passing between the two facilities is not sufficient to treat the facilities as separate operations. For example, a regional authority may have several regional WWTPs that are separated geographically. Operation of the WWTPs would be considered separate facilities. But if the facility that processed solids from all of the regional plants was located adjacent to one of the WWTPs, then that WWTP and the solids-processing facility would be considered one facility because they are under common ownership and are adjacent to one another.

The CAA Amendments of 1990 also group facilities by major industrial classification. "Sanitary services" are part of the U.S. Department of Labor Standard Industrial Classification (SIC) Major Group 4900 (for the SIC can be accessed at http://www.

osha.gov/pls/imis/sic_manual.html). The subgroup (Industrial Group 4952) includes wastewater systems. Thus, a municipality or regional authority that operates a WWTP and a solid-waste management facility on adjacent properties must consider both operations as part of the same facility.

5.1.3 *Potential Emissions*

Potential emissions are the emissions that would occur on an annual basis if the facility was operated at its design capacity continuously. Potential emissions can be reduced if there are physical limitations that constrain the process from operating at its design capacity on a continuous basis or if a federally enforceable operational limitation was adopted that limits operation (e.g., total quantity of solids to be processed annually). If the potential emissions from new source are greater than the emission thresholds for federal permit review, as presented in Table 17.2, then the facility is a new major emission source. A facility with emissions that are less than the major emission source thresholds is considered a minor source and is subject to state permit review requirements.

If the proposed action is a modification to an existing facility, then emissions from the existing facility need to be quantified to determine whether the existing source is a major or minor air emissions source and whether the proposed change will result in a net increase or decrease in air emissions. If the net change in emissions is greater than the emissions thresholds for a modification, then the proposed action is a major modification.

The process of determining what emissions credits apply when calculating a net change in emissions can be complicated and may involve the development of emission offset credits. Consultation with the governing regulatory authority may be needed to ensure that the changes in net emissions are being calculated appropriately.

The details of a proposed action may not be fully developed at the planning stage, yet estimating potential emissions is critical for future regulatory review. A preliminary estimate of emissions for an MHI or FBI can be made using emission factors. Estimating emission rates based solely on generic emission factors, however, is not advisable. Source-specific emission testing and vendor performance guarantees are preferred methods for setting emission limits.

State and local regulatory agencies also have established emission thresholds that determine whether the facility is subject to regulatory review. The appropriate regulatory authority should be consulted to identify emission thresholds and information needed to meet state and local permit review requirements.

Table 17.2 Major source thresholds and major modification thresholds.

Pollutant	Pollutant attainment status	Major source threshold[a] (ton/yr[b])	Major modification threshold[c] (ton/yr[b])
CO	CO attainment	100	100
	CO serious nonattainment	50	50
SO_2	SO_2 attainment or nonattainment	100	40
PM-10	PM-10 attainment	100	15
	PM-10 nonattainment	70	15
PM-2.5	PM-2.5 attainment or nonattainment	100	15
NO_2	NO_2 or ozone attainment	100	40
	Ozone attainment or marginal or moderate nonattainment or ozone transport region	100	25
	Serious ozone nonattainment	50	25
	Severe ozone nonattainment	25	25
	Extreme ozone nonattainment	10	Any increase in actual emissions
VOC	Ozone attainment	100	40
	Serious ozone nonattainment or ozone transport region	50	25
	Severe ozone nonattainment	25	25
	Extreme ozone nonattainment	10	Any increase in actual emissions
Lead	Pb attainment or nonattainment	100	0.6

[a] 40 CFR 51.165 (a) (1) (iv) major stationary source.

[b] ton/yr × 0.9072 = Mg/a.

[c] 40 CFR 51.165 (a) (1) (v) major modification.

Emission factors that can be used to estimate emission rates from proposed incinerator or to identify applicable emission limits can be found in the U.S. EPA's Technology Transfer Network (http://www.epa.gov/ttn). Chapter 2 of U.S. EPA's Compilation of Emission Factors contains a section on combustion wastewater solids. Emission factors for both MHIs and FBIs are provided. In most cases, the confidence in the reported data is questionable. As a result, U.S. EPA prefers the use of action emissions test data from similar incinerators. In addition, the RACT/BACT/LAER Clearinghouse is a database that contains emission limitations from across the county (http://cfpub.epa.gov/RBLC/).

5.2 Project Implementation and Permit to Install

Emissions limits for specific source categories are established under New Source Performance Standards (NSPS) (CAA Sec. 111, Standards of Performance for New Stationary Sources) and National Emission Standards for Hazardous Air Pollutant (NESHAP) regulations (CAA Sec. 112).

The Prevention of Significant Deterioration (PSD) provisions (CAA Sec. 165, Preconstruction Requirements) can result in more stringent limits as a result of the Best Available Control Technology (BACT) review process. The New Source Review (NSR) (CAA Sec. 173, Permit Requirements) review process includes an emission control evaluation that results in the Lowest Achievable Emission Rate (LAER).

To begin the air permit review process, a sufficient amount of information is needed to define process requirements to allow calculation of emission rates and evaluation of strategies. The permit process is an iterative process between the applicant and the reviewing authority where emission limits and performance criteria are evaluated. Thus, it is best to begin the permit process before final process design decisions have been made.

In a traditional design–bid–build construction process, the 30% design point is a good time to prepare and submit the air permit application. Enough technical information has been prepared to define the process, and it is early enough to make modifications to process equipment or air pollution control devices if necessary. Given that it can take anywhere from 12 to 24 months to obtain a permit, however, it is best to submit the permit-to-install (i.e., construct) application to the local air agency or state regulatory body as soon as possible.

5.2.1 *Permit Application Requirements*

The type of review and threshold levels are based on the attainment status of the region where the facility is located. Review requirements depend on the annual potential emission rate for either the total facility or the proposed modification. A facility is defined as a contiguous facility when it is within the same industrial classification code and under common ownership. Therefore, the incinerator for processing wastewater residuals would be considered part of the WWTP.

Adjacent landfills or solid-waste processing facilities, however, are also within the same industrial classification. If the regional authority or municipality also owns these facilities, then emissions from these facilities may be counted when determining the review status of a proposed solids thermal treatment facility.

Although the format and presentation will vary by the reviewing agency, permit applications typically should be bound in a document and submitted to the agency for review and would contain several elements:

- **Permit forms.** Each agency requires permit applicants to use a standard set of forms. Many agencies make the forms available electronically on their Web site either as downloadable documents or interactive online forms that feed directly into a database.
- **Process description.** A detailed process description that describes operation of the facility in sufficient detail to support and confirm emission calculations and control strategies. Often process flow diagrams and process equipment data sheets are included.
- **Emission estimates.** The basis for each pollutant emission rate should be presented with supporting information such as emission test results from similar units, vendor performance guarantees, or mass balance calculations.
- **Control technology assessment.** This may be a determination of the BACT or LAER, depending on the attainment status of the region.
- **Emission limitations.** This is an assessment to ensure that statutory emission limitations are achieved. Both federal and state emission limits should be identified.
- **Air quality compliance.** A dispersion modeling analysis may be necessary to determine whether the NAAQS or the PSD increments are exceeded.
- **Special issues.** State and local agencies may request additional demonstrations to show that hazardous air pollutants, noise, and odor are within acceptable limits.

Permit review times will vary depending on the complexity of the permit application, local conditions, and complexity of the project. To ensure that the permit is reviewed in the shortest amount of time, it is best to meet with the regulatory agency in advance, understand what information is needed for their review, and provide as complete a permit application as possible.

5.2.2 *Federal Regulatory Requirements*

The federal regulatory permit review requirements are defined in 40 CFR 51 and 52. The U.S. EPA has delegated review authority to many state and local regulatory agencies; state and local agencies can review the application and assess compliance with both the federal, state, and local permitting requirements. For state and local agencies over which U.S. EPA has retained review authority, the regional U.S EPA office will serve as the reviewing authority for compliance with federal regulations. The state or local agencies will still review and comment and state specific requirements.

5.2.3 *Nonattainment New Source Review*

New source review for pollutant nonattainment areas is conducted for facilities located in areas where pollutant concentrations are greater than the NAAQS. The degree to which the standards are not being met increases the requirements to provide emission reductions and emission offsets. Emission control strategies must demonstrate LAER.

5.2.4 *Prevention of Significant Deterioration*

For pollutants that are in attainment with the NAAQS, a PSD review is required (40 CFR 52.21). This review seeks to maintain ambient air concentrations below NAAQS by limiting air quality effects to incremental limits above the baseline concentrations. Emission control strategies must be demonstrated to represent BACT, considering environmental, energy and economic effects.

5.2.5 *New Source Performance Standards*

For combustion technologies, the New Source Performance Standard (NSPS) for Sewage Treatment Plants applies. This standard is defined in 40 CFR 60, Subpart O and established a particulate emission limit of 0.65 g/kg dry solids input (1.3 lb/ton). The rule also established opacity limits and associated monitoring, testing, recordkeeping, and reporting.

5.2.6 *National Emission Standards for Hazardous Air Pollutants*

The National Emission Standards for Hazardous Air Pollutants (NESHAP) are defined in 40 CFR 61. The limitations for beryllium (Subpart C) and mercury (Subpart E) are applicable to units that incinerate or dry WWTP solids. Emission of beryllium to the atmosphere from stationary sources shall not exceed 10 g (0.022 lb) of beryllium over a 24-hour period. Emission of mercury to the atmosphere from incineration plants, drying plants, or a combination of these that process wastewater residuals shall not exceed 3.2 kg (7.1 lb) of mercury per 24-hour period.

In addition to the criteria pollutants regulated by the NAAQS, there is another set of federally regulated air pollutants known as hazardous air pollutants (HAPs). Hazardous air pollutants are a set of 188 chemicals specifically regulated by the U.S. EPA that are known or believed to cause human health effects in excess of levels specified by the Agency.

Sources that emit more that 9 Mg/a (10 ton/yr) of an individual HAP or more than 23 Mg/a (25 ton/yr) of multiple HAPs are considered a major source of HAPs. Major sources of HAPs may be subject to the Maximum Achievable Control Technology regulations (40 CFR 63). Subpart VVV defined MACT standards for publicly owned treatment works. The standards in this subpart do not address solids-processing or thermal oxidation systems.

If a treatment plant treats industrial wastes and is considered a part of a miscellaneous organic chemical manufacturing process unit (MCPU), then emission control requirements may apply. These additional requirements are defined in 40 CFR 63 Subpart FFFF National Emission Standards for Hazardous Air Pollutants: Miscellaneous Organic Chemical Manufacturing.

5.2.7 *State and Local Regulatory Requirements*

Even if a thermal oxidation facility is relatively small, state preconstruction permitting requirements may apply. Although the review requirements may not be as stringent, the same emission limits may apply. Most states require a BACT analysis to determine the appropriate air pollution control equipment and emission limits.

Many states also have special regulations governing toxic air pollutant emissions. Some state rules define a control technology assessment for toxic air emissions (T-BACT). Other states require a dispersion modeling assessment to show toxic air pollutant emissions are in compliance with allowable ambient levels. State air toxics programs may also establish emission limits for some hazardous air pollutants.

Additional regulations may also apply for noise and odors. Special compliance demonstrations or preconstruction assessments of potential noise and odor may be required. Applicable standards may be developed on a case-by-case basis and include communication with the public and involvement of other interested parties.

Visual plume emissions are regulated by opacity limits as defined in many state regulations. Meeting these opacity limits may not prevent the formation of a yellow plume. Because of the negative response from the public that may result from the formation of a visible yellow plume, mitigation may be necessary.

5.3 Permit to Operate

Once the proposed facility is constructed, performance testing begins. The pollutant emissions to be tested are defined in the permit to construct. Testing procedures are defined in 40 CFR 60 Appendix A or methods agreed to by the reviewing authority. A testing protocol is developed that describes how testing will be conducted and emissions reported. If the measured emission rates are greater than the permitted emission limits established in the permit to construct, then immediate mitigation measures are implemented that may include modifications to air pollution control equipment or repermitting of the facility.

5.3.1 Federal Title V Operating Permit Program

Shortly after the facility begins operation, the Operating Permit program ensures continued compliance and reporting of actual emissions from major facilities. The Operating Permit program (Title V of the CAA) consolidates the emission limits established for the entire facility and defines a monitoring process to ensure compliance.

Once the facility is constructed, a performance test may be required to demonstrate that the emission limits established in the permit have been met by the operating facility. Satisfactory completion of the emission testing and submission of an operating permit application may be needed for continued operation of the facility.

If the facility is a major source, a Title V Operating Permit (40 CFR 70, State Operating Permit Programs) may be needed. The operating permit identifies all air emission sources present at the facility, summarizes the emission limitations and special conditions that have been established by the preconstruction permit, and outlines a process by which continued compliance with the emission limitations can be demonstrated.

5.3.2 State Operating Permit Program

The state operating permit program applies to major emissions sources that have accepted federally enforceable operating limits to restrict annual emission to levels below major source thresholds or to large minor emissions sources with potential emissions that are below the major source threshold but above the state operating permit program. The state operating permit program is similar to the federal operating permit program except that compliance with the program is administered through the state.

5.3.3 *Wastewater Solids Management*

The performance standards for the treatment and disposal of wastewater solids required by the Clean Water Act Amendments of 1987 are contained in Title 40, Part 503 of the Code of Federal Regulations (40 CFR 503), also referred to as Part 503 rules (promulgated 58 FR 9387, February 19, 1993). Subpart E defines the requirements for residuals fired in an incinerator.

For multiple hearth and FBIs, pollutant limits are established for each unit based on emission limits established elsewhere in the regulations, atmospheric dispersion, and control efficiency. Specific limits are derived for seven metals, total hydrocarbons, and/or carbon monoxide:

- Beryllium emission of 10 g per 24-hour period;
- Mercury emission limits of 3200 g per 24-hour period;
- Lead, arsenic, cadmium, chromium, and nickel feed cake concentrations limits based on site-specific conditions (i.e., feed rates, control efficiency of the scrubbing systems, dispersion factors, etc.);
- Total hydrocarbon monthly average concentration of 100 ppm by volume, as propane, corrected to 0% moisture and 7% oxygen; and
- As a surrogate for total hydrocarbon, a carbon monoxide monthly average concentration of 100 ppm by volume, corrected to 0% moisture and 7% oxygen may be used by the WWTP.

Sufficient monitoring equipment must be installed to ensure that performance standards are met. This includes a continuous emission monitor for total hydrocarbons or carbon monoxide and instruments to monitor oxygen and moisture concentrations in the stack. Combustion temperatures must be monitored at least daily to ensure that operating combustion temperatures established during the performance test is not exceeded by more than 20%.

Other parameters that monitor the performance of the air pollution control may also be required to ensure proper operation at a frequency that is appropriate for that device. Sampling and analysis for metal concentrations shall be at a frequency as determined by concentrations expected in wastewater solids. Records that define the performance of the incinerator and actual emissions rates must be reported annually and kept for 5 years.

5.3.4 *Air Quality Compliance*

Compliance Assurance Monitoring (CAM) requirements are defined in 40 CFR 64. The CAM rules applies to a pollutant specific emissions unit at a major source that is required to obtain a Title V Permit. The rule applies to an emissions unit subject to an emission standard or limitation, uses a control device to achieve compliance, and exceeds the uncontrolled emission criteria. Operation of an incinerator is required to achieve particulate control limits using a particulate control device. A CAM plan demonstrating that the scrubber is operated as intended and achieves the required emission limitations needs to be developed.

Permitting requirements under the CAA have evolved over the past 30 years. As the Clean Air Act is amended, new interpretations of existing regulations are made and additional requirements are added. Thus, it is necessary for owners of wastewater solids incinerators to review carefully the air quality permitting requirements as they apply to each facility location.

5.4 Maximum Achievable Control Technology Standards

On March 21, 2011, U.S. EPA published in the *Federal Register* a new set of air emission regulations for wastewater solids incinerators under 40 CFR Part 60 (U.S. EPA, 2011). These new regulations contain Standards of Performance for New Stationary Sources and Emissions Guidelines for Existing Sources. All wastewater solids incinerators in the United States have to be in compliance with the rule's nine emissions limits and other requirements by March 21, 2016. The new MACT standards for new MHIs and FBIs and existing MHIs and FBIs can be found in Table 17.3.

In May 2011, however, the National Association of Clean Water Agencies, Washington, D.C., formally requested that U.S. EPA reconsider the rule for a variety of reasons. As a result of NACWA's request and potential legal challenges to the rule, the reader is advised to check the rule's status with U.S. EPA personnel.

5.5 Recently Permitted New Fluidized Bed Incinerators

Detailed information on the permitting process can be obtained from agencies that have recently undergone permitting and installation of a new wastewater solids incineration system, like the Northeast Ohio Regional Sewer District, Cleveland, Ohio; the Metropolitan Sewer District of Greater Cincinnati, Cincinnati, Ohio; and the Metropolitan Council Environmental Services, St. Paul, Minnesota.

TABLE 17.3 Maximum achievable control technology standards for wastewater solids incinerators.

		Existing units		New units	
Pollutant	Units*	Multiple hearth	Fluidized bed	Multiple hearth	Fluidized bed
Particulate matter	mg/dscm	80	18	60	9.6
Hydrogen chloride	ppmvd	1.2	0.51	1.2	0.24
Carbon monoxide	ppmvd	3800	64	52	27
Dioxin/furans					
a. Total mass basis	ng/dscm	5.0	1.2	0.045	0.013
or					
b. Total equivalence basis	ng/dscm	0.32	0.10	0.0022	0.0044
Mercury	mg/dscm	0.28	0.037	0.15	0.0010
Oxides of nitrogen	ppmvd	220	150	210	30
Sulfur dioxide	ppmvd	26	15	26	5.3
Cadmium	mg/dscm	0.095	0.0016	0.0024	0.0011
Lead	mg/dscm	0.30	0.0074	0.0035	0.000 62
Visible fugitive emissions—ash handling	%	≤ 5% of the hourly observation period	≤ 5% of the hourly observation period	≤ 5% of the hourly observation period	≤ 5% of the hourly observation period

* All units are corrected to 7% oxygen.

6.0 REFERENCES AND RECOMMENDED READINGS

Clean Air Act. *U.S. Code*, Section 50–88, Title 40, 1963.

Clean Air Act. *U.S. Code*, Section 7401–7671, Title 42, 1990.

Clean Water Act. *U.S. Code*, Section 1251, Title 33, 1972.

Dangtran, K.; Butt, J. (2004) Minimization of CO and NOx Emissions by Optimization of Fluidized Bed Design/Operating Conditions and by Chemical Additives; *Proceedings of the Water Environment Federation Bioenergy Workshop*; Cincinnati, Ohio, Aug 11–12; Water Environment Federation, Alexandria, Virginia.

Dominak, R. P., et al. (2005) Long-Term Residuals Management Plan for the Northeast Ohio Regional Sewer District.

Harder, S.; Dominak, R. (2006) Biosolids Economics: A Case for Careful Analysis. *Proceedings of the Water Environment Federation Residuals and Biosolids Management Conference 2006* [CD-ROM]; Water Environment Federation: Alexandria, Virginia.

Lundberg, L. A. (2004) The Future of Fluidized Bed Incineration. *Proceedings of the Water Environment Federation Residuals and Biosolids Management Conference 2004* [CD-ROM]; Salt Lake City, Utah, Feb 22–25; Water Environment Federation: Alexandria, Virginia.

Stone, L. A.; Harder, S.; Lue-Hing, C.; Logan, T.; Schaefer, P.; Shea, T.; Welp, J. (2008) *Residuals Management Validation Panel—Summary of Findings and Recommendations.* Northeast Ohio Regional Sewer District.

U.S. Environmental Protection Agency (2005) *Guideline on Air Quality Models,* 40CFR Part 51 Appendix W; Office of Air Quality Planning and Standards: Research Triangle Park, North Carolina.

U.S. Environmental Protection Agency (1993) Standards for the Use or Disposal of Sewage Sludge. *Code of Federal Regulations,* 40CFR, Part 503, Subpart E.

U.S. Environmental Protection Agency (2009a) Standards of Performance for New Stationary Sources. *Code of Federal Regulations,* 40CFR, Part 60, Appendix A, Method 5.

U.S. Environmental Protection Agency (2009b) Standards of Performance for New Stationary Sources. *Code of Federal Regulations,* 40CFR, Part 60, Subpart O.

U.S. Environmental Protection Agency (2011) Standards of Performance for New Stationary Sources and Emission Guidelines for Existing Sources: Sewage Sludge Incinerators; Final Rule, *Code of Federal Regulations,* 40 CFR, Part 60, Subpart LLLL and Subpart MMMM.

Water Environment Federation (2009) *Wastewater Solids Incineration Systems,* Manual of Practice No. 30; McGraw Hill: New York.

Water Environment Federation; American Society of Civil Engineers; Environmental and Water Resources Institute (2009) *Design of Municipal Wastewater Treatment Plants,* 5th ed.; WEF Manual of Practice No. 8; ASCE Manuals and Reports on Engineering Practice No. 76; McGraw-Hill: New York.

Chapter 18

Pyrolysis and Gasification and Other Thermal Processes

1.0 INTRODUCTION

Thermal processes for sludge handling, once considered net energy users, have become capable of energy conservation and energy production. Improvements in mechanical dewatering devices and automated controls have helped bring about this change. Today, thermal technologies can cost-effectively use sludge in an environmentally responsible manor. Recent issues of energy-cost escalation and more-stringent environmental controls have affected all sludge handling and use technologies, including treatment, transportation, emissions control, and chemical manufacturing costs. Self-sustaining thermal processes can be less susceptible to influences of cost.

This chapter focuses primarily on new improvements to old technologies of pyrolysis and gasification. These processes use substoichiometeric oxygen process principals and may be one of the most technically convincing energy possibilities for a carbon-neutral economy. The more typically used direct-burn processes for sludge energy use are discussed in other chapters. Other thermal technologies for sludge processing will also be briefly discussed in this chapter. For purposes of this chapter, a thermal process is defined as one in which operating temperatures are in excess of drying-only technologies and that cause thermochemical alteration of the feedstock material. In this chapter, "biomass" includes any biodegradable waste that can be burnt as fuel. It excludes organic material that has been transformed by geological processes into substances such as coal or petroleum. Sludge is one type of biomass fuel.

New thermal processes are emerging. Caution is advised, however, in implementing any technology without an extensive track record. Although high-temperature processes have been used for combustion of wastewater solids since the early 1900s, newer thermal processes are subject to operational challenges that may differ appreciably from experience with direct-burn processes. Pilot testing is highly recommended. Thermal processes oxidize the volatile component of sludge and offer advantages and disadvantages. Several advantages of this process include

- **Environmentally responsible.** The process is carbon neutral and has proven air emissions compliance capabilities. Air emissions from processes can be tracked by continuous emissions monitoring systems on a minute-by-minute basis.
- **Volume reduction.** It significantly reduces the volume and weight of wet sludge cake. Ash-product use options are available and expanding.
- **Detoxification.** Thermal decomposition destroys pathogens, viruses, and other biological contaminants that may otherwise create adverse environmental impact.

- **Energy conservation.** The process offers the potential to recover energy by direct means (heat exchangers) or by the combustion of converted products; thereby reducing or eliminating the energy needed to operate the processes.
- **Beneficial use.** Pyrolysis and gasification byproduct, "char," is a low-grade fuel that shows promise as activated carbon additives, odor control media, filtration media, and soil additives. Heat-derived byproducts from syngas can be used for drying sludge and for other energy applications.

The process also has several disadvantages, including

- **Emerging.** Newer thermal processes present new design and operational technology challenges. Caution is advised in the implementation of any technology without an extensive track record.
- **Dry feedstock.** Economic viability of many thermal processes requires significantly less moisture in the feedstock than existing sludge incinerators require. The added complexity of separate-stage sludge dryer is required.
- **Cost.** Because of the lack of full-scale, long-term installations, operational and capital costs data is limited. Gasification systems have processed agricultural and forest waste and coal for more than 50 years; however, sludge applications differ enough to preclude a direct comparison of cost data.
- **Operating challenges.** High-temperature processes can create maintenance requirements that reduce steady-state reliability of operation.

Technological advancements in thermal energy conversion systems are rapidly occurring. These advancements, coupled with progress in other related emerging technologies, are likely to increase efficiency and lower capital cost. Economic incentives and pollution credits can reduce life-cycle cost. Biomass pyrolysis and gasification technologies hold promise to convert sludge and other biomass products into thermochemical-derived fuels, and other high-value chemicals.

2.0 PYROLYSIS AND GASIFICATION PRINCIPALS

Biomass gasification is a two-step or combined step process. Pyrolysis plus gasification principals are used to convert the energy potential from solid to gas. Gasification literally means to convert a solid or liquid substance into a gas by oxidization-reduction reactions that thermochemically convert biomass energy. Larger-molecule volatile solids and some fixed-carbon solids are converted into smaller-molecule combustible gas products, termed synthesis gas, or syngas.

Pyrolysis and gasification technologies are not new. Ancient Egyptians practiced wood distillation by collecting the tars and pyroligneous acid for use in their embalming processes. The process of making charcoal from wood dates back to 2000 B.C. In 1921 Ford Motor Company, Detroit, Michigan, used scrap wood in a wood carbonization and distillation plant to produce charcoal and commercial chemicals including methanol, acetates, acetone, alcohols, creosote, pitch, keytones, and heavy and light oil. In World War II, small downdraft woodchip-fired gasifiers produced fuel gas for motor vehicles (Skov and Papworth, 2006). Application of full-scale pyrolysis and gasification processes using sludge as the primary feedstock is limited. During the preparation of this manual, only several large-scale pilot or full-scale sludge-only feed gasifiers were in operation worldwide.

In a gasification energy conversion system, the first reaction, pyrolysis, chemically changes the volatile matter into carbon rich "char" by heat in an atmosphere devoid of oxygen. The pyrolysis-produced char and combustible gases then undergo gasification reactions. Conventional air/oxygen-blown gasification thermochemical reactions occur at temperatures of 600 to 900 °C (1100 to 1650 °F) and approximately 30% of the stoichiometeric oxygen required for complete combustion.

2.1 Gasification Reactions

Gasification systems regulate oxygen to control the rate of partial combustion to transfer energy from solid feedstock to syngas. As a result, the fuel demand associated with excess air required for direct combustion, higher fixed carbon oxidization processes is reduced. The energy transfer occurs as air/oxygen is introduced to glowing carbon in the gasifier. The oxygen reacts with the carbon; this semicombustion follows eq 18.1.

$$C + \tfrac{1}{2} O_2 \rightarrow CO \qquad 18.1$$

During the pyrolysis step, some complete combustion of the carbon takes place. Complete combustion or oxidization of carbon occurs per eq 18. 2

$$C + O_2 \rightarrow CO_2 \qquad 18.2$$

With the inclusion of nitrogen, carbon conversion occurs per eq 18.3.

$$C + O_2\,(+\,3.76\,N_2) \rightarrow CO_2\,(+\,3.76\,N_2) \qquad 18.3$$

Nitrogen aids in the gasification process because it acts to temper the gasification reactions. Without nitrogen, extremely high temperatures would prevail, and the gasification process would be difficult to control. If the temperature in the gasifier is maintained, then CO_2 produced by carbon combustion is immediately reduced to carbon monoxide per eq 18.4.

$$CO_2 + C \rightarrow 2\,CO \qquad 18.4$$

The carbon dioxide dissociation to carbon monoxide is temperature-dependent. Dissociation commences at approximately 400 °C; at less than 400 °C only inert smoke will be produced.

The composition (%N_2, CO_2, H, and CO) of the gasifier-produced syngas determines its calorific value. The calorific value of the syngas increases as CO_2 is reduced or dissociated in the gasification process into CO. Syngas at approximately 35% N_2:35% CO has a calorific value of approximately 5020 kJ/Nm^3 (350 Btu/cu ft). Typically, gasifiers can convert 65 to 85% of the energy in the solid feedstock to syngas energy (Chilson, 2003) A simplified heat balance is shown in Figure 18.1.

Unfortunately, the syngas energy produced by gasification often is needed to dry the sludge feedstock to support autogenous conditions. Separate-stage drying is necessary for overall performance viability. Efficiency of the system can be considered

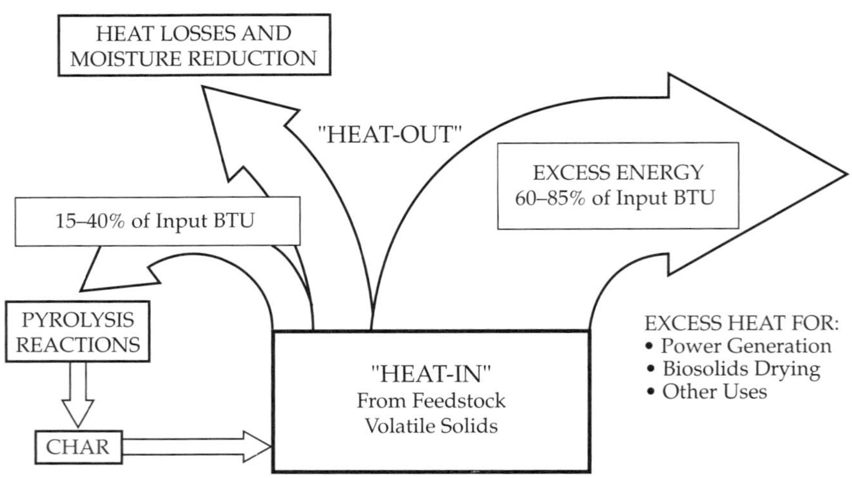

FIGURE 18.1 Simplified gasifier heat balance (1 Btu = 1.055 kJ).

as the ratio of the fuel calorific value to the produced syngas calorific value as illustrated in eq 18.5.

$$\% \text{ Efficiency} = (\text{Gas calorific value}/\text{Fuel calorific value}) \times 100 \qquad 18.5$$

Use of steam as the gasifying agent increases process efficiency and syngas calorific value. As steam contacts the glowing carbon mass it reacts per eq 18.6.

$$H_2O \rightarrow H_2 + 0.5\ O_2 \qquad 18.6$$

As shown, 1.0 Nm^3 of steam will produce 1.0 Nm^3 of hydrogen and 0.5 Nm^3 oxygen. The dissociation of steam is endothermic and requires 10 800 kJ/Nm^3. Hydrogen has a calorific value of 12 800 kJ/Nm^3. At this point in the process, the syngas released is primarily a mixture of CO (approximately 40%); hydrogen (15 to 20%); and nitrogen (approximately 40%). The hydrogen fraction can increase the calorific value of the syngas by as much as 60 to 65%.

The term "steam-reforming" is sometimes used when referring to gasification. When steam is used as the gasifing agent, steam-reforming converts the CO produced in the gasification step to hydrogen. This relationship, the water-gas shift reaction, is shown in eq 18.7. This reaction is slightly exothermic.

$$CO + H_2O \rightarrow CO_2 + H_2 \qquad 18.7$$

2.2 Pyrolysis

Detailed information on prolysis principals and processes are beyond the scope of this manual. The pyrolysis step has been used alone to convert biomass feedstock to liquid byproducts termed bio-oil. Pyrolysis reaction time and temperature relationship is critical to yield objectives, as indicated in Figure 18.2 (Barbooti, 1988). Flash heating of less than one second minimizes char production and maximizes gas or tar, depending on operating temperature. Higher temperature (± 900 °C) produce more gas, lower temperatures (± 500 °C) produce more tars and liquids. Slow heating of 100 seconds or more results in increased char production and less gas or tar production.

Fast pyrolysis, characterized by rapid heating (i.e., 0.1 to 2.0 seconds) to 450 to 900 °C maximizes yields of bioliquids and minimizes tar and charcoal or char production (González et al., 2003). The vibrations caused by this temperature range vibrate the biomass atoms apart into fragment molecules of bio-oil, volatile gas, and tar at random positions. Proficient rapid heating requires the biomass be ground into fine

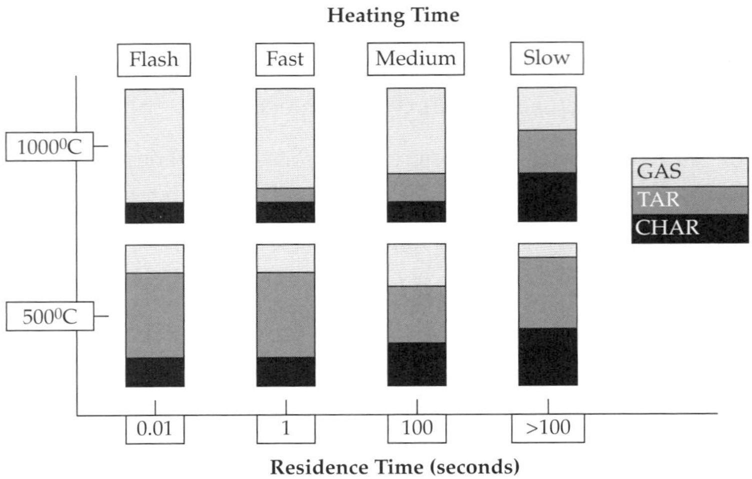

Figure 18.2 Pyrolysis byproduct variation by time and temperature (courtesy of European Biomass Industry Association).

particles to allow the insulating char layer that forms at the surface of the reacting particles to be continuously removed. Processing organic materials in a fast pyrolysis plant requires a feedstock particle size less than 6 mm and a moisture content of less than 10% on a weight basis. Extreme pyrolysis, which leaves only carbon as the residue, is called carbonization.

Pyrolysis has been used on a massive scale to turn coal into coke for steelmaking. The term pyrolysis is sometimes used to include thermolysis or hydrous pyrolysis (i.e., in the presence of water) used in steam cracking of oil. A flow schematic of a pyrolysis conversion of organic feedstock to oil is shown in Figure 18.3. Typically, 70 to 75% on a dry weight basis of feedstock volatiles can be converted into oil.

2.3 Gasifier Syngas Characteristics

Syngas from an air-blown gasifier have reported calorific value of 5 to 8 MJ/Nm^3 (150 to 250 Btu/cu ft). Syngas heating values from oxygen-blown gasifiers typically range 10 to 12 MJ/m^3 (300 to 350 Btu/cu ft) . Superheated steam gasifiers have reported calorific values of 12 to 20 MJ/Nm^3 (400 to 500 Btu/cu ft) (Lewis et al., 2002). Limited data on sludge-only gasification shows a syngas heating value of approximately 4 to

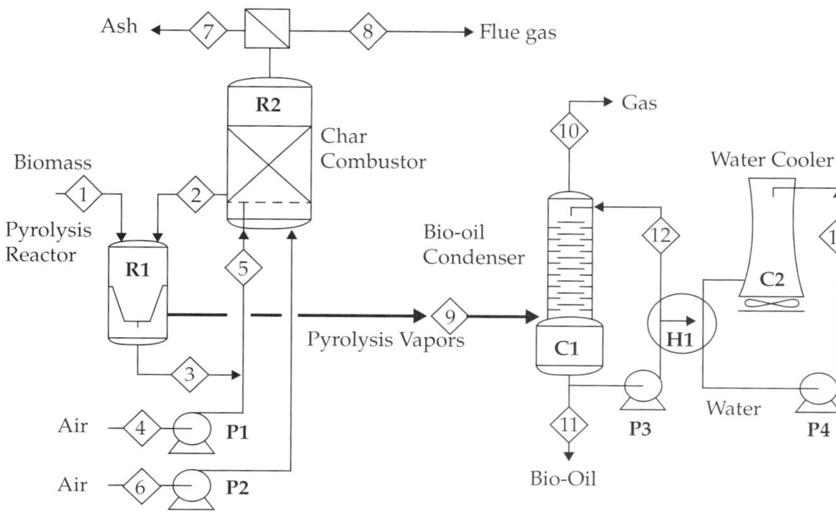

Figure 18.3 Pyrolysis to oil flow schematic (courtesy of BTG Biomass Technology Group).

12 MJ/m^3 (100 to 350 Btu/cu ft). Syngas is a mixture of varying amounts of nitrogen, carbon monoxide, and hydrogen. Small percentages of light hydrocarbons, oxygen, solid particles, tar, feed material, and other elements particular to the process may be present. The amount of light hydrocarbons and tars produced can be regulated by the pyrolysis reaction time. Control of particulate matter in the syngas requires special design attention.

Biomass gasifier-produced syngas typically contains 0 to 50% nitrogen, 10 to 50% hydrogen, 15 to 25% carbon monoxide, 5 to 25% carbon dioxide, and 2 to 10% methane, plus trace elements. The makeup of syngas can vary as a function of the feedstock and gasifier equipment used. Table 18.1 illustrates typical variation of syngas constituents from different gasification agents, using sludge as the feedstock. Table 18.2 shows air emissions were well below state allowable limits for the Hyperion Treatment Plant, Los Angeles, California, four-stage gasification system operated from 1986 to 1996 on digested sludge.

CIEMAT, the public research organization for energy and environment in Spain, tested fluid bed gasification on dried sludge (Sanchez et al., 2007). The trials were

Table 18.1 Syngas constituents from gasification of sludge biomass.

Syngas component	Gasification agent		
	Air	Steam	Air/steam
Nitrogen	44%	–	32%
Hydrogen	12%	49%	13%
Carbon monoxide	19%	17%	25%
Carbon dioxide	17%	24%	20%
Methane	6.5%	8%	10%
Heating value (MJ/Nm^3)	7.4%	11.9%	9.5%

Table 18.2 Hyperion air emissions control system performance (THC = tetrahydrocannabinol; NMHC = nonmethane hydrocarbon).

Air containment	Target emission limits dry at 5.5% oxygen	Actual emissions
NOx	75 ppm as NO_2	10–40 ppm as NO_2
SOx	20 ppm as SO_2	< 1 ppm as SO_2
CO	20 ppm as CO	< 10 ppm as CO
NMHC	10 ppm as CH_4	THC < 1 ppm as C
Particulate*	0.01 gr/dscf at 3% O_2	0.32 dscf at 3% O_2

*Front half plus back half, U.S. EPA Method 5.

conducted in a 2.4 Mg/d fluid bed gasifier fed by dried (92.5% total solids) sludge. The digested sludge characteristics were measured at 48.4% for volatile fraction; 46.1% ash; and 5.5% fixed carbon; although the individual trial runs were short (4 to 8 hours each). Table 18.3 shows stable syngas consistency and production rate during each trial.

Figure 18.4 illustrates the potential pathways and diversity of derivatives from syngas. In addition to the bio-oils produced by pyrolysis, Gonzalez et al. (2003) identified more than 200 potential hydrocarbon-based derivative compounds from gasifier syngas.

TABLE 18.3 Syngas from gasification of low-volatile solids sludge.

Trial ID	G1	G2	G3	G4	G5	G6
Syngas composition in (% volume, unless otherwise noted)						
N_2	45.94	42.44	37.27	40.31	43.82	47.85
H_2O	15.26	18.33	19.43	17.26	16.37	12.75
CO_2	21.34	20.78	25.46	24.46	23.52	23.21
CO	5.95	5.49	5.07	5.29	4.90	5.40
H_2	6.87	7.42	7.45	7.15	6.05	6.53
CH_4	2.37	2.72	2.32	3.02	2.79	2.54
C_2H_4	0.95	1.15	1.26	1.03	1.08	1.28
C_2H_6 (ppm)	230	246	441	469	538	353
C_2H_2	0.52	0.58	0.46	0.39	0.36	0.44
BTX (ppm)	3854	2605	3250	3322	3255	2953
CxHy (as C_3H_8 equiv.)	3.27	3.39	3.36	3.44	2.67	3.11
NH_3 (ppm)	7759	10618	12366	10217	10816	8320
H_2S (ppm)	832	–	–	–	513	–
HHV (MJ/Nm^3)	3.9	4.4	4.3	4.2	3.2	3.2

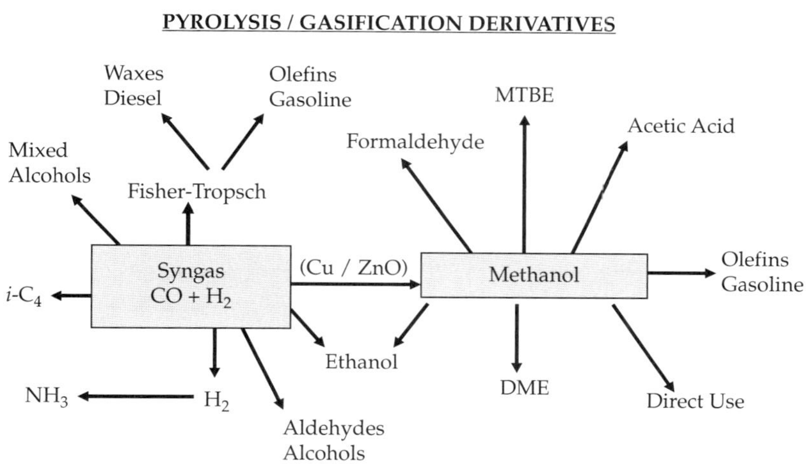

FIGURE 18.4 Pyrolysis and gasification derivatives (MTBE = methyl tertiary butyl ether; DME = dimethyl ether).

3.0 GASIFICATION PROCESS VARIATIONS

3.1 Gasifiers Design Variations

Gasifier designs can be subdivided by configuration, gasification agents, operating parameters, and function. Some of the more common variations include design according to heat; direct heat or autothermal gasifiers, indirect heat, or allothermal gasifiers; according to gasification agents; air-blown, oxygen-blown, steam gasifier/steam reformers; according to pressure; atmospheric, pressurized, according to configuration; fixed-bed updraft or downdraft gasifiers; fluidized, circulating, or bubbling-bed gasifiers; and entrained-flow gasifiers. The gasifier can be designed to combine pyrolysis and gasification reactions in a single reactor (direct-heated gasifiers) or isolate them in separate reactor vessels (indirectly heated gasifiers). Pyrolysis-only reactors and processes are also commercially available.

3.1.1 Direct-Heat Gasifiers

A direct-heat gasifier design combines pyrolysis and gasification reactions in a single reactor. Direct-heat gasifiers are more prevalent in the conversion of sludge-biomass feedstocks. In a direct-heat gasifier, a portion of the pyrolysis reactions-produced combustion products are oxidized to produce heat for the gasification step and subsequent pyrolysis of the new feedstock. Fixed-bed direct-heat gasifiers can be designed in a downdraft or updraft configuration as illustrated in Figure 18.5.

A fixed-bed, updraft gasifier is designed with a fixed bed of carbonaceous fuel through which the "gasification agent" (steam, oxygen, or air) flows counter-current to the bed. In this configuration, thermal efficiency is high, gas exit temperatures are low, and tar and methane production can be a design consideration. In a fixed-bed, downflow

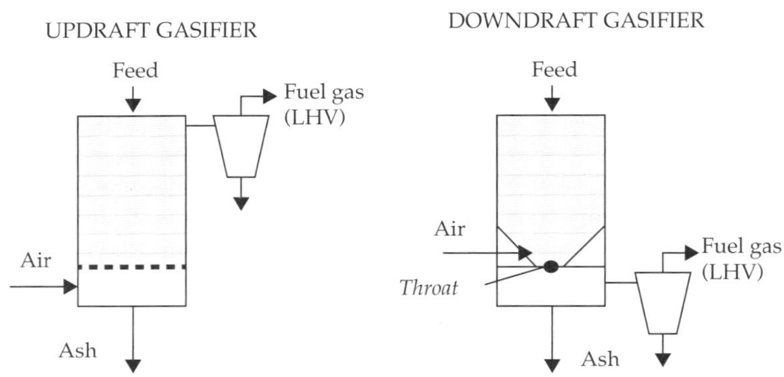

FIGURE 18.5 Direct-heat, fixed-bed gasifier configurations (LHV = lower-heating value).

gasifier, the gasification agent gas flows in a cocurrent pattern with the fuel through the bed. In this configuration, throughput is greater, gas exit temperature is high, energy efficiency is reported similar to upflow type, and tar levels tend to be lower. Heat typically is required in the upper part of the bed by char combustion or external heat sources.

Small downdraft gasifiers fired with wood chips were used during World War II to produce fuel gas that powered engine of motor vehicles (Skov and Papworth, 2006).

3.1.2 *Indirect-Heat Gasifier*

The thermochemical reactions in an indirect-heated gasifier are similar to direct-heat gasifiers. The fundamental difference is that in an indirect-heat gasifier, char combustion takes place in a separate char-combustor vessel.

3.1.3 *Moving-Bed Gasifiers*

Moving-bed gasifier reactor designs vary. The most typically used in the United States is the fluidized-bed gasifier. In a fluidized-bed gasifier, feedstocks are fluidized by air or oxygen and steam into a bed of hot sand. The ash is removed dry or as heavy agglomerates that defluidizes. Circulating fluidized-bed gasifiers are another variation in which particles are introduced into a circulating bed of hot sand.

3.1.4 *Direct-Heat, Entrained-Flow Gasifiers*

Entrained-flow gasifiers are pressurized, oxygen-blown reactors suited to large industrial applications. Dry pulverized solid or atomized liquid fuel is gasified with oxygen in cocurrent flow.

3.1.5 *Pyrolysis Reactors*

In-depth discussion of pyrolysis reactors is beyond the scope of this manual. Various pyrolysis technologies and configurations have been used to convert biomass feedstocks to produce bio-oils. Fixed beds were used for traditional charcoal production; however, slow heat transfer resulted in low liquid yields. The auger process adapted from the Lurgi process for coal gasification uses hot sand and biomass that are fed into one end of the screw. The screw mixes, conveys, and transfers heat. This process does not dilute the pyrolysis vapors; however, the sand must be reheated in a separate vessel, which adds mechanical complexity.

Other examples of pyrolysis systems are the EnerSludge pyrolysis operated in Perth, Australia. Unfortunately, the bio-oil was found to be unsuitable for fueling diesel engines, and the system economics reported were unfavorable.

The Pyrobustor process shown in Figure 18.6 is a two-stage, horizontal, rotary kiln provided by EISENMANN Corp. (Böblingen, Stüttgart, Germany) for the continuous

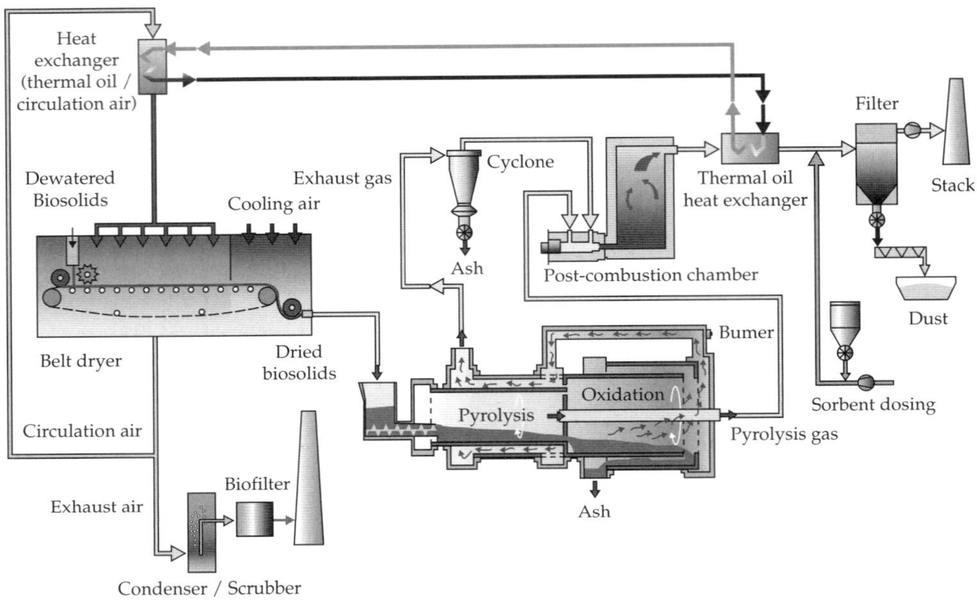

FIGURE 18.6 Pyrobustor process (St. Lorenzen, Trentino, Italy).

conversion of biosolids to recoverable energy and residual ash. Two, full-scale commercial units are operating in Europe that were commissioned in 2006 in Italy (St. Lorenzen, Trentino) and in 2008 in Germany (Crailsheim, Baden-Württemberg). These units have a combined operating history greater than 50 000 hours (exceeding 90% availability), and are rated for 550 to 650 kg/h (17 tons/day) dried solids, as fed, 88 to 94% dry matter.

In the first Pyrobustor stage, dried biosolids undergo slow pyrolysis to combustible organic gases and residual char. Heavy, condensable tars are minimized by retention time and residual moisture. In the second stage, the char is combusted to provide energy for the pyrolysis stage and to produce ash for reuse. The dryer is thermally "close-coupled" to the Pyrobustor to use recovered heat energy from the combustion of the pyrolysis gases, with any excess for beneficial use by the host facility. When the Pyrobustor is shutdown for maintenance, heat energy can be provided for dryer operation using auxiliary fossil or renewable fuels to produce Class-A equivalent biosolids. Excluding the dryer, the Pyrobustor plant in Italy cost \$5 million USD (\$3.5 million EUR).

4.0 ENVIRONMENTAL CONSIDERATIONS

Air emissions from biomass power production are regulated by federal and state standards. Air emissions permits are typically required. Pollution control technologies are proven to remove particulates and other air emissions compliant with regulatory standards. Typical control devices are wet and dry scrubbers including single or multiport, low-pressure Venturies, tray scrubbers, packed-bed scrubbers, and baghouses.

Sludge-derived energy from pyrolysis and gasification technologies offer the potential to reduce greenhouse gas emissions in two basic ways:

1. Burning sludge or biomass releases carbon dioxide that is largely balanced by the carbon dioxide captured in its own growth and use period. This relationship has been termed "carbon neutral."
2. Pyrolysis and gasification processes use only a portion of the fixed carbon in feedstock, compared to direct-burn processes to generate necessary heat; consequently, less carbon and nitrogen is converted to greenhouse gases, resulting in reduced carbon and nitrogen-based air emissions.

For gasification systems, a gas cooler in series with cyclones has proven effective not only to reduce particulate emissions but also to avoid de-novo-synthesis which is notorious for the formation of dioxins and furans (Sanchez et al., 2007).

Studies are ongoing to evaluate the beneficial use of pyrolysis byproduct char as a low-grade fuel, as an additive directly to dewatered sludge for odor control, as a filtration media for odor and volatile organic compound reduction. Other potential char use alternatives include

- Asphalt production because the mechanical consistency of the char allows its use as an admixture in the production of asphalt.
- Backfill and landscaping because of its high mineral content, primarily carbon, potassium, and phosphorus, allow consideration as additives for re-cultivation projects.
- Phosphorus recovery because the economics of phosphorus recovery is a function of feedstock phosphorous concentration.

Thermal decomposition of sludge produces energy. Anaerobic digesters for example, convert biomass energy or Volatile Solids into gas energy as methane. Unfortunately, the syngas energy produced by gasification is often needed to dry the sludge to support autogenous conditions, and the energy available for export is minimal.

5.0 SAFETY

Thermal processing of sludge carries with it certain mandatory code compliance, good engineering practices, and safety standards. These standards require sound design and attentive maintenance of equipment and devices.

Processes requiring separate-stage sludge drying require particular attention to safety-related equipment operation. Any process requiring a dried sludge feedstock must be evaluated for explosive potential. An explosive concentration of dry sludge dust in the presence of oxygen greater than 15% will explode when exposed to a spark at a temperature exceeding 354 °C (670 °F).

Minimizing the explosion potential in these areas requires added safety features to the systems. These include using inert purge systems to maintain the oxygen level in the dryer at less than 15%; eliminating objects from sludge that may cause sparks; applying water spray systems to produce steam to dilute oxygen concentration; and using rupture disks or explosion panels for reliving pressure buildup from explosions. Dry-product silos should be fitted with temperature, carbon monoxide, and explosive conditions gas sensors.

During the design phase, an overall safety review by qualified safety personnel is necessary. Safety review should include a list of applicable code and compliance statuses at the time of design transmitted to the owner for record. National Fire Protection Prevention Systems (NFPA) standards and other code compliance organizations should be reviewed. Efforts to develop more specific standards for sludge drying/gasification systems likely will continue.

6.0 OTHER THERMAL PROCESSES

Thermal processes give biomass/sludge tremendous flexibility in the way it can be used to produce energy. These combined power-and-heat and trigeneration (power, heating, and cooling) technologies use a variety of energy conversion devices to produce heat and electrical power. Devices include heat exchangers, gas and steam turbines, reciprocating engines, adsorption chillers, and fuel cells. Economic incentives, renewable energy credits, and carbon credits are often available. Positive environmental effects inherent to these systems could alter public perception for the handling of many of the organic residual wastes considered appropriate today. A brief overview descriptions of several processes follow. This list is limited, however, because further research is needed.

6.1 Supercritical Water Oxidization

Supercritical water oxidation (SCWO) is the oxidation of organics in a liquid or cake at supercritical temperature and pressure that are controlled to prevent vaporization of the liquid. The degree of oxidation is dependent on the temperature and pressure (Biomass Technology Group, 2011; World Student Community for Sustainable Development, 2011). These systems typically operate in the temperature range of 374 to 530 °C and pressure range of 22.1 to 27.6 MPa (3200 to 4000 psig) (Patterson et al., 2001).

Compressed air or oxygen is fed into the reaction vessel, and the reaction is exothermic. At supercritical temperature and pressure, solubility of organic substances and oxygen in water is greatly increased to completely decompose organic substances. SCWO can achieve high destruction efficiencies of greater than 99.99% of organics in short reactor residence times of less than one minute. Temperatures are sufficiently low to minimize formation of SOx and NOx gases (Patterson et al., 2001). Nonpolar organics (e.g., benzene and polychlorinated biphenyls) are completely miscible in SCWO and, in the presence of an oxidant, react to form (recoverable) carbon dioxide and water. Organic nitrogen is converted to molecular nitrogen.

Technology suppliers claim SCWO produces an inert, silty solid that settles out of the liquid stream. The volume of solids is significantly reduced. Metals are bound in a nonleachable form that passes the U.S. EPA's (2011) toxicity characteristic leaching procedure (TCLP). Air emissions are anticipated to be primarily carbon dioxide, oxygen, and nitrogen and to have a minimal odor.

The SCWO technology for sludge management is considered commercially unproven, although several demonstration systems have been built in Sweden and in Kobe, Japan. A variation of the SCWO has been installed in Rialto, California. A plant at Harlingen WaterWorks, Harlingen, Texas, was reported to have produced 9 Mg/d dry (10 tons/d) sludge from HydroSolids SCWO units, manufactured by HydroProcessing LCC, Austin, Texas. Corrosion in the heat exchangers occurred, however, and the system was no longer in service. The manufacturer suspected that an industrial user illegally discharged corrosive waste to the wastewater collection system, which caused damage to the system. The capital cost of the facility was $8 million, including $6 million for the two HyroSolids unit. The projected operating cost was reported at $180 per 0.9 Mg dry (1 dry ton). Revenue for waste-heat recovery and carbon dioxide projected a net operating cost of $60 per 0.9 Mg dry (1 dry ton).

6.2 Athos

The Athos process manufactured by Veolia Water, Chicago, Illinois, is based on the principal of SCWO or hydrothermal oxidization, an oxidation process in the liquid phase that involves heating thickened sludge to high temperatures of 250 to 300 °C (480 to 570 °F) and high pressures of 7000 to 15 000 kPa (1015 to 2175 psig) in the presence of an oxidizing gas of air or oxygen. A flow schematic of the Athos process is shown in Figure 18.7.

The manufacturer reports the process mineralizes the organic matter in the thickened sludge feedstock and is applicable to primary, biological, or digested sludges. The manufacturer reports the final product, "technosand," reduces the original sludge feedstock volume by 98 to 99%. Heavy metals are stabilized in a nonleachable form and malodorous organics (hydrogen sulfide, mercaptan, etc.) are destroyed.

6.3 SlurryCarb

Slurry carbonization is an emerging thermal treatment process to improve the dewaterability of biosolids. Manufacturers claim the process operates at lower temperature and pressure and enables a shorter reaction time than other SCWO pyrolysis processes (Vizier et al.). The process is reported to receive dewatered sludge at 15 to 30% total solids, macerated to particles of less than 12 mm (0.5 in.) and produces a product termed

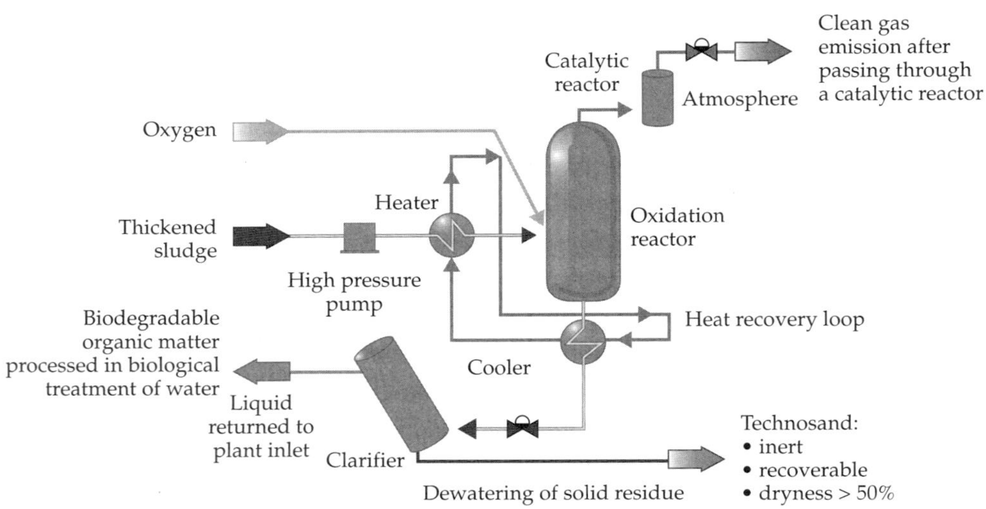

FIGURE 18.7 Athos process flow schematic (courtesy Veolia Water, Chicago, Illinois).

"E-fuel," that dewaters to 50 to 60% total solids with a heating value of approximately 15 000 to 19 000 kJ/kg (6500 to 8000 Btu/lb). In Realito, California, the SlurryCarb system is designed to process approximately 178 Mg/d (196 dry ton/d) of solids from five municipalities and generate approximately 150 Mg/d (167 ton/d) of product. According to the private operator, slurry carbonization can generate more than 1.98 times the energy it needs to treat undigested solids. The process is shown to generate a waste stream high in ammonia and organics. The carbon dioxide and reactor exhaust are typically directed to a regenerative thermal oxidizer (RTO) to oxidize any volatile organic compounds (VOCs) in the offgases. The project at Realito is still under development.

6.4 Blending of Sludge in Cement Manufacturing

Cement manufacturing requires large amounts of energy because its feedstock (limestone rock with small amounts of iron oxide and alumina) must be heated to greater than 1500 °C (2700 °F) in a pre-calciner tower and then burned in a rotating kiln to form a crystalline product called clinker—the essence of cement. Air emissions during cement manufacturing include carbon dioxide (CO_2), carbon monoxide (CO), and nitrous oxides (NOx). Some cement manufacturers use selective noncatalytic reduction (SNCR) to reduce the levels of NOx discharged within the kiln exhaust air. In SNCR, ammonia is injected into the kiln at temperatures of 870 to 930 °C. At this temperature range, the ammonia combines with NOx to form NO_2 and nitrogen (N_2) gas. The chemical reaction at the center of this process is shown in eq 18.8

$$\mathrm{NOx} + \mathrm{NH_3N}\ (\text{ammonia}) + \mathrm{O_2} + \mathrm{N_2} + \mathrm{H_2O} \qquad 18.8$$

Ammonia injection via the SNCR process has been shown to reduce NOx emissions. Based on the SNCR technique, a sludge injection process was developed by the Cement Industry Environmental Consortium (1996). The sludge injection process uses naturally occurring ammonia and moisture content of dewatered sludge as the reagent. The chemical reaction for NOx reduction through sludge injection is claimed to be the same as that for ammonia. The ammonia in the sludge combines with NOx in the exhaust air to form nitrogen gas. During the process, sludge is burned incidentally, and the inert materials (including metals) are bound into the cement.

The Mitsubishi Cement Corp.'s Cushenbury plant in Lucerne Valley, California, underwent long-term testing and adopted sludge injection in 1994 and 1995 and is still in operation today. The company estimates that 20 to 30% of NOx emissions are reduced when injecting sludge at the rate of approximately 50 Mg/d wet.

6.5 Vitrification

Vitrification is a process that has been used for many years to melt minerals into glass. The feedstock must contain 10% or less moisture and is then melted it in a furnace at high temperature of 1330 to 1500 °C (2400 to 2730 °F). The resulting liquid glass is quench-cooled to form a solid that is then crushed and sized to meet market specifications. Slagging is a concern for any thermal process operating at temperatures greater than ash softening. Offgas handling equipment may be susceptible to fouling/slagging and corrosion considerations. Vitrification of sludge is considered, at the time of publication, as a commercially unproven technology with limited application in facilities. A search of the technology found two: Lemar Environmental in New Zealand and Minergy in the United States. In 2008 to 2009, the Minergy vitrification process operation was discontinued because of reliability/online equipment issues that curtailed throughput and diminished system economics.

Discussion of other thermal processes is beyond the scope of this manual. In summary, thermal energy conversion technologies using sludge and other biomass feedstocks offer a clean, renewable energy source and an outstanding alternative to expensive and environmentally unfavorable disposal of organic residual wastes in landfills. Although bioenergy will never replace conventional fossil fuel-powered generating facilities, every kilowatt of fossil fuels replaced by these technologies reduces pollution and the effects of global warming. A promising future remains for innovations and implementation of projects using these environmentally sound technologies.

7.0 REFERENCES

Barbooti, M. M. (1988) Flash Pyrolysis for the Continuous Conversion of Reed into Hydrocarbons. *J. Anal. Appl. Pyrolysis*, **13** (3), 233–241.

Biomass Technology Group (2011) Supercritical Water Oxidation/Pyrolysis. www.btgworld.com (accessed August 2011).

Bridgwater, A.V. (2003) Renewable Fuels and Chemicals by Thermal Processing of Biomass. *Chem. Eng. J.*, **91** (2–3), 87–102.

Bridle, T. R.; Skrypski-Mantele, S. (2004) Experience and Lessons Learned from Sewage Sludge Pyrolysis in Australia. *Water Sci. Technol.*, **49** (10), 217–223.

Cement Industry Environmental Consortium (1996) Salt Precipitation and Scale Control in SCWO—Part B Commercial/Full-Scale Applications. U.S. Patent No. 5,586,510, December 24, 1996.

Chilson, S. J. (2003) *Conversion of Waste Products to Energy Biomass Gasification & Other High Temperature Technologies*; Water Environment Federation Bioenergy Technology Subcommittee: Alexandria, Virginia.

Encinar, J. M.; Beltrán, F. J.; Bernalte, A.; Ramiro, A.; González, J. F. (1996) Pyrolysis of Two Agricultural Residues: Olive and Grape Bagasse. Influence of Particle Size and Temperature. *Biomass Bioenergy*, **11** (5), 397–409.

González, J. F.; Encinar, J. M.; Canito, J. L.; Sabio, E.; Chacón, M. (2003) Pyrolysis: Energy Uses of the Different Fractions and Kinetic Study. *J. Anal. Appl. Pyrolysis*, **67** (1), 165–190.

Haug, R. T.; Lewis, F. M. (1999) *Four-Stage, Fluidized Bed Gasification Process Minimizes NOx. City of Los Angeles Hyperion WWTP*. Proceedings of the 15th International Conference on Fluidized Bed Combustion.

Lewis, F. M.; Swithenbank, J.; Russell, N. V.; Shabangu, S. V. (2002) High Temperature, Steam Only Gasification and Steam Reforming with Ultra-Superheated Steam, *Proceedings of the 5th International Symposium on High temperature Air Combustion and Gasification.*

Patterson, D. A.; Stenmark, L.; Hogan, F. (2001) Pilot-Scale Supercritical Water Oxidation of Sewage Sludge; *Proceedings of the 6th European Biosolids and Organic Residual Conference*, Nov.

Sanchez, J. M.; Murillo, J. M.; Otero, J.; Plaza, F.; Garralon, G. (2007) Thermal Valorisation of Dried Sewage Sludge Through Gasification. *Proceedings of the 10th International Conference on Environmental Science and Technology*, Kos Island, Greece, Sept.

Skov, N. A.; Papworth, M. L. (2006) *Driving on Wood: The lost Art of Driving without Gasoline*, 3rd ed.; Biomass Energy Press: Franktown, Colorado.

Vizier, P.; Riotte, M.; Duminy, D.; Perraud, S. *Degremount—Valenton, France, Agricultural and Energy Valorisation of Biosolids: The Paris 'Seine Amont' Waste Water Treatment Plant.*

U.S. Environmental Protection Agency (2011) TCLP Questions http://www.epa.gov/osw/hazard/testmethods/faq/faq_tclp.htm (accessed August 2011).

World Student Community for Sustainable Development (2011) Home page www.wscsd.org (accessed August 2011).

Chapter 19

Transport and Storage

(*continued*)

1.0 INTRODUCTION

This chapter discusses the transportation and storage of solids, the movement of sludge, scum, or other suspended or dissolved solids from point to point for treatment within the wastewater process, and transportation to offsite storage.

1.1 Need for Storage

Wastewater solids treatment requires effective storage. Storage that is compatible with the objectives of a treatment system must be incorporated into its design to enhance system reliability and efficiency.

The capacity of a treatment system to hold solids within its component processes influences system reliability during variable weather conditions, changes in staffing levels, repair schedules, and emergencies. For example, solids are generated from the wastewater treatment system 24 hours per day, but it may be most convenient or cost effective to operate dewatering equipment only during the day shift. Solids must, therefore, be stored during off-hours. Storage is necessary between interconnected treatment steps that operate at different rates such as, for example, when the centrifuge discharge and incinerator feeding have variable rates. Because the wastewater treatment plant (WWTP) generates solids all year, facilities that use land application as a beneficial reuse option must either provide onsite or offsite storage or have an alternative outlet for processed solids when sites are not accessible or available.

1.2 Storage within Wastewater Processes

Storage of solids within the treatment system is an aspect of plant operations that is at once both a process and a consequence of process. Solids may be stored for several hours in primary tanks to achieve gravity thickening before removal to digesters. The timing of primary solids removal influences a variety of subsequent processes, notably the thickening process. The thickness of feed to digesters influences plant capacity to accept extra flow during storms without a risk of washout with the effluent. Storage is also inherent in the treatment of solids through digestion; residence time of solids in a tank is a key operational control parameter. But storage can also be the consequence of problems in the processes, such as when grit chambers become loaded after storm events or when inorganic solids settle at the bottom of a digester. In these cases, timely removal of the solids from the system is necessary if treatment capacity is to be maintained.

This portion of Chapter 20 discusses the components of the treatment system where solids are collected in preparation for ultimate use or for temporary storage in the plant before transport offsite. This consists of three kinds of storage: liquid or dewatered biosolids storage in lagoons, tanks, or basins to achieve stabilization; liquid or dewatered biosolids storage before use season; and dewatered biosolids storage on pads, in silos, or other containers before transport offsite for final disposal or use.

In some treatment systems, storage requirements are a natural consequence of the choice of treatment processes. For instance, storage is an inherent element of process design for lagoon stabilization, air drying, and composting. For other solids processes, such as centrifugation and lime stabilization, design of postdewatering transport and temporary storage equipment involve a range of choices that will be influenced by factors such as plant size, climate, offsite land availability, and neighborhood proximity. Some of the principles that are described here will also apply to management of other waste streams, particularly grit, screenings, and scum. A key consideration of storage in the WWTP is minimizing odor impacts. Because solids largely consist of biological material, bacterial action on the organic matter can release compounds that are offensively odorous. Characteristics of the solids such as moisture content, age, and stabilization level will influence odor emission potential and the ability to achieve odor control. Sidestreams, odor management, and public outreach are covered in other chapters in this manual.

1.3 Transport of Solids

Transportation of solids is a necessity for WWTPs. Truck transportation remains the most common method of transport used in the United States. Transportation costs are typically the largest expense associated with any offsite solids management program. This trend will only increase as fuel prices continue to rise. Cost trends also are affected by regulations and safety issues. Transportation, more than any other solids management activity, simultaneously exposes biosolids managers to the highest number of uncontrolled variables and greatest exposure to the public. Trucks that travel on public roadways are visible to hundreds or thousands of other drivers with every trip. Operational risks largely beyond the control of the biosolids manager include collision, weather and roadway conditions, equipment failures, and traffic jams. It is the public nature of these situations that can turn a mundane equipment failure into a major media event. Later in this chapter, various transportation options such as truck, rail, and barge will be discussed. Each transportation option will include a discussion of equipment, costs, odor, and spill control practices.

1.4 Offsite Storage

Offsite storage, which is the focus of this chapter, can significantly facilitate the successful management of a biosolids recycling program. Offsite storage, which is storage either at the biosolids land application site or at an alternative location other than the WWTP, provides biosolids managers with important operational flexibility during periods of inclement weather or poor field conditions. This flexibility allows biosolids managers to make the best biosolids management decisions for a range of variable operational factors.

Odor typically is the key issue for storage facilities. Therefore, the location of a storage facility, along with other factors discussed in more detail later in the chapter, can significantly influence the efficiency of storage operations. Unlike transportation, offsite storage risk factors (apart from weather) *are* largely under the control of the biosolids manager. Proper management of the storage facility and surrounding area can determine the long-term success or failure of a storage site. Storage practices can range from short term, defined as a few hours to a few days, to "permanent" which can mean up to two years. Typically, the length of planned storage time influences both the type of storage facility and its level of management.

2.0 ONSITE STORAGE

2.1 Storage within Wastewater Treatment Processes

2.1.1 Screenings and Grit Removal

Wastewater facilities with screenings and grit removal should supply onsite storage until this material is transported offsite for disposal. Screening, the first step in wastewater treatment, removes large solids such as sticks, rags, and plastic bags that may clog piping and damage downstream equipment during the treatment process. Grit and larger solids removed from screens must be disposed of on a regular basis because of odor problems. A screw compactor may be used in conjunction with a screen to remove excess liquid from screenings and then discharge screenings to a container, such as a dumpster or bagging unit.

Influent flow will then typically enter grit removal facilities. The primary purpose of grit removal from influent wastewater is to remove abrasive inorganic material to reduce wear on downstream mechanical and process equipment. Grit removed from the wastewater flow is typically transferred to an onsite storage bin via conveyor or it may be pumped as a slurry to dewatering units that discharge directly into the storage hopper. The storage bin may be used for both grit and screenings. Some wastewater facilities include grit buildings to store grit and screenings until they are transported offsite for disposal. Additional information on grit removal and management process can be found in *Design of Municipal Wastewater Treatment Plants* (WEF et al., 2009).

2.1.2 *Aeration Tanks and Secondary Sedimentation*

Solids may be stored in the aeration tanks and secondary clarifiers of the WWTP. Storage in this area of the treatment process typically causes an increase in solids concentration and mean cell residence time (MCRT), with a resulting decrease in the food-to-microorganism ratio (F/M). With both of these conditions, careful operation of the secondary treatment process is required. Most WWTPs have a portion of storage capacity designed into the secondary process, which can be identified by reviewing the operation parameters for both the MCRT and F/M. In addition to tank size and process capacity, aeration capacity to maintain a healthy biological population and nitrification process must be considered. Operating within the plant's design ranges typically ensures that all components are able to support the increase in solids inventory.

Storing excess solids in the aeration basin and secondary clarifier can cause some undesirable consequences resulting in National Pollutant Discharge Elimination System (NPDES) permit compliance issues. As a rule, using the secondary system of the WWTP as a biomass storage area should be avoided if possible because this can significantly affect secondary operation. Filamentous bacteria and hydraulic or toxic overloads may develop. Operational precautions can reduce the likelihood of compliance issues, but caution must be taken when solids are stored in this portion of the treatment process.

Because redundancy of processes is often a requirement, isolation of a portion of the plant can be performed, allowing use of a tank to increase the storage capacity, often without increasing the likelihood of the operational issues described above. To control or reduce odors associated with this practice, constant aeration is recommended. For monitoring sludge blankets in tanks, different types of measuring devices are available. Sludge-blanket instrumentation can be as sophisticated as specialized electronic devices or as simple as a submerge-and-capture tube device.

2.1.3 *Stabilization Ponds*

Waste stabilization ponds, often referred to as oxidation ponds, store waste in large earthen basins. Used to store and treat domestic and industrial wastes, the pond's low construction and operating costs and reliable, easy-to-operate process can be arranged in various combinations to maximize treatment for small communities.

Ponds can be divided into four main categories: aerobic, anaerobic, facultative (aerobic-anaerobic), and maturation. The U.S. Environmental Protection Agency (U.S. EPA; 1979b) describes the ponds as follows:

> *Pure aerobic wastewater stabilization ponds provide only single-phase concentration type solids storage, whereas the more commonly used anaerobic and facultative ponds can provide for long-term, two-phase concentration type storage of removed settleable and created biological solids. Primary wastewater (usually anaerobic stabilization) ponds that receive raw sewage must be drained and cleaned approximately every five to ten years, depending on loadings. Secondary wastewater (usually facultative stabilization) ponds that are sufficiently deep (1.8–2.4 m) and that receive only those solids generated by biological activity probably never require cleaning.*

2.2 Storage within Solids Treatment Processes

Several units in the solids treatment processes can provide storage:

- Gravity thickeners;
- Dissolved air flotation;
- Blend tanks for primary and secondary solids (Qasim and Qasim, 1999);
- Anaerobic digesters (Malina and Pohland, 1992; Metcalf and Eddy, Inc., 2003; Speece, 2008; U.S. EPA, 1979b; Wang et al., 2006);
- Aerobic digesters (Liu et al., 1997; Metcalf and Eddy, Inc., 2003; U.S. EPA, 1979b; Wang et al., 2006); and
- Drying beds.

Storage within the listed treatment processes, with the exception of digesters, is incidental storage, and typically provides little capacity. Digesters not operated at full capacity, may be used to store solids for short periods.

2.3 Dedicated Onsite Storage Facilities

Liquid biosolids and wastewater solids can also be stored in several types of dedicated onsite storage facilities:

- Holding tanks;

- Facultative lagoons (U.S. EPA, 2002);
- Anaerobic liquid lagoons (U.S. EPA, 2002);
- Aerated storage basins;
- Drying lagoons;
- Drying beds;
- Storage silos; and
- Confined hoppers or bins.

2.4 Storage of Ancillary Treatment Residuals

This section discusses the storage of septage and fats, oils, and grease (collectively referred to as FOG). It also covers catch basins. Septage is defined in the federal regulations, 40 CFR Part 503, *Standards for the Use and Disposal of Sewage Sludge* as "either liquid or solid material removed from a septic tank, cesspool, portable toilet, Type III marine sanitation device, or similar treatment works that receives only domestic sewage" (U.S. EPA, 1993). Septage, not meeting this definition, must be handled and disposed of in accordance with 40 CFR Part 257, *Criteria for Classification of Solid Waste Disposal Facilities and Practices* (U.S. EPA, 1979a).

Three main management options for septage following collection are available:

- Treatment at a WWTP. Because of the strength of typical septage, as discussed above, WWTPs typically require that it be introduced to a collection system interceptor far upstream to allow for sufficient dilution by other wastes entering the system. Large WWTPs with sufficient capacity, however, may allow septage to be discharged to the plant headworks following screening.
- Treatment at a septage treatment facility. Following screening and lime treatment in a septage treatment facility, septage is dewatered. The dewatered solids may be composted, land applied, or landfilled. The filtrate from septage dewatering may be irrigated or land applied. With an industrial user discharge permit, filtrate may be discharged to a WWTP for further treatment. In rare cases, filtrate may be discharged to surface water with the applicable NPDES permit.
- Treatment followed by land application. In this case, the septage may be treated with lime while in the collection truck and then, preferably, screened before land application.

Septage presents challenging management issues. It contains five-day biochemical oxygen demand (BOD_5) and total suspended solids (TSS) that is 50 to 100 times that of typical domestic wastewater; higher concentrations of all other common domestic wastewater constituents; hair, grit, rags, plastics and other solids; and hydrogen sulfide and other odorous compounds, necessitating that odor control be part of any handling and storage plan.

Despite the above characteristics, septage can vary significantly from area to area and home to home (U.S. EPA, 1994; WEF, 1997). Several factors lead to this variability: homeowner habits, septic tank design and size, septic tank pumping frequency, use of garbage disposals and other home appliances, and climate.

Climate also affects septage tank pumping frequency, which contributes to another septage management issue—widely fluctuating volumes of trucked-in septage waste (i.e., daily septage production). For example, in colder portions of the United States, wintertime pumping of septic tanks typically occurs only in an emergency. Consequently, the volume of winter trucked-in septage may be approximately 10% of the volume of spring/summer trucked-in septage. Areas of the United States with more temperate climates may have more consistent rates of septage collection, which facilitates storage planning. Popular vacation locales, however, see dramatic increases in septage collection rates during tourist season. Storage design and planning must consider these factors. Additional information on the management of septage can be found in *Septage Handling* (WEF, 1997).

2.4.1 *Septage Receiving Facilities*

Septage receiving facilities may be located either at a WWTP or at a septage-only treatment facility. In either case, the receiving facility is critical to the smooth operation of all but the smallest septage handling enterprises. The storage volume provided by a septage receiving facility can be used to attenuate variations in septage characteristics and provide flow equalization and mixing and aeration of the septage before further treatment. Storage allows a controlled outflow of septage to additional treatment processes. Where treated septage is destined for land application, storage significantly increases the flexibility of operations. Because of the variability described above, a variety of storage devices may be used in septage receiving facilities.

Essential elements of receiving stations vary depending on septage volume, facility location, and septage management processes. A receiving station should include a concrete receiving pad; an inlet box, pipe, or quick-disconnect fitting to receive the septage; a trash rack to remove debris; and wash-down facilities. Other desirable features include holding tanks, fine-screening and grit-removal systems, metering, and odor control.

Septage must be screened before storage. This preliminary step ensures that grit and other foreign materials are removed and not passed on to other treatment processes. Removal of grit and other foreign materials protects both facility pump systems and septage quality. The grit and screenings may be stored in a concrete bunker or routed directly to a covered container for transfer to a landfill.

Septage receiving facilities typically use enclosed holding tanks for containment. The tanks control odors and provide management options for liquid decanting and solids settling and for solids removal or mixing, resulting in a more uniform product.

Enclosed storage tanks are subject to corrosion from the release of hydrogen sulfide and other odorous gasses. For this reason, most tanks have protective liners. Smaller tanks are often constructed of carbon steel with a suitable coating; epoxy-coated aluminum tanks might also be used. Steel tank interiors should have multiple coats of a high-build epoxy or coal tar. In addition, provisions should be made for interior cathodic protection and, where soil demands it, exterior protection. Special care should be used to coat the underside of the roof or exterior wall surfaces above the water line. Exposure to moisture and sulfide vapors occurs in these surfaces. Use of inorganic zinc as a heavy prime coating to provide sacrificial protection of the metal is an effective measure. Organic coatings are layered over the zinc undercoating. Aluminum or fiberglass covers are suitable alternatives for this. Large concrete tanks may be corroded by water that is saturated with calcium carbonate from lime treatment. Under such conditions, protective coatings for the concrete should be used. Aboveground, glass-lined metal storage systems are being increasingly used in septage treatment operations.

Water from any of the septage treatment processes is often stored in large, open lagoons or basins before final treatment or disposal. These basins typically are constructed of earth with clay or flexible membrane liners to prevent groundwater contamination. The purpose of the liner is to both hold the material inside and groundwater outside of the storage basin. Membrane liners are preferable to compacted soil liners. Bentonite linings (soil-cement linings) have been used where onsite clay is not available for compaction. Fabric-reinforced polyvinyl chloride membranes with a minimum thickness of 1 mm (0.4 in.) are popular selections. Hypalon is a common membrane lining material that is resistant to UV rays, which means it should be used where any part of the liner would be exposed to sunlight. Another group of materials sometimes used are EPDM rubber materials (i.e., rubber roofing products). The EPDM products tend to be a little higher priced but are more UV-resistant and easier to work with than other liner materials. These products meet the ASTM standard D-1418.

Necessary storage capacity of open-water lagoons is calculated using climate data and a detailed water balance equation. The equation balances the precipitation inputs versus evapo-transpiration of stored water to ensure enough storage capacity for both the natural inputs and facility water until discharge or land application.

Dewatered cake typically is stored in concrete bunkers or lined carbon steel tanks. Treated and dewatered cake for land application as a soil amendment is most often stored in a concrete bunker, allowing front-end loader access. Cake destined for a landfill may be stored in a truck or rolloff that will be used to deliver it for final disposal at a landfill.

2.4.2 *Fats, Oils, and Grease Facilities*

Often fats, oils, and grease (FOG) are treated or handled at the same location as septage and have similar storage requirements. As with septage, it is important to screen FOG before storage to prepare it for additional processing and treatment. It is also important that similar mixing and decanting ability is provided at the receiving facility.

The FOG volumes typically are smaller than septage volumes. Metal tanks are required and should be properly coated to avoid problems with hydrogen sulfide and tank deterioration. If the tanks are previously used, then they must be tested for water-tightness and structural soundness before use. This includes evaluation of the tanks based on their expected use in the processing of the product. For example, tanks that may have functioned well under normal operating conditions and temperatures may not maintain their soundness if subjected to high temperatures. Odors are a major problem with grease and biofuel facilities and require off-air filtration systems.

3.0 HAULING AND TRANSPORT OF SOLIDS

3.1 Truck Transportation

Transportation is a significant consideration in the management, marketing, and distribution of any commodity. Like many products, biosolids must be moved from where they are generated to where they can be used. Many factors should be considered when developing a biosolids trucking program; many of these are discussed below.

Individual states and localities may impose their own regulations on biosolids transportation. State and local rules typically require that biosolids loads have permits or permit numbers and load tickets (also called truck ticket, scale house ticket, weight ticket, trip ticket, manifest); and that transportation equipment be inspected, licensed, and marked with identification numbers.

The load ticket, which includes the date and time of biosolids pick up, the quantity shipped, and the biosolids source and destination, can have several functions. The load ticket is used as a manifest during transportation, can be referenced when invoicing, and is needed for reporting and recordkeeping for environmental compliance purposes.

3.1.1 *Advantages and Disadvantages*

Trucking provides operational flexibility not found with other modes of transport because terminal points and routes can be changed readily at low cost (Ettlich, 1976). This flexibility is valuable because locations of reuse or disposal can change quickly. If trucks are leased rather than purchased, truck hauling is not capital intensive and provides a viable option for transport of both liquid and dewatered sludge. For most small plants and some large plants, use of trucks is the best approach.

Some of the negatives associated with trucking can include difficulty in securing sufficient numbers of the right type of truck, fuel price fluctuations, risk of collisions and spills, administrative costs associated with labor management and Department of Transportation (DOT) compliance requirements. Comprehensive transportation requirements are promulgated and enforced by the DOT and are not addressed in this manual. Biosolids transportation regulations, which are typically an integral part of state or local requirements, must be followed in addition to Federal DOT regulations and are discussed below.

3.1.2 *Truck Types*

Biosolids are moved in tractor-trailer combination units; the percent total solids (%TS) of the biosolids being transported will largely determine the type of containment equipment used. Other important containment vessel criteria relates to the ease and speed of loading and unloading the biosolids.

Spillage or leakage from biosolids hauling operations is unacceptable and is a regulatory compliance issue. Spill and leak prevention begins with selection of appropriate transportation equipment. Liquid biosolids are best transported in waterproof, tanker-style trailers equipped with internal baffles to reduce load shifting and watertight seals on any openings or hose connection ports. Dewatered biosolids, sometimes called "cake," are typically transported in dump trailers with splashguards at the front and rear of the trailer and sealed tailgates. Splashguards typically must cover a certain percent, typically 50% total, of the open trailer top. In addition to splashguards, tarps may be required or recommended to further protect against accidental discharge of biosolids. Tarps that are stretched tightly, completely covering and overlapping the open trailer area between the splashguards, can effectively prevent both odor and spills.

Unless the top of the trailer is watertight, a minimum vertical distance may be required to be maintained between the top of the biosolids and the top edge of the trailer. This "freeboard" is typically required to be at least 0.9 m (2 ft). Freeboard allows space for movement of the biosolids without spilling over the edge during transportation.

Large tanker trailers can typically carry up to 25 000 L (6500 gal) of biosolids without exceeding road weight limits. Once at a land application site, biosolids transfer to an application vehicle can occur using a hose and a pump to prevent spills. Some application vehicles, such as Terra-Gators, are equipped with their own suction hoses to fill the 8000- to 15 000-L (2000- to 4000-gal) capacity tanks which they carry around the field on high floatation tires.

Dewatered biosolids are typically transported in tractor-trailer combination units using dump trailers. The dump trailers deposit cake biosolids on the field and front-end loaders scoop the biosolids into spreader boxes towed by tractors. Dump trailers have a biosolids capacity of 18 to 22 Mg (20 to 24 wet tons) without exceeding road weight limits.

All situations are different, but typically, larger trucks are more economical for biosolids transport except in the case of short, one-way haul distances and low annual sludge volumes. Diesel engines are used in larger trucks and are the economical choice for small trucks that are operated at high annual mileage (Ettlich, 1976). Where economic, environmental, and institutional considerations allow, dual-purpose trucks are available. These tank-trucks both transport and apply biosolids without the intermediate biosolids transfer step.

3.1.3 *Contracting versus Owning*

A WWTP has many different options from owning to leasing the trucking and from handling everything in-house to contracting-out the entire operation. Each WWTP situation is different and must be evaluated independently. Many decisions are made based on the availability of capital. Other important considerations include plant size, onsite storage capacity, frequency of haul, and personnel capabilities. For example, many small- to medium-sized plants choose to use their own trailers together with contracted tractor units and driver services. This choice can have two significant benefits. First, treatment plant staff assigned to sludge handling or dewatering operations are working in the immediate vicinity of the trailers and can reposition them under a loading conveyor belt as needed to ensure even truck loading. Second, this management option can result in lower bid prices because trucking companies that do not need to cover trailer amortization costs in their bid prices and can afford to charge less.

Written contracts with trucking companies or individual drivers can be used to document and enforce generator expectations regarding safe driving, spill response, and compliance with transportation permits and routes, truck maintenance, and appearance. Contracting out DOT compliance, which can include driver requirements relating to medical condition, age, eyesight, and driving time limits within 24-hour periods can ease a significant administrative burden (National Transportation and Safety Board, 2002).

3.1.4 *Route Selection and Haul Scheduling*

In developing a plan to manage biosolids recycling using trucks, several factors require evaluation. These factors relate to the quantity of biosolids, truck-loading time, travel time between the WWTP and the destination, and travel conditions along the haul route. All of these factors are discussed in more detail below.

First, the time it takes to load each truck (loading rate) will be affected by the %TS of the biosolids, the equipment and loading method, and the trailer type. For example, dewatered sludge will mound when loaded from a single point. Mounding will prevent effective use of the space inside the trailer. Efficient trucking operations use all available space within the transport vehicle without exceeding the weight limit allowed. Consequently, trailer loading rate may be affected by the need to re-spot trailers under a loading conveyor.

Additionally, the time required to travel roundtrip between the WWTP and the biosolids destination site, including unloading time at the site, will determine how many trips (how may loads of biosolids) a truck can deliver to a site each day. This number of "rounds" possible per day can help determine the number of truck units needed to move a given biosolids quantity within the time allotted. Some regulatory authorities may further restrict truck operations, by limiting operating times to certain hours or days (Ettlich, 1976).

Another major consideration is the transportation route itself. Carefully chosen transportation routes can promote a smooth and efficient operation while reducing risks, increasing safety, and saving money. Several issues should be considered when selecting a route:

- Daily and seasonal traffic patterns and congestion;
- Major road maintenance and construction projects;
- Road and bridge weight restrictions,
- Bridge and overpass height restrictions; and
- Road conditions related to road composition (pavement or gravel), which can relate to weight and seasonal restrictions.

Some state and local permitting authorities designate specific routes for transporting biosolids; others may require permit applicants to propose routes that regulatory authorities either approve or deny.

3.1.5 Costs

The cost of the trucking equipment, whether paid servicing a loan or through a trucking contract, will typically be the single greatest cost in using this method to transport biosolids. Fuel and preventative maintenance should be factored into budgets along with annual cost increases. Operational efficiency and, therefore, cost savings can be realized by:

- Moving biosolids from a WWTP directly to the biosolids land application site rather than to an offsite storage site can significantly reduce costs, because the biosolids material is transported only once rather than twice;
- Using WWTP onsite storage capacity whenever possible, including temporary storage of biosolids in tankers or trailers if operating permits and local regulations allow, to allow greater flexibility in haul scheduling;
- Using larger biosolids land application sites can reduce costs because it requires fewer equipment moves between sites;

- Minimizing distances between the WWTP and the land application site and between land application sites can also save on costs;
- Arranging for trucks to do a "back haul" on the return trip from delivering biosolids. A back haul means that the truck hauls cargo returning from its original destination. Carrying a payload both ways, rather than traveling full one way and empty on the return trip, can significantly cut transportation costs. As noted above, however, these opportunities are unusual because the locations, schedules, cargo characteristics, and other factors must all be compatible for it to work.

3.1.6 *Odor Control*

Trucks are large and noisy and, therefore, noticeable to the public and to neighbors of biosolids management sites and may be the only part of the entire biosolids transportation operation that is obvious to the public.

Consequently, the odor-free, safety-conscious operation of your trucks can significantly contribute to the reputation of your entire biosolids management project. Therefore, it is important that the trucks and trailers observe all rules of the road, operate in a courteous manner, and be kept clean looking. Covers over biosolids loads are also important. Clean trailers with covered loads look and smell are better than uncovered trailers with biosolids spilled down their sides. Inoffensive trailers raise fewer concerns than those that are obvious for the wrong reasons.

It is also important to choose transportation routes thoughtfully. When possible, major roadways should be used, rather than smaller routes, and areas with lots of pedestrians and traffic lights should be avoided.

As described above, transportation route selection should be more comprehensive than picking the shortest route from Point A to Point B. Carefully chosen transportation routes can

- Reduce the likelihood of odor complaints;
- Help avoid traffic delays;
- Minimize accident risk; and
- Promote a circumspect and successful biosolids management operation.

3.1.7 *Spill Control and Response*

Prevention of biosolids leaks and spills is a primary component of any successful biosolids management program; and, when necessary, prompt management and containment of a spill is equally important. Development of a comprehensive spill prevention and management plan that contains specific preventative measures, cleanup requirements, and personnel responsibilities, is important. Drivers and any other field personnel working with biosolids should be trained according to the spill management plan. Additionally, because the location of a spill may cause it to become a public event, a spokesperson should be designated to talk to the media or public if necessary.

As described in section 3.1.2, spill and leak prevention begins with selection of the appropriate transportation equipment. Tankers and trailers should be leak-proof, covered with a splashguard, and meet tarp and freeboard requirements.

Following biosolids loading and unloading operations, the exterior of the trailer should always be inspected for spills and leaks. Following off-loading of dewatered biosolids, dump trailer tires should be cleaned by scraping, brushing, or hosing before operating the truck on public roadways. As a preventative measure, a regular visual inspection of the exterior of the trailer can quickly reveal any developing equipment failures that could result in leakage of biosolids.

In the event of a spill, the following actions should be taken immediately:

- **Halt the source of spill.** This may mean turning a valve, righting a trailer, patching or plugging a hole, or other measures. Any faulty equipment should be repaired before resuming its use.
- **Contain the spill.** The area affected by the spill should be kept as small as possible. Straw bales or other portable barricades can be used to either form a barrier or to soak up liquid biosolids.
- **Remove spilled biosolids.** The type and quantity of biosolids spilled will determine the best cleanup and removal equipment to use. Common options include vacuum equipment from liquid biosolids applicator vehicles, front-end loaders, shovels, and brooms. Spilled biosolids may be spread on an approved application site, returned to the WWTP, or disposed of in an approved landfill.
- **Perform final cleanup.** In all cases, the final cleanup should be completed promptly to the satisfaction of the property owners where the spill occurred. Roadways may be swept or rinsed with water. Residuals from cleaned up spills can be allowed to dry and incorporated if on a tillable area.
- **Report the spill.** Because non-exceptional quality biosolids are regulated, any loss of material to a location other than a permitted site is a compliance issue.

Even distribution and marketing permits for exceptional quality biosolids issued by state regulatory agencies may require reporting some spills. To simplify the filing of complete spill reports, a form can be developed. The form can ensure that all required information is collected when a biosolids spill gets called in from field personnel. State and local agencies should be notified in accordance with requirements.

3.2 Rail Transport

Provided the right combination of access to loading and unloading facilities at origin and destination exists, rail transport can be a viable alternative to truck transport and is particularly appropriate for hauling distances greater than 240 km (150 mi).

3.2.1 *Advantages and Disadvantages*

Rail transportation can offer many advantages over other transportation methods. Transporting biosolids by rail is typically more cost effective than truck transportation, where distance limits truck transport to two roundtrips per day.

In the United States, there are seven major rail carriers, also called "Class I" operators: Burlington Northern Santa Fe (BNSF); CSX Transportation (CSX); Kansas City Southern (KCS); Norfolk Southern Corp. (NS); Union Pacific (UP). Also, there are two Canadian entities that operate within the United States and Canada under the respective laws of each country, Canadian National (CN) and Canadian Pacific (CP).

There are also hundreds of "short-line" rail haulers that can coordinate transportation with the seven major rail carriers. Working together, these large and small rail transporters can offer unique operating and marketing advantages for both shippers and receivers, including modest costs and customer service akin to relationships typically reserved for truckers. Both major rail carriers and short-line haulers can provide assistance in project development, for example, by finding real estate and establishing service and rates from the origin to the destination point.

Rail transport does have some disadvantages, including a substantial initial capital investment in property and rail equipment or the purchase of transportation services from an established rail transport company. For this reason, a shipper must have a biosolids contract lasting a minimum of five years, which is somewhat unusual in the industry, to amortize costs. Access to loading and unloading facilities at origin and destination can be another problem. If rail sidings are not close by, additional transportation will be required to get the biosolids to or from the rail, thereby adding to the overall costs. For the above reasons, rail haul is typically suited for higher volume (90 Mg/d or more) generators and long-term projects (five years or more).

3.2.2 *Routes, Haul Contracts, and Costs*

3.2.2.1 *Routes and Costs*

The volume and term of a contract dictate the choices the railroads make regarding rate and route. Typically, Class I railroads select origins and destinations under their own control or served by a short line spun from their railroad. Multiple railline routes can be more expensive, particularly if the individual carriers are "shorthauled" or not able to maximize the length of their haul, which affects revenue. Most rail carriers have "minimum line-haul revenue requirements" that dictate the minimum revenue they will accept as a company. Another alternative is working directly with a short line that can both serve the market's needs and advocate for the best rates and handling.

Although a shipper/receiver should be aware of all potential interchange locations between railroads or internal routing "probabilities" on any given railroad, attempts to dictate a particular route should be avoided. A unique aspect of rail transportation is that route options can be limited by property (rail line) ownership, necessitating management of the entire rail system, rather than a single piece of equipment. To ensure overall efficiency of their entire system, rail-roads make route adjustments by the minute.

3.2.2.2 *Haul Contracts and Costs*

Since deregulation, railroads have a variety of rate publication instruments at their disposal. A large-volume shipper may be accommodated with a comprehensive, long-term contract. A small-volume shipper may be provided with a deregulated instrument in the form of a confidential tariff, deregulated price quote, and confidential contract. No one document type is better or worse as long as it covers the essentials of the contract of carriage: commodity, price, route, conditions, and, in some cases, price adjustments. In today's world of escalating fuel prices, virtually all contracts will be subject to a flexible fuel surcharge. These fuel surcharges can be found on the Class I rail carrier websites.

Shippers are required by rail carriers to describe the commodity that will be transported by rail. Although various commodities can be used, two are typically applied to biosolids shipments using the standard transportation commodity code, or STCC (pronounced "stick"): Sludge, Sewage, Dry, Fertilizer Material STCC–40 291 27; or Sludge, Sewage, No Commercial Value STCC–40 291 91. The correct description is important in determining several factors when making a rail rate. Incorrect description or STCC or "shopping" can result in significant claims against the shipper/receiver if the rail carrier discovers that the commodity has been incorrectly identified.

3.2.3 *Railcar Types and Contracting*

In the case of Sludge, Sewage, Dry, Fertilizer Material STCC–40 291 27, shipments are typically placed in covered hopper cars that may be provided by the rail carrier, with the cost of their use calculated into the price offered. Shopping around for privately leased, covered hoppers is one way to determine if railroad's price is reasonable. Private leasing companies such as GE Railcar and GATX may offer a more competitive price on the same type of car. Additionally, boutique car-leasing agencies might offer deals for lease of anywhere from a few cars up to a couple of hundred. A quick search of the Internet will identify candidate companies in the railcar leasing business. Private leasing companies may also be able to offer solutions, including full maintenance, as opposed to triple-net leasing, which places the burden of repairs on the leasee of the equipment.

Railroads are obligated under Federal Railroad Administration (FRA) laws to inspect, cull, and repair any damage incurred on a railcar while on its line. The repairs may be as simple as a $25 fix to a grab iron/handhold or brake shoe to a multi-thousand dollar repair or change-out on wheels and axles. If equipment is leased, then full knowledge and understanding of these risks must be met by either learning it, a difficult task, or by hiring an expert to audit repair bills.

Rail and railcar access is controlled by the rail carriers according to the rules of demurrage. Rarely, if ever, will a railroad negotiate a deviation of a demurrage rule; as a result, the shipper/receiver should be familiar with the provisions. Railroad-owned equipment cannot sit indefinitely at any shipper or receiver's siding without incurring penalty charges. Likewise, customer-owned equipment is limited in the time it can sit on a railroad storage track. Demurrage is a tool by which the railroads manage their assets, both cars and track space.

The railroad does not provide equipment for the movement of Sludge, Sewage, No Commercial Value STCC–40 291 91, which is typically wet cake and not suitable for covered hoppers or other similar cars. Instead, they have written rules tariffs that outline their minimum requirements for accepting such sludges on their lines and point to these tariffs directly or indirectly in the rate instruments they publish to authorize rates, routes, and movements over their lines. It is incumbent on the shipper/receiver to find the equipment that meets the criteria set forth in these rules circulars or tariffs to successfully engage and ship this material with the railroads. The CSX railroad publishes a rules tariff for the waste industry that includes specifications for equipment in CSX 4048. The primary piece of equipment required for the rail haul of wet cake biosolids is a container of adequate construction to prevent leaks and mitigate odors.

3.2.4 *Odor Control*

Odor control for the shipment of Sludge, Sewage, Dry, Fertilizer Material STCC–40 291 27 is typically accomplished by using closed hopper cars, which largely prevent odors except during loading and unloading operations.

Odor control in shipment of wet biosolids cake, is typically managed through a rules tariff (see CSX 4048 at www.csx.com). Using proper containers, which are virtually watertight, covered with tarps of up to 60 mL thick, typically prevents odor issues on shipments that are moved promptly. As the train rolls down the track, a musty odor may develop; if that becomes a problem, the railroad reserves the right to require additional mitigation measures or to refuse service if determined necessary.

3.2.5 *Spill Control and Response*

A hauling contract with a rail carrier implies their responsibility to manage any spills on their property, should they occur. Their responsibility can be spelled out in an agreement if necessary, but as common carriers, they have obligations both express and implied when they accept shipments of all kinds into their system. It is for this reason that they set the rules requiring certain types of equipment and governing its use. Railroads routinely transport dangerous chemicals and materials including hazardous commodities; therefore, they have management teams—either employees or subcontractors—that respond when a spill occurs.

3.2.6 *Ancillary Facilities*

If the generator of dried biosolids or pellets has direct rail access, then a straight lease agreement may be negotiated and the freight may be loaded directly into the top of a covered hopper car. If the generator does not have direct rail access, then space may be rented from a railroad, but biosolids would need to be delivered there and elevated into the car for shipment. In certain industrial locations, the railroad may provide access to a "team track" for those clients who are able to get freight in and out of railcars independently. Whether a straight lease agreement or access to a team track, contract terms will vary widely based on local market conditions. In some cases, railroads may even offer sites owned by third parties to facilitate the movement of the freight.

Shipping wet cake, which requires containers (as opposed to a covered hopper cars) can present additional logistics issues. Once loaded, the container must be drayed to a facility that can lift the heavy, loaded container onto a railcar. Although railroads operate facilities that provide this service, they rarely extend that service to waste materials. In most cases, leasing property suited to this type of "intermodal" operation is required and will need to be fleeted with equipment that can lift containers on and off railcars at both ends of the operation. At a minimum, one top-pick container handler, such as those used in ports or rail intermodal yards, will be required to manage the loaded and empty containers. A fleet of tractors and chassis of sufficient load-bearing capability will be needed to shuttle containers to and from the WWTP. Subcontracted services may provide an alternative to owning and operating this type of transfer operation. Rail carriers should be consulted for their expertise before equipping and developing such an operation. Again, the railroads may even offer sites owned by third parties to facilitate movement of the freight.

3.3 Barge Transportation

3.3.1 *Advantages and Disadvantages*

Since the end of ocean disposal, barge transportation of wastewater solids is no longer widely practiced in the United States and, therefore, is covered briefly below. For port city utilities, relocating large volumes of solids either to other facilities or vessels, barge transportation remains a feasible and economical option. As of publication time, New York City and Philadelphia practice "in-city" barge transport of digested liquid biosolids from digester facilities to centralized dewatering stations. Barging provides the wastewater agency options between several vessel types and operation configurations at a low cost compared to many alternative transportation choices.

3.3.2 *Routes and Transit Times*

Overall barge turnaround time—the time it takes for a barge to return to the loading location after delivering its solids—is a key decision factor in determining the efficacy of barge transport for utility WWTP. The barging operation capacity must keep pace with both daily production and conservative estimates of barge return time. Tug and barge travel time can be affected by many factors such as tides, currents, drawbridges, and lock-opening schedules. Travel speeds for tugs and barges can range between 6 to 20 km/h (3 to 10 knots); loading and unloading times can also vary widely, from

90 minutes to 6 hours for liquefied solids that must be pumped off the barge. Logistics are also often affected by land side operations; if the biosolids are transferred to another transport mode, such as onto trucks, then the travel cycle time may be longer.

3.3.3 *Barge Types, Contracting versus Ownership*

The principal choice for most WWTPs is whether barge ownership makes sense for their solids transport needs. Authorities with large solids handling requirements may select barge ownership and contract separately for towing services. Full-service contracts may prove the best for smaller WWTPs with intermittent transport requirements.

Two types of barges are capable of handling wastewater solids: hopper barges for dewatered or dried biosolids and tanker barges for liquid biosolids. Barges are also classified by the type of waterway in which they may operate. The American Board of Shipping (ABS) sets standards for the construction, certification, and operation of barges in U.S. waterways. Agencies considering barge use should ensure that their barges operate under an ABS inspection and certification program because the latter affects insurance rates covering risks in operations.

The actual volume carried by a tanker barge may be approximately 85% the nominal capacity because of sediment buildup and foam formation during transport. Annual maintenance is usually required to remove sediment. Smaller barges offer greater operational flexibility; although higher speeds offered by a larger tug and barge combination can significantly cut transit time and, thus, reduce towing fees.

The overall flow capacity of a barging operation will reflect the anticipated travel time of a cycle in barge loading, movement, unloading and return, and onland capacity to transfer and hold biosolids. In Philadelphia, which has a long-term average transport of 3800 kL (1 mil. gal) per day five days per week; a barge system with two 3800-kL (1-mil. gal) barges and onland capacity to handle a total of 15 000 kL (4 mil. gal) has proven adequate, even during periods in which one vessel is out-of-service in dry-dock for inspection and repair.

Barges require comprehensive servicing in dry-dock once every five or six years, which takes the barge out of service for approximately one month. In addition, all pipes on the barge and on land require regular inspection for integrity and complete valve operation.

3.3.4 *Spill Control and Response*

One important element of a barge transportation system is a well-developed spill prevention and cleanup program. Careful attention to equipment and facility design and operator training can minimize the risk of spills. Spill containment on the barge can be located around hose connections or the entire deck perimeter. If the operation involves removal of dewatered biosolids from a barge hopper, then spill control and containment must consider shoreside slope and drainage to an area that can be swept or pumped clean of spilled biosolids to avoid drainage to the waterbody.

For electrical safety, all equipment must be designed to meet standards for operations in flammable atmosphere, as liquid biosolids will generate methane during transport. Barges are vented to the atmosphere with a gooseneck vent stack 76 mm (3 in.) off the deck floor.

3.3.5 *Ancillary Facilities*

An important factor in determining the feasibility of barging solids is the cost of onshore loading and offloading facilities. It may be desirable or necessary to locate a storage or transfer tank or lagoon near the barge loading or unloading dock. For liquid biosolids, it is desirable to load and meter the flow from a fixed pumping station located on a fixed wharf. Offloading typically is accomplished by a pump mounted on the barge itself. For dewatered biosolids, the dock needs to be of size and construction to support the weight and movement of crane and trucks.

An important operational and safety issue in designing barge operations is pier elevation in relation to tide and the vessel deck elevation when loaded and unloaded. Floating docks eliminate most of these problems but are typically more expensive than fixed wharfs because of the complexity of anchoring devices capable of sustaining the loads imposed by a large barge.

4.0 OFFSITE STORAGE

An in-depth discussion of offsite storage of wastewater solids is provided in U.S. EPA's *Guide to Field Storage of Biosolids* (2000).

4.1 Constructed Facilities

4.1.1 *Site Selection*

Possibly the single most important decision affecting the viability of offsite biosolids storage is site selection. The adage that the most important aspect to real estate is "location, location, location" applies equally well to storage site selection. The location of an offsite storage area will affect every aspect of storage site operation.

4.1.1.1 *Location*

It is best that the biosolids storage location be convenient to both the point of biosolids generation and the locations of reuse. After identifying a general geographic location, applicable regulatory and permitting requirements should be carefully examined. Different states and local governments typically have different requirements for biosolids storage. Assuming that legal and technical requirements make the constructed biosolids storage facility feasible, then a variety of logistical considerations must be evaluated.

4.1.1.2 *Access*

The accessibility of the storage site to construction equipment, biosolids trucks, and other ancillary equipment is another important consideration in site selection. The first consideration is whether the surrounding roadways have weight or other restrictions that could impede operation of the site. Daily travel routes for biosolids trucks and traffic patterns should be evaluated. Local ordinances may set requirements for the main access road to the storage facility. Site road length, surface, and foundation material and connection with public road may all be subject to review. Use of the storage site during all seasons and in all weather will be dependent on the site road conditions.

4.1.1.3 *Neighboring Land Use*

Typically, zoning criteria will ensure that a biosolids storage facility is not totally incongruous with other surrounding land uses. If zoning ordinances do not enforce this, however, then it is still wise to consider congruity because existing neighbors who have different land use needs could credibly argue that a biosolids storage site is an inappropriate use. Areas that are already home to various agricultural or industrial uses are seen as consistent land uses for a biosolids storage facility.

4.1.1.4 *Site Physical Characteristics*

Site slope and soil type will affect construction costs and potential complications. A soil survey should be consulted to determine whether the soil type is appropriate for the construction being considered. Size and orientation are important because minimum setback distances from property lines and neighbors are set by regulation, and the site may or may not allow for required buffer distances. A related consideration is whether the storage facility would be visible to neighbors and the public. The presence of flood plains and site proximity to sensitive water bodies can affect construction restrictions and costs.

4.1.2 *Design Considerations*

Many factors will influence design decisions for a constructed biosolids storage facility. These include, but are not limited to, volume and characteristics of the biosolids being stored, state and local regulations, cost, ownership, site life, neighboring uses, building and fire codes, and access to power and other utilities.

4.1.2.1 Materials

Initial decisions include determining the type of pad/pit and liner system to use and whether the facility will be covered. Pads and pits can range from compacted earth with clay subsoil to specialized, impervious asphalt or concrete. For constructed facilities, the trend is toward the latter. The working surface of the pad or pit should be designed to withstand daily operation of equipment and to be easily repaired if necessary. Covers can range from tarps to buildings with roofs. Again, the trend is to cover the stored biosolids if feasible because this can facilitate several other operational issues discussed in the next section. Pre-engineered or prefabricated structures are a popular cover choice because they provide many of the benefits of a building, at significantly lower cost. Cover materials should be resilient enough for the moist and corrosive environment of a biosolids storage site. Cover strength in high wind and heavy snow conditions should also be assessed.

4.1.2.2 Stormwater Management

Regulators increasingly are scrutinizing stormwater management practices. Acceptable technologies to manage stormwater are also increasing, given appropriate site characteristics. Therefore, it is recommended that stormwater management options are assessed as a part of facility design planning to identify the most effective, and possibly economical, option.

4.1.2.3 Odor Control

Odor control is discussed more fully in a later section but, as with stormwater management, it is recommended that odor control technologies be evaluated at the time of facility design planning. Sufficiently sized biofilters are economical to operate and effective in many biosolids applications, but the site must be designed to accommodate them.

4.1.2.4 Utilities

Depending on the proposed location of the biosolids storage site, access to utilities may be simple or prohibitively expensive. In the latter case, generators can typically provide enough power for lighting and ventilation, if necessary. Maximizing use of natural lighting and ventilation is another important consideration in facility design planning.

4.1.3 *Operational Practices*

Operational practices will likely be dictated by operating permits. Operating permit requirements may include tracking biosolids sources and volumes stored; maintaining facility staffing levels; not exceeding allowed days and hours of operation; controlling dust generated by equipment operation; controlling fly (and other vector) populations; managing leachate; and cleaning trucks and tires before site exit. It is likely that separate permits would govern groundwater monitoring and the operation of any odor control equipment on site.

Even if limited operating conditions apply, minimum recommended operational practices would include routine inspections, housekeeping, and maintenance to reduce the potential for odors, vector attraction, and unclean or unsightly conditions. All biosolids should be kept within the storage containment area and covered, if necessary. Appropriate supplies should be maintained onsite to manage flies and other vectors, odor, and biosolids tracked out of the storage facility.

4.1.4 *Security*

Site security can typically be accomplished fairly simply by using signs, locking gates and doors, and securing any valuable items or equipment that is onsite. If there will be periods when the facility is not in use, then routine inspections can be made to minimize the likelihood of unauthorized access. Fencing, security lighting, and remote monitoring are all additional options, but not typically warranted for this type of facility.

4.2 Field Storage/Stockpiles and Design

4.2.1 *Site Selection*

Short-term field storage options and requirements can vary from state to state. When storing biosolids for use on a single farm, the storage location needs to be agreed upon with the landowner/farmer. If multiple fields are scheduled for biosolids application, then a single, centrally located stockpile area may work best. Locating the storage site in a field along an existing field/farm lane is typically practiced. This option may require some surface work including grading and addition of stone base material. If a usable access road does not exist, then constructing one is the most costly option and may require local zoning approval.

When storing biosolids for use on multiple farms, storage access is equally important and will affect the ability to use the site year round in all weather conditions. Having a lengthy access road or a runway area for trucks to clean their tires after

digested cake that had been processed to achieve 38% VOC destruction (Easter et al., 2008). These data are from a municipal WWTP in southeastern Virginia using traditional anaerobic digestion.

This summary also allows a comparison between belt filter press and centrifuge dewatered cake. Centrifuge cake stored over time appeared to be more odorous than belt press cake. Odors in both cases were primarily H_2S, methyl mercaptan, dimethyl sulfide, carbonyl sulfide, carbon disulfide, dimethyl disulfide, tri-methyl amines, and ammonia. Out of 17 volatile acids analyzed, acetic acid was the only one always detected in the range of 16 to 130 parts per billion (ppb). Lower levels of propionic acid, isobutyric acid, methylbutanioc acid, isovaleric acid, and isocaproic acids were observed in some of the samples.

4.0 ODOR SAMPLING AND ANALYSIS METHODS

This section highlights basic odor-sampling and analysis methods typical for biosolids-handling processes.

4.1 Odor Sampling

Collecting representative samples of odorous air is an important element of characterizing the odorous emissions. The data resulting from the analysis of these emissions may be used to prioritize the source of odors via dispersion modeling or to develop design criteria for needed odor treatment process. Odor treatment technologies designed to remove hydrogen sulfide may not perform well on solids-handling odors unless the process is designed to handle these other compounds, such as RSCs and ammonia.

Sampling procedures differ depending on the type of source (point source or area source) and the type of analysis desired. A summary of sampling and analytical procedures is provided in Table 20.5. For sensory testing (odor concentration, odor intensity, hedonic tone) and reduced sulfur analyses, samples typically are collected in Tedlar bags. For VOC analyses, samples are collected in stainless-steel (SUMMA) canisters. The SUMMA canisters are placed under vacuum, and the vacuum is used to pull the sample into the canister.

In some instances, odorants must be concentrated onto sorption tubes and shipped to laboratories for desorption and analysis. In this case, a set amount of air is passed through a properly selected sorption tube specific to the odorous compound

TABLE 20.5 Odor sampling and analysis matrix (VOC = volatile organic compound; GC-FPD = gas chromatograph with flame photometric detector; GC-SCD = sulfur chemiluminescence detector; GC-MS = gas chromatograph-mass spectrophotometer; HPLC-UV = high-pressure liquid chromatography with UV)

Odor source	Type of analysis	Sampling device	Sample container	Analytical instrument	Method
Point	Sensory–odor concentration, odor intensity, hedonic tone	Vacuum chamber	10-L Tedlar bag	Olfactometer and odor panel	ASTM E-679, EN 13725, ASTM E-544
	Reduced sulfur compounds	Vacuum chamber	3-L Tedlar bag	GC-FPD, GC-SCD	EPA TO-14, ASTM D-5504
	VOCs	SUMMA canister	SUMMA canister	GC-MS	EPA TO-15
	Amines	Impinger	Impinger	–	–
	Organic acids (VFA)	Impinger	Impinger	HPLC-UV	EPA TO-5
Area	Sensory	Flux chamber/ vacuum chamber	10-L Tedlar bag	Olfactometer and odor panel	ASTM E-679, EN 13725, ASTM E-544
	Reduced-sulfur compounds	Flux chamber/ vacuum chamber	3-L Tedlar bag	GC-FPD, GC-SCD	EPA TO-14 ASTM D-5504
	VOCs	Flux chamber/ Summa canister	SUMMA canister	GC-MS	EPA TO-15
	Amines	Flux chamber/ Impinger	Impinger	–	–
	Organic acids (volatile fatty acids)	Impinger	Impinger	HPLC-UV	EPA TO-5

of interest. An example of this type of analysis would be to speciate the types of amine-based compounds or volatile acids.

For point sources such as stacks or exhaust fans, the sample is conveyed through Teflon tubing directly into the sample container. For odor panel testing or reduced sulfur analysis, a Tedlar bag is placed in a special vacuum chamber, and a sample pump is used to induce a vacuum, pulling the sample air from the source directly into the Tedlar bag. This avoids the potential for contamination by the sampling pump. For VOC analyses with a SUMMA canister, a valve under vacuum is opened, allowing sample air to enter the canister.

Area odor sources such as biosolids lagoons, compost piles, or sludge drying beds require use of an emissions isolation chamber (flux chamber) that floats or rests on the odor-emitting surface, preventing the effects of ambient wind currents that could dilute the sample. Additional information on the sampling of odorous air may be found in *Control of Odors and Emissions from Wastewater Treatment Plants* (Water Environment Federation, 2004).

4.2 Odor Analysis

Laboratory testing of odorous air samples can yield important information regarding the strength and offensiveness of the odor and the levels of odorous constituents. Laboratories that analyze the sensory properties of an odor report the following:

- Odor concentration, also referred to as the dilutions-to-threshold ratio (D/T), is an estimate of the number of dilutions of odor-free air required to make the odor undetectable by half of the odor panel. The test is conducted under controlled conditions using an olfactometer and a panel of six to eight trained panelists. A high D/T ratio indicates a strong odor, requiring many dilutions to render it undetectable.
- Odor intensity is the perceived strength of an odor above its recognition threshold. The odor panel compares the odorous sample with "standard" odors composed of various concentrations of *n*-butanol. Results are presented in equivalent parts per million butanol.
- Hedonic tone is a measure of the relative pleasantness or unpleasantness of an odorous air sample. A scale from −10 (most unpleasant) to +10 (most pleasant) is used.
- Odor character is assessed in response to the question, "What does it smell like?" A variety of descriptors is used, such as "fruity," "floral," "earthy," "musty," "rotten egg," and "decayed vegetables."

Laboratories that analyze air samples typically have the capability to measure the concentrations of RSCs and VOCs. Most solids-handling processes emit odors containing a variety of RSCs.

- Odorous sulfur compounds, RSCs, typically are measured with a gas chromatograph with flame photometric detector (GC-FPD) or sulfur chemiluminescence detector (GC-SCD). Depending on the specific equipment used, the instruments can measure to the parts-per-billion level. Many of these compounds are detectable by the human nose at concentrations of 1 ppb or less.
- Volatile organic compounds typically do not make a large contribution to odor emissions because their odor detection threshold is often in the parts-per-million (ppm) range versus the ppb range for RSCs. Analysis of these compounds may be justified where there is a high industrial input to the WWTP or the presence of air toxics is suspected. A gas chromatograph-mass spectrophotometer (GC-MS) typically is used.
- Nitrogen compounds such as ammonia and organic amines (trimethyl amine) are sometimes contributors to odors from biosolids, particularly in processes in which the pH is elevated (i.e., lime stabilization). The laboratory test for amines is high-pressure liquid chromatography.
- Organic acids such as butyric acid ("rancid butter") and acetic acid (vinegar) can contribute to biosolids odors, especially when fats, oils, or grease are included. The laboratory test for these compounds is high-pressure liquid chromatography with UV detector.

Hand-held devices may be used for screening of specific odorous constituents. Most electrochemical hydrogen sulfide meters measure in the parts-per-million range. Some devices have data-logging capability, allowing semicontinuous monitoring of H_2S for a period of days or weeks. These are useful to characterize diurnal variations in H_2S or spikes because of specific loading patterns. Color detector tubes, in which a known volume of gas is passed through a glass tube containing a color-reactive chemical, can be used to estimate the concentration of a variety of odorous compounds, including hydrogen sulfide, dimethyl sulfide, ammonia, and amines. The tubes do not provide highly accurate results, however, and are subject to interference from other compounds.

5.0 CONTROL METHODS AND TECHNOLOGIES

The following sections highlight liquid-phase chemical methods for reducing odors in liquid sludge, which can also lower odor effects in the dewatered biosolids cake.

The liquid-phase control discussion is followed by a section on capturing and treating the odorous emissions with gas-phase treatment systems.

5.1 Liquid-Phase Treatment Options

Although various organic and inorganic compounds can cause domestic wastewater odors, hydrogen sulfide is often the predominant odor-causing compound. As such, liquid-phase treatment for sludge streams typically focuses on removing hydrogen sulfide. Hydrogen sulfide is released into the gas phase when dissolved sulfide concentrations are high. This occurs under anaerobic conditions, and the bacteria primarily responsible for sulfide formation—"sulfate-reducing bacteria" (SRB)—use sulfate (SO_4^{-2}) as their terminal electron acceptor, reducing it to sulfide (H_2S, HS^-, and S^{-2}).

Liquid-phase control is a method of odor prevention by which chemicals are introduced into the sludge stream to react with the dissolved sulfides, thereby preventing the release of H_2S into the vapor phase. Various chemicals are potentially suitable for liquid phase odor control (Water Environment Research Foundation, 2003):

- Oxygenation and aeration—ambient air or pure oxygen is injected into oxidize sulfides or to maintain an aerobic environment;
- Chemical oxidation—chemical oxidizing agents are added to oxidize dissolved sulfide to sulfate, which remains in solution;
- Sulfur precipitation—metal salts are added to form a metal sulfide precipitate with the dissolved sulfide; and
- Biological treatment—process changes or microbial cultures are used to affect the ability of bacteria to produce H_2S.

5.2 Oxygenation and Aeration

If a sufficient amount of dissolved oxygen is present in the bulk liquid phase, then anaerobic odor-generating conditions can be avoided. This is typically possible only in the liquid sludge feed.

Oxygen reacts stoichiometrically at a ratio of 1.9 kg oxygen per kg hydrogen sulfide (1.9 lb O_2/1 lb H_2S) in accordance with the following reaction:

$$H_2S + 2O_2 \rightarrow 2H^+ + SO_4^{2-} \qquad 20.1$$

Oxygen demand can, however, be driven by more than just the need to react with sulfide, resulting in a higher "effective stoichiometric" demand. The specific oxygen uptake rate (SOUR) of active biosolids can be high.

5.3 Sodium Hypochlorite

Sodium hypochlorite (NaOCl) is the chemical found in bleach-containing home cleaning products but at triple the strength. Sodium hypochlorite is a strong oxidizing agent. It reacts stoichiometrically with hydrogen sulfide in wastewater at a theoretical ratio of 8.8 kg of sodium hypochlorite per kg hydrogen sulfide (8.8 lb NaOCl/lb H_2S):

$$4NaOCl + H_2S \rightarrow 4NaCl + H_2SO_4 \qquad 20.2$$

Because hypochlorite reacts indiscriminately with other non-H_2S constituents, the required effective dose could be much higher than stoichiometric.

5.4 Hydrogen Peroxide

Hydrogen peroxide (H_2O_2) is a strong oxidizer that reacts with most reduced compounds in wastewater, including sulfide. It is typically available in bulk concentrations of 35 to 50%. It has the advantage of having harmless reaction byproducts, and breakdown of excess H_2O_2 increases dissolved oxygen. Under typical wastewater pH ranges, reactions proceed according to the following equation:

$$pH < 8.5 \; H_2S + H_2O_2 \rightarrow S + 2H_2O \qquad 20.3$$

The reaction between hydrogen peroxide and sulfide is relatively slow (nearly complete in 30 minutes) compared to those of other oxidants, which take seconds to minutes. It reacts stoichiometrically with hydrogen sulfide in wastewater at a theoretical ratio of 1 kg/ kg H_2S (1 lb/1 lb). Peroxide can, however, be an indiscriminate oxidizer requiring higher overall dose rates because of other oxidation demands.

5.5 Permanganate

Sodium and potassium permanganate are strong oxidants that can be used to remove sulfides and other odor-causing compounds. Sodium permanganate is more soluble, whereas potassium permanganate is less expensive, but both are more expensive than other chemicals. Permanganate is used to treat sludge because it tends to react first "preferentially" with sulfide and can be cost effective in this type of application.

Potassium permanganate ($KMnO_4$) is a strong chemical oxidizing agent that reacts with hydrogen sulfide according to the following equations:

$$\text{Acidic pH: } 3H_2S + 2KMnO_4 \rightarrow 3S + 2H_2O + 2KOH + 2MnO_2 \qquad 20.4$$

$$\text{Basic pH: } 3H_2S + 8KMnO_4 \rightarrow 3K_2SO_4 + 2H_2O + 2KOH + 8MnO_2 \qquad 20.5$$

A major vendor reports that for sludge streams, typical dosing can be as high as 100 mg/L. This often equates to the demand for approximately 7 kg permanganate/kg sulfide (7 lb permanganate/1 lb sulfide).

5.6 Sulfur Precipitation Using Iron Salts

Metal salts can combine chemically with dissolved sulfide to form relatively insoluble metal-sulfide precipitates, thus preventing the release of H_2S gas. Metal-sulfide precipitates exist as soft black or reddish-brown flocs.

Ferrous iron salts such as ferrous chloride ($FeCl_2$) and ferrous sulfate ($FeSO_4$) are available for sulfur precipitation. Chloride species are typically used for sludge-stream dosing. Ferrous iron reacts with each sulfide species as shown below:

$$Fe^{2+} + S^{2-} \rightarrow FeS \qquad 20.6$$

$$Fe^{2+} + HS^{-} \rightarrow FeS + H^{+} \qquad 20.7$$

$$Fe^{2+} + H_2S \rightarrow FeS + 2H^{+} \qquad 20.8$$

These reactions indicate a stoichiometric dose of 13.8 kg $FeCl_2$/kg of sulfide (3.8 lb $FeCl_2$/1 lb of sulfide). Effective doses for reducing the sulfide concentration to less than 1.0 mg/L are reported from 4 to 12 kg of $FeCl_2$/kg of sulfide removed (4 to 12 lb/1 lb sulfide removed), depending on the target sulfide concentration.
Ferric iron reacts similarly to ferrous iron:

$$2Fe^{3+} + S2^{-} \rightarrow 2Fe^{2+} + S \qquad 20.9$$

$$2Fe^{3+} + 3HS^{-} \rightarrow Fe_2S_3 + 3H^{+} \qquad 20.10$$

These reactions indicate a "stoichiometric" dose of 3.3 to 4.9 kg $FeCl_3$/kg sulfide (3.3 to 4.9 lb $FeCl_3$/1 lb sulfide). Field and laboratory experiments, however, indicate that the typical required dose to remove sulfide can be higher.

6.0 GAS-PHASE TREATMENT

Although chemical addition to raw sludge or biosolids can often reduce odor concentration and odor intensity of the material, continuous chemical addition can be expensive and may not provide the degree of control required. Therefore, odors from sludge thickening, dewatering, stabilizing, and storage processes are often collected and treated.

6.1 Containment and Ventilation

The first step in gas-phase odor treatment is containment of the odorous air and conveyance to an odor treatment device. Containment alternatives vary widely in configuration and materials of construction. The objectives of containment are to

- Ensure worker safety;
- Prevent the escape of odorous gas;
- Minimize the volume of air to be treated; and
- Allow reasonable access for inspection and maintenance.

Because of the potential for hydrogen sulfide to be present in solids-handling processes, it is important that all materials of construction be resistant to corrosion. Cover and containment options include flat covers, full domes, low-profile domes, arches, equipment enclosures, and other options. The options chosen need to be balanced against need for access to the covered space and the total odorous air volume that needs to be treated.

Once enclosed, the space must be ventilated and the air conveyed to the odor control system. Ventilation rates may vary widely depending on the characteristics of the odorous air and the degree of conservatism in minimizing the escape of odors.

Criteria for establishing ventilation rates for covered processes include

- Maintaining a minimum negative pressure [e.g., 1.3 to 2.5 mm (0.05 to 0.1 in.) water column, or 12 to 25 Pa] to prevent the release of odors;
- Providing a safe working environment (e.g., H_2S less than 10 ppm) if the enclosure is designed to be entered routinely;
- Controlling the buildup of combustible gases such as methane; and
- Controlling H_2S levels and moisture to reduce corrosion.

Ventilation rates for enclosed areas typically are expressed by the number of air changes per hour (AC/h). An air exchange rate of at least 12 AC/h is recommended for continuously ventilated and occupied areas having exposed sewage or sludge surfaces. This criterion is suggested in the "Ten State Standards" (Water Supply Committee, 2003) and in the National Fire Protection Association's (2008) *Standard for Fire Protection in Wastewater Treatment and Collection Facilities*. Sometimes it can be cost-effective to enclose the primary odor-producing sources inside a larger complex isolating the odorous air and reducing the overall volume that has to be conveyed and treated. This can result in the larger volume of air to be treated as an HVAC ventilation issue without the need for added odor control.

For covered channels, storage tanks, and processes that are not routinely entered, the most important criterion is minimizing the escape of odors. The NFPA recommends a minimum negative pressure of 25 Pa (0.1 in. water column) under all operating conditions.

6.2 Odor Treatment

Technologies for treating odorous air are summarized in Table 20.6 and discussed below. More detail on gas-phase scrubber systems is provided by WEF (1998, 2009).

6.2.1 *Packed-Tower Scrubber*

Packed-tower scrubbers, also called wet scrubbers or chemical scrubbers, have been used to treat odorous air for many years. They offer high odor-removal efficiency, operating flexibility, and a small footprint. The main disadvantages are relatively high operations and maintenance requirements and operating cost and the need to handle hazardous chemicals.

Packed-tower scrubbers typically consist of a vertical, cylindrical, corrosion-resistant vessel containing 4 to 4 m (6 to 12 ft) of inert packing. Odorous air is passed up through the packing, and a chemical solution is sprayed down to contact the odorous air. The solution is collected in a sump and recirculated; makeup water and chemicals are added to maintain the proper conditions for contaminant removal. Empty bed velocities through a packed-tower scrubber are typically 100 to 165 m/min (300 to 500 ft/min).

TABLE 20.6 Summary of odorous air treatment alternatives.

Technique	Frequency of use	Cost factors	Advantages	Disadvantages
Packed tower wet scrubbers	High	Moderate capital, high operation and maintenance (O&M) costs	Effective and reliable; long track record; small footprint	High chemical consumption, high O&M
Activated carbon adsorbers	High	Cost-effectiveness depends on carbon replacement or regeneration frequency	Simple; few moving parts; effective; several media options	Best for relatively dilute airstreams to ensure long carbon life
Inground biofilters	High	Low to moderate capital; low O&M	Simple; low O&M; effective; no chemicals	Large footprint; design criteria varies; some failures from short-circuiting, overloading
Pre-engineered biofilters	Med	Moderate to high capital; low O&M	Low O&M, no chemicals; longer media life and smaller footprint than in-ground systems	Higher capital costs than inground biofilters
Bioscrubbers, biotrickling filters	Medium	Moderate capital; low O&M	Smaller footprint than biofilters, high H_2S loadings possible; little or no chemicals	Limited track record in United States; little long-term performance data for non-H_2S applications
Thermal oxidizers	Low	Very high capital and O&M (energy)	Effective for wide spectrum of odors and VOCs	Only economical for high-strength, difficult-to-treat airstreams
Diffusion into activated sludge basins	Low	Economical if use existing blowers or diffusers	Simple; low O&M; effective, reliable	Potential for corrosion of blower inlet components; additional blower cleaning may be required
Odor counteractants	High	Operating cost dependent on chemical usage	Low capital cost	Limited odor removal efficiency (< 40%); only applicable for dilute airstreams

Packed-tower scrubbers can include multiple stages designed for removal of specific contaminants. For example, composting odors may require a first-stage scrubber using an acidic scrubbant liquid to remove ammonia and amines. This may be followed by one or more stages using sodium hypochlorite to oxidize RSCs. For hydrogen sulfide removal, the traditional packed-tower scrubber uses both sodium hydroxide (to promote H_2S absorption into the liquid phase) and sodium hypochlorite (to oxidize the H_2S to elemental sulfur or sulfate).

6.2.2 *Carbon Adsorber*

Like wet scrubbers, carbon adsorbers have been a workhorse in the odor control industry for many years. Carbon treatment is simple and effective for a wide range of odorants. The most significant disadvantage is the potential need for frequent replacement or regeneration of the carbon because of higher-than-anticipated loads. Carbon adsorber vessels used for wastewater odor control applications are typically either deep-bed adsorbers, where odorous air is directed vertically through a minimum 1-m (3-ft) bed of media; or radial-flow adsorbers where the flow is directed horizontally through a bed of media and is collected at the outer annular space or the inner central column.

The major variable in selecting a carbon treatment system is the activated carbon media. "Catalytic" carbon significantly increases carbon's capacity to remove low-molecular-weight hydrogen sulfide, potentially increasing the lifetime of the media when H_2S is present. The new media has primarily replaced the caustic-impregnated carbon that was previously used for H_2S applications. The former system had significant disadvantages in handling and disposing of spent carbon and operating challenges. Carbon media capacities are summarized in Table 20.7. Face velocity (empty bed) through carbon adsorbers typically ranges from 10 to 20 m/min (30 to 60 ft/min).

TABLE 20.7 Carbon media capacities.

Carbon type	H_2S capacity, % by weight
1. Virgin	10
2. Caustic-impregnated	25
3. Water regenerable	16
4. High capacity	60

6.2.3 *Biofilters*

Biofilters are odor treatment devices in which odorous air is passed through a bed of organic or inorganic media that serves as a matrix for the attachment of biological growth. Odorous compounds such as hydrogen sulfide are biologically oxidized as they pass through the moist media. Biofilters can be custom, in-ground engineered units or pre-engineered modules with proprietary media.

The main design parameter for biofilters is the empty bed retention time (EBRT). The selection of EBRT is dependent primarily on the concentrations and types of odorous compounds to be removed. For example, airstreams containing only H_2S can be effectively treated at an EBRT of 30 seconds or less. However, airstreams containing more recalcitrant dimethyl disulfide may require an EBRT of 60 seconds or more.

Well-designed biofilters are capable of achieving high odor-removal efficiencies. For solids-handling odors, long detention times (EBRT of 60–90 seconds) may be necessary to remove non-H_2S sulfur compounds.

6.2.4 *Bioscrubbers or Biotrickling Filters*

Bioscrubbers and biotrickling filters are vertical vessels that use inert packing material to serve as sites to which biological growth can attach. The medium is continuously or intermittently irrigated with water containing nutrients to satisfy the biological population attached to the medium.

Bioscrubbers and biotrickling filters are highly effective for airstreams in which the predominant odorant is hydrogen sulfide. For H_2S, the EBRT may range from 10 to 20 seconds depending on loading, resulting in a small footprint and a reasonably sized vessel. For non-H_2S odors typically encountered in sludge-handling applications, however, such short detention times are likely to result in marginal odor-removal performance. For airstreams containing high amounts of H_2S and elevated levels of other sulfur compounds, a bioscrubber may be an appropriate first stage of a multistage odor control system.

6.2.5 *Activated Sludge Diffusion*

Diffusion of odorous air into aeration tanks has been successfully practiced at facilities around the world for at least 40 years. Testing has shown that diffusion of odorous air into the activated sludge process is highly effective for removing both H_2S and other RSCs. Odor panel testing has shown no effect on aeration tank emissions when foul air is sent to the blowers for treatment. Consideration needs to be given to the type of odor source and guidance provided under the National Fire Protection Association Standard 820 (2008).

6.2.6 *Thermal Oxidizers*

Thermal oxidizers are sometimes used for high-strength, difficult-to-treat airstreams. In most systems, the temperature of the airstream is increased to 760 to 815 ° C (1400 to 1500 °F) for one to two seconds in the presence of oxygen. Catalytic oxidizers operate at approximately 370 °C (700°F). The main disadvantage of the thermal oxidation process is the high operating costs. Four types of thermal oxidation systems are used:

- Direct combustors or flares;
- Regenerative thermal oxidizers;
- Recuperative thermal oxidizers; and
- Catalytic oxidizers.

Regenerative thermal oxidizers preheat the incoming air and are more thermally efficient than other options. These are more commonly used for odor/VOC destruction.

6.2.7 *Other Technologies*

There are a variety of other odor treatment technologies that are used. Successful performance data on these technologies for treatment of odors from sludge-processing operations, however, is limited. These include:

- Ionization—positive and negative ions are added to makeup air and introduced into an odorous room or process.
- Ultraviolet light—odorous air is subjected to UV radiation to destroy the odors.
- Counteractants—plant-derived extracts (essential oils) are sprayed into the air to react with the odorous compounds.

7.0 REFERENCES

Bowker, R. (1999) Overview of Odor Control Strategies for Wastewater Treatment. *Proceedings of the Virginia Water Environment Association Annual Meeting.*

Burrowes, P. A.; Witherspoon, J.; Quigley, C.; Easter, C. (2001) Biofilters-Wastewater Collection and Treatment Odour Experiences of a Sustainable

Technology in North America. *Proceedings of the First International Water Association International Conference on Odour and VOCs*. University of New South Wales: Sydney, Australia.

Clean Air Act. *U.S. Code*, Section 50–88, Title 40, 1963.

Clean Air Act. *U.S. Code*, Section 7401–7671, Title 42, 1990.

Easter, C.; Williams, T.; Witherspoon, J; Feltner, M. (2008) Odor Emissions from Anaerobically Digested Municipal Waste Water Treatment Plant Biosolids During Storage and Land Application: Impacts of Lime Dosing and the Type of Dewatering Process Used (Belt Presses versus Centrifuges). *Proceedings of the Water Environment Federation/Air & Waste management Association Odor and VOC Emissions Conference*, Phoenix, Arizona, April 6–9; Water Environment Federation: Alexandria, Virginia.

Epstein, E. (2007) *The Science of Composting*. CRC Press: Boca Raton, Florida.

Jubinville, R.; Stallings, R.; Bowen, W. (1997) Implementation of a comprehensive odor control program at the New London, Ct. Water Pollution Control Facility. *J. N. Engl. Water Environ. Assoc.* **31** (1), 40–48.

Koe, L.; Yang, F. (1999) The Effectiveness of the Cover and Treatment Scheme for Controlling Odorous Emissions at a Municipal Wastewater Treatment Facility. *Proceedings of the Chartered Institution of Water Environment Management and International Association on Water Quality Joint International Conference on Control and Prevention of Odors in the Water Industry*, London, England, United Kingdom; IWA Publishing: London, England, United Kingdom.

Melbourne Water (1989) *Hydrogen Sulphide Control Manual;* Melbourne Water: Melbourne, Victoria, Australia.

National Fire Protection Association (2008) *Standard for Fire Protection in Wastewater Treatment and Collection Facilities*, Publ. 820; National Fire Protection Association: Quincy, Massachusetts.

Sereno, D. J.; McGinley, C.; Harrison, D.; Haug, R. (1993) Dewatered Sludge Storage Emissions Control Using Multistage Wet Scrubbing. *Water Environ. Res.* **65** (1), 66–72.

U.S. Environmental Protection Agency (1974) *Sulfide Control in Sanitary Sewerage Systems*. EPA 625/1-74-005; U.S. Environmental Protection Agency: Washington, D.C.

U.S. Environmental Protection Agency (1985) *EPA Design Manual for Odor and Corrosion Control in Sanitary Sewerage Systems and Treatment Plants*, EPA/625/1-85/018; U.S. Environmental Protection Agency: Washington, D.C.

U.S. Environmental Protection Agency (1993) *Standards for the Use and Disposal of Sewage, Final Rule.* Code of Federal Regulations, Part 503, Title 40, 1993; *Fed. Regist.*, 58, 9248.

Water Environment Federation (1995) *Odor Control in Wastewater Treatment Plants*, Manual of Practice No. 22; Water Environment Federation: Alexandria, Virginia.

Water Environment Federation (1998) *Clarifier Design*, Manual of Practice No. 8; Water Environment Federation: Alexandria, Virginia.

Water Environment Federation (2004) *Control of Odors and Emissions from Wastewater Treatment Plants*, Manual of Practice No. 25; McGraw-Hill: New York.

Water Environment Federation; American Society of Civil Engineers; Environmental & Water Resources Institute (2009) *Design of Municipal Wastewater Treatment Plants*, 5th ed.; WEF Manual of Practice No. 8; ASCE Manuals and Reports on Engineering Practice No. 76; McGraw-Hill: New York.

Water Supply Committee of the Great Lakes—Upper Mississippi River Board of State and Provincial Public Health and Environmental Managers (2003) *Recommended Standards for Water Works* (aka "Ten States Standards"); Health Research Inc.: Albany, New York.

Water Environment Research Foundation (2003) *Identifying and Controlling Odor in the Municipal Wastewater Environment, Phase II: Literature Search and Review*, Report no. 00HHE5A; Water Environment Research Foundation: Alexandria, Virginia.

Water Environment Research Foundation (2004a) *Health Effects of Biosolids Odors: A Literature Review and Analysis*; Report no. 00HHE5C; Water Environment Research Foundation: Alexandria, Virginia.

Water Environment Research Foundation (2004b) *Identifying and Controlling Odor in the Municipal Wastewater Environment, Phase II: Impacts of In-plant Operational Parameters on Biosolids Odor Quality*, Report no. 00HHE5T; Water Environment Research Foundation: Alexandria, Virginia.

Water Environment Research Foundation (2007) *Minimization of Odors and Corrosion in Collection Systems,* Report no. 04CTS1; Water Environment Research Foundation: Alexandria, Virginia.

Witherspoon, J.; Sidhu, S.; Castleberry, J.; Coleman, L.; Reynolds, K.; Card, T.; Daigger, G. (1999) Case Study—Odor Emission Estimates and Control Strategies Using Models and Sampling for East Bay Municipal Utility Districts. *Proceedings of the Chartered Institution of Water Environment Management and International Association on Water Quality Joint International Conference on Control and Prevention of Odors in the Water Industry*, London, England, United Kingdom; IWA Publishing: London, England, United Kingdom.

Chapter 21

Sidestreams from Solids Treatment Processes

(continued)

1.0 INTRODUCTION

Solids processing at wastewater treatment plants (WWTPs) produce sidestreams and use thickening, dewatering, or decanting to separate liquid and solids. Because not all solids processing sidestreams are of equal concern, this chapter focuses primarily on those with significant nutrient loads (high nutrient-load sidestreams). These include dewatering recycle from anaerobic digestion, anaerobic lagoon decant, and fermenter overflow.

Sidestream of lesser concern (low nutrient-load sidestreams) are those from primary sludge and waste activated sludge (WAS) thickening or aerobic digested sludge dewatering or are returns from aerated lagoons. Even sidestreams with a low average load can have significant effects on the liquid process if they are returned in short increments (i.e., monthly lagoon decanting).

2.0 LOW NUTRIENT-LOAD SIDESTREAMS

2.1 Primary Sludge Thickening Return

Recycle composition (Table 21.1) from primary sludge thickening depends on primary clarifier solids capture and primary sludge thickening solids capture. In most cases, the primary sludge thickening return would be returned to the WWTP headworks, in which case it would go through primary clarification again. Recycle load as a percentage of the influent load is small, and its composition (Table 21.2) is similar to raw influent albeit with higher particulate concentrations. The numbers assume a typical raw influent composition of 5000 mg/L primary sludge total suspended solids (TSS); 60% primary clarifier solids capture; and 90% thickening solids capture.

Table 21.1 Typical composition of low nutrient load sidestreams (WAS = waste activated sludge; TSS = total suspended solids; BOD = biochemical oxygen demand; TKN = total Kjeldahl nitrogen).

Parameter	Unit	Primary sludge thickening return	WAS thickening return	WAS and primary sludge co-thickening return[b]	Fermenter overflow[c]
TSS	mg/L	400–600	1000–1400	650–750	700–900
BOD	mg/L	350–550	200–400	400–500	2000–2500
TKN	mg/L	40–70	70–90	60–70	80–120
NH_4-N	mg/L	20–40	1–20[a]	25	60–100
NOx	mg/L	0	0–15[a]	0	0
TP	mg/L	5–20	15–70[a]	3–5	10–20
PO_4-P	mg/L	4–10	0.1–3[a]	10–15	5–15

[a] Depending on secondary treatment process (i.e., BOD removal, biological nutrient removal, etc.).

[b] Assuming secondary process with BOD removal only, not recommended for facilities that nitrify.

[c] Volatile fatty acids = 500–700 mg/L.

TABLE 21.2 Typical recycle load as a fraction of the raw influent for low nutrient load side streams (WAS = waste activated sludge; TSS = total suspended solids; BOD = biochemical oxygen demand; TKN = total Kjeldahl nitrogen).

Parameter	Unit	Primary sludge thickening return	WAS thickening return	WAS and primary sludge co-thickening return[b]	Fermenter overflow
Flow	%	2.0–3.0	0.4–0.6	3.0–5.0	3–4
TSS	mg/L	4.0–8.0	2.5–3.5	10–15	8–13
BOD	mg/L	3.0–7.0	0.5–2.5	6–10	25–35
TKN	mg/L	4.0–8.0	1.0–2.5[a]	6–10	9–12
NH_4-N	mg/L	2.0–3.0	0–1[a]	3–5	9–12
NOx	mg/L	0	0–0.2[a]	0	0
TP	mg/L	3.0–5.0	2.0–8.0[a]	7–10	7–9
PO_4-P	mg/L	1.0–3.0	0.1–0.8[a]	3–4	6–8

[a] Depending on secondary treatment process (i.e., BOD removal, biological nutrient removal, etc.).

[b] Assuming secondary process with BOD removal only, not recommended for facilities that nitrify.

2.2 Waste Activated Sludge Thickening or Dewatering Return

Waste activated sludge thickening—such as with a dissolved air flotation thickener (DAFT), centrifuge, or rotary drum thickener—has capture rates ranging from 85 to 95%; thus, 5 to 15% of waste solids are returned. Dissolved compounds (i.e., ammonia, nitrate, etc.) would be present in similar concentration as found in the plant effluent. Table 21.1 shows a typical WAS thickening return compositions for a facility with nitrogen and phosphorus removal. The recycle load would be similar for thickening solids from integrated fixed film activated sludge (IFAS), moving bed bioreactors, or trickling filters.

The recycle load from WAS thickening has a relatively small effect on the secondary treatment process. This was illustrated by comparing the WAS thickening return load with the raw influent load (Table 21.2). Even in cases where the recycle load is compounded through shortened thickening schedules (i.e., eight hours per day on weekdays), the effect remains negligible in most cases.

In nitrifying facilities, operators should be aware of nitrifiers recycled with the WAS thickening return. When returned to the head of the plant, downstream biological oxygen demand (BOD) measurements may be skewed by generating oxygen demand from nitrification in the sample. In these cases, the carbonaceous BOD (cBOD) test method should be used instead.

2.3 Primary Sludge and Waste Activated Sludge Co-Thickening Return

Where WAS and primary sludge are co-thickened, the subsequent recycle load can be expected to equal those of WAS and primary sludge thickening combined, assuming identical solids capture rates

Co-thickening is generally not recommended, especially for nitrifying facilities or facilities with enhanced biological phosphorus removal. The recycle composition and fractural contribution are presented in Table 21.1 and Table 21.2 and assume WAS from a BOD removal-only process.

2.4 Dewatering Return from Aerobically Digested Sludge

Depending on the secondary treatment process and digester operation, the dewatering return may contain varying levels for ammonia, nitrate, and phosphate. Aerobic digesters can partially or fully nitrify and can remove nitrogen when operated in air on/off mode or with low dissolved oxygen. At WWTPs with enhanced biological phosphorus removal, some of the stored phosphorus is released during aerobic digestion.

2.5 Primary Sludge Fermentation Return

Some biological nutrient removal (BNR) facilities use primary sludge fermentation to generate volatile fatty acids (VFAs) for biological phosphorus removal or soluble carbon substrate for denitrification. Although this recycle load is desired, the recycle BOD load can be significant. When used as a denitrification carbon source, the ammonia load of the recycle can limit nitrogen removal performance. In most cases, the fermenter overflow is returned just upstream of the anaerobic zone to improve biological phosphorus removal. Tables 21.1 and 21.2 show examples of typical fermenter overflow composition and its fractural contribution to the influent load.

3.0 HIGH NUTRIENT-LOAD SIDESTREAMS

3.1 Dewatering Centrate from Anaerobic Sludge

Typically, the dewatering filtrate or centrate return is referred to as the solids process recycle load. Dewatering recycle loads from anaerobic digestion can add between 15 and 25% of ammonia to the plant influent load and between 5 and 100% of phosphorus. This amount depends on several parameters discussed later in this section.

At most facilities, the recycle load is further compounded by dewatering, which generates the daily recycle load in only a few hours, typically during the dayshift when influent loads are at their highest diurnal level. The instant recycle load can be calculated using eq 21.1. The average daily load may be expressed as a percentage of the influent load (kg/d or kg/h).

$$RL_{instant} = RL_{ave}(7/D \times 24/H) \qquad 21.1$$

where

$RL_{instant}$ = instant recycle load at time of return;
RL_{ave} = daily average recycle load;
D = number of dewatering days per week; and
H = number of dewatering hours per day.

The ammonia content is relatively consistent in the recycle from anaerobic digestion; phosphate, however, can vary significantly. The primary variable is the amount of phosphorus transferred to the digester via uptake by phosphorus accumulating organisms (PAOs). Typically, this uptake is intended in BNR systems, but may simply be an unintended consequence of the underlying mechanism of anaerobic selectors. Even anoxic selectors may produce different degrees of phosphorus uptake if they run low or out of nitrates at times.

The recycle phosphorus content is affected by metal (Ca, Mg, K) phosphate precipitation in the digester. Phosphorus removal may come from iron added for odor control, alum or iron in chemical sludge from the secondary process, additional calcium from lime addition, and from the release of accumulated Mg and K from the phosphorus uptake and release cycle (positive metal ions, predominantly Mg^{2+} and K^{+}, are found in enhanced biological phosphorus removal plants).

For facilities with seasonal effluent requirements and operations modes (i.e., chemical addition, BNR), the recycle composition will vary throughout the year. It may take several months after a change in operation for equilibrium to be achieved across the plant. Depending on the duration of the different treatment season, the WWTP may be in transition year around and show a sine curve-type trend over the years.

Table 21.3 shows the dewatering centrate composition from a conventional WWTP and a BNR plant with 90% phosphorus removal without chemical polishing or supplemental alkalinity addition. The concentrations in Table 21.3 are for a dewatering centrifuge. In the case of a belt filter press (BFP), spray and washwater will provide significant dilution, although the percentage load contribution (Table 21.4) will remain the same. Table 21.4 assumes 0.5% primary sludge TSS, 60% primary clarifier solids capture, and 90% thickening solids capture.

3.1.1 *Treatment Capacity and Performance*

Fully equalized, the recycle load from anaerobic digested sludge dewatering affects plant capacity at a level equal to its secondary treatment load contribution. Assuming that adequate capacity exists, typically nitrification should not affect effluent quality, unlike nitrogen and phosphorus removal. For nitrogen removal, depending on the removal goal, measures may be required to compensate for the additional load. This could mean higher internal recycle rates, higher dose of a carbon supplement, additional alkalinity demand, or higher demand for aeration.

If the conditions in the BNR system do not allow the additional phosphorus load to be removed biologically, chemical polishing or an increased dose may be required. When added to the secondary process, the side effect of chemical phosphorus removal

TABLE 21.3 Typical dewatering recycle from anaerobic digestion composition (BNR = biological nutrient removal; CAS = conventional activated sludge).*

Parameter	Unit	BNR	CAS-N
TSS	mg/L	1200–1600	
BOD	mg/L	400–500	
TKN	mg/L	1100–1300	
NH_4-N	mg/L	1000–1200	
NOx	mg/L	0	
Total phosphorus	mg/L	300–400	100–200
PO_4-P	mg/L	250–350	80–120

* Assuming 2.5% digested sludge total solids; 55% digester volatile solids reduction; and 95% dewatering solids capture. (Assumed influent composition summarized in Table 22.1.)

Table 21.4 Typical range of dewatering recycle for anaerobic digestions as fraction of raw influent load (based on 24 hours per day, 7 days per week return) (BNR = biological nutrient removal; CAS = conventional activated sludge; TSS = total suspended solids; BOD = biochemical oxygen demand; TKN = total Kjeldahl nitrogen).

Parameter	Unit	BNR	CAS-N
Flow	%	0.2–0.4	
TSS	%	2–3	2–3
BOD	%	0.7–1.0	0.7–1.0
TKN	%	12–18	13–19
NH_4-N	%	15–22	16–23
NOx	%	0	0
Total phosphorus	%	30–40	10–15
PO_4-P	%	15–30	8–12

(Assumed influent composition summarized in Table 22.1.)

is an increase in mixed liquor suspended solids (MLSS) and a potential reduction in overall capacity because of secondary clarifier load limitations.

Without equalization of sidestream treatment, the effect of the recycle load can be far more severe for secondary treatment capacity and performance. Figure 21.1 shows the aeration basin's ammonia load with and without a compounded recycle load from an eight-hour daytime dewatering operation, five days per week. This figure illustrates the load increase and shows that the daytime dewatering operation results in a recycle ammonia load return during the peak load time of the day.

For nitrogen removal, the effect described previously is amplified when the dewatering centrate is not equalized. The compounded nitrogen load return can result in a carbon substrate deficiency at the time of return, resulting in higher effluent nitrate and total nitrogen concentrations.

The compounded recycle phosphorus load affects the phosphorus removal process differently in facilities that practice BNR. It is likely that the compounded recycle load will exceed the biological removal capacity, making chemical polishing necessary. Although automatic chemical dosage based on real-time phosphate analysis is available, this control strategy is rarely used. Typically, the strategy is to feed continuously a higher coagulant dosage to capture the peak demand. When this operating strategy is combined with the compounded phosphorus return from the

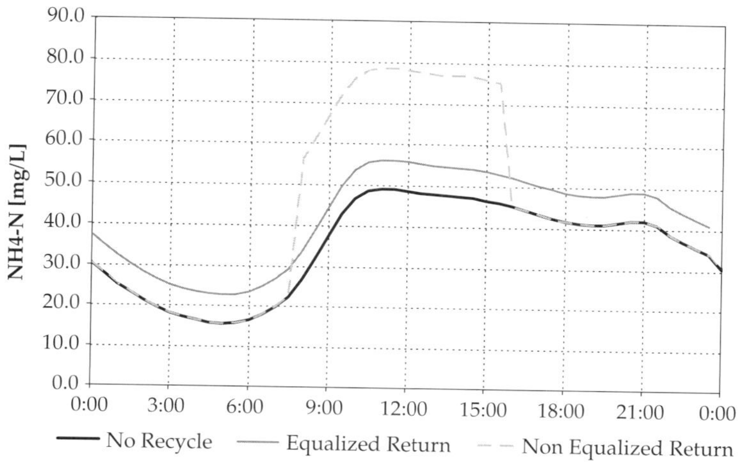

FIGURE 21.1 Illustration of dewatering recycle ammonia load effect on aeration basin influent ammonia concentration.

non-equalized dewatering recycle, it results in significantly higher coagulant use and chemical sludge production. The effect of a compounded recycle phosphorus load may be addressed easily by either installing online coagulant dosage control or adding coagulant to the sidestream.

For facilities with low influent alkalinity, the effect on alkalinity feed control is similar to coagulant dosage control. Unless the dosage is controlled online, overdosing for much of the day is required to capture the peak demand. Otherwise, the possibility exists that, during the dewatering recycle load return, the pH may drop because of alkalinity limitations. The latter can result in unstable nitrification.

3.1.2 *Process Control and Monitoring*

The level of the sidestream process control and monitoring related to the dewatering recycle return depends on the sidestream management strategy and discharge requirements. Regardless of its effect on secondary treatment process monitoring, the composition of the dewatering recycle reflects both anaerobic digestion and dewatering performance. For BNR facilities, the dewatering recycle composition can provide valuable feedback on biological phosphorus removal performance.

The composition of the dewatering recycle is consistent once past the dewatering equipment startup period. The composition of dewatering return flows would not change significantly from day to day because of the long retention time in the anaerobic digester. Thus, grab samples would be sufficient. At a minimum, flow-monitored grab samples should be analyzed for TSS to monitor dewatering performance (solids capture). Nitrifying plants should also monitor ammonia, and facilities with biological or chemical phosphorus removal should analyze phosphate concentration.

For facilities with anaerobic or anoxic selectors, dewatering equipment, including recycle conveyance pipes, pumps, and storage tanks, should be inspected routinely for struvite deposits (quarterly).

3.2 Dewatering Recycle or Decant from Anaerobic Lagoon Sludge

Anaerobic lagoons may be used in place of or in addition to conventional aerobic or anaerobic digesters. Anaerobic lagoons typically are used for digester sludge storage. Where the organic loading rate of the lagoon is very low, the subsequent recycle load is (on average) also low.

Because lagoons are used for sludge storage, decanting or dewatering of lagoon sludge occurs infrequently and can result in high recycle loads at the time of return. For instance, some facilities use the storage lagoon to hold digested sludge throughout the year until seasonal land application is possible. In such cases, the recycle load of an entire year may be returned to the main process over a period of several weeks or a few months. This load may be compounded further through the dewatering or decanting operation schedule.

Recycle from the lagoon decant can have a similar composition as the lagoon sludge dewatering recycle. Both significantly depend on the lagoon retention time, upstream sludge treatment, local climate, and decanting operation strategy.

3.2.1 *Effect on Treatment Process*

When returned continuously, recycle from an anaerobic storage lagoon that is storing previously digested sludge is low. It is, however, typical for such storage lagoons to be decanted, or emptied (dewatering the sludge), over a short time. This can compound a relatively small recycle load to such an extent that its effect becomes severe.

In addition to increasing the nitrogen and phosphorus load, the compounded return of anaerobic lagoon decant can result in an immediate decline in effluent quality, as fine particles and turbidity from the lagoon pass through the secondary treatment process. Facilities with a very low effluent discharger requirement or reuse programs should refrain from returning decant in slugs but should equalize it or recant more often in smaller increments. Facilities with effluent filtration risk blanketing filters with slugs of solids from the lagoon that did not settle out upstream.

This can be particularly troublesome for facilities that produce reuse waters; the fine particles returned from lagoons may not be retained by conventional filtration methods (i.e., sand or cloth filters).

Decant transfer lines from anaerobic lagoons have high struvite-formation potential. These lines should be inspected on a regular basis. Long transfer lines may require additional measures to prevent struvite scales, which can be difficult and costly to remove.

3.2.2 *Process Control and Monitoring*

Monitoring requirements for anaerobic storage lagoons are minimal for most facilities. Facilities with phosphorus or nitrogen limits or that have reuse programs should analyze the lagoon content before decant or dewatering and adjust the schedule to the treatment capacity of the secondary process.

3.3 Anaerobic Digester Decant and Recuperative Thickening Return

Some facilities use digester decanting to increase the digester solids content and its capacity. The decant composition is very similar to that of the anaerobic digester sludge dewatering centrate for the soluble components ammonia and phosphate (see Section 3.1). Compared to the dewatering return, the decant likely will have a higher concentration of suspended solids.

The effect of the decant on the secondary treatment process depends on the operation strategy: how much is decanted and how often. Thus, the recycle load cannot be quantified without knowledge of the operation strategy. Regardless, the effect can be minimized by avoiding decant operation during the peak-load hours of the day and by decanting frequently in small increments versus only a few large decants.

Recuperative thickening has the same objective as digester decanting but is typically a more continuous process. Therefore, its recycle load is distributed over a longer period. Similar to decanting, the composition is similar to dewatering centrate except with a higher concentration of suspended solids.

Both decanting and recuperative thickening can recycle a significant amount of suspended solids back to the liquid treatment system. At facilities without primary clarifiers, this recycle can increase aeration basin solids yield, MLSS, and secondary clarifier solids loading. This can result in a reduction of secondary treatment capacity. Operators will need to balance the need to increase digester capacity with the loss of secondary treatment capacity.

Monitoring requirements are also similar to those for dewatering return. The relevant recycle load (i.e., ammonia and phosphate) must be known to assess its effects. Flow and TSS monitoring is recommended for all facilities; ammonia should be included for nitrifying plants and phosphate for those that remove phosphorus.

4.0 SAMPLING AND MONITORING

Solids processing sidestreams can have significant effects on the capacity and performance of the liquid treatment process. The effect becomes more important as effluent quality requirements increase. Sidestream loads should be quantified to optimize the design and operation of WWTPs. In addition, sidestream characterization can help to determine proper handling and disposal methods. Unfortunately, sampling and monitoring of the sidestreams are performed rarely.

4.1 Sampling Locations

Sidestream sampling begins with identifying relevant sampling location, analytical methods, and equipment. A good sample location should be safe and easy to access and should provide a sample that is mixed well enough to represent the process flow.

Although it is preferable to sample sidestreams just before its return to the WWTP as a single stream, sometimes this is not possible. Therefore, several samples from several streams may be needed to produce a composite. For example, samples of streams from several individual sources that later combine downstream in a safe or accessible location.

Inline sample locations should be located near a floor drain to allow for sample-line purging. Whenever possible, sample taps should not be located on the bottom of the pipe; this prevents clogging. Pressurized lines should be equipped with slow-opening sample valves.

To quantify the recycle load, it is necessary to measure flow at the locations corresponding to the sampling point.

4.2 Sampling Schedules

Sampling requirements for solids processing sidestreams depend on the type of process from which the sidestream originates and whether the sample is collected for solids process monitoring (i.e., solids retention) or recycle-load monitoring.

For process monitoring, such as thickening or dewatering, daily samples of online monitoring are recommended, where feasible. For recycle nutrient loads from digested sludge dewatering, weekly sampling is sufficient because the composition does not change much from day to day.

Table 21.5 shows recommended sampling frequencies and analysis for the sidestreams covered in this chapter. Recycle nitrate and phosphorus measurements are only of relevance to facilities that nitrify or are required to remove phosphorus.

4.3 Sample Collection and Handling

Solids processing sidestream compositions are much more consistent compared to, for instance, plant influent samples. Therefore, grab samples are typically sufficient. Samples should be collected only when the process is actually running (i.e., dewatering), and sample lines should be thoroughly purged to ensure accuracy.

Samples collected for analysis of phosphate, ammonia, and nitrate may require special consideration depending on the process stream that is sampled. For process streams with active biomass, it is good practice to filter samples collected for analysis of soluble components such as ammonia, phosphate, or volatile fatty acid (VFA) in

Table 21.5 Recommended weekly sample frequencies for solids processing sidestream (VFA = volatile fatty acid; WAS = waste activated sludge; TSS = total suspended solids; BOD = biochemical oxygen demand).

Parameter	TSS[a]	VFA	BOD	NH_4-N	PO_4-P
Primary sludge thickening return	3		1	1	1
WAS thickening return	3				
Primary sludge and WAS co-thickening return	3				
Primary sludge fermenter return	3	3	1	1	1
Dewatering return from anaerobic sludge	5		1	1[b]	1[b]
Dewatering return from aerobic sludge	5		1	1[b]	1[b]
Lagoon decant/dewatering centrate	3		1	1[b]	1[b]
WAS dewatering return	5		1		

[a]TSS sample for monitoring dewatering or thickening performance.

[b]Facilities with NH_4-N, total nitrogen, and/or total phosphorus limits should monitor NH_4-N, and PO_4-P accordingly.

the field. This minimizes biological activity that could cause a change from the original sample composition. For instance, a sample collected from WAS thickener return can have higher phosphate concentrations after extended sample storage because of phosphate release.

Flow measurements using online measurement devices with totalizers are recommended for all sidestreams (see also Section 4.4).

4.4 Online Monitoring

Although technology improves daily, online monitoring equipment is not widely used in WWTPs. Online monitoring devices typically measure flows, solids, and nutrients. For stabilized sludge, Venturi flow tubes, magnetic meters, and positive-displacement pumps are used; for thickeners, magnetic meters and positive-displacement pumps are used; for dewatering belt presses, scales are used; and for dry composting thermal reduction, bulk container or truck scales are used. Flow should be totalized to account for variations in flow and noncontinuous operation.

The online monitoring of nutrients includes measurement of NH_4-N, NO_x-N, and PO_4-P. The three methods currently used to measure nutrients are colorimetric, UV absorption, and ion selectivity. Most analyzers can measure nutrients only in streams with low solids concentrations. The need for sample filtration and noncontinuous operation of many sidestream-generating processes makes the use of online nutrient analyzers less practical.

5.0 SIDESTREAM MANAGEMENT

5.1 Off peak or 24-Hour Operation

In principal, it is possible to avoid most issues associated with recycle loads through changes in operating strategy, such as the hours a given process is in operation. For most facilities, however, such changes can create staffing issues, especially for small and midsize facilities without 24 hours per day, 7 days per week staffing.

To avoid the compounded return of the dewatering recycle load, dewatering may be operated 24 hours per day, 7 days per week, provided the equipment is sized properly and is designed to operate largely unattended. Offpeak or extended-day dewatering can also significantly reduce the peak-load effect from the recycle stream.

For other scheduled recycle load-generating operations, such as digester or lagoon decanting, offpeak operation is easier to accommodate and is highly recommended.

In the case of the lagoon decant, its composition should be analyzed before decanting so that the rate of return can be limited to the secondary treatment capacity.

5.2 Sidestream Equalization

Sidestream equalization mitigates the effect of the sidestream load by managing the amount and timing of its return to the main process. The need for such management is typically limited to nutrient-rich recycle streams, such as dewatering centrate from anaerobically digested sludge or the decant from anaerobic lagoons or digesters.

Flow equalization is often the easiest and most cost-effective way to manage sidestreams. Although equalization does not eliminate the nutrient-cycle load, it enables operators to manage additional nutrient loads by returning sidestreams in a controlled manner. Equalizing the recycle load can be accomplished either operationally with a continuous dewatering schedule (24 hours per day, 7 days per week) or by incorporating an equalization basin where the sidestream is stored and returned gradually.

With an equalization basin, operator can choose to return more of the recycle load during offpeak hours of the day. In fact, the recycle load may be used to level out the plant load over the 24-hour period. When operating the offpeak return of high-ammonia recycle loads, changes in wastewater composition must be considered, including alkalinity for nitrification and available carbon for denitrification; unfavorable conditions for nitrification or nitrogen removal may occur.

Even partial equalization can reduce the peak load significantly. At most WWTPs, the diurnal peak influent load overlays with the dayshift dewatering schedule; storing the recycle load for the duration of the diurnal influent peak will significantly reduce the effect of the recycle load on liquid-stream process.

5.2.1 *Equalization Tank Sizing and Design*

The required volume for dewatering recycle equalization depends on the desired operation strategy. For full equalization, the storage tank has to hold enough recycle to return the average daily recycle flow each day between the two dewatering days, whichever are furthest apart (i.e. Friday of one week and Monday of the following week). For full equalization, the required volume for flow equalization can be calculated using equation.

$$V_{EQ} = D_{st} \times DR_{flow} \times d_{dw}/7 \qquad 21.2$$

where

V_{EQ} = equalization basin volume, ML;

D_{st} = largest number of consecutive days without dewatering per week;

d_{dw} = number of dewatering days per week; and
DR_{flow} = dewatering flow, ML/d (actual flows on dewatering days).

The daily average dewatering recycle flow can be determined through a mass balance or extrapolated from existing plant records. Depending on the dewatering equipment used, the wash or spray water volume needs to be added. For BFPs, this can easily double the recycle flow. A safety factor may be applied, depending on how critical the equalization is for overall plant performance and permit compliance. Redundancy is typically not required except for facilities with very low [limit of technology (LOT)] effluent nutrient limits.

Where site or financial constraints do not allow a tank large enough for 24 hours per day, 7 days per week equalization, smaller tanks can be nearly as effective in reducing the recycle load effect. Any tank large enough to store the daily dewatering recycle volume, thus permitting its equalization over a 24-hour period, would reduce load spike from a non-equalized dewatering recycle. The necessary volume may be minimized further by dewatering on as many days per week as possible.

The equalization tank design should include the following features:

- Mechanical mixing;
- Covered tank with foul-air treatment (odor control);
- Level sensor with high and low alarms;
- Sloped flow with drain;
- Overflow to drain; and
- Variable-speed return pumps or flow-control valves with flow metering.

The dewatering recycle equalization tank and conveyance system should be designed with consideration for potential of struvite formation. Struvite-formation potential will affect mixer design and control, pipe material or pipe liner selection, pump and valve selection, and the ability to remove scales mechanically or chemically (accessibility, cleaning loops). The prevention of harmful buildup of deposits requires routine inspections at least annually for systems with an operational history and monthly for new systems.

5.2.2 *Equalized Dewatering Recycle Control*

The simplest control strategy is 24 hours per day, 7 days per week constant return. The control system should enable the operator to enter a return flow based on anticipated dewatering return volume. High- or low-level alarms can notify the operator to adjust the return rate.

When load equalization is desired, the control system would have to allow the operator to enter a return-flow rate curve in 30- or 60-minute increments.

For either recycle strategy, flow monitoring and variable-speed pumps or flow-control valves are required. It is important that designers and operators anticipate the potential of struvite deposits, which can interfere with valve operation and flow metering and reduce the net pipe diameter. Dewatering recycle, particularly from centrifuges without dilution from washwater, has a high struvite-formation potential; even facilities without biological phosphorus removal can experience significant scaling problems in dewatering recycle storage and conveyance systems.

6.0 SIDESTREAM TREATMENT AND NUTRIENT RECOVERY

In cases where sidestream management alone is insufficient, several technologies may be considered. Sidestream treatment may be necessary or beneficial at facilities with lot requirements for nutrients that receive significant solids load transfers from other facilities or where the existing footprint requires a more compact solution. Sidestream technology selection may depend on whether the treatment is designed to remove or recover nitrogen, phosphorus, or both.

Sidestream treatment technologies described in this section are aimed at the most typical case of dewatering recycle from anaerobic digestion. Nevertheless, the technologies remain valid for sidestreams of similar composition but different sources.

6.1 Sidestream Nitrification—Return Activated Sludge Nitrification

Sidestream nitrification with return activated sludge (RAS) relies on the higher concentration of nitrifiers in RAS by sending dewatering recycle to an aerated tank that contains RAS (Figure 21.2). Return activated sludge nitrification is most effective when the dewatering recycle is equalized and only a fraction of the RAS is diverted to the RAS nitrification tank. At the full RAS rate, the net hydraulic retention time of RAS and dewatering recycle in the RAS nitrification tank is relatively short. Using only a small fraction of the RAS flow may require alkalinity addition to the RAS nitrification tank.

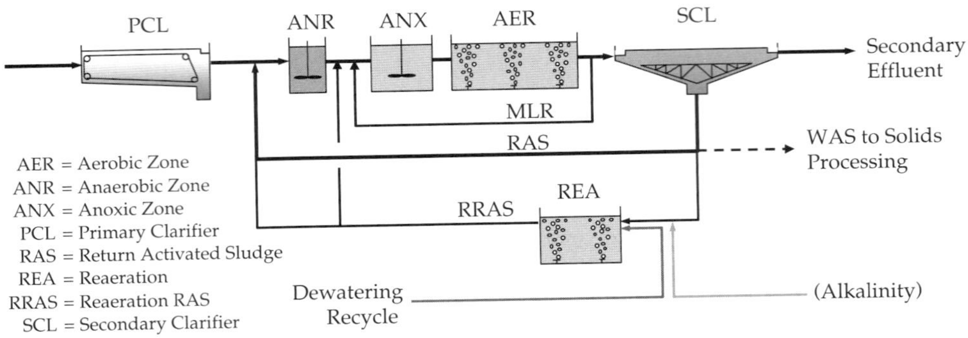

FIGURE 21.2 Reaeration process schematic (WAS = waste activated sludge; PCL = primary clarifier).

The design configuration, including RAS rate and tank size, is best determined with the aid of activated sludge modeling software. Preliminary sizing may be obtained by using the maximum oxygen uptake rate (OUR). The maximum attainable uptake rate depends on many factors but relies most on basin depth. Table 21.6 shows the maximum OURs based on basin depth that can be used to estimate the approximate size of the RAS nitrification basin or, conversely, which can be used to estimate how much ammonia can be nitrified given an existing volume. Two key assumptions are made in Table 21.6: sufficient nitrifiers and alkalinity are present.

Equation 21.3 may be used to estimate the size of the reaeration basin, assuming full nitrification and equalization of the dewatering recycle. For instance, at a 40 ML/d (10 mgd) plant with a recycle ammonia load of 230 kg/d (500 lb/d), the recycle ammonia oxygen demand for full nitrification would be 1050 kg/d (2300 lb/d) (4.6 kg O_2/kg ammonia nitrified). For a 6-m (20-ft) deep reaeration basin, the approximate maximum OUR is 100 mg/L·h or 2400 mg/L·d; and the necessary volume is 435 275 L (115 000 gal).

The nitrification of the recycle ammonia in the RAS nitrification basins would likely require an alkalinity supplement. In the previous examples, the alkalinity (eq 21.4) demand is 1620 kg/d (3570 lb/d) alkalinity as $CaCO_3$. For full nitrification, the alkalinity demand cannot be met without a supplement because dewatering

TABLE 21.6 Oxygen uptake rate (OUR) versus depth* in return activated sludge (RAS) nitrification basin (1 ft × 0.3048 = 1 m).

Reaeration basin depth	Unit	Maximum OUR
10 ft	mg/L·h	40–60
15 ft	mg/L·h	60–90
20 ft	mg/L·h	80–120
25 ft	mg/L·h	100–150

*Approximate values assuming fine-bubble aeration.

recycle from anaerobic digestion typically contains 50% of the alkalinity needed for nitrification.

$$V_{rea} = \frac{[NH_4 - N] \cdot Flow_{DR}\, 4.6\, \frac{lbO_2}{lb\,NH_4 - N}}{OUR_{\max} \cdot 24\, \frac{hr}{d} \cdot} \qquad 21.3$$

$$V_{rea} = [\Delta[NH_4\text{-}N]\ kg/d \times Flow_{DR}\ ML/d \times 4.6\ kg\ O_2/kg\ NH_4\text{-}N] / [OUR_{max}\ mg/L{\cdot}h \times 24\ h/d]$$

$$Alk\ [kg/d\ as\ CaCO_3] = \Delta[NH_4 - N] \times 7.14 \qquad 21.4$$

where

V_{rea} = reaeration basin volume;
$[NH_4\text{-}N]$ = dewatering recycle ammonia concentration, mg/L;
$Flow_{DR}$ = average daily dewatering flow, ML/d;
OUR_{max} = maximum oxygen uptake rate in reaeration basins, mg/L·h;
$\Delta[NH_4\text{-}N]$ = ammonia nitrified, kg/d.

The RAS nitrification system may be followed by an anoxic zone. Denitrification in the main anoxic zone, however, would be just as efficient and may not require supplemental carbon addition. If sufficient volume is available, the RAS nitrification basin may be operated in simultaneous nitrification and denitrification (SNDN) mode, air on/off nitrification, and denitrification (NDN).

6.2 Sidestream Nitrification—Activated Sludge

Sidestream activated sludge nitrification (Figure 21.3) can be independent from the secondary treatment system; providing some amount of WAS can improve flock forming This sidestream process benefits from the higher temperatures of anaerobic

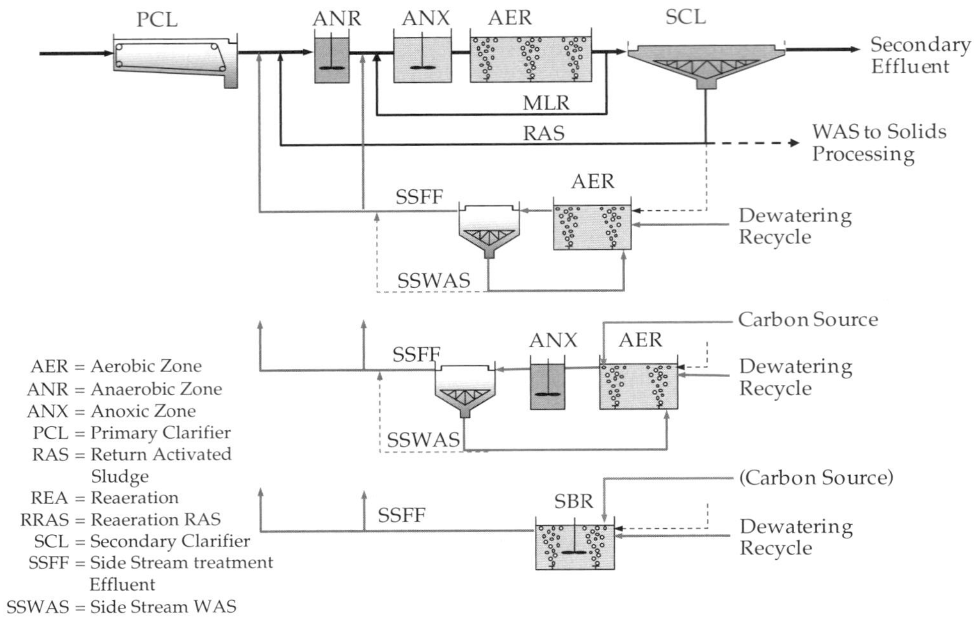

FIGURE 21.3 Sidestream activated sludge.

digester dewatering recycle, allowing nitrification at short solids retention times (SRTs), which reduces the reactor size. Because the sidestream treatment effluent is returned to the secondary process influent, complete nitrification is not as critical, which allows a more aggressive design with smaller safety factors and less redundancy. This is determined on a case-by-case basis.

The sidestream activated sludge system can provide seeding of the secondary treatment process with nitrifiers (if used to lower the mainstream nitrification SRT to less than two times the minimum SRT in a process known as INITRI). This seeding can potentially reduce the aerobic design SRT in the main activated sludge systems. Therefore, the sidestream system can reduce the size or increase the capacity of the secondary treatment system beyond the actual ammonia nitrified.

The design of the sidestream activated sludge system should follow the same principals as for the design of a secondary treatment nitrification system. For dewatering recycle from anaerobic digestion, the warmer temperature would allow for a short aerobic SRT. The minimum size of the aeration tank is driven by oxygen-transfer limitations. The size of the tank is estimated using the same approach, as outlined in Section 6.1 (eq 21.3 using low values from Table 21.6).

A sidestream activated sludge system should receive only equalized dewatering recycle. This maximizes sidestream activated sludge volume use and minimizes sidestream clarifier size by reducing peak flow.

The clarifier may be eliminated by using a sequencing batch reactor (SBR; Figure 21.4). This process is also known as the bioaugmentation batch enhanced (BABE) process..

For sidestream treatment, a single SBR reactor would be sufficient in combination with recycle equalization. Waste sludge from either system may be discharged directly to the secondary treatment system to maximize the beneficial effect of nitrifier seeding.

If full nitrification is desired, alkalinity addition is likely required. The amount of supplementation can be estimated using eq 21.4 from Section 6.1.

The sidestream activated sludge system would also lend itself to NDN and chemical phosphorus removal. Denitrification would likely require the addition of a supplemental carbon source. If added to the concentrated dewatering recycle from anaerobic digestion, then lime can precipitate phosphate while supplementing alkalinity.

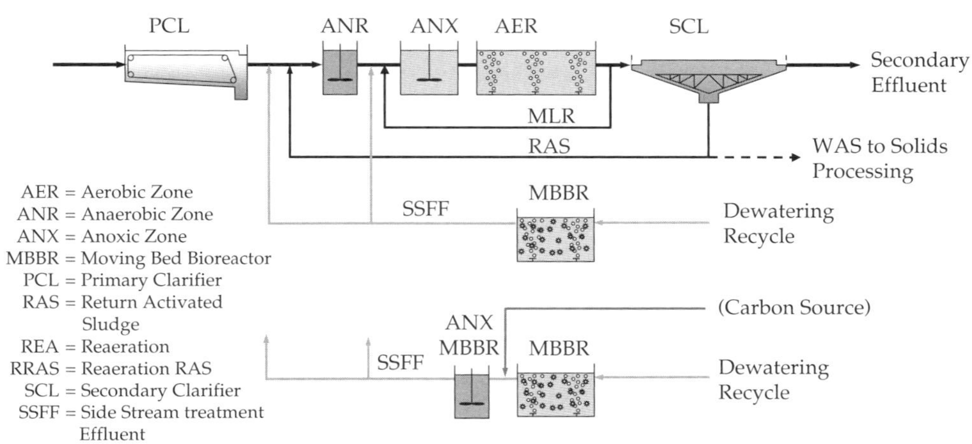

FIGURE 21.4 Sidestream moving bed bioreactor (MBBR) process schematic (nitrification and nitrification/gentrification; PCL = primary clarifier).

6.3 Sidestream Nitrification—Moving Bed Bioreactor

Moving bed bioreactors (Figure 21.4) may be used in a variety of municipal and industrial applications, including sidestream treatment. The main benefit of this over activated sludge (Section 6.2) is that biomass retention through clarification of SBR cycles is not required. The biomass is attached to suspended media, which are retained in the basin.

Pilot testing may be required to determine the best media solids processing sidestreams. Vendors offer different suspended media designs, some of which may be better suited for the solids processing sidestream environment. Because fouling or ragging is possible, prescreening of the dewatering recycle is recommended (e.g., prescreening with a 2-mm rotating drum screen). Whichever screening or straining technology is used, struvite formation should be anticipated.

Other MBBR design features, such as coarse-bubble aeration and retention screens, are the same for secondary treatment.

6.4 SHARON-Anammox Nitrogen Removal Technologies

The biological nitrogen removal processes, single-reactor system for high-activity ammonium removal over nitrate (SHARON) and anaerobic ammonia oxidation (Anammox) are innovative processes that offer new alternative to conventional nitrogen removal. These processes were developed at the Technical University of Delft, The Netherlands, and other European universities in the 1990s and are licensed by Paques BV, The Netherlands.

Unlike conventional activated sludge NDN, these processes perform partial nitrification of ammonia to nitrite. Once (partially) nitrified, nitrogen removal can occur by either (1) reducing nitrite under anoxic conditions to nitrogen gas by heterotrophic denitrifiers seen in activated sludge NDN (SHARON) or (2) oxidizing the residual ammonium under anaerobic conditions to nitrogen gas, with nitrite as the electron acceptor (Anammox).

The primary benefits of the SHARON and Anammox processes compared to conventional activated sludge NDN, is the reduced power consumption (up to 60% less), little or no required external carbon source, low yield, and significantly reduced CO_2 emissions. The elevated design temperature (30 to 36°C) and potentially long hydraulic retention time (HRT; >1 d) limits their application, however, to process streams with both high ammonia concentrations and temperatures. Dewatering recycle from anaerobic digestion is an ideal candidate for these processes because it is at or close to the required design temperature (> 27°C). Dewatering recycle flow equalization is recommended for all SHARON and Anammox processes.

6.4.1 *SHARON*

The SHARON process is a high-rate NDN process with minimal SRT and smaller reactor volume compared to conventional NDN. In addition, the process enables both a savings of 25% in oxygen transfer energy and 40% in carbon requirements for denitrifying bacterial growth compared to conventional NDN.

Production of alkalinity during the denitrification process controls reactor pH. Ammonia removal efficiencies of greater than 90% have been found, which translates to a 90% reduction in recycle flow comprised of 600 to 1000 mg/L ammonia-nitrogen. Ammonia removal of 85 to 95% is feasible through optimization of SRT, pH, and dissolved oxygen levels. The SHARON reactor accomplishes this at a low SRT and an elevated temperature to select for ammonia-oxidizing bacteria (AOB) over nitrite-oxidizing bacteria (NOB). The process occurs within a completely mixed reactor in which both phases of nitrification and denitrification take place simultaneously. Hence, there is no need to promote large microbial flocs to settle or retain biomass as the reactor operates with small microbial flocs.

The reactor may be run as an intermittent aerated SBR or a continuous-flow staged reactor. By using intermittent aeration, NDN may be accomplished in a single reactor; here, separate-stage reactors are used with partial nitrification in the first followed by denitrification to reduce nitrite to nitrogen gas in the second.

6.4.2 *Anammox*

The Anammox reaction consists of nitritation of ammonia and anaerobic oxidation of ammonia using nitrite as an electron donor (see eq 21.5). Depending on the process design solution, these two reactions can occur in series, simultaneously or in a batch sequence. Equation 21.6 is an expansion of Equation 21.5 that incorporates alkalinity and pH during Anammox oxidation step.

$$NH_4^+ + NO_2^- \rightarrow N_2 + 2H_2O \qquad 21.5$$

$$NH_4^+ + 1.32\,NO_2^- + 0.066\,HCO_3^- + 0.13\,H^+ \rightarrow 1.02\,N_2 + 0.26\,NO_3^- + 0.066\,CH_2O_{0.5}N_{0.15} + 2.03\,H_2O \qquad 21.6$$

The balance between aerobic and anaerobic ammonia oxidation is managed through control of dissolved oxygen and pH. Growth of NOB is limited by maintaining a short aerobic SRT that, along with the higher reactor temperature, favors AOB; Anammox bacteria form a natural granula through extracellular polymeric substance (EPS) excretion. These granules settle rapidly, which allows selection of Anammox over other bacteria by separation through sedimentation/overflow zones, hydrocyclones, or batch settling.

In addition to energy savings and reduced carbon demand, Anammox has a significantly higher rate process than conventional activated sludge NDN. Suggested loading rates for Anammox range from (8 to 26 kg total N/d/m^3) (Jetten et al., 1999), which is more desirable than rates typically found in conventional activated sludge systems of (1 kg total N/d m^3) (STOWA, 1996a and 1966b).

One downside of the Anammox process is the slow growth of the granule. Therefore, it is recommended that new facilities be seeded before startup. With adequate seeding, full capacity may be reached in three to six months. Experience from full-scale installations shows that once Anammox granular media are established subsequent reseeding is not required.

6.4.2.1 SHARON-Anammox

The SHARON-Anammox process uses two separate stages in series for nitritation and Anammox. Some of the ammonia-rich sidestream is fed to the first station (SHARON) where the ammonia is oxidized to nitrite. The remaining flow and first-stage (SHARON) effluent are combined and fed to the second stage (Anammox), where anaerobic ammonia oxidation takes place. This process solution offers separate control of the SHARON and Anammox biomass and reactions. This two-stage design is used rarely and has been replaced by single-stage process solutions.

6.4.2.2 One-Step Anammox

The one-step Anammox (OSA) process is based on partial nitrification to nitrite, followed by ammonium nitrification to nitrogen gas by nitrite, all of which occurs in a single-stage. The two reactions, nitritation and Anammox, are controlled primarily through pH, dissolved oxygen, and controlled growth of AOB while continuously wasting nitrite oxidizing organisms from the reactor. This is accomplished by controlling the design aerobic HRT at 1 to 1.5 days.

With the reactor volume constant, the aerobic HRT is varied by turning the air on and off as required. Anammox granular media are retained by taking advantage of its good settling characteristics in a specially designed settling and overflow zone where the suspended, nongranular biomass is wasted with the effluent via launders.

6.4.3 *DEMON*

Similar to the one-step Anammox, the DEMON process uses only a single reactor operating in batch mode much like an SBR. The growth of AOB and NOB also is controlled through the aerobic HRT. Granular Anammox media are retained by separating other biomass and granular media with a hydrocyclone.

6.5 Chemical Phosphorus Removal

Chemical phosphorus removal relies on precipitation of soluble (reactive) phosphorus, mostly orthophosphate (PO_4^{3-}), followed by solids removal from the liquid stream. In the case of sidestream phosphorus removal, the goal is to transform soluble reactive phosphorus to a particulate, nonreactive form. In this form, the returned phosphorus load will have limited effect on the secondary treatment process. It is advantageous to precipitate soluble phosphorus upstream of dewatering such to allow the precipitated phosphorus to be removed with the dewatered biosolids. This reduces the overall treatment system phosphorus inventory, which directly translates into lower effluent phosphorus concentrations.

Returning the chemical sludge from alum or ferric to the process may have additional removal benefits because of longer chemical sludge retention time, which improves chemical utilization through adsorption and complexation of phosphate. However, the additional chemical sludge increases the secondary clarifier solids load and may reduce the secondary treatment capacity.

If an alkalinity supplement is needed for the secondary process, then a chemical such as lime or magnesium hydroxide [only in sidestreams with high ammonia where magnesium ammonium phosphate (struvite) can form] may be preferred because it provides a dual benefit of phosphorus removal and alkalinity supplementation.

Typically, the following chemicals are used for the precipitation of phosphorus:

- Aluminum sulfate (alum);
- Ferric chloride;
- Polyaluminum chloride; and
- Lime.

6.6 Air Ammonia Stripping

Air-stripping systems can use either hot or cold air depending on the desired removal efficiency and incoming water temperature. Increasing the pH will also further improve ammonia removal efficiency.

Typically, ammonia-stripping systems consist of stripping towers filled with packing media. Ammonia-rich waste stream is distributed over the packing media while air is blown up through the tower (Figure 21.5). This disperses the flow and increases the total liquid/air contact surface for the gas transfer to occur. The larger the airflow—thus, the lower the ammonium concentration in the air, the more efficient the ammonia stripping. Because larger airflows require more energy, however, there are economical limits to improving stripping performance through increased airflow.

Because of the necessary pH increase for efficient ammonia stripping, the treated sidestream can have a high pH and may require neutralization or at pH reduction before being returned to the secondary treatment process. On the other hand, the waste stream now carries a significant amount of alkalinity that could be used in secondary treatment. It is recommended that the treated sidestream return from an ammonia stripping unit returns to the plant influent to permit mixing and dilution.

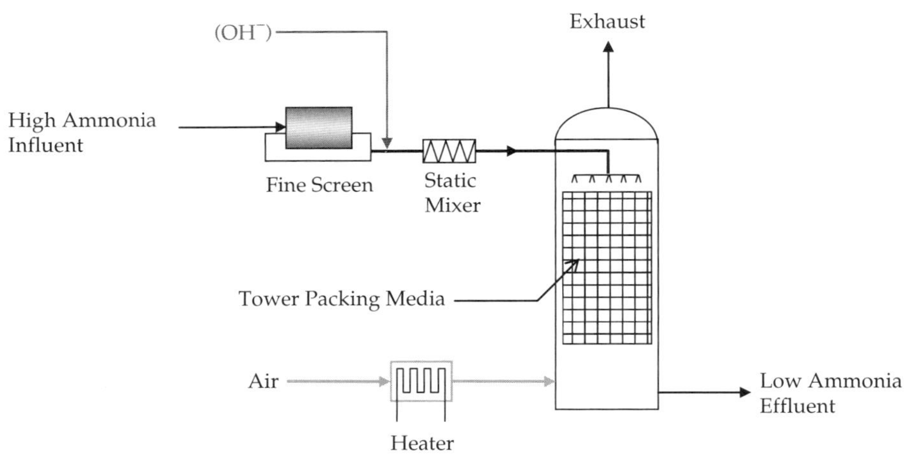

FIGURE 21.5 Example schematic of an ammonia hot air stripping unit with fine screening and optional pH adjustment.

The basic air-stripping process simply discharges the stripped ammonia into the air. Many locations, however, require air discharge permits and odor control equipment, which make implementation of basic air stripping more difficult.

The stripped ammonia may be captured in a closed-loop system (Figure 21.6). In this process, the stripped ammonia is removed in the absorption tower where concentrated ammonia sulfate is produced after sulfuric acid addition.

6.7 Steam Stripping

By introducing steam in the stripping reactor, followed by steam condensation, the stripped ammonia may be recovered as aqueous ammonia.

Low-grade stripping steam (approximately 1030 kPa or 150 psig) is introduced near the bottom of the column (Figure 21.7). Steam flowrates may be adjusted relative to feed rates to conserve energy because they are not providing any transfer or mixing support in the selective catalytic reduction (SCR) process. A condensing section is added to the tower top to concentrate the stripped ammonia, lowering the water dew point and reducing the ammonia pipe size. Concentrations of greater than 97% ammonia may be produced readily; however, the condensing section requires a cooling medium, such as water or ambient air.

Using steam, the stripper will operate hotter and, therefore, at a higher pressure than with air, which reduces the amount of stripping media necessary to remove the

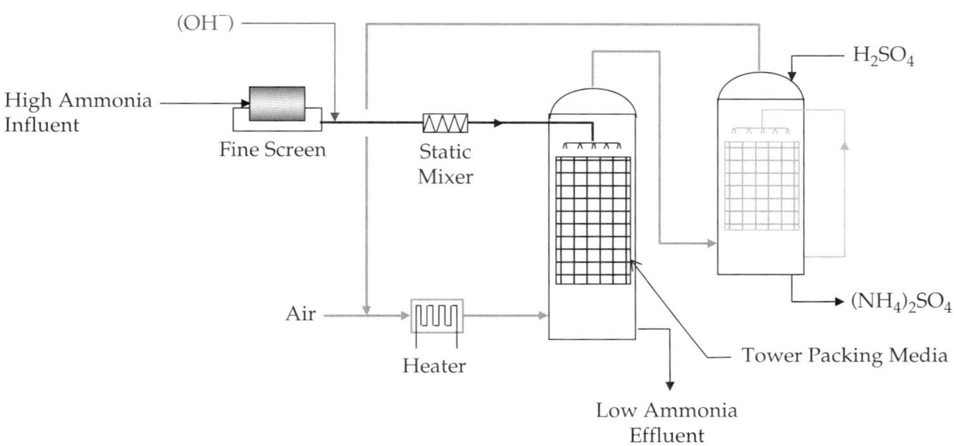

FIGURE 21.6 Example schematic of a closed-loop ammonia air stripping unit

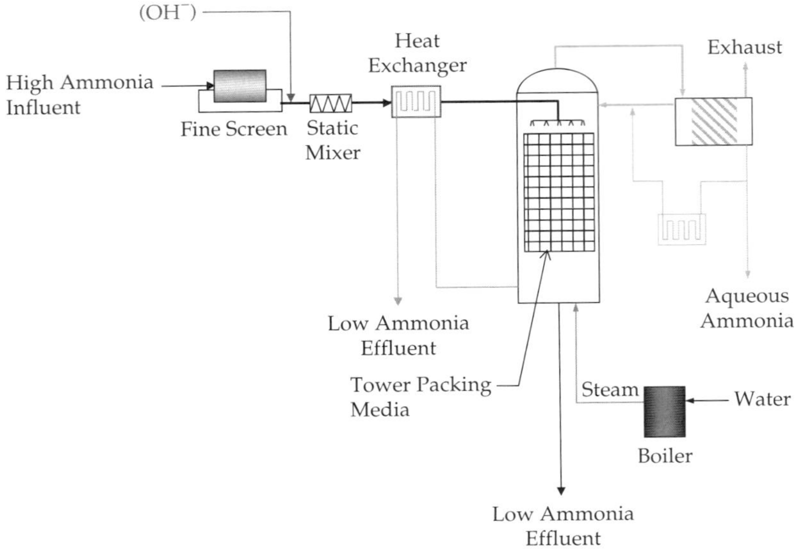

FIGURE 21.7 Example schematic of an ammonia steam stripping unit.

ammonia. The increased temperature at the top of the column also makes water or ambient air feasible as a cooling medium.

The concentrated top product is essentially anhydrous ammonia vapor and needs to be handled with great care. Residual water will collect at the bottom of the tower and should be recycled back to the column or disposed of as needed.

A significant benefit of steam stripping is that no offgas treatment is required because the offgas vapors may be condensed into a small amount of concentrated liquid stream. In the steam stripper, the stripped ammonia from the tower is sent to a distillation column where the ammonia is cooled and condensed. A pure ammonia solution is formed as a result. The ammonia water solution may be used to form ammonium salts, such as ammonium sulfate, with the addition of sulfuric acid.

Ammonia removal rates of 80% or more can be accomplished using steam stripping with alkali addition. Solids processing recycles containing high ammonia concentrations typically will require pretreatment before ammonia stripping. Otherwise, the stripping tower packing media will quickly foul. The recommended pretreatment consists of fine screening such as a rotating drum screen to retain rags and hair.

The pretreated sidestream also requires flow equalization to allow the stripping reactor to be fed at constant rate. For the equalization reactor, any necessary reheating or pH adjustments would occur between the storage tank and the ammonia stripper.

7.0 PHOSPHORUS RECOVERY

Processes to recover phosphorus from solids treatment process sidestreams have been under development since the 1990s. These processes have focused primarily on recovery of phosphorus as either calcium phosphate (hydroxyl apatite) or magnesium ammonium phosphate (struvite) compounds. Phosphate recovery processes operate on similar principles as chemical phosphate removal processes, except that they typically produce solid crystalline products with low water content rather than amorphous flocculent sludges.

Phosphate recovery from wastewater is regarded as a way to recover a portion of the phosphate used in food production and detergents, thus, reducing the rate of depletion of global phosphate reserves. Phosphate recovery systems tend to be more complex than chemical phosphate removal. These systems are, however, able to remove phosphorus from the sidestream and produce a useable byproduct that is easily dried, stored, and transported. This byproduct may be sold to offset system costs.

7.1 Calcium Phosphate

Calcium phosphate recovery focuses on production of hydroxyapatite or $Ca_5(PO_4)_3(OH)$. Hydroxyapatite is one of the main components of bones and dental enamel, and is part of the family of apatites that are mined as a source of most of the world's phosphate ore. Hydroxyapatite recovered from wastewater can, therefore, be used as an input to fertilizer production or used directly as a substitute for rock phosphate. Typically, calcium phosphate recovery is carried out by adding lime to digested sludge dewatering liquors. The process can require pretreatment to remove carbonate by CO_2 stripping to avoid the formation of calcium carbonates. Lime addition provides a source of calcium ions and increases the pH of the liquors. Hydroxyapatite solubility is pH-dependent and governed by the solubility product or K_{sp}. As with many phosphate compounds, a range of solubility products are reported for hydroxyapatite in wastewater. This is likely because of the variety of calcium phosphate compounds that can precipitate and competing chemical reactions with magnesium, carbonate, and other compounds in dewatering liquors.

Phosphate levels may be reduced to approximately 1 mg/L PO_4-P through calcium phosphate-based recovery processes.

7.2 Magnesium Ammonium Phosphate

Magnesium ammonium phosphate or struvite recovery is similar to calcium phosphate recovery in that both processes produce crystalline product. Struvite recovery offers the added advantage of contributing to ammonia removal and recovery in addition to phosphorus recovery. Ammonia removal through struvite recovery is typically between 5 and 20% of sidestream loads. This is because of the relatively high ammonia concentration in sludge dewatering liquors relative to phosphate levels. Phosphate removal levels can vary depending on treatment objectives and operating conditions; removal rates of 80 to 90% are typical.

Struvite deposits are found often in piping, valves, and heat exchangers and in mixers in anaerobic digesters and digested sludge handling and dewatering systems. Struvite recovery processes enhance this naturally occurring process by dosing supplemental magnesium to the sludge dewatering liquors and adjusting the solution pH. Magnesium is typically dosed as either magnesium chloride or magnesium hydroxide, and the pH is adjusted using either sodium hydroxide or by air stripping of CO_2.

Struvite precipitation occurs when the activity product of Mg^{2+}, NH_4^+, and PO_4^{3-} is greater than the solubility product of struvite. The reaction of these ions is controlled by system pH; total concentrations of dissolved magnesium, ammonia, and phosphorus species; and the ionic strength of the wastewater. Because pH dictates the speciation of these components, struvite solubility also changes with pH. The solubility product of struvite is expressed as follows in eq 21.7, which shows the activity of each ion.

$$K_{sp} = -\text{Log}([MG^{2+}] \times [NH_4^+] \times [PO_4^{3-}]) \qquad 21.7$$

where K_{sp} = solubility product.

Struvite recovery processes typically occur in either fluidized bed reactors or mixed tanks, depending on the process vendor. Recovered struvite ranges from fine powders to solid pellets of approximately 3 mm in diameter. These products typically may be used directly as fertilizers. Figure 21.8 shows a completely mixed reactor tank based-recovery process.

Fluidized bed reactor-type facilities assist crystal growth through long retention times and internal reticulation with the goal or producing a reusable product directly

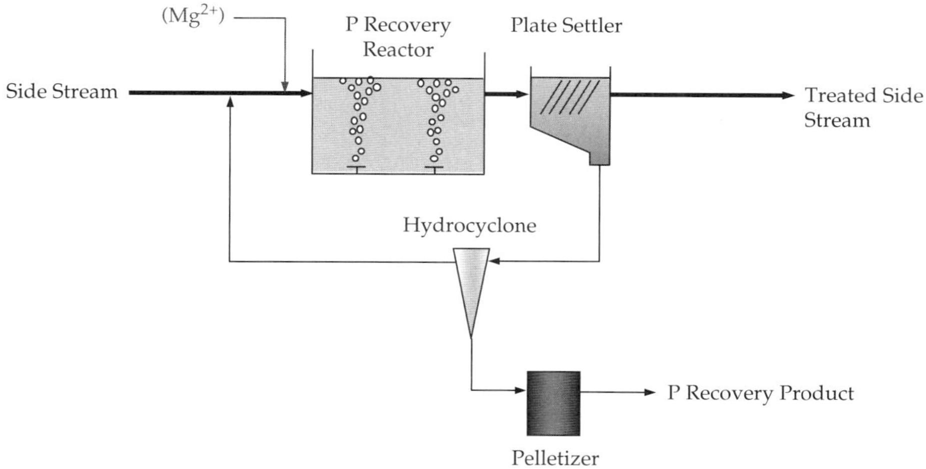

FIGURE 21.8 Completely mixed reactor based phosphorus recovery process.

in the reactor. Complete mixed reactor design is more similar to chemical phosphorus removal because they have a short retention time, and the struvite crystals are small. They can, however, be removed with a hydrocyclone because of their high specific gravity. Recovered struvite requires further processing to generate a reusable product.

Struvite recovery processes can reduce phosphate levels of less than 5 mg/L PO_4-P, but typically are used to produce effluent in the range of 20 to 40 mg/L PO_4-P to improve the economics of operation.

7.3 Economics of Phosphorus Recovery

The primary difference in evaluating the economics of phosphate recovery relative to chemical precipitation is that, in many cases, selling the recovered product can result in a significant cost offset compared with the cost of disposing of chemical precipitation sludges. It is expected that, over time, as phosphate reserves diminish, the value of the recovered phosphate products will increase.

In addition to the product itself, phosphorus recovery offers additional indirect savings, which can become a significant economic factor. Phosphorus removal by chemical precipitation produces chemical sludge that, typically, is carried through at least part of the process, thus occupying treatment capacity not available for biological treatment. Chemical sludge also generates direct costs associated with the chemicals, dewatering, hauling, and disposal.

7.4 Phosphorus Recovery Technology

Phosphorus recovery is an emerging technology. Many experimental facilities have been in operation around the world with varying degrees of success, both operationally and economically. Full-scale experience, however, is limited.

Various industry drivers, such as regulations, nutrient trading, commodity prices, hauling costs, or interest in sustainable designs, are improving the economical validity of phosphorus recovery. It is expected that the quantity and quality of available phosphorus recovery technologies will increase.

As of 2010, examples of commercially available phosphate recovery systems are the DHV's Crystalactor (Netherlands), Paques' PHOSPAQ (Netherlands), and Ostara Nutrient Recovery Technologies' PEARL process (Canada).

8.0 REFERENCES AND SUGGESTED READINGS

Carrier, S.; Tarre, M.; Beliavski, N.; Denekamp, A.; Gieseke, D.; de Beer, D.; Green, M. (2004) High Nitrification Rate at Low pH in a Fluidized Bed Reactor with Chalk as the Biofilm. *Water Sci. Technol.*, **49** (11–12) 99–105.

Cheremisinoff, P. N. (1995) *Waste Minimization and Cost Reduction for the Process Industries*; Noyes Publications: Park Ridge, New Jersey.

Cornwell, D. A. (1990) *Air Stripping and Aeration.* In *Water Quality and Treatment: A Handbook of Community Water Supplies*; 4th ed.; Pontius, F. W., Ed.; McGraw-Hill: New York.

Culp, R. L.; Wesner, G. M.; Culp, G. L. (1978) *Handbook of Advanced Wastewater Treatment*, 2nd ed.; Van Nostrand Reinhold Co.: New York.

De Boer, W.; Klein Gunnewiek, P. J. A.; Veenhuis, M.; Bock, E.; Laanbroek, H. J. (1991) Nitrification at Low pH by Aggregated Chemolithotrophic Bacteria. *Appl. Environ. Microbiol.*, **57** (12), 3600–3604.

Gabriella Eliasson Samtek, V. A. (1994) *Supernatant Treatment in Eslv, Sweden; Results and Experiences of Ammonia Stripping Method 1993–1994.*

Hellinga, C.; Schellen, A. A. J. C.; Mulder, J. W.; Van Loosdrecht, M. C. M.; Heijnen, J. J. (1998) The SHARON Process: An Innovative Method for Nitrogen

Removal from Ammonium-Rich Waste Water. *Water Sci. Technol.*, **37** (9), 135–142.

Hunik, J. H. (1993) Engineering Aspects of Nitrification with Immobilized Cells. PhD Thesis. Wageningen University, The Netherlands.

Jetten, M. S. M.; Strous, M.; Van de Pas-Schoonen, K. T.; Schalk, J.; Van Dongen, L. G. J. M.; Van de Graaf, A. A.; Logemann, S.; Muyzer, G.; Van Loosdrecht, M. C. M.; Kuenen, J. G. (1999) The Anaerobic Oxidation of Ammonium. *FEMS Microbiol. Rev.* **22** (5), 421–437.

Kim, S. H.; Kim, W. J.; Chung, T. H. (2002) Release Characteristics of Nitrogen and Phosphorus in Aerobic and Intermittent Aerobic Sludge Digestion. *Kor. J. Chem. Eng.*, **19** (3), 439–444.

KMPS Pilot Plant Services Group. Steam Stripping Systems. http://www.modular-process.com/steam-stripping.htm (accessed August 2011).

Logemann, S.; Schantl, J.; Bijvank, S.; Van Loosdrecht, M. C. M.; Kuenen, J. G.; Jetten, M. S. M. (1998) Molecular Microbial Diversity in a Nitrifying Reactor System without Sludge Retention. *FEMS Microbiol. Ecol.* **27** (3), 239–249.

Lu-Kwang, J.; Shah, H. K.; Porteous, J. (2005) Phosphorus Release in Aerobic Sludge Digestion. *Water Environ. Res.*, **77** (5), 553–559.

New York City Department of Environmental Protection (1998) *Nitrogen Control Feasibility Plan*; New York City Department of Environmental Protection: New York.

Pennsylvania Department of Environmental Correction (2002) The SHARON® High-Rate Nitrogen Removal System: An Innovative Wastewater Treatment Process http://www.dep.state.pa.us/dep/DEPUTATE/Watermgt/WSM/WSM_TAO/InnovTech/ProjReviews/SharonHiRate.htm (accessed August 2011).

Sagberg, P.; Ryrfors, P.; Grundnes Berg, K. (1997) *The Mass Balances of Nitrogen and Carbon in a Compact Nitrogen Phosphorous WWTP.*

STOWA (1996a) *One-Reactor System for Ammonium Removal via Nitrite*, Report no. 96-01.; STOWA: Utrecht, The Netherlands.

STOWA (1996b) *Removal of Ammonium from Sludge Water with the Anammox Process: Feasibility Study*, Report no. 96–21; STOWA: Utrecht, The Netherlands.

STOWA (2001) *The Combined Sharon/Anammox Process*, Report no. 2000-25; STOWA: Utrecht, The Netherlands.

Tarre, S.; Green, M. (2004) High-Rate Nitrification at Low pH in Suspended- and Attached-Biomass Reactors. *Appl. Environ. Microbiol.*, **70** (11), 6481–6487.

van der Star, W. R. L.; Abma, W. R.; Blommers, D.; Mulder, J. -W.; Tokutomi, T.; Strous, M.; Picioreanu, C.; van Loosdrecht, M. C. M. (2007) Start-Up of Reactors for Anoxic Ammonium Oxidation: Experiences from the First Full-Scale Anammox Reactor in Rotterdam. *Water Res.*, **41** (18), 4149–4163.

van der Star, W. R. L.; Miclea, A. I.; van Dongen, U. G. J. M.; Muyzer, G.; Picioreanu, C.; van Loosdrecht, M. C. M. (2008) The Membrane Bioreactor: A Novel Tool to Grow Anammox Bacteria as Free Cell. *Biotechnol. Bioeng.*, **101** (2), 286–294.

Chapter 22

Instrumentation and Monitoring

(*continued*)

1.0 INTRODUCTION

Wastewater solids processing and management systems are using increasingly complex processes for treatment and disposal. Concurrently, regulatory requirements are increasing process performance requirements. These systems require sufficient monitoring, alarming, and reporting to maintain the processes within operational parameters while meeting regulatory permit requirements. Properly applied instrumentation provides operations, management, and engineering with the data necessary to manage, operate, and design solids processing facilities.

This chapter focuses primarily on instrumentation information for utility management, facility managers, and project managers and provides an overview of the instruments and their application for solids processing design and management. It reviews application of process control and measuring for unit solids technologies and processes. Table 22.1 provides a summary of the instrumentation typically used for solids processing applications. Automation system designers and professionals desiring an in-depth discussion on automation of wastewater treatment facilities can find more detailed information in *Automation of Wastewater Treatment Facilities* (WEF, 2006).

2.0 THE NEED FOR INSTRUMENTATION

Properly applied instrumentation in solids processing systems provide several benefits including monitoring for regulatory compliance and reporting requirements; enhancing personnel safety and providing notification of process upsets; improving treatment process efficiency; leveraging and optimizing use of operational resources including labor, chemicals, and energy; and providing information for future planning.

2.1 Monitoring and Reporting Regulatory Compliance

State and federal regulatory permits include reporting requirements for solids processing compliance. Properly applied instrumentation allows for automatic unattended collection of this information.

2.2 Process Alarming

Application of instruments to monitor the solids process can provide operators with alarms before an anomaly becomes a critical issue. Implementing instrumentation systems to provide protective or preventative alarms can provide a significant cost savings through early detection, which can help utilities avoid costly equipment maintenance.

2.3 Improving Process Efficiency

Many of the solids processes used in treatment systems require consistent monitoring and timely process control adjustments for efficient treatment. The use of instruments to measure critical process data for automatic adjustment provides repeatable process results. For example, total suspended solids (TSS) analyzers used to measure solids content in the return and waste activated sludge flows provide information to optimize the solids retention times within the process leading to improved liquid and solids treatment efficiency.

TABLE 22.1 Instrumentation matrix.

Solids Process Design and Management **Chapter**	**Flow**						**Pressure**						**Temperature**										
	Magnetic	**Differential pressure**	**Mechanical (rotary or variable-area)**	**Area velocity (Doppler, ultrasonic, or magnetic)**	**Coriolis**	**Thermal dispersion**	**Pressure**	**Gauges**	**Capacitance and impedance**	**Differential-pressure**	**Sonic and ultrasonic**	**Microwave (radar)**	**Thermocouple**	**Resistance**	**Thermister**	**Polymer**	**Nitrogen**	**pH**	**Phosphorus and orthophosphate**	**Gas analyzers (oxygen, combustible, hydrogen sulfide)**	**Solids—microwave density**	**Solids—near infrared**	**Sludge blanket level**
Chapter 01																							
Chapter 02																							
Chapter 03																							
Chapter 04																							
Chapter 05																							
Chapter 06	X						X	X			X	X											

Chapter 07	X		X		X	X	X	X	X	X	X	X			X		X		X		
Chapter 08	X		X	X	X	X	X	X	X		X		X		X		X			X	X
Chapter 09																					
Chapter 10	X		X			X	X	X		X	X			X					X		
Chapter 11	X		X			X	X	X		X	X			X					X		
Chapter 12	X			X			X	X		X	X		X	X	X		X		X	X	X
Chapter 13	X		X			X	X	X			X			X					X		
Chapter 14	X						X	X			X	X	X				X				
Chapter 15	X		X			X	X	X		X	X	X	X	X					X		
Chapter 16	X	X	X			X	X	X					X	X					X		
Chapter 17	X	X					X	X		X	X		X	X					X		
Chapter 18	X	X					X	X		X	X		X	X					X		
Chapter 19	X	X	X	X		X	X	X		X	X		X	X		X	X		X		X
Chapter 20	X		X			X	X	X		X	X			X		X		X	X		
Chapter 21	X						X	X		X	X					X	X		X		X
Chapter 22																					
Chapter 23																					
Chapter 24																					
Chapter 25																					
Chapter 26	X	X	X	X		X	X	X		X			X	X		X	X		X		

2.4 Leveraging Operational Resources

Solids processing facilities must collect process data for multiple purposes including for regulatory compliance, monitoring of plant processes, and improving operating efficiency as discussed above. The collection of this data can help to leverage the owner's limited operational resources including labor, chemicals, and energy. The consolidation of the data into one control and monitoring system allows operators—via computer-based workstations located strategically on or off the plant site—to monitor, review, inspect, and adjustment process operations without physically going to each individual process. This reduces the amount of time operators spend monitoring and controlling a system, allowing them to take on additional duties and responsibilities. Control systems can automate tedious tasks, such as chemical dosing, to save operational resources.

3.0 JUSTIFICATION AND DESIGN CONSIDERATIONS

Instrumentation that can help reduce labor needs or chemical or energy consumption are justified primarily from an economics viewpoint. Economics may, however, be a secondary consideration in decisions to install instrumentation. For instance, instrumentation used to verify compliance with discharge requirements may be justified for noneconomic reasons. The information provided by the instrumentation may be essential for planning new facilities or for improving existing facilities. Such information may also be required for monitoring treatment results for reports to various government agencies. Economic considerations will be secondary for those systems requiring continuous monitoring to protect operating personnel. Economic analyses of instrumentation, when required, must include both capital and operation and maintenance (O&M) costs. These O&M costs can be high, especially in solids management, where the materials being measured are typically solids-laden and possibly corrosive. Effective application of instruments for solids processing and management can be challenging because of the harsh environment associated with these processes. In some instances, regulatory requirements, such as those associated with *NFPA 820 Standard for Fire Protection in Wastewater Treatment and Collection Facilities*, will determine the need for instrumentation devices (NFPA, 2008).

4.0 AUTOMATION AND PROCESS CONTROLS

Virtually every wastewater management conference now includes presentations documenting how instrumentation and automation has improved plant process performance. As solids processes become more complex and require constant monitoring to remain within designed operational parameters, automated control systems are becoming a necessity.

Control systems are becoming more reliable, cost effective, and easier to use as technologies progress. These systems can oversee repetitive tasks such as batching of chemical day tanks, monitoring process flows, and controlling chemical pumps to maintain chemical dosage. This control reduces the amount of time operators spend monitoring and adjusting the process to maintain set points, thus, allowing them more time for other responsibilities. Automated control also reduces process variability by eliminating individualized control by each operator. Adding instrumentation, especially sophisticated analytical instrumentation, however, may require additional technicians for device maintenance.

5.0 SOLIDS PROCESSING AND MANAGEMENT

5.1 Flow Meters

Several types of flow meters are available: magnetic, ultrasonic, flume and weir, differential-pressure, mechanical, and mass-flow. Flow meters should be sized to handle the flow that will exist at wastewater treatment plant (WWTP) startup or the first few years of operation. If they are sized to handle future maximum flows, then they may be too large to accurately register lower initial flowrates. Flow meters have specific installation criteria for accurate operation; therefore, it is important that manufacturer's installation requirements and constraints be incorporated to the design.

Flow meters should not be installed immediately downstream of pumps or throttling valves.

Flow meters designed for closed conduits need to be continuously full of liquid unless designed specifically for partially filled applications. Mounting the flow meter in a vertical pipe section with upward flow helps to ensure that the pipe will remain full during operation. Horizontally mounted flow meters may require air-bleed valves if there is a tendency to trap air in the pipe. Pumps, pipe fittings, and other flow-disturbing obstructions must be located far enough away from the meter to provide an even profile.

All flow meters need enough surrounding space for calibration, maintenance, or removal. Considerations should be given during design to minimize disruptions to the process if the meter body needs to be removed for service. Consideration should also be given to providing a means of future calibration and maintenance.

5.1.1 *Magnetic*

Magnetic flow meters ("magmeters") are the most popular method of measuring full-pipe flow in WWTP applications. They typically have a long life, are cost-effective, and are accurate and precise when properly installed.

5.1.1.1 *Operating Principles*

Magnetic flow meters are based on Faraday's principle of electromagnetic induction, in which the voltage induced by an electrical conductor moving through a magnetic field is proportional to the conductor's velocity (Figure 22.1).

Another type of magnetic flow meter is the insertion probe. These operate on the same principle as the full-body flow-tube type but do not actually sample the full flow path through the pipe. The insertion probe can sample multiple points across the diameter of a pipe or as a single sample on the tip of the probe. The probes typically

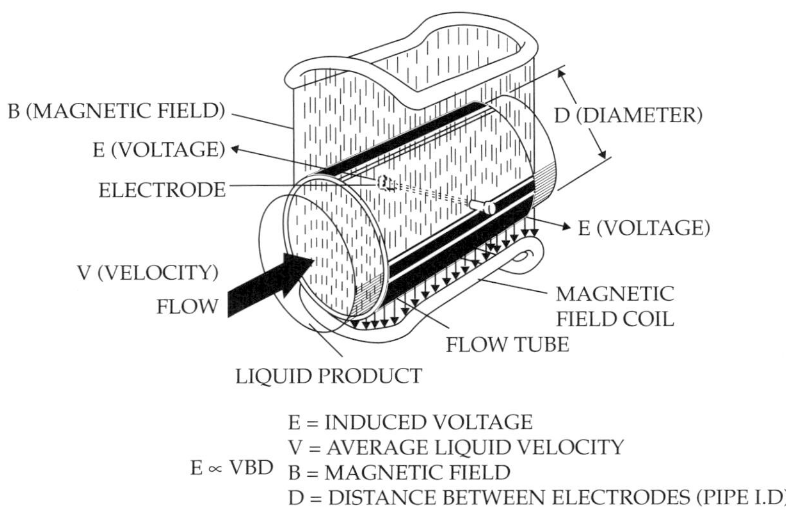

FIGURE 22.1 Faraday's principle of electromagnetic induction (© Copyright 2006 Omega Engineering, Inc. All rights reserved. Reproduced with the permission of Omega Engineering, Inc., Stamford, CT 06907).

have an alternating current coil-and-electrode design and are inserted into the process pipe through a fully ported ball-valve tap (insert meters may not be suitable for WWTP measurement where the process fluid has suspended solids or stringy material that could foul the sensor).

5.1.1.2 *Construction Materials*

Magnetic flow meter electrodes are available in platinum, Alloy 20, Hastelloy C-22, Monel, nickel, platinum-alumina ceramic, platinum-rhodium alloy, 316 stainless steel, Tantalum, titanium, and tungsten carbide. Some manufacturers offer optional electrode shapes (Figure 22.2) and self-cleaning versions (via heat, ultrasound, or polarity switching) to help avoid fouling. Field-replaceable electrodes are also available from some manufacturers. The meter liners are available in ceramic, fiber glass, hard rubber, Kynar, perfluoroalkoxy (PFA), polyurethane, soft rubber (EPDM), Teflon (PTFE), and vitreous enamel; the choice depends on the process fluid in contact with the meter.

Depending on piping material and application, grounding electrodes or grounding rings may be required. Cathodically protected piping systems may require special powering and grounding practices.

5.1.1.3 *Accuracy and Precision*

Magnetic flow meters are available in both an AC- and DC-coil type. Under typical field conditions, pulsed direct current flow meters are accurate to within 0.2% of full

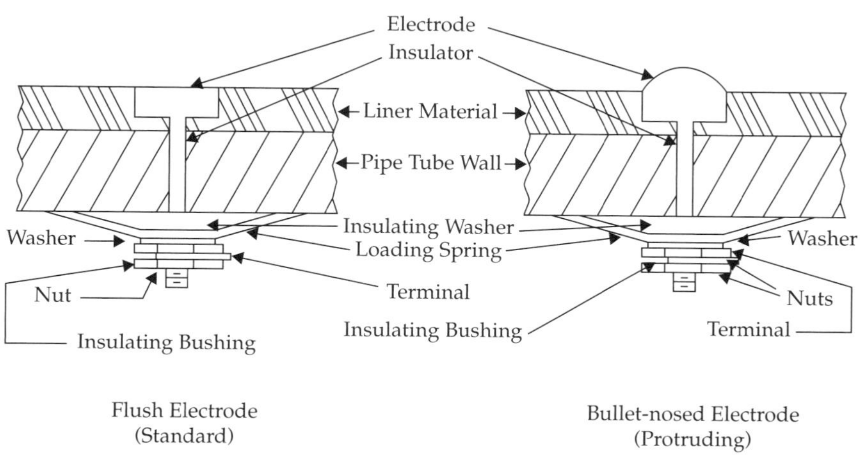

FIGURE 22.2 Optional electrode shapes (WEF, 2006).

scale. The alternating current flow meters are typically accurate to within 1% of full scale. Most full-body magnetic meters can read flow accurately in either direction. Figure 22.2 illustrates optional electrode shapes.

Magnetic flow-meter performance can be degraded if misapplied. For example, flow-disturbing obstructions such as elbows, reducers, and valves that are located too close to the meter inlet and outlet may add some error or uncertainty to the measured flow. Accuracy can also be affected by pipe occlusions/narrowing because of grease deposition or solids buildup in the meter.

Performance can be severely degraded if a flow meter is improperly oriented so the pipe only partially fills (trapping gases) or if material builds up on the electrodes. When this happens, some flow meters may have irregular readings, and others may not report flow at all.

Full-body magnetic flow meters are best suited for applications where the process fluid has a conductivity greater than 5 μS/cm (5 μmho/cm) and solids concentration is less than 10% (by weight). Special low-conductivity meters are available which will function with as little as 0.05 or 0.1 μS/cm.

Magnetic flow meters are not suitable for nonconductive liquids, gas streams, streams with powdered or granular dry chemicals, or partially full pipes (unless the meter is specifically designed for such applications).

5.1.2 Differential Pressure

Differential-pressure flow meters are based on the Bernoulli principle, which states that when a fluid passes through a flow-tube restriction, it causes a predictable pressure drop that is proportional to the square of the flowrate. Differential-pressure flow transmitters used with Venturi flow tubes or orifice plates typically are used for measuring flow in clean water applications such as WWTPs. They are not suitable for most wastewater plant sludge and solids metering applications. Differential flow meters are discussed in the *Instrument Engineers Handbook* (Liptak, 1999) and *Automation of Wastewater Treatment Facilities* (WEF, 2006).

5.1.3 Mechanical

Mechanical flow meters, such as rotary-element (e.g., turbine or propeller) and positive-displacement meters, operate on the principle that an object moved by water or air can be correlated to the flowrate of the fluid or gas.

Many mechanical flow meters are being replaced by magnetic flow meters, which are less expensive and easier to maintain. Positive-displacement flow meters are still used for billing-related applications because they are highly accurate, even if their headloss is high. This meter basically works by letting its internal cavity fill with fluid, rotating to discharge this fluid, and repeating the process.

5.1.4 *Variable Area*

Variable-area flow meters (rotameters) are small floats often used in manifolds to verify that throttling valves are properly dividing a chemical among various injection points and that the proper amount of chemical is being delivered to each location. They are popular for measuring low flows because they are inexpensive and allow operators to visually inspect the status of a process.

A variable-area flow meter typically consists of a tapered glass flow tube and a stainless-steel weight (float). To function properly, the tube must be vertical with the flow rising through it. The more flow through the instrument, the higher the float rises in the tube. A calibration plate mounted next to the tube allows operators to read the flowrate.

Because the float must be visible in the fluid, the meters can only be used in clean fluids such as air, drinking water, or some chemicals. Also, because glass and stainless steel have different thermal-expansion properties, the meters are limited to applications where the temperature is less than 93 °C (200 °F).

Most variable-area flow meters are factory precalibrated, assuming water or air with a specific gravity of 1.0 and a temperature of 20 °C (68 °F). To calculate actual flows of a substance with a different density, conversion equations are required.

5.1.5 *Area Velocity*

The area-velocity flow meter continuously measures both level and velocity to calculate flow volume in an open channel or pipe. An ultrasonic or pressure sensor is installed at the bottom of the pipe or channel to measure the liquid level. The sensor transmits ultrasonic pulses that travel through the water and reflect off the liquid surface. The instrument measures the time it takes for echoes to return to the sensor. Based on the speed of sound in water, the liquid level depth is measured. For pressure-type level measurement, the sensor converts the sensed head pressure into level. Flow velocity is measured with an ultrasonic Doppler signal continuously injected into the water. This high-frequency wave is reflected back to the sensor from particles or bubbles suspended in the liquid. If the fluid is in motion, then echoes return at an altered frequency proportionate to flow velocity.

Alternately, a separate down-looking ultrasonic level sensor can be used for highly aerated or turbulent-flow applications. It measures level by transmitting ultrasonic pulses through the air to the liquid surface with accuracy of ±0.25% of range. Along with the level sensor, a submerged Doppler velocity sensor is used to measure the water velocity. Most area-velocity flow meters will work in partially full and surcharged pipes, and in rectangular, trapezoid, and egg-shaped channels.

5.1.6 *Liquid Mass*

Liquid-mass flow meters directly measure the liquid-mass flowrate of a material (i.e., kilograms or pounds delivered in a unit of time). Some use density information to convert the output signal to a volumetric flowrate.

Liquid mass flow meters are not typically used in the wastewater industry because the instruments are sensitive to process conditions and do not handle fouling well. They can monitor difficult-to-measure process streams, however, such as chemical feed systems.

5.1.7 *Coriolis*

A coriolis flow meter is a small-diameter meter [up to 254 mm (10 in.) in diameter] that can measure liquids or gases by inducing a vibration and comparing it to the measured vibration of the material flowing through the meter to determine the mass of the material. The meters typically can measure multiple variables [e.g., mass flow, temperature, viscosity, and density (and, therefore, solids concentration or flow volume)] can be used in applications where temperatures are as high as 350 °C. The small tubes inside the meters result in a large pressure drop and may be prone to plugging. The meters may work well when measuring chemical process feed lines. Coriolis flow meters can measure the flow of nonconductive liquids or gases that cannot be measured with a magnetic flow meter. They can also measure gases, such as chlorine or ozone.

5.1.8 *Thermal-Dispersion Mass*

Thermal-dispersion mass meters use heat transfer and known characteristics of the fluid being measured to determine the mass of material flowing through the meter per unit of time.

A thermal-dispersion mass meter consists of a heater element and temperature transducer on two separate probes on the meter's head. It works by adding a known amount of energy to the heater element, and then measuring how much heat is dissipated as the process fluid flows over the temperature transducer mounted downstream from the heater element (the actual temperature of the fluid is relatively unimportant). If the fluid's thermodynamic constants are known, then the mass flowrate can be calculated. If the fluid's physical characteristics are reasonably constant, then the meter can also report volumetric flow. It is worthwhile to note that thermal dispersion flow meters are not sensitive to flow direction. In applications where flow direction may reverse, erroneous values will result. In addition, moisture content of the gas being measured can have an effect on the accuracy of thermal dispersion meters. In applications with high moisture content, erroneous values will result.

5.2 Level Measurement Devices

Level measurement instruments are available for continuous level measurement and single point sensing. Figure 22.3 shows some typical level instruments of each type.

5.2.1 *Continuous Level Measurement*

Continuous level-measurement instruments include bubbler, capacitance and impedance, differential-pressure, sonic, ultrasonic, and microwave (radar) (see Figure 22.3). Because the sonic, ultrasonic, and microwave devices do not involve direct contact with the product being measured, they typically are used to measure corrosive liquids or sludge. In addition to analog signals, most continuous level-measurement devices have options for alarm outputs and on–off control.

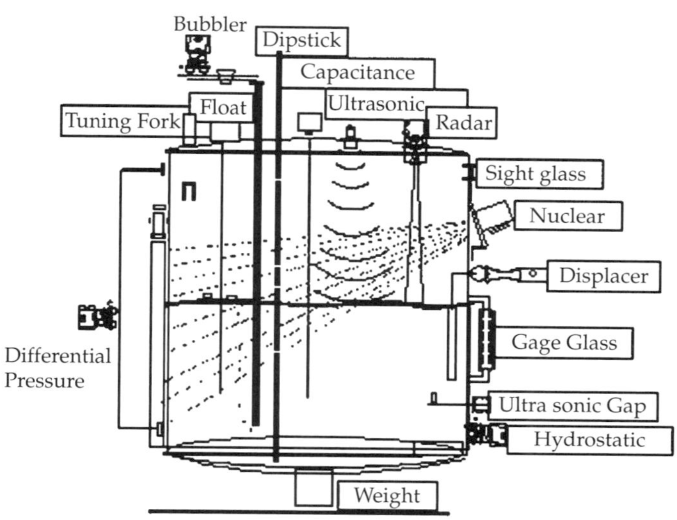

FIGURE 22.3 An example of one of the technologies available for level measurement (courtesy of the City of Roseville, California).

5.2.1.1 *Capacitance and Impedance*

Capacitance and impedance probes can be used to measure the level of liquids and some slurries and dry materials. They also can be configured to distinguish between interface levels in WWTPs. This is useful when a water surface level that requires monitoring is covered by a foam layer.

Some transmitters are characterized to produce a signal proportional to flow in flumes and weirs via electronic conversions.

5.2.1.2 *Differential-Pressure*

Differential-pressure (head-level sensing) instruments are typically used in WWTPs to measure levels of liquids and slurries; they are not effective for solids measurement.

Pressure-sensing instruments typically use solid-state electronic pressure sensors isolated from the process fluid being measured by a diaphragm. Diaphragms are flat or concentrically corrugated metal or ceramic disks that compress in response to pressure increases. As the diaphragm compresses, a connected electrical component (e.g., capacitor, strain gauge, resistor, or inductor) produces an electrical signal proportional to the differential pressure.

Diaphragm seal systems consist of a differential-pressure transmitter, a diaphragm seal, fill fluid, and direct-mount or capillary tubing (Figure 22.4). When process pressure is applied, the diaphragm is displaced, transferring the measured pressure through the fluid-filled system and the capillary tubing to the sensor.

Submersible pressure-sensing transducers are available for submersion directly into the tank to be measured. This assembly includes a breather tube that acts as the transducer's low-pressure atmospheric reference leg. It is routed above the maximum height of the measured substance and either vented into the enclosure or routed to a breather bag. It is typical to have a desiccant filter in the enclosure at the end of the vent tube to ensure that humidity cannot condense into the tube and make level readings inaccurate. Figure 22.4 is an example of a differential-pressure sensor with seal and capillary system.

5.2.1.3 *Sonic and Ultrasonic*

Sonic and ultrasonic level transducers do not touch the process fluid and can be used in many different WWTP processes.

Sonic and ultrasonic level sensors typically are based on the time-of-flight principle. A sensor sends pulses that travel through air and are reflected back by the surface of the measured substance. The instrument is calibrated to calculate the difference between the distance from the sensor to the level of the measured substance. Figure 22.5 shows an ultrasonic transducer mounting.

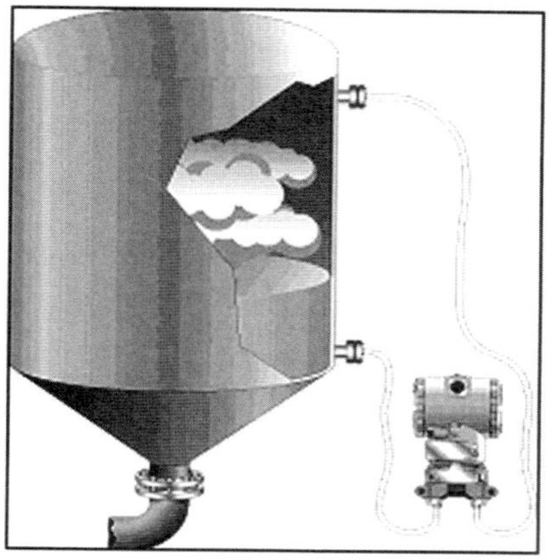

FIGURE 22.4 An example of a differential-pressure sensor with seal and capillary system (courtesy of the City of Roseville, California).

FIGURE 22.5 Typical ultrasonic level transducer mounting (courtesy of the City of Roseville, California).

Sonic and ultrasonic wave velocity depends on temperature, pressure (somewhat), and humidity (to a minor extent). If changing conditions are anticipated, then automatic temperature compensation can be provided (compensation is unnecessary for the other two factors). Sonic and ultrasonic level sensors can measure distances from 152 mm to 61 m (6 in. to 200 ft), depending on the sensor. Figure 22.5 shows a typical ultrasonic level transducer mounting.

The signal from an ultrasonic sensor can be attenuated via absorption in the vapor space, reflection away from receiver's sensing area, and absorption by foam on top of the measured substance. Foam on liquid surfaces or carbon-dioxide-rich vapor spaces (e.g., anaerobic digesters) may completely absorb ultrasonic waves. Ultrasonic sensors require a fairly quiescent (flat) surface at the measured material to obtain a reliable signal. In turbulent applications, a stilling well may be required.

Corrosion and freezing can also be problematic. Sensor heaters are available to prevent freezing. Proper material selection can reduce corrosion.

5.2.1.4 *Microwave (Radar)*

Although sonic, ultrasonic, and microwave level sensors may seem identical, microwaves measure the distance differently.

Two types of microwave level sensors are in use: frequency-modulated and guided-wave. The frequency-modulated level sensor involves a frequency-modulating, continuous-wave transmitter and antenna assembly. The device is mounted on top of the process vessel. The transmitter sends radar signals into the tank, and the antenna detects both those and the signals that reflect off the measured substance. The transmitted signal has a different frequency than the reflected one, and this difference is proportional to the level of the measured substance.

5.2.2 *Point Level Measurement*

Point level monitors include ultrasonic gap, float, frequency-shift tuning fork, thermal dispersion, hydrostatic, conductivity, capacitance, and inductance. Point level monitors can be divided into two groups: direct-acting switch devices and inference-based level switches.

Direct-switch devices (e.g., float and some hydrostatic) directly operate discrete switches via interaction with the process medium. For example, as the process level increases to the point of contact with a ball float, its buoyancy will cause it to raise, which will open or close a switch (depending on the ball-float configuration).

Inference-based level switches use electronic circuitry to sense a change when a sensing element touches the measured substance. Thermal dispersion technology measures the heat loss (cooling effect) of a fluid flowing across a heated sensor. A typical sensor uses two resistance temperature detectors (RTDs) that are sheathed in thermowells and separated by a gap. Heat is applied internally to one RTD, creating a differential temperature between the two. Immersing the measuring element into a liquid directly affects the extent to which heat is dissipated and, in turn, the magnitude of the temperature differential between the RTDs. This differential is converted into an electrical signal that can be used to trip a relay in level or interface-level switch applications.

5.3 Pressure Measuring Devices

Typical uses for pressure sensors include monitoring pressure in compressed-air distribution systems, discharge of suction pumps, and pressure vessels. With diaphragm isolation systems, pressure sensors can be used in many WWTP processes. Diaphragm seals can be connected to capillary tubing so the instrument can be installed at a convenient location for maintenance.

Pressure-sensing instruments typically use some type of solid-state electronic pressure sensor isolated from the process fluid being measured by a diaphragm. Diaphragms are flat or concentrically corrugated metal or ceramic disks that compress in response to pressure increases. As the diaphragm compresses, an electrical component (e.g., capacitor, strain gauge, resistor, or inductor) connected to it produces an electrical signal proportional to the differential pressure.

Diaphragm seal systems consist of a pressure transmitter, a diaphragm seal, fill fluid, and direct-mount or capillary tubing (Figure 22.4). When process pressure is applied, the diaphragm is displaced, transferring the measured pressure through the fluid-filled system and the capillary tubing to the sensor.

5.4 TEMPERATURE MEASURING DEVICES

Temperature sensors typically include thermocouples and RTDs (see Table 22.2). Thermocouples are rugged and inexpensive and are typically used for high-temperature measurements such as in sludge incinerators. Resistance temperature detectors are used for lower temperature ranges and when accuracy is important.

5.4.1 Thermocouple

Thermocouples operate on the Seebeck effect, which describes the flow of current caused by heating a circuit junction composed of two dissimilar metals. The open-

Table 22.2 Temperature measurement application guidelines.

Measurement type	Temperature measurement application	Application comments
Thermocouple	All	Good range; less cost
Resistance temperature detector	0 to 300 °C	Stable; more expensive and more accurate than thermocouple

circuit voltage is a nonlinear function of the junction's temperature and the metal compositions. All junctions of dissimilar wires produce the Seebeck effect; junctions of dissimilar metals in the measuring circuit, other than those at the point of interest, must be compensated for or prevented. For example, thermocouple extension wires may be made of the same metal as the thermocouple or maintained at a known temperature so unwanted voltage can be taken into account. All thermocouple circuits have at least two junctions: one at the point of measurement and the other at the point of connection to copper leads, which are then routed to a device that measures the voltage caused by the junction of interest. Refer to Table 22.2 for temperature measurement application guidelines.

5.4.2 *Resistance*

Resistance temperature detectors operate on the principle that the resistivity of a metal changes with its temperature. By measuring the resistance at a point of interest, a temperature can be determined.

Resistance temperature detectors consist of a thin wire or film supported by a substrate with matching thermal-expansion characteristics. The thin metal element is necessary to obtain adequately high resistance [typically, approximately 100 ohms at 0 °C (32 °F)]. The element is supported so the RTD is rugged enough for process measurement. The temperature-detecting elements may be placed in glass or ceramic bobbins, embedded in glass, or mounted on ceramic substrates of various shapes.

5.5 Online Analyzers

Online analyzers are instruments that, when properly calibrated and installed, allow continuous measurement of a parameter that can otherwise only be measured by a time-consuming grab sample and laboratory test. Instrument output is typically calibrated to match a laboratory result.

5.5.1 *Chlorine Residual Analyzers*

The disinfection potential of chlorine can be measured in one of two ways: chlorine residual analyzers and oxidation reduction potential analyzers. Chlorine residual analyzers are available to measure either free chlorine, combined chlorine, or total chlorine. The analyzers for these three variables are similar but may use different chemical reagents to measure the desired parameter. For most WWTP applications, use of sample conditioning (filtering) system is required to remove suspended solids and other particles that could foul the analyzer and result in unreliable readings. Depending on brand, chlorine analyzers use different technologies to detect chlorine values. These technologies include amperometric/polarographic sensors, colorimetric sensors, and gas stripping sensors.

5.5.2 *Oxidation–Reduction Potential*

Oxidation–reduction potential (ORP) analyzers measure the water's potential to oxidize (or reduce) organic matter or compounds in the water. The potential is generated by the oxidizing or reducing agents in the sample (chlorine or sulfite).

Oxidation–reduction potential is a potentiometric measurement in which the potential (tendency) of a medium to transfer electrons is sensed by an inert metal electrode and read relative to a reference electrode immersed in the same medium, according to *Standard Methods for the Examination of Water and Wastewater* (APHA et al., 1998). It also is called a "redox" (reduction–oxidation) measurement. The sensor's readout is a voltage (relative to the reference electrode), with positive values indicating an oxidizing environment (ability to receive electrons) and negative values (e.g., –300 mV) indicating a reducing environment (ability to lose electrons).

Knowing ORP is useful in waters with a relatively high concentration of reducing or oxidizing substances [e.g., chlorine (a strong oxidant) and sulfite (a strong reductant)]. However, ORP is a nonspecific measurement (the measurement notes the effect of all dissolved species in the medium) that is affected by multiple parameters (e.g., temperature, nitrification, and denitrification).

The ORP reading must be related to some known value (e.g., a known oxidant residual). Use of the signal for chlorination and dechlorination requires measuring the ORP value before chlorine is added and after the dechlorination chemical is added. Currently, continuing interest and research is occurring for using ORP to control nutrient removal processes at WWTPs.

5.5.3 *Streaming Current Analyzers*

When the correct polymer dose is added to a treatment process, the zeta potential of the water changes, destabilizing colloids and allowing flocculation to allow easier removal of suspended solids. It is not possible for operators to measure residual polymer in wastewater to optimize a treatment process. A streaming current monitor (SCM), however, can monitor the charge characteristics (zeta potential) of a process flow and provide an output that can be used to control the coagulant dose.

Streaming current monitors have been used successfully in WWTPs, where tests have shown a strong correlation between polymer dose and centrate or filtrate charge characteristics (zeta potential). Monitors have been used in thickening and dewatering processes. Pilot testing is recommended to verify suitability for specific applications.

Streaming current monitors measure the electrokinetic charge in the wastestream. They respond to changes in wastewater characteristics (TSS, percent solids, blends, pH, conductivity, etc.) and flowrates. A typical SCM consists of a sensor, a monitor, a sensor maintenance module, and a controller (Figure 22.6).

The sensor consists of a plastic reciprocating piston in a plastic body with metal electrodes. When the sample is pulled into the measurement chamber (i.e., where the probe is located), colloidal particles attach themselves to the measurement probe. As the piston moves the water forward, the ions surrounding these particles are sheared

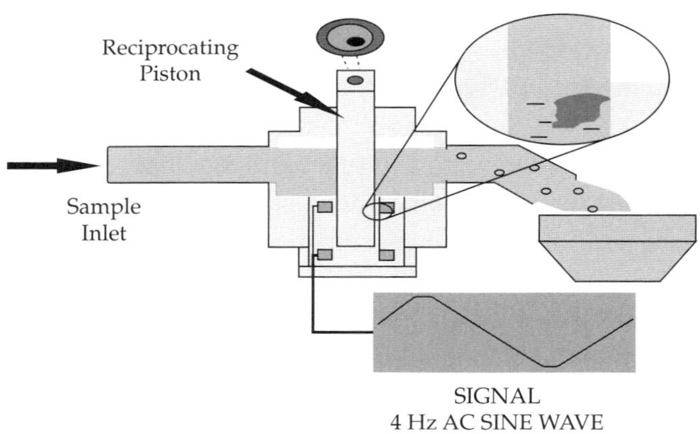

FIGURE 22.6 A diagram of a streaming current monitor unit (courtesy of Chemtrac Systems, Inc., Norcross, Georgia).

away by the water. Movement of the ions creates a voltage that is detected by the electrodes in the probe. The transmitter tracks the charge of the ions and the amplitude to convert the signal to usable information that indicates whether coagulant feed is sufficient, underfed, or overfed.

The sensor maintenance module operates the cleaning water solenoids, which are connected to clean plant water or potable water. The solenoids flush both the instrument and the sample supply line when required. Flushing them every 15 minutes helps prevent the accumulation of debris that could lead to blockages. Cleaning agents, acids, or bases may be added to the flush water to prevent precipitate or grease from accumulating on the electrodes. Figure 22.6 shows a diagram of a streaming current monitor unit. In areas where pH or conductivity changes are known to occur or sludge is fed from multiple sources, SCM field trials should be performed before plans for a permanent installation are made.

5.5.4 *Nitrogen*

Most WWTPs must comply with a permit limit typically expressed in terms of ammonia as nitrogen (NH_3-N). Ammonia is a gas, however; when dissolved in wastewater, it exists in equilibrium with the hydrated ion, ammonium (NH_4).

The relationship between dissolved ammonia and ammonium is pH-dependent. As pH increases, the equilibrium is driven toward dissolved ammonia. At greater than a pH of 11, all of the ammonium is converted to dissolved ammonia. Under normal circumstances, however, almost all of the dissolved ammonia will escape from the liquid, leaving only ammonium ion. In a WWTP, the ammonium concentration is critical.

Depending on the measuring principle used, manufacturers may express results as either ammonia or ammonium. An ion-selective electrode (ISE)-based instrument, for example, measures ammonium. A gas-sensitive electrode (GSE)-based instrument, however, actually measures ammonia gas (after raising the sample's pH to convert all the ammonium to ammonia). The best way to avoid confusion is to express the results in terms of nitrogen (e.g., NH_3-N or NH_4-N). Ammonia is typically monitored in processes such as plant influent, aeration tanks, plant effluent, and supernatant of dewatered biosolids.

Three principal methods are used for measuring ammonia with an online instrument: ISE, GSE, and colorimetry. Ammonium ISEs are similar in principle to a pH electrode, can measure over a wide range, and have a fast response time. However, they are subject to interferences and operate over a limited pH range.

Gas-sensitive electrodes are more selective than standard ammonium ISEs and typically are used in WWTP applications. An ammonia GSE is effectively a pH electrode behind a gas-permeable membrane. Sodium hydroxide is added to the sample to raise the pH above 11 and drive all the ammonium to dissolved ammonia. The ammonia gas is allowed to pass across the GSE membrane and redissolve in an electrolyte (typically NH_4Cl) surrounding a pH electrode. The ammonia gas decreases the electrolyte's pH. The analyzer then converts this change in pH to an ammonia reading.

Two colorimetric methods are used for measuring ammonia:

1. Indophenol blue method; and
2. Monochloramine-F method, in which hypochlorite is added to the sample to convert the ammonia to monochloramine. A second reagent containing an indicator is then added to the sample, and the resulting color development is measured via a colorimeter.

The method is similar to GSE. Sodium hydroxide is added to the sample to raise the pH above 11. The ammonia gas is then stripped from the sample and redissolved in a liquid pH indicator solution. The change in the indicator's color is measured via a colorimeter and converted to an ammonia reading. Some instruments measure at multiple wavelengths, providing compensation for light scattering by particles in the sample and variations in the sample color.

5.5.5 *pH*

In WWTPs, pH sensors are used to monitor plant conditions, track biological treatment process conditions, and control acid–base additions for pH adjustment. Regulators require the pH of WWTP influent and effluent to be measured and reported. Specific industrial discharges also may require pH monitoring to provide advance warning of possible toxic conditions. Although some biological processes (e.g., activated sludge systems) can tolerate a pH variance of 5 to 9, others are pH-sensitive. The anaerobic digestion process, for example, requires pH to be between 6.6 and 7.6; the process fails when pH is less than 6.2. Therefore, the pH of anaerobic digester liquor should be monitored. Online monitoring of digester pH is not recommended, though, because the solids will continually foul the sensor; periodic sampling is preferred.

An online pH monitor can provide feedback in other processes requiring pH adjustment. For example, pH adjustment may be required to neutralize low-pH industrial wastes, enhance phosphorus removal by alum addition, or optimize nitrification or denitrification processes.

Figure 22.7 shows a typical pH sensor arrangement. The heart of the sensor is a glass membrane, across which an electrical potential is generated. This potential varies as pH changes. The difference between this potential and a reference electrode is measured and amplified by an electronic signal conditioner. The complete electric circuit includes the glass electrode wire, glass membrane, process fluid, reference electrode fill solution, and reference electrode wire.

5.5.6 *Phosphorus and Orthophosphate*

Most National Pollutant Discharge Elimination System (NPDES) permits note phosphorus limits in terms of total phosphorus. Online total phosphorus monitors are available but are relatively expensive and require a lot of maintenance. Therefore, orthophosphate (PO_4) monitors are more widely accepted. An analyzer measuring 3 mg of PO_4 is measuring the equivalent of roughly 1 mg of phosphorus; an orthophosphate analyzer measuring 0.3 mg/L PO_4 is equivalent to a total phosphorus analyzer measuring 0.1 mg/L of phosphorus. Expressing orthophosphate values as PO_4-P avoids confusion.

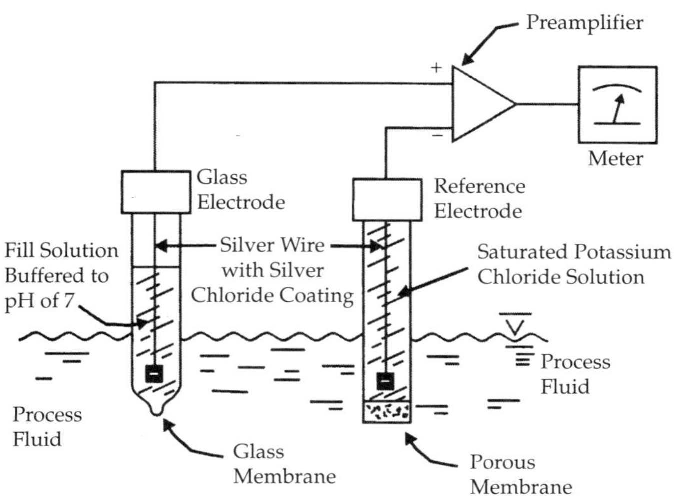

FIGURE 22.7 Typical pH sensor arrangement (WEF, 2006).

Online orthophosphate instruments are used to monitor plant influent for feed-forward control of chemical dosing and flow balancing; aeration basin effluent to ensure that phosphorus uptake was adequate; and final plant effluent as a surrogate for total phosphorus. Orthophosphate instruments also are used to monitor anaerobic effluent to ensure adequate release of phosphorus from phosphorous accumulating organisms (PAOs). In addition, they can be used to monitor and control phosphate release by controlling mixed-liquor nitrate recirculation back to the anaerobic zone.

All online orthophosphate instruments use colorimetry. The two colorimetric methods typically used are the molybdovanadate method (also called the yellow method) and the ascorbic acid method (also called the blue method or the molybdenum blue method). The yellow method is typically used because of the relative simplicity of the instrument and related reagents. When very low detection levels (less than 0.1 mg/L of PO_4-P) are required, the blue method is used.

Because of their inherent complexity (and corresponding expense), online total phosphorus analyzers typically are only used to monitor the final effluent at WWTPs. Many plants prefer to use orthophosphate analyzers as a surrogate parameter in final effluent.

Online total phosphorus instruments first convert polyphosphate and organic phosphorus compounds to orthophosphate. They then measure orthophosphate colorimetrically (typically using the molybdenum blue method). The conversion step requires high pressures and temperatures, which is why online total phosphorus instruments are relatively expensive and maintenance-intensive.

5.5.7 *Gas Detector/Analyzers*

Gas detectors frequently are used in WWTPs to monitor the atmosphere for potentially hazardous gases, and thus provide an alarm to warn plant staff of potential danger. Gas detectors typically are connected to a visual alarm light, audible alarm horn, or both located in the area being monitored. Simple detectors typically provide an alarm output when a preset gas concentration is detected in the atmosphere. More elaborate detector/analyzers can also transmit a signal showing the actual gas concentration being monitored. Most manufacturers recommend that gas detector/sensors be calibrated or checked on a routine basis to ensure proper operation. Several types of detectors typically found in WWTPs are discussed below.

5.5.7.1 *Combustible Gas*

Combustible gases that can be found in the atmosphere at a wastewater treatment plant include both methane and gasoline. Methane is produced by the decomposing or digesting of wastewater solids and can occur at various locations throughout a wastewater plant, especially in areas such as the influent headworks and solids dewatering/treatment areas. Methane is lighter than air and, as a result, detectors/sensors are typically located at a point high in the area being monitored. Gasoline can sometimes be present in the incoming wastewater to a plant and may be a concern for monitoring in the influent headworks and primary treatment areas. Gasoline is heavier than air, and, as a result, detectors/sensors are typically located low near the floor of the area being monitored. In areas where both methane and gasoline are monitored, multiple sensors at high and low points in the area are required.

Electro-catalytic diffusion-type sensors are typically used for detecting combustible gases; however, these sensors can be poisoned or damaged by hydrogen sulfide gases in the atmosphere. Thus, when high concentrations of hydrogen sulfide are anticipated, a type of sensors that uses infrared sensing technology is recommended. Infrared radiation (IR) sensors are not affected by hydrogen sulfide in the atmosphere. These sensors can be either a single-point detector or a "path"-type sensor that will monitor the entire area between a pair of sensors/transmitters.

5.5.7.2 Hydrogen Sulfide

Hydrogen-sulfide gases can be present in the atmosphere at most wastewater treatment plants, especially in the influent headworks and solids processing area. Even small amounts of hydrogen sulfide gas can be harmful to humans, especially if exposure occurs for long periods of time.

Sensors typically are one of two types: electro-chemical or metal oxide semiconductor. Because sensors are primarily intended to protect people, they typically are mounted at a height where an average height person would breathe, such as 1 to 2 m (4 to 6 ft) above the floor.

5.5.7.3 Oxygen Deficiency

Oxygen deficiency is a safety concern in areas with limited air flow or uncertain ventilation, such as the lower levels of pumping stations or equipment vaults and manholes. Sensors are typically electro-chemical type and will set off an alarm if the oxygen concentration drops approximately 1% below the normal expected ambient concentration.

5.5.7.4 Chlorine Gas

In areas where chlorine gas is stored or used, a gas detector is typically provided to detect a possible leak. Chlorine gas is heavier than air and detectors are located approximately 0.3 to 0.6 m (1 to 2 ft) above the floor. Detectors are typically located in the chlorine storage and feed rooms and anywhere that chlorine gas is piped or routed in pressurized piping. Chlorine detector systems are frequently connected to a safety valve system on the chlorine storage cylinders that will automatically close if a leak is detected. It is typical to design a chlorine-gas distribution piping system that operates under a vacuum. An eductor or similar device must produce enough vacuum to pull the chlorine gas from a vacuum regulator located near the storage cylinder to the point of use. This scheme significantly reduces the risk of a chlorine gas leak because a break in the piping will cause a loss of vacuum, causing the chlorine vacuum regulator to shut off the supply.

Sensors are typically the electro-chemical type. In addition to activating alarm lights and horns, the alarm condition may cause the chlorine supply valves to close, may start or stop a ventilation system, and may start a chlorine gas scrubber/neutralization system.

5.5.8 *Online Solids Measurement*

Online solids analyzers are sometimes used to monitor solids-bearing fluids in WWTP processes. Many instruments use near-infrared technologies, although instruments based on microwave, nuclear, or ultrasonic technologies are also available. Online solids analyzers work by developing an electronic correlation between a sludge's characteristics and actual solids concentrations. This correlation is established during calibration, which compensates for some variation in solids characteristics. If solids characteristics vary significantly, however, then so do the correlation constants, and recalibration typically will be required. Most optic-based instruments, for example, are sensitive to size distribution and significant color changes. Other instruments may be sensitive to air bubbles, conductivity, or precipitate buildup. This section describes near-infrared and microwave analyzers. Nuclear and ultrasonic solids analyzers have been used at WWTPs but are rare and, therefore, not discussed.

5.5.9 *Near-Infrared Analyzers*

Optical suspended solids analyzers typically consist of a controller or interface device that communicates with a sensing head that is either inserted into a pipe or suspended in a process tank. Two types of optical solids analyzers have been developed for use in at WWTPs: (1) transmittance (two- and four-beam alternating light); and (2) reflectance (single- and multiple-angle backscatter). Transmittance analyzers are suitable for light-colored (brown) and light-concentrated sludge (typically return activated sludge or waste activated sludge). The transmittance meter measures the relative loss of light and converts the signal to a solids concentration. The four-beam version uses two emitters and two detectors, creating an array of light paths that produce a matrix of results. A ratio-metric comparison of these signals produces a reading that effectively eliminates interferences, compensates for fouling, and allows measurement in harsh environments.

Reflectance analyzers can accurately measure a wider range and up to 5% solids concentration depending on sludge type. The multiple-angle version consists of a light source and two detectors positioned with their optical axes at 90 and 140 deg from each other (Figure 22.8).

The emitted light is backscattered by solids particles and received by the detectors. Detectors have filters to reduce interferences from ambient light sources and to compensate for color changes. Sensor cleaning systems should be an integral part of the optical solids meters. Water or air purging, ultrasonic vibration, and wipers are methods used to keep the light source and detector windows clean. Multiple outputs from the detectors are used by the instrument processor (transmitter) to determine solids concentration.

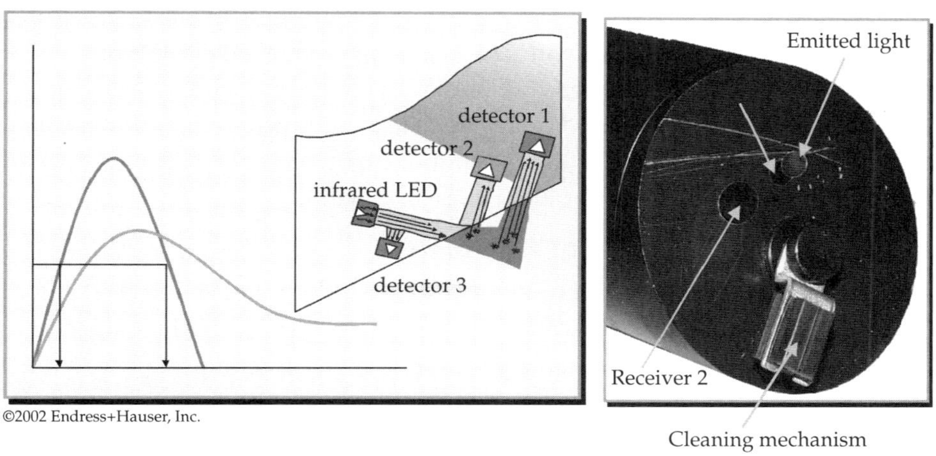

FIGURE 22.8 Illustration of how backscattered light principle instrument works (courtesy of Endress + Hauser, Greenwood, Indiana).

All optical instruments are sensitive to environmental conditions that may interfere with output over time. Possible problems include interference from entrained air, grease or fiber accumulation, changes in water or sludge color, and changes in sludge particle size. Most optical instruments have internal software to compensate for sensor head fouling or changes in sludge color or size, but some manufacturers are better at it than others. Unfortunately, these types of analyzers may not be good fit for streams that are black (i.e., primary and digested sludge).

5.5.10 *Microwave Analyzers*

Online microwave analyzers typically consist of a flanged spool piece installed in the solids/sludge pipeline. They measure total solids based on the measured phase difference of microwaves passing through the sludge. They operate on the principle that when microwaves go through a substance, the physical density can be determined by measuring the phase lag of the waves. The phase difference correlates to density, which can be directly related to the total solids concentration after compensation for ambient conditions.

Online microwave analyzers are linear and easy to calibrate. They are also typically expensive ($20,000 USD). Correction factors for span, conductivity, and slope and intercept of the solids density are used to calibrate the instrument based on grab-sample percent solids results. Online microwave analyzers work well even without a formal calibration, however, and are not overly sensitive to upstream or downstream fittings. Color and color changes do not affect the reading. The analyzers are sensitive to conductivity, changes in conductivity and temperature, air entrainment, and partially filled pipes. To prevent gas bubbles from interfering with measurement, the sensing element must be full and under pressure at all times. Onboard instrumentation can compensate for many of these parameters, but engineers should advise the manufacturer of process characteristics before installation.

5.5.11 *Interface/Sludge Blanket Level Analyzers*

An interface/sludge blanket level (ISBL) analyzer (sludge blanket detector) automatically identifies and reports the depth of the sludge blanket in a clarifier and a gravity thickener. It can replace the manual core sampler and, more importantly, eliminate the variability of the manual readings.

Ultrasonic ISBL analyzers are similar to the level sensors used for tanks or channels. In other applications, the sound waves travel through air and bounce off the top of the liquid. In clarifiers, the sound waves travel through water and bounce off the top of the sludge blanket.

The transmitter electronics typically are rail-mounted over the point where the blanket needs to be measured on a rigid or hinged mount with its face submerged. The distance between the sensor face and the tank floor is measured immediately after installation and keyed into the analyzer. The blanket height is calculated as the difference between this distance and the echo depth to the top of the blanket, similar to a level transmitter.

An ISBL analyzer typically is used to measure the height of a variable sludge blanket, develop a suspended solids profile in addition to measuring blanket height, and sound an alarm when a sludge blanket reaches a predetermined depth. The choice of technology depends on the use and the plant's control strategy.

5.5.12 Vibration

Vibration sensors are used to monitor the mechanical condition of pumps, motors, compressors, blowers, centrifuges, and other rotating and reciprocating devices. They can detect gradual or sudden mechanical failures and abnormal process conditions that are stressing the machine.

Vibration sensors typically consist of two or three main components: a transducer (sensor), signal conditioners, and a computer and software (if necessary). The transducers are mounted on the case or inside the monitored machine, typically as close as possible to the anticipated source of vibration.

The three types of vibration transducers are: velocity, acceleration, and displacement. Their outputs are oscillating voltage or current signals that are proportional to the vibration's velocity, acceleration, or displacement characteristics. The choice of transducer depends on the monitored machine's construction, the speed of its moving components, the mechanical conditions to be detected, and the conditions' vibration characteristics.

6.0 REFERENCES

American Public Health Association; American Water Works Association; Water Environment Federation (1998) *Standard Methods for the Examination of Water and Wastewater*, 20th edition; American Public Health Association: Washington, D.C.; American Water Works Association: Denver, Colorado; Water Environment Federation: Alexandria, Virginia.

Endress and Hauser Consult AG (1998) *Document TIF*, 008 3.98; Endress and Hauser Consult AG: Reinach, France.

Instrumentation Testing Association (2002) *Online Interface/Sludge Blanket Level Analyzers Performance Evaluation Report*; Instrument Testing Association: Henderson, Nevada.

Liptak, B. G. (1999) *Instrument Engineers Handbook*; CRC Press: Boca Raton, Florida.

National Fire Protection Association (2008) *Standard for Fire Protection in Wastewater Treatment and Collection Facilities, NFPA 820*; National Fire Protection Association: Quincy, Massachusetts.

New York Department of Environmental Conservation (2006) *Biosolids Land Application Facility Annual Report* www.dec.ny.gov/docs/materials_minerals_pdf/arla.pdf (accessed August 2011).

Water Environment Federation (2006) *Automation of Wastewater Treatment Facilities*, 3rd ed.; Manual of Practice No. 21; McGraw-Hill: New York.

Water Environment Federation (2008) *Operation of Municipal Wastewater Treatment Plants*, 6th ed.; Manual of Practice No. 11; McGraw-Hill: New York.

Chapter 23

Land Application and Product Distribution

(*continued*)

(*continued*)

1.0 INTRODUCTION

1.1 Overview

Land application takes advantage of the soil conditioning and fertilizing properties of biosolids by spreading or spraying the biosolids on the soil surface, or incorporating or injecting biosolids to soil. Land application, which has been practiced in many countries for centuries, uses the nutrients (e.g., nitrogen, phosphorus) and organic matter in these materials to grow crops or other vegetation and to improve soil quality.

Over the years, land application has been increasingly managed to protect human health and the environment from various potentially harmful constituents typically found in biosolids such as bacteria, viruses, and other pathogens; metals (e.g., cadmium and lead); toxic organic chemicals (e.g., polychlorinated biphenyls); and nutrients (e.g., nitrogen as nitrate). Management approaches have included regulatory measures; voluntary and mandatory pretreatment of wastewater or solids to improve quality (e.g., lower pollutant levels); and use of good management practices at land application sites (e.g., buffer zones, slope restrictions).

Biosolids are defined as wastewater solids that meet all regulatory requirements for beneficial use through land application (40 CFR 503, *Standards for the Use and Disposal of Sewage Sludge*; U.S. EPA, 1993). Increasingly, other beneficial uses are being found for biosolids including heat or energy recovery via incineration. This reflects an expanded view of biosolids as a renewable resource; these other beneficial uses are not addressed in this chapter.

The beneficial use of biosolids through land application is a dominant practice for solids use in the United States. According to a recent survey by the New England Biosolids and Residuals Association (2007), 55% of solids generated nationwide in

2004 were "applied to soils for agronomic, silvicultural, and/or land restoration purposes, or were likely stored for those purposes." Further, 74% of the material applied to soils was used for agricultural purposes.

This chapter reflects design information for the land application of biosolids and biosolids marketing/distribution. This information is an abridgement and update of the comprehensive U.S. Environmental Protection Agency's (U.S. EPA's) *Process Design Manual: Land Application of Sewage Sludge and Domestic Septage* (1995). Other key U.S. EPA documents that serve as useful supplements to this design manual include

- *Environmental Regulations and Technology: Control of Pathogens and Vector Attraction in Sewage Sludge* (1992);
- *Standards for the Use or Disposal of Sewage Sludge*, 40 CFR Part 503 (1993); and
- *A Plain English Guide to the EPA Part 503 Biosolids Rule* (1994a).

Additionally, documents developed by the National Biosolids Partnership (NBP)—a collaboration of U.S. EPA, the Water Environment Federation (WEF), and the National Association of Clean Water Agencies (NACWA)—provide solid guidance for the development and implementation of biosolids land application programs. Key among these is the *National Manual of Good Practice for Biosolids* (NBP, 2005). Both the manual and Chapter 3 of this manual address public outreach, a critical component of any successful land application program.

1.2 Regulations

The 40 CFR 503 *Standards for the Use and Disposal of Sewage Sludge* (discussed extensively in Chapter 2, Appendix A) set forth requirements for the land application of biosolids, but the rule allows states to impose more stringent requirements. Many states and local jurisdictions have restrictions on biosolids use beyond those specified in Part 503, including

- Requirements for plans that specify where and how biosolids will be applied;
- Requirements for site-specific land application permits. On a state level, this may require obtaining a permit to meet State Environmental Protection Act (SEPA) requirements, if applicable;
- More stringent pollutant limits, setback requirements, or other measures.

Many states also have regulations requiring nutrient management plans (NMPs) for the land application of biosolids. These requirements vary widely, with some states focusing on management of the nutrient nitrogen, and others incorporating comprehensive management requirements for phosphorus. Because state and local requirements can vary widely, it is important to identify regulatory requirements at these levels.

1.3 Nutrient Management Planning

As discussed in previous sections, nitrogen-based agronomic application rates are required under Part 503 rule (the approach for calculating nitrogen-based application rates is discussed in subsequent sections of this chapter). The more recent focus on phosphorus-based management is described below.

Approaches to phosphorus management vary nationwide. Some states specify in National Pollutant Discharge Elimination System (NPDES) permits that biosolids application rates may not exceed agronomic rates for phosphorus. Other states (for example Kansas) require that a certified crop advisor or other agricultural professional be consulted when soil test phosphorus (the soil phosphorus concentration) reaches a specified threshold (typically 150 to 200 ppm). But many states are moving to the use of a phosphorus index to manage phosphorus loadings on land application sites.

The phosphorus index is a risk management-based approach that considers transport and source factors in estimating the potential for offsite movement of phosphorus from a given land application site. The Natural Resources Conservation Service (NRCS) has developed phosphorus indices for various geographical areas. These tools were developed to help land managers assess the risk of offsite phosphorus migration for an individual site, but were not designed to determine compliance with water quality regulations. A phosphorus index typically generates a site rating based on transport and source factors, and uses the site rating to assign a vulnerability class (low, medium, high, very high) indicating the potential for offsite transport. This process uses the transport and source factors shown in Table 23.1.

Agronomic phosphorus loading limitations have the potential to increase land requirements two to three times beyond that required based on agronomic nitrogen loadings. The agronomic phosphorus loading rates typically place more severe restrictions on wastewater treatment plants (WWTPs) that use phosphorus removal because significant amounts of phosphorus leave the plant site as stored phosphorus in biosolids or plants that land apply biosolids in phosphorus-limited watersheds.

TABLE 23.1 Typical phosphorus index transport and source factors.

Transport factors	Source factors
Subsurface drainage	Soil test phosphorus concentration
Soil erosion (sheet and rill, wind)	Commercial phosphorus fertilizer application method
Runoff class	Commercial phosphorus fertilizer application rate
Irrigation-induced erosion	Organic phosphorus source application rate
Flooding frequency	Organic phosphorus source application method
Distance to surface waters/buffer width	

Some states that require P-based management will incorporate those requirements into a more encompassing nutrient management plan (NMP). An NMP describes the rates, timing, and placement of nutrients sources (including nitrogen and phosphorus) at a land application site. In many states, NRCS policies and standards determine NMP requirements. The NRCS *Title 190 General Manual Part 402* establishes a framework for state conservationists to develop their own practice standards, including NMP contents.

The development of such plans, if they are performed in accordance with NRCS guidance, is detailed in the NRCS' *Conservation Practice Standard: Nutrient Management, Code 590* (2006). Each state is responsible for supplementing Part 402 requirements to adjust for local conditions in its own 590 standard. Consequently, specific NMP elements will vary from state to state. Each state includes their Code 590 requirements in their individual field office technical guide (FOTG), which is available in electronic form and is known as the eFOTG. Access to individual state eFOTG documents is available through http://efotg.sc.egov.usda.gov//efotg_locator.aspx.

2.0 LAND APPLICATION AND PRODUCT DISTRIBUTION PRACTICES

The term "land application" typically refers to use of biosolids on agricultural fields; however, silviculture, reclamation, and product distribution are alternative land application practices that can be pursued. The following sections provide an overview of these practices.

2.1 Application to Agricultural Lands

Agricultural land application of biosolids is practiced in nearly every state. Hundreds of communities, both large and small, have developed successful agricultural land application programs.

Biosolids applied to agricultural land must be applied at a rate that is equal to or less than the "agronomic rate," which is defined in Part 503 as the rate designed to provide the amount of nitrogen needed by the crop or vegetation while minimizing the amount of nitrogen in the biosolids that will pass below the root zone of the crop or vegetation to the groundwater. The amount of available nitrogen applied to the site is based on that required by the crop. This amount of nitrogen might otherwise be applied to the site as commercial fertilizer by the farmer. By limiting nitrogen loadings to fertilizer recommendations, the effect on groundwater should be no greater than in agricultural operations using commercial fertilizers or manure; groundwater effects may even be less because of the agronomic rate requirement. Note that some states have also established phosphorus-based management requirements in addition to nitrogen limitations.

Section 23.5 of this chapter details both nitrogen- and phosphorus-based agronomic rate calculations for agricultural sites.

2.2 Application to Forest Lands

Except for certain areas in the Great Plains and the Southwest, forested lands are abundant and well distributed through most of the United States. Many WWTPs are located near forests; in fact, it is estimated that approximately one-third of the land within standard metropolitan areas is forested. Although 1% of land-applied biosolids are applied to forest sites, the application of biosolids to forest soils has the potential to become a significant use Biosolids have been land applied at forest sites in more than ten states on an experimental, field-scale level. The most extensive experience with this practice is in the Pacific Northwest.

Three categories of forest land may be available for biosolids application: recently cleared land before planting; newly established plantations (3 to 10 years old); and established forests. The availability of sites and application considerations for each type of site will determine which type of site or combination of sites is best for a forest land application program. The *Process Design Manual: Land Application of Sewage Sludge and Domestic Septage* (U.S. EPA, 1995) provides detailed guidance on the development of forest land application programs.

2.3 Land Application at Reclamation Sites

Surface mining of coal, exploration for minerals, generation of mine spoils from underground mines, and tailings from mining operations have created more than 1.5 mil. ha (3.7 mil. ac) of significantly disturbed land. The properties of these drastically disturbed and marginal lands vary considerably from site to site.

Biosolids can be used to return barren land to productivity or to provide the vegetative cover necessary for controlling soil erosion. A relatively large amount of biosolids (7 to 450 Mg/ha or 3 to 200 ton/ac) must be applied to a reclamation site to provide sufficient organic matter and nutrients capable of supporting vegetation until a self-sustaining ecosystem can be established. Because of these typically large, one-time applications of biosolids at such sites, the Part 503 rule allows biosolids application at reclamation sites to exceed agronomic rates for nitrogen if approved by the permitting authority. Surface water or groundwater monitoring may be required as a condition for biosolids application, however, if deemed necessary.

Because biosolids typically are applied only once to land reclamation sites, an ongoing program of biosolids application to disturbed lands requires that a planned sequence of additional sites be available for the life of the program. This objective may be achieved through arrangements with land owners and mining firms active in the area or through planned sequential rehabilitation of existing disturbed land areas. Once a site is reclaimed, biosolids can be applied to the site in compliance with the requirements for the type of site it may become (e.g., agricultural or forest land or a public site). For example, reclaimed areas may be used for crop production using agronomic rates of biosolids application. As with forest land application, *Process Design Manual: Land Application of Sewage Sludge and Domestic Septage* should be consulted when developing a land reclamation program using biosolids (U.S. EPA, 1995).

2.4 Distribution for Public Use

As indicated by NEBRA (2007), approximately 22% of biosolids that are beneficially used are distributed as a Class A "EQ" (exceptional quality) product in the United States. Class A "value-added" materials, such as compost and heat-dried product, typically are sold or given away for public use in bagged or bulk form.

Compost is often used as a top dressing by homeowners or turf managers, who typically spread the product at a depth of 6 to 13 mm (0.25 to 0.5 in.) on established lawns. For this application, the compost provides nutrients and organic matter to the soil as it degrades. Compost can also be mixed directly into planting beds to improve poor soils or mixed with other materials (such as sand) to create a top dressing or manufactured soil.

Like compost, heat-dried product can be used as a standalone product or mixed with other materials. Because of its typically higher nutrient content, however, it is often used as a fertilizer (where compost is used as a soil conditioner). Alone or mixed with other nutrients by fertilizer blenders, heat-dried product can be spread with conventional fertilizer spreaders on turf or agricultural lands.

Uses and product distribution considerations for these and other Class A products distributed to the public are further detailed in Section 6.0.

3.0 BIOSOLIDS CHARACTERISTICS

Determining the suitability of biosolids for land application or distribution is a necessary first step in planning and designing programs for those uses. A key component of that assessment is defining biosolids characteristics.

This section describes the properties of biosolids that might be land-applied or distributed to the public (additional data important to product marketing are discussed in Section 6 of this chapter). The information provided is intended, however, primarily for illustrative purposes. Although data are useful in preliminary planning and design, analysis of the actual biosolids to be land-applied or distributed is necessary for design purposes because biosolids characteristics vary significantly.

3.1 Biosolids Quantities

The amount of biosolids to be land applied will affect site evaluation and design in several important ways, including land area needs, size of transportation equipment and storage facilities, and cost. Quantities of biosolids available also will affect the selection of land application practices (i.e., application at agricultural, forest, reclamation, or public sites) and application rates and operating schedules.

Biosolids quantity can be measured in two ways: the volume of the wet biosolids, which includes the water and solids content; or the mass of dry solids. Biosolids volume is expressed as liters, cubic meters, or gallons; biosolids mass typically is expressed in terms of weight, in units of dry megagrams or dry tons (a U.S. measure also known as a short ton, equivalent to 2000 lb). Because the water content of biosolids can be high and variable, the mass of dry solids is typically used to compare biosolids with different proportions of water.

3.2 Physical Characteristics

Biosolids' physical textures and particle sizes are as varied as the processes that are used to create them. Heat-dried biosolids can be granular and irregular or spherical and similar in size to a small ball bearing. Composts are similar in texture to peat, but depending on manufacturing process, may contain some wood fibers or chips. Alkaline-stabilized materials can have a clay- or soil-like texture, depending upon moisture content and other factors. Digested biosolids can be applied to land in liquid form or as a clay-like dewatered material. Table 23.2 presents typical physical characteristics for these materials.

3.3 Chemical Characteristics

Historically, chemical characteristics of interest focused on metals and nutrients. Concern is growing, however, regarding the potential presence of pharmaceutical and other compounds. This section focuses on chemicals that affect biosolids management and briefly addresses chemicals of emerging concern.

3.3.1 Nutrients

Biosolids contain varying concentrations of macronutrients and micronutrients. Table 23.3 summarizes the typical primary nutrient (nitrogen, phosphorus, and potassium) contents of various types of biosolids as a percentage of dry weight.

TABLE 23.2 Typical biosolids physical characteristics (Lue-Hing et al., 1998; Logan, personal communication).

Property	Class A alkaline-stabilized	Compost	Digested liquid	Digested cake	Heat-dried
Total solids (%)	30–65	58	4	22	95
Volatile solids (% of total solids)	12–20	60	61	60	67
Bulk density					
(kg/m^3)	1040–1200	720	960	880	530–740
(lb/cu ft)	(65–75)12–20	(45)	(60)	(55)	(33–46)
Organic content (%)	–	50–75	60–75	60–75	78
pH					
	12–12.5	–	–	–	6.4–8.0

Table 23.3 Biosolids primary nutrient contents.

Nutrient content (as % of dry weight)	Class A alkaline-stabilized	Compost	Liquid	Cake	Heat-dried
Nitrogen	1.0	2.8	5.3	4.1	6.0
Total phosphorus	0.4	1.7	2.2	2.2	3.1
Potassium	0.3	0.3	0.3	0.2	0.3

As shown in Table 23.3, biosolids contain total nitrogen ranging from 1.0% of dry weight for Class A alkaline-stabilized biosolids and up to 6.0% for heat-dried biosolids. Primary and waste activated biosolids contain mostly organic nitrogen, in the form of protein; digested biosolids may have as much as one-third of the total nitrogen as ammonia (NH_3). Biosolids contain little if any nitrate (NO_3^-).

Biosolids typically contain total phosphorus contents ranging from 0.9 to 3.1%. The phosphorus in biosolids includes organic phosphorus, and various forms of inorganic phosphorus. Biosolids from treatment plants that use iron and aluminum salts for phosphorus removal will contain inorganic phosphorus in the form of iron and aluminum phosphates; those that use lime for dewatering or for disinfection will have inorganic phosphorus in the form of calcium phosphates. Iron, aluminum, and calcium phosphates are relatively insoluble, which will affect the buildup of available phosphorus in soils receiving repeated applications of biosolids.

In some states, the amount of water extractable phosphorus (WEP) is used to determine phosphorus availability and is incorporated into phosphorus Indices. Because of the interest in WEP in biosolids, this parameter was included in U.S. EPA's *Biosolids: Targeted National Sewage Sludge Survey* (2009). The survey indicates that WEP can range from 0.07 to 34% of total phosphorus.

Biosolids contain small amounts of potassium, typically 0.1 to 0.3% dry weight, that are not a significant source of potassium for crops. Most of the potassium in biosolids is in a water-soluble form. Biosolids also contain variable concentrations of other macronutrients and micronutrients. These include calcium, magnesium, sulfur, iron, manganese, and the regulated elements, copper, molybdenum, selenium, and zinc. Some plant species require nickel for growth. Plant availability of these nutrient elements depends on the significant chemical phases in the biosolids. Elements like copper and zinc are known to be complexed with the biosolids organic matter, and most sulfur in biosolids is in the form of protein. On the other hand, iron and manganese exist as relatively insoluble oxides.

3.3.2 *Metals*

At the time of promulgation of the 503 rules, biosolids from several WWTPs had failed to meet the most stringent pollutant limits (40 CFR 503.13, Table 3) for one or more elements (U.S. EPA, 1993). Most, however, met the ceiling concentrations and, therefore, their biosolids could still be land-applied, but with a limit on the cumulative loading to a given site. By the late 1990s, industrial pretreatment was so successful that most WWTPs were able to meet the Table 3 pollutant concentrations limits. Table 23.4 summarizes average biosolids metals concentrations based on product type.

3.3.3 *Other Constituents*

Biosolids contain many organic and inorganic chemical compounds. Some are discharged to collection systems as household chemicals that are then concentrated in biosolids after treatment. These include surfactants and conditioning agents used in detergents, and very low levels of various drugs, such as birth-control. Biosolids also contain natural materials such as silica (SiO_2) and aluminosilicate clays that enter the system as sediments from combined sewers.

The largest source of data for the concentration of some of these chemicals (and others of interest) comes from U.S. EPA's *Biosolids: Targeted National Sewage Sludge Survey* (2009).

Table 23.4 Average biosolids metal concentrations (mg/dry kg) (WWTP = wastewater treatment plant) (Lue-Hing et al., 1998).

Metal	Class A alkaline-stabilized	Compost	Digested liquid	Digested cake	Heat-dried	40CFR 503 Table 3 limits
Number of WWTPs in sample	7	10	48	117	5	NA
Arsenic	5.79	5.39	7.78	13.86	5.58	41
Cadmium	3.07	4.64	4.90	7.03	8.94	39
Chromium	50.3	72.7	62.0	119.4	152.6	NA
Copper	176	317	448	559	472	1,500
Lead	62.0	80.4	74.8	128.6	93.3	300
Mercury	0.73	1.94	2.60	2.01	1.60	17
Molybdenum	8.20	13.8	11.0	15.3	17.9	75
Nickel	38.8	26.5	33.5	69.3	35.0	420
Selenium	2.43	3.73	5.86	6.06	8.32	100
Zinc	878	878	807	886	906	2,800

3.4 Biological Characteristics

Biosolids contain varying concentrations of pathogens depending on the level of treatment they have received. Raw primary sludge contains the highest levels of pathogens. Biosolids treated to 40 CFR Part 503 Class A standards have pathogen concentrations that are below detection limits. Concentrations of specific pathogens, such as helminths, vary with specific sources in a watershed and also geographically as a consequence of climate. Table 23.5 summarizes the results from a study of indicator pathogens in stabilized biosolids.

4.0 SITE EVALUATION AND SELECTION PROCESS

The process of planning a biosolids land application project begins with the collection and assessment of basic data on biosolids characteristics. Biosolids characteristics in conjunction with estimated application rates can then be compared to applicable federal, state, and local regulations for an initial assessment of biosolids suitability for any of the land-application practices discussed in Section 2.0 of this chapter. Public perception and acceptance of such a proposed project and land availability,

TABLE 23.5 Typical biosolids indicator pathogen contents (WWTP = wastewater treatment plant) (Lue-Hing et al., 1998).

Pathogen	Class A alkaline-stabilized	Class A composted	Class A heat-dried	40 CFR 503 Class A Standards
Number of WWTPs in sample	5	4	5	NA
Fecal coliform (MPN/g mean)	3	76	6	< 1000
Fecal coliform (MPN/g median)	1	506	8	NA
Salmonella sp. (MPN/4g mean)	2	2	0	< 3
Salmonella sp. (MPN/4g median)	2	2	0	NA
Pathogen	**Class B alkaline-stabilized**	**Class B composted**	**Class B heat-dried**	**40 CFR 503 Class B Standards**
Number of WWTPs in sample	55	26	41	NA
Fecal coliform (MPN/g mean)	60	104 600	6 521	< 2 000 000
Fecal coliform (MPN/g median)	9 600	472 600	16 071	NA
Salmonella sp. (MPN/4g mean)	2 070	NA	NA	NA
Salmonella sp. (MPN/4g median)	4 000	NA	NA	NA

transportation modes, and climatic conditions, must all be considered and evaluated to determine the feasibility of the proposed program.

Careful identification, evaluation, and ultimate selection of land application sites can prevent future environmental problems, reduce monitoring requirements, minimize overall program costs, and moderate or eliminate adverse public reaction. Poor site selection and management practices in the past have resulted in environmental problems and public resistance.

Figure 23.1 provides an overview of a six-step process for identifying the best biosolids land application practice and the best sites for land application of biosolids. These steps are further discussed below.

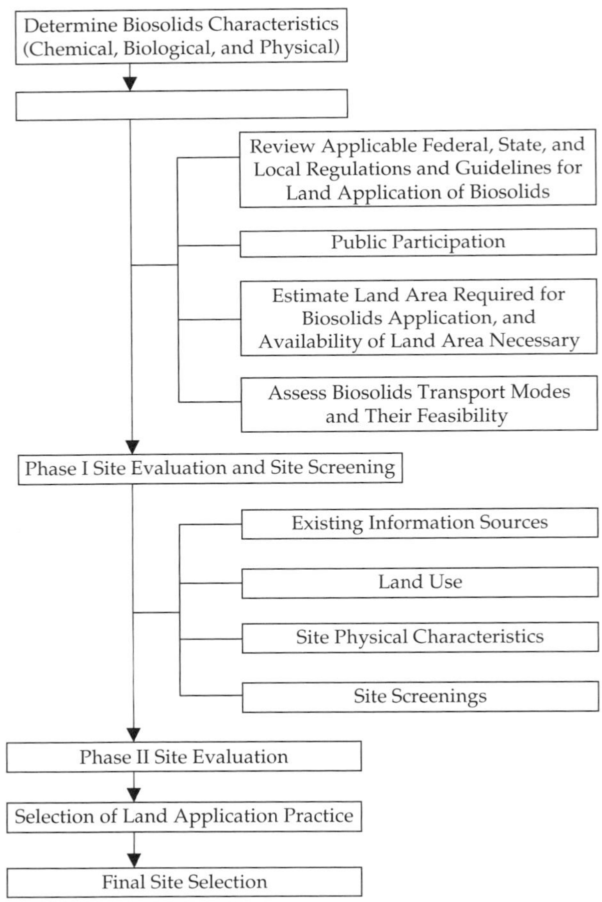

FIGURE 23.1 Simplified planning steps for biosolids land application site selection.

4.1 Preliminary Planning

As shown in Figure 23.1, the first step in the site selection process is preliminary planning. Key elements include regulatory assessments, public participation, land use requirements, and biosolids transport. Each of these elements is further addressed below.

4.1.1 *Institutional and Regulatory Framework*

All current federal, state, county, and municipal regulations and guidelines should be reviewed during the preliminary planning process. Depending on local procedure, permits may be required from both state and local regulatory agencies. Consideration must be given to protection of endangered species. Figure 23.2 shows the agencies that have jurisdiction over land application of biosolids. Federal requirements under the Part 503 rule are discussed further in Chapter 2 of this manual.

Other federal regulations also may apply to biosolids application in wetlands:

Clean Water Act, Sections 401, 402, and 404 (33 U.S.C. § 1251);

Rivers and Harbors Act of 1899 (33 U.S.C. § 407);

Executive Order 11990, Protection of Wetlands (42 FR 26961);

National Environmental Policy Act (42 U.S.C. § 4321 et seq.);

Migratory Bird Conservation Act (16 U.S.C. § 715 to 715s);

Fish and Wildlife Coordination Act (16 U.S.C. § 661 et seq.);

Coastal Zone Management Act (16 U.S.C. § 1451–1464);

Wild and Scenic Rivers Act (16 U.S.C. § 1271 et seq.); and

National Historic Preservation Act (16 U.S.C. § 470 et seq.).

Published information that may be useful in assessing those requirements includes U.S. Geological Survey topographic maps, National Wetland Inventory maps, NRCS soil maps, and wetland inventory maps prepared locally. District offices of the U.S. Army Corps can provide a wetland delineation to indicate whether all or some portion of a potential or actual land application site is in a wetland. The state agency regulating activities in wetlands should also be asked to inspect the area in question. The definition of a wetland and the regulatory requirements for activities in wetlands may be different at the state level.

4.1.2 *Public Participation*

Public participation is critical during the early stages of planning a land application project. Most involvement should come at the beginning of the planning process when

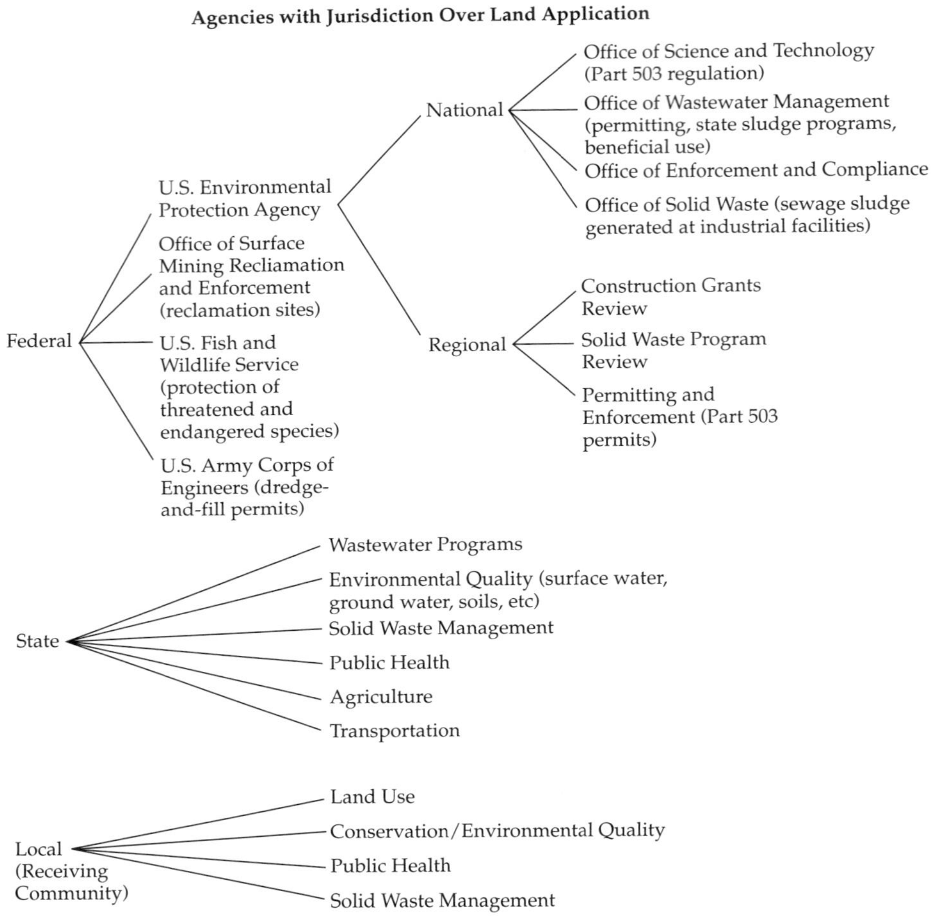

FIGURE 23.2 Institutional framework (adapted from U.S. EPA, 1981).

public input has the greatest potential to shape the final plan. This early involvement helps determine the limits to public and political acceptability of the project. During this phase, the public plays a constructive, as opposed to a reactive, role in decision-making.

Most public interest and involvement—including the most vocal and organized protests—occur during the site-selection stage. Therefore, the major thrust of the public participation program should come during this stage, with a particular emphasis on two-way communication using such avenues as public meetings, workshops, and radio talk shows. Chapter 3 of this manual provides a detailed discussion of how to design and implement a public participation program for a land application project. The *National Manual of Good Practice for Biosolids* also provides excellent guidance on this issue (NBP, 2005).

4.1.3 *Preliminary Land Area Requirements*

A precise estimate of the land area required for biosolids agricultural land application should be based on design calculations provided in Section 5 of this chapter. For preliminary planning, a rough estimate of the land application area can be obtained from Table 23.6 (different practices may not necessarily involve repeated annual applications).

For example, a project that intends to land apply 1000 Mg (1100 ton) dry weight of biosolids annually can use the typical rates shown in Table 23.6 to create a rough estimate. In this case, the land required for agricultural application would be 90 ha (220 ac), plus additional area required, if any, for buffer zones, biosolids storage, and other uses. For a one-time application of 1000 Mg of biosolids at a land reclamation site, the typical values shown in Table 23.6 indicate that 9 ha (22 ac) would be required.

4.1.4 *Biosolids Transport Assessment*

Transport can be a significant cost of a land application project and requires a thorough analysis. Biosolids can be transported by truck, pipeline, or rail. In certain instances, combined transport methods (e.g., pipeline-truck) are also used. The choice of a transportation method depends on the type of land selected, the volume and solids content of the biosolids, and the distance to and number of destination points. Chapter 19 of this publication provides additional information on transportation assessments and design.

TABLE 23.6 Preliminary estimates of biosolids applications (dry weight) for different types of land.

		Reported range of application rates		Typical rate	
Land type	**Period of application**	**Mg/ha**	**ton/ac**	**Mg/ha**	**ton/ac**
Agricultural land	Annual or twice annually	2–70	1–30	10	5
Forest land	Annually, or at 3- to 5-year intervals	10–220	4–100	18	8
Land reclamation site	One time	7–450	3–200	112	50

4.2 Phase I Site Evaluation and Site Screening

A Phase I site evaluation uses the information obtained during preliminary planning to identify potential land application sites. Existing information sources are used to identify multiple sites considering land use and physical characteristics within the area that biosolids can feasibly be transported. Site screening allows elimination of unsuitable areas because of physical, environmental, social, or political reasons and identification of sites for more detailed Phase II site evaluation.

4.2.1 *Existing Information Sources*

Sources of information on land characteristics, cropping patterns, and other relevant data in the geographic search area include

- U.S. Geological Survey;
- U.S. Department of Agriculture—NRCS, U.S. Forest Service, and Cooperative State Research, Education, and Extension Service;
- U.S. EPA;
- U.S. Army Corps of Engineers;
- Private photogrammetry and mapping companies;
- State agricultural mining and geologic agencies;
- State water resource agencies;
- State universities and local colleges;
- Local planning and health departments;
- Local water conservation districts;
- Groundwater uses (municipalities, water companies, individuals, etc.); and
- State land grant universities and water resource centers.

4.2.2 *Land Use and Availability*

Prevailing or projected land use significant influences site selection and acceptance of a particular biosolids land-application practice. It is necessary to determine both current and future land use in assessing the land area that is potentially suitable or available for biosolids application. Other important considerations include zoning compliance, aesthetics, access, and site acquisition.

4.2.3 *Physical Characteristics of Potential Sites*

Site physical characteristics of concern are

- Topography;
- Soil permeability, infiltration, and drainage patterns;
- Depth to groundwater; and
- Proximity to surface water.

The planner/designer should review federal and state regulations or guidelines that place limits on these physical characteristics of application sites. Summarized below are physical characteristics that might result in a site being deemed unsuitable for biosolids applications:

- Areas bordered by ponds, lakes, rivers, and streams without appropriate buffer areas;
- Wetlands and marshes without a permit;
- Steep areas with sharp relief;
- Undesirable geology (karst, fractured bedrock; if not covered by a sufficiently thick soil column);
- Undesirable soil conditions (rocky, shallow);
- Areas of historical or archeological significance;
- Other environmentally sensitive areas such as flood plains or intermittent streams, ponds, or other as specified in Part 503 regulation.

This section focuses on information that can typically be obtained from existing topographic and soil maps to identify sites where more detailed investigations may be justified. Additional information on the effect of site physical characteristics on land application can be found in U.S. EPA's *Process Design Manual: Land Application of Sewage Sludge and Domestic Septage* (1995) and the *National Manual of Good Practice of Biosolids* (National Biosolids Partnership, 2005).

4.2.4 *Site Screening*

Site screening is an integral part of the Phase I site evaluation process. Initially, development of land area, transportation distance, topographic, hydrologic, soils, and other site screening criteria helps focus efforts on collecting relevant information.

When potential sites are identified, ownership should be determined. Often city hall, county courthouse, or a real estate broker will have community or areawide maps indicating the tracts of land, present owners, and property boundaries. Much of this information is also available online now through sources such as county tax assessor and other Web sites. The county recorder and title insurance companies are also useful sources of information on property ownership, size of tracts, and related information. Contacting landowners prematurely without adequate preparation may result in an initial negative reaction, which is difficult to reverse. A public information program should be prepared and local political support secured.

Initial contacts concerning the proposed project should be made with the prospective landowners/site managers through personal interviews. Initial contacts via telephone are not recommended to avoid misunderstandings regarding the benefits of a land application program. The individuals involved in making the initial owner contacts should be knowledgeable about potential program benefits and constraints.

Ideally, the Phase I site evaluation and screening process will identify two or three sites that merit a more detailed Phase II site evaluation, discussed below.

4.3 Phase II Site Evaluation: Field Investigation

The Phase II site evaluation step involves field investigations of one or more sites to determine whether soil survey and other map information used in the Phase I site evaluation and screening process is accurate and to obtain additional, more detailed information required for final selection of the land application practice and final site selection. Key elements of Phase II investigations include

- Preliminary field site surveys;
- Procedures for detailed site investigations;
- Special considerations for detailed site evaluations for different land application practices: agricultural, forest land, reclamation sites, and public sites.

The following list provides a summary of site-specific information required:

- Property ownership;
- Physical dimensions of site;
- Overall boundaries;
- Portion usable for biosolids land application under constraints of topography, buffer zones, and other restrictions;

- Current land use;
- Planned future land use;
- If agricultural crops are to be grown
    - Cropping patterns;
    - Typical yields;
    - Methods and quantity of fertilizer application;
    - Methods of soil tillage;
    - Irrigation practices, if any;
    - Final use of crop grown (animal/human consumption, nonfood chain, etc.); and
    - Vehicular access within site;
- If forest land
    - Age of trees;
    - Species of trees;
    - Commercial or recreational operation;
    - Current fertilizer application;
    - Irrigation practices; and
    - Vehicular access within site;
- If reclamation site
    - Existing vegetation;
    - Historical causes of disturbance (e.g., strip mining of coal, dumping of mine tailings, etc.);
    - Previous attempts at reclamation, if any;
    - Need for terrain modification; and
    - Surface/groundwater conditions and characteristics including location and depth of wells, if any; location of surface water (occasional and permanent); history of flooding and drainage problems; seasonal fluctuation of groundwater level; and quality and users of groundwater.

A field site survey should be conducted after potential sites have been identified in the map study performed during the Phase I site evaluation. A drive or walk through the candidate areas should verify or provide additional information as included in the field survey form shown in Figure 23.3.

4.4 Selection of Land Application Practice

When the most feasible land application practices have been identified (e.g., application to agricultural land, forests, reclamation sites, public sites, or lawns and home gardens), preliminary estimates of site life and costs (capital and operation and maintenance) for the individual practices should be made. Potential social and environmental effects resulting from each practice also should be assessed. Comparing these data should reveal the most suitable land application practice that fits both the needs of the WWTP and local conditions. The facility might also consider adopting more than one land application practice (e.g., agricultural and forest land) if the combined practices appear to be cost-effective.

A checklist of relevant design features for each land application site is typically helpful for compiling information and provides baseline data for cost estimates (Table 23.7).

Individual practices can be compared and evaluated based on both quantitative and qualitative factors:

- Estimated costs;
- Reliability;
- Flexibility;
- Land area requirements and availability;
- Land use effects;
- Public acceptance; and
- Regulatory requirements (federal, state, and local).

A qualitative comparison of each land application practice is based on the experience and judgment of the project planners and designers. This is more difficult than a cost comparison because the level of each effect is more ambiguous and subject to differences of opinion.

A. Property Location ______________________ Property Owner ______________________

______________________ ______________________

______________________ ______________________

______________________ ______________________

B. Topography
Relief (sharp, flat, etc.) ______________________

Slope Estimate ______________________

Drainage Patterns
-Open/Closed ______________________

-Drainage Class No.* ______________________

-Any Underdrains ______________________

C. Distance from Site Boundary to
Surface Water ______________________

Water Supply Well ______________________

D. Estimate of Site Dimensions
Area ______________________

Natural Boundaries ______________________

Fences ______________________

E. Available Access
Road Types ______________________

Other ______________________

F. Existing Vegetation/Crops and Commonly Used Crop Rotations

Onsite ______________________

Neighboring Properties ______________________

G. Soil
Texture ______________________

Variability ______________________

*Refer to Soil Conservation Service drainage classes.

FIGURE 23.3 Sample form for preliminary field site survey.

Table 23.7 Example design features checklist/comparison of candidate land application practices (WWTP = wastewater treatment plant).

Subject	Candidate practice or combination of practices		
	Agriculture	**Forest**	**Reclamation**
1. Distance and travel time from WWTP to the candidate site			
2. Distance and travel time from the storage facility to the candidate site			
3. Distance from the nearest existing development, neighbors, etc., to the candidate site			
4. Biosolids modification requirements (e.g., dewatering)			
5. Mode of biosolids transportation			
6. Land area required			
7. Site preparation/construction needs:			
None			
Clearing and grading			
Access roads (onsite and offsite)			
Buildings (e.g., equipment storage)			
Fences			
Biosolids storage and transfer facilities			
Onsite drainage control structures			
Offsite runoff diversion structures			
Onsite runoff storage			
Flood control structure			
Ground water pollution control structure (e.g., subsurface drain system)			
Soil modification requirements (e.g., lime addition, etc.)			
8. Equipment needs:			
Biosolids transport vehicle			
Dredge			
Pumps			
Crawler tractor			
Subsurface injection unit			
Tillage tractor			

(*continued*)

TABLE 23.7 Example design features checklist/comparison of candidate land application practices (WWTP = wastewater treatment plant) (*Continued.*)

Subject	Candidate practice or combination of practices		
	Agriculture	Forest	Reclamation
Biosolids application vehicle			
Nurse tanks or trucks			
Road sweeper			
Washing trucks			
Irrigation equipment			
Appurtenant equipment			
9. Monitoring requirements:			
Soil			
Biosolids analysis			
10. Operational needs:			
Labor			
Management			
Energy			
Repair			

4.5 Final Site Selection

The final selection of the site is often a simple decision based on the availability of the best sites. This is frequently the case for small communities. If, however, the site-selection process is complex, involving many potential sites or several biosolids use or disposal practices, a weighted scoring system may be useful. Additionally, decisions should consider:

- Compatibility of biosolids quantity and quality with the specific land application practice selected;
- Public acceptance of both the practices and sites selected;
- Anticipated design life, based on assumed application rate; land availability (capacity); projected heavy metal loading rates (if Part 503 cumulative pollutant loading rates are being met); and soil properties.

A preliminary estimate of relative costs should be made as part of the site-selection process. These estimates are necessary for comparing alternative sites or land application practices. Proximity of the biosolids land application site to the WWTP is important in the decision-making process because of transportation costs. Further, the cost of biosolids dewatering equipment should be evaluated in view of estimated fuel savings through decreased total loads or shorter haul distances. For ease of comparison, all costs can be expressed in dollars per dry weight of biosolids. Capital costs should be estimated over the life of the site; whereas operating costs should be estimated annually. Cost factors that are of prime importance are summarized in Table 23.7. These assessments should be based on experience and best engineering judgment.

The Phase I and Phase II site evaluation process should result in detailed information on two or more sites that have been identified as suitable for the selected land application practice. This information, combined with the preliminary cost analysis, should provide a basis for selecting the most cost-effective site or sites.

4.6 Special Considerations for Reclamation Sites

At reclamation sites involving abandoned mine land, field investigations to characterize ground-water distribution and quality are typically required. The detailed site investigation should determine:

- Depth to groundwater, including seasonal variations;
- Quality of existing groundwater;
- Present and potential future use of groundwater;
- Existence of perched water; and
- Direction of groundwater flow.

5.0 PROCESS DESIGN FOR AGRICULTURAL LAND APPLICATION SITES

Agricultural land is the type of land most typically used for the application of biosolids.

This section presents detailed design information for the application of biosolids to agricultural land, placing primary emphasis on the growing of crops such as corn, soybeans, small grains, cotton, sorghum, and forages. The design example presented at the end of this section assumes that the designer has (1) selected agricultural land application; (2) completed preliminary planning; and (3) chosen a transportation system to convey biosolids to the application site.

Information must also be available on the types of crops to be grown, attainable yield level, and the relationship between soil fertility tests and recommended fertilizer application rates. The overall goal is to develop a nutrient management plan for the use of biosolids and fertilizer to meet the nutrient needs of the crop to be grown.

5.1 Regulatory Requirements and Other Considerations

Chapter 2 presents the requirements specified by the federal Part 503 regulation. When designing a land application system, state and local agencies must be consulted for any other requirements that must be met. Several criteria should be considered including nitrogen and other nutrients, soil pH and requirements for pH adjustment, and application methods and scheduling. Detailed information, including guidance for biosolids application on arid land, is provided in U.S. EPA's *Process Design Manual: Land Application of Sewage Sludge and Domestic Septage* (1995).

5.2 Determining Biosolids Application Rates for Agricultural Sites

Biosolids application rates are calculated from data on biosolids composition, soil test information, nitrogen and phosphorus fertilizer needs of the crop grown, liming requirements if applying lime-stabilized biosolids, and concentrations of trace elements. In essence, this approach views biosolids as a substitute for conventional nitrogen or phosphorus fertilizers or lime in crop production. The number of years that biosolids may be applied to a given field may be limited, based on Part 503 cumulative pollutant loading rate limits for metals.

The approach for determining biosolids application rates on agricultural cropland can be summarized as follows:

- Nutrient requirements for the crop selected are based on yield level and soil test data. If biosolids have been applied in previous years, then fertilizer recommendations are corrected for carryover of nutrients added by previous biosolids additions.
- Annual biosolids application rates are calculated based on nitrogen and phosphorus crop needs, liming requirements if lime-stabilized biosolids are applied, and Part 503 annual pollutant loading rate limits, where applicable.

- Supplemental fertilizer is determined from nitrogen, phosphorus, and potassium needed by the crop and amounts of nitrogen, phosphorus, and potassium by biosolids application.
- Biosolids applications are terminated when a Part 503 cumulative pollutant loading rate limit is reached if applicable.

The nitrogen and phosphorus content of biosolids is unbalanced with respect to plant needs, typically supplying more phosphorus than needed when applied at a rate to meet crop nitrogen needs. In addition, phosphorus applications with biosolids accumulate and progressively increase the concentration of plant-available phosphorus in the soil. High levels of plant-available phosphorus have been correlated with increased losses of phosphorus in surface runoff. Some states require biosolids application rates to be based on crop needs for phosphorus and on background plant-available phosphorus levels. Supplemental nitrogen is than applied to make up the difference between nitrogen supplied by the biosolids and crop needs.

The agronomic rate typically is the limiting factor for biosolids application rates, rather than Part 503 pollutant limits. Only when the cumulative pollutant loading rate (CPLR) at a site is being met will Part 503 become the limiting factor for the biosolids application rate. U.S. EPA's *Process Design Manual: Land Application of Sewage Sludge and Domestic Septage* (1995) should be consulted for guidance on determining CPLR and annual pollutant loading rates.

Developing an accurate mass balance for nitrogen in the biosolids and soil-crop system is critical in calculating biosolids application rates. Equations required for calculating a nitrogen mass balance are relatively simple; choosing reasonable input values for calculations, however, is more challenging. For initial calculations, "typical" and "suggested" values for all necessary parameters are provided in tables throughout this chapter. Site-specific data or the best judgment of individuals familiar with the nitrogen dynamics of the soil-crop system at the site should always be used in preference to "typical" values.

5.2.1 *Crop Selection and Nutrient Requirements*

The crops grown in an area will influence the scheduling and methods of biosolids application. Using the cropping systems already present will typically be advantageous because the crops have evolved because of local soil, climatic, and economic conditions. Because biosolids applications typically are limited by the nitrogen requirements of the crop, high nitrogen-use crops, such as forages, corn, and soybeans, will minimize the amount of land needed and the costs associated with biosolids transportation and application. Applying biosolids to meet nitrogen needs of crops will add excess phosphorus, however, and eventually rates may need to be reduced to manage biosolids phosphorus additions. Therefore, not only is it good practice to use fields with a mixture of crops, but the prudent manager of a land application program will continue to identify additional land areas that can be held in reserve.

Fertilizer recommendations for crops are based on the nutrients needed to achieve the desired yield of the crop to be grown and the capacity of the soil to provide the recommended plant-available nutrients. Accurate measurement of plant-available nitrogen (PAN) in soil is difficult and also dependent on climate. As a result, fertilizer nitrogen recommendations for a particular locality are typically developed using a combination of (1) guidelines developed by state agricultural experiment stations and the cooperative extension service, based on historical experience with crop yields on different soil types using different management practices; (2) soil test data; and (3) estimates of residual nitrogen carryover from previous applications of biosolids, animal manures, or nitrogen-fixing crops, such as alfalfa and soybeans.

Data can be used to estimate the amount of plant-available nitrogen that will be mineralized from biosolids organic nitrogen applied initially and from organic nitrogen remaining from previous applications (Table 23.8). These estimates can then be used to adjust the fertilizer nitrogen recommendations for the crops to be grown and residual soil nitrogen.

5.2.2 Calculation of Annual Application Rates

Recommended annual rates of biosolids application on cropland are based on nitrogen or phosphorus requirements of the crop grown, nitrogen and phosphorus levels in the biosolids, and metal concentrations in the biosolids for which pollutant limits have been set in the Part 503 rule. For lime-stabilized biosolids, the lime requirement of the soil should also be considered.

5.3 Design Example of Nutrient- and Metals-Based Land Application

This design example demonstrates the calculation of nitrogen-, phosphorus-, and metal-based application rates. For calculations of other application rates or for more complicated land application scenarios, the U.S. EPA's *Process Design Manual* (1995) should be consulted. This example involves a wastewater treatment facility that plans

TABLE 23.8 Mineralization factor (*f*) for nitrogen in various types of biosolids (U.S. EPA, 1994b).

Time after sludge application (year)	Percent of organic nitrogen mineralized from stabilized primary solids and waste activated sludges	Percent of organic nitrogen mineralized from aerobically digested biosolids	Percent of organic nitrogen mineralized from anaerobically digested biosolids	Percent of organic nitrogen mineralized from composted biosolids
0–1	40	30	20	10
1–2	20	15	10	5
2–3	10	8	5	3
3–4	5	4	3	3
4–5	3	3	3	3

to surface apply biosolids to hay crops. The facility produces 3 dry metric ton/d of liquid, aerobically digested biosolids. The biosolids will not be incorporated. The fields have not received soil amendments or fertilizers in the past. The digested biosolids have the following characteristics:

- Total nitrogen content = 3%;
- Ammonia-nitrogen content = 2%;
- Nitrate content = 0%;
- Total phosphorus content = 1%; and
- The limiting metal is copper at 400 mg/dry kg solids.

1. Determine the land-application rates and required acreages on (a) nitrogen-limiting basis and (b) phosphorus-limiting basis. Assume that the annual crop requirement is 235 kg nitrogen/ha·yr (210 lb nitrogen/ac/yr) and 73 kg phosphorus/ha·yr (65 lb phosphorus/ac/yr).
2. Determine the metal-limiting land application rate.
3. Which is more limiting: crop nutrient requirements or metals limits?

5.3.1 Nitrogen-Limiting Rate Calculation

First, the plant-available nitrogen (N_{PA}) in the biosolids can be calculated as follows:

$$N_{PA} = 1000[(\mathrm{NH_3})K + \mathrm{NO_3} + \mathrm{N_0}f]$$

where

N_{PA} = plant-available nitrogen (kg/metric ton dry solids);
NH_3 = percent ammonia-nitrogen in biosolids (as a decimal);
K = volatilization factor for ammonia;
NO_3 = percent nitrate in biosolids (as a decimal);
N_0 = percent organic nitrogen in biosolids (as a decimal); and
f = mineralization factor (conversion of organic nitrogen to ammonium-nitrogen).

K depends on the application method used (see Table 23.8). Because the biosolids will not be incorporated, K is assumed to be 0.5 (50% loss of NH_3-N). The mineralization factor f can be determined via Table 23.9. Because this is the first year of application and digested biosolids are being applied, f can be assumed to be 0.3. Therefore,

N_{PA} = <u>13 kg nitrogen/metric ton dry solids (26 lb nitrogen/ton dry solids)</u>

Next, the nitrogen-limiting biosolids loading (R_N) can be calculated as follows:

$$R_N = \frac{U_N}{N_{PA} + N_{PM}}$$

where

U_N = annual crop requirement for nitrogen;
N_{PA} = plant-available nitrogen from this year's sludge application; and
N_{PM} = plant-available nitrogen from mineralization of all previous applications.

Table 23.9 Volatilization factor (K) for ammonia (U.S. EPA, 1994b).

If solids are	K factor is
Liquid and surface applied	0.5
Liquid and injected into the soil	1.0
Dewatered and applied in any manner	1.0

As mentioned above, U_N is 235 kg N/ha/yr (210 lb N/ac/yr). Also, N_{PM} is zero since it is the first year of application. Therefore,

$$R_N = \underline{18 \text{ metric ton dry solids}}/\text{ha·yr } \underline{(8.0 \text{ tons dry solids/ac/yr})}$$

The total land required for application on a nitrogen-limiting basis is then

$$= (3 \text{ metric ton solids/d}) \times (365 \text{ days})/(18 \text{ metric tons solids/ha·yr})$$
$$= 61 \text{ ha } (150 \text{ ac})$$

5.3.2 *Phosphorus-Limiting Rate Calculation*

First, the total phosphorus percentage must be converted to the percentage of plant-available phosphorus, (P_2O_5). This is done as follows:

$$P_2O_5 = \text{total phosphorus } (\%) \times 2.29$$
$$= 22.9 \text{ kg } P_2O_5/\text{dry metric ton solids } (45.8 \text{ lb } P_2O_5/\text{dry ton solids})$$

The phosphorus-limiting biosolids loading (R_P) can be calculated using a formula analogous to that for nitrogen. In this case, the phosphorus crop requirement (U_P) is 73 kg phosphorus/ha·yr (65 lb phosphorus/ac/yr), all P_2O_5 is plant-available (P_{PA}), and the plant-available form from previous applications is zero because this is the first year of biosolids application. Thus,

$$R_P = 3.2 \text{ metric ton dry solids/ha·yr } (1.4 \text{ ton dry solids/ac/yr})$$

The total land required for application on a phosphorus-limiting basis is then

$$= (3 \text{ metric ton/day}) \times (365 \text{ days})/(3.2 \text{ metric ton/ha/yr})$$

$$= 344 \text{ ha } (851 \text{ ac})$$

5.3.3 *Metal-Limiting Rate Calculation*

Because copper is the limiting metal, the lifetime biosolids application rate is calculated based on the maximum lifetime cumulative loading for copper [1500 kg/ha (1340 lb/ac), listed in 40 CFR 503 Table 2. The biosolids contain 400 mg copper/kg solids, which is equivalent to 0.4 kg copper/metric ton dry solids (0.8 lb copper/ton dry solids). Therefore, the metal-limiting solids loading rate is 3750 metric ton dry solids/ha (1675 ton dry solids/ac).

5.3.4 *Rate Calculation Comparison*

A comparison of the nitrogen-, phosphorus-, and metal-limiting solids loading rates calculated above shows that crop nutrient requirements are the limiting factors for determining appropriate solids loading rates:

- Nitrogen-limiting rate = 18 metric ton dry solids/ha·yr (8.0 ton dry solids/ac/yr);
- Phosphorus-limiting rate = 3.2 metric ton dry solids/ha·yr (1.4 ton dry solids/ac/yr); and
- Metal-limiting rate = 3570 metric ton dry solids/ha (1675 ton dry solids/ac).

5.4 Additional Considerations for Land Application Program Planning

To simplify the design example, only two fields with different crops were considered. For most land application programs, however, biosolids will be applied to more than two crops and many individual fields. Application rates should be calculated for each field receiving biosolids as a detailed plan is developed. For this design example, additional crops could be oats, soybeans, and forages for hay and pasture. Crop rotations and relative hectares of each crop will vary. Farms with livestock will be producing animal manure that also will be used to meet plant nutrient requirements on land available for growing crops.

6.0 MARKETING BIOSOLIDS-DERIVED PRODUCTS

Biosolids treatment processes generate several marketable products:

- Compost is a typically accepted product marketed as a soil conditioner.
- Heat-dried biosolids are marketed as a fertilizer because they retain the input plant nutrients. Dried biosolids also have value as a renewable fuel.
- Alkaline-stabilized biosolids are used agriculturally for plant nutrients or liming. Some efforts exist to use this material as a renewable fuel.
- Ash is produced by biosolids incinerators and can be used in brick manufacturing, cement production, and road construction and in production of concrete pipes and culverts.

Merely producing biosolids that meet Class A pathogen reduction standards, vector attraction reduction requirements, and low pollutant concentrations ("exceptional quality" or "EQ biosolids") does not guarantee a marketable product or that there will be a market for the product. Managers deciding whether to invest in the infrastructure to produce a biosolids-derived product must consider marketability factors, some of which are discussed below. The section also discusses strategies for developing and maintaining markets. Further discussion is provided in Chapter 13 for composted product and Chapter 16 for thermal-dried product.

6.1 Marketability Criteria

Not all Class A biosolids are marketable (the terms "marketable" or "marketability" imply that distribution of these products may generate revenue). Some Class A biosolids may have an unfavorable appearance or poor handling properties and, thus, are not marketable without further processing. It is important to be aware that although biosolids-derived product characteristics typically exhibit some variability, inconsistent product quality may erode consumer confidence and inhibit marketability. Biosolids-derived product manufacturers should test the product frequently and strive to achieve the greatest consistency possible.

6.1.1 *Chemical Properties*

Chemical properties to be considered when assessing the value of any biosolids-derived product may include plant nutrients (primarily nitrogen, phosphorus, and potassium), salinity, liming value (an expression of an amendment's capacity to increase soil pH), and heating value (to meet increasing interest in using biosolids as fuel).

6.1.2 *Physical Characteristics*

The physical characteristics affecting a product's marketability are dependent on the type of biosolids-derived product; particle size (can determine target markets); bulk density (affects transportation costs); dust content (should be limited); stability (reflects a reduced potential for putrefaction); inert content (textiles, glass, plastic, and metal could be considered problematic in high quantities); maturity (applies to compost); viable weed seeds in finished compost (can be problematic); and durability (applies to heat-dried biosolids; i.e., resistance to degradation).

6.1.3 *Odor*

Product odor typically can be minimized through proper operations and diligent controls. Improper field storage (e.g., in a poorly drained area or close to sensitive receptors) may create anaerobic conditions in composts or alkaline-stabilized products resulting in odor issues. Heat-dried biosolids should be kept dry to prevent product degradation and to minimize odors. Alkaline dosage should be controlled to minimize ammonia-based odors in alkaline-stabilized products. Thorough mixing of alkaline admixture and biosolids allows for dissipation of potentially odorous compounds.

6.2 Other Considerations

In addition to marketability characteristics, a successful biosolids distribution program must consider a variety of other market factors including a good understanding of the target market, seasonality, consistency of product supply, price, public perception, bulk versus bag sales, and proper labeling.

6.3 Distribution Mechanisms

Biosolids producers can consider direct distribution to users, contracting with private distributors, or contracting with private brokers.

Direct distribution requires product marketing to be conducted by the generator. This option may be attractive if the target markets consist of local users, and the generator is committed to support the marketplace. Direct selling to a few large users can be a good way to develop an initial customer base before expanding production and marketing operations. Generators that choose this method may need to educate staff or hire new staff to carry out marketing. Some distributors of biosolids-derived products offer application services to facilitate distribution. One variation of this approach is selling directly to a blender who undertakes marketing, selling, and distribution of the final product. A generator should provide product quality data and information on use restrictions to the blender. This option is most typical with producers of heat-dried biosolids. Composts and alkaline-stabilized products can also be distributed to soil blenders who incorporate these materials in manufactured topsoil or materials for specific applications (e.g., turf mix for golf courses). In some cases, one or more biosolids-derived products may be blended to create a third product (e.g., compost and alkaline-stabilized).

Private contractors may be a person or company that will develop and service markets. In some cases, the contractor may prefer to operate the processing facilities so that they have greater control over the final product quality. A contractor typically signs an agreement with the WWTP that requires routine removal of product from the manufacturing facility and management of the product in accordance with regulatory standards and any restrictions enumerated in the agreement. Most contractors are paid on a per weight basis for providing routine services. Typically, a contractor will offer some sharing of revenues to entice the WWTP to agree to the arrangement but, most often, will not guarantee any revenue-sharing. Contractors may or may not accept full responsibilities for regulatory requirements.

Brokers take on some or all of the distributing and marketing tasks and typically have access to broader regional or nationwide markets. Often, brokers market a variety of products and offer their "best efforts" to place product into the marketplace. Ancillary activities such as billing and collections and transportation remain the responsibilities of the generators. Brokers typically work on a commission basis, so their incentive to move product is a function of their potential to earn commission.

6.4 Market Development

Successful biosolids product marketing programs with strong market demand have evolved only with proactive marketing efforts initiated early in the planning stages of the biosolids program.

Market planning should begin before process selection; it makes no sense to commit capital and operating expenses to a process for which there is no final product outlet. A limited assessment of local and regional markets helps qualify processes and products offering the best potential. Including plant personnel in the planning process helps ensure the facility can be operated and maintained in alignment with the biosolids program goals.

Communicating with potential users should be undertaken early in the process because the product will be viewed as a "new" to the local marketplace. Prospective customers can be sent periodic facility progress updates. Approved product labels can be provided. A material safety data sheet (MSDS) that provides information on product characteristics and safe handling and information on how to handle spilled product and who to contact in the event of an issue should be developed. Introduction of the product can include a broad range of activities such as tours of the processing facility, demonstration plantings, appearances on gardening radio shows, presentations to civic organizations, and communications with ratepayers via newsletters and bill stuffers. Continued attentiveness of a marketing program to communications is critical to ensure reliable outlets for the product. Any issues with the product or its application must be quickly addressed to the complete satisfaction of the customer because most segments of the "green industry" are fairly small, and word of mouth can ruin a market.

A dynamic sales model or business plan can help guide biosolids distribution and marketing as operating or market conditions change. It can also be used to aid in development and support of new market opportunities. The model should consider the range of market sectors identified and assess how the product can fit into those sectors. Some opportunities may require additional capital or labor, and the value of directing product to those opportunities must be weighed against cost. Considerations include market sector size (i.e., how many megagrams per year); seasonal constraints; delivery method (e.g., bulk in dump trailers, in bags); and distance to market. Over time, the range of opportunities may change. The model should be structured to account for changes, and marketing efforts should be redirected as needed. A key component should be a commitment to customer service. Especially with a new product, coordination of delivery, consultation on applications, and advice on proper handling and storage will be critical to the customer. Fact sheets and frequently-asked-question guides can also support customers.

Research to support marketing is important for comparing the performance of a product with competing products. Work closely with potential customers can demonstrate product efficacy. Generators can provide small quantities of product for users and cooperate in setting up field trials, demonstration plantings, and similar efforts.

Partnerships can be formed with entities that can support marketing efforts. Partners may include transporters, regulators, researchers, marketers/brokers, and companies that bag products, among others. These entities will handle and help promote the product, so it is important to work closely with them as the markets are developed and serviced. The chief partnership, however, will be with the customers. It is critical that the generator work closely with customers to make sure the product performs up to or beyond expectations. These relationships are critical because the user always has a choice—many competing products are on the market. It is important to service the customer efficiently and expeditiously. Little things, like incorrect billing, or late product delivery can present negative impression. Properly treated customers will become long-term partners, providing reliable outlets for the product.

7.0 OTHER DESIGN CONSIDERATIONS

This section provides guidance for designing the components of a land application system, including methods, site preparation, and supporting facilities. Two other key components of land application systems—transport and storage—are addressed in Chapter

19. The designer should take into consideration each component's effect on overall system efficiency, reliability, and cost when selecting and designing each component.

7.1 Land Application Methods

The technique used to apply biosolids to land can be influenced by the means used to transport the biosolids from the treatment works to the land application site. Typically used methods include

- The same transport vehicle hauls the biosolids from the WWTP to the application site and applies the biosolids to land.
- One type of transport vehicle, typically with a large-volume capacity, hauls biosolids from the WWTP to the application site. At the application site, the biosolids is transferred either to an application vehicle, a storage facility, or both.
- Biosolids are pumped and transported by pipeline from WWTP to a storage facility at the application site. Biosolids are subsequently transferred from the storage facility to an application vehicle.
- Irrigation equipment is used to spray biosolids onto fields.

Biosolids land application methods involve either surface or subsurface application. Subsurface application may be accomplished with subsurface injection of liquid biosolids, or with disking or other subsequent incorporation of liquid or cake applied biosolids into the soil. Each has advantages and disadvantages that are discussed in the following subsections. With all of the application techniques, the biosolids eventually become incorporated into the soil, either immediately by mechanical means or over time by natural means. Biosolids are applied either in liquid or dewatered form. The methods and equipment used are different for land application of these two types of biosolids, and each has advantages and disadvantages that are highlighted below.

7.1.1 Application of Liquid Biosolids

Application of biosolids to land in liquid form is relatively simple and includes surface application, subsurface application, and irrigation application. Dewatering processes are not required, and the liquid biosolids can be readily pumped. Liquid biosolids application systems include

- Vehicular application by tank truck or wagon spreading or an umbilical system spreading; and
- Irrigation application by spraying or flooding (gravity flooding).

7.1.2 *Application of Dewatered Biosolids*

Dewatered biosolids are applied to land by surface application techniques. The principal advantages of using dewatered biosolids are reduced biosolids hauling and storage costs and the ability to apply biosolids at higher application rates with one pass of the equipment. A potential disadvantage is that additional equipment may be needed. Dewatered biosolids typically cannot be directly off-loaded from transportation to application equipment or directly injected to help mitigate potential nuisance conditions.

Spreading dewatered biosolids is similar to surface application of solid or semisolid fertilizers, lime, or animal manure. Dewatered biosolids cannot be pumped or sprayed. Spreading is done by box spreaders, bulldozers, loaders, or graders. The biosolids may then be plowed or disked into the soil. The box spreader is most typically used; builders, loaders, and graders are used for sites with high biosolids application rates.

Specially designed equipment is used to spread dewatered biosolids. Surface spreading of dewatered biosolids may be followed by incorporation of the biosolids into the soil. It is not typically necessary to incorporate dewatered biosolids into the soil when the biosolids are applied to existing pasture or hay fields. Standard agricultural disks or other tillage equipment pulled by a tractor can incorporate the dewatered biosolids into the soil.

7.2 Site Preparation

For agricultural land application systems where biosolids are applied to privately owned farms at low agronomic application rates, site modifications are not typically cost-effective. More forest land typically is available within the local area than is needed for biosolids land application, so unsuitable land can be avoided rather than modified. For reclamation sites, extensive site grading and soil preparation is necessary. These site preparation costs, however, are typically borne by the landowner (e.g., mining company, ore processor, etc.).

7.2.1 *Protection of Groundwater and Surface Water Quality*

One of the significant environmental tasks at a biosolids land application site is the prevention of surface and subsurface water contamination by constituents in biosolids. In addition to proper application rates, erosion control can be used to prevent water contamination.

7.2.2 *Erosion Control*

Every land application site should be designed to minimize soil erosion. The measures used to prevent soil erosion include strip-cropping, terraces, grassed waterways, and reduced tillage systems (e.g., chisel plowing, no-till planting). The presence of vegetation or crop residues on the soil surface is effective in reducing runoff from steeply sloping soils. For many cropping systems (e.g., corn, soybeans, small grains), liquid biosolids applied to the surface are incorporated into the soil by plowing or disking before planting, further reducing the potential for loss of biosolids constituents via surface runoff.

Selecting proper biosolids application method (surface or incorporation) and adopting recommended practices for control of soil erosion will eliminate potential contamination of surface waters. Information on proper slopes, effective conservation tillage, and wind erosion techniques for agricultural lands can be obtained from the NRCS.

7.2.3 *Access Roads*

A permanent road should be provided from the public road system to the land application site. For large sites, the roadway should be 6.5 to 8 m (20 to 24 ft) wide to allow for two-way traffic; for smaller sites, a 5-m (15 ft) wide road should suffice. To provide all-weather access, the roadway, at a minimum, should be gravel-surfaced, preferably with asphalt pavement capable of supporting heavyweight transport vehicles. Grades should not exceed equipment limitations. For loaded vehicles, uphill grades should be less than 7%.

7.2.4 *Public Access: Site Fencing and Security*

Under the Part 503 rule, public access restrictions apply to land application sites where biosolids meeting Class B pathogen requirements are applied (see Chapter 2). Access to lands with a high potential for public exposure, such as parks or ballfields, must be restricted for one year after application. Examples of restricted access include ensuring remoteness, posting with "no trespassing" signs, or fencing. Access to land with low potential for public exposure (e.g., private farmland) must be restricted for 30 days after biosolids application.

Depending on the topography and vegetation of the site and adjoining areas, entrance gates may suffice to prevent unauthorized vehicular access. At some sites, it will be necessary to construct peripheral fences to restrict trespassers and livestock. Fencing requirements are influenced by the relative isolation of the site. Sites close to residences will require fencing. Facilities that are in relatively isolated, rural areas may require a less sophisticated type of fence or only fencing at the entrance and other select places to keep unauthorized vehicles out.

7.2.5 *Lighting and Other Utilities*

If land application operations occur at night, then portable lighting should be provided at the operating area. Lights may be affixed to haul vehicles and onsite equipment. These lights should be situated to provide illumination to areas not covered by the regular vehicle headlights. If the facility has structures such as employee facilities, office buildings, equipment repair, or storage sheds or if the access road is in continuous use, then permanent security lighting may be needed.

Larger sites may need electrical, water, communication, and sanitary services. Remote sites may have to extend existing services or use acceptable substitutes. Portable chemical toilets can be used to avoid the high cost of extending sewer lines; potable water may be trucked in; and an electrical generator may be used instead of having power lines run onsite.

Water should be available for drinking, dust control, washing mud from haul vehicles before entering public roads, and employee sanitary facilities. Telephone or radio communications may be necessary because accidents or spills can occur that necessitate the ability to respond to calls for assistance.

8.0 REFERENCES AND SUGGESTED READINGS

California Integrated Waste Management Board (2003) The Importance of Compost Maturity. http://www.ciwmb.ca.gov/publications/organics/44303007.pdf (accessed August 2011).

California Integrated Waste Management Board (2003) Guidelines for Writing Compost Procurement Specifications http://www.ciwmb.ca.gov/organics/products/CMSpecs/Guide.htm (accessed August 2011).

Clean Water Act. *U.S. Code*, Section 1251, Title 33, 1972.

Coastal Zone Management Act. *U.S. Code*, Section 1451–1464, Title 16, 1972.

Executive Order 11990, Protection of Wetlands. *Fed. Regist.*, **1977**, 26961, 3 CFR, 1977 Comp., p. 121.

Fish and Wildlife Coordination Act. *U.S. Code*, Section 661–et seq., Title 16, 1934.

Logan, T. (2001) Personal communication.

Lue-Hing, C.; Tata, P.; Granato, T. A.; Sustich, R.; Johnson, R.; Pietz, R. I. (1998) *Sewage Sludge Survey*. Association of Metropolitan Sewerage Agencies: Washington, D.C.

Migratory Bird Conservation Act. *U.S. Code*, Section 715–715s, Title 16, 1929.

National Biosolids Partnership (2005) *National Manual of Good Practice for Biosolids*. http://biosolids.org/docs/MGP_National_Manual_of_Good_Practice_0105.pdf (accessed August 2011).

National Environmental Policy Act. *U.S. Code*, Section 4321–et seq., Title 42, 1966.

National Historic Preservation Act. *U.S. Code*, Section 470–et seq., Title 16, 1966.

Natural Resources Conservation Service (1999) *General Manual Part 402*; U.S. Department of Agriculture, Natural Resources Conservation Service: Washington, D.C. http://www.nrcs.usda.gov/Internet/FSE_DOCUMENTS/stelprdb1043143.pdf (accessed November 2011).

Natural Resources Conservation Service (2006) *Conservation Practice Standard: Nutrient Management, Code 590*; U.S. Department of Agriculture, Natural Resources Conservation Service: Washington, D.C. http://www.nrcs.usda.gov/Internet/FSE_DOCUMENTS/stelprdb1043135.pdf (accessed November 2011).

New England Biosolids and Residuals Association (2007) *National Trends in Biosolids Management: A Survey Overview*. New England Biosolids and Residuals Association: Tamworth, New Hampshire.

Rivers and Harbors Act of 1899. *U.S. Code*, Section 407, Title 33, 1899.

Sommers, L.; Parker, C.; Meyers, G. (1981) *Volatilization, Plant Uptake and Mineralization of Nitrogen in Soils Treated with Sewage Sludge*, Technical Report 133; Water Resources Research Center, Purdue University: West Lafayette, Indiana.

Sommers, L. E.; Van Volk, V.; Giordano, P. M.; Sopper, W. E.; Bastian, R. (1987) Effects of Soil Properties on Accumulation of Trace Elements by Crops. In *Land Application of Sludge—Food Chain Implications*; Page, A. L.; Logan, T. J.; Ryan, J. A., Eds.; Lewis Publishers: Chelsea, Michigan.

Switzenbaum, M. S.; Moss, L. H.; Epstein, E.; Pincince, A. B.; Donovan, J. F. (1997) Defining Biosolids Stability. *J. of Environ. Eng.*, **1123** (12), 1178–1184.

Texas Department of Transportation. Compost Specification Item 161. ftp://ftp.dot.state.tx.us/pub/txdot-info/cmd/cserve/specs/2004/standard/s161.pdf (accessed August 2011).

U.S. Environmental Protection Agency (1981) *Institutional Constraints and Public Acceptance Barriers to Utilization of Municipal Wastewater and Sludge for Land Reclamation and Biomass Production: A Report to the President's Council on Environmental Quality;* EPA-430/9-81-013; U.S. Environmental Protection Agency: Washington, D.C.

U.S. Environmental Protection Agency (1992) *Environmental Regulations and Technology: Control of Pathogens and Vector Attraction in Sewage Sludge,* EPA/625/R-92/013; U.S. Environmental Protection Agency, Office of Research and Development: Cincinnati, Ohio.

U.S. Environmental Protection Agency (1993) *Standards for the Use and Disposal of Sewage, Final Rule.* Code of Federal Regulations, Part 503, Title 40, 1993; *Fed. Regist.*, 58, 9248.

U.S. Environmental Protection Agency (1994a) *A Plain English Guide to the EPA Part 503 Biosolids Rule,* EPA/832/R-93/003; U.S. Environmental Protection Agency, Office of Wastewater Management: Washington, D.C.

U.S. Environmental Protection Agency (1994b) *Guidance for Writing Permits for the Use or Disposal of Sewage Sludge;* U.S Environmental Protection Agency: Washington, D.C.

U.S. Environmental Protection Agency (1995) *Process Design Manual: Land Application of Sewage Sludge and Domestic Septage;* U.S. Environmental Protection Agency: Washington, D.C.

U.S. Environmental Protection Agency (2007) *Alkaline Treatment of Municipal Wastewater Treatment Plant Sludge Technical Guide,* Draft; U.S. Environmental Protection Agency: Washington, D.C.

U.S. Environmental Protection Agency (2009) *Biosolids: Targeted National Sewage Sludge Survey Report;* EPA 822-R-08-014; U.S. Environmental Protection Agency: Washington, D.C.

Water Environment Federation; American Society of Civil Engineers; Environmental and Water Resources Institute (2010) *Design of Municipal Wastewater Treatment Plants,* 5th ed., WEF Manual of Practice No. 8; ASCE Manual and Report on Engineering Practice No. 76; McGraw-Hill: New York.

Water Environment Research Foundation (1993) *Document Long-Term Experience of Biosolids Land Application Programs*, Project no. 91-ISP-4; Water Environment Research Foundation: Alexandria, Virginia.

Water Environment Research Foundation (2002) *Evaluating Risks and Benefits of Soil Amendments Used in Agriculture*, Project no. 99-PUM-1; Water Environment Research Foundation: Alexandria, Virginia.

Wild and Scenic Rivers Act. *U.S. Code*, Section 1271–et seq., Title 16, 1968.

Chapter 24

Landfill Management Systems

(*continued*)

(continued)

1.0 INTRODUCTION

This chapter discusses land disposal options for when wastewater treatment solids are unsuited for beneficial reuse or land application sites are not available. Two principal land disposal methods—landfilling and dedicated land disposal—are approved methods for land disposal. The methods differ both in application method and rate. Landfill operations involve subsurface application of dewatered wastewater solids to depths of several meters. Dedicated land disposal operations, however, typically involve repetitive liquid–solids applications, which may only raise the land surface by less than 25 mm (less than 1 in.) per year.

In this Chapter, the term biosolids refers to material that meets Part 503 regulations; otherwise the material will be referred to as sludge (U.S. EPA, 1993).

1.1 Regulatory Agency Guidance

Development of formalized methods for biosolids disposal to land has been encouraged and funded by the U.S. Environmental Protection Agency (U.S. EPA) since 1974. Chapter 2 discusses guidance and regulatory documents, which deal with federal laws to control the disposal of wastewater solids.

1.1.1 *Environmental Protection Agency Regulations*

As required by the Clean Water Act Amendments of 1987 (33 U.S.C. § 1251), the U.S. EPA developed a new regulation to protect public health and the environment from any reasonably anticipated adverse effects of certain pollutants that might be present in sewage sludge. This regulation, Standards for the Use or Disposal of Sewage Sludge (Title 40 of the *Code of Federal Regulations* [CFR], Part 503), was published in the *Federal Register* (58FR 9248 to 9404) on February 19, 1993, and became initially effective on March 22, 1993 (U.S. EPA, 1993, 1994).

The Part 503 rules address biosolids monofills. The 40 CFR 258 rules for municipal solid waste landfills address co-disposal landfills that receive municipal solid waste (MSW) or refuse and biosolids. These rules are typically referred to as the Subtitle D landfill rules. They were promulgated in the *Federal Register* on October 9, 1991, and most of the rule provisions became effective on October 9, 1993 (U.S. EPA, 1991).

The details of Subtitle D regulations are not covered here. This chapter assumes that anyone considering biosolids disposal will either develop their own biosolids monofill or will dispose solids with a third party at an existing refuse landfill in a co-disposal application. Under those circumstances, there is less need for the reader to understand the full Subtitle D rules. The U.S. EPA and other parties, however, have produced several excellent publications that present, describe, and summarize the Subtitle D regulations.

1.1.2 *State/Local*

Most states have adopted the Part 503 and Subtitle D rules and have primacy (authority) in permitting and enforcing these regulations. Some states have developed regulations more restrictive than the corresponding federal standard. Any state authorized to run its Clean Water Act (CWA) Part 503 program and Resource Conservation and Recovery Act (42 U.S.C. § 6901) Subtitle D program likely issues its own specific permits for such biosolids landfills within their jurisdiction .

2.0 LANDFILL

2.1 Definition

Landfill disposal of biosolids is the planned burial of wastewater treatment solids, including biosolids, screenings, grit, and ash at a designated facility. The solids are placed into a prepared site or excavated trench, and covered with a layer of soil. The soil cover must be deeper than the depth of the plow zone, 20 to 25 cm (8 to 10 in.) to be landfilling and not land application. Screenings, grit, and ash are disposed of in a landfill by similar methods.

2.2 Suitability of Solids for Landfilling

Table 24.1 summarizes the suitability of biosolids of varying solids contents and other factors for different types of landfilling.

2.3 Biosolids Landfilling Methods

Solids landfill methods can be grouped into three categories: biosolids-only monofill trench fill, biosolids-only monofill area fill, and co-disposal with refuse. Site and design criteria are discussed under these categories. A detailed discussion of biosolids landfills is presented in the *Process Design Manual, Municipal Sludge Landfills* (U.S. EPA, 1978b). The remaining parts of the landfill portion of this chapter summarize the information in this design manual.

2.3.1 *Monofill*

Monofills are active sites where sludge with a solids content of at least 15% (or more depending on the type of monofill) are disposed and covered periodically (U.S. EPA, 1995).

2.3.1.1 *Trench Fill*

The biosolids-only trench method involves excavating trenches so that dewatered biosolids may be entirely buried below the original ground surface. Disposal of screenings, grit, and ash can also occur in such trenches, if the physical management of these various waste types together is demonstrated to work in a given application. Physical testing of such co-disposal operations should be performed in advance to demonstrate compatibility of materials. In this method, the biosolids are deposited directly into the trench from a haul vehicle. Typical operations procedure requires daily soil cover. Trench disposal is appropriate for dewatered, unstabilized sludge, because the immediate application of cover material reduces associated odors. Vector control requires daily cover, except during very cold weather.

TABLE 24.1 Suitability of sludge and biosolids for landfilling (U.S. EPA, 1979).

	Sludge/biosolids only landfilling		Co-disposal landfilling	
Type of sludge/biosolids	**Suitability**	**Reason**	**Suitability**	**Reason**
Liquid–unstabilized				
- Gravity-thickened primary, WAS and primary, and WAS	NS	OD, OP	NS	OD, OP
- Flotation thickened primary and WAS and WAS without chemicals	NS	OD, OP	NS	OD, OP
- Flotation-thickened WAS with chemicals	NS	OP	NS	OD, OP
- Thermal-conditioned primary or WAS	NS	OD, OP	MS	OD, OP
Liquid–stabilized				
- Thickened anaerobic digested primary and primary, and WAS	NS	OP	MS	OP
- Thickened aerobic digested primary and primary, and WAS	NS	OP	MS	OP
- Thickened lime stabilized primary and primary, and WAS	NS	OP	MS	OP
Dewatered–unstabilized				
- Vacuum filtered, lime conditioned primary	S	–	S	–
Dewatered–stabilized				
- Drying bed digested and lime-stabilized	S	–	S	–
- Vacuum filtered, lime-conditioned digested	S	–	S	–
- Pressure filtered, lime-conditioned digested	S	–	S	–
- Centrifuged, digested and lime-conditioned digested	S	–	S	–
Heat-dried				
- Heat-dried digested	S	–	S	–
High-temperature processed				
- Incinerated dewatered primary and primary and WAS	S	–	S	–
- Wet-air oxidized primary and primary and WAS	NS	OD, OP	MS	OD, OP

WAS = waste activated sludge; OD = odor problem
NS = not suitable; OP = operational problems
MS = marginally suitable
S = suitable

2.3.1.2 Area Fill

In the biosolids-only area fill method (including area fill mound, area fill layer, and dike containment), biosolids are mixed with soil, and the mixture is placed on the original ground surface. This method requires substantial amounts of imported soil but may be suitable in areas where groundwater is shallow (liners can be easily installed) or bedrock prevails (that is, where excavation is neither possible nor required). Adequate drainage and runoff control are necessary to prevent contamination of nearby surface waters.

2.3.2 Co-Disposal with Refuse

The term co-disposal is used when biosolids are disposed of at a refuse landfill. This method presents distinct trade-offs compared with only biosolids. Biosolids can be disposed of in this manner if it is mixed with refuse or with soil. Mixing techniques are discussed in detail in the *Disposal of Sewage Sludge into a Sanitary Landfill* (U.S. EPA, 1974). This method can include biosolids/sludge/refuse mixture or biosolids/soil mixture.

2.3.3 Bioreactor Landfills

A bioreactor landfill is designed to promote decomposition of waste solids at an accelerated rate. It achieves physical and chemical stabilization of the waste mass at an earlier time than would occur in a landfill. Bioreactor landfills have been an initiative of the research and landfill operations community and, since the late 1990s, the regulatory community. The addition of biosolids to such sites has had a significant role in the function of bioreactor landfills.

The value of biosolids application (particularly liquid residuals) is that it tends to fill voids in the waste mass and not consume landfill air space at the same rate as refuse. In addition, through the addition of water and other biosolids content, it accelerates the chemical and biological decomposition of the landfill, thus providing physical stabilization. This decomposition creates additional air space throughout the landfill life, puts less stress on the landfill during the 30-year postclosure period, and leaves a lower risk profile for continued decomposition after the 30-year postclosure period.

2.4 Preliminary Planning

The purpose of preliminary planning is to select a disposal site and suitable method or methods of disposal. Preliminary planning is followed by detailed design, initial site development, site operation and maintenance, and final site closure.

2.4.1 *Public Participation*

Federal requirement do not require a public participation program (PPP) in the development of biosolids landfills under either the Part 503 rules for biosolids monofills or the Subtitle D rules for refuse landfills and co-disposal sites; although some states may have requirements. (Subtitle D does have a requirement for public participation if the site has an environmental problem that triggers "Corrective Action.")

The U.S. EPA does, however, strongly encourage public participation programs in the development and operation of biosolids disposal sites under both the Clean Water Act (CWA) including its Part 503 facilities, and the Resource Conservation and Recovery Act (RCRA) and its Subtitle D program. Chapter 2 of this report provides a detailed discussion of public participation in this and other phases of a project.

2.4.2 *Solids Characterization*

Biosolids quantity and quality must be characterized. Table 24.1 describes the suitability of various types of biosolids for landfilling operations.

An estimate of the average biosolids quantity is necessary to establish landfill area requirements and the probable life of the disposal site. Data on minimum and maximum biosolids quantities are important for developing daily operating requirements including equipment needs, storage facility sizing, and daily operating schedules.

The character of the biosolids is directly related to the choice of a landfill method. Biosolids quality and the corresponding leachate can be roughly correlated; design of leachate treatment facilities is more effective if biosolids or sludge quality is known. Parameters that are required or may be required to be analyzed by the Part 503 rules or the Subtitle D rules for biosolids landfills include solids content, paint filter test, metals, hazardous waste, pathogen, and polychlorinated biphenyls (PCBs). Design, operation, and monitoring of facilities may be aided by further characterization of biosolids to include volatiles content, nitrogen, inorganic ions, toxic organic compounds, and pH.

2.4.3 *Selection of a Landfilling Method*

The landfilling method should be compatible with the characteristics of the biosolids or sludge and landfill site and should be considered in preliminary planning.

2.4.4 *Site Selection*

Site selection is a critical process that is directly related to the method of ultimate disposal. The site must be suitable for the type of biosolids to be disposed of and situated in a convenient, yet unobtrusive, location. Chapter 4 of *Process Design Manual:*

Municipal Sludge Landfill provides an in-depth approach to site selection (U.S. EPA, 1978b).

2.4.4.1 *Site Considerations*

The following factors must be considered during the evaluation of possible landfill sites: site life and size; topography; surface water; soils and geology; groundwater; vegetation; meteorology; environmental sensitivity; use (i.e., grazing or cropland); archaeological and historical significance; site access and haul distances; zoning restrictions; and costs. Information on these factors should be collected and assessed in before decision-making.

2.4.4.2 *Site Selection Methodology*

The site selection procedure can be roughly divided into three phases: initial inventory and assessment, screening, and final selection. Initial inventory and assessment is designed to develop a list of potential sites that can be evaluated and rapidly screened to produce a manageable number of candidate sites. Information used in this phase is typically readily accessible. Investigation of each option becomes more detailed as the selection procedure progresses.

2.4.4.3 *Vectors*

Vectors include airborne insects such as flies and mosquitoes and burrowing animals that may come in contact with the waste. To control for vectors, a site should not be near areas such as swamps and woodlands that have significant colonies of airborne insects and burrowing animals.

2.4.4.4 *Odors*

Odors can be a significant issue for any biosolids operation including landfills. As with vectors, it is best to maximize the buffer distance between waste deposits and local populations. Prevailing wind speed and direction related to local populations should also be evaluated, including seasonal changes.

2.4.4.5 *Water Resources*

Typically, sites should be located as far as possible (horizontally and vertically) from surface water and groundwater resources. Existence of natural conditions that impede or restrict exposures between waste deposits and the surface or groundwater resources can help significantly. Although natural exposures can be manipulated to some degree by engineered control mechanisms, natural protections dictate higher suitability and ranking in any evaluation of candidate sites. Sensitivity or use of the water resource and its potential to affect receptors or users should be considered.

2.4.4.6 *Road Access*

Good road access from the source of the biosolids to the candidate disposal sites is a key consideration in site selection. Typically, a shorter haul on high-traffic, high-capacity road systems is best. Short haul distances save hauling costs and time and reduce the environmental effects of transportation. Better roads on the haul routing are essential. Use of federal, state, and superior county roads ensures that roads can readily handle haul vehicles. The number and proximity of population receptors along the haul route, and increased traffic, noise, safety hazards, mud, and odors are all relevant in this process. Chapter 19 provides a more detailed discussion of hauling and transport.

2.4.4.7 *Zoning*

If zoning is not in place, then it is unlikely to be a restriction. If zoning is in place and applies to the proposed site, then it is ideal if the existing zoning and planning designations are considered a "compatible use." Compatible use means that a proposed use is considered to be explicitly defined in, or at least plausibly related to, the zoning designation and description. For example, it is unlikely that a zoning classification will specifically be named or scoped to include biosolids or sludge disposal; however, it may be specific for waste disposal, or wastewater and pollution control activities, or even "public services." If the zoning is an incompatible use, then a special use exemption must be granted by the local zoning authority.

2.4.4.8 *Distance from Receptors*

Distances from receptors and buffers areas, either on the site property or with additional un-owned but undeveloped adjacent properties, is a key consideration in site selection. The Part 503 rules specifically require additional buffers be incorporated into the site property, consistent with the type of biosolids to be disposed. These required buffers should be considered minimums.

Considering only distance may not be adequate because topography, existing natural and manufactured features, prevailing wind direction, and aesthetics may require increased separation or possibly even allow less distance than typically would be acceptable.

2.4.4.9 *Geology and Soils*

Ease of excavation and soil suitability for cover application and protection of the groundwater at the site need to be considered. Excavation typically occurs at the disposal site, which is the case with most landfills. Shallow bedrock or groundwater may preclude excavation, or at least make some excavations quite expensive. Depth to bedrock and groundwater should be part of the evaluation criteria for site selection.

The type of soil on site is also a consideration. Less permeable, tighter soils (clay, for example) are best for groundwater protection, if the site will remain unlined. If liner systems will be installed for groundwater protection, then in-situ availability of such soils at a candidate site is less relevant.

The type of soil also has affects its use for cover. Workable but cohesive soils of medium permeability are most desirable because they are workable even when wet, allowing site operation in wet, muddy conditions. These soils can easily be used as cover and provide resistance to infiltration, reducing the opportunity for surface water infiltration and leachate generation.

2.4.4.10 Threatened and Endangered Species

A survey should be performed on each candidate site to determine if critical habitats of threatened or endangered species are present. Such habitats are to be avoided. If candidate sites are so restricted, then the site becomes relatively less desirable and may be ruled out.

2.4.4.11 Proximity to Wetlands or Waterways

Location in wetlands or waterways is specifically called out under "locations restrictions" of the Part 503 rules. Although work-arounds are available, it may be best to eliminate sites in such designated areas. Failing that, these sites should be ranked significantly lower as a site candidate because permits will be required from the U.S. Army Corps of Engineers or other entities.

2.5 Facility Design

2.5.1 Regulations and Standards

Local, state, and federal regulations and standards must be fully understood before the landfill is designed. Consideration must be given to requirements governing the degree of biosolids stabilization, loading rates, frequency and depth of cover, monitoring, and reporting. Design should conform to all building codes and should include adequate buffer zones to protect public roads, private structures, and surface waters. Obtaining permits for construction and operation of biosolids landfills can be a long and costly process. To minimize delays associated with this task, permit application should be initiated early in the design stage. A sound regulatory-consultant relationship and a mutual understanding should be developed. The following is a partial list of permits that may be required:

- National Pollutant Discharge Elimination System (NPDES) permit if the landfill is in wetlands;
- U.S. Army Corps of Engineers permits for construction of levees, dikes, or containment structures to be placed in the water in a wetlands area;
- Office of Endangered Species permit if the landfill is located in a critical habitat of an endangered species;
- Solid waste management permit;
- Special use permit; highway department permit; construction permit;
- Building permit; and
- Drainage or flood plain alteration permit.

2.5.2 *Site Characteristics*

Site characteristics should be clearly described and analyzed to ensure the suitability of the landfill site and the method of landfilling. Design work will build upon planning data, but will include a higher level of detail and working drawings. Site characteristics include: site plan (e.g., fill area, buffer zone, topographic features, surface water locations, roads, utilities, structures, etc.); soils; groundwater; subsurface geology; climate; and land use.

2.5.3 *Landfill Type and Design*

More than one biosolids landfill method may be suitable for the selected site. If this is the case, then a method must be selected before the final design is begun. Maximizing use of the site is an important consideration in method selection. Additional design criteria are summarized in Table 24.2.

2.5.4 *Ancillary Facilities*

Ancillary facilities may be needed at the landfill site: leachate controls, gas control, roads, soil stockpiles, structures, utilities, fencing, lighting, wash racks, monitoring wells, and landscaping.

2.5.5 *Environmental Effects*

Environmental effects will vary among landfill locations. Crucial areas include traffic, land use, air quality, surface and groundwater quality, public health, aesthetics,

TABLE 24.2 **Landfill design criteria (U.S. EPA, 1979).**

Method	Biosolids solids content (%)	Trench width (m)	Bulking required	Bulking agent	Bulking ratio[a]	Cover thickness (m) Interim	Cover thickness (m) Final	Imported soil required	Biosolids application rate (m^3/ha)[b]	Equipment
Monofill–trench fill										
Narrow trench	15–20 [c]	0.6–0.9	No	–	–	–	0.6–0.9	No	2 267–10 580	Backhoe with loader, excavator, trenching machine
	20–28 [c]	0.9–3.0	No	–	–	–	0.9–1.2			
Wide trench	20–28 [c]	3.0	No	–	–	–	0.9–1.2	No	6 046–27 395	Track loader, dragline, scraper, track dozer
	> 28 [d]	3.0	No	–	–	–	1.2–1.5			
Monofill–area fill										
Area fill mound	> 20[c, d]	–	Yes	Soil	0.5–2 soil: 1 biosolids	1	0.9–1.5	Yes	5 668–26 450	Track loader, backhoe with loader, track dozer
Area fill layer	> 15[d]	–	Yes	Soil	0.25–1 soil: 1 biosolids	0.2–0.3	0.6–1.2	Yes	3 779–17 004	Track dozer, grader, track loader
Diked containment	20–28[c]	–	No	Soil	0.25–0.5 soil:		0.9–1.2	Yes	9 069–28 340	Dragline, track dozer, scraper
	> 28[d]		No	Soil	1 biosolids					
Codisposal with Refuse										
Biosolids/refuse mixture	> 3[d]	–	Yes	Refuse	4–7 tons refuse: 1 wet ton biosolids	0.2–0.3	0.6	No	945–7 935	Dragline, track dozer
Biosolids/soil mixture	> 20[d]	–	Yes	Soil	1 soil: 1 biosolids	0.2–0.3	0.6	No	3 023	Tractor with disc, grader, track loader

[a]Volume basis unless otherwise noted.
[b]In actual fill areas.
[c]Land-based equipment.
[d]Biosolids-based equipment.
[e]Sometimes used.

1 meter = 3.28 ft
1 m^3 = 1.3 cu yd
1 ha = 2.5ac
1 ton = 0.9 Mg

wildlife, and habitats of endangered species. Adverse effects should be mitigated during the site selection process or by specific measures in the design.

2.5.6 *Gas Control and Gas to Energy*

When sludge or biosolids is landfilled, it decomposes over time anaerobically and generates landfill gas. Landfill gas consists of methane, carbon dioxide, and many trace gases. Methane gas is combustible at levels greater than the lower explosive limit (LEL), which is 5% by volume. Landfill gas can migrate laterally from disposal areas through adjacent soils to onsite and offsite structures, where gas can accumulate and become a fire or explosion hazard. Trace gases can be malodorous and create an odor nuisance for onsite workers and offsite neighbors.

The first requirement of gas control is to prevent the lateral migration of methane gas at combustible levels beyond the property line or into onsite or offsite structures. Passive venting of such gases onsite or impermeable blockage of migrating gases adjacent to disposal areas can be used to address these hazards. Comprehensive active gas extraction for control purposes at biosolids monofills is typically not required or appropriate.

2.5.7 *Access Roads, Fencing, and Scales*

Access roads into and around the site, for haul vehicles to the dumping area, and for site operational vehicles are a key consideration addressed under facility design. Access roadways need to be addressed continuously and as a priority during site operations. Under the Part 503 rules, site access must be controlled, primarily to prevent unauthorized site access and use or open dumping.

Site scales are installed at virtually all MSW landfills and co-disposal sites; however, scales at monofill landfills are not typically provided because the weight or volume of waste hauled to the site is determined at the wastewater treatment plant (WWTP).

2.5.8 *Leak/Explosive Gas Detection*

Gas detection can be done in onsite or offsite structures and in natural soil or ambient air at the property boundary. The appropriate standards for gas levels that signify an exceedance requiring mitigative action, which may be an emergency, include 100% of the LEL or 5% methane (these values are equivalent) in natural soils at the property line. The standard for occupied air space in onsite and offsite structures is 1.25% methane, which is 25% of the LEL.

A handheld, combustible-gas meter is best for monitoring from subsurface probes or in the air space of occupied structures. It can also be used to test ambient air at the property boundary as prescribed under the Part 503 rules. As an alternative to handheld gas meters, permanent gas detectors with integrated alarm systems can constantly scan for gas exceedances.

2.5.9 *Odor Control Equipment/Structures*

Biosolids disposal sites can have odor emissions, perceived as a nuisance to both off site receptors and on site workers. Probably the best odor control mechanism is sludge treatment, applied offsite. The most common and effective sludge treatments are aerobic and anaerobic digestion and lime application. At the site itself, there can be multiple sources of odor. The largest source is typically the recent loads of biosolids disposed at the site. Odor controls applicable here include daily cover and then tighter and thicker covers such as clay and even plastic, as intermediate and final cover.

2.5.10 *Landfill Liners and Leachate Management*

Biosolids monofills may or may not be required to have liner systems installed. Refuse co-disposal sites, however, are required to have liners under the Subtitle D criteria. Once a liner system is installed, a leachate collection system must be installed to remove the wastewaters (known as leachate) that percolate to the bottom of the fill and collect atop the liner. These collection systems must be operated to remove leachate for subsequent treatment and discharge.

2.5.11 *Communications/Alarms*

If the site is staffed full time, or even part time, mobile or hardwired phones should be available in all cases. Security and access control alarms can be installed for site security and environmental control.

2.5.12 *Radioactivity Monitoring System*

These systems have become popular at some large refuse disposal sites to detect if any prohibited radioactive wastes (typically from medical applications) might find their way into the MSW waste stream. Some WWTPs have also installed systems to monitor for the discharge of radioactive wastes into municipal wastewaters.

2.5.13 *Water Supply/Fire Protection*

A potable water supply may be needed for uses by onsite workers and for nonpotable uses such as controlling dust, cleaning equipment and haul vehicles, cleaning tires, and extinguishing fires. Municipal water supply to the site should be provided if feasible. But in many cases, the remote nature of such disposal sites dictates that alternative water supply will be needed.

Onsite water supply for extinguishing fires in the waste is a significant need at conventional refuse landfills and co-disposal sites of biosolids and refuse. Extinguishing fires with water is less of a need at biosolids monofills because the waste is typically noncombustible. A water supply for extinguishing onsite fires on equipment, in facilities, and in brush is advisable.

2.6 Operation and Maintenance

A biosolids landfill should be viewed as an ongoing construction site. Unlike conventional construction, however, the operating parameters of a biosolids landfill often change and may require innovative alterations and contingency plans. An effective landfill requires a detailed operational plan. Equipment selection should be compatible with biosolids characteristics, site conditions, and landfill method.

Operational procedures can be separated into those specific to the landfill method and those applicable to biosolids landfills in general. Method-specific procedures include site preparation and biosolids unloading, management, and covering. These procedures are discussed in detail in *Process Design Manual: Municipal Sludge Landfill* (U.S. EPA, 1978b). General procedures include scheduling, equipment selection and maintenance, management and reporting, safety, and environmental controls. These items are also discussed in the U.S. EPA manual.

2.6.1 *Operations Plan*

As with any construction activity, biosolids landfilling must proceed according to detailed plans and operating schedules. The operation plan should address all relevant method-specific or general operating procedures for the landfill, including hours of operation; measuring procedures; traffic flow and unloading procedures; special wastes handling; cover excavation, stockpiling and placement; maintenance procedures and schedules; inclement weather operations; environmental monitoring and control practices. An operations plan is an important tool for providing continuity of activities, monitoring and control of progress, and personnel training.

2.6.2 *Operating Schedule*

Features of the operating schedule include hours of operation, availability of qualified personnel, site preparation schedules, and equipment maintenance schedules. The hours of operation must be arranged so that the site is open when biosolids are to be received. If variations in the rate of receipt are expected during the day, then it may be desirable to schedule for equipment and personnel accordingly. The schedule may also need to provide for the application of daily soil cover.

2.6.3 *Equipment Selection and Maintenance*

Equipment selection depends upon landfill method, design dimensions, and biosolids quantity. Selection must be based upon the functions to be performed and the cost of individual equipment models.

2.6.4 *Biosolids Characterization*

Solids characterization during ongoing landfill operations must be considered. The purpose of ongoing characterization is to ensure the nature of the biosolids remains consistent and within the limits of the rules and permit conditions and to note any changes that may affect the ability of the site to operate properly. If the biosolids characterization alters over time, then changes in operational procedure may be required.

2.6.5 *Management and Reporting*

Management and reporting activities include maintenance of activity records, performance records, required regulatory reports, and cost records and onsite supervision and public relations activities. Performance records may be required as part of the regulatory process. Regulatory agencies may perform periodic inspections on a scheduled or an unscheduled basis.

2.6.6 *Safety/Emergency Procedures*

Providing a safe working environment at the landfill site should be a part of general operation and maintenance. Safety features should be built into the design, and certain practices must be followed daily to ensure safe working conditions. The operations plan should have a separate safety section and specific safety guidelines for each operation and feature of the landfill. Also, initial and ongoing training programs are appropriate.

2.6.7 *Environmental Controls*

Protection of the environment and of public health is the most important aspect of a landfill's operation. The operations plan should contain guidelines for providing this protection and actual operations should conform to these guidelines without compromise.

2.6.8 *Solids Handling*

The solids content of the delivered biosolids significantly affects the ability of the disposal site to handle the waste, bury it, and maintain its buried and managed condition once covered. Regulations do not require a minimum solids content at a biosolids monofill. Experience with most wastewater solids shows that a minimum of 15% solids content is required to maintain the cover system securely atop the waste without excessive leachate breakouts or soft spots in the landfill cover thereafter. If the waste has a solids content of less than 15%, then it can be mixed with soil or other stabilizers to increase its solids content and make it manageable.

2.6.9 *Vectors and Odors*

Vectors are the mechanism by which diseases and other agents deleterious to human health and comfort can be carried from the waste disposal site to receptors. At landfills, vectors typically include flies, mosquitoes, and other such flying insects and birds. Vectors also include burrowing and scavenging animals. In the context of the Part 503 regulations, grazing animals are prohibited from biosolids disposal sites during operation and after closure. Such grazing animals would also have the potential to serve as vector carriers.

Vector control is a specific intent of the Part 503 regulations. These regulations require application of daily cover to a site when the waste has not been treated in a manner to achieve certain reduced-pathogen levels. Daily cover is the best control for airborne insects and burrowing animals.

2.6.10 *Leak/Spill Prevention and Cleanup Procedures*

Leaks and spills of biosolids in transit from the source to a disposal site are a common risk and sometimes occur in any disposal operation. Every step should be taken to minimize and preferably prevent this. Contingencies must be developed so that when spills do occur, they are cleaned up promptly to avoid affecting humans or the environment.

2.7 Site Closure

When closing a biosolids landfill site, certain criteria must be met to make the site publicly acceptable. These criteria are established according to the type of landfill and the location, size, and ultimate use of the site. The procedures for closure should be included in the operations manual and updated or modified as the original landfill plan is altered.

2.7.1 *Ultimate Land Use*

The ultimate use of the site should be described and illustrated in the operation and maintenance manual or in a separate document describing site closure. The actual work involved in completing the site will depend on its ultimate use and on the care taken in day-to-day fill operations.

2.7.2 *Grading at Completion of Filling*

When each section of the landfill is completed, the final cover should be applied and fine-graded according to a predetermined plan. It is imperative that no biosolids become or remain exposed after the grading is completed.

2.7.3 *Suitability of Material for Final Grading*

Final grading of the site is to be performed after sufficient time has elapsed to allow for initial settlement. The final grading plan should be designed in accordance with the intended ultimate use of the landfill site. It is important that all biosolids be completely covered to the specified depth with cover material.

2.7.4 *Sedimentation and Erosion Control*

Sound practices for sedimentation and erosion control at biosolids disposal sites can start with local regulations for good management practices. Local regulations typically will dictate the design and performance standards for the site during its use. Next, the Part 503 and the Subtitle D rules stipulate additional standards that apply to biosolids landfills. Both sets of rules require design of surface water management systems onsite to accommodate a 25-year, 24-hour peak storm event for that particular locale.

2.7.5 *Landscaping*

The landscaping plan should reflect the intended ultimate use of the landfill site. Where practical, landscaping may be done on completed sections before the entire fill project is completed.

2.7.6 *Continued Leachate and Gas Control*

Because decomposition of the organics in the biosolids may continue even after the landfill has been completed, an ongoing monitoring and control program must be maintained. Leachate and gas must be controlled even after filling operations have stopped. Completion plans should clearly outline this program.

2.7.7 *Suitability of Biosolids for Cover Material*

Application of biosolids in cover materials at MSW landfills is permitted and practiced at many sites in many states. These MSW landfills may be refuse-only or co-disposal with biosolids sites. Mixing with cover soils can be done with daily, intermediate, and even final cover. Approval of such biosolids mixing in landfill covers is state by state, with many states accepting this practice, but a few not recognizing or allowing it.

2.7.8 *Application of Cover Material*

Application of cover soil at biosolids co-disposal sites with refuse follows the routine of a refuse-only landfill. By rule, daily cover is applied at a rate of 0.15 m (6 in.) of soil over the refuse and sludge working face at the end of each working day. Next, intermediate cover is applied when intermediate waste grades are achieved, which is typically when no additional waste deposits will be applied for 30 days or more. Intermediate covers typically are 0.3 m (12 in.) of soil. Final cover is applied when final grade is achieved. Final cover typically has at least 0.46 m (18 in.) of clay soil, may have a plastic membrane added, and then has additional layers of drainage zone and vegetation zone above that. The cover scheme above is as prescribed by the Subtitle D rules, and is sometimes enhanced under state regulation.

Application of cover soils at biosolids monofills follows a different scheme. As a practical matter, every biosolids monofill must be covered at some point to meet the definition of a landfill as having buried waste (in this case, biosolids only). Although Part 503 requires daily cover for any sludge or treated biosolids that has not met the pathogen reduction requirements, it does not prescribe a design specification for daily cover and makes no mention of intermediate or final cover.

Final covers should provide all the benefits of daily cover and should achieve the heightened goal of lasting for the minimum three-year postclosure period as stipulated in the Part 503 rules for maintenance and operation of the leachate management system and monitoring of landfill gas at monofill landfills. (The post-closure care period for co-disposal landfills is 30 years under Subtitle D.)

3.0 DEDICATED SURFACE DISPOSAL

3.1 Definition

Dedicated surface disposal (also referred to as dedicated land disposal) is the process in which biosolids are placed on land by injecting it below the surface or incorporating it into the soil after being sprayed or spread on the surface. Dedicated surface disposal sites receive repeated applications of typically liquid biosolids for the sole purpose of final disposal (U.S. EPA, 1995). Although application of dewatered biosolids is possible, it is not typical.

As with other land disposal techniques, dedicated surface disposal requires that wastewater sludge be stabilized before application. Once the sludge has been stabilized, however, it can be applied to dedicated land in either the liquid or dewatered state.

Because biosolids are placed on dedicated surface disposal sites at higher rates than are allowed when biosolids are used as a soil amendment, dedicated sites do not qualify as land application sites under Subpart B of the Part 503 regulations. Rather, Part 503 Subpart C specifically addresses active biosolids units (monofills, dedicated surface disposal sites, piles and mounds, and impoundments).

3.2 Suitability of Solids for Dedicated Surface Disposal

Part 503 regulates three characteristics of biosolids: the content of certain heavy metals, the level of pathogens, and the attractiveness of the biosolids to vectors. Stabilization of biosolids will typically be necessary to meet pathogen and vector-attraction reduction requirements of any site where biosolids are not covered at the end of each operating day.

3.3 Background

Dedicated surface disposal was first developed as an informal practice before Part 503 regulations to reduce high operational costs associated with biosolids disposal. Groundwater contamination, odor production, and aesthetic concerns typically were not addressed.

A more sophisticated approach to dedicated surface disposal was necessary as biosolids quantities increased with enhanced wastewater treatment, and onsite biosolids disposal was perceived as associated with environmental problems. Part 503 Subpart C regulations address dedicated surface disposal including siting restrictions and management practices to protect human health and the environment.

3.3.1 Rule 503 Subpart C

Dedicated surface disposal sites are covered by the Part 503 Subpart C regulation including management practices to help protect human health and the environment from potential adverse effects of pollutants in biosolids. Several management practices are required under Subpart C that pertain to dedicated surface disposal sites:

- Runoff from an active biosolids unit must be collected and disposed of properly;
- Runoff collection systems must be capable of handling a 25-year, 24-hour storm event;

- When an active biosolids unit has a liner, leachate must be collected and disposed of properly; and
- Biosolids placed in an active biosolids unit must not contaminate an aquifer.

More details are provided in Process Design Manual, Surface Disposal of Sewage Sludge and Domestic Septage (U.S. EPA, 1995).

Subpart C also requires the owner/operator of surface disposal sites to restrict public access. To protect groundwater, a liner for a dedicated land disposal unit may be necessary. A liner is a layer of relatively impervious soil, such as clay, or a synthetic material that covers the bottom of an active biosolids unit with a hydraulic conductivity of 1×10^{-9} m/s (3.281×10^{-9} ft/s) or less. The liner prevents the liquid in the disposal unit from seeping into the groundwater below. A leachate collection system is a system or device installed immediately above a liner that collects and removes leachate (liquid waste from rainfall that seeps through the contents of the active biosolids unit).

Biosolids placed on a dedicated surface disposal unit without a liner and leachate collection system must meet pollutant limits for arsenic, chromium, and nickel established under the Part 503 Subpart C regulation. In addition to pollutant limits, for units without a liner, the permittee must demonstrate that groundwater will not be contaminated from other constituents associated with biosolids including nitrates.

3.4 Site Selection

Selection of a dedicated surface disposal site must consider regulatory, technical, and economic requirements and constraints and public participation and acceptance. Several additional important considerations are addressed in Part 503 Subpart C: groundwater, erosion control and runoff management, topography, soil types, availability of sufficient land, water resources, community infrastructure, zoning, distance from receptors, geology, endangered species, proximity to wetlands or watersheds, and archaeological or historic significance.

3.5 Storage

Storage of biosolids is typically necessary at dedicated surface disposal sites because of adverse weather and other factors that prevent the continuous spreading of biosolids. Liquid biosolids are typically stored in lined lagoons or metal tanks. Dewatered biosolids is typically stored by mounding in areas protected from runoff. Methods for calculating biosolids storage requirements at dedicated land disposal sites are presented in *Surface Disposal of Sewage Sludge and Domestic Septage* (U.S. EPA, 1995). *Guide to Field Storage of Biosolids* provides discusses recommended practices for odor prevention and mitigation, spill prevention and response, and runoff management practices (U.S. EPA, 2000).

3.6 Operational Methods and Equipment

The methods used at dedicated surface disposal sites are dictated by biosolids characteristics, cost, and aesthetics (e.g., odors or other community concerns) (U.S. EPA, 1995). Methods for biosolids disposal for a dedicated surface disposal site include

- Subsurface methods for liquid biosolids, including subsurface injection or plow or disc covering;
- Surface spreading of liquid biosolids by tank trucks or tank wagons;
- Spraying of liquid biosolids; and
- Surface spreading of dewatered biosolids.

With all of these disposal techniques, the biosolids eventually becomes incorporated into the soil, either immediately by mechanical means or over time by natural means.

The technique used to apply biosolids to the land can be influenced by the means used to transport the biosolids from the WWTP to the dedicated surface disposal site. These methods include (U.S. EPA, 1995)

- One transport vehicle both hauls biosolids from the WWTP to the site and spreads it on land.
- One type of transport vehicle, typically with a large-volume capacity, hauls biosolids from the WWTP to the dedicated surface disposal site. At the site, the biosolids is transferred to an application vehicle, into a storage facility, or both;
- Biosolids are pumped and transported by pipeline from the WWTP to a storage facility at the site. Biosolids are then transferred from the storage facility to the biosolids application vehicle.

Land spreading of biosolids by gravity irrigation/flooding is not recommended because of difficulty achieving uniform spreading rates, clogging of soil pores, and tendency of biosolids to turn septic with resulting odors. *Surface Disposal of Sewage*

Sludge and Domestic Septage provides more details on liquid biosolids spreading methods (U.S. EPA, 1995).

3.6.1 *Dewatered Biosolids*

Application of dewatered biosolids (20% solids or more) is similar to application of solid or semisolid fertilizers, lime, or animal manure. Dewatered biosolids can be spread with bulldozers, loaders, graders, or box spreaders and then plowed or disked in.

The principal advantage of using dewatered biosolids is that conventional equipment for application of fertilizer and lime and for tillage can be used. The disadvantage is the higher energy and operational costs associated with biosolids dewatering and the treatment required for the resulting sidestream. The disadvantages appear to outweigh the advantages, because dewatered biosolids are infrequently used for dedicated surface disposal units.

3.6.2 *Solids Application Rates*

The disposal rate of biosolids to dedicated surface disposal sites is limited based on Part 503 requirements. Part 503 limits pollutants in biosolids for surface disposal sites if the site does not have a liner and leachate collection system. Representative samples of biosolids must be tested for arsenic, chromium, and nickel if the site does not have a liner and leachate collection system.

If the biosolids meet Part 503 pollutant limits, or if a liner and leachate collection system is used, then other factors to consider for controlling application rates include (U.S. EPA, 1995):

- Rate of biosolids that can be disposed during each spreading activity while still maintaining aerobic conditions in the soil;
- Method of biosolids disposal, soil drainage, soil characteristics;
- Number of days during the year when biosolids can be disposed, as dictated by weather conditions, ability of the biosolids spreading equipment to operate with existing soil conditions, and equipment breakdown and maintenance requirements; and
- Evaporation rates of biosolids liquids.

3.7 Environmental Controls

Control of transport mechanisms for potential pollutants via surface and groundwater and through aerosols and odors is required. Environmental controls for dedicated surface disposal include leachate control if the site has a liner, runon/runoff controls, possible groundwater pumping, and management practices for controlling odors (e.g., drying periods). If the biosolids are stabilized, then vector controls should be negligible (U.S. EPA, 1995).

3.7.1 *Site Layout*

Good site planning is the key to environmental pollution control for dedicated surface disposal sites. Division of the dedicated surface disposal site into several disposal fields is desirable for operational and environmental controls. Individual fields ranging of 4 to 40 ha (10 to 100 ac), with average of 20 ha (50 ac), are typical.

3.7.2 *Groundwater Controls*

As described previously, Part 503 requires proof that the biosolids placed on an active biosolids unit is not contaminating an aquifer. This proof must be either the results of a groundwater monitoring program or use of a liner and leachate collection system. It is difficult to demonstrate that groundwater will not be contaminated in the absence of a liner, unless groundwater is very deep and protected by a natural clay layer.

In cases where there is no liner and leachate collection system (e.g., the site has a natural protected layer), groundwater monitoring is typically necessary to evaluate migration patterns from the site and for ensuring that the quality of external waters are not reduced.

3.7.3 *Surface Water Runoff Controls*

Dedicated surface disposal sites should be graded so that all surface runoff drains toward one point near the edge or corner of the field. Each site should be surrounded by a berm to keep uncontaminated surface run-on out and to contain contaminated disposal site runoff. A center drain should either direct the contaminated runoff back to a biosolids storage area or to a pump that directs the runoff back to the treatment plant or separate onsite treatment system. Temporary holding of the runoff to permit settling of settleable solids and monitoring may be desirable. If the stored water is found to be of sufficient quality to meet discharge standards, then it can be released without any treatment, although an NPDES permit would be required if the discharge is to surface water.

3.7.4 *Air Pollution Control*

The two air pollution concerns are aerosol transport and odor. There must be adequate buffer zones around the dedicated surface disposal site to mitigate the offsite transport of aerosols and odors. Operationally, systems that minimize the length of time biosolids are directly exposed to the air are preferred. It is possible to incorporate special design features for air pollution control, for example, vacuum stripping of the digested biosolids to remove before land application.

3.7.5 *Site Monitoring*

Monitoring requirements for dedicated surface disposal sites are relatively straightforward. Groundwater monitoring is essential, unless the site is lined and has a leachate collection system, and should be conducted from a pattern of groundwater wells located primarily at the downstream boundary of the site.

In addition to site monitoring, if the site does not have a liner and leachate collection system, then the Part 503 pollutant limits for surface disposal must be met. The owner/operator must then monitor the biosolids for the regulated pollutants and ensure that certain pathogens and vector-attraction reduction requirements are being met.

3.7.6 *Control of Pathogens, Vectors, Odors, and Dust*

Biosolids for dedicated surface disposal sites must meet the Part 503 operational standards for pathogen and vector-attraction reduction. Part 503 requires that the pathogen density be reduced through certain processes. The rule also contains management practices that help ensure that pathogens will not regrow.

Odors can be managed by preventing anaerobic conditions of biosolids by using drying cycles at the dedicated surface disposal site. Lime or chemical masking agents can be applied to reduce odor problems. An effective means of reducing odors is to limit storage of the biosolids. To minimize dust, access roads should be graveled or water can be applied.

3.7.7 *Noise Control*

Noise sources at surface disposal sites include operating equipment and haul vehicles. These noises typically are similar to that generated by any heavy construction activity and are confined to the site and the streets used to bring biosolids to the site. To minimize this effect, every effort should be made to route traffic through the least populated areas. Further, the site can be isolated so that the noise cannot carry to nearby neighborhoods (U.S. EPA, 1995). The use of earthen berms and trees as noise barriers may be effective. On the site, existing Occupational Safety and Health Act (OSHA) standards will govern noise protection for employees.

4.0 REFERENCES

Clean Water Act. *U.S. Code*, Section 1251, Title 33, 1972.

Resource Conservation and Recovery Act. *U.S. Code*, Section 42, Title 6901, 1976.

U.S. Environmental Protection Agency (1974) *Disposal of Sewage Sludge into a Sanitary Landfill*; SW-71d; U.S. Environmental Protection Agency, Office of Solid Wastes, Washington, D.C.

U.S. Environmental Protection Agency (1978a) *Principals and Design Criteria for Sewage Sludge Application on Land*; Sludge Treatment and Disposal, Part 2; EPA-625/4-78-012; U.S. Environmental Protection Agency, Environmental Research Information Center: Cincinnati, Ohio.

U.S. Environmental Protection Agency (1978b) *Process Design Manual: Municipal Sludge Landfill*; EPA-625/1-78-010, SW-705; U.S. Environmental Protection Agency, Environmental Research Information Center, Office of Solid Wastes: Cincinnati, Ohio.

U.S. Environmental Protection Agency (1978c) *Subsurface Disposal of Municipal Wastewater Treatment Sludge*; SW-167c; U.S. Environmental Protection Agency: Office of Solid Wastes, Washington, D.C.

U.S. Environmental Protection Agency (1979a) *Process Design Manual for Sludge Treatment and Disposal*; EPA-625/1-79-011; U.S. Environmental Protection Agency, Center for Environmental Research Information: Cincinnati, Ohio.

U.S. Environmental Protection Agency (1979b) *Regulations on Public Participation in Programs Under the Resource Conservation and Recovery Act, The Safe Drinking Water Act, and The Clean Water Act*; 40 CFR 25, 44 CFR 10292; U.S. Environmental Protection Agency, Office of Waste and Hazardous Materials: Washington, D.C.

U.S. Environmental Protection Agency (1991) *Criteria for Municipal Solid Waste Landfills*, Code of Federal Regulations, Part 258, Title 40, 1991; *Fed. Regist.*, **56**, 51016.

U.S. Environmental Protection Agency (1993) *Standards for the Use and Disposal of Sewage, Final Rule*. Code of Federal Regulations, Part 503, Title 40, 1993; *Fed. Regist.*, **58**, 9248.

U.S. Environmental Protection Agency (1994) *A Plain English Guide to the EPA Part 503 Biosolids Rule*; EPA/832/R-93/003; U.S. Environmental Protection Agency, Office of Wastewater Management: Washington D.C.

U.S. Environmental Protection Agency (1995) *Surface Disposal of Sewage Sludge and Domestic Septage*; EPA-625/K-95/002; U.S. Environmental Protection Agency, Office of Research and Development, Washington, D.C.

U.S. Environmental Protection Agency (2000) *Guide to Field Storage of Biosolids*; EPA/832-B-00-007; U.S. Environmental Protection Agency, Office of Wastewater Management: Washington, D.C.

Chapter 25

Emerging Technologies

(continued)

(*continued*)

1.0 INTRODUCTION

New technologies for treating sludge and biosolids are continuously being developed. Some of these technologies arise from the need for a technology to address changing regulatory, economic, or social issues affecting current solids management practices. Others arise from modifications or enhancements of existing technologies. Most take years of research, development, and testing to be ready for the municipal market, and often, several more years before they are accepted as proven processes within the industry. The purpose of this chapter is to provide an introduction to some of these emerging technologies along with sufficient information to determine the potential merits of a technology for specific applications.

2.0 APPROACH

More technologies are being introduced to the market than can be reasonably addressed in this chapter. Therefore, for the purposes of this document, an emerging technology was defined as one that meets one of the following criteria:

- Pilot-scale testing at municipal wastewater treatment plants (WWTP) in North America or Europe over the last five years;

- Limited full-scale applications at North American WWTPs;
- Full-scale application at European facilities but not at North American facilities.

New solids processing technologies are identified and summarized in *Emerging Technologies for Biosolids Management* (U.S. EPA, 2006). The technologies were screened to identify those that were considered "innovative," which is similar to the emerging classification in this document. It also included technologies that were considered "embryonic" (i.e., still in their infancy). Many of the innovative technologies identified in the 2006 manual are expanded on in this chapter, but embryonic technologies are not included.

3.0 THICKENING AND DEWATERING PROCESSES

Thickening and dewatering processes remove water from sludge and biosolids, making them easier and less expensive to further process and transport. This chapter focuses on emerging technologies that may offer unique advantages related to thickening and dewatering.

3.1 Membrane Thickening

3.1.1 Identification of Process and Scientific Basis

Membrane thickening is the result of the development and implementation of membrane bioreactors (MBRs) for secondary wastewater treatment. Thickening basins are filled with waste activated sludge (WAS), which is concentrated by the withdrawal of water through the pores of the membranes. The WAS can be thickened to 4% total solids or greater.

The tank is typically aerated to maintain aerobic conditions and reduce membrane clogging and blinding. Modules of membranes may either be tubular or sheet, and the thickener can be operated as a batch or continuous-flow process.

3.1.2 Process Description

Membranes are composed of two basic materials: organic polymers and inorganic materials such as ceramics. Organic polymer-based membranes are typically used for municipal wastewater treatment and are formed from either modified natural cellulose acetate materials or synthetic materials. Modules are of the following types:

tubular, hollow-fiber, spiral-wound, plate-and-frame, and pleated cartridge filters. The type of module selected will depend on the desired application.

Some membrane thickeners are designed with the membranes immersed in the reactors. Other applications locate the membranes in a separate stage or compartment. Any type of membrane can be used in either application.

One issue with this system is the potential for membrane fouling, which is the systematic accumulation of suspended solids, colloids, precipitates, and macromolecules on the membrane surface or inside the pores. Fouling will reduce membrane permeability. Techniques to control membrane fouling include chemical washing and cleaning and air-scour and permeate back-pulsing to prevent cake layer formation.

Membrane thickening systems can require less space than other systems. Membrane units are flexible and can handle varying flow rates and addition or subtraction of units as necessary. Sludge levels must be maintained so that membranes remain immersed. Throughput limits are set by the physical properties of the membrane. Typically, peak design flows should be no more than 1.5 to 2.0 times the average design flow.

Membranes should be used in an aerobic environment to achieve separation of liquid from biomass. Anaerobic environments have plugged membranes too quickly in the past. Therefore, aerobic environments are needed for oxygen mixing.

Some typical design parameters for hollow fiber and flat sheet membranes are presented in Table 25.1.

3.1.3 *Status of Development*

By December 2009, membrane thickeners had been installed in two locations in the United States. Full-scale facilities were in use at the Village of Dundee WWTP in

TABLE 25.1 Typical design parameters for membrane thickeners.

Parameter	Hollow-fiber membranes	Flat sheet membranes
Design flux		
Average	325–490 $L/m^2 \cdot d$	490–650 $L/m^2 \cdot d$
Maximum	815–1140 $L/m^2 \cdot d$	980–1300 $L/m^2 \cdot d$
Peak	< 1467 $L/m^2 \cdot d$	1630 $L/m^2 \cdot d$
Scouring air	0.23–0.29 $m^3/m^2 \cdot h$	0.51 – 0.54 $m^3/m^2 \cdot h$

Dundee, Michigan, and at the Cauley Creek Water Reclamation Facility in Fulton County, Georgia.

3.1.4 Pre- and Post-Treatment

Waste activated sludge needs to be free of objects that could clog or damage the membranes. Therefore, fine screening at 2 mm at some point in the solids processing before membrane thickening is required. Also, the WAS should be aerobic to avoid any septicity problems in the thickener that could also damage the membranes or cause odor or foaming problems. Post-treatment will depend on the ultimate use or disposal method.

3.1.5 Environmental Effects

This technology is not anticipated to have adverse environmental effects, and it should improve the efficiency of downstream processing and transportation, potentially reducing environmental effects. The compact nature of this process should facilitate covering of the thickeners for odor control if necessary.

3.1.6 Technology Prognosis

The potential for this technology appears good. Although the number of installations at the time of development of this manual were limited, it is likely to grow as the technology becomes more widely known. Some of the manufacturers supplying this technology include Enviroquip, Inc., Infilco Degremont, Mitsubishi International Corp., US Filter/MEMCOR, Veolia Water Solutions and Technologies, and GE Power and Water. Inc.

3.2 Electro-Dewatering

3.2.1 Identification of Process and Scientific Basis

The objective of electro-dewatering is to enhance conventional dewatering performance using an electric current. The principle of electro-dewatering is the movement of a polar liquid through a membrane or other porous structure under the influence of an applied electric field. When applied to dewatered sludge cake, a flow of ions migrates from the anode to the cathode. This movement drags water molecules out of the biosolids. Electro-dewatering is used in conjunction with conventional dewatering to remove surface and interstitial water around and inside the floc, thereby enhancing the removal of free water.

Electro-dewatering is based on the principle of electro-osmosis, which describes the movement of a polar liquid through a porous media. With electro-osmosis, the rate of flow of water through a porous media is proportional to the voltage gradient and cross-sectional area. This is similar to Darcy's law for flow through a porous medium under the influence of a hydraulic gradient, but hydraulic permeability is a function of the size of the pores in the material and of type of particles forming the porous material.

Electro-osmotic permeability is not dependant on particle size and remains constant for different materials that support electro-osmosis. Therefore, electro-osmosis offers an advantage over hydraulic flow (mechanical dewatering) when applied to materials such as wastewater solids, which are typically comprised of colloidal sized particles (Decker et al., 2006). During the application of direct current, cations in the sludge are attracted to the cathode. Movement of the cations helps carry water towards the cathode (i.e., electrophoresis), where it can be released from the sludge (McKay et al., 2007).

3.2.2 *Process Description*

The point of application of the electric current varies between system suppliers. Some require a thickening or dewatering step before electro-dewatering. Once a cake is formed, the direct current will cause electro-osmosis to occur. Ions will migrate to the appropriate electrode to compensate for particle charges. Electro-dewatering can be combined with conventional belt filter presses or other dewatering technologies. According to a recent study by the Water Environment Research Foundation (2006), the cost–benefit of electro-dewatering is likely to be greatest for sludge not responding well to traditional pressure filtration. The study demonstrated that the electro-dewatering technique is applicable to a wide range of sludge, and performance might be limited for sludge with high conductivities.

3.2.3 *Status of Development*

This technology was categorized as innovative by the U.S. EPA (2006) in its recent manual *Emerging Technologies for Biosolids Management*. Electro-dewatering has been attempted by a variety of groups since the 1920s with limited success. Recent bench-scale tests have provided favorable results. For example, a bench-scale demonstration was conducted by the Electric Power Research Institute (2003) involving an electroacoustic dewatering process using electric and ultrasonic fields. The test demonstrated improved water removal over a conventional belt filter press by approximately 8%.

Several manufacturers have developed full-scale machines that have been used for large-scale tests or full-scale applications, such as the units tested at the Achille Gagnon WWTP in Victoriaville, Quebec, Canada, from March 2005 through

December 2006 (McKay et al., 2007). Testing at Achille Gagnon WWTP compared mechanical dewatering to electro-dewatering after mechanical dewatering. The WAS was mechanically dewatered using screw presses to approximately 16% total solids before electro-dewatering and total cake concentrations ranging from 25 to 45% total after electro-dewatering. Dryer feed solids to electro-dewatering were thought to reduce the final product total solids. The higher polymer dosage associated with the higher cake total from the screw press was suspected of interfering with the electro-dewatering process.

The tests at Victoriaville also indicated that electro-osmosis affects pathogen levels in wastewater. This is consistent with other studies where pathogenic bacteria and viruses were reduced to below detectable levels in electro-dewatered biosolids. Tables 25.2 and 25.3 contain the data from Victoriaville's testing of electro-dewatering.

3.2.4 *Pre- and Post-Treatment*

Depending on the equipment manufacturer, sludge may require some thickening or dewatering before application of the electric current. Post-treatment requirements will be dependent on the final use or disposal practices.

3.2.5 *Environmental Effects*

No environmental effects are associated with using this technology at this time.

3.2.6 *Technology Prognosis*

This technology should have a good opportunity to continue to progress and develop. It has demonstrated favorable results at the Achille Gagnon WWTP since 2005. As the number of installations increases and equipment and system improve, it should be

TABLE 25.2 Bacterial pathogen destruction following electro-dewatering.

Sample Source	E. coli MPN/g[a]	Salmonella[b] MPN/g
Mechanically dewatered samples	94 000	71
	1 900 000	< 5
	33 000	5
Electro-dewatered samples	< 7	< 3
	< 7	< 3
	< 8	< 3

[a] MPN/g = most probable number/gram of dry solids.

[b] Salmonella analysis = MA.700-Sal-tm 1.0.

TABLE 25.3 Viral pathogen destruction following electro-dewatering.

Sample source	Enteric viruses (plaques) [PFU/4 g][a,b]	Enteric viruses (HISG-immunoperoxidase) [MPNIU/4 g][c]	Somatic coliphages [PFU/4 g]	F-specific coliphages [PFU/4 g]
Mechanically dewatered	< 0.08	< 0.16	608	5000
	< 0.08	0.28	280	1024
	0.08	0.08	10 120	3920
Electro-dewatered	< 0.08	< 0.08	< 0.08	< 0.08
	< 0.08	< 0.08	< 0.08	< 0.08
	< 0.08	< 0.08	< 0.08	< 0.08

[a] All results are expressed as liter per 4 g of humid sample as submitted to the laboratory.
[b] PFU = plaque-forming units on MA-104 cells.
[c] MPNIU = most probable number of infectious units on MA-104 cells.

a candidate for consideration in enhancing dewatering performance. Manufacturers include Elcotech Technologies, Inc., Waste Technologies of Australia, and Ashbrook-Simon Hartley, Inc.

3.3 Geotextile Tube Dewatering

3.3.1 Identification of Process and Scientific Basis

Geotextile tube dewatering is a low-maintenance, cost-effective dewatering solution using tubes made from a high-strength polypropylene fabric similar to that used for belt filter presses. The concept is for free water in sludge to filter through the fabric, leaving the dewatered biosolids in the tube. Dewatered solids within the bag can be allowed to sit for optimum dewatering before transporting the solids to beneficial use or disposal. Hydrostatic loading of the biosolids and constriction of the fabric apply pressure to the biosolids. The tubes need to be placed in a location where the drained water can be collected for treatment.

3.3.2 Process Description

As the large geotextile tube is filled by pumping, the tube retains the floc and allows effluent water to permeate through the tube wall. Polymer conditioning is needed to improve the filtration process. Volume reduction within the tube allows for repeated filling. After the final cycle of filling and dewatering, dewatered solids continue to consolidate by desiccation because water vapor escapes through the fabric.

A rapid dewatering test can be done by placing 75 to 150 L of biosolids to be dewatered into a small hanging bag and observing the ability of water to pass through the fabric over time. Based upon test results, the number of bags required to dewater a specific volume of biosolids within acceptable time limits can be calculated. For continuous dewatering, previously filled bags would need to be emptied as new bags are filling. If this is optimized, the amount of space needed can be minimized.

One advantage of a geotextile system is the limited labor required for operation. In addition, odor containment is because the process is enclosed. Disadvantages include the space required and susceptibility to climate conditions. Freezing temperatures can be a problem for this type of system, especially for the polymer conditioning system.

3.3.3 *Status of Development*

This technology was categorized as innovative by the U.S. EPA (2006) in its recent manual *Emerging Technologies for Biosolids Management*. Studies in California, Georgia, Ohio, and New Hampshire have shown improved dewatering of sludge, successful containment of odors, reduction in effluent suspended solids, and cost savings resulting from the use of the geotextile containers (Mastin and Lebster, 2006). The technology has also been used by a regional contractor at multiple locations in Iowa including Clinton, Wapheton, Royal, and West Liberty (Wienands, S. 2010).

3.3.4 *Pre- and Post-Treatment*

Biosolids should be relatively free of large particles that could cut or damage the fabric. Depending on the source of the solids (e.g., a lagoon), some screening or grinding may be required before filling the geotextile tube. In addition, polymer conditioning tests will be needed to identify the best polymer for the application. Filtrate will need to be collected for treatment. Post-treatment requirements for the dewatered solids will be dependent on the final use or disposal practices.

3.3.5 *Environmental Effects*

Environmental effects appear to be minimized by enclosing the solids within the fabric bags.

3.3.6 *Technology Prognosis*

This technology appears to be best suited for dewatering applications at small treatment facilities. Dewatering of solids stored in lagoons may be a niche application as well. Use medium and large facilities could prove difficult because of logistics and space requirements of this technology. Increased use of this technology is expected in the future.

4.0 THERMAL PROCESSES

4.1 Supercritical Water Oxidation

4.1.1 Identification of Process and Scientific Basis

Supercritical water oxidation (SCWO) is a process for thermal oxidation of the organics in wastes. It has been used to treat a broad range of aqueous organic wastes such as polychlorinated biphenyls (PCBs), pesticides, cyanide wastewater, and wastewater sludge. The process involves increasing the temperature and pressure of the aqueous waste in a reactor to oxidize the organic compounds. The process is also known as supercritical wet-air oxidation, supercritical wet oxidation, and hydrothermal oxidation.

The natural solvent properties of water are enhanced at supercritical conditions, which are achieved by raising the temperature and pressure of the water to above its critical point of 374 °C and 22 100 kPa (705 °F and 3204 psi). At the supercritical point, the liquid and gas phases are indistinguishable. Under these conditions, gaseous oxidants such as oxygen become completely miscible in the supercritical water, and the solubility of organics in the water is significantly increased. This results in rapid oxidation of the organic compounds, with times ranging from a few seconds to a few minutes (Yesodharan, 2002). Tests with wastewater sludge have shown chemical oxygen demand (COD) reductions of greater than 99.9% (Griffith et al., 2002; Stendahl and Jafrerstrom, 2003).

An oxidizing agent such as oxygen (air) is typically added to promote the oxidation process. Organic compounds are oxidized to carbon dioxide (CO_2) and water, and nitrogen compounds are oxidized primarily to nitrogen gas with minimal formation of nitrogen oxides. Inorganic salts and some minerals are almost insoluble and form precipitates. Phosphorous forms phosphoric acid and sulfur forms sulfuric acid. The residue for disposal is a wet inert material consisting primarily of silt, sand, and clay particles, which dewaters well (Boulding, 1996; Svanström et al., 2004).

4.1.2 Process Description

Although the thermodynamic principles used to achieve SCWO are the same, the method to achieve supercritical conditions will vary with the equipment suppliers. Typically, the sludge is thickened to increase COD concentrations to desired levels and to reduce equipment size. Some pilot tests have been conducted with sludge dewatered to as much as 35% total solids and diluted to 17% total solids for pumping. Following thickening, the sludge is ground to improve homogeneity and then pressurized to approximately 25 000 kPa (3626 psi). The pressurized sludge is heated to 400 to 600 °C (752 to 1112 °F) and conveyed to a reactor where oxygen is added.

The oxidation process is exothermic. Heat can be recovered for use in the process to reduce energy needs. An external heat source is required for initiation; however, once operational temperatures have been achieved, the process may become self-sustaining if the waste contains sufficient organic concentrations. Chemical oxygen demand concentrations of 15 000 mg/L or greater are required for the process to be self-sustaining (Boulding, 1996). One supplier of a proprietary SCWO system for wastewater sludge has indicated that organic concentrations of 3% or higher are needed for energy recovery to be possible (O'Regan, 2008).

Following the oxidation process, treated flow can be sent through heat recovery systems followed by solids/liquid separation processes. The separated inert solids can be disposed in a landfill or possibly beneficially used. Preliminary testing done at a full-scale demonstration in Harlingen, Texas, indicated that the residual could comply with all the metals limitations of 40 CFR Part 503 (58FR 9248 to 9404) for beneficial use (Griffith et al., 2002; U.S. EPA, 1993). Some researchers have demonstrated recovery of phosphorous from the process (Stendahl and Jafrerstrom, 2003). Reported effluent quality has varied. Effluent at Harlingen contained relatively high ammonia levels, warranting additional treatment (Griffith et al., 2002).

4.1.3 *Status of Development*

Testing of SCWO with wastewater sludge has been conducted since the 1980s. Several large pilot facilities have been constructed and operated, but no full-scale systems have been operated for extended periods. The first large demonstration facility was constructed in Harlingen, Texas, in 2000, and test operations began in 2001. The facility included two 2725 L/h (12.5 gpm) trains that ran periodically for several months. The feed sludge was thickened anaerobically digested sludge. Sludge was thickened to approximately 4.5% total solids, but the goal was to ultimately feed 6% total solids to increase the COD concentration. Chemical oxygen demand reductions of greater than 99.9% were reported. Organic nitrogen conversion to elemental nitrogen was not as high as expected, and ammonia concentrations ranged from 410 to 2075 mg/L in the underflow (Griffith et al., 2002). One of the goals of the project was to recover carbon dioxide and sell it to a neighboring industry, but the industry closed soon after the start of the pilot program. The facility had numerous start-up mechanical and process issues associated with grit and conveyance of thickened sludge. The plant was shut down in 2001 and the developer is no longer actively pursuing marketing of the system.

In 2006, the city of Orlando, Florida, began development of a full-scale demonstration project at the Iron Bridge Regional Water Reclamation Facility using a system similar to that demonstrated at Harlingen. The facility was scheduled to begin operation by mid-2009, but as of early 2011 it had not achieved continuous operation. Energy recovery from the exothermic process is expected to provide the heat needed to sustain the process (Sloan et al., 2008).

A European company is working to commercialize the SCWO process. SCFI Group, located in Cork, Ireland, United Kingdom, recently acquired a proprietary SCWO technology from Chematur Engineering of Sweden. The process is modular with a capacity of 5 dry Mg/h (5.5 dry ton/hr). The process includes heat recovery and proposes to generate electricity using excess heat. SCFI Group has a demonstration plant operating in Cork with a capacity of 250 L/hr at 22% total solids (O'Regan, 2008).

4.1.4 *Pre- and Post-Treatment*

The SCWO process is well-suited for processing dilute organic wastes that are not easily incinerated. Therefore, minimal preprocessing of wastewater sludge is required. Some thickening may be required to achieve the organic loading needed for the process to be self-sustaining.

The inorganic residual may need to be mechanically dewatered before disposal. Available literature indicates the residue separates easily from the water, but no information was found discussing dewatering systems and performance. It is expected that the material will dewater similar to the ash slurry from a conventional incineration process. The Harlingen, Texas, demonstration project disposed of the residual on a dedicated land disposal site without any treatment (Svanström et al., 2004).

4.1.5 *Environmental Effects*

The emissions consist primarily of CO_2 with low levels of nitrous oxide (N_2O) and volatile organic compounds (VOCs). Recovery of CO_2 has been a feature of several research and demonstration projects. Treatment requirements will depend on the location and governing agency.

No testing of the residual for pathogen reduction was reported in the reviewed literature; however, it is anticipated that the residuals should be pathogen-free because of the temperatures and pressures of the process. Destruction of the organics should render the product nonputrescible and, as a result, should not be an attraction to vectors. Griffith and Raymond (2002) indicated that feasibility tests showed the Harlingen residuals should be able to comply with Part 503 metals limitations for land application, but test results from the full-scale demonstration project had not been completed (U.S. EPA, 1993).

The primary contaminant of the liquid sidestream is ammonia. Ammonia concentrations as great as 2075 mg/L were observed in the return effluent at Harlingen, Texas (Griffith, 2002). The need for treatment of the sidestream before returning to the liquid treatment process will be site-specific.

4.1.6 Technology Prognosis

Several challenges are associated with implementation of a full-scale SCWO system. The high temperatures and pressure warrant specialized equipment and safety procedures. In addition, the equipment must be able to withstand the corrosive conditions created by the process. Special alloys may be required to address this issue. Salt precipitation can also create scale buildup that can impede flow through the reactor vessel and heat exchanger (Yesodharan, 2002). Research into SCWO has demonstrated the potential for the process, but additional work with larger scale systems will be needed. As additional projects are brought online, solutions to these obstacles will likely be identified, and long-term viability of the process can be assessed.

4.2 Sludge-to-Fuel

4.2.1 Identification of Process and Scientific Basis

Several different proprietary configurations have been developed and tested for converting municipal sludge to a carbon-rich fuel. These processes use high pressure and heat to convert the organic matter in sludge into a carbon-rich material (carbonization). The carbonized sludge can be readily dewatered or dried to generate a product that can be used as a fuel.

The processes involve pressurizing the solids above the saturated steam pressure to prevent boiling when heated above the normal boiling point. This eliminates the loss of thermal energy through evaporation and minimizes the additional thermal energy requirement. The solids are then heated to alter the molecular structure of the solids slurry, and carbon dioxide gas is removed from the solids. This reduces the mass of solids by approximately 40% and reduces the affinity of the solids to water. The resulting "carbonized" waste is then cooled, depressurized, and dewatered. The product has an energy value ranging from approximately 14 to 21. MJ/kg (6000 to 9000 Btu/lb), depending on post-processing.

4.2.2 *Process Description*

The biosolids are received from the WWTP at 15 to 30% total solids. The sludge is macerated and then pressurized above the saturated steam pressure to prevent boiling. Under this pressure, the slurry remains a liquid throughout processing, and thermal energy inputs for the evaporation of water are minimized.

Following pressurization, the slurry is moved through heat exchangers to raise its temperature. The actual temperature depends on the proprietary process, but varies from approximately 200 °C to more than 260 °C (400 to 500 °F). This causes the cellular structure of the biological sludge to rupture and the carboxyl groups of the organic molecules to break off and release as carbon dioxide gas. This reaction is known as decarboxylation. Decarboxylation improves the morphology of the solids, causing them to become hydrophobic with a negative surface charge.

Following decarboxylation, the slurry undergoes partial cooling and depressurization. The slurry is then dewatered. Centrifuges have been able to attain at least 50% total solids in demonstration facilities. The dewatered product can be used as is, or further dried.

4.2.3 *Status of Development*

The first commercial-scale project to use this technology is EnerTech Environmental's SlurryCarb facility at the Rialto, California, WWTP. EnerTech began startup testing of the 800 wet Mg/d (883 wet ton/d) facility in October 2008 with full-scale operation starting at the end of 2009. As of January 2010, the facility had not achieved continuous operation.

ThermoEnergy Corp. has developed a sludge-to-fuel process called ThermoFuel. This system is based on the sludge-to-oil-reactor-system (STORS) developed by Batelle Memorial Institute. A STORS system was demonstrated by ThermoEnergy at Colton, California, from early 1999 to late 2000. The testing was sponsored by U.S. EPA.

4.2.4 *Pre- and Post-Treatment*

The level of pretreatment required will depend on the proprietary system used. One system is configured to process dewatered sludge; the other is designed to process thickened sludge. Sludge should be free of large trash and debris to prevent fouling of the process.

Post-treatment will be required for emissions and sidestreams. The level of treatment will be dependent on the system. Sidestreams are expected to be high in organics and solids, and may require pretreatment before being returned to the liquid treatment train. One supplier uses an anaerobic/aerobic process before discharging the treated sidestream. Odor control may be required for the sidestream treatment.

4.2.5 Environmental Effects

Emissions control will be required to deal with the volatile organics released during the thermal treatment and drying processes. Emissions control should mitigate potential environmental effects. Energy recovery from the process will minimize natural gas needs and help reduce the overall carbon footprint of the process. Use of the product to replace coal for cement production will further reduce the carbon footprint.

4.2.6 Technology Prognosis

The first full-scale facility has experienced difficulties during startup and has not been able to operate continuously. Plant modifications were being made at the time this document was prepared. It will take time before it is known if this process is viable. The complexity of the process will likely require specialized expertise to implement and operate the system. If full-scale operation proves to be successful, the prognosis for this technology should be good, especially for areas with high population density that generate the large quantities of biosolids needed for the process to be cost-effective.

4.3 Plasma-Assisted Thermal Oxidation

4.3.1 Identification of Process and Scientific Basis

Plasma-assisted thermal oxidation is a proprietary process that uses a low-power plasma arc torch to sustain the oxidation process in a rotary kiln. Plasma arc technology is designed to oxidize biosolids at temperatures lower than that of conventional combustion systems, which is projected to improve the energy efficiency of the combustion process. Hydro Québec, Canada's largest electric utility, developed and patented the plasma-assisted sludge oxidation (PASO) process in 1997 for treating solid wastes from pulp and paper mills. In 2001, Hydro Québec licensed the production and marketing of the system to Fabgroup Technologies, Inc., a Montreal-based custom-metal equipment manufacturer (Metro Wastewater Reclamation District, 2006; Mulhern, 2008). Fabgroup Technologies has three manufacturing plants and has been in operation more than 50 years (Mulhern, 2008).

Plasma oxidation uses electricity to highly ionize a gas (air or inert gas) to produce a high-temperature plasma. Plasma oxidation is similar to pyrolisys in that it requires little excess oxygen. The process has been used to generate high temperatures (> 1600 °C) for the destruction of specialty wastes, such as medical and military wastes. Plasma melting has been used in Japan since 1993 to solidify ash from the combustion of municipal solid waste.

The PASO process differs from other applications of plasma for waste destruction. The oxidation process occurs in a rotary kiln maintained at a temperature of approximately 650 to 700 °C (1202 to 1292 °F) and at slightly subatmospheric pressure. In contrast, conventional sludge combustion in a fluid-bed furnace operates at approximately 850 °C (1562 °F).

4.3.2 Process Description

In PASO, the biosolids and preheated air are introduced continuously and concurrently into the rotary kiln, which rotates at a rate of one revolution per hour. A plasma torch is located at the opposite end of the kiln, as shown in Figure 25.1. According to Fabgroup Technologies, the total solids concentration in the biosolids can vary from 20 to 65%. The oxidation process is sustained by a plasma arc torch that generates ultraviolet radiation and ionic radicals that catalyze the oxidation process. The

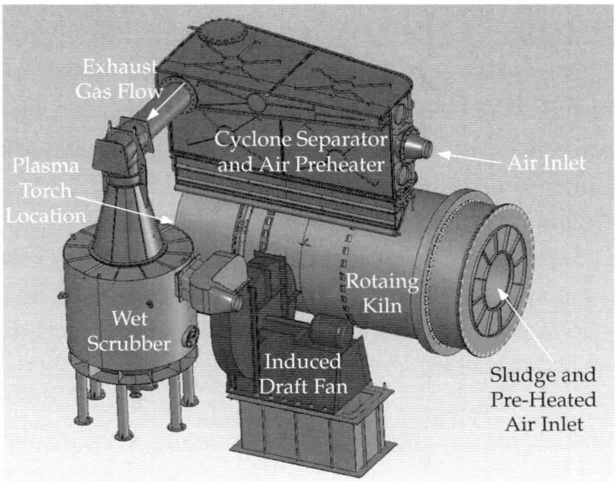

FIGURE 25.1 Plasma-assisted sludge oxidation process main components (courtesy of Fabgroup Technologies, Inc.).

energy released during the oxidation process evaporates water from the solids and heats the reaction products.

The residual ash from the oxidation process collects in the kiln and serves as a heating medium for the freshly introduced solids. Excess ash is continuously removed by passing the exhaust gasses through a cyclone separator. The cyclone separator plenum is double-walled. Feed air for the kiln flows through the gap between the walls to preheat the feed solids and air entering the kiln. After the cyclone separator, the exhaust gas is passed through a wet scrubber for cooling and further removal of particulates and other constituents. An induced draft fan is used to draw air through the process and exhaust the gas through a stack. The process achieves a volume reduction of approximately 95% (Bacon, 2002; Metro Wastewater Reclamation District, 2006; Mulhern, 2008). Fabgroup Technologies indicates the feed solids should have an organic content of 20% or greater by mass (Mulhern, 2008).

As with other oxidation processes, the potential exists for recovery of waste heat from PASO for uses in power generation, building heating, or other beneficial uses.

4.3.3 *Status of Development*

In September 2008, Fabgroup Technologies began operation of a full-scale PASO system at Valleyfield, Québec, Canada, for biosolids treatment. The unit has a design capacity of approximately 3.6 wet Mg/h (4.0 wet ton/hr) and is processing dewatered solids at 30% total solids concentration and 65% organic content by mass (Mulhern, 2008).

4.3.4 *Pre- and Post-Treatment*

The only pretreatment required for the PASO process is mechanical dewatering. Fabgroup Technologies has indicated the process will operate with feed total solids concentrations ranging from 20 to 65%.

Post-treatment is comprised of ash disposal and exhaust gas treatment. Ash is removed from the system in two ways: over a weir in the kiln and through the exhaust gases, as previously described. Fabgroup Technologies indicates the bottom ash from the kiln has low metals concentration and is suitable for beneficial uses, but the fly ash has elevated metals levels and must be disposed separately by Canada regulation (Mulhern, 2008). Scrubber blow-down water may require clarification for particulate removal before being returned to the wastewater treatment process.

As previously noted, exhaust gas is treated via a cyclone separator and wet scrubber before discharge. Fabgroup Technologies has observed particulate concentrations close to regulatory limits in the exhaust and plans to add mist eliminators to reduce moisture carryover, which is believed to be the source of the particulates (Mulhern, 2008).

4.3.5 Environmental Effects

Limited information was readily available on the emissions characteristics; however, based on the temperatures of the oxidation process and the limited oxygen concentration, it is anticipated that the emissions will be comprised primarily of CO_2 and low levels of traditional pollutants, such as NOx and SOx. Traditional emissions controls practices should enable the process to comply with regulatory requirements.

Ash characteristics will depend on the concentration of metals in the feed solids. It is anticipated the ash from this process should be similar to that produced by other oxidation processes, such as traditional incineration.

4.3.6 Technology Prognosis

Application of an emerging technology to the treatment of municipal biosolids typically requires demonstration or full-scale facilities to identify process, design, and cost issues that must be addressed in the future. Fabgroup Technologies put into operation a large-scale PASO process in Valleyfield Québec in late 2008. This large facility will be invaluable in identifying any design and operational issues and potential system costs.

4.4 Vitrification

4.4.1 Identification of Process and Scientific Basis

Vitrification is the melting of minerals within the biosolids to create a glass-like product that can be beneficially used as an aggregate. It has been used to process and stabilize industrial and hazardous wastes and municipal sludge. The melting process sequesters metals and destroys most volatile organic compounds. The U.S. EPA has used soil vitrification as part of a Superfund site cleanup (U.S. EPA, 2001).

During the process, organic matter is thermally destructed at temperatures of approximately 1400 °C (2550 °F) or greater. Synthetic combustion air, which is a mixture of 90% pure oxygen and recycled exhaust gas, is added to support the high temperature process (Dorn et al., 2007). In contrast, conventional sludge combustion in a fluid-bed furnace operates at approximately 850 °C (1562 °F). The resulting molten glass is cooled rapidly in a water quench chamber to create the glass aggregate. Exhaust gas is separated and subject to emissions control requirements that will vary based on local regulations and the specific process implemented.

4.4.2 Process Description

Minergy Corp., a Wisconsin energy company, has developed and is marketing a vitrification system for biosolids processing. It combines thermal drying with vitrification in a modular system. The heat energy for drying is recovered from the vitrification process, and the energy for vitrification is provided primarily from the organic content of the dried biosolids.

Minergy developed a patented closed-loop combustion air system that uses oxygen-enriched synthetic combustion air to develop the high temperatures needed for vitrification of the biosolids (1315 to 1482 °C). The high oxygen content in the synthetic combustion air also aids with control of NOx emissions.

In a second zone, the inert fraction of the solids is melted, and the molten material is dropped into a water quench tank for rapid cooling and formation of the glass aggregate. Following separation of the hot combustion gases from the molten material in the second zone, the hot gases are cooled in a third zone to between 370 and 870 °C with lower temperature dilution gases collected from other points in the system.

The hot exhaust gas from the furnace is conveyed to a gas-to-oil heat exchanger that recovers heat for the fluid-bed drying system. The heated oil is then used in an oil-to-gas heat exchanger within the fluid-bed dryer to provide the heat energy required for drying the solids to 90% total solids or greater.

Although both the furnace and drying components have closed-loop gas circulation systems, some exhaust is discharged from each. Emissions control technologies used include a fabric filter, packed tower scrubber, and additional NOx control as needed.

4.4.3 Status of Development

Vitrification was first applied to municipal wastewater in Japan in the early 1980s (U.S. EPA, 1985). Limitations on land available for land application or landfill disposal practices in Japan was a key driver in the development and application of the process. The resulting slag is crushed and screened for use as an aggregate in construction. Some of the furnaces used by the Japanese for vitrification include coke bed, electric arc, and cyclone furnaces (Lue-Hing et al., 1992).

Vitrification was first used in the United States at Monticello, New York, in 1991. A subsidiary of Kubota Corp. constructed a 23 wet Mg/d (25 wet ton/d) demonstration facility that was operated for several years. The vitrified product was approved by the New York State Department of Transportation for use as a fine aggregate substitute (U.S. Department of Transportation).

In the mid-1990s, Minergy developed a vitrification process to treat pulp and paper sludge. The process uses the organic content of the sludge as a fuel for vitrification in a cyclone furnace. Waste heat is recovered and used to produce steam that is used for electric power generation. The Fox Valley Glass Aggregate Plant was constructed in Neenah, Wisconsin, using the Minergy technology, and began operation in 1998. The facility was designed to process approximately 1180 wet Mg/d (1300 wet ton/d) at 40% total solids. Minergy operated the facility until 2006, when it was sold to Thermagen Power Group. Glass aggregate from the plant has been used in the production of asphalt, cement, construction fill, drainage material, and blasting media (Minergy, www.minergy.com).

Following the development of the Fox Valley plant, Minergy developed a second-generation vitrification system focused on biosolids processing. The first biosolids system was installed by the North Shore Sanitation District (NSSD) near Waukegan, Illinois. The facility began startup operations in early 2007. The NSSD facility was designed to process 170 wet Mg/d (187 wet ton/d) of sludge at 17% total solids from three different plants. Approximately 20% of the sludge is anaerobically digested; the remainder was undigested (Dorn et al., 2007).

After approximately one year of startup service, the vitrification system at the NSSD facility was out of service for Minergy to work through process and maintenance issues that were preventing continuous operation (Dorn, 2008).

4.4.4 *Pre- and Post-Treatment*

Vitrification requires that the sludge be dewatered before the process. The NSSD Minergy system was designed to process dewatered solids at approximately 17% total solids. Minergy's patented process takes the pretreatment one step further by incorporating drying to 90% total solids or greater into the overall process. Emissions will require treatment. Condensate also requires treatment before being returned to the plant.

4.4.5 *Environmental Effects*

Emissions control is required, but permit limits will be site-specific. The NSSD uses a fabric filter for particulate control and a packed tower condenser with sodium hydroxide addition for removal of sulfur dioxide. Following the condenser, the gas is reheated to prevent condensation and the exhaust is treated using a prefilter and impregnated carbon media for mercury removal (Dorn et al., 2007). Emissions limits for mercury were set at 0.80 kg/yr (1.77 lb/yr) from the cyclone furnace. Table 25.4 shows the permitted emissions limits for other pollutants based on the various components at the NSSD facility.

Table 25.4 Permitted emissions limits for the North Shore Sanitation District, Waukegan, Illinois, vitrification process (Dorn et al., 2007).

	Cyclone furnace		Auxilary heater		Two-stage packed tower scrubber	
Pollutant	**Emissions kg/h (lb/h)**	**Emissions kg/yr (ton/yr)**	**Emissions kg/h (lb/h)**	**Emissions kg/yr (ton/yr)**	**Emissions kg/h (lb/h)**	**Emissions kg/a (ton/yr)**
Volatile organic matter	0.13 (0.28)	1134 (1.25)	0.05 (0.11)	435 (0.48)	0.75 (1.65)	6550 (7.22)
Particulate matter	0.23 (0.51)	2041 (2.25)	0.07 (0.15)	608 (0.67)	0.14 (0.30)	1179 (1.30)
NOx	8.86 (19.54)	77 664 (85.61)	0.91 (2.00)	7947 (8.76)	0.07 (0.16)	653 (0.72)
CO	0.22 (0.49)	1950 (2.15)	0.76 (1.68)	6677 (7.36)	0.18 (0.40)	1569 (1.73)
SO_2	3.44 (7.58)	30 137 (33.22)	0.005 (0.01)	45.4 (0.05)	NA	NA

Condensate must be treated to remove mercury before being returned to the sanitary sewer. The NSSD uses impregnated carbon for this treatment step (Dorn et al., 2007).

4.4.6 *Technology Prognosis*

Vitrification has proven to be a feasible treatment process for biosolids. Japan has successfully used the technology since the 1980s. The complexity of the system and its high energy use are a few factors that have adversely affected acceptance and implementation in the United States. Minergy's proprietary GlassPack system was designed to address the high energy consumption through the use of extensive energy recovery systems. A full-scale system has been constructed and the owner and manufacturer are working to resolve operational and maintenance issues that have prevented the system from operating reliably during its first year of operation.

4.5 Chemical Drying for Higher Quality Fertilizer Production

4.5.1 *Identification of Process and Scientific Basis*

Chemical drying is a process that incorporates wastewater biosolids in the production of an organic/inorganic ammonium sulfate fertilizer. Chemical drying produces a value-added biosolids product that is high in nutrient value. In chemical drying, biosolids are added to other traditional fertilizer components and, because of chemical exothermic reactions, a significant percentage of the contained water is removed before later physical thermal drying. In this process, the biosolids are only a small fraction (5 to 30%) of the dry weight of the resulting fertilizer product.

Chemical drying occurs when dewatered biosolids, sulfuric acid, and ammonia are mixed in a pressure vessel. The exothermic reaction drives off water, and an organic ammonium sulfate fertilizer is produced. The reaction occurs at 135 to 150 °C (275 to 302 °F) at pressures of 207 to 480 kPa (30 to 70 psi). The high temperature/pressure process meets the standards of the U.S. EPA for Class A pathogen reduction.

4.5.2 *Process Description*

Two proprietary chemical drying processes are being marketed. Although they are similar, the equipment used to accomplish the chemical drying process is different. Both processes require that the biosolids first be dewatered to 13 to 30% total solids. The quantity of water in the solids will affect the sizing of equipment. Chemicals, such as sulfuric acid and anhydrous ammonia, are added to the solids in a hydrolysis step. The chemical reaction causes the mixture to heat to temperatures in from 135 to 150 °C. The hydrolyzed blend is then blended with previously processed product before being granulated. Granules approximately 2 to 3 mm diameter are then dried and screened. The fertilizer product has the following characteristics:

- Sterile and chemically modified;
- Hardness of 2.3 to 2.7 kg (5 to 6 lb);
- Dry (less than 1% moisture);
- High nitrogen content (16%);
- pH of 6.8;
- Density of 880 kg/m^3 (55 lb/cu ft);
- Indefinite shelf life; and
- Nonflammable.

4.5.3 *Status of Development*

Two U.S.-based companies are offering chemical drying: Unity Envirotech, Clearwater, Florida, and VitAg Corp., Beech Island, South Carolina. Both were built based on experience at an ammonium sulfate fertilizer production facility operated from 1994 to 2004 in Helena, Arizona.

Cypress Chemical Co., Helena, Arizona, produced ammonium sulfate fertilizer using the pipe-cross reactor process that was first developed by the Tennessee Valley Authority in the 1960s. The 40-year-old plant had a daily capacity of 544 Mg (600 ton) of fertilizer product. Cypress Chemical was successful in obtaining anaerobically digested dewatered biosolids produced by the New York City Department of Environmental Protection (NYCDEP).

Between 1999 and 2004 the facility produced approximately 120 000 Mg (130 000 ton) of organic/inorganic ammonium sulfate fertilizer using biosolids. The product was sold to distributors throughout the United States, primarily within an 800-km (500-mile) radius of the site. Crops using the product included wheat, soybeans, corn, cotton, rice, sod, and citrus.

As the slow-release characteristics of the product became better understood, the selling price and demand for the product increased dramatically. By late 2004, transportation costs of biosolids from New York to Arizona exceeded the tipping fee revenues and it was no longer cost-effective to continue to manage New York biosolids at the Helena facility. Attempts to locate other large, closer supplies of biosolids also failed, and the company ceased operations.

Unity Envirotech reported it is developing a regional facility in Polk County, Florida, to serve utilities in central and south Florida. VitAg Corp. is reported to be developing a facility in New Jersey.

4.5.4 *Pre- and Post- Treatment*

The only pretreatment required for the process is mechanical dewatering with a target solids content of approximately 20 to 30%. Solids concentrations less than 20% can be processed; however, energy and chemical costs will be proportionately higher.

Post-treatment is comprised of sidestream treatment and emissions and odor control. Most of the wastewater (approximately 90%) will be reintroduced as process makeup water.

4.5.5 *Environmental Effects*

Emissions control is required, but permit limits will be site-specific. Chemicals used in the process, including acid and anhydrous ammonia, should be used carefully and stored properly. System suppliers indicate the processes should be environmentally sustainable because it converts municipal biosolids into a fertilizer product that has a lower carbon footprint than typical chemical fertilizers.

4.5.6 *Technology Prognosis*

Chemical drying has been proven to be a feasible treatment process for biosolids. The complexity of the technology and the need to market the product as a high-end fertilizer will likely require the system suppliers to also operate facilities and market product. The technology was used for an extended period by the Cypress Chemical to process large quantities of biosolids from the NYCDEP from 1999 to 2004. Previous experience with product sales from the Cypress Chemical facility indicates the sale price of the fertilizer is expected to be more than ten times higher than traditional biosolids products. Current system suppliers indicated that additional facilities were under development.

5.0 CHEMICAL PROCESSES

5.1 Neutralizer Process

5.1.1 *Identification of Process and Scientific Basis*

Chemical stabilization of wastewater treatment plant sludge has been used since the 1970s. One emerging chemical stabilization process is the Neutralizer process, a proprietary process supplied by BCR Environmental (BCR). The process, which uses chlorine dioxide for stabilization, is a chemical batch process that can treat raw, aerobically digested, or anaerobically digested WAS to produce Class A biosolids. It requires thickening to approximately 4% total solids to minimize the volume entering the chemical process tanks. Tests carried out at Tulane University, New Orleans, Louisiana, showed that this process provides complete disinfection of the sludge is effective against helminth eggs, bacteria, bacterial spores, and viruses (Reimers et al., 2006).

5.1.2 *Process Description*

The acid-oxidative process works in two stages. Before chemical treatment, however, the WAS should be thickened to approximately 4% total solids. Then in the first stage of the process, chlorine dioxide, at a dosage of 50 to 75 mg/L, is used to partially disinfect the thickened WAS and increase the ORP above 100 mV. The chlorine dioxide is created by reaction of sulfuric acid and sodium chlorite in-pipe by injection through the chlorine dioxide generator. This stage takes approximately one to two hours to destroy viruses and bacteria. This stage of treatment can produce Class B stabilized biosolids, if the pH is adjusted to neutral.

In the second stage, the pH is lowered to 2.3 by adding sulfuric acid, which is followed by introduction of 1800 mg/L of sodium nitrite. Under these oxidation reduction potential (ORP) and pH conditions, nitrite converts to nitrous acid (900 mg/L). The second stage takes approximately six hours to destroy helminth ova and other parasitic pathogens (Reimers et al., 2006). At the end of the batch operation, sodium hydroxide is added to elevate pH to neutral. Once the pH is stabilized, the disinfected biosolids can be beneficially used.

5.1.3 *Status of Development*

The Neutralizer process has been tested at the bench-, pilot-, and full-scale levels. A full-scale system began operation in March 2007 at the Spencer WWTP, Clay County, Florida. Two additional facilities have since been constructed at other Clay County plants, and a new facility was under construction in late 2009 at the City of Starke, Florida, WWTP.

Bench- and pilot-scale testing and data from the Spencer WWTP were used to secure site-specific equivalency as a process to further reduce pathogens (PFRP) for Class A biosolids as defined under Title 40 Code of Federal Regulations Part 503 (U.S. EPA, 1993). BCR Environmental reportedly was conducting additional monitoring and testing to obtain national equivalency.

5.1.4 *Pre- and Post-Treatment*

Thickening to approximately 4% total solids is suggested before the Neutralizer process. Odor control requirements should be considered during implementation.

5.1.5 *Environmental Effects*

Three sidestreams are generated from the Neutralizer process: two filtrates produced during the thickening and dewatering processes and gas (primarily NO_2/NO_3) vented from the process tanks after chemical stabilization. During full-scale monitoring, no adverse effect was observed on the Spencer WWTP's liquid treatment processes. Chlorine oxidation of residuals may result in increased levels of toxic chlorinated organics, however. Additionally, acidic conditions during the chemical stabilization process may enhance the release of heavy metals from sludge. Therefore, measures should be taken to deal with the filtrate, centrate, or decant from the process.

Five chemicals that are either oxidants or corrosives are used for the Neutralizer process. These chemicals should be used carefully and stored properly. In addition, the pH of the final product should be increased to neutral before it is used for land application. Otherwise, the low pH product may adversely affect crops.

5.1.6 *Technology Prognosis*

The Neutralizer process has been demonstrated to disinfect sludge to produce Class A biosolids on a full-scale basis at small WWTPs. The final product can be land applied or used as a commercial fertilizer. Currently, BCR is in the process of obtaining national PFRP equivalency from U.S. EPA's Pathogen Equivalency Committee.

6.0 REFERENCES AND SUGGESTED READINGS

Bacon, M. (2002) Plasma-Assisted Sludge Oxidation. *Proceedings of the Water Environment Federation 16th Annual Residuals and Biosolids Management Conference*, Austin, Texas, March; Water Environment Federation: Alexandria, Virginia.

Boulding, J. R. (1996) *EPA Environmental Engineering Sourcebook.* CRC Press: Boca Raton, Florida.

Decker, W.; Sieger, R. B.; Lamont-Black, J. (2006) A New, Real Technology for Solids Dewatering. *Proceedings of the 79th Annual Water Environment Federation Technical Exposition and Conference* [CD-ROM]; Dallas, Texas, Oct 21–25; Water Environment Federation: Alexandria, Virginia.

Dorn, B.; Speth, D.; Jensen, B. (2007) Biosolids Recycling Facility—Sludge to Glass Aggregate. *Proceedings of the Water Environment Federation/American Water Works Association Joint Residuals and Biosolids Management Conference* [CD-ROM]; Denver, Colorado; Water Environment Federation: Alexandria, Virginia.

Dorn, B. (2008) Personal communication. North Shore Sanitary District, Waukegan, Illinois.

Electric Power Research Institute (2003) *Emerging Environmental Technologies: An Analysis of New Treatment Technologies for the California Energy Commission, Palo Alto, California, California Energy Commission, Sacramento, California.* Electric Power Research Institute: Palo Alto, California.

Environmental Resources Management (2008) *Life Cycle Analysis of Greenhouse Gas Emissions for EnerTech's E-Fuel Product*; Environmental Resources Management: Washington, D.C.

Environmental Resources Management (2009) *Comparison of Greenhouse Gas Emissions for EnerTech's E-Fuel Product Versus Conventional Means for Managing*

Biosolids from Wastewater Treatment; Environmental Resources Management: Washington, D.C.

Gidner, A.; Stenmark, L. (2011) *Supercritical Water Oxidation of Sewage Sludge—State of the Art.* Chematur Engineering AB: Karlskoga, Sweden.

Griffith, J. W.; Raymond, D. H. (2002) The First Commercial Supercritical Water Oxidation Sludge Processing Plant. *Waste Manage.*, **22**, 453–459.

Kearney, R.; Bolin, K.; Dooley, B. (2008) New Technology for Converting Biosolids to a Renewable Fuel Is Starting Up. *Proceedings of the 13th European Biosolids and Organic Resources Conference*, Manchester, England, United Kingdom.

Kearney, R.; Bolin, K.; Dooley, B. (2008) New Thermal Treatment Technology Will Energy Efficiently Convert Biosolids to a Renewable Fuel; *Proceedings of the 27th Annual International Conference on Thermal Treatment Technologies*; Montreal, Quebec, Canada.

Kearney, R.; Bolin, K. (2008) The New SlurryCarb Process, Under Construction in Rialto, CA, Will Convert Biosolids to a Renewable Fuel; *Proceedings of the Water Environment Residuals and Biosolids Management Conference* [CD-ROM]; Philadelphia, Pennsylvania; Water Environment Federation: Alexandria, Virginia.

Lue-Hing, C.; Zenz, D. R.; Kuchenrither, R. (1992) *Municipal Sewage Sludge Management: Processing, Utilization, and Disposal*, Water Quality Management Library, Vol 4.; Technomic Publishing Co.: Lancaster, Pennsylvania.

Mastin, B. J.; Lebster, G. E. (2006) Dewatering with Geotube Containers: A Good Fit for a Midwest Wastewater Facility? *Proceedings of the Water Environment Federation/American Water Works Association Joint Residuals and Biosolids Management Conference* [CD-ROM]; Covington, Kentucky; Water Environment Federation: Alexandria, Virginia.

McKay, S. F.; Paradis, R.; Blanchette, M. (2007) Reducing Disposal Costs Through Advanced Electro-Dewatering; *Proceedings of the Water Environment Federation/American Water Works Association Joint Residuals and Biosolids Management Conference* [CD-ROM]; Denver, Colorado; Water Environment Federation: Alexandria, Virginia.

Metro Wastewater Reclamation District (2006) *Biosolids Management Program/Facility Study* PAR880; Metro Wastewater Reclamation District: Denver, Colorado.

Mulhern, T. (2008) Personal communication. Fabgroup Technologies, Inc., Montreal, Québec, Canada.

O'Regan, J. (2008) Personal communication. SCFI Group Limited, Cork, Ireland. May 26, 2008 personal e-mail to Scott Carr.

Reimers, R. S.; Pratt-Ward, L. S.; Bradford, H. B.; Mussari, F. P.; Schmitz, W. (2006) Development of the Neutralizer Process for Disinfection and Stabilization of Municipal Wastewater Residuals; *Proceedings of the Water Environment Federation Residuals and Biosolids Management Conference* [CD-ROM]; Covington, Kentucky; Water Environment Federation: Alexandria, Virginia.

Sloan, D.; Pelletier, R.; Modell, M. (2008) Sludge Management in the City of Orlando—It's Supercritical. *Fl. Water Resour. J.*, **60** (6), 46–54.

Stendahl, K.; Jäfverström, S. (2003) Phosphate Recovery from Sewage Sludge in Combination with Supercritical Water Oxidation; *Water Sci. Technol.*, **48** (1), 185–191.

Svanström, M.; Fröling, M.; Modell, M.; Peters, W.; Tester, J. (2004) Environmental Assessment of Supercritical Water Oxidation of Sewage Sludge. *Resour. Conserv. Recycl.*, **41**, 321–338.

U.S. Department of Transportation. Federal Highway Administration. User Guideline for Waste and Byproduct Materials in Pavement Construction http://www.fhwa.dot.gov/publications/research/infrastructure/structures/97148/index.cfm (accessed August 2011).

U.S. Environmental Protection Agency (1985) *Seminar Publication: Municipal Wastewater Sludge Combustion Technology*; EPA 625/4-85/015; U.S. Environmental Protection Agency, Center for Environmental Research Information: Cincinnati, Ohio.

U.S. Environmental Protection Agency (1993) *Standards for the Use and Disposal of Sewage, Final Rule*. Code of Federal Regulations, Part 503, Title 40, 1993; *Fed. Regist.*, **58**, 9248.

U.S. Environmental Protection Agency (2001) *A Citizen's Guide to Vitrification*; EPA-542-F-01-017; U.S. Environmental Protection Agency: Washington, D.C.

U.S. Environmental Protection Agency (2006) *Emerging Technologies for Biosolids Management*; EPA-832-R-06-005; U.S. Environmental Protection Agency: Washington, D.C.

U.S. Environmental Protection Agency (2008) *Emerging Technologies for Wastewater Treatment and In-Plant Wet Weather Management*; EPA-832-R-06-006; U.S. Environmental Protection Agency: Washington, D.C.

Water Environment Research Foundation (2006) Demystifying the Dewatering Process: New Techniques and Technologies Shed Light on a Complex Process. http://www.werf.org/AM/Template.cfm?Section=Progress1&Template=/CM/HTMLDisplay.cfm&ContentID=3052 (accessed August 2011).

Wienands, S. (2010) Personal communication. Nutri-Ject Systems, Inc., Hudson, Iowa.

Yesodharan, S. (2002) Supercritical Water Oxidation: An Environmentally Safe Method for the Disposal of Organic Wastes. *Current Sci.*, **82** (9–10), 1112–1122.

Chapter 26

Treatment and Use of Green Gases

(*continued*)

(continued)

1.0 INTRODUCTION TO GREEN GASES

This chapter provides technical background information and a practical and helpful guide for the effective treatment and use of biogas. Green gas or biogas, sometimes called biomethane, is the collective name for gas produced from the biological decomposition of organic material typical in wastewater biosolids or biodegradable solid waste.

1.1 Natural Gas Comparison

Natural gas is a naturally occurring mixture consisting primarily of methane and small amounts of nitrogen, propane, butane, and other hydrocarbon gases. Raw, untreated natural gas (wellhead gas) frequently contains many of the same contaminants as biogas and is often extensively treated before being used in commercial pipelines. The exact composition of pipeline quality gas at a given point depends on many factors (Hansen, 2008).

1.2 Digester Gas

Digester gas, also known as sludge gas, sewage gas, or biogas, is a combustible methane-rich gas produced as a byproduct of microbial decomposition of organic material during the anaerobic digestion process. The principal constituents of digester gas are methane and carbon dioxide.

Most municipal wastewater anaerobic digesters today operate in the 34 to 37 °C (94 to 99°F) mesophilic temperature range. Typical characteristics of municipal wastewater treatment plant (WWTP) mesophilic digester gas, compared to natural gas, are included in Table 26.1.

Table 26.1 Typical digester gas and a comparison to typical natural gas.

Item or parametera	Digester gas		Pipeline quality natural gas
	Range	Common value	
Methane, CH_4, % (dry basis)	50–70	60	80–98
Carbon dioxide, CO_2, % (dry basis)	30–45	39	0–2
Nitrogen, N_2, % (dry basis)	0.2–2.5	0.5	0.2–10
Hydrogen, H_2, % (dry basis)	0–0.5	0.2	~0
Water vapor, H_2O, %	5.9[b]	6	dry
Hydrogen sulfide, H_2S, ppm_v (dry basis)	200–3 500	500	< 16
Propane, C_3H_8, % (dry basis)	0	0	0.6–5
Butane, C_4H_{10}, % (dry basis)	0	0	0.5–3
Ammonia, NH_3, ppm_v	0–600	5	0
Siloxanes, ppb_v	200–10 000	800	0
Specific gravity (based on air = 1.0)	0.8–1.0	0.92	0.59
Wobbe Index, per the LHV	400–690	570	1336
Higher heating value, HHV, MJ/m^3	20–28	24	41
Higher heating value, HHV, Btu/cu ft	500–700	610	1030–1040
Lower heating value, LHV, MJ/m^3	18–25	21	36.5
Lower heating value, LHV, Btu/cu ft	450–640	550	930–950

[a]Percents are by volume; ppm_v = parts per million by volume; ppb_v = parts per billion by volume; HHV includes the heat of the water of vaporization; LHV does not include the heat of the water of vaporization.

[b]Mesophilic digester gas at 37 °C (98 °F); thermophilic digester gas at 55 °C (131°F) contains approximately 15% water vapor.

A few digesters in the United States operate in the 52 to 60 °C (126 to 140 °F) thermophilic temperature range.

Acid-gas phase anaerobic digesters produce a specialized digester gas. Gas from the first stage of acid phase digesters will typically be lower in caloric or heating value and much higher in hydrogen sulfide (H_2S). Because of the challenging nature and limited quantity of this acid-phase digester gas, it might be better off being flared and not beneficially used. Acid-gas can be blended with raw digester gas and flared.

1.3 Landfill Gas

Landfill gas (LFG) is a biologically produced methane-rich gas that forms by the microbial decomposition of organic material in municipal solid waste. Although similar to wastewater digester gas, LFG is typically produced in larger quantities and with a more variable composition than digester gas. Landfill gas is noted here because many treatment technologies and gas handling equipment now used for digester gas were originally used for LFG. Landfill gas is composed primarily of carbon dioxide, methane, and nitrogen. Typical characteristics of LFG, compared to natural gas, are included in Table 26.2.

1.4 Other Biogas

Other biogases include manure digestion gas, grease and food waste digestion gas, upward anaerobic sludge bed and similar anaerobic process gases.

1.4.1 Manure Digestion

Digester gas from the anaerobic digestion of animal manure is typically similar in composition to gas from municipal wastewater sludge digestion. The Inland Empire Utilities Agency's (Chino, California) Regional Plant No. 5 Solids Handling Facility (RP-5 SHF) is designed to receive and process animal manure and food waste in anaerobic digester units and has co-digested manure for several years (Inland Empire Utilities Agency, 2011).

1.4.2 Grease and Food Waste Digestion

Municipal WWTPs have begun adding grease—also known as fats, oils, and grease (FOG)—and similar easily biodegradable food waste to existing digesters for disposal, to reduce grease in the collection system, and to increase methane gas production. The composition of FOG and food waste digestion gas is typically similar to sludge digestion gas, with slightly more methane and less carbon dioxide.

Wastewater treatment plants that add FOG to the anaerobic digestion process include Pinellas County, Florida; West Lafayette, Indiana; Des Moines, Iowa; and East Bay Municipal Utility District, Oakland, California. Riverside, California, has successfully added FOG to anaerobic digesters since 2005, resulting in increased

Table 26.2 Typical landfill gas and a comparison to typical natural gas.

Item or parameter*	LFG Range	LFG Common value	Pipeline quality natural gas
Methane, CH_4, % (dry basis)	35–65	45–50	80–98
Carbon dioxide, CO_2, % (dry basis)	30–60	45	0–2
Nitrogen, N_2, % (dry basis)	0.2–40	5	0.2–10
Oxygen, O_2, % (dry basis)	0–2.5	0.5	0
Hydrogen, H_2, % (dry basis)	0–0.5	0.2	~0
Water vapor, H_2O, %	4–12	6	Dry
Hydrogen sulfide, H_2S, ppm_v (dry basis)	200–3500	500	<16
Halogens or halides, ppm_v	0–250	130	~0
Ammonia, NH_3, ppm_v	0–500	100	~0
Propane, C_3H_8, % (dry basis)	0	0	0.6–5
Non-methane organic compounds (NMOC), ppm_v	200–12 000	2000	NA
Siloxanes, ppbv	400–12 000	1000	0
Wobbe index, per LHV	12–25 (300–630)	18 (455)	48 (1220)
Specific gravity (based on air = 1.0)	0.87–1.13	0.91	0.59
Higher heating value, HHV, MJ/m^3	16–26	19	41
Higher heating value, HHV, Btu per cubic foot	350–650	480	1030–1040
Lower heating value, LHV, MJ/m^3	13–23	17	37
Lower heating value, LHV, Btu per cubic foot	320–590	440	930–950

*Percents are by volume; ppm_v = parts per million by volume; ppb_v = parts per billion by volume; HHV includes the heat of the water of vaporization; LHV does not include the heat of the water of vaporization.; NA = not available.

gas production, methane concentrations, and average kilojoule values; and reduced biosolids production (Bailey, 2007).

1.4.3 *Upward Anaerobic Sludge Bed and Similar Anaerobic*

High strength industrial waste can be treated by similar anaerobic treatment processes such as the upward anaerobic sludge bed (UASB) process. The biogas from this anaerobic process is similar to municipal anaerobic digester gas, but the methane content on some high loads beds can be as high as 70 to 80% by volume. Another similar anaerobic process is the bulk volume fermenter (BVF) process sometimes used for anaerobically treating high strength dairy or brewery wastes.

1.5 Why Use Green Gas?

Green gases are an important renewable energy resource. Reasons for using green gas include process optimization, economic benefits and incentives, renewable energy usage goals, and carbon balance and sustainability.

Biogas formation is an indication that the anaerobic process is performing as desired for process optimization. Economic benefits and incentives can result from using biogas, particularly because energy represents one of the largest WWTP operating costs. Renewable energy goals or targets have been set by many states, cities, municipalities, and institutions. Biogas is a sustainable resource and carbon is balanced during electricity and heat generation. Methane (CH_4) combustion produces carbon dioxide (CO_2) and water vapor (H_2O). Each molecule of CH_4 forms one molecule of CO_2:

$$CH_4 + 2O_2 \rightarrow CO_2 + 2H_2O \qquad 26.1$$

For complete combustion, each 1 mol of CH_4 produces 1 mol of CO_2. Because CH_4 has a molecular weight of 16, and CO_2 has a molecular weight of 44, each kilogram of CH_4 will produce 2.75 kilograms of CO_2 when completely combusted (44 / 16 = 2.75). By comparison, when a more complex but common substance such as cellulose ($C_6H_{12}O_6$) is combusted, as in burning wood or other biomass, each molecule of cellulose forms six molecules of CO_2 as follows:

$$C_6H_{12}O_6 + 6O_2 \rightarrow 6CO_2 + 6H_2O \qquad 26.2$$

1.6 Biogas Use

The majority of biogas use applications involve producing electricity or heat.

Landfill gas is widely used for electric power generation; and LFG-to-electricity (LFGE) facilities represent one of the highest beneficial uses of biogas. More than 200 facilities generate more than 1000 megawatt (MW) of electric power in the United States. By far the most common LFG energy recovery application, comprising approximately 75% of the LFGE projects, is onsite electric power generation for sale to the electric utility. (Yazdani et al., 2006). Most LFGE technology is mature and well proven.

Combined heat and power (CHP) is the production of electricity and useful heat simultaneously or sequentially from a single fuel source. Many digester gas-to-energy applications are based on the concept that heat from power generating equipment can be captured and used to warm anaerobic digesters. The terminology "combined heat and power" is synonymous with the word cogeneration; CHP is used more widely outside of the United States.

2.0 ELECTRIC AND COMBINED HEAT-AND-POWER EQUIPMENT

Electric power and CHP generation technologies for biogas applications include internal combustion engines, combustion gas turbines, microturbines, steam boilers with steam turbine generators, Stirling cycle engines, and fuel cells.

2.1 Internal Combustion Engines

Digester gas has been used in WWTP internal combustion engines in the United States since the 1920s and 1930s. Internationally, treatment plants in Bombay, India, in 1907 (small installation) and Birmingham, England, United Kingdom, in 1921 were among the first treatment plants to use sludge gas for engine fuel (Greeley and Velzy, 1936). In the United States the first facilities included Sugar Creek, Charlotte, North Carolina, in 1928; Cedar Rapids, Iowa, in 1935; and Durham, North Carolina, in 1937.

Internal combustion engines produce heat that can be captured or recovered as hot water, hot oil, or low-pressure steam for a CHP application. Only a portion of the total heat available from the engine can be used to make high-pressure steam or high-temperature hot oil in CHP applications because of thermodynamic temperature limitations.

2.1.1 Reciprocating Engines

Reciprocating engines are the most typical and widely accepted technology for digester-gas fueled CHP applications. Reciprocating engines include both spark-ignition and compression-ignition dual-fuel or gas/diesel engines.

2.1.1.1 *Spark-Ignition Engines*

Spark-ignition reciprocating engines comprise the vast majority of all LFGE and digester gas-to-energy applications. Spark ignition reciprocating engines are available in sizes less than 10 kW to more than 20 000 kW; some can use both biogas and natural gas in a mixture. Spark-ignition Otto cycle gas engines are much more typical in sizes less than approximately 1000 to 2000 kW power (kWp) or 1500 to 2700 hp (1 kWp = 1.341 hp) because of specific engine models configured for use with low-kilojoule biogas. Most spark-ignition engines constructed today are called lean-burn engines because their air-to-fuel ratio is much leaner than required for stoichiometric combustion.

2.1.1.2 *Dual-Fuel Engines*

Dual-fuel or gas/diesel reciprocating engines are compression ignition engines. They simultaneously burn gas and a small amount of diesel fuel pilot oil. This pilot oil ignites the fuel in lieu of a spark plug. The dual-fuel capability of these engines always requires using some diesel fuel, but engine controls allow automatic switchover to straight diesel fuel operation without changing load if the gas fuel supply is interrupted. Typically, dual-fuel engines use 1 to 5% diesel fuel oil, but many can operate with anywhere from 1 to 100% diesel fuel. This capability is a beneficial feature for a standby unit because it can operate even if the plant experiences a complete power failure. Diesel fuel storage and handling equipment and gas compressors delivering 520 kPa (75 psig) pressurized gas are required to supply fuel to this type of engine.

2.1.2 *Advanced Reciprocating Engine Systems Engines*

The U.S. Department of Energy initiated a multiyear manufacturing and development program in 2001 to develop more technically advanced spark-ignition lean-burn engines with improved fuel economy and reduced exhaust emissions. The new engines are called advanced reciprocating engine systems (ARES) and have an electrical output of approximately 1000 to 2600 kW. Engines with similar performance qualities are available from a few European manufacturers. The increased fuel efficiency of these advanced engines means that the added power is produced from a smaller quantity of digester gas than older, less-efficient engines now in service. Efficiency is important for biogas because although the fuel is "free," it is available in limited quantity and thus should be used wisely.

2.2 Combustion Gas Turbines

Combustion gas turbines burn a mixture of compressed air and fuel and capture the energy of the expansion of the gases from the fuel combustion. Gas turbines can be attractive prime movers for generating electricity and have a track record of low maintenance and high availability. High-temperature exhaust gases can be used to produce hot water, hot thermal oil, or steam for heat-recovery applications. In most gas turbines, high-pressure steam, typically greater than 22 kPa (150 psig), can be produced from the hot exhaust gases.

Combustion gas turbines are widely used in new, large commercial electric power plants coupled with a steam turbine powered by steam made from the hot exhaust gases. Gas turbines are adaptable to combusting low-kilojoule fuel such as digester gas, and are successfully in use at several WWTPs, including King County South, Seattle, Washington; Metro Denver, Denver, Colorado; and Joint Water Pollution Control Plant, Carson, California. Some form of exhaust emissions control will likely be required for any proposed gas turbine. Gas turbine emissions reduction technologies include

- **Water or steam injection.** For these wet technologies, pressurized water or steam is injected directly into the combustion zone to minimize nitrogen oxides (NOx).
- **Dry/low-NOx combustors.** This new emissions control technology may be the only technology appropriate for digester gas operation. Not all gas turbine manufacturers offer this technology, although dry-NOx units are becoming more proven. Most recent advanced gas turbines are available with low-NOx combustors.
- **Catalytic converters.** This technology is expensive and requires careful and complete gas treatment to remove all siloxanes and sulfur contaminants.

2.3 Microturbines

Microturbines are small, high-speed recuperated combustion gas turbines that are available in sizes from 30 to 250 kW for distributed generation and cogeneration. Microturbines use new technologies including extended-surface recuperators, air bearings, and ultra-fast operating speeds. More than 40 WWTPs have purchased microturbines in the United States, including Lewiston, New York (60 kW);

Durango, Colorado (65 kW); Albert Lea, Minnesota (120 kW); Lancaster, California (250 kW); Sheboygan, Wisconsin (300 kW); and Allentown, Pennsylvania (360 kW). Several California WWTPs installed microturbine without sufficient digester gas treatment and have experienced operating difficulties. Several plants no longer use their microturbines because of operational, maintenance, and fuel quality issues.

All gas turbines, including microturbines, generate less power when installed at high elevations and when inlet air temperatures to the machines exceed their rated 15 °C (59 °F) performance temperature. For example, at the Salt Lake City, Utah, site elevation of 1300 m (4250 ft) above sea level, a 15 to 20% reduction in performance could be expected compared to a similar unit operating at sea level.

2.4 Steam Boilers with Steam Turbine Generators

Although typical in large, commercial power plants, the use of steam boilers with steam turbine generators does not scale down efficiently for biogas applications. Only a few large WWTP facilities in the United States, such as Deer Island in Boston, Massachusetts, have sufficient capacity to burn digester gas in large steam boilers and generate high-pressure steam, producing electricity with a steam-driven rotating steam turbine generator. In smaller sizes, the steam boiler/turbine technology is inefficient for power generation. This inefficiency, along with the requirement to continuously staff a licensed steam boiler operator onsite, limits their attractiveness.

2.5 Stirling Cycle Engines

Stirling cycle engines are an emerging power generation technology that uses an external combustion process to convert heat to mechanical power based on the expansion and contraction of a closed cycle gaseous working fluid.

2.6 Fuel Cells

A fuel cell is an electrochemical device that combines hydrogen with oxygen to continuously produce electricity. The hydrogen is extracted from the fuel delivered to the unit, and the oxygen is obtained from the air.

In the United States, fuel cells have been converting WWTP digester gas to electricity since approximately 1997. Fuel cells have evolved and progressed from being a curiosity to an accepted technology used in a growing number of municipal and industrial WWTPs, including El Estero, Santa Barbara, California; Fairfield, Connecticut; and Riverside, California.

Four different types of fuel cells are now in active commercial service or undergoing development: phosphoric acid, carbonate, solid oxide, and proton exchange membrane. Other fuel cell types are not being considered here. Stationary fuel cells are available as fully modularized units in sizes of 200 kW and larger. Table 26.3 provides a summary of phosphoric acid and molten carbonate fuel cell technologies.

TABLE 26.3 Comparison of wastewater treatment plant digester gas fuel cells.

Item	Phosphoric acid	Molten carbonate	Remarks
Representative fuel cell manufacture	United Technology Corporation (UTC)	FuelCell Energy, Inc. (FCE)	UTC purchased the ONSI fuel cell business
Modular sizes, electrical output	200 kW	300 kW and 1400 kW units	FCE recently increased their unit capacity
Electrical efficiency (LHV), %	36–40	45–47	New equipment performance, typical
Operating electrolyte temperature	190 °C (375 °F)	650 °C (1200 °F)	Operating temperature affects startup time
Recoverable heat output, at 82–93 °C (180–200 °F)	0.45 mil. kJ (0.425 mil. Btu/hr) for 1 MW	1.57 mil. kJ (1.49 mil. Btu/hr) for 1.2 MW	Per manufacturer's performance claims
Expected performance degradation, %	Approximately 2% yearly capacity decrease	2–3% yearly capacity decrease	Capacity drops off as the cell stack fouls
Water consumption, Continuous, gpm	6 L/min (1.6 gpm) for 1 MW	7.6 L/min (2 gpm) for 1.2 MW	Water is used to make steam in the reformer
Required digester gas fuel pressure	140–210 kPa (20–30 psi)	170 kPa (25 psi)	Gas compressors are required
Supplemental natural gas to the digester gas	Not required	Manufacturer originally recommended 10% natural gas	Originally was recommended by the fuel cell supplier; no longer needed

2.6.1 *Phosphoric Acid*

The first commercial fuel cells were the phosphoric acid-type fuel cells, which are an established, proven technology. Several early digester gas phosphoric acid fuel cells, such as installations at Deer Island (Boston, Massachusetts) and Columbia Boulevard (Portland, Oregon) are best characterized as developmental or experimental applications and might not accurately characterize current fuel cell equipment offerings. United Technologies Corp. (UTC) now offers a 200 kW PureCell 200 standard phosphoric acid fuel cell product. At least 10 municipal WWTPs have installed digester gas phosphoric acid fuel cells, including some installations with more than seven years of operational experience.

2.6.2 *Molten Carbonate*

Most new fuel cell installations use the molten-carbonate type fuel cell, also called carbonate or direct-carbonate type. Some molten-carbonate fuel cell technology, such as the reformer and inverter, are extremely similar to those in phosphoric acid fuel cells. An important difference is the electrolyte solution within the unit.

Within the fuel cell unit there are four main process modules: the gas cleanup unit, reformer, cell stack, and inverter. The gas cleanup unit purifies the digester gas or natural gas, producing pure, clean methane gas (CH_4). If not removed, then even minute quantities of sulfur compounds will quickly damage the fuel cell's cell stack catalyst. The reformer produces high-temperature steam that reacts with the pure CH_4 react to produce hydrogen gas (H_2) essential to fuel cell operation. The cell stack produces electricity from the H_2 and O_2 from the air. The inverter converts the fuel cell's direct current (DC) electric power into alternating current (AC). Water is both created by the chemical reactions and consumed in the steam reformer to disassociate the methane gas into hydrogen (Electric Power Research Institute, 2006).

2.6.3 *Solid Oxide and Proton Exchange Membrane*

Other fuel cell technologies, including the solid-oxide and the proton-exchange-membrane (PEM) type fuel cells, are under development. These technologies are not ready for long-term digester gas applications and no known WWTP digester gas installations are in service.

2.6.4 *Fuel Cell Heat*

Chemical reactions in the current generation of fuel cells are exothermic and generate heat as a part of these reactions. This heat is sufficient to vaporize water that is chemically produced by the same reactions. Excess fuel cell heat is often captured and productively used. A 1400-kW (1.4 MW) fuel cell assembly produces approximately 8300 kg (18 300 lb) per hour of 343 to 399 °C (650 to 750 °F) exhaust, consisting of steam and clean hot gases. This hot exhaust, when passed through a heat-recovery heat exchanger, is sufficient to produce 3.94 mil. kJ/h (3.73 mil. Btu/h) by cooling the exhaust gases to approximately 120 °C (248 °F) (FuelCell Energy, Inc., 2010).

3.0 COMBUSTION AND HEAT GENERATION EQUIPMENT

Technologies for combusting or producing heat energy and disposing of excess biogas include boilers and flares.

3.1 Boilers

Boilers are energy conversion devices used to combust fuel and extract usable heat energy via steam or high-temperature water from the hot combustion gases. Boilers are historically the most typically used technique for capturing digester gas fuel energy. Plants that use engine-generators or combustion gas turbines also use boilers as standby or supplemental heating equipment. One of the most typical uses for heat generated from the boilers is for heating anaerobic digesters.

3.1.1 Fire-Tube

Packaged fire-tube boilers are the most typically used boilers in WWTPs because of their wide availability, low initial cost, and ease of installation. Fire-tube boilers are boilers in which the flames and combustion products are sealed inside airtight, closed steel tubes surrounded by a water bath. Fire-tube boilers include packaged Scotch marine boilers popular in sizes from approximately 2 to 53 mil. kJ/h (2 to 50 mil. Btu/hr).

3.1.2 Firebox

Firebox boilers are a specialized type of fire-tube boiler with an oversized combustion chamber that may be helpful for proper combustion of relatively low kilojoule-content digester gas. Firebox boilers are limited in size from approximately 2 to 11 mil. kJ/h (2 to 10 mil. Btu/hr).

3.1.3 Water-Tube

Water-tube boilers are boilers in which the flames and the product of combustion are in a chamber outside of multiple water-filled steel tubes. Water-tube boilers include small packaged boilers of less than 1 mil. kJ/h (1 mil. Btu/hr); large package boilers are up to 106 mil. kJ/h (100 mil. Btu/hr); and very large field-erected boilers are constructed onsite in sizes of up to approximately 1 mil. MJ/h (1 bil. Btu/hr). Water-tube boilers offer several advantages for many biogas applications:

- No formation of acid-laden exhaust gas condensate that can corrode tubes.
- Flexible, serpentine tubes accommodate thermal shock better than rigid tubes.
- Flexible tubes that flex when heated and cooled allow any solid deposits to flake off.
- Significantly larger outer surface area of tube allows silica deposits to be easily noted and removed.

3.2 Waste Gas Burners or Flares

Waste gas burners, or flares, are gas safety devices used to burn excess digester gas not used by the electricity generation system or by the boilers. Because biogas is continuously generated in the digesters by a biological process, waste gas burners are needed to reduce the possibility of odor and gas explosions resulting from directly venting excess digester gas at pressure relief valves. Any significant modifications to the digester gas use system should include a review of these waste gas burners. There are two types of waste gas burners now in service at WWTPs: conventional, or open, flares and enclosed combustion flares.

Conventional, or open, combustion waste gas burners are simple devices that include basic gas combustion and safety equipment. Most flares in service at WWTPs, and virtually all flares installed before approximately 1990, were conventional flares. Flare systems typically include gas pipeline safety equipment, ignition systems, burner heads, and pilot lights.

Enclosed, or controlled, combustion flare systems include more precise combustion controls to carefully optimize the gas and air proportions and combustion control to better ensure a more complete combustion process. Careful control and metering of combustion parameters, including the air-to-fuel ratio, residence time, and mixing characteristics, result in increased combustion efficiency while producing fewer exhaust emissions than conventional flares. Conversely, enclosed combustion flares often have smaller operating ranges than conventional flares. Enclosed combustion flare technologies include both low-NOx (typically capable of achieving 20 to 30 ppm NOx) and ultra-low-NOx (9 to 15 ppm) types.

4.0 OTHER BIOGAS-TO-ENERGY TECHNOLOGIES

Other biogas-to-energy technologies center on using the energy contained in biogas offsite through exporting the gas itself. Because these technologies do not produce electricity or heat at the same place the gas is produced, this is typically an LFG application where little or no requirement exists for onsite power or heat.

4.1 Direct Gas Pipeline to Industry

High market prices for commercial natural gas typically prompt increased production of pipeline quality biomethane from biogas. Medium-kilojoule LFG direct use applications are the second most typical LFGE projects, and are often most popular where electricity prices are relatively low. One of the simplest applications for LFG is to pipe the medium-kilojoule gas, with a minimum of treatment, directly to an adjacent industrial user. This application is extremely site specific, and the applicability depends on locating an appropriate industrial user and securing a long-term purchase agreement throughout the year.

4.2 Pipeline-Quality Natural Gas Substitute

One biogas use technology that avoids the inherent thermodynamic conversion efficiency drawbacks of electric power generation or onsite combustion issues is cleaning the biogas and selling it directly to the local gas utility. For some locations, direct-pipeline injection to a nearby natural gas pipeline may be the most appropriate application. The first biogas-to-pipeline-quality natural gas project was probably at the Palos Verde Landfill, in Rolling Hills, California in 1975 treatment plants that have implemented pipeline-quality high-kilojoule biogas scrubber systems include Renton South Treatment, Seattle, Washington; Dos Rios, San Antonio, Texas; and Hale Avenue, Escondido, California. Natural gas utilities can accept whatever volume of cleaned biogas is available for delivery. The limitations on cleaning and selling the biogas to the natural gas utility are sometimes political and economical, not technical.

Treated biogas is required to be cleaned to at least 95 to 98% pure methane and compressed to the high pressure needed for the gas utility. The cleaned biomethane typically should contain less than 3% CO_2 (Cairns and Pincince, 1984); less than 0.1 mg/L (7 lb/mil. cu ft) water; less than 2.7 ppm H_2S; and a higher heating value (HHV) of at least 38.3 MJ/m^3 (975 Btu/cu ft). Several different cleaning techniques have been successfully used, including membrane separation, amine treatment, cryogenic refrigeration, pressure-swing adsorption, Selexol treatment, temperature-swing adsorption, and water scrubbers.

4.3 Compressed Biogas Used as Fleet Vehicle Fuel

This biogas application uses highly compressed and cleaned biogas as a substitute for compressed natural gas or diesel fuel in fleet vehicles. This application has seen more widespread use in Europe, especially Sweden, compared with the United States (Hagen, 2001). High conventional vehicle fuel costs can make this application a timely and economical option. To convert a truck to use compressed biogas fuel can cost $20,000 to $55,000 per vehicle. For this technology to be practical, suitable fleet vehicle users in the immediate proximity of the fueling equipment must be located, and arrangements must be made to supply fuel to their vehicles at an attractive cost.

5.0 GAS SAFETY

Several regulations and guidelines and important gas safety measures for biogas applications must be considered, such as pressure control, confined space safety, and fire safety.

5.1 Regulations and Guidelines

Gas safety regulations and guidelines include NFPA 70: National Electrical Code; NFPA 820: Fire Protection in Wastewater Treatment and Collection Facilities; NFPA 54: National Fuel Gas Code (NFPA, 2008, 2009, 2011); and the Canadian Code for Digester Gas and Landfill Gas Installations (CAN CGA-B105-M2007). The NFPA 820 provides safety standards for tanks, vessels, piping, and equipment containing digester gas; NFPA 54 provides safety requirements for combustible gas handling piping, and valving. Although CAN CGA-B105 is not specific to the United States, it does offer safety guidelines for designers to consider.

5.2 Pressure Control

Pressure control equipment includes vacuum- and pressure-relief valves, pressure safety valves, and pressure regulating valves. These devices should be used where appropriate in biogas use applications.

5.3 Confined-Space Safety

Confined-space safety procedures and equipment are required for entering a digester for inspection and cleaning. All systems must be designed to ensure maximum safety, such as specifying the appropriate personal safety equipment and providing proper enclosures and ventilation. Additional safety guidelines are noted by the *NFPA 70: National Electrical Code*; *NFPA 820: Fire Protection in Wastewater Treatment and Collection Facilities*; *NFPA 54: National Fuel Gas Code* (NFPA, 2008, 2009); *Recommended Standards for Wastewater Facilities* (GLUMRB, 2004); and *Safety and Health in Wastewater Systems* (WEF, 1994).

5.4 Fire Safety

Biogas is a combustible fuel. Flame traps and automatic thermal shutoff valves must be included in the design of the gas collection and piping systems to combustion equipment. Biogas fire safety equipment includes devices for limiting the spread of a flame front or the propagation of a flame, including flame arresters, thermal shutoff valves, and flame checks. Other similar fire safety features or equipment include manual shutoff valves and nonflammable piping.

5.5 Stainless-Steel Gas Piping

Some plants were originally designed with carbon-steel gas piping, but this practice changed because of corrosion issues and deposits. Stainless-steel piping has proven to be effective with biogas use. Some of the added cost of the stainless steel can be avoided by using piping with thinner walls than traditional piping. Both Type 305 and Type 316 stainless steel is used, but the latter provides better protection against corrosion (see Chapter 10).

6.0 BIOGAS TREATMENT

Raw biogas constituents can negatively affect the performance or efficiency of equipment or fail to meet the requirements of other uses (such as quality and emission).

6.1 Biogas Constituents Removed by Treatment

Biogas constituents typically removed by treatment include water vapor, hydrogen sulfide, siloxanes, and carbon dioxide, occasionally along with other volatile organic compounds, halides, ammonia, nitrogen, and mercaptans.

6.1.1 *Water Vapor*

Because it is produced below liquid level, biogas is typically fully saturated with water vapor. Fully saturated means the percent water in a biogas is the ratio of the partial pressure of water vapor at gas temperature to the absolute pressure of the biogas. As an example, in a mesophilic digester operating at 37 °C (98 °F), the vapor pressure of water is 6.160 kPa (0.8935 psia). At an elevation of 150 m (500 ft) above sea level, the atmospheric air pressure is approximately 99.495 kPa (14.43 psia). If in this case, the digester gas is at a typical pressure of 1.99 kPa (8 in. H_2O, or 0.289 psig), then digester gas absolute pressure is

$$99.495 + 1.99 = 101.485 \text{ kPa, absolute } (14.43 + 0.289 \text{ psia} = 14.719 \text{ psia}) \tag{26.3}$$

Thus, if the gas from the digester is fully water-saturated, then the percent water vapor in the wet gas as it is produced within the digester is

$$6.160 \text{ kPa}/101.495 \text{ kPa} = 6.07\%, \text{ by weight} \tag{26.4}$$

Cooling this gas to 4 °C (39 °F) means less than 1% (0.78) of the gas will then be water vapor.

In a thermophilic digester, because of the higher process temperature, the gas produced will contain approximately 2.5 times as much saturated water as a mesophilic digester. For thermophilic digestion at 131 °F (55 °C), the vapor pressure of water is 15.738 kPa (2.2826 psia), resulting in a calculated percent water vapor of 15.51%, by weight. The same principal holds true for various other volatile compounds. At higher process temperatures in the digestion vessels, more of the organics typically will volatize into their gaseous phase.

As the water-saturated gas cools even slightly, some water vapor condenses. Strategically located and properly sized sediment traps (accumulators) should be installed in the biogas piping system to collect moisture and sediment and to remove particles in the gas piping from the digester. Appropriate locations for these devices are downstream of the digesters, at the end of long pipe runs, and anywhere the gas is cooled or compressed, resulting in condensation. All horizontal piping in the system should slope toward traps. Designer should consider a protective coating for fabricated carbon-steel sediment traps. Steel sediment traps may be galvanized by the manufacturer or painted with corrosion-resistant coating. Fabricated stainless-steel sediment traps are also available at significantly higher cost than protected steel.

It is recommended that drop traps be installed on each sediment trap and at all low spots in the gas system for convenient, safe removal of accumulated condensate. Low-pressure drip traps are typically manufactured from low copper-aluminum castings. High-pressure devices require fabricated steel or stainless steel. One typical flaw in the design of a gas-collection system is the lack of sufficient drip traps to remove the condensate from the gas piping. Corrosion and freezing of the collected condensate in the gas system can also be serious problems.

Manually operated drip traps should be used in cases of indoor installation. Float-controlled, automatic drip traps require frequent maintenance to keep the valve from sticking open. These should be used for outdoor installations only (when local codes and safety considerations permit).

6.1.2 *Hydrogen Sulfide*

Hydrogen sulfide (H_2S) is a poisonous, combustible, odorous gas that is one of the most typical and troublesome gaseous impurities in biogas. Unless controlled, the concentration of H_2S in digester gas can range from 150 to 3500 ppm or more for municipal applications; and up to 20 000 ppm for industrial applications, depending on the influent wastewater composition. Significant sources of sulfur compounds are potable water supply and industrial discharge. Sulfates occur naturally in water from the decomposition of urine and protein and result from alum treatment in the water supply system. The H_2S in digester gas is formed by the reduction of sulfates by anaerobic bacteria within the digester.

Removing digester gas H_2S may be necessary to reduce corrosion in boilers and engines. In addition, H_2S is a toxic air pollutant that can create a safety hazard and odor, even in minute concentrations; H_2S is a hazardous, possibly deadly gas and proper safety precautions must be followed when it is present. Gas safety sensors or equipment should be used where appropriate for personnel safety.

Fuel gas with H_2S contributes to air pollution. Therefore, the H_2S might have to be removed to satisfy local air quality standards. Regulatory agencies have long studied the effect of burning gas containing small amounts of H_2S on emissions. In the refinery industry, federal source performance standards limit fuel gas for newly constructed burners and combustion units to 160 ppm or less of H_2S Although flaring solves the odor problem of the gas, H_2S produces sulfur dioxide (SO_2), which is a major component of acid rain. The H_2S also must be removed to protect some gas compression and use equipment, such as internal combustion engines and gas compressors.

6.1.3 *Siloxanes*

Siloxanes are a family of manufactured volatile organic compounds containing silicon that are becoming increasingly common in many household products. Approximately 100 siloxane compounds are used in consumer products such as deodorants, cosmetics, shampoos, dyes, lubricants, dry cleaning fluids, and waterproofing compounds (Schneider, 2008). As a result, siloxanes are found in wastewater and landfill solid waste and volatilized in digester and landfill gases. Siloxanes have become an increasing concern in biogas use. Siloxanes in biogas are often only a few parts per million or less, near their detection limits (Wheless and Pierce, 2004). This concentration makes them difficult to measure and, therefore, control. Properties of siloxane contaminants commonly found in biogas are included in Table 26.4.

Siloxanes are considered to be biologically stable compounds, but there are some indications that octamethylcyclotetrasiloxane (D4) might slowly decay under biological conditions (Arnold, 2009). When combusted, siloxane compounds form tough,

TABLE 26.4 Properties of common siloxanes in biogas (Narros et al., 2009; Schneider, 2008; Wheless and Pierce, 2004).

Chemical name	Common abbreviation	Chemical formula	Molecular weight (g/mol)	Silicon Atoms per Molecule	Water solubility, mg/L at 25°C (77 °F)	Silicon, %
Hexamethyldisiloxane	MM or L_2	$C_6H_{18}OSi_2$	162.4	2	0.93	34.5
Octamethyltri-Siloxane	MDM or L_3	$C_8H_{24}O_2Si_3$	236.5	3	0.035	35.5
Decamethyl-tetrasiloxane	MD_2M or L_4	$C_{10}H_{30}O_3Si_4$	310.7	4	0.007	36.1
Hexamethylclylo-trisiloxane	D_3	$C_{12}H_{18}O_3Si_3$	222.4	3	1.56	37.8
Octamethylclylo-tetrasiloxane	D_4	$C_8H_{24}O_4Si_4$	296.6	4	0.056	37.8
Decamethylclylo-pentasiloxane	D_5	$C_{10}H_{30}O_5Si_5$	370.8	5	0.017	37.8
Dodecamethyl cyclohexasiloxane	D_6	$C_{12}H_{36}O_6Si_6$	444.9	6	0.005	37.9

often abrasive silicon dioxide (SiO_2) deposits (the chemical name for ordinary beach sand). Typically, this stimulates coformation of other chemical deposits including those with calcium, sulfur, iron, and zinc. Rock-hard SiO_2 and other chemical deposits clog engine heads, foul exhaust and intake valves, and coat combustors and fuel injectors. They also coat exhaust catalysts, boilers surfaces, and exhaust heat recovery equipment tubes downstream of the engine or gas turbine. Siloxanes are best controlled in their gaseous form before combustion.

6.1.4 *Carbon Dioxide*

Carbon dioxide is an inert stable gas, technically not an impurity or contaminant, and thus might be considered a biogas dilutant. It is a naturally occurring component of anaerobic biogas that mainly reduces the heating value, or the caloric valve, of the biogas. Dry digester gas is typically 30 to 45% CO_2 by volume; but the same dry digester gas is approximately 55 to 62% CO_2 on a mass basis. Physical qualities of biogas based on mass should carefully consider this fact. As an example, the CO_2 effect of digester gas might be greater than expected because carbon emissions are determined on a mass basis not on a volumetric basis.

As the global concern for climate change increases, some WWTPs have begun to implement systems for monitoring or removing CO_2. Technologies that remove CO_2 from biogas or sequester the carbon, might be considered a first step in a more comprehensive carbon-capture and reduction process.

6.1.5 *Other Volatile Organic Compounds*

Biogases contain smaller amounts of other volatile organic compounds (VOCs), also called non-methane organic carbon compounds (NMOC). These VOCs include benzene, ethyl-benzene, toluene, and xylenes (BETX) and are found in trace quantities in LFG but are rarely noted in digester gas.

6.1.6 *Halides*

Halides are important constituents in LFG, but are typically rare in digester gas. Halides typically include chlorides and fluorine compounds and are called halogenated hydrocarbons. Halogenated hydrocarbons are typically produced from industrial refrigerants called chlorofluorocarbons or CFCs. Halides can be damaging to gas-using equipment such as large engines because when combusted, they form acidic hydrogen chloride (HCl) and hydrogen fluoride (HF) compounds (Caterpillar, 2009).

6.1.7 *Ammonia*

Ammonia (NH_3) is typically not present in biogas in quantities that are detectable or of concern. Digester gas contains up to 500 ppm of ammonia (NH_3) (Chambers and Potter, 2002).

6.1.8 *Nitrogen*

Nitrogen (N_2) comprises approximately 78% of the Earth's atmosphere and typically is not considered a contaminant or a concern with biogas. In some biogas use application, such as pipeline-quality natural gas substitute, N_2 must be removed from the gas to achieve the required gas quality.

6.1.9 *Mercaptans*

Mercaptans are a family of sulfur-based organic compounds characterized by a disagreeable odor. Trace amounts (less than 1 ppm) have been detected in some digester gas analysis; however, they are not typically problematic. Other sulfur compounds sometimes found in digester gas include carbon disulfide (C_2S); carbonyl sulfide (COS); and dimethyl sulfide (CH_3OS).

6.2 Gas Use Equipment Fuel Quality Requirements

Gas use equipment such as combustion engines, combustion gas turbines, microturbines, and fuel cells have specific fuel quality requirements.

6.2.1 *Internal Combustion Engines*

Internal combustion engines have been combusting low-kilojoule fuels, such as digester gas, for more than 80 years. Historically, engines have been used to extract energy from biogas because they can easily combust biogas with a caloric value or an energy content as low as 13 MJ/m^3 (350 Btu/cu ft), and with fuels that contain the traditional biogas impurities.

Internal combustion engines have evolved over the last 30 years. Engine manufactures have begun requiring increasingly cleaner fuel for newer engine designs, particularly for the newer generation of low-emissions lean-burn engines. This has, in many cases, created the need to treat biogas proper to use in the engine to meet the manufacturer's fuel quality specifications.

A summary of fuel quality requirements from engine manufacturers with biogas experience is included in Table 26.5.

Table 26.5 Internal combustion engine fuel quality requirements (Caterpillar, 2009; Cummins, 2005; GE Jenbacher, 2008; Waukesha, 2010).

	Maximum quantity allowed in the fuel				
Fuel contaminant	**Caterpillar**	**Cummins**	**GE Jenbacher**	**Waukesha**	**Remarks**
Caloric value (LHV*)	15.7–23 MJ/m^3	15–30 MJ/m^3	13.8–34.4 MJ/m^3	15.73 MJ/m^3	Varies by model
Variations in Caloric value	0.16–0.28 MJ/m^3 (7–11 Btu/cu ft)	4 MJ/m^3 (based on 16 MJ/m^3)	1% by volume per 30 seconds	Less than 5% of LHV per minute	
Water	80% relative humidity No liquid water	80% relative humidity No droplets	80% relative humidity	No liquid water	
Sulfur compounds (primarily H_2S)	57 µg/MJ (60 µg/Btu)	100 mg/m^3	2 000 mg/m^3	81 µg/MJ	Lower is better
Halogens as Cl	19 µg/MJ (20 µg/Btu)	1 mg/m^3	100 mg/m^3	8.5 µg/Btu	
Ammonia (NH_3)	2.61 µg/MJ (2.96 µg/Btu)	30 mg/m^3	55 mg/m^3	1.4 µg/Btu	
Silicon in fuel[3] (Si)	56 mg/MJ (60 mg/Btu)	1 mg/m^3	10 mg/m^3	0.7 µg/ Btu total organic	Typically siloxanes
Particulates	0.80 µg/MJ (0.84 µg/Btu)	30 mg/m^3	50 mg/m^3	Not given	
Particle size	1 micron	1 µm	3 µm	0.3 microns	
Oil residual	1.19 µg/MJ (1.25 µg/Btu)	1 mg/m^3 CH_4	5 mg/m^3 CH_4	0.3 microns	

* LHV= lower heating value, or net heating value.

6.2.2 *Fuel Cells*

Fuel cells have the most stringent fuel quality requirements of all biogas utilization equipment. Although several different types of fuel cells have been developed, the most common types of fuel cells with biogas experience are the phosphoric acid type and molten carbon type. A summary of fuel quality requirements from fuel cell manufacturers with biogas experience is included in Table 26.6.

TABLE 26.6 Fuel cell fuel quality requirements.

	Maximum quantity allowed		
Fuel contaminant	Phosphoric acid (UTC)	Molten carbonate (FCE)	Remarks
Oxygen, volume %	<4%	0.05%	Increases bed temperature
Water, volume %	remove	0.13 %	Damages valves
Hydrogen sulfide, H_2S	4 ppm maximum	10 ppmv	
Carbonyl sulfide, COS	N/A	0.1 ppmv	
Carbon disulfide, CS_2	N/A	0.05 ppmv	
Sulfur, total	N/A	10 ppmv	
Chlorides	See halogen, total	50 ppbv	
Fluorides		10 ppbv	
Halogens, total	< 4 ppm	< 100 ppbv	Corrodes fuel processors
Siloxanes	NA	< 1 ppmv	
Particulates	0.7 microns	< 10 ppmv	< 10 micron

6.2.3 *Combustion Gas Turbines*

Fuel requirements include an inlet pressure of 1000 to 1700 kPa (150 to 250 psig), gaseous fuel free of particles greater than 5 μm, and gas that is free of liquid condensate. To meet fuel requirements, a high-pressure gas compressor and a moisture separator/filter is required.

The gas turbine package requires a fuel gas booster compressor to supply the required 1400 kPa (200 psig) fuel to the combustion chamber. Some turbine models are available in a "dry low emissions" (DLE) version. Other types of gas turbines burning clean fuels such as natural gas can meet severe NOx emission standards with a post combustion selective catalytic reduction (SCR) system.

Catalysts are not suitable for use with digester gas unless the gas has been thoroughly and reliably cleaned of impurities. Various minute contaminants in the digester gas quickly poison the noble metals in the catalyst. This point has been proven convincingly many times at many WWTPs. A summary of gas turbine fuel quality requirements from a manufacturer with biogas experience is included in Table 26.7.

TABLE 26.7 Gas turbine fuel quality requirements (Solar Turbines, 2009).

Fuel contaminant	Maximum quantity allowed	Remarks
Water	No free water	Liquid water
Hydrogen, by volume	< 4 %	By volume
Hydrogen sulfide	3000 ppm max	On a mass basis
Siloxanes	0.1 mg/m^3	Total
Particulates	< 30 ppm mass × LHV*/21 500	Total
Particle size	10 μm	

*LHV= lower heating value, or net heating value.

6.2.4 *Microturbines*

Suppliers initially indicated that microturbines could use a wide range of fuels with minimal treatment. Experience has shown, and microturbine suppliers now recognize, that some fuel conditioning is required. Biogas fuel cleanup and gas pressurization packaged modules are now readily available for LFG and digester gas microturbine applications from the vendors. A summary of microturbine fuel quality requirements from manufacturers with biogas experience is included in Table 26.8.

6.3 Gas Compression Equipment

Gas is typically produced in digesters at a pressure of 1 to 2.5 kPa (4 to 10 in. WC or 0.15 to 0.36 psig), and LFG is typically at a pressure less than 7 kPa (1 psig). This pressure is typically not sufficient for the applicable biogas use equipment, necessitating gas compression equipment such as low-pressure gas blowers or medium- and high-pressure gas compressors.

6.3.1 *Low-Pressure Gas Blowers*

Low pressure for gas blowers is limited to less than 48 kPa (7 psig) for this discussion and include centrifugal and positive displacement types. A summary of low-pressure gas blower technologies used for biogas is provided in Table 26.9. One rule is that biogas use equipment needing low pressures will typically result in a simpler, more reliable, and less expensive facility.

6.3.1.1 *Centrifugal*

Low-pressure centrifugal blowers have been used successfully with biogas in numerous projects. Centrifugal blowers are popular for several reasons, including their lack

TABLE 26.8 Microturbine fuel quality requirements (Capstone, 2005).

	Maximum quantity allowed		
Fuel contaminant	**Capstone**	**Ingersoll-Rand**	**Remarks**
Water	0 %	–	Per ASTM D5454
Hydrogen sulfide	400 ppm_v low-kj gas 3.5% sour, low-kj gas	25 ppm_v	Varies by model. See gas compressor spec limitations.
Hydrogen	5 %	5 %	
Chlorine or chlorides	1500 ppm, mass	200 ppm	
Fluorides	250 ppm_v	--	
Ammonia	200 ppm_v	--	
Halogenated organics	–	200 ppm_v	
Alkali metal (Na, K, Li)	0.51 ppm, mass	0.6 ppm, mass	Per ASTM D3605
Siloxanes	5 ppb_v	10 ppm_v	
Particulates	20 ppm, mass $< 10\ \mu m$	3 microns	
Oil	2 ppm, mass	–	

of direct metal-to-metal contact (low wear), reliability, and ability to produce oil-free compressed gas. Many boilers, flares, and some beneficial gas use equipment only require an inlet gas pressure of 9 to 14 kPa (2 to 7 psig) and may be able to use low-pressure centrifugal gas blowers.

Centrifugal gas blowers often operate at 3000 to 3550 rpm and use flow-control valves or inlet guide vanes to modulate their output. Because capacity drops off in proportion to speed squared, variable-speed drives are rarely a good option for centrifugal blower flow modulation. Construction material is important in any application handling a wet gas such a LFG or digester gas. Centrifugal gas blowers are available in cast iron and stainless steel.

6.3.1.2 *Positive Displacement*

Positive displacement blowers have been used successfully in low-pressure biogas applications, particularly in small gas flow applications. Positive displacement blowers typically produce constant discharge pressure. Because output varies proportionally with speed, they are often equipped with variable-speed drives for flow modulation.

Table 26.9 Low-pressure gas blowers summary.

Equipment type	Pressure range kPa (psig)	Flow range, Nm3/h (cu ft/min)	Advantages	Disadvantages
Hermetically sealed centrifugal blowers	Up to 41 (6)	Up to 4000 (up to 7000)	• Low gas leakage • Low maintenance • Oil-free gas • Available in stainless steel	• Limited discharge pressure • Made by a single supplier
Open multistage centrifugal blowers	Up to 120 (18)	1200–8000 (2000–14 000)	• Low maintenance • Made by several vendors • Oil-free gas	• Limited discharge pressure • Requires special seals to minimize gas leakage
Roots-type positive displacement blowers	Up to 80 (12)	1200–8000 (2000–14000)	• Inexpensive • Made by several vendors	• Produces pulsations or gas pressure fluctuations
Rotary lobe positive displacement	Up to 140 (20)	Up to 2900 (5000)	• Higher efficiency than most compressors • Oil-free gas	• Produces pulsations or gas pressure fluctuations

6.3.2 *Medium- and High-Pressure Gas Compressors*

Medium pressure for gas compressors typically ranges from 340 to 1700 kPa (50 to 250 psig). High pressure for gas compressors is typically greater than 1700 kPa (250 psig).

For many applications, medium-pressure gas compression equipment is needed to supply the needs of the ultimate gas consuming device. Some boilers, engines, turbines, and fuel cells need medium-pressure gas to function. Equipment originally designed to combust natural gas will typically require a slightly higher inlet pressure when used with biogas because of the lower heating value of the gas. The type of compressor frequently depends on gas flow rate, required discharge pressure, and the need for clean, oil-free compressed gas.

Specialized compressors and their accessories are required for high pressure. For discharge pressure in excess of approximately 1700 kPa (250 psig), two or three stages of gas compression typically are required. High-pressure gas compressors also require added cooling; the compressed gas should be cooled between each stage of compression.

6.4 Compressed Gas Storage

Compressed gas storage equipment includes low-, medium-, and high-pressure applications.

6.4.1 *Low-Pressure Storage*

Within an operating digester gas system, biogas is constantly being created and used. This is a dynamically active system within essentially constant volume gas containment piping and vessels. Gas is produced in the digesters at a highly variable rate that depends on how recently each individual digester was fed with raw sludge. Meanwhile, the digester gas using devices, such as engines and boilers, typically have either variable or relatively constant gas consumption rates.

Storage is a common technique for equalizing gas production and use. Only a small amount of digester gas is contained within the large-diameter digester gas piping and beneath the fixed covers of the digesters. Therefore, many plants use some form of onsite gas storage. Low-pressure gas storage systems include several different containment concepts including floating, deep-skirted digester gas storage covers, dry-seal cylindrical steel gas holder tanks, and flexible membrane covers.

The floating-type digester covers were typical many years ago but have lost favor in some areas because of annular space gas leakage and air quality concerns. Virtually all digester gas covers in California are now sealed or fixed covers.

The dry-seal type or piston type gas holder is a vertical steel tank within a tank gas pressurization device. The ample steel vessel weight is used to maintain digester gas pressure at a virtually constant value while production or usage varies. The weighted, movable piston helps to maintain a constant digester gas pressure within the low-pressure gas piping.

Flexible membrane covers include the Dystor cover; the similar, flexible dual-membrane Walker Process GasHolder; the Westech DuoSphere; and the JDV Double Membrane gas holder flexible cover system. All of these flexible-membrane type gas holders are relatively new compared to other available systems. The added capacity of these flexible membrane digester gas holders is limited by their low-pressure rating. These covers do not have a demonstrated reliable method for reporting the status of percent storage used.

6.4.2 *Medium- and High-Pressure Storage*

Medium or high compressed gas pressure allows a smaller storage vessel to contain a specific volume of gas. Compressing digester gas to a high pressure consumes more electricity and requires a more expensive gas compressor system. Some WWTPs in California and Wisconsin effectively use a higher percentage of their digester gas via a system of medium- or high-pressure gas compressors and gas storage spheres or horizontal storage tanks (pressure vessels). These plants operate their digester gas systems at pressures of approximately 140 to 1030 kPa (20 to 150 psig).

6.5 Gas Treatment Technologies

Gas treatment technologies include the removal of water, hydrogen sulfide, siloxanes, carbon dioxide, halides, ammonia, nitrogen, and other volatile organic compounds.

6.5.1 *Water Removal*

Water is a constant problem with biogas, and its removal is an important first step in many biogas cleansing technologies. Digester gas can be dried via several techniques, including refrigerated dryers, desiccant dryers, coalescent filters, and glycol adsorption.

6.5.1.1 Refrigerated Dryers

Refrigerated dryers are a simple and successful technique typically used for drying digester gas. Refrigerated dryers use mechanical chillers and gas heat exchangers to cool the gas and condense the liquid water where it is easily removed. To conserve energy, many refrigerated dryers use gas-to-gas corrugated plate heat exchangers to use cooled and dried gas to partially pre-cool the raw gas entering the machine. Refrigerated dryers for biogas need to be robust, continuous-duty industrial machines designed for combustible and corrosive gases; light-duty compressed-air dryers should not be used for biogas.

6.5.1.2 Desiccant Dryers

Desiccant dryers are sometimes used to remove moisture in packaged compressed air drying systems. Desiccant dryers use hydroscopic material to adsorb moisture from the gas to be treated. The saturated desiccant media is then regenerated, often via thermal means, to drive off the moisture. Digester gas typically contains more water and other contaminants than compressed air applications; thus, the use of desiccant dryers for biogas has not achieved wide acceptance.

6.5.1.3 Coalescent Filters

Coalescent filters can also be used to remove minute liquid water droplets from digester gas or LFG but not molecular water vapor.

6.5.1.4 Glycol Adsorption

Glycol is sometimes added to raw natural gas to remove moisture. The hydrated glycol is then separated from the gas, and the glycol solution is dried and recycled. One form of glycol treatment is via the Selexol process, used at the Mountain Gate landfill near Los Angeles, California (Environment Agency, 2004). The Selexol process typically is used for acid gas (CO_2 and H_2S) removal and has been successfully applied for more than 30 years.

6.5.2 Hydrogen-Sulfide Removal

Hydrogen sulfide can cause significant corrosion and flow problems in piping and equipment, and its removal is a constant challenge with most anaerobic digester gas and LFG. Many of the techniques used for H_2S removal were developed by the natural gas industry because of high H_2S concentrations in some raw natural gas wells. One method used successfully to remove many contaminants from biogas streams, including H_2S, is scrubbers; H_2S is removed using various chemistries in dry or wet scrubber processes.

6.5.2.1 Dry Scrubbers

Dry-scrubber processes use an iron-oxide media to remove H_2S gas. Dry scrubbers include iron sponges and SulfaTreat. Iron-sponge technology, a dry-scrubber process, is one of the oldest gas treatment technologies widely used today. In the process, moist gas flows through iron oxide-impregnated wood chip media in a low-pressure vessel. The H_2S reacts with iron oxide to form elemental iron, elemental sulfur, and water. The iron sponge is periodically regenerated by oxidizing the iron sulfide to form iron oxide and elemental sulfur. The chemical equations in a dry scrubber process are:

$$FeO + H_2S \rightarrow Fe + S\downarrow + H_2O$$

$$2Fe_2S_3 + 3O_2 \rightarrow 2Fe_2O_3 + 6S\downarrow \qquad 26.5$$

The iron sponge method typically is best-suited to relatively small gas flows.

The SulfaTreat dry scrubber process, in use at many WWTPs, is similar to iron sponges. The granular media contained in the vessel is a proprietary product.

6.5.2.2 *Wet Scrubbers*

Conventional wet scrubbers use a liquid containing caustic (NaOH) at a high pH to enhance H_2S absorption. An oxidizing agent may also be used to reduce absorbent disposal problems and increase useful life. The overall chemical equation in a wet scrubber process is indicated below:

$$H_2S + 4NaOCl + 2NaOH \rightarrow Na_2SO_4 + 4NaCl + 2H_2O \qquad 26.6$$

Gas leaving the wet scrubber is saturated with moisture and requires condensate removal downstream. Headloss is typically too high for low-pressure digester gas systems; therefore, compression is required before the wet scrubbers. Wet scrubbers, not common for digester gas, are used for odor control at Eagle River Water and Sanitation District (Avon and Vail, Colorado) and Hyperion (Los Angeles, California) WWTPs.

6.5.2.3 *In Situ Control via Iron Salts*

As an alternative to gas scrubbing, some plants add iron salts directly to the digester or plant influent. Iron reacts with H_2S to form insoluble iron sulfide as indicated in the equations below:

$$\text{(ferrous chloride) } FeCl_2 + H_2S \rightarrow FeS\downarrow + 2\,HCl \qquad 26.7$$

$$\text{(ferric chloride) } 2FeCl_3 + 3\,H_2S \rightarrow Fe_2S_3\downarrow + 6\,HCl \qquad 26.8$$

Equipment requirements include a bulk-storage tank, chemical feed pumps, piping, and monitoring equipment. Although the operating cost is typically low, this method requires a higher level of operator skill than the iron-sponge method. Adding iron salts to heated sludge lines should be avoided. Operators at several WWTPs have found this can result in rapid buildup of vivianite (ferrous phosphate) scale on sludge lines. Use of iron salts can reduce digester alkalinity. A designer must monitor and control the solution strength and dosing rate to avoid lowering the pH in the digester.

6.5.2.4 *Adsorption Systems*

Adsorption systems are available that use activated carbon treated with an alkaline material to absorb H_2S from the gas stream. These types of adsorption systems are only suitable for treating low flows of digester gas (such as those from digester pressure-relief valves). Spontaneous combustion of these systems is possible under the certain conditions.

6.5.2.5 *Catalytic Scrubbers*

Catalytic scrubbers are available that use an aqueous chelated iron catalyst to treat the gas. With this method, elemental sulfur is produced. The catalyst is later reactivated with air in oxidizer vessels.

6.5.3 *Siloxane Removal*

Several WWTPs and landfill-to-energy sites have achieved success removing siloxanes from biogas for several years or more. Many siloxane removal systems involve more than one process step.

Most siloxane removal systems are based on activated carbon adsorption systems. Moisture removal typically occurs upstream of the activated carbon, thus both water removal and H_2S treatment should be done for the longest interval between servicing the carbon media.

Because siloxanes are both organic gases and compounds dissolved in wastewater, their tendency to volatize depends on the digester gas pressure and the siloxane's vapor pressure at that temperature. For higher operating temperatures of thermophilic digestion, siloxane vapor pressures increase with temperature, causing a higher percentage to volatize.

Specialized silica gel adsorption media systems are one of the newer concepts for siloxane removal from digester and landfill gas. These systems include automatic regeneration equipment to periodically backflow heated air thru their media vessels. The off-gassed siloxanes from the regeneration are then combusted in a small dedicated flare. Packaged continuously regenerative systems are now available. Regenerative silica gel media siloxane removal systems are likely to be more cost effective in larger-capacity systems greater than approximately 850 m^3/h (500 cu ft/min). Initially, regenerative silica gel media systems were developed for LFGE applications. A regenerative silica gel media system is being used at the Bay View WWTP in Toledo, Ohio.

6.5.4 *Carbon-Dioxide Removal*

Carbon dioxide is a significant component of biogas. It is a relatively inert and stable gas in biogas that mainly acts as dilutant and, thus, for most technologies only serves to reduce the heating value or the caloric value of biogas. Removing CO_2 from biogas allows a much more energetic fuel to be transported from production to the final point of gas use. Most biogas use applications do not remove CO_2 because of added complexity and operational cost. Moreover, retaining CO_2 in the gas may allow combustion-based biogas use technology to achieve lower levels of NOx emissions because the inert CO_2 in the biogas fuel helps to reduce the combustion zone flame temperature and, thus, initial NOx formation.

Techniques available for removing CO_2 from biogas include amine treatment, cryogenic refrigeration, membrane separation, pressure-swing adsorption, temperature-swing adsorption, and water scrubber systems. A summary of typical characteristics of CO_2 removal systems is included in Table 26.10.

6.5.4.1 *Amine Treatment*

Amines are a class of substances derived from NH_3 with a specific affinity for CO_2, H_2S, or other acid gases. Common amines include monoethanolamine (MEA) and diethanolamine (DEA). Amines are frequently used in removing both H_2S and CO_2 from raw or sour natural gas wells, for which most amine systems were developed.

6.5.4.2 *Cryogenic Refrigeration*

Cryogenic, or low-temperature, refrigeration involves cooling gas to liquefy and separate the gas components, producing liquid CH_4 or liquid natural gas (LNG) suitable for vehicle fuel. Although cryogenic systems can work well, they require a significant amount of mechanical energy input to refrigerate the gas to the required temperatures. Cryogenic refrigeration technologies take advantage of the fact that CO_2 gas will liquefy at −79 °C (−109 °F), a higher temperature than CH_4 gas at −162 °C (−259 °F). Cryogenic processes typically liquefy or freeze the CO_2 to separate gaseous CH_4.

The Frank R. Bowerman Landfill in Irvine, California, uses a cryogenic process. The gas liquefaction plant produces approximately 19 m^3 (5000 gal) of LNG per day,

Table 26.10 Carbon-dioxide removal systems summary (PSA = pressure-swing adsorption treatment system; TSA = temperature-swing adsorption treatment system).

	Amines	Cryogenic	Membrane	PSA or TSA	Water Scrubber
Low Initial cost	✓				✓
Low operating cost	✓		✓		
Proven technology	✓✓	✓	✓	✓	✓
Highest CH_4 yield	✓	✓	✓	✓	
Accommodates high contaminates levels		✓			✓

Note: ✓ = an applicable parameter or quality; ✓✓ = most appropriate technology.

and has been operating successfully since 2006. The LNG fuel is used by Orange County (California) Transit Authority buses. The plant is permitted for future expansion to produce 150 m^3 (39 600 gal) of LNG per day, the equivalent of the approximately 283 000 m^3 (10 mil. cu ft) of gas flared daily (Clarkson, 2007).

6.5.4.3 *Membrane Separation*

Membrane separation is used on raw natural gas well applications and some LFG installations. Membrane separation works by preferentially permeating the gaseous molecules to be removed and rejecting the methane and vice versa. In late 2010, the San Antonio Water System Dos Rios WWTP (San Antonio, Texas) commissioned a 42 000 m^3/day (1.5 mil. cu ft/d) membrane separation process for cleaning their digester gas.

6.5.4.4 *Pressure-Swing Adsorption*

Pressure swing adsorption (PSA) processes are used to separate gases, such as CO_2 and CH_4, from the gas stream by selective adsorption on a bed of granular material. The technology is based on the capacity of specialized solid materials to adsorb gases as the pressure is elevated and selectively desorb the gases as the pressure is lowered. The PSA technology was introduced commercially in the 1960s and is used extensively today for industrial production of oxygen, nitrogen, and hydrogen, and for removal of H_2S in refineries.

The molecular sieve processes and activated carbon beds are two types of PSA used in treatment of biogas for CO_2 removal to increase CH_4 concentration in the product gas. The difference between these types is the solid adsorbent material used; molecular sieve processes typically use zeolites rather than activated carbon in the bed. A continuous product is achieved through a cyclic process requiring multiple connected process vessels, with some adsorbing under pressure while others are desorbing. The Palos Verde landfill near Los Angeles, California uses a molecular sieve PSA process to purify landfill gas (Environment Agency, 2004).

6.5.4.5 *Temperature-Swing Adsorption*

Temperature-swing adsorption (TSA) might be the most typically used technique for removing CO_2. In the TSA process, CO_2 molecules are initially adsorbed onto a permeable media such as activated carbon at a relatively low temperature, typically near ambient temperatures of 10 to 27 °C (50 to 80 °F). Once the adsorption media is saturated, CO_2 is expelled from the media surface at a desired regeneration location when, by heating the media to much higher temperatures of approximately 150 to 200 °C (300 to 400 °F), the CO_2 molecules are rejected.

6.5.4.6 *Water Scrubber Systems*

Water scrubbers are typically used in chemical processing and gas separation industries. In the system, the gas to be removed is captured by the scrubbing liquid in a spray or packed tower. This technology relies on CO_2 being more readily soluble in water under high pressures than CH_4. The chief disadvantage of the water scrubber process is the large electric power requirement and cost of compressing the raw gas. Water scrubber systems range in size from 300 to 1800 m^3/h (10 000 to 64 000 cu ft/hr).

One proprietary water scrubber system for CO_2 removal is the Binax process. In this process, the biogas is compressed to 300 psig (U.S. EPA, 1995). The Binax process has successfully cleaned biogas to 98% CH_4 and only 2% CO_2, by volume. King County (Washington) WWTP has operated a successful installation of Binax since 1983; the system was enlarged in 1990.

6.5.5 *Halide Removal*

Halides are based on the halogen elements chlorine, fluorine, bromine, and iodine. Chlorine and fluorine are not common in significant quantities in digester gas, and measurable bromines and iodides are even less common. Chlorine and fluorine are water-soluble and halogens can be removed from biogas by removing water.

6.5.6 *Ammonia Removal*

Ammonia (NH_3) concentrations in digester gas are typically fairly small and not a concern with biogas. Ammonia can be present in thermophilic digester gas or in biogas produced by a large amount of protein in the feed material (Arnold, 2009). Ammonia is both combustible and water soluble. It can be removed by several techniques and typically is removed by one of the processes used to treat the other gas contaminants.

6.5.7 *Nitrogen Removal*

Nitrogen (N_2) concentrations in digester gas are typically less than 2% by volume, not sufficient to warrant removal, especially because the presence of free molecular nitrogen does not affect common biogas use applications. Nitrogen can be removed by the membrane separation process or cryogenic treatment. Removal from LFG can be a challenge because of the physical size of the N_2 molecule.

6.5.8 *Removal of Other Volatile Organic Compounds*

Typical gas cleanup systems, including activated carbon used for siloxane removal, which are effective for removing other VOCs, including water-soluble BETX compounds.

7.0 REFERENCES

Arnold, M. (2009) *Reduction and Monitoring of Biogas Trace Compounds*; VTT Technical Center Research: Wartsila, Finland.

Bailey, R. (2007) City of Riverside Grease to Gas to Power Anaerobic Digestion of Restaurant Grease Interceptor Wastewater. *Santa Ana River Basin Section Clarifier*, **58** (2), 1–5. California Water Environment Association: Las Flores, California.

Cairns, C. A.; Pincince, A. B. (1984) *Sale of Surplus Digester and Landfill Gas to Public Utilities*; EPA/600-S2-84-039; U.S. Environmental Protection Agency: Washington, D.C.

Capstone (2005) *Capstone Microturbine Fuel Requirements*; Technical Reference 410002-001. Capstone Turbine Corporation: Chatsworth, California.

Caterpillar (2009) Caterpillar Gas Engine Lubricant, Fuel, and Coolant Recommendations; SEBU6400-04. Caterpillar: Peoria, Illinois.

Chambers, A. K.; Potter, I. (2002) *Gas Utilization from Sewage Waste*. Carbon and Energy Management, Alberta Research Council: Alberta, Canada.

Clarkson, D. (2007) Fuelling New Ideas: Conversion of Biogas into LNG Becomes Reality. *Waste Manage. World*, **8** (3).

Cummins (2005) *Fuels Requirements*, Service Bulletin 3379001. Cummins: Columbus, Indiana.

Doorn, M.; Pacey, J.; Augenstein, D. (1995) *Landfill Gas Energy Utilization Experience: Discussion of Technical and Non-Technical Issues, Solutions, and Trends*, EPA/600-SR-95-035; U.S. Environmental Protection Agency: Washington, D.C.

Doris, E.; McLaren, J.; Healey, V.; Hockett, S. (2009) *State of the States 2009: Renewable Energy Development and the Role of Policy*; NREL/TP-6A2-46667; National Renewable Energy Laboratory, U.S. Department of Energy: Golden, Colorado.

Electric Power Research Institute (2006) *Assessment of Fuel Gas Cleanup Systems for Waste Gas Fueled Power Generation*; 1012763; Electric Power Research Institute: Palo Alto, California.

Environment Agency (2004) *Guidance on Gas Treatment Technologies for Landfill Gas Engines*; Environment Agency: Almondsbury, Bristol, United Kingdom.

FuelCell Energy, Inc. (2010) *DFC1500 Specifications*. FuelCell Energy, Inc.: Danbury, Connecticut.

GE Jenbacher (2008) *Fuel Gas Quality*, Technical Instruction No. (TI) 1000-0300; GE Jenbacker: Houston, Texas.

Great Lakes–Upper Mississippi River Board (2004) *Recommended Standards for Wastewater Facilities*; Great Lakes–Upper Mississippi River Board of State and Provincial Public Health and Environmental Managers: Albany, New York.

Greeley, S.; Velzy, C. R. (1936) Operation of Sludge Gas Engines. *Sewage Works Journal*, **8** (1), 57–62.

Hagen, M.; Polman, E.; Myken, A.; Jensen, J.; Jönsson, O.; Dahl, E. (2001) *Adding Gas from Biomass to the Gas Grid*, Report SGC 118; Swedish Gas Center: Malmö, Sweden.

Hansen, T. (2008) Meeting Gas Quality Challenges. *Power Eng.*, **112** (5), 46–47.

Inland Empire Utilities Agency Regional Solids Plant No. 5 Solids Handling Facility (RP-5 SHF) http://www.ieua.org/facilities/rp5solid.html (accessed August 2011).

King County (2009) *King County Fuel Cell Demonstration Project*, Final Report. Prepared for U.S. Environmental Protection Agency. King County: Seattle, Washington.

Leicht, R. K.; Regan, J. T.; Troy, D. A. (1986) Refinery Hydrogen Sulfide Emissions Cut 99.9% with Chalated Catalyst. *Chem. Proc.*, **Aug**.

Literary Digest (1929) *Engines Run on Sewage Gas*, July 6, 1929.

Narros, A.; Del Peso, M. I.; Miele, G.; Vinot, M.; Fernandez, E.; Rodriguez, M. E. (2009) Determination of Siloxanes in Landfill Gas by Adsorption on Tenax Tubes and TD-GC-MS. *Proceedings of Sardinia 2009, 12th International Waste Management and Landfill Symposium*; Cagliari, Italy, October 5–9; CISA Publisher: Padova, Italy.

National Fire Protection Association (2008) *NFPA 820: Fire Protection in Wastewater Treatment and Collection Facilities*, 2008 ed.; National Fire Protection Association: Quincy, Massachusetts.

National Fire Protection Association (2009) *NFPA 54: National Fuel Gas Code*, 2009 ed.; National Fire Protection Association: Quincy, Massachusetts.

National Fire Protection Association (2011) *NFPA 70: National Electrical Code,* 2011 ed.; National Fire Protection Association: Quincy, Massachusetts.

Schneider, M. (2008) Gas Engine O&M Issues: Strategies for Protecting Gas Engines that Run on Nonconventional Gas. *Power Engineering,* **112** (6), 64–66.

Solar Turbines (2009) *Fuel Requirements;* Solar Turbines: San Diego, California.

U.S. Environmental Protection Agency (1979) Process Design Manual: Sludge Treatment and Disposal; EPA/625-1-74-006; U.S. Environmental Protection Agency: Washington, D.C.

U.S. Environmental Protection Agency (1995) *Case Studies in Residual Use and Energy Conservation at Wastewater Treatment Plants;* EPA/832-R-95-003; U.S. Environmental Protection Agency: Washington, D.C.

Water Environment Federation (1994) *Safety and Health in Wastewater Systems,* Manual of Practice No. 1; Water Environment Federation: Alexandria, Virginia.

Waukesha (2010) *Gaseous Fuel Specification for Waukesha Engines,* S-7884-7.

Wheless, E.; Pierce, J. (2004) Siloxanes in Landfill and Digester Gas. *Proceedings of the 27th Annual SWANA Landfill Gas Symposium;* San Antonio, Texas, March 22–25; Solid Waste Association of North America: Silver Spring, Maryland.

Yazdani, R.; Kieffer, J.; Sananikone, K.; Augenstein, D. (2006) *Full-Scale Bioreactor Landfill for Carbon Sequestration and Greenhouse Gas Emissions Control;* DE/FC26-01NT41152; U.S. Department of Energy: Washington, D.C.

Index

A